XII International Symposium on

MULTIPARTICLE DYNAMICS 1981

Edited by W.D. SHEPHARD
V.P. KENNEY

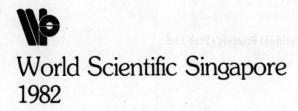

World Scientific Singapore
1982

World Scientific Publishing Co Pte Ltd
P. O. Box 128
Farrer Road
Singapore 9128

Editorial Advisory Committee

ISBN 9971-950-30-8
Printed by Singapore National Printers (Pte) Ltd

PROCEEDINGS OF THE
XII INTERNATIONAL
SYMPOSIUM
ON
MULTIPARTICLE
DYNAMICS

NOTRE DAME
JUNE 21–26, 1981

Editors: W. D. Shephard
V. P. Kenney

ORGANIZING COMMITTEE

W. D. Shephard (Chairman)
N. N. Biswas
N. M. Cason
V. P. Kenney
R. C. Ruchti

INTERNATIONAL ADVISORY COMMITTEE

A. Białas	Jagellonian University, Krakow, Poland
W. Kittel	University of Nijmegen, Netherlands
D. R. O. Morrison	CERN, Geneva, Switzerland
S. Ratti	IFN, Pavia, Italy
F. Verbeure	Univ. Instelling Antwerpen, Belgium
A. Wroblewski	University of Warsaw, Poland

SPONSORS

University of Notre Dame
U.S. National Science Foundation
U.S. Department of Energy
George I. Alden Trust

PREFACE

The XII International Symposium on Multiparticle Dynamics
was held at the University of Notre Dame, Notre Dame, Indiana
USA from June 21 to June 26, 1981. About 80 physicists partici-
pated in the Symposium which focussed on recent developments in
a variety of areas involving multiparticle final states in elementary
particle interactions.

The Symposium was the 12th in a series started in Paris in
1970 to provide an opportunity for intense discussion of the special
area of multiparticle production. The Symposia have been held
annually since then at various locations in Western and Eastern
Europe and India. This XII Symposium was the first of the series
to be held in the United States. Over the years the topics covered at
the symposia have been modified extensively as the field of elementary particle physics has grown and evolved, but the emphasis has
continued to be on understanding the various aspects of multiparticle final states as they are produced in a variety of interactions and on using the insights gained from these studies to
throw light on the basic processes of elementary particle interactions. In recent years, emphasis had tended to shift toward
consideration of hard processes in which hadrons are produced and
of developments in perturbative QCD, in the hope that these
processes could be completely understood. While these were still
topics of major interest at the 1981 Symposium, a shift back
toward increased interest in soft hadronic production may be
detected. Prof. L. Van Hove, in the concluding lecture of the
Symposium, notes: "I would like to remark on the renewed
importance of soft hadronic processes, and of the soft phases
involved in all hard hadronic processes, for the study of strong
interaction dynamics, i.e., of QCD. Nature does not provide us
with any situation controlled by perturbative QCD alone, nor does
theory provide us yet with the tools required for non-perturbative
QCD. It would be overoptimistic to expect that non-perturbative
QCD will be elucidated without the guidance of experiment." Thus
the Proceedings include the contents of extensive sessions on soft
hadron theory and experiment and on heavy-nucleus interactions
and quark-gluon plasma, in addition to recent work on lepton-
hadron interactions, high-p_T hadronic interactions, e^+e^- interactions, and the production and decay of systems involving heavy
quarks.

These Symposia have been traditionally limited in attendance
to a relatively small number of active workers in the field of multi-

particle production. At Notre Dame as many of these participants as possible were encouraged to present their latest results in experiment and theory. In addition, there were longer review talks summarizing developments in various subfields and invited talks on topics of particular interest. As at previous meetings in this series, extensive discussion was encouraged, both inside and outside the formal sessions. While the outside discussion, although a major component of the Symposium, cannot be included in the Proceedings, an effort has been made to retain the spirit of the meeting by including as much as possible of the discussion following the individual talks in as close to the original form as possible. The discussion was transcribed by the editors and has not, in general, been reviewed by the participants. All papers received from speakers in the actual sessions have been included. We wish to thank the participants for their efforts.

All sessions were held at the Center for Continuing Education on the campus of the University of Notre Dame. The staffs of the Center and of the Morris Inn deserve much credit for providing an environment conducive to fruitful scientific discussion. We are most grateful to the sponsors of the Symposium: University of Notre Dame, U.S. National Science Foundation and Department of Energy, and the George I. Alden Trust for financial assistance. Members of the Notre Dame administration, including Prof. W.C. Miller, Chairman of the Department of Physics. Prof. F. J. Castellino, Dean of the College of Science, and Prof. T. O'Meara, Provost, provided invaluable support, encouragement, and assistance.

Major credit for the success of the Symposium goes to the staff and graduate students of the Notre Dame High Energy Physics group and to other members of the Physics Department and its staff for their enthusiastic efforts in a variety of essential roles. Special thanks are due to S. Clark for her work on the Proceedings, Dr. J. M. Bishop and K. Perry for keeping track of participants and their travel plans, and to R. Erichsen and B. Baumbaugh. The advice and aid of our International Advisory Committee is gratefully acknowledged. The hard work and good ideas of the members of the Organizing Committee were essential, as were the efforts and enthusiasm of Barbara Shephard and the wives of the organizers.

The Editors want to thank those who served as Chairmen and Secretaries at the various sessions, the speakers for their interesting and timely contributions, and, most of all, all the participants of the Symposium, whose enthusiasm and cooperation made the

Symposium a pleasure even for the organizers. We hope that you
will find these Proceedings a valuable record of recent developments
in multiparticle dynamics and that they will be of help in the
continued growth of our understanding of elementary particle
interactions.

W. D. Shephard
V. P. Kenney

MULTI
XII

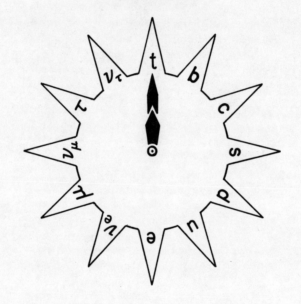

NOTRE DAME
JUNE 21-26, 1981

CONTENTS

Session A1: Soft-Hadron Physics: Experiment I   1

Monday, June 22, a.m.

        Chairman: J. Whitmore
        Secretary: V. P. Kenney

Review Talk:

     W. Kittel: Review of Developments in Soft Hadronic
            Multiparticle Physics   3

Invited Talk:

     A. Wroblewski: Multiplicities in e^+e^-, Lepton-
          Hadron, and Hadron-Hadron Collisions
                (text not received)

Contributed Talks:

     J. Beaufays: Jet-like Properties of Multiparticle
          Systems Produced in K^+p Interactions
          at 70 GeV/c   49

     V. Kistiakowsky: Comparison of 147 GeV/c π^-p
          Low Transverse Momentum Hadron
          Production with Deep-Inelastic Lepto-
          production of Hadrons   59

     S. Ratti: Planar Multiparticle Production in
          Hadron-Proton Collisions at 147 GeV/c   67

     M. Bardadin-Otwinowska: Comparison of Jets
          in Leptonic and Hadronic Interactions   75

     M. Spyropoulou-Stassinaki: Forward Distributions
          of Identified Charged Particles and Net
          Charge and Strangeness Distributions
          in K^+p Interactions at 70 GeV/c   89

     H. Frehse: Neutral Strange Particle Production in
          Proton-Proton Collisions at $\sqrt{s} = 63$ GeV   105

Session A2:   Soft-Hadron Physics: Experiment II   117

<u>Monday, June 22, p.m.</u>

      Chairman: A. Białas
      Secretary: V. P. Kenney

Invited Talk:

    D.R.O. Morrison: Diffraction Dissociation Revisited
                           (text not received)

Contributed Talks:

    J. Pernegr: A New 0^-S Meson and New Results
          on the 1^+S State in the 3π System
          Coherently Produced on Nuclei   119

    B. Buschbeck: Fragmentation of u-Quark,
          d-Quark, and the (ud)-Diquark
          System - Can It Be Measured in K^-p
          Reactions by Selection of Fast $\bar{K}^o$'s
          or Λ's?   129

    A.M. Touchard: Fragmentation of Quark and
          Diquark Jets into Pions in K^-p
          Interactions at 70 GeV/c   137

    W.A. Mann: Study of Reactions $pp \rightarrow (\omega^o, \rho^o, \eta^o)$ +
          Anything at 300 GeV/c   145

    J. Poirier: The Search for Resonant States from
          1 to 5 GeV   153

    M. Kalelkar: Inclusive Strange Resonance Production
          in pp, π^+p, and K^+p Interactions at
          147 GeV/c   161

    R. Ross: Inclusive Strange Particle Production and
          Vector and Tensor Meson Production in
          K^+p Interactions at 70 GeV/c   171

    E. De Wolf: Multiparticle Fragmentation in K^+p
          Interactions at 32 GeV/c   189

M. Suk: Observation of ρ^0 - Meson Spin Alignment
in $\overline{p}p$ Interactions at 22.4 and 5.7 GeV/c   201

D. Minette: Charged and Neutral Strange Particle
Production in 300 GeV/c Proton-Neon
Interactions   209

P. Seyboth: Multiparticle Production on Hydrogen,
Argon, and Xenon Targets in a Streamer
Chamber   217

W.D. Walker: Neutral and Charged Pion
Production in π^-p and π^-Ne
Collisions   225

Session B1: Soft-Hadron Physics: Theory I   245

Tuesday, June 23, a.m.

Chairman: F. Verbeure
Secretary: N. M. Cason

Invited Talks:

R. Hwa: Central Production and Small-Angle
Elastic Scattering in the Valon Model   247

I. Dremin: Multiparticle Correlations   261

Contributed Talks:

B. Andersson: Soft Gluons, Baryon-Antibaryon
Production and Low $p_\perp$-Physics in
the Lund Model   285

T. Kanki: Dual Topological Unitarization
Approach to Single Particle Spectra   305

Xie Qu-Bing: Inclusive Production of Meson
Resonances and the Sea-Quark
Distributions in the Proton   319

Session B2: Soft-Hadron Physics: Theory II   329

Tuesday, June 23, p.m.

      Chairman: W. Kittel
      Secretary: N. M. Cason

Invited Talk:

    J. Gunion: Perturbative QCD for Hadronic
           Processes at Low-p_T 331

Review Talk:

    T. DeGrand: Prejudices and Scenarios in
           Small Transverse Momentum Physics   361

Session C: Heavy Nucleus Interactions and Quark-Gluon
          Plasma   381

Tuesday, June 23, p.m.

      Chairman: D. R. O. Morrison
      Secretary: N. M. Cason

Invited Talks:

    L. McLerran: The Quark-Gluon Plasma and the
           Little Bang   383

    H. Pugh: Relativistic Heavy Ion Experiments   399

Session D: Lepton-Hadron Interactions   429

Wednesday, June 24, a.m.

      Chairman: J. A. Poirier
      Secretary: N. N. Biswas

Invited Talk:

    R. Orava: Structure of Quark Jets   431

Contributed Talks:

U. Sukhatme: Diquark Jets   465

N. Schmitz: Recent Results on Hadron Production
        in a Neutrino-Hydrogen Experiment with
        BEBC   481

D. R. O. Morrison: Charge Distributions in
        νp and $\overline{\nu}$p Interactions   505

D. Zieminska: Production of Hadrons in High
        Energy νD Interactions   519

W. A. Mann: Fragmentation of u-Quark Jets in
        High Energy νD Interactions   527

H. Abramowicz: (text not received)

B. Pietrzyk: Measurement of the Fragmentation
        of the Hadronic System in Different
        Flavour and Color States   539

Session E: High Transverse Momentum Hadronic
            Interactions   551

Thursday, June 25, a.m.

Chairman: A. Wroblewski
Secretary: J. M. Bishop

Review Talk:

J. Owens: Theoretical Developments in Large
        Transverse Momentum Hadronic
        Reactions   553

Contributed Talks:

P. Seyboth: Deep Inelastic Hadron Scattering
        with a 2_π Calorimeter Trigger   571

R. Gokieli: Study of Events with an Identified High
Transverse Momentum Particle at the
ISR at $\sqrt{s}$ = 63 GeV   583

M. Mestayer: High p_T Hadron Production   593

Invited Talk:

L. Resvanis: Review of Direct Photons and
Associated Particle Production at
the ISR   603

Contributed Talks:

P. Garbincius: Preliminary Results for High
X_t π^o Production and Charged Particle
Correlations for Hadronic Interactions
with Nuclear Targets   631

A. Contogouris: K-Factors in Large-p_T Direct
γ and $\ell^+\ell^-$ Production   639

Session F: e^+e^- Interactions and Comparisons with
Hadronic Interactions   647

Thursday, June 25, p.m.

Chairman: S. Ratti
Secretary: J. M. Bishop

Invited Talk:

S.-L. Wu: Recent Developments at PETRA   649

Contributed Talks:

G. Valenti: Low-p_T Proton-Proton Collisions
at the CERN Intersecting Storage Rings   717

W. T. Meyer: Study of Hadronic Events in pp
Collisions at $\sqrt{s}$ = 62 GeV and Comparison
with Hadronic Events in e^+e^- Collisions   733

Session G1:   Production and Decay of Systems Involving
               Heavy Quarks I   743

Thursday, June 25, p.m.

        Chairman: S. Ratti
        Secretary: J. M. Bishop

Invited Talk:

    M. S. Alam: Results from C.E.S.R.   745

Contributed Talks:

    A. Fridman: Results on e^+e^- Interactions in the
          $\Upsilon(9.46)$ and $\Upsilon'(10.01)$ Energy Region   791

    A. Contin: Charm Production at the CERN
          Intersecting Storage Rings Split-Field
          Magnet   799

    P. Giusti: Beauty Production at the CERN
          Intersecting Storage Rings Split-Field
          Magnet   815

    E.-E. Kluge: Was a Beauty Baryon Observed
          at the ISR?   843

Session G2:   Production and Decay of Systems Involving
               Heavy Quarks II   859

Friday, June 26, a.m.

        Chairman: V. P. Kenney
        Secretary: L. Dauwe

Review Talk:

    C. Peterson: Theory of Hadronic Production of
          Heavy Quarks   861

Invited Talk:

    R. Raja: χ Meson Production in 225 GeV/c
           π^-p Interactions   887

Review Talk:

    R. C. Ruchti: Hadronic Charm Production   897

Session H: Closing Session   925

Friday, June 26, a.m.

    Chairman: V. P. Kenney
    Secretary: L. Dauwe

Invited Talk:

    L. Van Hove: Recent Developments in
              Multiparticle Dynamics   927

List of Submitted Papers   947

List of Participants   955

SOFT-HADRON PHYSICS: EXPERIMENT I

Monday, June 22, a.m.

Chairman: J. Whitmore
Secretary: V. P. Kenney

REVIEW OF DEVELOPMENTS IN SOFT HADRONIC MULTIPARTICLE PHYSICS

W. Kittel
University of Nijmegen, Nijmegen, The Netherlands

ABSTRACT

Either partons are responsible for particle production in soft hadronic collisions or we shall have to reconsider deep inelastic lepton-hadron and e^+e^- collisions. This is shown in a comparison of data on soft hadron-hadron collisions to each other and to deep inelastic lepton-hadron and e^+e^- collisions in the light of quark-parton pictures. Three basic observations can be made. They are the success of quark combinatorics to explain hadron yields of various quantum numbers, the reflection of the valence quark structure function in hadronic pion production, and the similarity of jets in hadron-hadron, deep inelastic lepton-hadron and e^+e^- collisions. In terms of kinematical variables, the jet similarity can be accounted for by equal multiplicity and transverse momentum distributions. The rest follows from longitudinal phase space.   The models evolving from these observations help to illuminate various aspects of the data. Quark and diquark fragmentation functions extracted from hadron-hadron collisions agree with those from neutrino-hadron collisions. Kaon and pion structure functions agree with those obtained from muon-pair production. Hyperon polarization at large energies, not understandable from reggeology, follows naturally from quark-fragmentation-recombination. While a new field has silently evolved in these observations over the last years, the field will have to gain in depth. In a second approximation, deviations from similarity will have to be shown to exist where expected.

Review given at the XIIth International Symposium on Multiparticle Dynmaics, Notre Dame, Indiana, June 21-26, 1981.

3

Solid indication for the existence of quarks now comes from hadron spectroscopy as well as from deep inelastic lepton-hadron scattering and from e^+e^- annihilation into jets. Apart from the discovery of the Ω^-, mainly the states involving c- and b-quarks have led to an increased interest in quark bound state spectroscopy, at a time when in deep inelastic scattering leptons appeared to interact with point-like fermions (partons) of fractional charge and when the distribution of jets in e^+e^- collisions turned out to be consistent with that expected for spin 1/2 parent particles.

Furthermore, good evidence now exists for the existence of field particles (gluons) expected to be responsible for the interaction between quarks. From deep inelastic scattering it is known that quarks only carry about half of the nucleon momentum, the remainder is attributed to gluons. In e^+e^- annihilation planar events with a three jet structure are generally interpreted in terms of one of the produced quarks radiating a fast gluon. Finally, the experimental data on $\Upsilon(9.46)$ decay are in agreement with a three gluon decay mode.

In the context of quarks and gluons (partons), three observations have recently been drawing increased attention to soft hadronic collisions:
1. Resonance and particle yields in central and fragmentation regions can be understood from quark combinatorics.
2. Pion production in the nucleon fragmentation region of soft hadron-hadron collisions seems to reflect the valence quark distribution in the nucleon as observed in moderately deep inelastic lepton-nucleon collisions.
3. Quark fragmentation jets from e^+e^- annihilation and deep inelastic collisions seem to resemble soft hadronic particle production in longitudinal, transverse and multiplicity behaviour of the produced hadrons.

These observations lead to the expectation that the parton structure of hadrons also governs soft hadron-hadron collisions. To test this unifying concept and to use its far-reaching consequences to illuminate not only the complicated hadron-hadron collisions, but the (soft) quark fragmentation and recombination in general, is the basis of present experimental and theoretical effort in this field. Before discussing the status of this effort in the later section, we shall first review the three basic observations.

I. BASIC OBSERVATIONS

I.1. Quark Combinatorics

Early data[1] give evidence for a two component picture of inclusive particle and resonance production. The fragmentation component depends on the produced particle or resonance and on the fragmenting incoming particle. The shape of the central component is universal, i.e. does neither depend on the incoming particles nor on the produced particle or resonance.

In particular, the following experimental results have been

obtained:
- suppression of mesons in the proton fragmentation region;
- a ratio $\rho^+ : \rho^0 : \rho^- \approx 2:1:0$ for $\pi^+ p \to \rho$ in the π^+ fragmentation region;
- suppression of $\bar{K}^{*0}$ in π^- fragmentation, suppression of K^{*-} in π^+ fragmentation;
- suppression of ϕ in proton or pion fragmentation;
- more $\pi \to \rho$ than $K \to \rho$;
- a ratio of $(3 + 3\lambda) : (3\lambda) \approx 4-8$ for K^{*+} produced from K^+ and π^+ beams (or charge conjugate).

Furthermore, it has been noted that only of the order of 10% of the pions are produced directly. The largest pion sources are $\rho^{\pm 0}$ and ω^0.

An interesting approach to understand these observations is an extension of the additive quark model[2a] for multiparticle production to inclusive resonance production by Anisovich and Shekhter[2b] and Bjorken and Farrar[2c]. According to this probably too simple model, one quark of the beam particle reacts with one quark of the target to produce a large quark-antiquark ($q\bar{q}$) sea in the center. One or more quarks (or antiquarks) from the center join respectively the beam spectator and the target spectator quarks to produce hadrons in the fragmentation regions. The remaining quarks and antiquarks join to produce hadrons in the central region. SU(6) symmetry is assumed to be broken only by a relative suppression ($\lambda \approx 1/3$) of strange quark production. The probability of quark production in the sea does not depend on spin projection and charge. Furthermore, due to the large quark multiplicity at high energy, the probability of central quark production does not depend on the nature of the incoming quarks. In meson production in the fragmentation region, on the other hand, one spectator is joined by one quark from the sea, so that there the distribution depends on the nature of the spectator.

The quark suppression factor can be measured from K^*/ρ or K^*/ϕ ratios[3]. Values lie between $\lambda \approx 0.15$ at the lower energies to $\lambda = 0.35 \pm 0.6$ from a simultaneous fit (fig.1) to all meson resonances observed at ISR energies[3f].

One of the main assumptions is that a "gas" of quarks and antiquarks with non-correlated spin projections is formed. As a consequence, the number of $q\bar{q}$ pairs with total spin s is proportional to the statistical weight $2s + 1$. This gives an expected ratio 3:1 for ρ/π or $K^*(890)/K$, more generally for mesons containing the same quarks and belonging to the same SU(6) multiplet. The actually observed ratio depends on the contamination from decays of unobserved resonances. The ratio 3:1 has been verified[4] on K, $K^*(890)$ and $K^{**}(1420)$ mesons (which appear in decay processes to a smaller extent and in a better controlled way than π and ρ) from pp and $K^- p$ data.

The probability for a spectator quark to pick up an antiquark (rather than a quark) from the sea has been estimated[5] to $\beta \approx 80\%$, consistently from $\bar{\Lambda}/K_s^0$ and Λ/K_s^0 multiplicity ratios in π^- and K^- fragmentation.

The most detailed application of quark combinatorics is a systematic study of baryon production in the proton fragmentation region[6]. The 240 GeV/c data[3d] suggest that antibaryons can be attributed dominantly to the non-fragmentation type $B\bar{B}$ pair production with local compensation of quantum numbers. In that case, each valence quark of the incident proton either recombines with an antiquark to give an outgoing meson or recombines with other quarks, to give an outgoing baryon. From a consistent description of proton, Λ, Σ, Ξ and Ω^- production in proton fragmentation[3d], these two possibilities turn out to occur in an uncorrelated way for each valence quark of the incident proton, with the respective probabilities of about 40% and 60%[6b].

Furthermore, the probabilities a_i (i= 0, ...3) for the produced baryon to contain i valence spectator quarks can be consistently deduced from the same data constrained by the assumption of uncorrelated valence behaviour to lie in the regions

$$a_0 \lesssim 0.1 \qquad a_1 \lesssim 0.2$$
$$a_2 \approx 0.4 \qquad a_3 \approx 0.4 \ .$$

A strange quark suppression of $\lambda \approx 0.22$ is needed for these data.

We can conclude that pure quark combinatorics can give a consistent description of relative particle yields in central and fragmentation regions. The approach of ref.2b,c) is particularly well suited for the central and meson fragmentation regions , the approach of ref.6) to target fragmentation. The description is global, but it can be taken as a first hint that quarks may play a role also in soft hadronic collisions. If we want to know how, we have to go to more differential distributions.

I.2. Reflection of the Valence Quark Distribution

The analysis of lepton induced reactions has shown that only about half of the momentum of the proton is carried by quarks (q) and antiquarks ($\bar{q}$) and a small fraction by antiquarks alone. As $q(x)=\bar{q}(x)$ at x=0, the antiquark distribution is concentrated at small x (say x $\lesssim$ 0.2, the "sea" region) and the same is expected for gluons which could dissociate into a $q\bar{q}$ pair. The presence of an $\bar{q}$ component in the proton structure function implies that the proton, which primarily consists of three quarks, has dissociated into a state with $q\bar{q}$ pairs if hit by the electron.

The suggestion of Ochs[7] is that the same process of dissociation into partons occurs if the proton fragments into hadrons in the field of another hadron at high energies. The longitudinal momenta of partons will be approximately conserved during the interaction and there is a correspondence between the hadron state and the parton state in the following way: A quark in the parton state outside the sea-region (x $\gtrsim$ 0.3) is a valence quark of a hadron of the final hadron state, and the momentum of the quark contributes to the momentum of the hadron.

In the fragmentation region of the proton, the π^+ can be assumed to be composed of a u valence and a $\bar{d}$ sea quark. Since the latter carries very little momentum, we expect to find a $\pi^+ = |u\bar{d}\rangle$ with similar momentum to that of the u-quark. The same holds for a

6

$\pi^- = |d\bar{u}\rangle$ and the d-quark. As a consequence, the x-distribution of a
pion in the fragmentation region of an incident proton is expected
to be similar to that of the valence quark which it shares with the
proton. Fig.2a-b recalls[8] that the x-distribution of π^+ is indeed
similar to the u-quark distribution u(x) derived from SLAC data on
electron-nucleon deep inelastic scattering. The π^- distribution
agrees with d(x) up to x~ 0.7, and is only slightly above d(x) for
larger x-values.

We conclude that the quantum numbers and the distribution of
the target proton valence quarks can be found back in the particles
produced in the target fragmentation region.

What is the role of the beam quantum numbers? Fig.2c gives a
comparison[9] of the ratio of the inclusive invariant cross sections
for π^+ and π^- production as a function of x in the target fragmentation region for the reactions $\pi^{\pm} p \rightarrow \pi^{\pm} X$ at 16 GeV/c to that
for $p \rightarrow \pi^{\pm} X$ (scaling between 19 and 2000 GeV/c). The ratio R is
considerably larger for 16 GeV/c π^+p than for pp reactions. On the
other hand, R falls below the proton curve for π^-p collisions.

At higher energies, however, the $\pi^{\pm} p$ ratios tend to converge.
The approach of R to a high energy limit is given in fig.2d for
$\pi^{\pm} p$ collisions[10], as a function of $s^{-1/2}$, for several x intervals. The asymptotic limit is consistent with the ratio R for pp
reactions. A similar conclusion can be drawn for $K^{\pm}p$ collisions.
This means that at $p_{lab} \gtrsim 200$ GeV/c, the influence of the proton
valence quarks alone is observed to govern proton fragmentation in
soft hadronic collisions.

A particular significance[7] may be attributed to the value of
the ratio R for $x \rightarrow 1$. In a model in which quarks interact by the
exchange of vector gluons, the helicity of a fast quark (x~1) is
the same as the helicity of the proton. Therefore, by examining
the wave function of the proton
$$|p(\uparrow)\rangle = 2|u(\uparrow)u(\uparrow)d(\downarrow)\rangle - |u(\uparrow)d(\uparrow)u(\downarrow)\rangle + \text{perm.}$$
one finds u/d $\rightarrow$ 5 for $x \rightarrow 1$. Early pp data[8a] were in agreement
with this expectation. More recent data[8b] show (fig.2e) that R≈5
is reached from below at x≈0.8, but that R then suddenly decreases
again at highest x values, so that another regime seems to hold
there.

The second basic observation of valence quark reflection in
proton fragmentation supports the quark recombination picture[11] of
hadron production in the fragmentation region. In the framework of
this picture, a fast meson with small p_t is formed by a two step
process

   i) a quark-antiquark pair is picked out from a proton,
   ii) this pair recombines to form the meson.

Neglecting many-body recombination (expected to contribute at $x \rightarrow$
1, see fig.2e), the inclusive x distribution of a meson M is therefore given by

$$\frac{x}{\sigma_t} \frac{d\sigma}{dx} = \int f_{v\bar{s}} (x_v,x_s) R(x_v,x_s) \delta(\frac{x_v}{x}+\frac{x_s}{x}-1) dx_v dx_s$$

where $f_{v\bar{s}} (x_v,x_s)$ is the probability density of finding simultaneously within the proton two partons v and $\bar{s}$ with respective momentum fractions x_v and x_s (the two-quark structure function) and $R(x_v, x_s)$ is the recombination probability for the v and $\bar{s}$ quarks to form the meson M. The δ-function assures momentum conservation for the process.

Justification[12] for the particular form of $f_{v\bar{s}}$ and $R(x_v, x_s)$ comes from the so-called "valons" (kind of dressed quarks). The distribution of quarks in a valon and of the valons in the proton can be completely determined in shape and normalization by ensuring that they reproduce the quark distribution from not too high Q^2 lepton-nucleon collisions.

In the same quark-picture, the influence of the beam quantum numbers at lower energies and the convergence to a common π^+/π^- ratio above $p_{lab} \sim 200$ GeV/c can be understood from valence quark-antiquark annihilation followed by recombination[9].

A problem that remains is the p_t dependence of the π^+/π^- ratio. In the proton, the two u quarks are expected to be more separated on the average than the u and d quarks. As a consequence, on the average the u quark could be more peripheral than the d quark and the π^+/π^- ratio could fall with p_t at not too large p_t values. Such a behaviour is indeed reported for π^-n reactions[13a] (in a charge symmetric way) at 21 GeV/c, while at 205 and 360 GeV/c the same reactions are consistent with a constant p_t behaviour (see fig.3a). However, for K^+p reactions[13b] at 32 GeV/c, the π^+/π^- ratio increases with p_t, especially for the larger x-values (fig.3b). Indication for such an increase, rather than decrease, is also found in $\pi^\pm p$ data at 147 GeV/c. A systematic study of the p_t dependence would therefore be very important.

I.3 Jet Universality

The decrease in average sphericity $\langle S \rangle$, thrust $\langle 1-T \rangle$ and spherocity $\langle S' \rangle$ with increasing energy, generally interpreted as evidence for jet production, is not only a feature of reactions in which single quark effects are expected to dominate[14], but also of low p_t hadron-hadron collisions[15]. As is shown in fig.4a-c, the average shape of the hadronic system is the same in all three types of collision at given hadronic energy, as is its change with energy.

The energy dependence of the shape of the sphericity distribution itself[15a,c] is shown in fig.4d,e. As for e^+e^- collisions, one observes a change from a dip at S=0 at low energy to a sharp peak at S=0 at higher energies. Very good agreement is observed between the S distribution in K^-p at 110 GeV/c ($\sqrt{s} = 14.8$ GeV) with the PLUTO result at 17 GeV, and the K^+p data at 70 GeV/c ($\sqrt{s} = 11.5$ GeV) with the TASSO distribution at 13 GeV (for the curves LPS and FF see below). Diffractive events contribute only at the most "jet-like" values, i.e. low S. The agreement between the shape of the K^-p and the leptonic distributions is improved when the diffractive events are removed. The only difference in the shapes of S distributions in K^-p and in leptonic data is a peaking at lower S values in the K^-p data at 16 GeV/c, when compared with the SPEAR data. This probably results from the problems in defining the sphericity axes in low $\sqrt{s}$ e^+e^- events.

The normalized 70 GeV/c K^+p rapidity distribution, evaluated with respect to the thrust axis[15c] is compared to e^+e^- results at 13 GeV/c (TASSO)[16] in fig.5a. At similar energy, the two rapidity distributions agree remarkably well and it can be expected from the insert (and from old results on hadronic rapidity plateaus) that hadron-hadron and e^+e^- data show a similar $\ln\sqrt{s}$ increase of the plateau hight and, above all, the same plateau hight at the same energy!

Also the normalized p_t^2 distributions relative to the sphericity axis (fig.5b) show agreement at 13 GeV/c. There may be a small indication of a larger cross section in the high p_t tail for the e^+e^- results, but this can be understood from the higher e^+e^- energy. (For LPS and FF see below.)

The conclusion of jet universality is further supported from a comparison[15] of the energy dependence of $\langle n \rangle$, as well as of $\langle p_t \rangle$ and $\langle p_{//} \rangle$ relative to the thrust axis (fig. 5c). Again, the hp and e^+e^- data have essentially the same values and the same energy behaviour. In particular, the rise in $\langle p_t \rangle$ with E_{cm} felt to be a characteristic feature of single quark jets is in fact also a feature of hadronic low p_t particle production.

At this stage we may ask whether the agreement between the e^+e^-, lh and hh data is e.g. due to longitudinal phase space, or has any more fundamental dynamical origin. In ref.15c the sphericity, thrust, spherocity and other distributions are compared to mere Longitudinal Phase Space (LPS)[17a] and the Field and Feynman (FF) parametrization of quark-parton jets[17b]. In all cases, both the FF parametrization as well as LPS more or less describe the data. Furthermore, for hadronic reactions sphericity, thrust and spherocity axes agree with the beam direction. One may conclude that

(i) FF is a somewhat complicated way to parametrize longitudinal phase space,

(ii) jet universality turns out to reduce to the equal p_t distributions shown here and to the equal average multiplicities[18], the rest follows from independent emission + conservation laws.

The point is, that this holds equally well for hadronic, deeply inelastic and e^+e^- hadron production. In all cases the same mechanism is at work, governing multiplicities and transverse momenta in an identical way.

What is even more surprising, is that the p_t development of e^+e^- and $\nu(\bar\nu)N$ multiparticle production with energy can be understood from transverse cut phase space up to $\sqrt{s} \approx 16$ GeV. Fig.6a.b shows[19] the $\sqrt{s}$ dependence of $\langle p_t \rangle$ for e^+e^- collisions and the W dependence of $\langle p_t^2 \rangle$ for current fragments of $\nu(\bar\nu)N$ collisions. The curves are from Monte Carlo calculations with a matrix element

$$M \propto \prod_{i=1}^{N} \exp(-Ap_{ti}) \quad \text{with } A=3.5$$

and with a Gaussian multiplicity distribution centered at

$$\langle N \rangle = 3.69 + 0.06 \exp[1.92(\ln 4s)^{1/2}].$$

Up to about 16 GeV there is little need for anything else.

As shown by M. Bardadin-Otwinowska[20], hadron-hadron data also show the increase with energy of the tail of the distribution in $p_{t,in}$ (p_t component in direction corresponding to the second largest eigenvalue of the momentum tensor). I find it important that this increase can be observed also in hadron-hadron colli-sions, the fact that it is not as fast there as in e^+e^- and νN col-lisions is not surprising. If two chains contribute to hadronic particle production, according to fig.6a p_t distributions have to be compared at corresponding chain energies and not at the total cms energy.

Fig.6b shows $\langle p_t^2 \rangle$ for lepton-hadron and hadron-hadron colli-sions as a function of $W^2=s$ and $W^2=s/3$, respectively. Clearly, the hadron data follow the lepton data when plotted as a function of $W^2=s/3$. Such a shift in energy is not necessary for $\langle n \rangle$, $\langle p_{//} \rangle$ or sphericity distributions, since the two chains add in the longitudi-nal direction.

If quark fragmentation is indeed governed by longitudinal phase space, it will be interesting to look for the trivial differ-ences expected from the quantum numbers or excitation modes of the fragmenting quarks, as well as for (non-trivial) flavour correla-tions within the fragmentation jet.

One difference of the hadron system produced in e^+e^- annihila-tion and pp collisions is that the baryon number is zero in the first case and two in the second. In other words, there are more interacting valence quarks in pp collisions than in e^+e^- hadron production. A CERN-Bologna-Frascati Collaboration[21] using the CERN Split-Field Magnet at $\sqrt{s}=62$ GeV has attempted to remove the expected effects of baryon-number conservation by selecting events with a leading proton and redefining the fractional variables of the particle after removal of the proton energy.

The pp data (normalized to the mean charge multiplicity) are compared to (normalized) e^+e^- data from the Tasso Collaboration[22] at three different energies E_{had}. The agreement between the two distributions at all three E_{had} values, both in shape and absolute value (mean charged multiplicity), suggests that the mechanism for transforming energy into particles in these two processes must be the same. This would be surprising if the fragmentation process would not be governed by longitudinal phase space, since exclud-ing the leading proton means excluding the leading parton plus its com-panion. To be fair, one would have to exclude the leading quark and its companion in e^+e^- annihilation, as well.

The 147 GeV/c h^+p collaboration studied the effect on $\langle S \rangle$ and $\langle T \rangle$ when removing the leading particle. As expected, $\langle S \rangle$ and $1-\langle T \rangle$ increase, but the available hadron energy E_{had} decreases. Both values are compared to the e^+e^- and νN data at the corresponding values of E_{had}. Removing the proton does not improve the agree-ment. I would suggest to forget the approach of excluding the pro-ton energy and to look for expected differences (possibly in corre-lations) rather than to exclude them before they are shown to exist.

As shown by V. Kistiakowsky[23] on π^-p data at 147 GeV/c, the z distribution of the fastest and second fastest hadrons in the beam hemisphere are strikingly similar to those for hadrons produced in deep inelastic lepton interactions and to FF (and therefore also to LPS) predictions (fig.7a,b). The same holds for the second fastest particle and the fastest negative particle in target fragmentation. As expected, the distribution for the protons in the target hemisphere (fig.7c) are significantly different, however.

Another possible approach seems to me to select certain events where the jet is expected to derive from $q\bar{q}$ separation as in e^+e^- jet production. A possible laboratory for this study is meson or photon diffraction dissociation[24]. The diffractive dissociation can be assumed to be a two-step process. First, the meson dissociates into a $q\bar{q}$ pair, then this $q\bar{q}$ system hadronizes in the same manner as hadronization takes place of neutral quark-antiquark pairs during e^+e^- annihilation to two jets of hadrons. Diffraction dissociation may have the enormous advantage over e^+e^- annihilation that the $q\bar{q}$ pair is known and not a mixture of $u\bar{u}$, $d\bar{d}$, $s\bar{s}$, $c\bar{c}$ or $b\bar{b}$. This may provide the unique opportunity to investigate jets of known flavour!

We conclude that in first approximation, fragmentation jets are equal in e^+e^-, νN and soft hadronic multiparticle production. The jet universality can be reduced to equal p_t and multiplicity distributions and can further be described by transverse cut phase space (LPS). As a second approximation, one has to continue to look for differences from LPS in flavour correlations and for differences due to different type parent partons (diquark, quark, strange quark, glue).

The fact that jets produced in e^+e^- annihilation and deep inelastic lepto-production are very similar to those produced along the beam direction in ordinary hadron-hadron collisions has led to the assumption of parton fragmentation as a common underlying dynamical mechanism. In the quark fragmentation view[25], inelastic (non-diffractive) scattering is dominated by events where one valence quark of the baryon with low momentum interacts with the other hadron and is fixed in the central region. The remaining diquark system will carry almost all the original baryon momentum and subsequently fragment into hadrons in the baryon fragmentation region. The rapidity density of the fragmentation chain is universal, i.e. it depends only on the nature of the system at the chain end and not on the nature (hard or soft) of the process which produces them. The inclusive one-particle distribution of hadron H in the fragmentation region of hadron h is given by the convolution integral

$$\frac{1}{\sigma_t} \frac{d\sigma}{dx} = \sum_q \int_x^1 dydz \, f_q^h(y) \, D_q^H(z) \, \delta(yz-x)$$

between the probability distribution $f_q^h(x)$ of the constituents of h to carry momentum fraction x and the fragmentation $D_q^H(z)$ of constituent q into the hadron H, as obtained from leptonic reactions.

11

The important assumption is, that the mechanism responsible for hadronic interactions favours configurations of valence quarks different from those measured in lepto-production. Only configurations in which the diquark carries almost all the momentum are assumed to be responsible for pion production in the proton fragmentation region. This amounts to

$$f_q^h(x) = k_q^h \, \delta(x-x_o)$$

with k being a statistical factor corresponding to the additive quark model and $x_o \approx 1$. So, hadronic spectra are given by the fragmentation functions $D_q^H(x)$ in a parameter free form.

II. SELECTED APPLICATIONS

The recombination and fragmentation pictures have been designed to describe single particle distributions from two different points of view. Attempts to show the duality or complementarity of the two views have been started and should be continued. Rather than trying to prove one of the pictures and disprove the other, we shall here use their complementarity to illuminate different aspects of the data.

II.1 Target Fragmentation

II.1a. Factorization

Of particular interest in connection with the influence of the valence spectators are various types of quantum number correlations between the two different fragmentation regions and within one fragmentation region itself. Long range correlation of pions, each coming from a different one of the two incident protons, has been measured[26,86] in the form of the two pion correlation function

$$Q(x_1,x_2) = \frac{N_{12}(x_1,x_2) \cdot \sum_{x_1}\sum_{x_2} N_{12}(x_1,x_2)}{\sum_{x_1} N_{12}(x_1,x_2) \cdot \sum_{x_2} N_{12}(x_1,x_2)}$$

for pp $\rightarrow \pi\pi X$ at $\sqrt{s}$ = 44.7 and 62.3 GeV (fig.8). Over most of the x range, the $\pi^+\pi^+$, $\pi^+\pi^-$ and $\pi^-\pi^-$ data are essentially uncorrelated (Q$\approx$1), in agreement with factorization of the two fragmentation processes. This is expected for gluon exchange (dashed lines), but not for valence or sea-quark exchange (solid lines).

II.1b. Correlations

For comparison to single particle production in the proton fragmentation region, the ratio R of π^+ to π^- production can be studied in association with various triggers. With a π^+ trigger, the spectator system ($u_v d_v d_s$) should produce equal amounts of π^+'s and π^-'s at fixed $\tilde{x}_\pi = x_\pi/(1-x_{tr})$. This is confirmed by the data[27] in fig.9: Whereas the untriggered π^+/π^- ratio rises with increasing x, the ratio for a π^+ trigger is compatible with unity for $x \gtrsim 0.4$. In comparing this associated ratio to the π^+/π^- ratio measured in charged current $\bar{\nu}p$ collisions where the spectator system is the same, the agreement is indeed striking (see insert fig.9a).[28]

As shown by B. Buschbeck[28], a strongly enriched ud valence system can also be obtained when requireing a fast $\bar{K}^0$ in relatively low energy K^-p collisions. In this case, one of the proton u quarks is absorbed through annihilation by the $\bar{u}$ quark in the K^- beam. Fig.9d shows that indeed, the associated π^+/π^- ratio is unity. A similar conclusion can be drawn[28] from forward Λ's.

For a π^- trigger, the spectator system is $(u_v u_v u_s)$ and a strong increase of the ratio R is expected with increasing x. Also this increase is indeed observed in fig.6a,b, as well as in νp interactions[29] having a similar spectator system. While these results naturally follow from the quark recombination picture, calculations[30] show in fig.9c that the same trend is expected from the quark fragmentation view.

II.1c. Diquark Fragmentation

According to the fragmentation model[25], pion production in the proton fragmentation region can be written in terms of the diquark fragmentation functions D_{qq}^{π} as

$$\frac{1}{\sigma_t} \frac{d\sigma^{p \to \pi^{\pm}}}{dx}(x) \approx \frac{1}{3} D_{uu}^{\pi^{\pm}}(x) + \frac{2}{3} D_{ud}^{\pi^{\pm}}(x).$$

From isospin invariance follows $D_{ud}^{\pi^-} = D_{ud}^{\pi^+}$, so that three independent functions remain:

$$H_1 = D_{uu}^{\pi^+}, \quad H_2 = D_{ud}^{\pi^+} = D_{ud}^{\pi^-} \quad \text{and} \quad H_3 = D_{uu}^{\pi^-}.$$

One can expect $H_1 > H_2 > H_3$ for large x since in H_1 both u-quarks can contribute to the creation of a π^+, while in H_2 only one quark contributes, and in H_3 none of the two u-quarks contribute to π^- production.

Furthermore, if baryon production is dominant in the diquark fragmentation, the approximate relation $2H_2 \sim H_1 + H_3$ is valid[25]. For the pion cross section in proton fragmentation then follows

$$\frac{1}{\sigma_t} \frac{d\sigma^{p \to \pi^+}}{dx}(x) \approx \frac{2}{3} H_1(x) + \frac{1}{3} H_3(x) \quad \text{and} \quad \frac{1}{\sigma_t} \frac{d\sigma^{p \to \pi^-}}{dx}(x) \approx \frac{1}{3} H_1(x) + \frac{2}{3} H_3(x).$$

As an example, K^-p data at 70 GeV/c[31] are compared to νp and $\bar{\nu}p$ data[32] in fig.10. The following observations can be made:
- $H_1(x)$ from νp data indeed agrees with $H_1(x)$ from K^-p data.
- $2H_2(x)$ from $\bar{\nu}p$ data agrees with $H_1(x)$, giving support to the above approximation.
- $H_3(x)$ from νp data is much smaller than $H_1(x)$. It is compatible with zero for the K^-p data (not shown).

Good agreement between diquark fragmentation functions from $\nu(\bar{\nu})N$ and soft hadronic collisions is also shown by N. Schmitz[29] and B. Buschbeck[28].

II.1d. At what Q^2?

Recently, the standard recombination model was significantly improved by the introduction of valence quark clusters, called "valons"[12]. The parton distribution in the hadron is understood as a convolution of the Q^2 independent distribution of valons in the hadron and the Q^2 dependent internal structure of valons. With this concept a Q^2 dependent continuous link is created between the valence quarks seen at high Q^2 and the constituents seen at low Q^2.

The (flavour dependent) valon distribution was obtained[33] from high Q^2 deep inelastic data. Provisional internal structure functions for the (non-strange) valons were estimated by Hwa[12] using not too deeply inelastic DIS data (Q^2=9 GeV^2). These functions where shown to describe existing hadronic data without any further parameter, but the question remained at what effective Q^2 deep inelastic data should be used to predict soft hadronic behaviour.

In a recent paper[34], $K^+p \rightarrow \pi^-$ data were used to extract the internal structure of valons from hadronic collisions directly (fig.11a). The same parameters can be used to predict $K^+p \rightarrow \pi^+$ correctly (fig.11b). The prediction for νW_2 is shown in fig.11c and follows the trend of deep inelastic data very well, indicating an effective Q^2 of somewhat smaller than 9 GeV^2.

We conclude that proton fragmentation is well behaved and understood, so that one can go further and apply the same ideas to extract new information from meson fragmentation.

II.2 Meson Fragmentation

II.2a. Single Quark Fragmentation Functions

The 70 GeV/c K^-p data were used to extract $D_u^{\pi^+}$ and $D_u^{\pi^-}$ from the π^+, π^- and π^0 distribution in the K^- fragmentation region[31] (see fig.12). The contamination from charged kaons has been estimated from K^n production and has been removed. The study of π^- production is restricted to the region $x \leq 0.7$ to avoid contamination from leading K^-. The authors claim rather good agreement with the $\nu(\bar{\nu})p$ data[17b] with W>4 GeV and with the Field and Feynman D_u^π functions[17b]. This is certainly true for the comparison of $D_u^{h^-}$. Note that the difference between νp and hp data for $D_u^{h^-}$ is smaller than between (the isospin symmetric) $D_u^{h^-}$ and $D_u^{h^+}$ from $\nu(\bar{\nu})p$ data. $D_u^{\pi^+}$ seems consistently flatter for hp collisions, but one has to take into account that
- in the $\nu(\bar{\nu})p$ data it is only at high values of z (where the contribution of the sea is negligible) that the functions D_q^h represent the fragmentation of a pure u(d) quark (in addition they represent production of hadrons $h^\pm$ rather than just $\pi^\pm$);
- it is not clear that diffraction dissociation (Q, L ..) has been sufficiently removed. There may be a contribution $K^- \rightarrow \pi^-$ ...below x=0.7.

Similar results have been obtained for 110 GeV/c K^-p and 70 GeV/c K^+p collisions[35].

With the above limitations in mind, one can conclude that one can extract the single quark fragmentation function to a very good approximation from hadronic collisions.

II.2b. Meson Structure Functions

Alternatively, the recombination picture[11,12] can be used to determine the valence quark distribution in mesons[36], for which there is no direct information from deep inelastic lepton interactions. The results can be given in terms of the power n of the (1-x) distribution of the valence quarks. For a pion it follows from charge conjugation and isospin invariance that the quark distribution function is the same for both valence quarks. For a kaon the situation is expected to be non-symmetric. A value of n=1.0±0.1 has been obtained for the pion structure function. Furthermore, the power n is indeed larger than unity for the non-strange valence quark in the kaon, while it is smaller than unity for the strange valence quark. These results are compatible with those extracted via the Drell-Yan model from μ-pair production[37]. One can conclude that meson valence quarks are harder than those in the nucleon and that strange valence quarks are harder than non-strange ones.

II.3 Meson Resonances

Of large interest in connection with partons is inclusive production of resonances. One can assume that resonances are more abundantly and more directly produced than pions and kaons, represent a larger variety of quantum numbers and allow for conclusions about their production from their decay density matrix.

II.3a. The φ Meson

In the quark picture, the φ is of central interest because of its hidden strangeness quantum numbers. In particular, it has a valence quark in common with an incoming kaon, but not with an incoming pion, proton or antiproton. The x-distribution is therefore flatter for kaon induced φ production than in the other cases[38].

Furthermore, $K^-p \rightarrow \overline{K}^{*0}$, $K^+p \rightarrow K^{*0}$ and $Kp \rightarrow \phi$ all have the same x dependence[39] and are definitely more forward peaked than Kp $\rightarrow \rho_0$. This is a direct consequence of the fact that the strange quark in the incident kaon is harder than the nonstrange quark. The ratio of φ and ρ_0 is plotted in fig.13a for K^-p at 32 GeV/c[38b] and compares well to the ratio of s and $\bar{u}$ distribution functions.

II.3b. Comparison to Models

Fig.13b gives the forward ρ^0 distribution from $\pi^{\pm}$ p collisions at 147 GeV/c[40]. The curves shown correspond to a power law fit[41] (DCR), a recombination fit (QRM)[17b] and fragmentation predictions (QFM) with Field and Feynman[17b] (dashed) and Lund[25] (dot-dashed) parametrizations of the fragmentation functions. The Lund prediction works very well, so do DCR and QRM. In the latter two cases, the distribution function comes out consistently slightly flatter than for pions. This observation does not change after exclusion of diffractive events and should be checked with good statistics.

II.3c. Multiparticle Fragmentation vs Resonances

In fig.14, the x dependence is shown for the production of $K^n \pi^{\pm}$ pairs, for three effective mass intervals of the pairs[42]. The solid line is drawn on the $K^n \pi^-$ data of fig.14b and repeated in fig.14a. It coincides with $K^n \pi^+$ in the backward hemisphere and even describes low mass $K^n \pi^+$ pairs up to $x \approx 0.7$. However, pairs containing the $K^{*+}(890)$ resonance are produced with a much flatter x dependence than those with lower or higher mass. This observation is confirmed by a resonance to background ratio strongly increasing with x even within the K^* mass interval $0.85 < M(K^n \pi) < 0.93$ GeV (not shown), and by the difference of the invariant cross sections for $K^{*+}(890)$ and background in the same mass interval in fig.14c. While a power $n \approx 1.5$ is obtained for $K^n \pi^{\pm}$ in the same mass interval, $n = 0.30 \pm 0.06$ for $K^{*+}(890)$ (after exclusion of quasi-two-body and diffractive channels).

The observation of $n(K^+ \rightarrow K^{*+}) < n(K^+ \rightarrow K^n \pi^+)$ is in disagreement[43] with the Kuti-Weisskopf (or longitudinal phase space[17a]) model of the valence quarks in a hadron being uncorrelated except for energy-momentum correlation. From uncorrelated valence quarks, powers $n_1 > n_2$ are expected for systems inheriting one or two valence quarks from the incident hadron, respectively. For our case this would mean

$$n(K^+ \rightarrow K^{*+}) \approx n(K^+ \rightarrow K^0 \pi^+) < n(K^+ \rightarrow K^{*0}) \approx n(K^+ \rightarrow K^0 \pi^-).$$

The observation of fig.14 is in clear contradiction to this expectation, as is the above mentioned (sect.II.3a) similarity of K^{*+} and K^{*0} production. Both these features can be understood if only one valence quark (the $\bar{s}$ quark in the K^+) is responsible for K^0 or K^{*0} production, the remaining pion being produced from the sea. This correlation result would fit well into the quark fragmentation picture, where one of the valence quarks is held back at small $|x|$ (see ref.44).

This result gets support from observations for proton fragmentation in the same experiment[42]. It is important enough to be checked with good statistics at higher energies. Inspite of the exoticity of the K^+p state, a lab momentum of 32 GeV/c cannot yet be considered asymptotic, since (for a unknown reason) the π^+/π^- ratio in the proton fragmentation region[9] is still different from that of pp collisions.

II.3d. The Density Matrix

Fig.15a,b shows experimental results for the density matrix in the Gottfried-Jackson frame for $K^+p \rightarrow K^{*0}X$ at 32 and 70 GeV/c[45] as a function of t and M^2/s. For both reactions, there is a tendency for ρ_{oo} to decrease with $|t|$ and increase with M^2/s. Distributions like these should be a challenge to the quark pictures.

Model calculations based on SU(6) in the framework of the recombination picture[46] lead to a prediction of $\rho_{oo} \approx 1/3$ and $\rho_{1-1} \approx 0$ in the transversity frame, for intermediate x values. The data are compared to this prediction in fig.15c. Even there may be indication for small systematic deviations and for a p_t depedence not expected from the calculations, rough agreement can be observed.

II.4 Hyperon Polarization

Non-zero Λ polarization is known to persist up to high energies since 1976 from pBe collisions at 400 GeV/c[47a] and has been observed in $\pi^- p$ collisions at NAL[47b] and pp collisions at ISR[47c]. This observation cannot be explained from the Regge model since at these energies RPR terms dominate and these terms do not give rise to polarization[48]. Non-zero polarization has also been observed in ep scattering[48]. In fig.16a the pp and ep points show an increasing polarization for $p_t > 0$. Furthermore, from scattering off deuterium and other nuclei, no difference is found in polarization between proton and neutron targets[49].

The best recent data come from the Michigan-Minnesota-Rutgers-Wisconsin Collaboration[50] (see fig.16b,c). One can conclude that:
- Λ^o's produced in hh, hA and ep scattering are polarized transverse to the production plane, along the $(\vec{p}_\Lambda \times \vec{p}_p)$ axis.
- This polarization is probably independent of the beam energy, the projectile type and weakly dependent on the x value of the Λ^o.
- The polarization increases linearly with the transverse momentum of the Λ^o.
- $\bar{\Lambda}$ and protons are not polarized.
- Ξ^o, Ξ^- have the same polarization as the Λ^o.
- Σ^+ has the same polarization as Λ^o in magnitude but with opposite sign.

A semi-classical model for basically soft Λ^o-production able to explain the observed polarization effect is suggested in the quark fragmentation picture[51]. In this picture, a diquark continues forward as a unit after the collision and a string-shaped colour dipole field is stretched between the diquark and the central collision region. This field can break up by the production of quark-antiquark pairs (as in e^+e^- hadron production). A Λ^o-particle can be formed if an $s\bar{s}$-pair is produced in the field of a (ud)-diquark (of isospin and spin I=S=0), so that the spin of the Λ^o is determined by the spin of the s-quark.

The transverse momentum p_t of the Λ^o with respect to the beam direction is made up of two contributions, the transverse momentum $\vec{q}_t$ of the diquark (the direction of the field string) and the (locally conserved) transverse momentum $\vec{k}_t$ of the s quark with respect to the string direction. A pair of massless quark-antiquarks can be produced point-like, but massive quarks have to be produced at a certain distance from each other. Therefore, the pair will obtain an orbital angular momentum perpendicular to the string; this is assumed to be compensated by the spin of the $s\bar{s}$-pair. In a Λ sample of definite p_t, we obtain an enhanced number of events where $\vec{k}_t$ and $\vec{p}_t$ point in the same direction. So the observed effect is explained by a sort of trigger bias. The curves in fig.16a show the model prediction and its upper and lower limits (without inclusion of the effect of Λ production via Σ^o and Σ^*).

A somewhat similar picture has recently been developed[52] in the framework of the recombination model. Data for Λ, Σ^+, Ξ and $\bar{\Lambda}$ are consistent with the observation that slow partons preferentially recombine with their spins down in the scattering plane, while fast partons recombine with their spins up. Here, the polarization arises via Thomas precession of the quarks' spin in the recombination process. A prediction for the Λ polarization at fixed x vs p_t and at fixed p_t vs x is shown to be in good agreement with the data of ref.50) in figs.17a,b. Also this picture can be extended to e^+e^- annihilation and deep inelastic scattering.

A particularly important test is $K^- \rightarrow \Lambda$ (or $K^+ \rightarrow \bar{\Lambda}$) in the forward direction. In this case, all the asymmetry resides in the leading s (or $\bar{s}$) quark and a positive asymmetry is expected. As shown in fig.17a, a non-zero polarization is indeed observed in the forward direction for $K^-p \rightarrow \Lambda$ up to 32 GeV/c[53] and for $K^+p \rightarrow \bar{\Lambda}$ up to 70 GeV/c.[54] Since the sign convention is opposite to that of fig.16b,c, the asymmetry is indeed opposite to that of the Λ there. What is surprising, is the large absolute value of P. Again, a high statistics K^+p experiment at higher energy will be welcome.

Of special interest in this context is, furthermore, the prediction for polarization of baryons (and vector mesons) in e^+e^- annihilation by Bartl et al.[55] and the explanation of hyperon polarization by multiple scattering of the strange quark by Szwed[56].

We conclude that hyperon polarization, which cannot be explained by the triple Regge model, may find an explanation from the quark composition of incident and produced particles.

III. CONCLUSIONS

With the help of quark combinatorics one can get to a consistent understanding of relative particle yields. This first basic observation is taken as a hint that partons may play a role also in soft hadronic reactions.

The similarity of pion production in the nucleon fragmentation region to the valence quark distribution in the nucleon as measured from deep inelastic collisions suggests that valence quarks are governing pion production in nucleon fragmentation. This observation has led to a recombination and valon picture of hadron-hadron collisions. Transverse momentum dependence of π^+/π^- ratio is expected to give information on the geometrical correlation between valence quarks, but needs further study.

Particle production in soft hadronic collisions shows jet features very similar to those in e^+e^- and lepton-hadron collisions. This third observation has led to the quark fragmentation picture. The jet similarity boils down to equal transverse momentum and multiplicity distribution plus independent emission (LPS). Future work will have to aim at flavour correlations and at the study of differences where they are expected to show up.

Particle production and correlation in proton fragmentation is well understood in terms of these two pictures. Extracted di-quark fragmentation functions are identical to those from deep inelastic neutrino scattering. The effective Q^2 for soft hadronic collisions can be extracted from valon model fits and is below 9 GeV^2.

Application of the two pictures to meson fragmentation allows to extract meson structure functions similar to those obtained from μ-pair production on one hand, and quark fragmentation functions similar to those obtained from deep inelastic neutrino scattering on the other. Differences in the x distribution for K^* and non-resonant $K\pi$ background point in the direction of a valence quark correlation as postulated by the fragmentation models.

Differential distribution for resonances start to become available at high energies. They allow to check the results on more directly produced particles, to compare a larger variety of quantum numbers and to challenge models on their decay density matrix.

Hyperon polarization at high energies, not understood from triple Regge exchanges, seems to be explained from both the recombination and fragmentation pictures of hadron production.

We conclude that a new field has evolved in the last 3 years and that partons either do play a role in soft hadronic reactions or we have to reconsider the conclusions drawn from e^+e^- annihilation and neutrino reactions. Work to be done on the theoretical side is on a solid foundation of the two pictures as well as on an explanation of the apparent complementarity of the two views. On the experimental side, high statistics is needed on hydrogen and on higher nuclei at p_{lab} > 200 GeV/c, in particular from strange meson beams in combination, with good momentum resolution and particle identification. This will allow to study flavor correlations and resonance production more conclusively than was possible until now and, in particular, to isolate the influence of longitudinal phase space in quark fragmentation.

REFERENCES

1) K. Böckmann, "Inclusive Vector Meson Production and Hadron Structure", Symp. on Hadron Structure and Multiparticle Production, Kazimierz 1977.
W. Kittel, "Inclusive Resonance Production and Fireballs", VIII Int. Symp. on Multiparticle Dynamics, Kaysersberg 1977 and "Soft Hadronic Interactions-Review of Three Basic Observations", Lectures given at the Europhysics Study Conf. on Partons in Soft Hadronic Processes, Erice 1981.
A. Zieminski, "Multiparticle Production", EPS Conf. on Particle Physics, Budapest 1977.

2a) J.J.J. Kokkedee and L. Van Hove, Nuovo Cimento 42A (1966) 711.
H. Satz, Phys.Letters 25B (1967) 220 and Phys.Rev.Letters 19 (1967) 1453.

b) V.V. Anisovich and V.M. Shekhter, Nucl.Phys. B55 (1973) 455.

c) J.D. Bjorken and G.R. Farrar, Phys.Rev. D9 (1974) 1449.

3a) K. Böckmann et al., Nucl. Phys. B166 (1980) 284;

b) V. Blobel et al., Phys. Letters 48B (1974) 73 and Phys. Letters 59B (1975) 88.

c) V.V. Ammosov et al., Yad. Fiz. 24 (1976) 59;

d) M. Bourquin et al., Z. Physik C5 (1980) 275;

e) A. Suzuki et al., Nuovo Cim. Letters 24 (1979) 449.

f) D. Drijard et al., "Production of Vector and Tensor Mesons in Proton-Proton Collisions at √s=52.5 GeV", CERN/EP 81-12;

g) B. Ghidini et al., Phys. Rev. Letters 68B (1977) 186.

4) V.V. Anisovich, M.N. Kobrinsky and J. Nyiri, Phys. Letters 102B (1981) 357.

5) Yu.M. Shabelski, "Production of Projectile Fragments in Additive Quark Model; Inelasticity Coefficients and Probabilities of Inelastic Charge Exchange", Leningrad 1980. Data from R.T. Edwards et al., Phys.Rev. D18 (1978) 76.

6a) J. Kalinowski, S. Pokorski and L. Van Hove, Z. Physik C2 (1979) 85.

b) L. Van Hove, "The Role of Valence Quarks in Proton Fragmentation", Ref. TH 2997-CERN, 21 Nov. 1980.

7) W. Ochs, Nucl.Phys. B118 (1977) 397.

8a) J. Singh et al., Nucl.Phys. B140 (1978) 189.

b) G.J. Bobbink, "Correlations Between High Momentum Particles in Proton-Proton Collisions at High Energies", PhD Thesis, Univ. Utrecht 1981.

9) B. Buschbeck, H. Dibon, H.R. Gerhold and W. Kittel, Z. Physik C3 (1979) 97.
B. Buschbeck, H. Dibon and H.R. Gerhold, Z. Physik C7 (1980) 73.

10) M.M. Schouten et al., "Approach to Scaling of π^+/π^- Ratios in Target Fragmentation from $\pi^+/K^+/pp$ Interactions at 147 GeV/c", Nijmegen preprint 1981.

11) K.P. Das and R.C. Hwa, Phys.Letters 68B (1977) 459.
J. Ranft, Phys.Rev. D18 (1978) 1491.

L. Van Hove, Lectures at the XVIII. Int. Universitaetswochen fuer Kernphysik, Schladming, Acta Physica Austriaca, Suppl. XXI (1979) 621.

12) R.C. Hwa, Phys.Rev. D22 (1980) 759 and 1593.

13a) H. Abramowicz et al., Z. Physik C7 (1981) 199;
  b) I.V. Ajinenko et al., Z. Physik C4 (1980) 181.

14) Ch. Berger et al., Phys.Letters 78B (1978) 176 and 81B (1979) 410.
P.C. Bossetti et al., Nucl.Phys. B149 (1979) 13.
K.W.J. Barnham et al., Phys.Letters 85B (1979) 300.

15a) R. Göttgens et al., Nucl.Phys. B178 (1981) 392.
  b) Ph. Herquet et al., "Properties of Jet-Like Systems Observed in π^+p, K^+p and pp Interactions", Proc. XVth Renc. de Moriond 1980, Vol. I, edited by J. Tran Than Van, R.M.I.E.M. Orsay, p215.
  c) M. Barth et al., "Jet-Like Properties of Multiparticle Systems Produced in K^+p Interactions at 70 GeV/c", CERN/EP 81-44 and J. Beaufays, this Symposium.
  d) J.M. Laffaille et al., "Energy Dependence of Transverse Momentum Invariant Distribution of Pions and Neutral Kaons in K^-p Interactions between 14.3 and 70 GeV/c", DPh PE 80-16.

16) R. Brandelik et al., Z. Physik C4 (1980) 87.

17a) L. Van Hove, Rev. Mod. Phys. 36 (1964) 655;
E.H. de Groot and T.W. Ruijgrok, Nucl. Phys. B101 (1975) 95.
  b) R.D. Field and R.P. Feynman, Nucl.Phys. B136 (1978) 1.

18) A. Wróblewski, "Multiplicities in e^+e^-, lh and hh Collisions", this Symposium.

19) A.B. Clegg and A. Donnachie, "A Description of Jet Structure by p_t-limited Phase Space", M/C TH 81/11.

20) M. Bardadin-Otwinowska, "Comparison of p_t Distributions of Hadrons in e^+e^-, νN and K^-p Collisions", this Symposium.

21) M. Basile et al., Phys.Letters 92B (1980) 367, Phys. Letters 95B (1980) 311, Nuovo Cimento 58A (1980) and Nuovo Cim. Letters 29 (1980) 491.

22) R. Brandelik et al., Phys.Letters 83B (1979) 261 and 89B (1980) 418.

23) V. Kistiakowsky et al., "Comparison of 147 GeV/c π^-p Low Transverse Momentum Hadron Production with Deep-Inelastic Leptoproduction", this Symposium.

24) S.P. Misra, A.R. Panda and B.K. Parida, Phys.Rev. Letters 45 (1980) 322.
J. Randa, Phys. Rev. D23 (1981) 1662.
J. Gunion, "Unification of Low-p_t Models and their Relation with other Approaches", Europhysics Study Conf. Partons in Soft-Hadronic Processes, Erice 1981.

25) B. Andersson, G. Gustafson and C.P. Peterson, Phys.Letters 69B (1977) 221 and Phys.Letters 71B (1977) 337.
B. Andersson et al., Nucl.Phys. B135 (1978) 273.
A. Capella et al., Phys.Letters 81B (1979) 68.
A. Capella, U. Sukhatme, J. Tran Thanh Van, Z. Physik C3 (1980) 329.

A. Capella and J. Tran Thanh Van, Phys. Letters 93B (1980) 146.

G. Cohen-Tannoudji, A. El Hassouni, J. Kalinowski and R. Peschanski, Phys.Rev. D19 (1979) 3397.

G. Cohen-Tannoudji, A. El Hassouni, J. Kalinowski, O. Napoly and R. Peschanski, Phys.Rev. D21 (1980) 2699.

26) G.J. Bobbink et al., Phys.Rev.Letters 44 (1980) 118.

27a) E.A. De Wolf et al., "Two-Particle Production in K⁻p Interactions at 32 GeV/c. Tests of the Quark-Parton Model?" EPS Conf. on High Energy Physics, Geneva 1979.

b) M. Barth et al., Z.Physik C7 (1981) 187

c) F. Erné, "Correlations Between Fragmentation Baryons and Mesons Produced in High-Energy pp Interactions" XVIth Rencontre de Moriond, Les Arcs 1981.

28) B. Buschbeck and H. Dibon, "Fragmentation of u-Quark, d-Quark and the (ud)-Diquark System", this Symposium.

29) N. Schmitz, "Hadron Production in a Neutrino-Hydrogen Experiment with BEBC", this Symposium.

30) B. Andersson, G. Gustafson, I. Holgersson and O. Mansson, Nucl. Phys. B178 (1981) 242.

31) L. de Billy et al., "Study of the Longitudinal Distribution in the Fragmentation Regions Using the Quark Parton Model in K⁻p Interactions at 70 GeV/c", Paris preprint LPNHEP/81-01 and A.M. Touchard, this Symposium.

32) J. Bell et al., Phys. Rev. D19 (1978) 1;
M. Derrick et al., Phys. Rev. D17 (1978) 1.

33) R.C. Hwa and M.S. Zahir, Phys. Rev. D23 (1981) 2539.

34) M. Barth et al., "Determination of the Internal Structure of Valons", paper submitted to this Symposium and private communication from L. Gatignon.

35a) R. Göttgens et al., "Fragmentation Spectra in K⁻p Interactions at 110 GeV/c", CERN/EP 80-102 ReV;

b) M. Spyropoulou- Stassinaki, this Symposium.

36) R.C. Hwa and R.G. Roberts, Z.Physik C1 (1979) 81.
Aachen-Berlin-CERN-Cracow-London-Vienna-Warsaw Collaboration, "The Fragmentation Spectra in K⁻p Interactions at 110 GeV/c", EPS Conf. 1979.
W. Aitkenhead et al., Phys.Rev.Letters 45 (1980) 157.
D. Denegri et al., Phys.Letters 98B (1981) 127.
M. Barth et al., "Inclusive Neutral Kaon Production in 70 GeV/c K⁺p Interactions", Nijmegen preprint HEN 202, 1981.

37) J. Badier et al., Phys. Letters 93B (1980) 354.

38a) D.R.O. Morrison, "Inclusive φ-Meson Production-Relation to J/ψ and Production", EPS Conf. Geneva 1979.

b) Yu. Arestov et al., Z. Physik 8 (1981).

c) C. Daum et al., "Inclusiv φ-Meson Production in 93 and 63 GeV Hadron Interaction", Amsterdam preprint NIKHEF-H/81-1.

39) P.V. Chliapnikov, V.G. Kartvelishvili, V.V. Kniazev and A.K. Likhoded, Nucl.Phys. B148 (1979) 400.
C. Cochet et al., Nucl.Phys. B155 (1979) 333.

40) M.M. Schouten et al., International Hybrid Spectrometer Consortium, "Inclusive and Semi-Inclusive ρ⁰ Production in

$\pi^+/\pi^-/K^+/pp$ Interactions at 147 GeV/c", Nijmegen preprint HEN187, 1981.

41) J.F. Gunion, Phys.Letters 88B (1979) 150.

42) E.A. De Wolf et al., "A Study of Multiparticle Fragmentation in K^+p Interactions at 32 GeV/c", and E.A. De Wolf at this Symposium.

43) J. Kuti and V.F. Weisskopf, Phys. Rev. D4 (1971) 3418;
K. Fialkowski, Acta Phys. Polonica, B11 (1980) 659;
E. Takasugi and X. Tata, "Particle Correlations in the Recombination Model Associated with Modified Kuti-Weisskopf Structure Functions", Austin preprint DOE-ER-03992-337, 1981.

44) L. Van Hove, this Symposium.

45) M. Barth et al., "Production of Vector and Tensor Mesons in K^+p Interactions at 70 GeV/c, a Preliminary Analysis", and R. Contri at this Symposium.

46) H. Miettinen, private communication.

47a) G. Bunce et al., Phys.Rev.Letters 36 (1976) 1113.
  b) N.N. Biswas et al., Nucl.Phys. B168 (1980) 4.
  c) S. Erhan et al., Phys.Letters 82B (1979) 301.

48) H. Pressner, Proc. Xth Int. Symp. on Multiparticle Dynamics, Goa 1979.

49) K. Raychaudhuri et al., Phys.Letters 90B (1980) 319.
S. Dado et al., Phys.Rev. D22 (1980) 2656.

50) K. Heller et al., Int. Conf. on High Energy Physics, Madison 1980.
C. Wilkinson et al., Phys.Rev.Letters 46 (1981) 803.

51) B. Andersson, G. Gustafson and G. Ingelman, Phys.Letters 85B (1979) 417.

52) T.A. De Grand and H.I. Miettinen, Phys. Rev. D23 (1981) 1227.
T.A. De Grand and H.I. Miettinen, "Models for Polarization in Inclusive Hadron Production", Santa Barbara preprint UCSB TH-27, 1981.

53) M.L. Faccini-Turluer et al., Z. Physik 1 (1979) 19.

54) M. Barth et al., "Inclusive Λ and $\bar{\Lambda}$ Production in K^+p Interactions at 70 GeV/c", CERN/EP 81-51 and R. Ross at this Symposium.

55) A. Bartl, A. Fraas and W. Majerotto, Z. Physik C6 (1980) 335.

56) J. Szwed, "Hyperon Polarization at High Energies", Cracow preprint TPJU-8/81.

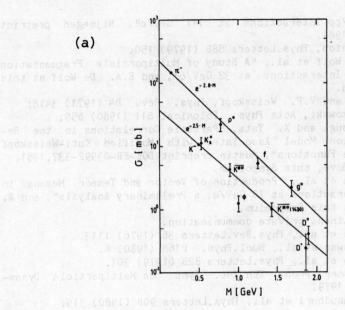

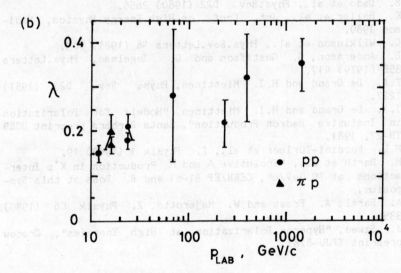

Fig.1a) Resonance cross section from pp collisions at √s̄=52.5 GeV
as a function of the resonance mass[3f].
b) Strange quark suppression factor λ as a function of the lab momentum[3].

24

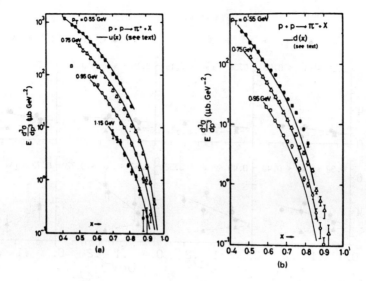

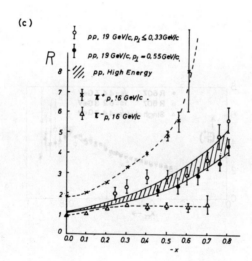

Fig.2a-b) Comparison of the invariant π^+ and π^- cross section from pp collisions at $\sqrt{s}$=45 GeV to the u and d quark distribution functions u(x) and d(x), respectively[8],
c) π^+/π^- ratio[9] in the proton fragmentation region of 16 GeV/c π^+p collisions (crosses) 16 GeV/c π^-p collisions (triangles) and pp collisions (shaded area and circles).

25

(d)

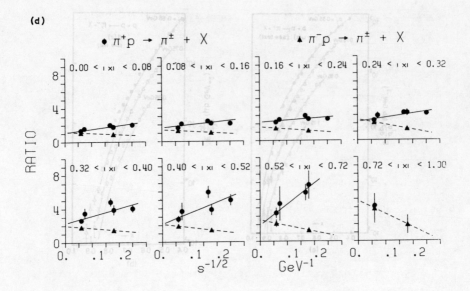

(e)

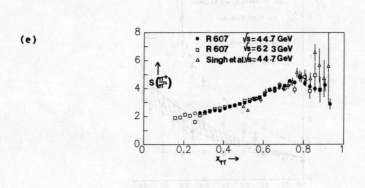

Fig.2d) The π^+/π^- ratio in the proton fragmentation region for $\pi^\pm p$ collisions, as a function of $s^{-1/2}$, for different x-intervals, as indicated[10],

e) π^+/π^- ratio at fixed polar angle in pp collisions at $\sqrt{s}=44.7$ and 62.3 GeV[8b].

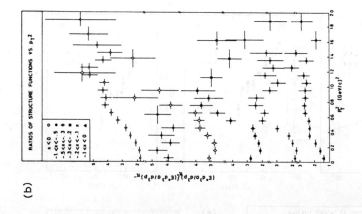

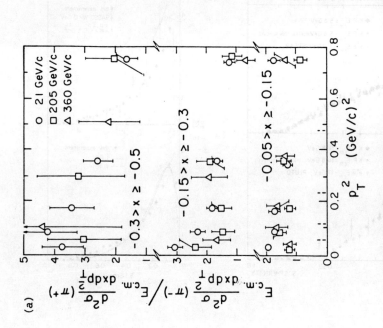

Fig.3a) The p_t^2 dependence of the π^-/π^+ ratio in the neutron fragmentation region of $\pi^- n$ collisions at 21, 205 and 360 GeV/c, for the three x intervals indicated [13a].

b) The p_t^2 dependence of the π^+/π^- ratio in the proton fragmentation region of $K^+ p$ collisions at 32 GeV/c, for the x intervals indicated [13b].

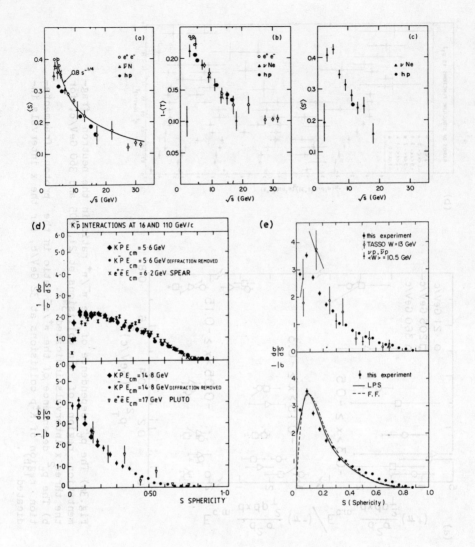

Fig.4a-c) Average sphericity, thrust and spherocity as a function of cms energy for e^+e^-, neutrino and hadron-hadron interactions.
d) Sphericity distribution for K^-p at 16 and 110 GeV/c compared to e^+e^- at corresponding energies[15a].
e) Same for K^+p at 70 GeV/c[15c], compared to e^+e^- and $\nu(\bar{\nu})p$ data at the corresponding energy and to predictions from longitudinal phase space (LPS) and from Field and Feynman (FF).

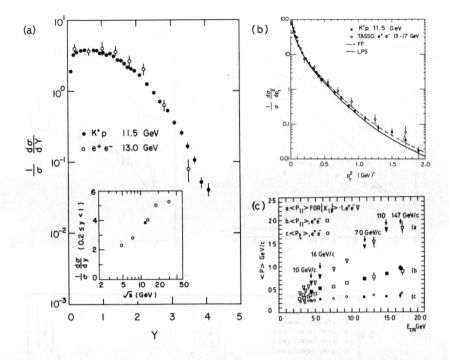

Fig.5a) Rapidity distribution $1/\sigma_t \cdot d\sigma/dy$ for non-diffractive K^+p events at 70 GeV/c (using thrust axis). Only charged particles produced in the c.m. forward hemisphere are used. In the insert, rapidity distribution $1/\sigma_t \cdot d\sigma/dy$ averaged over $0.2<y<1$ as a function of the c.m. energy [15c].

b) $1/\sigma_t \cdot d\sigma/dp_t^2$ as a function of p_t^2 relative to the sphericity axis for K^+p collisions at 70 GeV/c. The superimposed curves are the p_t^2 distributions for FF (dashed line) and for LPS (continuous line) [15c].

c) Energy dependence of $\langle p_t \rangle$ and $\langle p_{//} \rangle$ relative to the thrust axis for hadron-proton and e^+e^- reactions [15].

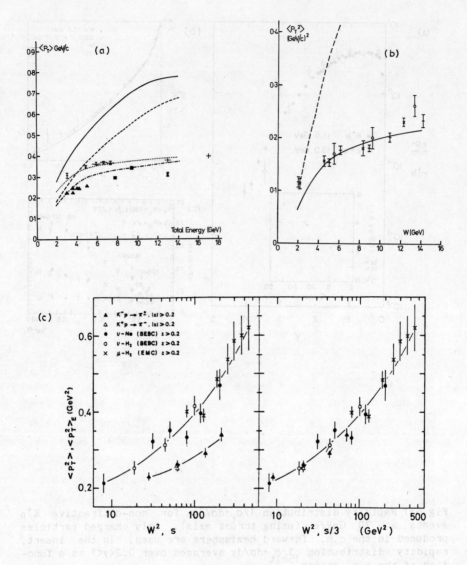

Fig.6a) $\langle p_t \rangle$ with respect to sphericity or thrust axis $(\_,\_,+,\ )$, compared to phase space with A=0 (dashed line) and A=3.5 (dot-dashed), respectively corrected to the true axis $(\mp,\ \text{full}\ \text{and}\ \text{dotted lines})$ [19].

b) $\langle p_t^2 \rangle$ in the current fragmentation region of $\nu(\bar{\nu})N$ collisions compared to phase space, with A=0 (dashed) and A=3.5 (full) [19].

c) $\langle p_t^2 \rangle$ dependence of hadron-hadron collisions compared to that of lepton-hadron collisions in terms of W^2 and s, respectively W^2 and s/3 [15d].

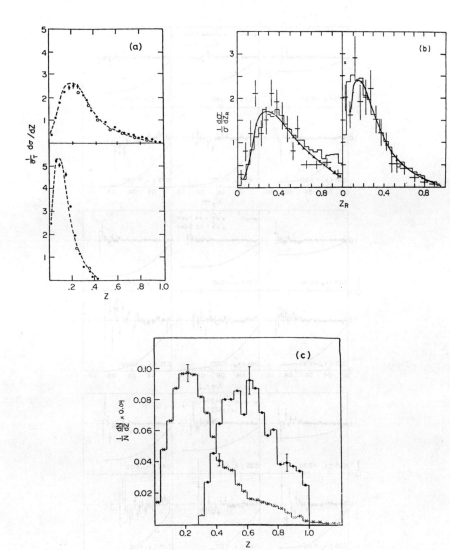

Fig.7a) z-Distribution for the fastest and second fastest charged forward particle of π^-p collisions at 147 GeV/c ($\bullet$) compared to deep inelastic ep data ($\circ$) and to FF predictions.
b) z_R-Distribution for the fastest negative forward particle in the same experiment (histogram) compared to the fastest positive particle in νN collisions (+) and FF predictions. The same for the fastest positive in the same experiment, compared to the fastest negative in νN collisions and FF predictions.
c) z-Distribution for the fastest charged forward particle (x) and for identified protons ($\bullet$) for π^-p collisions at 147 GeV/c[23].

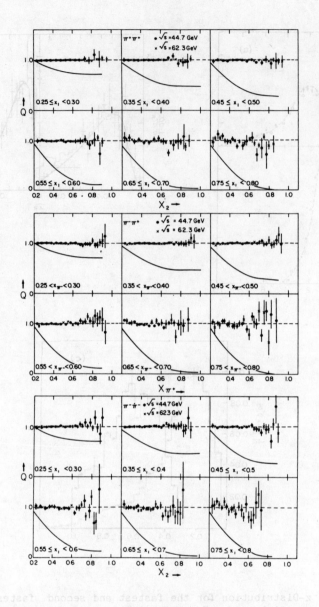

Fig.8) The correlation quotient Q for the $\pi^+\pi^+$, $\pi^+\pi^-$ and $\pi^-\pi^-$ combinations in pp collisions at $\sqrt{s}=44.7$ GeV and $\sqrt{s}=62.3$ GeV. The data are shown as a function of the Feynman x of one pion at several fixed values of x of the other pion. The lines are predictions from the counting rules[8b].

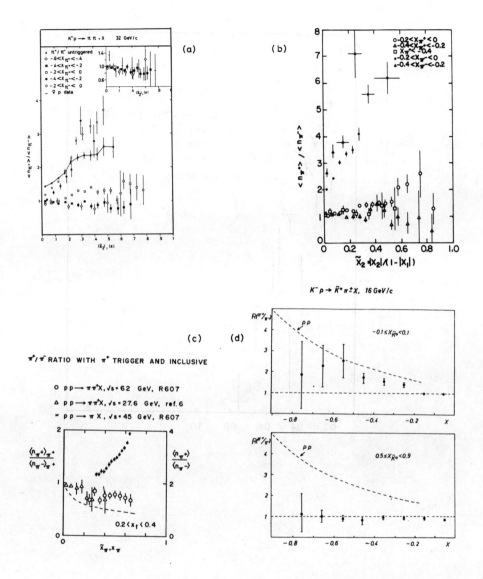

Fig.9a–b) The π^+/π^- ratio as a function of $\tilde{x}_2$ in the proton fragmentation region of 32 and 70 GeV/c K^+p collisions, associated with a π^- or π^+ trigger, for several trigger momentum intervals[27a,b].
c) The π^+/π^- ratio as a function of $\tilde{x}_\pi$ (respectively x) for pp collisions, associated with a π^+ trigger (l.h. scale) and no trigger (r.h. scale)[27c].
d) The π^+/π^- ratio as a function of x for $K^-p \rightarrow \bar{K}^0\pi^\pm X$ at 16 GeV/c, with an $\bar{K}^0$ in the given x intervals, compared to the π^+/π^- from pp collisions.[28]

33

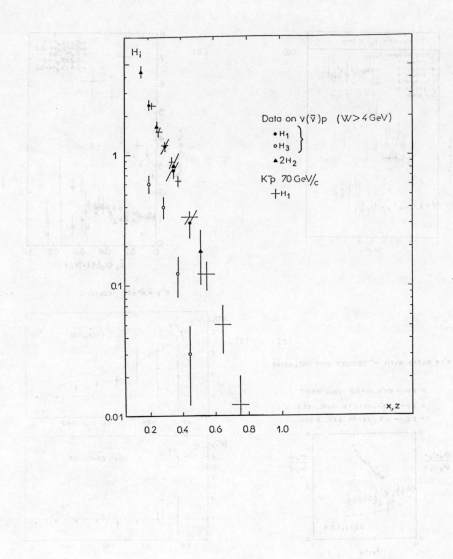

Fig.10) Diquark fragmentation function $H_1(x)=D_{uu}^{\pi^+}(x)$ obtained in K^-p collisions at 70 GeV/c[31]. Data on $H_1(x)$, $H_2(x)=D_{ud}^{\pi^+}(x)$, $H_3(x)=D_{uu}^{h^-}(x)$ from $\nu(\bar\nu)p$ experiments are also reported.

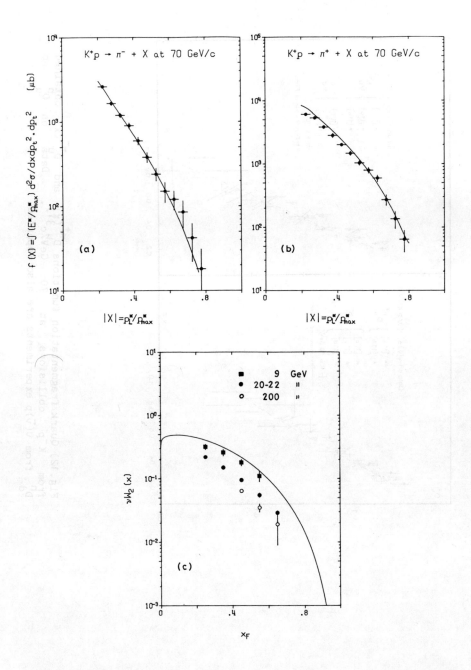

Fig. 11) Invariant x distribution for $K^+p \to \pi^{\pm}X$ at 70 GeV/c and νW_2 from deep inelastic scattering data at Q^2=9, 20-22 and 200 GeV^2. The full lines are from a valon model fit to the $K^+p \to \pi^-x$ data[34].

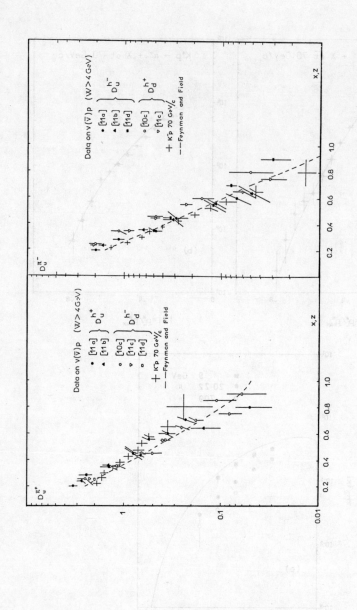

Fig.12) Quark fragmentation functions $D_u^{\pi^+}(x)$ and $D_u^{\pi^-}(x)$ obtained from K^-p collisions at 70 GeV/c. Data on D_u^h and D_d^h from $\nu(\bar\nu)p$ experiments are also reported.

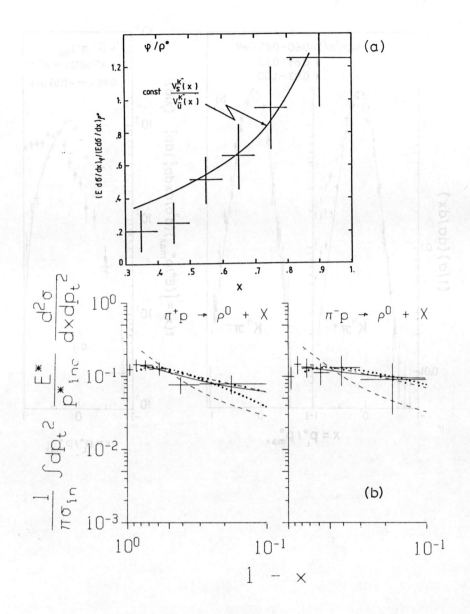

Fig.13a) The ϕ/ρ^0 ratio as a function of x for K^-p collisions at 32 GeV/c, compared to the ratio of the strange to non-strange valence distribution functions [38b].
b) the (1-x) dependence of ρ^0 production in the pion fragmentation region of $\pi^\pm p$ collisions at 147 GeV/c [40]. The curves are explained in the text.

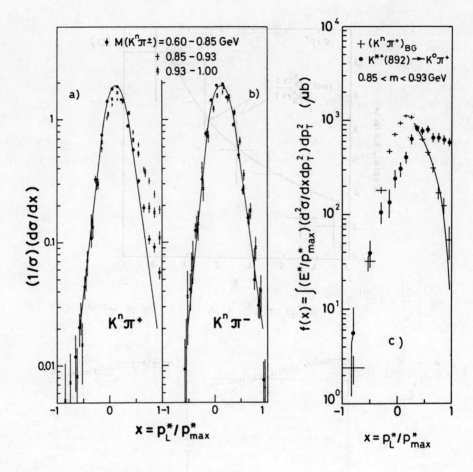

Fig.14a-b) Feynman x distribution of $K^n \pi^\pm$ systems from $K^+ p \to$ [42] $K^n \pi^\pm X$ at 70 GeV/c, in intervals of effective mass as indicated [42]. The curve corresponds to the data of fig.14b).
c) Invariant x distribution for $K^{*+}(892)$ and background in the corresponding mass interval [42]. The curve is a fit to $(1-x)^n$.

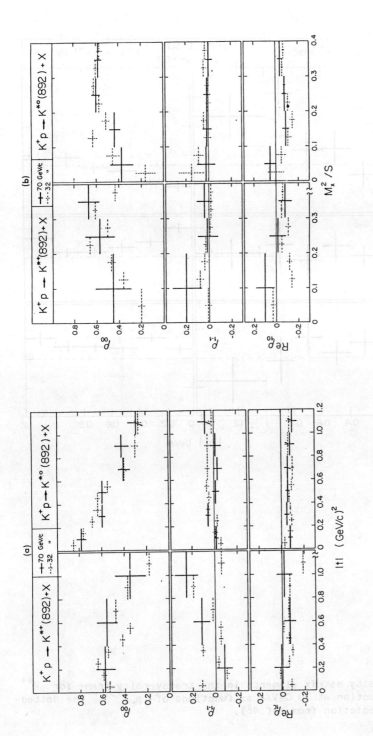

Fig.15a-b) Density matrix elements in the t-channel for K^{*+} and K^{*0} production from K^+p collisions at 32 and 70 GeV/c, as functions of t and M^2/s.

39

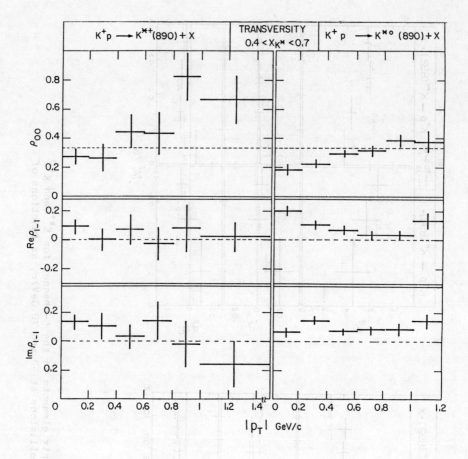

Fig.15c) Density matrix elements in the transversity frame for K^{*+} and K^{*0} production at 70 GeV/c as functions of $|p_t|$ [45]. The dotted line is a prediction from ref.46).

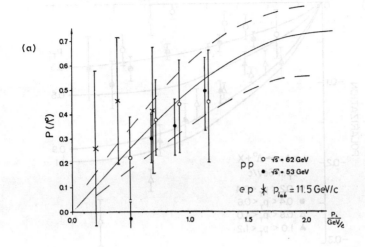

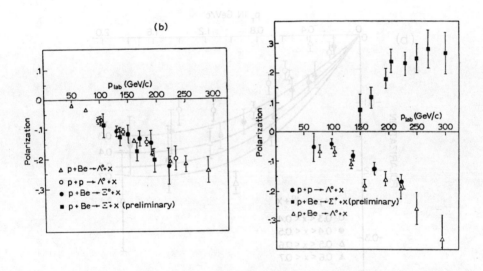

Fig.16a) Λ polarization as a function of p_t for pp collisions at $\sqrt{s}$=53 and 62 GeV[47c] and ep collisions at p_{lab}=11.5 GeV[48]. The lines are the expectation and its limits from the fragmentation model[51].

b) Hyperon polarization for pp and pBe collisions at 400 GeV/c, as a function of the hyperon lab momentum[50].

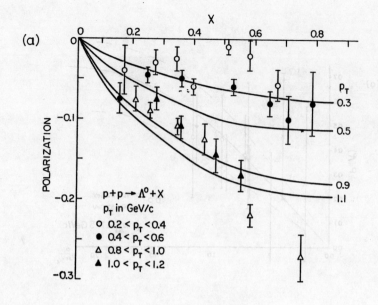

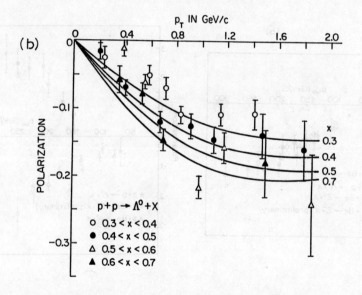

Fig.17a) Polarization of Λ^0 from pp collisions at 400 GeV/c as a function of x of Λ, for the p_t intervals indicated[50].
b) Same as a), but as a function of p_t of Λ for the x intervals indicated. The lines are described in the text[52].

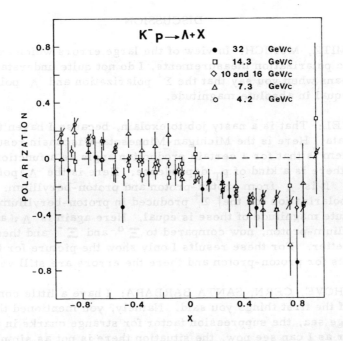

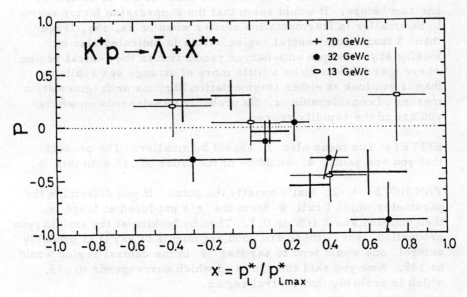

Fig.18) Λ polarization for $K^-p \to \Lambda + X$ up to 32 GeV/c[53] and $\bar{\Lambda}$ polarization for $K^+p \to \bar{\Lambda} + X$ up to 70 GeV/c[54] as a functions of x.

43

DISCUSSION

SCHMITZ, MUNICH: In view of the large errors which one has
· on the polarization measurements, I do not quite understand what
it means when you say that the Σ^+ polarization and Λ polarization
are equal in absolute magnitude.

KITTEL: That is a nasty job to explain, because I haven't shown
the data. Here is the Michigan-Minnesota-Wisconsin result. One
problem is that for a certain production angle as a function of
p_{lab} there is a kind of p_T dependence. Here is the Λ [polariza-
tion]. It is Λ from proton-proton and proton-beryllium. Here is
the [polarization for the] Σ^+ produced in proton-beryllium. The
absolute magnitude of these is equal. Here again is Λ, again
beryllium-proton, now compared to Ξ^0 and Ξ^- and these results
are better. For these results I only show the picture for the ISR
results for proton-proton and there the errors are still very large.

VAN HOVE, CERN, SANTA BARBARA: I have a little comment on
one of the first things you said. Namely, you mentioned the
strange sea, the suppression factor for strange quarks in the sea.
As far as I can see now, the situation there is not as simple as
one saw before. It would seem that the suppression factor seems
to be smaller in fragmentation studies when x is, say, larger
than .5 than in the central region. This is indicated also by
looking at the baryon anti-baryon production in the central region
where there seems to be a little more of strange sea available
than if you look at either fragmentation baryons or fragmentation
mesons of considerable x. So probably this depends on where
you are in the rapidity range.

KITTEL: You mean also λ should be smaller. The product
that you are using, Ψ, would be of the order of .15 with this λ.

VAN HOVE: Well, that's exactly the point. If you determine the
parameter which I call Ψ from the Λ's produced at large x,
then you find about 10% or 0.1. Then by looking at the anti-baryon
production, where there are good arguments to say it is probably
central, one would tend to say that Ψ in the central region would
be .15. Now you said the s/u is .3 which corresponds to .15,
which is probably that central region.

KITTEL: But this 1/3 is taken from a recent fit of ISR data. I
should say that at lower energies, up to 200 GeV, their λ is of

44

the order of .2, in agreement with what you said.

VAN HOVE: Yes. Ah, but what you say is that ISR data at large
x, which then presumably are K/π ratios, have changed?

KITTEL: Yes, they have changed in the sense that . . . Here is
the cross section as a function of the mass of the produced
resonance. If you take the K here and the π here, there may be
also this mass effect. The K/π ratio is, in any case, different
than the K/ρ ratio or K*/f because here the masses move closer
to each other. So they have treated simply all strange resonances
and the non-strange resonances [here you have the ϕ] and fitting
all these together that would lead to a λ of .35.

VAN HOVE: Yes, but this is the Chliapnikov type of fit that needs
t: be corrected with very large mass differences. This is not a
direct determination in a recombination model. Maybe we had
better talk that over privately.

HWA, OREGON: Concerning the jet universality, while it gives
support to the fragmentation picture it does not preclude the
recombination picture. It works there also.

KITTEL: No, I didn't want to say that. Did I say that?

KALELKAR, RUTGERS: On the Λ polarization, is there any
evidence for a Feynman-x dependence of the polarization?

KITTEL: Very little. One can say it is essentially not depending
on x and not depending on s.

MORRISON, CERN: I would like to ask you about diquarks. The
evidence on polarization being different (opposite) in Λ and Σ^+
is very suggestive about diquarks, because in one case the s
quark is fixed (on the outside) and in the case of the diquark it is
not. But is there any absolutely compelling evidence which says
you must believe in diquarks, and if you don't believe in diquarks
you can't explain the results?

KITTEL: I think this difference that Kistiakowski will present
of the fastest particle in a proton fragmentation jet compared to
that in a quark fragmentation jet may be some evidence.

DEGRAND, SANTA BARBARA: Theorists don't need diquarks
to explain the polarization asymmetry, but it is a convenient

shorthand.

CONTOGOURIS, MCGILL: There is one place I know where one cannot explain the situation in terms of the usual parton model and the usual QCD, and this is large p_T baryon production in proton proton collisions. It seems that there is a case where one needs a diquark.

ANDERSSON, LUND: In connection with the x dependence on Λ polarization, I get the impression that the people who are working at the FNAL hyperon beam did have an essentially smaller x than Peter Schlein had when he was working at the ISR. And if you would plot both sets of data you would find that the polarizations for the hyperon beam, that is the hyperon data done on a hydrogen target are of the order of at least 30 or 40% below Peter Schlein's ISR data. I take that as an indication that smaller x actually corresponds to smaller polarization. There is also evidence that if you are just in the center of phase space you do not see any polarization. So if you get to small x, the dominating Λ-polarization process is probably different.

KITTEL: Okay, so I should limit my statement a little bit then, as to the large-x region. In the central region there is certainly a different mechanism.

BIAŁAS, KRAKOW: I have two comments. One about x dependence of the parameters, any parameter in these models. I think that, for example, in this valon model you have to distinguish that there are two valons which are spectators and one valon which interacts. There is, I think, no reason to believe that they will fragment or behave afterwards in a simple way. The one which interacts probably behaves differently. At least you can guess it behaves differently. Furthermore, there is quite good evidence from analyses of scattering on heavy nuclei with indeed a quark which interacts. I mean if you interpret the data in terms of the valon model then the one which interacted emits particles and behaves very differently from the two spectators. This was already, I think, proven beyond any doubt. So I think one does expect differences as you go along the rapidity axis because in the forward region only spectators contribute and in the central region mostly the interacting quark contributes. So I'm glad to hear that, for example, this strange suppression factor changes and maybe also some other things change. Also this problem of having different behavior of fastest and slow pion, to my mind, reflects this phenomenon. So that's one comment. Second is

46

that this Λ polarization can be also explained if you assume that there is some effect of double scattering of the quark. That was done by Leo Stodolsky (Munich) and Szwed (Krakow).

CERN/EP/0469R/JB/ed
27 July 1981

JET-LIKE PROPERTIES OF MULTIPARTICLE SYSTEMS PRODUCED

IN K$^+$p INTERACTIONS AT 70 GeV/c

Brussels-CERN-Genova-Mons-Nijmegen-Serpukhov Collaboration

Presented by J. Beaufays
University of Mons, Belgium

We present results on the jet-like properties of multiparticle systems produced in K$^+$p interactions at 70 GeV/c. The data are analysed in terms of several variables commonly used to study the jet structure of an event. An extensive comparison is made with jets found in electron-positron annihilations and in deep inelastic νN interactions at comparable energy. Many similarities are found between low-p_T jets in this experiment and jets observed in leptonic interactions. Our data are very well reproduced by the Field-Feynman quark fragmentation parametrisation but equally well by a longitudinal phase space model, suggesting that these similarities do not prove or disprove the universal character of the jet fragmentation.

Recently striking similarities have been observed between jets in non-diffractive low-p_T interactions (initiated by different particles) and hadronic jets produced in e^+e^- annihilations and deep inelastic $\nu(\bar{\nu})$ nucleon scattering, when the data are expressed in terms of collective variables (like sphericity, thrust or spherocity) commonly used to study the jet structure [1]. In this paper [2], we present a comparison of the jet-like properties of low-p_T K^+p interactions with leptonic induced interactions. We also compare our results to the predictions of a longitudinal phase space model and to the Field and Feynman quark fragmentation parametrisation. These comparisons allow us to draw some conclusions on the significance of the observed "universality" of jet properties.

The data come from an exposure of BEBC, filled with hydrogen, exposed to an RF separated K^+ beam of nominal momentum 70 GeV/c, corresponding to a c.m. energy $\sqrt{s}$ = 11.5 GeV/c. This analysis is based on a partial sample of 9561 complete well measured events, having at least four charged outgoing tracks.

To reduce the effect of the diffractive component in our data, we removed from our sample events with at least one leading particle with $|x| > .8$ where x is the Feynman variable ($x = p_L^* | p_{max}^*$).

Our results are compared with the predictions of the parametrisation of quark fragmentation as proposed by Field and Feynman [3] and with those which follow from the minimal assumptions of longitudinal phase space [4].

In the Field-Feynman model (FF), two back-to-back jets were generated, allowing one valence quark of each incident particle to fragment into hadrons. In the fragmentation process, the longitudinal momenta are assigned to each quark in the jet cascade according to a prescription, which was originally determined by data from deep inelastic lepton scattering and e^+e^- annihilation. The quark transverse momenta are generated with a gaussian distribution $\exp[-p_T^2/2\sigma_q^2]$: the σ_q parameter was fixed to 300 MeV/c in order to reproduce our experimental p_T^2 distribution relative to the jet-axis. We also note that the Monte-Carlo reproduces quite well the measured charged multiplicity.

In the Longitudinal Phase Space model (LPS), exclusive final states were generated according to the following matrix element:

$$|M|^2 = \prod_{i=1}^{n} \exp[-B(y_i)m_T^i]$$

where m_T^i is the transverse mass of particle i and $B(y_i)$ is a rapidity dependent parameter (taken from the experimental data). The various exclusive channels were properly weighted using our measured average charged and neutral multiplicities.

In fig. 1, we compare our multiplicity distribution in a KNO plot to e^+e^- results [5]. We observe a good agreement with these data, when the diffractive component is removed, whereas the corresponding distribution for the complete sample (solid line on the figure) shows a different shape. This emphasizes that the diffractive events must be removed from the sample for comparison with the leptonic data.

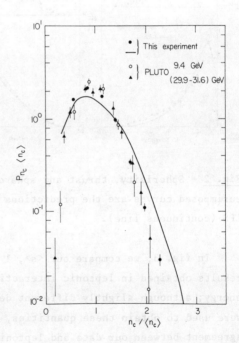

Fig. 1 - KNO plot for non diffractive events. The solid curve corresponds to the full sample.

In fig. 2, the normalized sphericity, thrust and spherocity distributions are shown and compared to data obtained in e^+e^- annihilation [6], and νN deep inelastic scattering [7,8] at similar hadronic energies. There is good overall agreement between our data (with diffraction removed) and the corresponding distributions from leptonic interactions. We note also that the two models (FF and LPS) describe the data very well and yield very similar predictions.

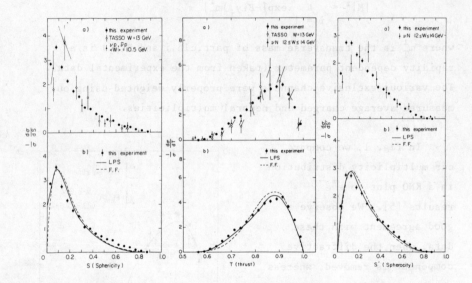

Fig. 2 - Sphericity, thrust and spherocity distributions. The surimposed curves are the predictions of FF (dashed line) and from LPS (continuous line).

In fig. 3, we compare our <s>, 1 - <T> and <s'> with selected results obtained in leptonic interactions over a wide range of energy; although slightly different definitions and reference frame were used to obtain these quantities, we observe a remarkable agreement between our data and leptonic results.

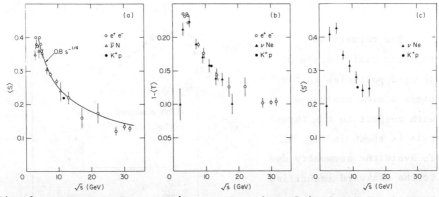

Fig. 3 - $\langle s \rangle$, $1 - \langle T \rangle$ and $\langle s' \rangle$ as a function of $\sqrt{s}$.

Characteristic features of hadronic jets, as well as of hadron-hadron collisions are the limitation of $\langle p_T \rangle$, the mean particle momentum transverse to the jet (beam) axis and the increase of $\langle p_L \rangle$, the mean longitudinal momentum, with increasing energy. The $\langle p_T \rangle$ and $\langle p_L \rangle$ relative to the thrust axis are shown in fig. 4 together with PLUTO data [9]. To suppress the non-scaling part of the single particle distribution, we show also $\langle p_L \rangle$ with $x = 2p_\parallel^*/E_{cm} > .1$. This quantity shows a linear rise with c.m. energy while the transverse momentum shows little change with energy.

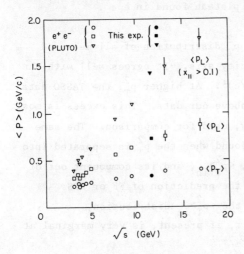

Fig. 4 - $\langle p_L \rangle$ and $\langle p_T \rangle$ with respect to the thrust axis

53

The normalized
rapidity distribution of
charged particles in the
overall c.m. evaluated
with respect to the thrust
axis is shown in fig. 5.
To avoid the asymmetry due
to the backward identified
protons, we use here only
particles produced forward
in the overall c.m. The
TASSO results at c.m.
energy 13 GeV, to which
our data are compared,
refer only to charged
particles, assumed to be
pions. The plateau

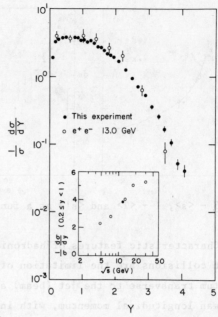

Fig. 5 - Rapidity distribution of charged
particles with respect to the thrust axis

at $0.2 \leqslant y \leqslant 0$ is in good agreement with e^+e^-. Moreover, we note
(insert of fig. 5) that our K^+p data follow the linear increase
with $\ell n E_{cm}$ for the height of the plateau found in e^+e^-.

Fig. 6 shows the normalized p_T^2 distribution of all measured
particles relative to the sphericity axis. It agrees well with the
TASSO data [10] for $p_T^2 < 0.5(GeV/c)^2$. At higher p_T^2, the TASSO data
are slightly but systematically above our data. This excess is most
probably due to the higher energy, used for comparison. The same
agreement with e^+e^- data [6] is found when the p_T^2 is separated into
its component in the event plane, $<p_T^2>_{in}$, and its component out of
the plane, $<p_T^2>_{out}$. Furthermore the prediction of FF or LPS
accounts for most of the tail of the $<p_T^2>_{in}$ distribution,
indicating that a planarity effect, if present, is very marginal at
our energy.

54

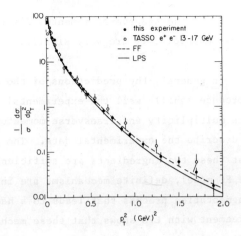

Fig. 6 - p_T^2 distribution
relative to the sphericity
axis. The surimposed curves
are the predictions from FF
(dashed line) and from LPS
(continuous line)

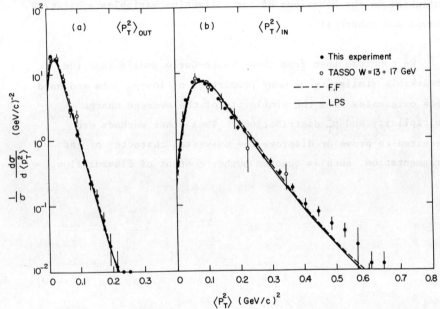

Fig. 7 - $<p_T^2>_{out}$ and $<p_T^2>_{in}$ distribution. The surimposed curves are the
predictions from FF (dashed line) and LPS (continuous curve).

In conclusion, we find striking similarities between many
properties of jets in low-p_T non diffractive K^+p interactions
and in leptonic induced reactions at the same c.m. energy: in

particular the transverse momentum distribution and the average charge multiplicity are very similar.

In general, the predictions of the two models (LPS and FF) reproduce equally well the experimental results. In the LPS model, both multiplicity and transverse momentum distributions are chosen to describe the experimental data. The success of this model shows that these two ingredients are sufficient to describe the data. In the FF model, definite mechanisms are incorporated in the fragmentation process that leads to a hadronic jet. However, its agreement with LPS shows that these mechanisms have essentially no influence on the behaviour of the collective variables sphericity, thrust and spherocity.

We conclude then from these Monte-Carlo models that the remarkable similarity of many properties of low-p_T jets and hard jets originates from the similarity of the average charge multiplicity and p_T^2 distributions. Thus other methods are required to prove or disprove the universal character of jet fragmentation, such as quantum number content of flavour-flow.

REFERENCES

[1] R. Göttgens et al., Nucl. Phys. B178 (1981) 392;

     D. Brick et al., International Hybrid Spectrometer Consortium,
     IFNUP/AE01/81.

[2] For more details see M. Barth et al., CERN/EP 81-44.

[3] R.D. Field and R.P. Feynman, Nucl. Phys. B136 (1981) 1.

[4] L. Van Hove, Rev. Mod. Phys. 36 (1964) 655.

[5] Ch. Berger et al., PLUTO Collaboration, Phys. Lett. 95B (1980)
     313.

[6] R. Brandelik et al., TASSO Collaboration, Zeitschrift für
     Physik C, Particles and Fields 4 (1980) 87.

[7] M. Derrick et al., Phys. Lett. 88B (1979) 177.

[8] K.W.J. Barnham et al., Phys. Lett. 85B (1979) 300.

[9] Ch. Berger et al., PLUTO Collaboration, Phys. Lett. 81B (1979)
     410.

[10] R. Brandelik et al., TASSO Collaboration, Phys. Lett. 86B
     (1979) 243.

DISCUSSION

RATTI, PAVIA: In spite of the fact that I agree completely with your conclusions, does it make any difference whether you include an extra flavor in the Field-Feynman predictions for thrust?

BEAUFAYS: We have not tried it. I don't think so.

RATTI: How many flavors are included?

BEAUFAYS: uu, u$\bar{\text{u}}$, d$\bar{\text{d}}$, $\bar{\text{s}}$ for the hadronization process and only the valence quark for the original.

RATTI: No c and b?

BEAUFAYS: No, not c and b.

RATTI: Because the e^+e^- data are compared with the Feynman-Field prediction which includes also charm and bottom; charm in the low energy at 13 GeV, and bottom in the 35 and 37 GeV. The second question I wanted to ask is if you have, by any chance, the average values of $< p_T >_{in}$ and $< p_T >_{out}$.

BEAUFAYS: No, not here.

COMPARISON OF 147 GeV/c π⁻p LOW TRANSVERSE MOMENTUM HADRON PRODUCTION WITH DEEP-INELASTIC LEPTOPRODUCTION OF HADRONS

V. Kistiakowsky[a], E. D. Alyea, Jr.[b], D. Brick[c], E. B. Brucker[d], W. M. Bugg[e], H. O. Cohn[f], E. S. Hafen[a], R. I. Hulsizer[a], M. Kalelkar[g], E. L. Koller[d], A. Levy[a], T. Ludlam[h], P. Lutz[a], S. H. Oh[a], R. J. Plano[g], I. A. Pless[a], H. Rudnicka[c], A. Shapiro[c], J. P. Silverman[a], P. E. Stamer[g], T. B. Stoughton[a], H. D. Taft[h], P. C. Trepagnier[a], T. L. Watts[g], M. Widgoff[c], R. K. Yamamoto[a].

ABSTRACT

Distributions with respect to z and z_R for the fastest and second fastest particles produced in both beam and target hemispheres from 147 GeV/c π⁻p interactions are compared with each other and with the distributions for the fastest and second fastest particles from deep-inelastic leptoproduction and Field-Feynman curves. The distributions and curves for the fastest particles are very similar with the exception of those for the identified protons. Those for the second fastest are in agreement. Distributions with respect to z' for the first through fourth fastest particles (excepting the protons) are all similar.

INTRODUCTION

In this paper, we present a comparison of our data for hadron production of charged hadrons (π⁻ + p → hadrons) in regions of low transverse momentum with data from studies of leptoproduction of charged hadrons (ℓ + p → ℓ + hadrons) at high transverse momentum[1,2]. We compare the distributions with respect to the Field-Feynman variables, z and z_R , for the fastest and second fastest charged hadrons in both the beam and the target hemispheres with both the leptoproduction results and with curves from the Field-Feynman quark model[3]. There is a striking similarity between the hadron production results in the beam hemisphere and both the leptoproduction data and the Field-Feynman curves. However, in the target hemisphere, the distributions for the fastest charged particles - the identified protons in events with identified protons - are found to be significantly different. There is no a priori understanding for the similarities between the hadro- and lepto-produced hadrons, but the difference observed for the proton may be qualitatively

a Massachusetts Institute of Technology, Cambridge, Massachusetts 02139
b Indiana University, Bloomington, Indiana 47401
c Brown University, Providence, Rhode Island 02912
d Stevens Institute of Technology, Hoboken, New Jersey 07030
e University of Tennessee, Knoxville, Tennessee 37916
f Oak Ridge National Laboratory, Oak Ridge, Tennessee 37830
g Rutgers University, New Brunswick, New Jersey 08903
h Yale University, New Haven, Connecticut 06520

understood in terms of the differences in the valence quark content of hadrons produced from pions and protons. The distributions for the first through fourth fastest particle with respect to a scaled energy variable, z_R', are very similar, which may point to a dependence of the distributions on the phase space available to the particles.

ANALYSIS

The data presented in this paper come from an experiment performed at the Fermi National Accelerator Laboratory using the Fermilab Hybrid Spectrometer exposed to a beam of 147 GeV/c negative pions. The experimental setup consists of a 30-inch bubble chamber, preceded and followed by spectrometer elements, which has been described elsewhere[4]. The data were derived from the analysis of 100,000 pictures together with downstream spectrometer information.

The Field-Feynman variables, z and z_R, were used for this analysis to permit comparison with deep-inelastic leptoproduction. In the latter case, the momentum fraction of the produced hadron is $z = p_{lab}/\upsilon$ where υ is the laboratory energy of the exchanged boson, and the energy fraction is $z_R = E_{lab}/E_{tot}$ where E_{tot} is the sum of the energies of the charged hadrons in a jet. For the low $p_\perp$ hadron production case, $z = (E^* + p_\parallel^*) / \sqrt{s}$ where E^* and $p_\parallel^*$ are the energy and longitudinal momentum of the hadron in the center of mass system, and $z_R = E_{lab}/E_{tot}$ where E_{tot} is the sum of the energies of the charged hadrons in either the forward of the energies of the charged hadrons in either the forward or backward hemisphere.

In order to exclude quasi-elastic events from our sample, only events with six or more charged particles were included. From previously published cross sections[5], it can be shown that the single and double diffraction dissociation component of these data is less than 5%.

RESULTS

Figure 1 gives the z distributions for the fastest (largest z) and second fastest charged particles, and the z_R distributions for the fastest like charge and unlike charge particles in the beam and target hemispheres in our data. Since for events with identified protons, the protons are the fastest particle in 99% of the events, the fastest charged particles in events without an identified proton are included in the second fastest distributions. It can be seen that the distributions for the second fastest charged particles in the two hemispheres are very similar, but that the proton distributions are significantly different from the distributions for the fastest charged particle in the beam hemisphere.

There are two sources of possible systematic differences between the beam and target hemisphere distributions. First, because the bubble chamber determination of ionization only permits identification of protons up to $p_{lab} = 1.4$ GeV/c, protons are missing from the distributions for $z < 0.6$ and for all values of z_R, and unidentified protons are included in the pion distributions for $z < 0.1$ and $z_R < 0.4$. The second possible source of difference between the beam and target

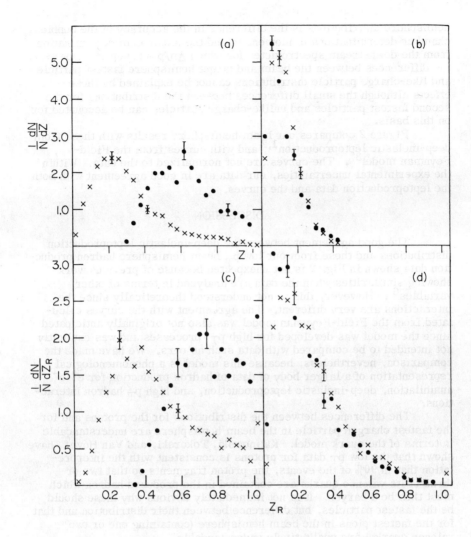

Figure 1. Distributions with respect to z for (a) the fastest charged particle in the beam hemisphere (x) and the identified protons (•) and (b) the second fastest charged particle in the beam (x) and target (•) hemispheres, and with respect to z_R for (c) the fastest negative particle in the beam hemisphere (x) and the identified protons (•) and (d) the fastest positive particle in the beam hemisphere (X) and the fastest negative particle in the target hemisphere (•).

hemisphere distributions is the difference in the accuracy of the bubble chamber determinations of momentum and the momentum determinations from the downstream spectrometer for which $\Delta p/p = 0.06p\%$[4]. However, the differences between the beam and target hemisphere fastest particle and like-charge particle distributions cannot be explained by these effects although the small differences between the distributions for the second fastest particles and unlike-charge particles can be accounted for on this basis.

Figure 2 compares our beam hemisphere results with those from deep-inelastic leptoproduction[1,2] and with curves from the Field-Feynman model[3]. The curves are not normalized to the data. Within the experimental uncertainties, our data are in good agreement with both the leptoproduction data and the curves.

DISCUSSION

The good agreement between the deep-inelastic leptoproduction distributions and those from our low p_T beam hemisphere hadron production data shown in Fig. 2 is not unexpected because of previous work showing similarities when the data are analyzed in terms of other variables[6]. However, this is not understood theoretically since the interactions are very different. The agreement with the curves calculated from the Field-Feynman model was also not originally anticipated since the model was developed for high p_T processes and was explicitly not intended to be compared with data such as ours. We have made the comparison, nevertheless, because this model is a phenomenological representation of a larger body of data of hadron production for e^+ e^- annihilation, deep-inelastic leptoproduction, and high p_T hadron interactions.

The differences between the distributions for the protons and for the fastest charged particle in the beam hemisphere are understandable in terms of the quark model. Kalinowski, Pokorski, and Van Hove[7] have shown that the low p_T data for protons is consistent with the interpretation that in 90% of the events, the proton fragments so that two or three of its valence quarks are contained in the produced hadron which must thus be a baryon. It is not immediately obvious why these should be the fastest particles, but difference between their distribution and that for the fastest pions in the beam hemisphere (containing one or two valence quarks) are qualitatively understandable.

It is interesting to pursue this analysis deeper into the central region of hadron production. One may compare the z_R' distribution of the second fastest charged particle in the beam hemisphere with that of the fastest charged particle in the target hemisphere, which is not an

identified proton. Here z_R' is defined to be $z' = \dfrac{E_{had}}{E_{tot}'}$ where E_{tot}' is

the sum of the energies of the charged hadrons in the particular hemisphere, not including those of the particles which have been removed (the fastest particle in the case of the beam hemisphere and the identified proton in the case of the target hemisphere). These distributions are

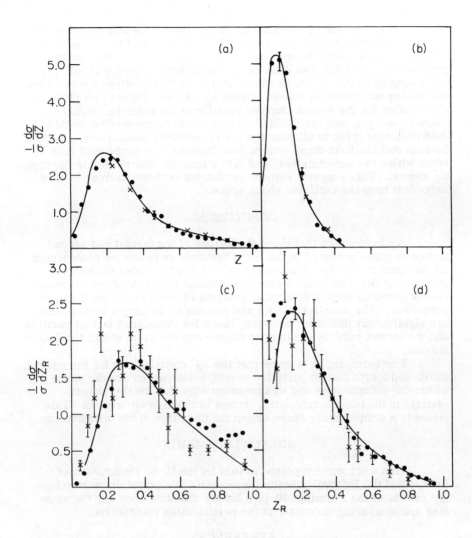

Figure 2. Distributions with respect to z for (a) the fastest charged particle in the beam hemisphere in this experiment (•) and from deep-inelastic electron scattering [1] (x) and (b) the second fastest charged particle in the beam hemisphere in this experiment (•) and from deep-inelastic electron scattering [1] (X), and with respect to z_R for (c) the fastest negative particle in the beam hemisphere from this experiment (•) and the fastest positive particle from neutrino interactions [2] (X) and (d) the fastest positive particle in the beam hemisphere from this experiment (•) and the fastest negative particle from neutrino interactions [2]. The smooth curves are taken from Field and Feynman [3].

shown in fig. 3(a) and the agreement is excellent. Figures 3(b) and 3(c) give the distributions for the cases where the fastest of the remaining charged particles have been removed from the events in 3(a) and 3(b), respectively, and E_{tot} has been recalculated for the charged particles remaining after this has been done. Again the distributions agree within increasing uncertainties except at high z_R' values. Figure 3(c) gives the distribution for the fastest charged particle in the beam hemisphere (here $z_R' = z_R$), and the Field-Feynman curve corresponding to this distribution is given in all four figures to facilitate comparison. It can be seen that the first three sets of distributions also agree with each other within the uncertainties, and differ from the fourth only in the high z_R' region. This suggests that the production of these hadrons is only dependent upon the available phase space.

CONCLUSIONS

We have shown that the distributions of the fastest and second fastest hadrons produced in the beam hemisphere in low p_T pion-proton interactions at 147 GeV/c are strikingly similar to those for hadrons produced in deep-inelastic lepton interactions and to the Field-Feynman curves based on phenomenological analysis of those and other high p_T processes. The distributions for the protons in the target hemisphere are significantly different; however, those for the second fastest particle and the fastest particle of opposite charge are the same within the uncertainties.

Furthermore, it is found that the z_R' distributions for the second, third, and fourth fastest particles in both hemispheres are the same within the uncertainties and in agreement with that for the fastest particle in the beam hemisphere, except in the high z_R' region. This points to a simple phase space origin for the shape of the distributions.

ACKNOWLEDGMENTS

This work was supported in part by the U. S. Department of Energy and the National Science Foundation. We gratefully acknowlege the efforts of the Fermilab 30-inch bubble chamber crew and the scanning and measuring personnel at the participating institutions.

REFERENCES

1. G. Drews et al., Phys. Rev. Lett. 41, 1433 (1978).
2. P. C. Bosetti et al., Nucl. Phys. B149, 13 (1979).
3. R. D. Field and R. P. Feynman, Phys. Rev. D15, 2590 (1977); Nucl. Phys. B163, 1 (1978).
4. D. Fong et al., Nucl. Phys. B102, 386 (1976).
5. D. Brick et al., Phys. Rev. D21, 1726 (1980).
6. B. Andersson, G. Gustafson, and C. Peterson, Phys. Lett. 69B, 221 (1977); M. Deutschmann et al., Nucl. Phys. B155, 307 (1979); K. W. J. Barnham et al., Phys. Lett. 85B, 300 (1979); P. H. Herquet, Mons preprint PNPE, 191 (25 July 1979); Proceedings of the X Inter-

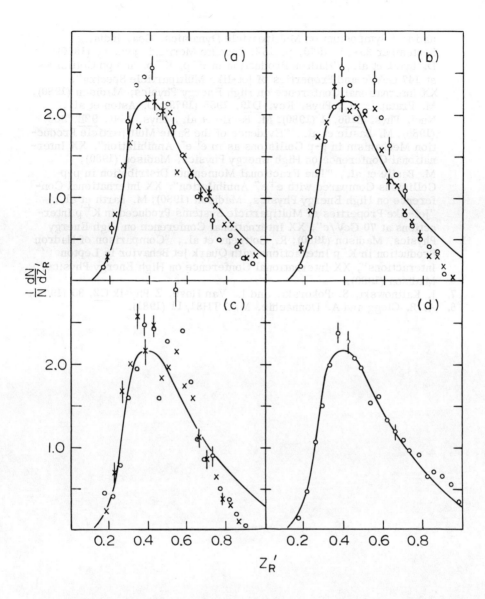

Figure 3.   Distributions with respect to z_R' for (a) the second fastest, (b) the third fastest, (c) the fourth fastest, and (d) the fastest particle in the beam (O) and target (X) hemispheres. The smooth curves are taken from Field and Feynman[3] .

national Symposium on Multiparticle Dynamics, Goa, India, September 25-29, 1979, p. 152; Reconte Moriond, January (1980); D. Brick et al., "Hadron Production in $\pi^+ p$, $K^+ p$, and pp Collisions at 147 GeV/c and Properties of Jet-like Multiparticle Spectra", XX International Conference on High Energy Physics, Madison (1980); M. Pratap et al., Phys. Rev. D19, 1955 (1979); D. Aston et al., Nucl. Phys. B166, 1 (1980); M. Basile et al., Phys. Lett. 92B, 368 (1980); M. Basile et al., "Evidence of the Same Multiparticle Production Mechanism in p-p Collisions as in $e^+ e^-$ Annihilation", XX International Conference on High Energy Physics, Madison (1980); M. Basile et al., "The Fractional Momentum Distribution in p-p Collisions Compared with $e^+ e^-$ Annihilation", XX International Conference on High Energy Physics, Madison (1980); M. Barth et al., "Jet-like Properties of Multiparticle Systems Produced in $K^+ p$ Interactions at 70 GeV/c", XX International Conference on High Energy Physics, Madison (1980); R. Gottgens et al., "Comparison of Hadron Production in $K^- p$ Interactions with Quark Jet Behavior in Lepton Interactions", XX International Conference on High Energy Physics, Madison (1980).

7. J. Kalinowski, S. Pokorski, and L. Van Hove, Z Physik C2, 85 (1979).
8. A. B. Clegg and A. Donnachie, M/C TH81/11 (1981).

PLANAR MULTIPARTICLE PRODUCTION IN HADRON-PROTON COLLISIONS AT 147 GeV/c.

(presented by S.P.Ratti)

D.H.Brick, H.Rudnicka, A.M.Shapiro, M.Widgoff - Brown University., R.E.Ansorge, W.W.Neale, D.R.Ward - Univ. of Cambridbe, R.A.Burnstein, H.A.Rubin - Illinois Institute of Technology, E.D.Alyea Jr. - Indiana Univ., L.Bachman, C.Y. Chien, P.Lucas, A.Pevsner - Johns Hopkins Univ., J.T.Bober, M.Elahy, T.Frank, E.S.Hafen, P.Haridas, D.Huang, R.I. Hulsizer, V.Kistiakowsky, P.Lutz, S.H.Oh, I.A.Pless, T.B. Stoughton, V.Suchorebrow, S.Tether, P.C.Trepagnier, Y.Wu, R.K.Yamamoto - Massachusetts Inst. of Tech., F.Grard, J. Hanton, V.Henri, P.Herquet, J.M.Lesceux, P.Pilette, R.Windmolders - Univ. de l'Etat, Mons; F.Criins, H.de Bock, W. Kittel, W.Metzger, C.Pols, M.Schouten, R.Van de Walle - Univ. of Nijmegen, H.O.Cohn - Oak Ridge Nat. Lab., F.Carminati, C.Castoldi, R.Dolfini, S.Ratti - Istituto di Fisica Nucleare and Sezione I.N.F.N. Pavia, R.Di Marco, P.F.Jacques, M.Kalelkar, R.J.Plano, P.E.Stamer, T.L.Watts - Rutgers Univ., E.B.Brucker, E.L.Koller, S.Taylor - Stevens Institute of Technology, Hoboken, L.Berny, O.Benary, J.Grunhaus, R.Heifetz, A.Levy - Tel Aviv. Univ., W.M.Bugg, G.T. Condo, T.Handler, E.L.Hart, A.H.Rogers - Univ. of Tennessee, Y.Eisenberg, D.Hochman, U.Karshon, E.E.Ronat, A.Shapira, R.Yaari, G.Yekutieli - Weizman Institute, T.Ludlam, R. Steiner, H.Taft - Yale Univ.

In a recent paper[1] we have reported evidence for planar events detected in an hydrogen bubble chamber, associated to a down-stream magnetic spectrometer and a lead glass gamma detector, exposed at FNAL to a tagged 147 GeV/c beam made of 51% π^+, 9% K^+, 40% protons.
In this paper we expand our analysis and investigate in more details the possible planar events with a statistics improved by about 1/4, using 15346 events having at least 6 charged prongs in the final state. Using the notations suggested by M.J.Counihan[2] and having described the events in momentum space[1], adopting a proper principal axes reference frame (see later for its definition), the beam direction is given by two Euler angles: the polar angle θ_B and the azymuthal angle ϕ_B. In ref. (1) evidence for planar events was given by the azymuthal angle distribution which shows a striking deviation from isotropy at $\phi_B \simeq 0$. For the purpose of this paper the principal axes reference frame has a slightly different definition which is made following the assumptions of D.P. Barber et al.[3]. The first principal axis $\vec{u}_1$ is the Thrust axis, that is $\vec{u}_1$ is chosen so that

67

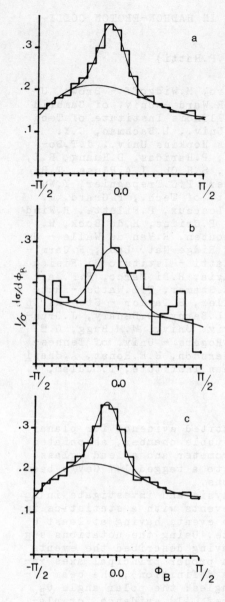

$$T = \frac{\Sigma|\vec{P}_i \cdot \vec{u}_1|}{\Sigma|\vec{P}_i|}$$

is maximized and the sum extends over the charged particles.
The <u>Major</u> axis $\vec{u}_2$ is defined perpendicular to $\vec{u}_1$ and such that

$$F_{major} = \frac{\Sigma|\vec{P}_i \cdot \vec{u}_2|}{\Sigma|\vec{P}_i|}$$

is maximized. Finally the Minor axis is defined as the third axis $\vec{u}_3$ which makes a right handed reference frame ($\vec{u}_3 = \vec{u}_1 \times \vec{u}_2$). <u>Fig. 1a</u> shows the azymuthat angle distribution for all 15346 events producing at least six charged particles. <u>Fig. 1b</u> shows the same distribution for 2668 "diffrac<u>c</u>tive-like events" selected, <u>a</u> mong the 6-20 prong topologies according to the rules given in Sect. 2 of ref. (1). <u>Fig. 1c</u> shows the same distribution for the remaining non-diffrac<u>c</u>tive events which account for ∿85% of the whole sample. Although diffractive-like events show a broader accumulation of events with vanishing ϕ_B, the two distributions of figs. 1b and 1c are essentially similar. The different structure of diffractive-like and non-diffractive events can be well seen by inspecting the momentum flow within the three pla<u>a</u>nes defined by the 3 principal axes $\vec{u}_1, \vec{u}_2, \vec{u}_3$. Fig.s 2 show polar displays of the mo<u>o</u>mentum flows (in 10° bin) the Thrust-Major plane (Fig. 2a); in the Thrust-minor plane (Fig. 2b) and in the "tran<u>n</u>sversal" Major-Minor plane (Fig. 2c) for all events. In Fig. 2a and 2b the maximum mo<u>o</u>

<u>Fig. 1</u> - Distributions of the azymuthal angle ϕ_B; a - 15346 events producing at least 6 charged particles; b - 2668 diffraction-like events; c - 12678 non diffractive events.

68

mentum is 1.9 GeV/c flowing in the first 10° degrees next
to the direction of the Thrust axis $\vec{u}_1$. Fig. 2c on the o-
ther hand shows a "cilindrical" shape in which the Major
axis has been systematically privileged against the minor
axis even in the case in which nature does not indicate a
ny clear preference between the two "transversal" axes.
It is relevant to point out the fact that the maximum tran

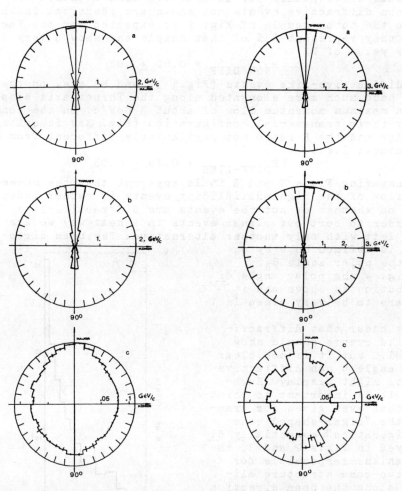

Fig.2 - Momentum flow di
stribution in a polar
diagram for 15346 events.
a- Thrust-Major plane (Ra
dius=1.9 GeV/c); b-Thrust
-Minor plane (Radius=1.9
GeV/c); c-Major-Minor pla
ne (Radius=0.11 GeV/c).

Fig.3 - Momentum flow distributions in a polar diagram
for 2668 diffractive-like e-
vents. a- Thrust-Major plane
(Radius = 3.03 GeV/c); b-
Thrust-Minor plane (Radius =
2.97 GeV/c); c- Major-minor
plane (Radius = 0.11 GeV/c).

sversal momentum flow (in Fig. 2c) along the Major axis is 110 MeV/c and that the ratio R = (momentum flow along the Major ($\pm$ 10°) axis/momentum flow along the Minor axis $\pm$ 10°) is

$$R_{ALL} = 0.80 \pm .02$$

Fig.s 3 show the polar displays of the momentum flows in the same three planes for diffractive-like events.
The non diffractive events not shown are identical in shape to the total sample of Fig. 2 as expected for the fact that they represent 83% of that sample and show a very similar value of R:

$$R_{NON\ DIFF} = 0.80 \pm .03$$

The diffractive-like events (Fig.s 3a and b) are, on the other hand much more elongated along the Thrust axis displaying a maximum momentum flow of about 3 GeV/c. On the contrary their transversal configuration (Fig. 3c) is quite similar and the R ratio not particularly different from the above; i.e.:

$$R_{DIFF-LIKE} = 0.79 \pm 0.08$$

By comparing Fig.s 2 and 3 it is apparent that the momentum flow of the high multiplicity events is rather independent on whether or not the events are diffractive-like. In order to sort out planar events in a neat way we are then left with only another alternative. The beam direction is defined both by ϕ_B and the polar angle θ_B.
In Fig. 4 the polar angle distribution is shown and it appears to be very steep in $\cos\theta$.
It is clear that diffractive-like events should show $\cos\theta \approx 1.0$ but it is not clear what angle θ non diffractive events might display. However, it is important to point out that even given for granted the "cigar-shaped cone configuration" essentially displayed in Fig.s 2 and 3, the mechanisms responsible for the two-cone structure aligned along the beam direction might be different from those responsible for the "non-aligned" two-cone structures.
Fig. 5a shows the azymuthal angle ϕ_B distribution for events "aligned along the beam" having $0.9 < \cos\theta_B \leqslant 1.0$. The distribution contains

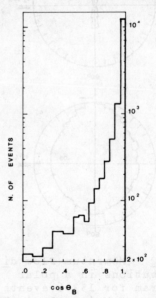

Fig.4 - Distribution of the polar angle θ_B formed by the beam direction with the thrust axis.

13808 events, that is ~90% of the total sample. As expected, it is identical to Fig. 1c.
On the other hand Fig. 5b, containing 1537 "non aligned" ($\cos\theta_B < 0.9$) events is strikingly different.
A single cut on the polar angle θ_B of the beam direction with respect to the Thrust axis, indeed selects a sample of events almost perfectly lying in a plane close to the one identified by the Thrust and the Major axis. The distribution of Fig. 5b is reproduced by a cubic background plus a Gaussian.
The values of $<\phi>$ and σ_ϕ for the distributions shown in fig.s 1 and 5 are collected in TABLE I.

TABLE I

		$<\phi>$ (rad)	σ_ϕ(rad)	χ^2/ndf
total sample	fig. 1a	-0.21±0.02	0.24±0.03	5.92/13
diff.-like events	" 1b	-0.17±0.10	0.34±0.09	5.15/13
non diff.event	" 1c	-0.21±0.03	0.24±0.03	3.16/13
aligned events	" 5a	-0.24±0.09	0.29±0.09	0.97/13
non. al.events	" 5b	-0.19±0.01	0.20±0.01	30/13

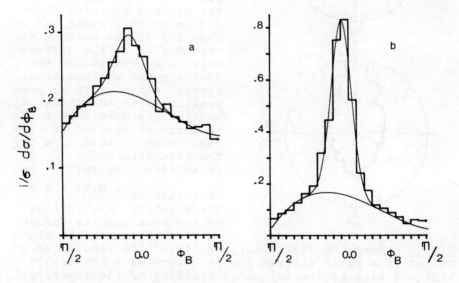

Fig. 5 - Distributions of the azymuthal angle ϕ_B;
a - 13809 aligned events ($0.9 < \cos\theta_B < 1.0$);
b - 1537 non aligned events ($\cos\theta_B < 0.9$).

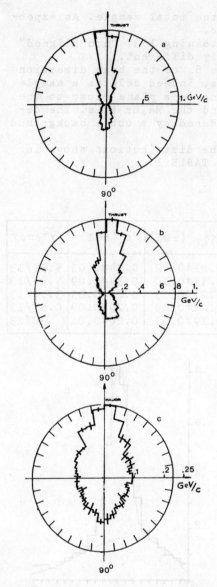

The investigation of the momentum flows for the two sets of events selected above is very instructive. The momentum flows for the aligned events do coincide with those displayed in Fig.s 2.

The maximum momentum flows in the two longitudinal planes are 2.02 GeV/c and 2.04 GeV/c respectively. The maximum transversal flow along the Major axis, is 100 MeV/c and the ratio R is

$$R_{ALIGNED} = 0.84 \pm 0.04$$

Quite different is the picture displayed in Fig.s 6 by 1537 "non aligned events". For these events the maximum momentum flow along the Thrust axis (1.05 GeV/c in Fig. 6a and 0.8 GeV/c in Fig. 6b) is about 1/2 of the maximum momentum flow for an average event and about 1/3 of the maximum momentum flow of the diffractive like events. On the contrary the maximum transversal momentum flow, perpendicular to the Thrust direction, is more than twice as large as that of an average event or that of a diffractive like event.
In addition the ratio R is

$$R_{NON\ ALIGNED} = 0.51 \pm 0.02$$

The relative orientations of the beam and the Thrust axis do influence the planarity of the events: we are observing a noticeable shrinking of the tranversal dimension of the non aligned events which appear to be very planar.

Fig. 6 - Momentum flow distribution in a polar diagram for 1537 non aligned events $(\cos\theta_B <0.9)$; a - Thrust-Major plane (Radius = 1.05 GeV/c); b - Thrust Minor plane (Radius = 0.80 GeV/c); c - Major-Minor plane (Radius = 0.23 GeV/c).

References

1. - D.Brick et al., I.H.S. Consortium Report INFN/AE 01.81
     (submitted to Nuovo Cimento)
2. - M.J.Counihan, P.L. $\underline{59B}$, 367 (1975)
3. - D.P.Barber et al., P.R.L. $\underline{43}$, 12 (1979)

EUROPEAN ORGANIZATION FOR NUCLEAR RESEARCH

CERN/EP 81-Draft
17 August 1981

COMPARISON OF JETS IN LEPTONIC AND HADRONIC INTERACTIONS

Maria Bardadin-Otwinowska

CERN, Geneva, Switzerland
and
University of Warsaw, Poland

ABSTRACT

Jet structure of hadronic final states produced in $K^{\pm}p$
and pp interactions is compared with recent data on e^+e^-
annihilation and on lepton-hadron scattering (νN, $\bar{\nu}N$ and μp).
The variables discussed are the sphericity, thrust, the hadron
transverse momentum with respect to the jet axis and the components of the transverse momentum vector in and out of the event
plane. Striking similarities are found between the data for
various reactions for energies of the hadronic system up to
$\sim$ 15 GeV. At the highest Petra energies ($\sim$ 30 GeV), significant
differences are observed between the transverse momentum distributions for e^+e^- annihilations and for the mesonic subsystem obtained by removing leading protons from the final state of the pp
interactions.

1. INTRODUCTION

The subject of this report is a phenomenological comparison
of the properties of hadronic final states produced in low p_T
hadron-hadron collisions, e^+e^- annihilations and lepton-hadron
interactions at high energies. A common feature of all the final
states is their two-jet structure. The comparison of low p_T
jets with the quark or diquark jets produced in leptonic interactions may bring information on the mechanism of the hadron-
hadron interaction at the quark-parton level.

Similarities between jets produced in hadronic and leptonic
reactions are predicted by the fragmentation model [1] sketched
diagramatically in fig. 1. In the extreme version of the model,
all final particles come from the hadronization of a fast
spectator quark or diquark emerging from the initial interaction
between wee valence quarks.

Talk given at the XII International Symposium
on Multiparticle Dynamics, 21-26 June 1981, Notre Dame, USA

EP/0426P/MBO/ef

This version of the fragmentation model is used here as a quide-line in presenting the experimental data.

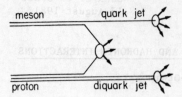

Fig. 1 – The digram representing the fragmentation model. In the extreme version of the model, the contribution from the central part is neglected.

In most of the published papers on jet analysis in hadronic reactions, the shape of the final state in the c.m. momentum space is studied and compared with the data on reactions $e^+e^- \to$ hadrons and lepton + proton $\to$ lepton + hadrons in the rest frame of the hadronic system. The elongation is measured by the global variables such as sphericity or thrust which refer to the whole hadronic system, or by the transverse momentum of produced hadrons with respect to the jet axis. Planar features are investigated in terms of two components of the hadron transverse momentum vector, defined to measure the width and the flatness of the final state.

It should be stressed that the present comparison is limited to the jet shape and neglects differences between the quantum numbers of the jets produced in various reactions.

2. EXPERIMENTAL DATA

The first evidence for striking similarities between hadronic and leptonic interactions comes from the comparison of the sphericity and thrust distributions in K^-p interactions at 10, 16 and 110 GeV/c with the data for the reactions $e^+e^- \to$ hadrons and $\nu N \to \mu^- +$ hadrons at the same total energies of the hadronic system [2(a)]. However, some differences are observed; the values of $\langle p_T^2 \rangle$ tend to be lower in K^-p than in the ν interactions.

The jet analysis of the K^-p data is now extended [2(b)] to the search for planar effects and results are available for K^+p interactions at 70 GeV/c [3] and for h^+p interactions at 147 GeV/c [4]. In all the experiments the diffractive component is removed (because no corresponding process exists in leptonic interactions) and all the remaining non-diffractive events are analyzed.

A different approach is taken by two ISR groups who apply more complex selection criteria in order to eliminate the effect of the leading proton in the pp collisions before comparing with the e^+e^- data. Meyer et al. [5] select events with two leading protons, remove the protons and analyze the remaining system in its own rest frame. Basile et al. [6] perform the analysis for

the mesonic subsystem in one hemisphere only. In the framework
of the fragmentation model (as in fig. 1 but for pp), the leading
proton can be interpreted as the first-rank particle in the fragmentation chain of the initial diquark. An antiquark left after
the proton emission continues fragmenting into hadrons and this
process is analoguous to the hadronization of a quark created in
the e^+e^- annihilation.

3. RESULTS

The results obtained in various hadronic and leptonic
experiments are collected in figs 2 to 8.

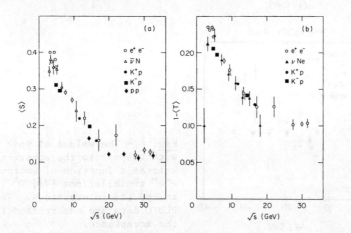

Fig. 2 - The variation of the average sphericity <s>
and thrust, 1 - <T>, with the energy of the hadronic
system [2-5].

The fig. 2 shows the variation of the average values of the
sphericity and thrust as a function of energy. The figure is
taken from ref. [3] and supplemented with the data of refs [2,4]
and [5]. The results from $K^{\pm}p$, pp, νN and e^+e^- reactions follow
the same trend. Similar conclusions can be drawn from fig. 3,
which shows the energy variation of the longitudinal and transverse components of the average hadron momentum with respect to
the thrust axis, for c.m. energies below 20 GeV [2,3,4]. On the
other hand, the values of $<p_T^2>$ when extended to energies of
∿ 30 GeV, show significant differences between leptonic and
hadronic interactions: the values of $<p_T^2>$ relative to the
sphericity axis, plotted in fig. 4 rise rapidly with energy in e^+e^-
annihilations [7] while no energy dependence is observed in
hadronic reactions above 20 GeV [5].

77

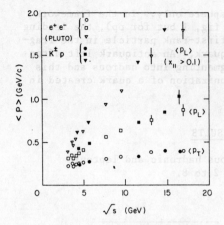

Fig. 3 - Energy variation of the longitudinal and transverse momentum with respect to the sphericity axis [2-4].

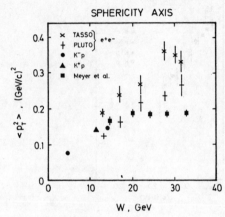

Fig. 4 - The values of $\langle p_T^2 \rangle$ with respect to the sphericity axis as a function of energy for e^+e^- annihilations [7], $K^{\pm}$ [2,3] and pp [5] interactions. The PLUTO data are uncorrected for the acceptance.

The same tendency is seen in fig. 5 where the values of $\langle p_T^2 \rangle$ in $K^{\pm}p$ interactions are compared with the recent data for ν [8], $\bar{\nu}$ [9] and μ^- [10] interactions; here $\langle p_T^2 \rangle$ is defined with respect to the weak current direction in leptonic interactions and with respect to the incident beam direction for $K^{\pm}p$. At high energies, the lepton-hadron data tend to lie above the $K^{\pm}p$ points.

Figs 3-5 indicate that the variable p_T^2 is more sensitive than p_T in detecting differences between various reactions. The distributions of p_T^2 relative to the sphericity axis are plotted in fig. 6 for reactions e^+e^-, K^-p and pp at several energies. In K^-p interactions, the tail of the p_T^2 distribution increases rapidly with the energy. The distributions for K^-p at 110 GeV/c, corresponding to $\sqrt{s}$ = 14.4 GeV and for K^+p at 70 GeV/c [3] (not plotted) are consistent with the e^+e^- data at similar energy (13-17 GeV). From there to 30 GeV, the tail of the e^+e^- distribution shows a considerable increase interpreted as an evidence for the gluon jets [7].

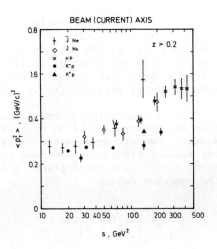

Fig. 5 – The values of $\langle p_T^2 \rangle$ with respect to the current axis for νNe [9], νNe [8] and $\mu^- p$ [10] interactions and $\langle p_T^2 \rangle$ with respect to the beam axis for $K^- p$ [2] and $K^+ p$ [3] interactions, plotted as a function of the energy squared of the hadronic system.

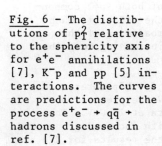

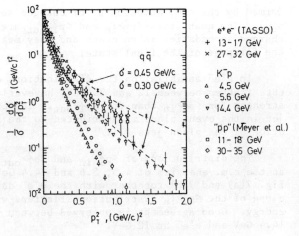

Fig. 6 – The distributions of p_T^2 relative to the sphericity axis for $e^+ e^-$ annihilations [7], $K^- p$ and pp [5] interactions. The curves are predictions for the process $e^+ e^- \rightarrow q\bar{q} \rightarrow$ hadrons discussed in ref. [7].

The behaviour of hadron-hadron data in the energy range 15-30 GeV is very interesting because no hard scattering effects are expected to occur there and consequently the hadronic data can be used as an estimate of the contribution due to soft processes alone in $e^+ e^-$ annihilations.

The agreement of the p_T^2 distributions in $K^\pm p$ and $e^+ e^-$ interactions at $\sim$ 15 GeV indicates that at this energy non-perturbative effects dominate in $e^+ e^-$ annihilation. At $\sim$ 30 GeV, the only hadronic data available are those obtained indirectly from the pp interactions. They show very little energy dependence and remain significantly below the $e^+ e^-$ data at high values of p_T^2, indicating that the increase of the tail of the p_T^2 distribution in hadronic interactions saturates below 30 GeV. However, it should be noted that the properties of a mesonic subsystem

extracted from a pp subsample could be different from the properties of a non-diffractive hadron-hadron final state. This is indeed suggested by the incompatibility of the p_T^2 distributions
for K^-p at 14.4 GeV and pp at 11-17 GeV. More data are needed
before a conclusive statement can be made on the energy dependence of the p_T^2 distribution in hadron-hadron interactions.

The evidence for gluon jets in e^+e^- annihilation comes also
from the observation of non-planar features of the hadronic final
state. They are seen in the distributions of $<p_T^2>_{in}$ and $<p_T^2>_{out}$,
the squared transverse momentum components in and out of the
event plane. The event plane is defined by the eigenvalues of
the momentum tensor

$$M_{ij} = \sum_{k=1}^{n_{ch}} p_i^{(k)} p_j^{(k)}$$

formed by the c.m. momentum p_i of charged secondaries (k=1, ...
n_{ch}). The quantities $<p_T^2>_{in}$ and $<p_T^2>_{out}$ are averaged over the
charged secondaries in an event and they measure the width and
the flatness of the final state.

In e^+e^- annihilation the distributions of both $<p_T^2>$ components flatten between 12 and 30 GeV, however the effect is much
stronger for $<p_T^2>_{in}$ than for $<p_T^2>_{out}$. The broadening of the
jet in the event plane is attributed to the increasing contribution of the gluon jets [7].

The distributions of $<p_T^2>_{in}$ and $<p_T^2>_{out}$ for K^-p interactions
at the c.m. energies of 4.5, 5.6 and 14.4 GeV are plotted in
fig. 7(a) and 7(b) together with the e^+e^- data at 12 GeV. The
slope of the $<p_T^2>_{in}$ distributions flattens with increasing
energy. Good agreement is observed between the results for K^-p at
14.4 GeV and e^+e^- at 12 GeV.

Fig. 7(c) shows the distributions of $<p_T^2>_{in}$ at higher
energies: 27-32 GeV and 35-36.6 GeV for e^+e^- [7], 11-18 GeV and
30-35 GeV for pp [5] and the K^-p data at 14.4 GeV presented already
in fig. 7(b).

The $<p_T^2>_{in}$ distributions become less steep with increasing
energy up to 14.4 GeV, whereas the pp data indicate that a saturation occurs at energies < 30 GeV. The observed different behaviour of the pp and K^-p reactions can be at least partly due to
the different methods of analysis and therefore the question of
the energy dependence of the $<p_T^2>_{in}$ distriburions remains
open. Nevertheless, one may conclude from fig. 7(c) that the
highest energy hadronic data available are significantly below
the e^+e^- points, as stressed in ref. [5].

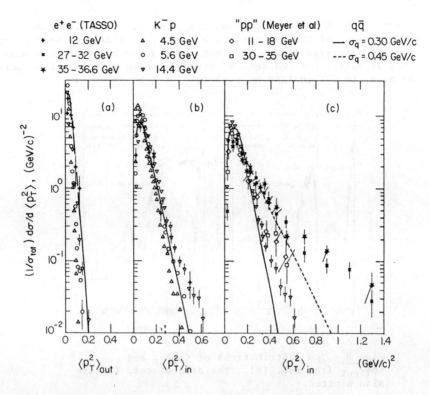

e$^+$e$^-$ (TASSO)   K$^-$p   "pp" (Meyer et al)   q$\bar{q}$

+   12 GeV   △   4.5 GeV   ◇   11 – 18 GeV   — σ_q = 0.30 GeV/c

×   27 – 32 GeV   ○   5.6 GeV   □   30 – 35 GeV   --- σ_q = 0.45 GeV/c

✱   35 – 36.6 GeV   ▽   14.4 GeV

Fig. 7 – The distributions of $\langle p_T^2 \rangle_{in}$ and $\langle p_T^2 \rangle_{out}$ for e$^+$e$^-$ annihilations [7], K$^-$p and pp [5] interactions. The curves are predictions for the process e$^+$e$^-$ → q$\bar{q}$ → hadrons discussed in ref. [7].

This conclusion is in contradiction with the statement made in ref. [6] that the pp data are perfectly consistent with the e$^+$e$^-$ data at similar energies. The comparison of the experimental data from refs [5] and [6] is presented in fig. 8. The $\langle p_T^2 \rangle_{in}$ distributions from the two experiments are compatible, both at lower energies (around 15 GeV) and at high energies (∼ 30 GeV). Therefore, the contradiction is merely due to a different inter-pretation of the experimental data.

We now compare the K$^-$p data at 110 GeV/c with the recent results obtained by the European Muon Collaboration who found evidence for planar events in the forward going hadrons produced in the reaction μ$^-$p → μ$^-$ + hadrons [11]. The authors define the plane which contains the current direction and which maximizes the sum $\sum_h p_{Tin}^2$, where the transverse momentum of each hadron is decomposed in components p$_{Tout}$ and p$_{Tin}$, normal and parellel to this plane, and the index h numbers charged secondaries in an event. The variables $\Sigma p_{T_{in}}^2$ and $\Sigma p_{T_{out}}^2$ measure the width and the flatness of the hadronic final state, similarly to $<p_T^2>_{in}$ and $<p_T^2>_{out}$ discussed previously, but relative to a different longitudinal axis. The two sets of variables also differ by the factor equal to the charged multiplicity of a given event.

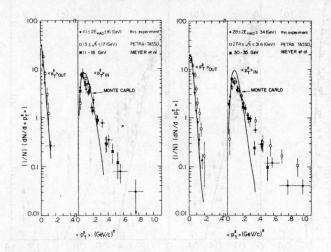

Fig. 8 - The distributions of $<p_T^2>_{in}$ and $<p_T^2>_{out}$ from ref. [6]. The data of ref. [5] are also plotted.

The distributions of $\Sigma p_{T_{in}}^2$ and $\Sigma p_{T_{out}}^2$ for $\mu^- p$ events with total hadronic energy $100 < W^2 < 460$ GeV² are presented in fig. 9 and compared with the $K^- p$ results at $W^2 = 208$ GeV². All selection criteria and acceptance cuts in the muon data are applied also to the sample of the $K^- p$ events. The current direction is replaced by the incident K^- direction. The distributions of $\Sigma p_{T_{in}}^2$ and $\Sigma p_{T_{out}}^2$ have remarkably similar shapes for $K^- p$ and $\mu^- p$ interactions. The $K^- p$ points are systematically lower than the muon data, however, the value of W^2 for $K^- p$ interactions, $W^2 = 208$ GeV², is considerably smaller than the average value of the total hadronic energy squared for μp, $<W^2> = 310$ GeV².

CONCLUSIONS

Hadronic final states produced in νN, $\bar{\nu} N$, μN, $e^+ e^-$ and low-p_T hadronic interactions are compared at the total hadronic energies from 5 to 35 GeV.

No significant differences are found between hadronic and leptonic reactions in the energy variation of the global variables such as thrust of sphericity neither in the behaviour of the average momentum components of produced hadrons with respect to the jet axis. However, the average p_T^2 increases with energy faster for $e^+ e^-$ and lepton-proton than for hadron-hadron interactions.

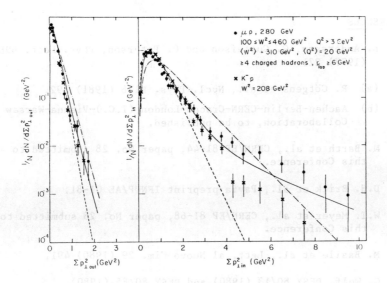

Fig. 9 - The distributions of $\Sigma p_{T_{in}}^2$ and $\Sigma p_{T_{out}}^2$ relative to the current axis in reaction $\mu^- p \to \mu^-$ + hadrons [11] and relative to the beam direction in reaction $K^- p \to$ hadrons, for forward going hadrons.

The p_T^2 distribution for $e^+ e^-$ annihilation shows a marked increase with energy between 13 and 30 GeV. For $K^- p$ interactions a fast increase is found between 5 and 15 GeV, whereas the pp data indicate a saturation below 30 GeV.

Similarly, a strong energy variation of $\langle p_T^2 \rangle_{in}$ distribution is found for the $e^+ e^-$ between 12 and 35 GeV and the $K^- p$ below 15 GeV, but not for the pp interactions above this energy.

A different behaviour of the K^- and pp data and their disagreement at $\sim$ 15 GeV could be due to different methods of analysis. More hadronic data are needed to make the comparison with leptonic data more conclusive in the interesting energy region of $\sim$ 30 GeV.

Acknowledgements
The author gratefully acknowledges helpful discussions with Drs K.W.J. Barnham, R.T. Ross and other colleagues from the 110 GeV/c $K^- p$ Collaboration.

83

REFERENCES

[1] B. Andersson, G. Gustafson and C. Peterson, Phys. Lett. 69B
     (1977) 221.

[2] (a) R. Götgens et al., Nucl. Phys. B178 (1981) 392.

     (b) Aachen-Berlin-CERN-Cracow-London (I.C.)-Vienna-Warsaw
          Collaboration, to be published.

[3] M. Barth et al., CERN/EP 81-44, paper No. 28 submitted to
     this Conference.

[4] D.H. Brick et al., Pavia preprint IFNUP/AE 01-81.

[5] W.T. Meyer et al., CERN/EP 81-68, paper No. 23 submitted to
     this Conference.

[6] M. Basile et al., Lett. al Nuovo Cim. 29 (1980) 491.

[7] G. Wolf, DESY 80/13 (1980) and DESY 80/85 (1980).

[8] H. Deden et al., Nucl. Phys. B181 (1981) 375.

[9] V.V. Ammosov et al., Serpukhov preprint HIEP-80-125 (1980).

[10] J.J. Aubert et al., Phys. Lett. 95B (1980) 306.

[11] J.J. Aubert et al., Phys. Lett. 100B (1981) 433.

DISCUSSION

De GRAND, SANTA BARBARA: My comment is that no theorist imagines that the large-transverse-momentum behavior of hadron-hadron interactions and e^+e^- interactions will be similar other than in very gross features and, in fact, I don't think any experimentalist will argue the same thing either, because you know that in proton-proton collisions large-transverse-momentum events are four-jet events, two small-p_T and two large-p_T, and people don't see three-jet events, which is the sort of thing you see in e^+e^- annihilation. That's common. So I think most of us tend not to pay much attention to what happens at very large energy because we know it's different. At small p_T my question would be: Is there any deviation in the data from what you would expect (and this is for all data, e^+e^- versus pp, say, from simple Lorentz, simple limited-transverse-momentum-type phase-space arguments? Is there any deviation at all?

BARDADIN-OTWINOWSKA: It depends on what you call low energy. As far as I know, at energies of up to 15 GeV in the center of mass system there is an overall agreement with the simple two-quark jet model. We have an example of a distribution which I had no time to show where, at our energy, we seem to have an effect which is in disagreement with the model you mentioned. Namely, that is again the p_T^2 distribution from the muon interactions, which is now compared with our 110 GeV K^-p data which corresponds to a slightly lower center of mass energy than the muon data. And now, the curves plotted are the simply quark anti-quark models, the Feynman-Field parameterization which is more or less the same as the longitudinal phase space, and the curve (which) comes from the assumption of a gluon jet. So that is an example of the hadron-hadron distributions where the hadron-hadron data seem to deviate from the simpler quark anti-quark model.

RATTI, PAVIA: I want to make a comment. It's a matter of reference frames. In comparing p_{in} and p_{out} there are essentially two different ways of defining the principal axis planes. One is using the eigenvectors and the eigenvalues of the tensors, one is using the thrust axis. They only coincide as an average and not always are they the same. I'm afraid that some of the equality or the discrepancies might come from this slightly different definition. MARK J uses a different reference plane to define the event plane because it maximizes the thrust and then maximizes the fake thrust in the perpendicular plane. Are these reference frames always compatible in the different experiments?

BARDADIN-OTWINOWSKA: Yes, they are. I mean, in all the hadronic experiments I was extremely careful to use the exactly same variables and definitions and to compare . . . .

RATTI: In your experiment, not . . . . .

BARDADIN-OTWINOWSKA: No, no, in other places as well. I mean, if that's what people say. For example the $< p_T^2 >$ I presented was summed with respect to the sphericity axis, some others were with respect to the thrust axis, and care was taken to put together things that are comparable.

KITTEL, NIJMEGEN: I'm asking myself if I should really compare these p_T averages at the same energy. I had shown these s predictions by Clegg and Donnachie who show that all neutrino p_T dependence on energy and all e^+e^- p_T dependence on energy can be explained just from the same transverse-cut phase space and for all energies the transverse cut is the same. Now, if you say that the hadron data have the same tendency to increase but do that slower, then I would say we just have to divide the energy by some constant and get an effective energy per jet which should be smaller, in fact, for hadron-hadron interactions than in e^+e^- because they are probably two jets, two chains. I have the feeling we don't have to divide by very much. We don't have to divide the square root of s by 2, but 30% decrease in the effective energy in the center of mass in enough to get your p_T dependence again on the same increasing curve.

BARDADIN-OTWINOWSKA: But in this case I'm not sure whether we get agreement with other variables. For example, if you look at the variation of the sphericity with energy that I show, there seems to be no need to reduce the hadron-hadron energy that you compare with the e^+e^- data. The distribution you talked about is the p_T^2 relative to the beam axis, and I think that it's not a question of shifting the points, the hadron points relative to the lepton points. I mean, for hadron-hadron this variation is rather flat. I'm not sure whether you can achieve something, whether you can get the points from two experiments to agree simply by shifting the scale.

KITTEL: It has been shown last year in Bruges by Laffaille that you get the same increase when you divide s by 2 or 3. Now for the sphericity . . . . .

86

BARDADIN-OTWINOWSKA: I don't think so, because if you divide the energy by 2 you just . . . . Well, I don't know.

KITTEL: We can look at that plot. But now for the sphericity the form is different. If you have two chains which are partially overlapping then the total event looks much longer than each chain itself. Imagine two chains, one going from the first, from the fastest, to the middle, and then the second from the middle to the slowest particle. The total chain would look long [but leave the p_T distribution equal to that of individual chains] and therefore the sphericity is larger than for each chain individually. So that effect would cancel in the case of sphericity.

BARDADIN-OTWINOWSKA: Well, I mean, whatever you say about the detailed mechanism, the data is just the data, and I really don't see that you could achieve it by shifting the energy scale, that you could get a better agreement with e^+e^- in the sphericity.

SCHMITZ, MUNICH: If I understood your plots correctly, when you showed $< p_T^2 >$ versus energy, you always took all hadrons in an event. In most of the plots you didn't separate between forward-going and backward-going. I think it would be quite interesting to make that separation, to look at the backward and forward hemisphere separately, because from the lepton production one knows that quark fragmentation [forward hemisphere] and diquark fragmentation [backward hemisphere] behave differently [with energy] as far as p_T^2 is concerned. So I think it is a little bit dangerous for you to compare the p_T^2 distribution for your case, meson proton scattering, and e^+e^-. In one case you have quark anti-quark fragmenting and in the other case you have a quark and a di-quark fragmenting. So I would propose to make a separation between forward- and backward-going [hadrons] and then carry out the comparison.

BARDADIN-OTWINOWSKA: Well, for example, this plot is for forward hemisphere only. Now in this one on p_T relative to the sphericity axis as a function of energy, you have here the points from the ISR experiment where the effect of the leading proton was removed. So, in a way it perhaps partly satisfies [your request].

SCHMITZ: Well, I mean, your quark diagram is correct. Then in proton-proton you have diquark fragmentation in both hemispheres and, of course, in e^+e^- you have a different fragmentation.

BARDADIN-OTWINOWSKA: No, in the proton-proton the leading proton contribution is removed, so there is only the mesonic system that remains. That's all that is plotted.

SCHMITZ: No, but it is still diquark fragmenting into mesons. If you remove the leading proton . . . . Oh, you say the leading proton . . . .

BARDADIN-OTWINOWSKA: Was removed in the ISR experiments and only mesons were plotted.

ANDERSSON, LUND: In regard to Tom DeGrand's question whether we should expect that there should be a broadening [in p_T] in connection with the hadronic interactions in a similar way as in e^+e^- and other hard processes, I think that evidently we should expect a broadening. If you think about it, I mean, when you throw in more and more energy, evidently, if we have scaling [structure functions] you would have more and more of those "soft gluons" [i.e. with small Bjorken x] [with energies in the subsystems sufficiently large (e.g. a few GeV) to cause maybe not independent jets but transverse "bulbs" on the force fields] hitting each other. That means that you will evidently get, in a confined force field or whatever, which you are going to break up into final state hadrons, all kinds of small wrinkles. I mean you will get it a little bit excited here and there. It will be a little bit like a snake, or something like that, rather than something which is completely straight from the point of view of transverse momentum. However, certainly that would be very similar to what you would expect in a hard process where you need a certain number of soft gluons in bremsstrahlung. I think that it would be much more interesting to look upon particular correlations. For instance, if you look upon baryon anti-baryon production in different processes and you look upon what kind of z spectrum as compared to p_T you would have, then you would see things which are different in baryon production, for instance, than in meson production, etc. I'm not going to talk any more about this but I think that such structureless things which we have seen now do not say any more than that evidently there is a certain amount of general straggling on these essentially flat force fields.

CERN/EP/0475R/MSS/ed
11 September 1981

FORWARD DISTRIBUTIONS OF IDENTIFIED CHARGED PARTICLES AND NET CHARGE AND STRANGENESS DISTRIBUTIONS IN K^+p INTERACTIONS AT 70 GeV/c

Brussels-CERN-Genova-Mons-Nijmegen-Serpukhov Collaboration

Presented by M. Spyropoulou-Stassinaki
CERN, Geneva, Switzerland

ABSTRACT

We present preliminary results from 70 GeV/c K^+p interactions in BEBC filled with hydrogen, using the External Particle Identifier (EPI) to yield a separation of π^+ and K^+ mesons in the forward region. The single charged particle (π^+, π^-, K^+) longitudinal distributions are studied and compared to the quark counting rules. The (π^+/π^-) ratio is given for the K^+ fragmentation region. From linear combinations of the x_F distributions, the charged pion fragmentation functions are extracted. A comparison of the net charge and net strangeness distributions of the beam fragments as function of the c.m. rapidity y, gives an estimate of the charge and strangeness correlation lengths.

INTRODUCTION

Results from the CERN Big European Bubble Chamber (BEBC) used together with the CERN External Particle Identifier (EPI) to yield a separation of π^+'s and K^+'s in the momentum range of 10-70 GeV/c are presented. The longitudinal momentum distributions ($d\sigma/dx_F$) for inclusive π^+, K^+ and π^- production in the forward region are given together with π^+/π^- ratios as a function of x_F. Fits of the

Talk given at the XII International Symposium
on Multiparticle Dynamics, 21-26 June 1981, Notre Dame, USA

invariant structure functions to the form $(1 - x)^n$ are made and their results are compared with QCR predictions. The $D_u^{\pi^\pm}$ fragmentation functions have been extracted from the data in the context of the Lund fragmentation model [15(c)].

The beam and target fragmentation characteristics are investigated by studying the net charge and net strangeness distributions as functions of the c.m. rapidity.

EXPERIMENTAL SAMPLE

The present analysis is based on a sample of 25 000 events (corresponding to a sensitivity of 1.4 events/microbarn) obtained from an exposure of BEBC, filled with hydrogen to an r.f. separated beam of positive kaons of nominal momentum 70 GeV/c. Details of data taking, run conditions and cross sections can be found elsewhere [1,2].

In order to identify the positive charged particles with momentum higher than 10 GeV/c, in addition to the BEBC data, the EPI information is used. The EPI is a system of 4096 proportional counters arranged in 128 layers of 32 cells each, positioned 11 m behind BEBC (dimensions 8.5 x 1.9 x .9 m). It was constructed [11] in order to distinguish pions and kaons up to 70 GeV/c, by multiple sampling of the ionization in the region of the relativistic rise [11]. Each track can be sampled in up to 128 cells, but the background (crossing tracks, etc.) limits the number of usable cells. The resolution in the measurement of ionization was calibrated in a separate run [11]. The resolution during the experiment was found to be close enough to the calibration values to permit particle identification. The EPI software is written in order to follow the fast tracks from BEBC down stream to the EPI, to find the strings of cells corresponding to this track, to filter out the background and crossing tracks and using the sampled ionization to attempt to assign a mass to the track.

For the last step, namely the assignment of a mass to an individual track the selection criteria have been formed by taken into account two probability functions: (a) a probability derived from the measurement of the ionization and (b) a probability reflecting the momentum dependent population of kaons and pions. The latter was derived from the distribution of ionization of all EPI measured tracks studied in successive momentum bins. A paper describing these procedures is in preparation [16]. We estimate that our identified samples of π^+'s and K^+'s contain less than 15% contamination. The preliminary results presented here are not corrected for this contamination.

In table 1 the EPI acceptance is presented as a function of x_F and p_T^2. It is defined as the ratio of the number of positive outgoing tracks with good EPI information (cleaned tracks with more than 50 cells) over the number of all the positive outgoing tracks in the same phase space region (x_F, p_T^2). Approximately 23% of positive tracks with $x_F > .2$ have good EPI information.

π^+, K^+ AND π^- LONGITUDINAL DISTRIBUTIONS

In fig. 1 we present the $d\sigma/dx$ distributions of π^+'s, K^+'s and π^-'s and the corresponding structure functions for the reactions

$$K^+ p \rightarrow \pi^+ + X \tag{1}$$

$$K^+ p \rightarrow K^+ + X \tag{2}$$

and

$$K^+ p \rightarrow \pi^- + X \quad . \tag{3}$$

For details on reaction (3) see ref. [2]. It is clear that the $d\sigma/dx$ distribution for reaction (1) falls more steeply than that of reaction (2) in the region $x_F(0.2 - 0.7)$. The π^+'s and K^+'s structure functions show similar differences. In order to obtain a sample containing the non-diffractive events, we have introduced a first "diffractive cut", removing from the sample all events having one track with $x_F < -.80$. It is known that this cut is closely

Table 1 EPI acceptance, as it is defined in sect. 2.

P_T^2 \ x	0.2	0.3	0.4	0.5	0.6	0.7	0.8	0.9	1.0
1.6			.285	.182	.40		.60	.25	
1.4		.063	.280	.077	.256	.375	.143	.286	
1.2	.018	.068	.025	.320	.167	.25	.25	.667	.158
1.0	.0465	.016	.100	.205	.25	.143	.125	.133	.273
0.8	.0227	.104	.139	.239	.261	.222	.50	.182	.267
0.6	.021	.046	.178	.310	.348	.235	.227	.164	.312
0.4	.0122	.0835	.226	.323	.297	.266	.333	.208	.20
0.2	.015	.119	.300	.366	.283	.296	.213	.224	.235
0	.001	.145	.314	.320	.330	.225	.176	.171	.124

$$x = P_L^* / P_{max}^*$$

Table 2 Fits of the single particle structure functions by the expression $A(1 - |x|)^n$.

	Exp. n	x^2/ND	x_F range
$K^+ p \to K^+ + X$	.57 ± .13	10.4/9	.3 < x < .75
$K^+ p \to \pi^- + X$	2.7 ± .2	5.4/10	.3 < x < .85
$K^+ p \to \pi^+ + X$	1.3 ± .12	17/10	.3 < x < .85

Table 3 Fits of the net charge and net strangeness distributions by the expression $\frac{dN}{dy} = \exp[-(\frac{y - \bar{y}}{L_{Q(S)}})]$

L_S = .63 ± .37	χ^2/ND = .09/8,	-1.6 < y < 1.6
L_Q = 1.8 ± .5	χ^2/ND = 1.4/13,	-4.4 < y < .8
L_Q = 1.5 ± .5	χ^2/ND = 3/15,	-4.4 < y < 1.6

related to other methods e.g. rapidity gap and gives similar results. The cut removes the K^+ diffraction. The proton diffraction remains and its contribution is clearly seen in the K^+ distribution above $x_F > .85$, together with the elastic peak. Thus the removal from the sample of all diffractive events is obtained by the combined "diffractive cuts" $x_F < -.80$, $x_F > .85$.

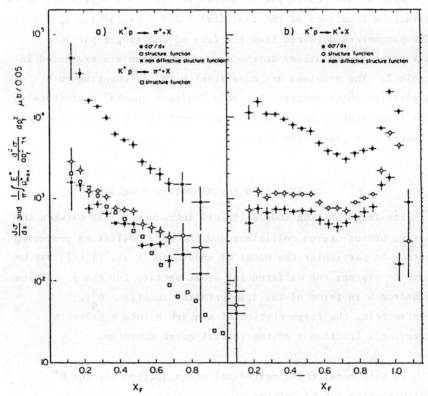

Fig. 1   (a) Forward longitudinal distributions for the reaction
$K^+ p \to \pi^+ + X$:
        $d\sigma/dx$,
        Structure function,
        "Non diffractive" structure function,
        Structure function for the reaction $K^+ p \to \pi^- + X$.

        (b) Forward distributions for the reaction $K^+ p \to K^+ + X$:
        $d\sigma/dx$,
        Structure function,
        "Non diffractive" structure function.

The ratio of the structure functions for reactions (1) and (3) is increasing as a function of x_F (fig. 2). According to the quark/parton model ideas the tendency of an increase for $x_F \to 1$ is attributed to the fact that the π^+'s carry a valence quark in common with the beam while π^-'s do not.

Several soft quark/parton models predict an x-dependence of the structure functions of the type $f(x) = A(1 - |x_F|)^n$ for $x_F \to 1$. The parameters obtained from the fits of the single particle invariant distributions to the above expression are presented in table 2. The n-values are consistent with the counting rule prediction which suggest that only "valence quarks" contribute in the fragmentation region and in agreement with previous results on charged pion production [3].

CHARGED PION FRAGMENTATION FUNCTIONS

The fragmentation models [15(c)] introduce a link between the low p_T hadron-hadron collisions and the hard collisions producing jets. In particular the model of Andersson et al. [15(c)] may be used to express the differential cross section for the production of a hadron h in terms of the fragmentation functions $D_q^h(x)$, representing the fragmentation of a quark q into a hadron h, carrying a fraction x of the initial quark momentum.

In this model the longitudinal cross section for the K^+ fragmentation can be written as

$$\frac{1}{\sigma} \frac{d\sigma^{K^+ \to h}(x)}{dx} = \frac{1}{2}[D_s^h(x) + D_u^h(x)]$$

where $\bar{s}$ and u are the valence quarks of the K^+. For the production of charged pions the above relation becomes

$$\frac{1}{\sigma} \frac{d\sigma^{K^+ \to \pi^\pm}(x)}{dx} = \frac{1}{2}[D_s^{\pi^\pm} + D_u^{\pi^\pm}(x)] \quad .$$

94

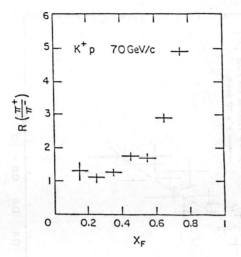

Fig. 2 π^+/π^- ratio in the forward region.

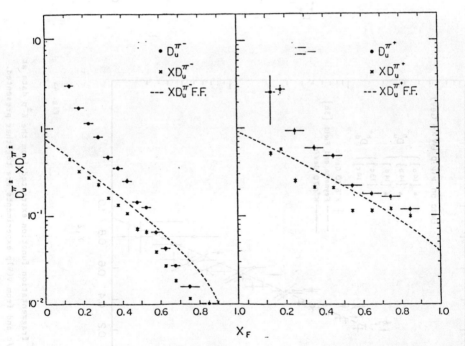

Fig. 3   The $D_u^{\pi^\pm}$ and $xD_u^{\pi^\pm}$ fragmentation functions extracted from K^+ data at 70 GeV/c. The dashed line is the $D_u^{\pi^\pm}$ derived from the FF model.

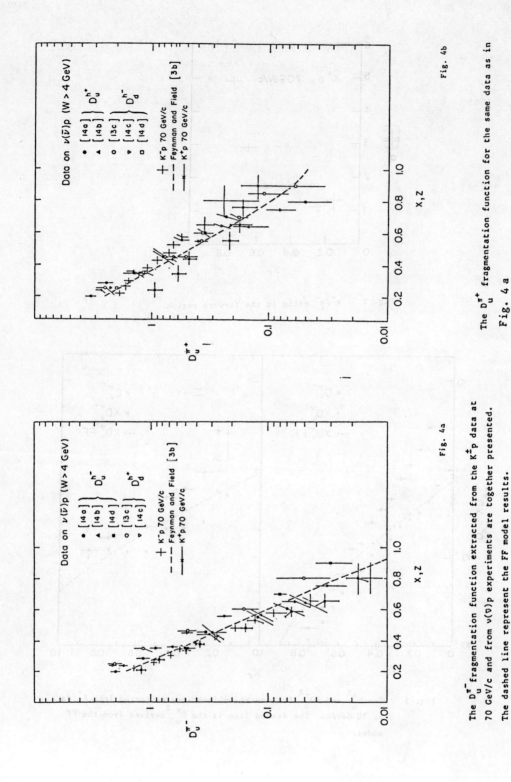

The $D_u^{\pi^-}$ fragmentation function extracted from the $K^{\pm}p$ data at 70 GeV/c and from $\nu(\bar{\nu})p$ experiments are together presented. The dashed line represent the FF model results.

Fig. 4a

The $D_u^{\pi^+}$ fragmentation function for the same data as in Fig. 4 a

Fig. 4b

Isospin and charged invariance reduce the number of independent D_q^h functions to the three $D_u^{\pi^+}$, $D_u^{\pi^-}$ and D_s^π. According to the Field and Feynman [15] prescription the approximation $D_u^{\pi^+} = D_s^{\pi^-}$ can be used, since the creation of $\pi^-(\bar{u}d)$ from a u quark or the production of π from s quark requires the creation of the same pairs of new quarks. Conversely the fragmentation functions can be derived from the differential cross sections. So for the fragmentation functions $D_u^{\pi^\pm}$ we obtain the following relations:

$$D_u^{\pi^-}(x) = \frac{1}{\sigma_{n.d.}} \frac{d\sigma^{K^+ \to \pi^-}(x)}{dx}$$

$$D_u^{\pi^+}(X) = \frac{1}{\sigma_{n.d.}} [2 \frac{d\sigma^{K^+ \to \pi^+}(x)}{dx} - \frac{d\sigma^{K^+ \to \pi^-}}{dx}]$$

In Fig. 3 we present the $D_u^{\pi^+}$ and $D_u^{n^-}$ obtained from our data together with the invariant fragmentation functions $xD_u^{\pi^-}$, $xD_u^{\pi^+}$.

The dashed lines correspond to the xD_u^π as they are derived from the Field and Feynman model [15]. The data and the model predictions are not in disagreement although there is a deviation at high x_F. Also a comparison of our D_u^π function with those of ref. [8] for K^- data and $\nu(\bar{\nu})$ data (fig. 4) show a satisfactory agreement. The symbol $\sigma_{n.d.}$ denotes the non-diffractive cross section which has been estimated by subtracting from the inelastic cross section the cross section corresponding to the events removed by the "diffractive cuts". We note that in our data the pions are identified individually by using the EPI information while in the K^-p data additional assumptions are used, based on an analogy between K^- and $\bar{K}^0$ production. The K^- data at 70 GeV/c are contained in a paper also presented in this conference [8].

NET CHARGE AND NET STRANGENESS STUDY

The net charge distribution [9,12] as function of the c.m. rapidity gives information about the separation of the beam and target fragments and is related to the average correlation length of the charged particles. Similar distributions for the net strangeness as function of c.m. rapidity allows the estimation of the average correlation length between the produced strange particles. Theoretical predictions [17] based on the Regge pole model suggest that the correlation lengths, $L_{Q(S)}$, obtained by fitting the above net Q and net S c.m. rapidity distributions in the central region by the expression $\sim e^{-(Y-y)/L_{Q(S)}}$ (1) will have values $LQ = 2$ and $L_S = 1$ respectively.

In this analysis the net charged distribution for kaon beam and proton target have been extracted using the following procedure.

We plotted the difference between the number of positive and negative particles by a function of the c.m. rapidity y for the inelastic K^+p reactions at 70 GeV/c, normalized to the inelastic cross section, namely (fig. 5 continuous line)

$$\text{Net } Q = \frac{dQ}{dy} = \frac{1}{s_{inel}} (\frac{d\sigma^+}{dy} - \frac{d\sigma^-}{dy})$$

The integral under this curve is 2.

We used also the corresponding distributions obtained from K^- data at 70 GeV/c [9] (fig. 5(a) dashed line). The integral of this distributrion in the forward region is 0.66.

In order to obtain the distribution for the kaon fragments we have taken the linear combination $1/2[dQ/dy(K^+p) - dQ/dy(K^-p)]$ (fig. 5(b)). Taking the combination $1/2(dQ/dy(K^+p) + dQ/dy(K^-p)]$ we obtain the distributions of the charged fragments of the proton (fig. 5(c)). The integral under the kaon curve for y > 0 is .78 ± .08. It is clear that the beam and target charge fragments are not completely separated at 70 GeV/c. This conclusion is in agreement with a similar comparison of $\pi^\pm p$ interaction at 100 GeV/c [12(a)] and in K^-p interactions study at the beam momentum range 10–110 GeV/c [12(b),9].

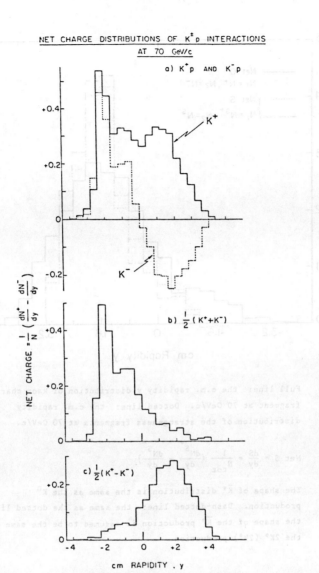

a) K$^+$ p AND K$^-$ p

K$^+$

K$^-$

NET CHARGE $= \frac{1}{N} \left(\frac{dN^+}{dy} - \frac{dN^-}{dy} \right)$

b) $\frac{1}{2}$ (K$^+$+K$^-$)

c) $\frac{1}{2}$ (K$^+$- K$^-$)

cm RAPIDITY , y

Fig. 5   (a) Net charge distributions of K$^+$p interactions at 70 GeV/c:

Net Q $= \frac{dQ}{dy} = \frac{1}{N} \left(\frac{dN^+}{dy} - \frac{dN^-}{dy} \right)$.

(b) c.m. rapidity distribution of the charged fragments of
target proton

$\frac{1}{2} \left(\frac{dQ}{dy}(\text{K}^+\text{p}) - \frac{dQ}{dy}(\text{K}^-\text{p}) \right)$ at 70 GeV/c.

99

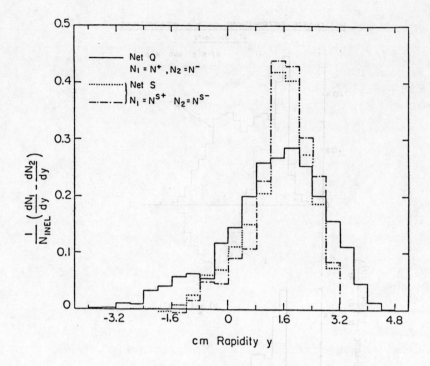

Fig. 6　Full line: the c.m. rapidity y distribution of kaon charged
fragment at 70 GeV/c. Dotted line: the c.m. rapidity
distribution of the strangeness fragments at 70 GeV/c.

$$\text{Net } S = \frac{dS}{dy} = \frac{1}{N_{tot}} \left(\frac{dN^{S^+}}{dy} - \frac{dN^{S^-}}{dy} \right)$$

The shape of $\bar{K}^0$ distribution is the same as the K^n
production. Dash-dotted line: the same as the dotted line but
the shape of the $\bar{K}^0$ production is assumed to be the same with
the $2K^0$ ($2V^0$) production.

For the net strangeness distributions we worked as follows. We plotted the c.m. rapidity distributions for the neutral strange particles K^n, Λ, $\bar{\Lambda}$ where K^n denotes a neutral kaon (K^0 or $\bar{K}^0$) (properly weighted for their detection probability) [5,6]. The K^+ sample was obtained from the EPI data by identification of K^+'s corrected for the EPI acceptance.

In fig. 6 (dotted line) we present the difference

$$\text{Net } S = \frac{dS}{dy} = \frac{1}{\sigma_{inel}} \left(\frac{d\sigma^{S^+}}{dy} - \frac{d\sigma^{S^-}}{dy} \right)$$

where σ^{S^+} is the cross section for strange particles produced with the same strangeness as that of the beam and σ^{S^-} is the cross section for strange particles with opposite than the beam's strangeness. In our case Λ and $\bar{K}^0$ are carrying the negative strangeness S^-. For the $\bar{K}^0$ contribution we made two extreme estimations. The first assumes that the $\bar{K}^0$ distribution has the same shape as the K^n distribution with an integral normalized to the measured $\bar{K}^0$ cross section. This assumption tends to overestimate the forward $\bar{K}^0$ production. The second assumes that the $\bar{K}^0$ has the shape of the K^n distribution, as observed for events with only two V^0's in the final state reconstructed as K^n by the kinematics. In this case the estimate of the $\bar{K}^0$ production tends toward an overestimate of a central component. Again the integral is normalized to the calculated $\bar{K}^0$ cross section of 1.6 mb. The two results obtained by using these assumptions for $\bar{K}^0$ distributions are indicated by the overall net S curves (first assumption by dotted line, second assumption by dash-dotted line).

We fitted the above dQ/dy and dS/dy distributions in the region $y < 2$ by the expression (1) in order to compare with the theoretical predictions. The results are included in table 3. They show values which are consistent, within errors, with these predictions and in particular that $L_Q > 2L_S$.

SUMMARY

We present the first preliminary results using the hybrid system
BEBC with EPI to identify the high momentum tracks (10-70 GeV/c).
From the study of the single particle longitudinal distributions of
the identified positive tracks (π^+ and K^+) and π^- in the beam
fragmentation region we observe that:

(a) The π^+'s structure function is flatter than the K^+'s one. The
parametrization with the expression $A(1 - x)^n$ yields values for
the exponent n which are consistent with the "valence quarks"
counting rules. The ratio (π^+/π^-) is increasing rapidly at
$x \to 1$, supporting the quark/parton model ideas that when the
outgoing particle has a valence quark in common with the beam
the x_F distribution falls less steeply.

(b) The charged pion fragmentation functions, $D_u^{\pi^\pm}$, extracted from
our data are consistent with those extracted from other hadron data
(K^-p at 70 GeV/c) and ν, $\bar{\nu}$ data and also in general agreement
with predictions from the FF fragmentation model.

(c) The net charge distributions of the beam and target charge
fragments at 70 GeV/c plotted as a function of the c.m.
rapidity, indicate that the beam and target charge fragments
are not completely separated at 70 GeV/c confirming the
conclusions from previous studies [12,9].

(d) For the K^+p interactions the net strangeness distribution has
different shapes from that of the net charge distribution of
the beam fragments as a function of the c.m. rapidity. By
fitting these two distributions with the formula
$\sim \exp[-(Y-y)/(L_{Q(S)})]$ we get an estimate of the average
correlation lengths $L_Q \simeq 1$ and $L_S \simeq 2$ for the net charge
and net strangeness respectively in agreement with theoretical
predictions.

Acknowledgements

It is a pleasure to thank our scanning and measurement staff at our respective laboratories and the operating crews of BEBC and the SPS accelerator and members of the EF Division for their help with the RF beam. In particular we thank the EF members who designed, constructed and tested the EPI counter. Finally, we acknowledge the work of the physicists who have been involved in the development of the EPI software.

REFERENCES

[1]   M. Barth et al., Zeitschrift für Physik C 2 (1979) 285.

[2]   M. Barth et al., Zeitschrift für Physik C 7 (1979) 187.

[3]   M. Barth et al., Zeitschrift für Physik C 7 (1981) 89.

[4]   L. Gerdyukov, CERN/BEBC-Bug 80-4, BEBC-BUG/PR13, 17 July 1980.

[5]   IFGE/II-81A, June 1981.

[6]   CERN/EP 81-51, 2 June 1981.

[7]   CERN/EP 81-44, 6 May 1981.

[8]   Study of longitudinal distributions in the fragmentation regions using the quark parton model in K⁻p interactions at 70 GeV/c, Sicily, 9 March 1981, Paris-Rutherford-Saclay Collaboration.

[9]   R. Göttgens et al., Zeitschrift für Physik 9 (1981) 13.

[10] D.Ph.PE 79-05, April 1979.

[11] M. Aderholz, P. Lazeyras, I. Lehraus, R. Matthewson and W. Tejessy:

      (a) CERN/EF/BEAM 77-3, 8 November 1977;

      (b) CERN/D.Ph.II/BEAM 74-5, 14 August 1974 and

      (c) Nucl. Instr. & Methods 153 (1978) 347.

[12] (a) J. Whitmore et al., Phys. Rev. vol. 16 No. 11 (1977) 3137;

      (b) P. Bosetti et al., Nucl. Phys. B62 (1973) 46-60;

      (c) Dennis Sivers, Phys. Rev. D, vol. 8, No. 7 (1973) 2272.

REFERENCES (Cont'd)

[13] (a) B. Musgrave, Proceedings of the Int. Conf. on
         Neutrinos, Bergen (1979).

     (b) N. Schmitz Int. Symposium on Lepton and Photon
         Interactionsat High Energy, Fermilab 1979;

     (c) M. Derrick et al., Phys. Lett. 91B (1980) 470.

[14] (a) J. Bell et al., Phys. Rev. D19 (1978) 1;

     (b) P.C. Bosetti et al., Nucl. Phys. B149 (1979) 13.

     (c) J.P. Berge et al., Proceedings of the Int. Conf. on
         Neutrinos, Bergen 1979;

     (d) H.J. Lubatti, id. Bergen 1979 (data on $\nu\bar{\text{He}} \rightarrow h\ X^-$ and
         $\nu\text{He} \rightarrow h\ X$).

[15] (a) R.D. Field and R.P. Feynman, Phys. Rev. D15 (1977) 2590;

     (b) R.D. Field and Feynman, Nucl. Phys. B136 (1978) 1;

     (c) B. Andersson et al., Phys. Lett. 69B (1977) 221 and
         refs in it.

[16] Forthcoming technical report on EPI.

[17] T. Inami and H. Miettinen, Phys. Lett., vol. 49B No. 1 (1974) 67;

     P. Grassberger, C. Michael and H. Miettinen, Phys. Lett.
     vol. 52B, No. 1 (1974) 60 and private communication.

CERN/EP/0473R/HF/ef
28 July 1981

NEUTRAL STRANGE PARTICLE PRODUCTION IN PROTON-PROTON COLLISIONS

AT $\sqrt{s}$ = 63 GeV

CERN-Dortmund-Heidelberg-Warsaw Collaboration

D. Drijard, H.G. Fischer, H. Frehse, P. Hanke, P.G. Innocenti,
J.W. Lamsa[*], W.T. Meyer[*], G. Mornacchi, A. Norton, O. Ullaland
and H. Wahl[**]
CERN, European Organization for Nucl. Research, Geneva, Switzerland

W. Hofmann, T. Lohse, M. Panter, K. Rauschnabel, J. Spengler and
D. Wegener
Institut für Physik der Universität Dortmund, Germany

W. Geist, M. Heiden, W. Herr, E.E. Kluge, T. Nakada, A. Putzer and
W. Rensch
Institut für Hochenergiephysik der Universität Heidelberg, Germany

K. Doroba, R. Gokieli and R. Sosnowski
Institute of Experimental Physics of Warsaw, University and Institute
for Nuclear Research, Warsaw, Poland

Presented by H. Frehse

ABSTRACT

The inclusive cross section for the production of K_s^o mesons and
Λ^o particles in proton-proton interactions at $\sqrt{s}$ = 63 GeV is presented. The produced particles have been detected in the full phase
space. Behaviour of the longitudinal and transversal dependences of
the cross sections are discussed. The total production cross
sections for K_s^o mesons and Λ^o particles was determined to $\sigma_{K_s^o}$
= (13 ± 0.38)mb and σ_{Λ^o} = (4.06 ± 0.55)mb respectively.

Presented at the XII Intern. Symposium on Multiparticle Dynamics
21-26 July 1981, Notre Dame, Indiana, USA

(*) Visitor from Ames Laboratory, Iowa State University, USA.
(**) Now at Institut für Hochenergiephysik, Vienna, Austria.

EP/0473R/HF/ef

1. INTRODUCTION

The investigation of strange particle production in hadronic interactions at high energies is of particular interest, since a new quantum number is created.

Up to now detailed measurement for the reactions

$$pp \rightarrow K^0_s + X \tag{1a}$$
$$pp \rightarrow \Lambda^0 + X \tag{1b}$$
$$pp \rightarrow \bar{\Lambda}^0 + X \tag{1c}$$

only exist at lower energies [1-13]. In the region of energies accessible at the CERN proton-proton Intersecting Storage Rings (ISR), the inclusive cross section for neutral strange particle production ($V^0 = K^0_s$, Λ^0, $\bar{\Lambda}^0$) production is determined up to now only for limited parts of the phase space [14-16]. The present paper describes the results of an experiment in which the inclusive cross section for V^0 particle production has been measured in the full phase space (4π steradian) at $\sqrt{s}$ = 63 GeV.

2. EXPERIMENT AND DATA REDUCTION

2.1 Experimental set-up and trigger

The experiment was performed at the CERN ISR using the upgraded Split Field Magnet detector (SFM) [17], which allows to measure the momenta of charged particles in 4π steradian. In previous publications, the experimental set-up [17-18] and the trigger (minimum bias trigger), have been described [19,20]. The cross section seen by the trigger amounts to $\sim$ 95% of the inelastic cross section.

The raw data were processed with the SFM off-line program chain to reconstruct those tracks which result from the primary interaction vertex.

2.2 V^0 particle reconstruction

In a second stage of the track finding, the secondary vertices from V^0 decays were reconstructed. A special fit procedure was designed to determine simultaneously the geometrical and
kinematical variables of a V^0 decay, taking into account the
relations between these variables due to the magnetic field. A
detailed description of this procedure is given in ref. [21]. In
fig. 1(a-c) the distributions of the invariant mass after the
vertex fit is shown for those V^0 particles, where momentum is
determined with a precision of $\Delta p/p < 0.1$ and where the decay
length ℓ is five times larger than the error on its decay length,
determined in the fit. The mass resolution is 40 MeV FWHM and
10 MeV FWHM for K^0_s mesons and Λ^0 particles respectively.

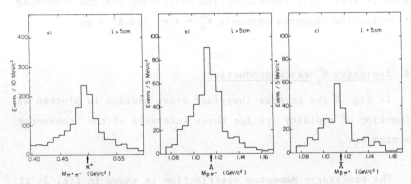

Fig. 1 Invariant mass spectrum after the secondary vertex
fit for good V^0's [21]: (a) $K^0_s \to \pi^+\pi^-$, (b) $\Lambda^0 \to p\pi^-$ and
(c) $\bar{\Lambda}^0 \to p\pi^+$.

A careful study of the sources of background, losses and
ambiguities has led us to apply the following cuts to get a
sample of uniquely reconstructed V^0 particles:

- V^0 fit with $\chi^2 < 18$,
- momentum uncertainty $\Delta p/p < 0.1$,
- decay length $\ell \geqslant 1$ cm,
- $\ell \geqslant 5 *$ uncertainty of decay length,
- $\cos\theta^*$ cut to remove e^+e^- pairs in $\Lambda^0(\bar{\Lambda}^0)$ sample.
- $P_{lab} \geqslant 0.1$ GeV/c.

After these cuts the probability to confuse V^0 particles is smaller than 5%.

3. EXPERIMENTAL RESULTS

A sample of 58 200 minimum bias events was used to determine the differential and total cross section for the reaction (1). The sample of uniquely reconstructed V^0 particles analyzed in this experiment, consists of 2773 K_s^0 mesons and 758 Λ^0 particles. Only data in phase space regions, where the acceptance is larger than 5% have been used for cross section calculations. This acceptance includes the condition of uniqueness. The data have been corrected for detector acceptances including the cuts discussed in sect. 2.2, reconstruction efficiency and the branching ratios for the observed channels $K_s^0 \rightarrow \pi^+\pi^-$ and $\Lambda^0 \rightarrow p\pi^-$.

3.1 Inclusive K_s^0 meson production

In fig. 2 the Lorentz invariant cross section is plotted as a function of rapidity $|y|$ for three intervals of the transverse momentum p_T.

The transverse momentum distribution is shown in fig. 3; it is compatible with an exponential dependence on p_T. A fit to our data gives a slope of $b = (4.68 \pm 0.08) \text{GeV}^{-1}$. This value is compatible with the values found for the production of K^+ and K^- mesons [22,23]. The p_T dependence of the Lorentz invariant cross section at $y = 0$ is shown in fig. 4; for comparison the results of Büsser et al., [15] at lower ISR energies (averaged over the energy interval 30 GeV $< \sqrt{s} <$ 53 GeV) are included. The mean value

$$E \frac{d^3\sigma}{d^3p} = \frac{1}{2} \left[E \frac{d^3\sigma}{d^3p} (pp \rightarrow K^+ + X) + E \frac{d^3\sigma}{d^2p} (pp \rightarrow K^- + X) \right] (2)$$

of ref. [23], calculated from the fit to this data, is in good agreement with the cross section of reaction (1a) as measured in this experiment (solid line in fig. 2).

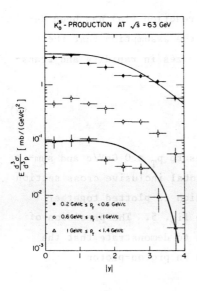

Fig. 2 Lorentz invariant cross section for K_S^0 mesons. The full lines represent interpolated values from ref. [23] at $\sqrt{s}$ = 53 GeV.

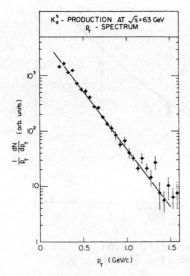

Fig. 3 Transverse momentum distribution for K_S^0 mesons.

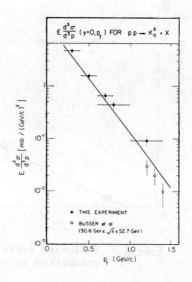

Fig. 4 Lorentz invariant cross section at y = 0 as a function of p_T including data from Büsser et al., [14].

The total cross section for the reaction (1a) has been determined from the present data under the assumption that the Lorentz invariant cross section factorizes in rapidity and transverse momentum

$$E \frac{d^3\sigma}{d^3p} = \frac{d\sigma}{dy} e^{-bp_T} \,. \qquad (3)$$

Extrapolating the measured values to $p_T = 0$ GeV/c and summing over all rapidities one gets a total inclusive cross section of $\sigma(pp \to K_s^o + X) = (13 \pm 0.38)$mb, which is plotted together with the results at lower energies in fig. 5. The precision of the present experiment is good enough to demonstrate that the cross section for K_s^o meson production in proton-proton collisions rises as ℓn^2 s.

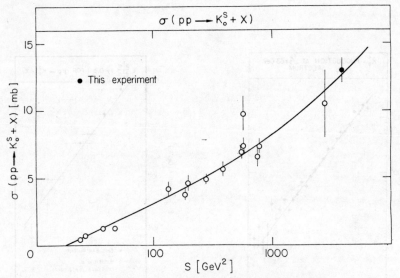

Fig. 5 Inclusive cross section for K_s^o meson production as a function of s [1-16].

3.2 Lambda particle production

The Lorentz invariant cross section for the reaction (1b) is plotted as a function of the Feynman variable $|x|$ in fig. 6 for

110

three intervals of the
transverse momentum. For
comparison, the results of
ref. [15] in the fragmentation
region are included (full
line). The two experiments
are in good agreement in the
region of the overlap.

In fig. 7 the transverse
momentum distribution for Λ^0
particle production is shown.
The shape of the distribution
is similar to that one for
proton production [22]

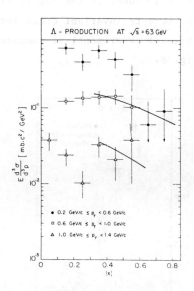

$$E \frac{d^3\sigma}{d^3p} \sim e^{-4p_T} , \qquad (4)$$

<u>Fig. 6</u> Lorentz invariant cross
section for Λ^0 production
including the results of
ref. [15] as full lines.

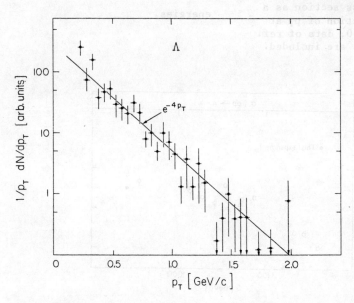

<u>Fig. 7</u> Transverse momentum distribution for Λ^0 particles. The
full line represents the shape measured for proton
production [23].

which is included in the figure (full line). The p_T dependence
of the Lorentz invariant cross section at x = 0 as determined in
the present experiment, is compared with the results of ref. [14]
in fig. 8.

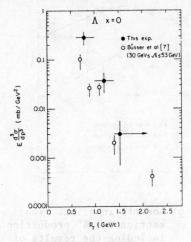

Fig. 8 Lorentz invariant
cross section as a
function of p_T at
x = 0, data of ref.
[14] are included.

Assuming again that factorization holds (formula (3)),
the total inclusive cross section for Λ^0 production has been
derived from the measured
differential cross section.
The value obtained is
$\sigma(pp \rightarrow \Lambda^0 + X) = (4.06 \pm 0.55)$mb.
It is shown together with the
results of other experiments in
fig. 9. The inclusive cross
section for Λ^0 production
seems to level off at higher
energies.

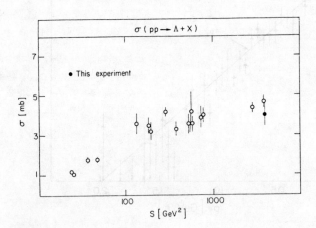

Fig. 9 Inclusive cross section for Λ^0 production as a function
of s [1-16].

4. DISCUSSION OF THE RESULTS

The production cross section for K_s^o mesons in
proton-proton collisions shows the same rapidity and transverse
momentum dependence as the cross section for K^+/K^- production
(figs 2 and 4). This behaviour is expected in constituent models
of particle production [24], since the quark content of the
detected K_s^o mesons is comparable to the quark composition of
a mixture of charged K mesons. The observed $\ln^2 s$ dependence
of the inclusive cross section (fig. 5) is compatible with a
dominant central production mechanism [16].

The Λ^0 production shows a similar behaviour as the
production of neutrons. The similarity is demonstrated by
fig. 10, where the X dependence of the Lorentz invariant cross
section for the reaction [25]

$$pp \rightarrow n + X$$

and the result of the present experiment for the reaction (1b)
are compared.

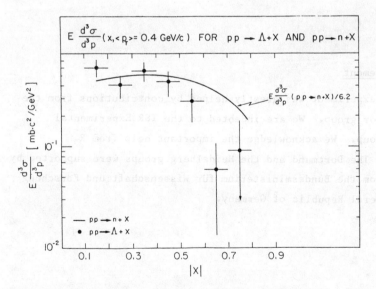

<u>Fig. 10</u> Comparison of the Lorentzinvariant cross section at
$p_T = 0.4$ GeV for $pp \rightarrow \Lambda^0 + X$ as measured in this
experiment and for $pp \rightarrow n + X$ [25].

5. CONCLUSIONS

The inclusive cross section for the production of K^o_s mesons and Λ^0 particles have been measured at the highest presently accessible energies:

(i) The K^o_s meson production is compatible to the mean value of the K^+ and K^- meson productions.
$\sigma(pp \rightarrow K^o_s + X)$ increases with ℓn^2 s.

(ii) The inclusive cross section for Λ^0 particle production is similar to that one for neutrons. $\sigma(pp \rightarrow \Lambda^0 + X)$ shows only a weak energy dependence, if at all.

The observed features are in concordance with the expectations of constituent models for particle production.

Acknowledgement

This experiment was greatly helped by contributions from the SFM detector group. We are indebted to the ISR Experimental Support group. We acknowledge the important help from R. Messerli. The Dortmund and the Heidelberg groups were supported by a grant from the Bundesministerium für Wissenschaft und Forschung of the Federal Republic of Germany.

REFERENCES

[1]   K. Jaeger et al., Phys. Rev. D11 (1975) 1756.

[2]   V. Blobel et al., Nucl. Phys. B69 (1974) 454;
      H. Fesefeldt et al., Nucl. Phys. B135 (1978) 379.

[3]   H. Bøggild et al., Nucl. Phys. B57 (1973) 77;
      P. Achlin et al., Physica Scripta 21 (1980) 12.

[4]   H. Blumenfeld et al., Phys. Lett. 45B (1973) 528.

[5]   M. Alston-Garnjost, Phys. Rev. Lett. 35 (1975) 142.

[6]   J.W. Chapman et al., Phys. Lett. 47B (1973) 465.

[7]   D. Brick et al., Nucl. Phys. B164 (1980) 1.

[8]   K. Jaeger et al., Phys. Rev. D11 (1975) 2405.

[9]   A. Steng et al., Phys. Rev. D11 (1975) 1733.

[10] F.T. Dao et al., Phys. Rev. Lett. 30 (1973) 1151.

[11] F. Lopinto et al., Phys. Rev. D22 (1980) 573.

[12] R.D. Kass et al., Phys. Rev. D20 (1979) 605.

[13] H. Kichimi et al., Phys. Rev. D20 (1979) 37.

[14] F.W. Büsser et al., Phys. Lett. 61B (1976) 309.

[15] S. Erhan et al., Phys. Lett. 85B (1979) 447.

[16] ACCDHW Collaboration, D. Drijard et al., CERN/EP 81-12 and
      Zeitschr. für Physik C (in print).

[17] W. Bell et al., Nucl. Instr. & Meth. 156 (1978) 111.

[18] CCHK Collaboration, M. Della Negra et al., Nucl. Phys. B128
      (1977) 1.

[19] W. Bell et al., Nucl. Instr. & Meth. 125 (1975) 437.

[20] CCHK Collaboration, D. Drijard et al., Nucl. Phys. B155
      (1979) 269.

[21] K. Rauschnabel, V^0 reconstruction in the Split Field Magnet
      Spectrometer, CERN/EP/Int. Report 81-1, April 1981;

      K. Rauschnabel, Thesis University of Dortmund 1981
      (unpublished).

[22] J. Singh et al., Nucl. Phys. B140 (1978) 189.

[23] B. Alper et al., Nucl. Phys. B87 (1975) 19.

[24] V.M. Shekhter and L.M. Skcheklova, Sov. J. Nucl. Phys. 27
      (1978) 567.

[25] J. Engler et al., Nucl. Phys. B84 (1975) 70.

[26] L. Van Hove, CERN/TH 2997 (1980);
      Yu V. Fisjak and E.P. Kistenev, CERN/EP 80-209.

SESSION A2

SOFT-HADRON PHYSICS: EXPERIMENT II

Monday, June 22, p.m.

Chairman: A. Białas
Secretary: V. P. Kenney

I) A NEW O^-S MESON AND NEW RESULTS ON THE 1^+S STATE IN THE 3π SYSTEM COHERENTLY PRODUCED ON NUCLEI.-

II) 5π COHERENT PRODUCTION ON NUCLEI.-

G. Bellini , M. di Corato , P.L. Frabetti , Yu I. Ivanshin ,
L.K. Litkin , D. Menasce , S. Otwinowski , F. Palombo , J. Pernegr ,
A. Sala , S. Sala , S.I. Sychkov , A.A. Tjapkin , I.M. Vassilievski ,
G. Vegni , V.V. Visniakov , O.A. Zaimidoroga .

Dubna - Milano Collaboration

Presented by J. Pernegr

XII INTERNATIONAL SYMPOSIUM ON MULTIPARTICLE
DYNAMICS

Notre Dame , INDIANA   JUNE 21 - 26 1981

I - The behaviour of the hadronic matter immediately after it has been produced can be studied in principle of the production takes place inside the nucleus. The new-born hadronic system can interact with the nucleons before it reaches asymptotic conditions.

A systematic study of the hadronic states going into 3π and 5π channels has been carried out by our collaboration. The channel :

$$\pi^- A \to \pi^- \pi^- \pi^- \pi^- A \qquad (1)$$

with A = Be, C, Al, Si, Ti, Cu, Ag, Ta, Pb has been studied at 40 Gev at the Serpukhov PS Accelerator, with a statistics of $\sim$ 120,000 events.

Details of the experimental set-up can be found in ref. 1[1].

The coherence mechanism is strongly present in the data in the full mass range, as shown by the $t'=|t-t_{min}|$ distributions (see as an example in fig. 1 the t' distributions of the 3π system in the low mass region).

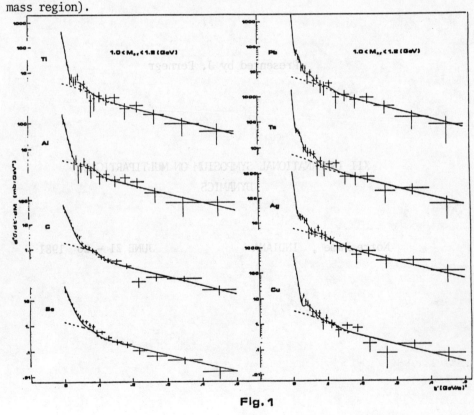

Fig. 1

Coherent samples are defined for every target selecting the events with t' smaller than the first diffractive minimum. Total cross sections for the coherent production in the full t' range were also measured by subtracting the incoherent background.

The data of the channel (1) were analyzed using the program PWA (Partial Wave Analysis)[2]. The set of important waves in the coherent sample consists of eight contributions: 1^+S, 1^+P, 1^+D; 0^-S, 0^-P; 2^-S, 2^-P, 2^+D. The spin flip amplitudes have been found negligible. The behaviour of the contributions and of the phase differences for the more important waves has been investigated as a function of $M_{3\pi}$, t' and of the atomic weight (A) of the nuclear target.

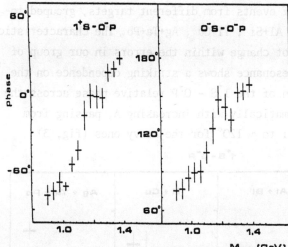

First of all we have performed the 3π mass dependent PWA in the sample of the events of all targets together (fig. 2). The mass shapes and the phase variations show not only for 1^+S, but also for 0^-S amplitudes, a resonance behaviour in the A_1 region.

The parameters of the 0^-S resonance have the following values:

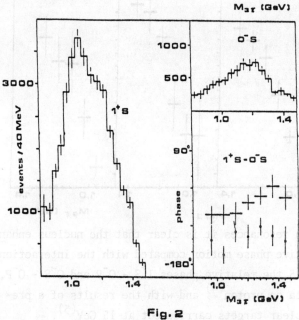

Fig. 2

121

$M_{3\pi} \simeq 0.20 \pm 0.03$ GeV and $\Gamma \simeq 0.330 \pm 0.040$ GeV; its phase variation is $\sim 80°$ (fig. 2).

The quantum numbers, the mass and the strong decay of this state correspond, in terms of the quark model, to the first radial excitation of the pion (π') [3].

The contribution and the phase variations of the amplitudes have been studied in samples of events from different targets, grouped together as follows: Be+C , Al+Si , Ti+Cu , Ag+Ta+Pb. The characteristics of the 0^-S resonance do not change within the errors in our group of elements, while the 1^+S resonance shows a striking dependence on the nuclear target. The motion of the 1^+S - 0^-P relative phase across the A_1 region increases systematically with increasing A, passing from $\sim 60°$ for the light nuclei to $\sim 120°$ for the heavy ones (fig, 3)

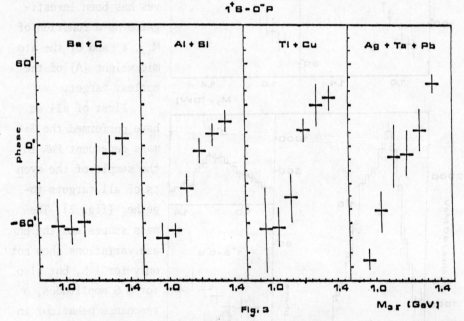

Fig. 3

In any case for both resonances it is clear that the nucleus enhances drastically the relative phase motion compared with the interaction on proton. In figs 4 and 5 the relative phases 1^+S - 0^-P and 0^-S - 0^-P, are compared with the data on proton[4] and with the results of a previously experiment on nuclear targets carried out at 15 GeV[5].

The phase motions of 1^+S and 0^-S seem to increase with the incident energy in the coherent interactions on nuclei.

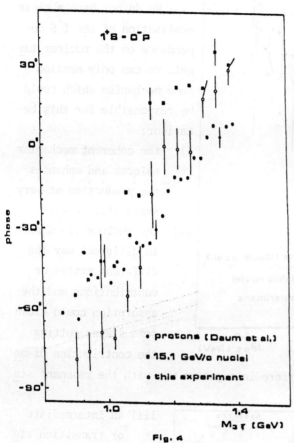

$1^+s - 0^-p$

phase

30°

0°

-30°

-60°

-90°

• protons (Daum et al.)
■ 15.1 GeV/c nuclei
● this experiment

1.0   1.4

$M_{3\pi}$ (GeV)

Fig. 4

A different contribution of 1^+S and 0^-S amplitudes as a function of the target atomic weight has been found analyzing the samples on the different targets. In the coherent region and in the full mass range the percentages of the 1^+ state increases with increasing A, while the 0^- state percentages defini tely decreases. The increase of the 1^+ state is due mostly to the 1^+S wave; the decrease of the 0^- to 0^-S wave. These effects are cle arly shown by figs. 6a and 6b (see next page).

While in the light nuclei the 1^+ and 0^- waves exhibit a flat dependence on t', in the heavy targets the 1^+ waves have a clear enhancement and the 0^- waves a strong depression in the coherent region.

A different dependence on t' for the different nuclear targets is shown also by the relative phase $1^+S - 0^-P$, which is nearly constant in the light nuclei, while changes fastly at small t' in the heavy tar gets (fig. 7).

From the previous results two sound conclusions can be drawn:

i)   the nucleus is a powerful mean in selecting and enhancing resonan-
     ces, as 1^+S and 0^-S;

ii) the characteristics of the 0^-S resonances do not change with the
     nuclear target, while the 1^+S moves and it is more pronounced as

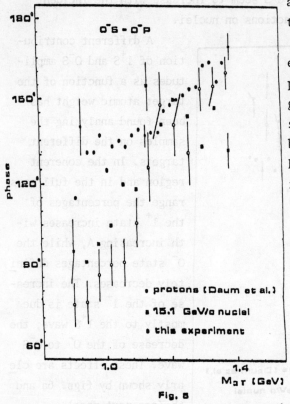

phase

180°

O⁻S - O⁻P

150°

120°

90°

● protons (Daum et al.)
■ 15.1 GeV/c nuclei
○ this experiment

60°

1.0   1.4
M₃ᵣ (GeV)

Fig. 5

as larger as the nuclear a
tomic weight;

We do not have a clear
explanation of the 1⁺S de
pendance on the nuclear tar
get. We can only mention
some mechanism which could
be responsible for this be
haviour:

i)   the coherent mechanism
     selects and enhances
     the production at very
     small t'.

ii) the nucleus can absorb
     in different way the
     different states or
     contributions and the
     absorption could clean
     some states cutting
     the contribution of me

chanism, which normally interfere in negative way with the resonant sta
tes.

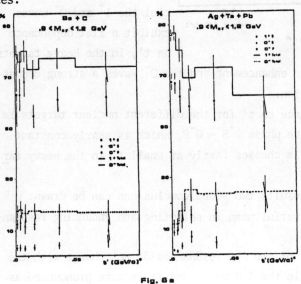

%   Be + C
         .8 < M₃ᵣ < 1.2 GeV

%   Ag + Ta + Pb
         .8 < M₃ᵣ < 1.2 GeV

t' (GeV/c)²   t' (GeV/c)²

Fig. 6 a

iii) an intermediate
     or transition sta
     te after the ha
     dron-hadron colli
     sion but before
     fixed final states
     have been reached,
     probably exists;
     during this tran
     sition time the
     new-born hadronic
     matter could inter
     act with the nu
     cleons and its cha

124

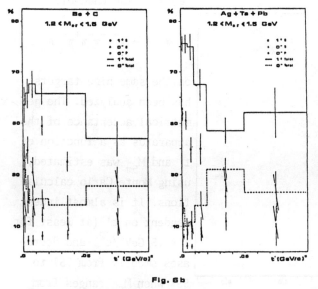

Fig. 6b

racteristics can be cha
nged, as it is found
for 1^+S.

A more detailed a-
nalysis of the nuclear
effects on the contribu
tion of the different
waves can be carried out
using a specific model.
To this purpose the Köl-
big-Margolis-Glauber mo
del has been used, even
if in its approach the
interaction is assumed
to be instantaneous and point-like. As well known, the nuclear absorption is measured by the parameter σ_2 , which in the frame of this model
is interpreted as the collision cross-section between the state under
investigation and the bound nucleons.

The best fit value obtained for σ_2 is of the order of 15 mb, with
small fluctuation for the different mass regions, has been evaluated
also for the different part-
ial waves. 0^- and 1^+ give
for σ_2 the following ranges:
21 - 30 and 11 - 16 mb, respectively, which reflects the
different contributions of
these waves in the coherent
region as shown in figs. 6a
and 6b.

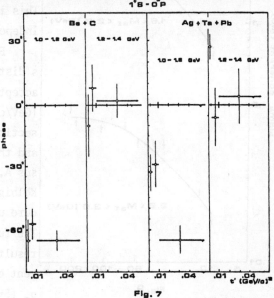

Fig. 7

II. 5π coherent production on nuclei.

A sample of ∿ 15,000 e-

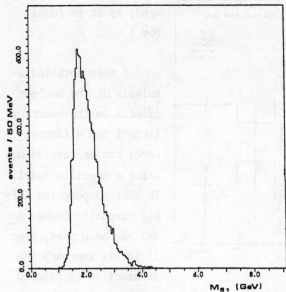

Fig. 8

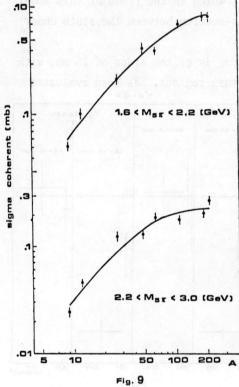

1.6 < $M_{5\pi}$ < 2.2 (GeV)

2.2 < $M_{5\pi}$ < 3.0 (GeV)

Fig. 9

vents of the channel:

$$\pi^- A \to \pi^- \pi^- \pi^- \pi^+ \pi^+ A$$

on the same nine targets has been analyzed. The geometrical acceptance of the apparatus as a function of t' and $M_{5\pi}$ was estimated, using Monte Carlo calculations. It is almost independent on t' (at least for t' $\leq$ .5 GeV^2/c^2) and decreases smoothly from 75% to 55% when $M_{5\pi}$ ranges from 1.8 to 3.0 GeV/c^2.

The inefficiency of the reconstruction programs was calculated recovering the lost events by means of an interactive graphic system[6]. This inefficiency is $\sim$ 25%, almost independent of t' and $M_{5\pi}$ values.

Fig. 8 shows the five-pion mass distribution, not corrected for acceptance and with t' cut at t'=.5 $(GeV/c)^2$. The differantial cross-section $d\sigma/dt'$ versus t' versus A and the total cross-section σ_c versus A, have been fitted using the Kölbig-Margolis formula. The procedure used was the same as in ref. 1. The full line drawn in fig. 9 is the result of the fit on the total coherent cross-section. The parameter σ_2 (5π) is always definitely smaller than 10 mb and tends to decrease with increasing $M_{5\pi}$

REFERENCES

1) G. Bellini et al., The (3π)-nucleon collision in coherent production on nuclei at 40 GeV/c; Preprint CERN-EP/81-40 (1981). Submitted to Nuclear Physics.
2) G. Ascoli et al.,Phys. Lett. 25 (1970) 962
3) For example: D.P. Stanley and D. Robson; Phys. Rev. 21 (1980) 3180
4) C.Daum et al., Diffractive production of 3π states at 63 and 94 GeV. CERN-EP/80-219 (1980)
5) J. Pernegr et al., Nucl. Phys. B134 (1978) 436
6) D. Menasce, F. Palombo, S. Sala; Comp. Phys. Comun. 22 (1981) 317

FRAGMENTATION OF u-QUARK, d-QUARK AND THE (ud)-DIQUARK SYSTEM -
CAN IT BE MEASURED IN K⁻p REACTIONS BY SELECTION OF FAST $\bar{K}^{0'}$ s OR Λ's ?

B. Buschbeck, H. Dibon

Institut für Hochenergiephysik, Vienna, Austria

ABSTRACT

A π^+/π^- ratio of 1 is observed in the proton fragmentation region in K⁻p reactions at 16 GeV/c with a fast $\bar{K}^0$ or Λ. This leads to the conjecture that in both cases a u-quark of the proton is removed by annihilation against the rather slow $\bar{u}$-quark of the incoming K⁻. A u- and d-quark are left in the fragmentation region yielding equal numbers of π^+ and π^-. In this region pion production in reactions with a fast $\bar{K}^0$ resembles very much the ud-diquark fragmentation found in $\bar{\nu}$ p-reactions, whereas in reactions with a fast Λ more pions are produced and their inclusive x-distributions are similar to those from single quark fragmentation.

1. INTRODUCTION

Recently it has been shown [1] that scaling deviations of the π^+/π^- ratio in the proton fragmentation region, observed in $\pi^\pm p$ and $K^\pm p$ reactions at 1o-1oo GeV/c can be explained by valence quark - valence antiquark annihilation with subsequent recombination of the remaining quarks. This mechanism, exploited quantitatively in the framework of the quark recombination picture [2,3] corresponds to the planar term in DTU models [4].

In the following the question is considered, whether a sample of hadronic reactions can be found where the annihilation contribution is enriched and dominant. To answer the above question, π-production in the proton fragmentation region in the reactions $K^- p \rightarrow K^0_{fast} \pi^\pm X$ and $K^- p \rightarrow \Lambda_{fast} \pi^\pm X$ at 16 GeV/c incident momentum [5] is investigated and compared to that in $\bar{\nu} p \rightarrow h^\pm \mu^+ X$, $K^+ p \rightarrow \pi^\pm X$ reactions and $p\bar{p}$ annihilation reactions. Shown are data on correlations which can be regarded as new experimental information, whereas the quark-model arguments (mainly following the philosophy of DTU-models) should serve as a guideline.

2. PION PRODUCTION IN THE REACTION $K^- p \rightarrow K^0 \pi^\pm X$

We selected a sample of K⁻p reactions where only one V^0 has been observed. At an incident energy of 16 GeV/c we expect that more than 80% of the events with $V^0 = K^0_n$ have only one strange particle [6], hence in most cases $K^0_n = \bar{K}^0$ and contains the valence s-quarks of the incident K⁻.

Fig.1 shows the π^+/π^- ratio

$$R = \int_0^\infty dp_\perp^2 \cdot E^x \cdot \frac{d^2\sigma(\pi^+)}{dp_\perp^2 dx} \bigg/ \int_0^\infty dp_\perp^2 \cdot E^x \cdot \frac{d^2\sigma(\pi^-)}{dp_\perp^2 dx} \qquad (1)$$

in the region of Feynman-$x = x_F < 0$ for $K^- p \to \bar{K}^0_{fast} \pi^\pm X$ (Fig.1a) and $K^- p \to \bar{K}^0_{slow} \pi^\pm X$ (Fig.1b).

We observe a ratio comparable to 1, if the $\bar{K}^0$ is fast, but it approaches a value greater than 2 at $x_F \approx -0.6$ if the $\bar{K}^0$ is slow.

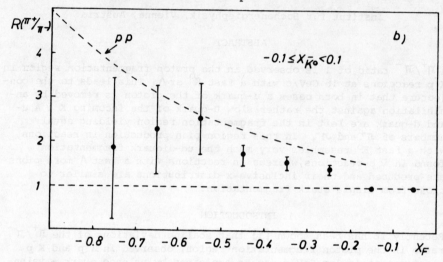

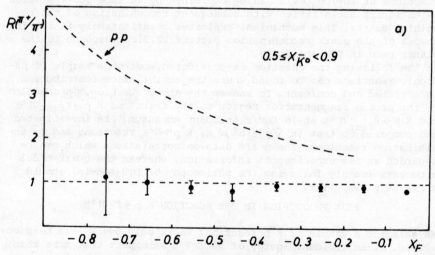

Fig.1: π^+/π^- ratios for $K^- p \to \bar{K}^0 \pi^\pm X$, 16 GeV/c

We try to explain the behaviour observed in fig.1 as follows: selecting a fast $\bar{K}^0$ in the final state means it contains a fast s-quark identical with the incoming s-quark. Therefore the $\bar{u}$-quark of the K^- beam is slow and the annihilation contribution (fig.2a) is enriched. The remaining u and d-quark of the proton yield to $R=1$

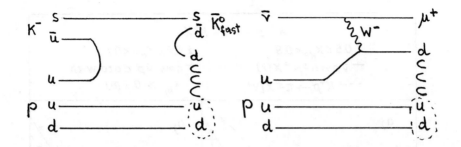

Fig.2a the annihilation
contribution in
$K^- p \rightarrow \bar{K}^0_{fast} \pi^{\pm} x$

Fig.2b hadron production
in $\bar{\nu} p \rightarrow h^{\pm} \mu^+ x$

in the proton fragmentation region. On the other hand, if the $\bar{K}^0$ is
slow, the $\bar{u}$ needs not to be slow, one observes an admixture of annihilation with non-annihilation reactions (planar contribution and
2-sheet contribution) where the latter cause the observed rise of
the ratio with $|x_F|$.

The quark configuration of fig.2a is similar to that in $\bar{\nu} p \rightarrow h^{\pm} \mu^+ x$
reactions shown in fig.2b and if the interpretation given above is
correct one might expect also the normalized inclusive distributions

$$F_n^{\pi^{\pm}} (x_F) = \frac{1}{\sigma} \cdot \frac{2}{\pi \sqrt{s}} \int_0^\infty E^* \frac{d^2\sigma \ (\pi^{\pm})}{dx_F \cdot d\, p_\perp^2} \, d\, p_\perp^2 \tag{2}$$

for $x_F < 0$ to be similar or even to be the same in $K^- p \rightarrow \bar{K}^0_{fast} \pi^{\pm} x$
and $\bar{\nu} p \rightarrow \mu^+ \pi^{\pm} x$ (with $x_{Bj} > 0.2$). In both cases the remaining (ud)-
system is expected to fragment to give the observed hadrons [*].

Fig.3 shows a comparison of F_n functions. We observe good agreement
of both π^+ and π^- production in the region $x_F < -0.2$ in $K^- p \rightarrow \bar{K}^0_{fast} \pi^{\pm} x$
(full circles) with that in $\bar{\nu} p$ reactions (full line [7] and big cros-
ses [8]). Little difference is found for all $F_n^{\pi,+}$ functions considered
(Fig.3a) in the fragmentation region but $F_n^{\pi^-}$ (Fig.3b) has a definitely steeper x-dependence in the reaction $K^- p \rightarrow K^0_{slow} \pi^{\pm} x$ (small
crosses) than $F_n^{\pi^-}$ in $\bar{\nu} p$ reactions and resembles much more the $F_n^{\pi^-}$
observed in $K^+ p \rightarrow \pi^{\pm} x$ [9] (eye-guide=dashed line) where valence quark
annihilation is not possible [**] .

We want to stress especially the tendency of the slope parameter n
given in table I to be smaller in the case of $K^- p \rightarrow \bar{K}^0_{fast} \pi^{\pm} x$ than
in $K^- p \rightarrow K^0_{slow} \pi^{\pm} x$ and $K^+ p$ reactions. This is expected [4] if in the
first case 1-sheet topology is enriched where the (ud)-system carries
the whole incoming momentum.

---

[*] x_F is defined in the hadronical system in $\bar{\nu} p$ reactions [7,8] and
is taken here in the overall CMS in the case of hadron-hadron reactions.

[**] A similar difference has been found comparing $F_n^{\pi^-}$ in $\bar{\nu} p$ reactions with that in pp reactions [8,10].

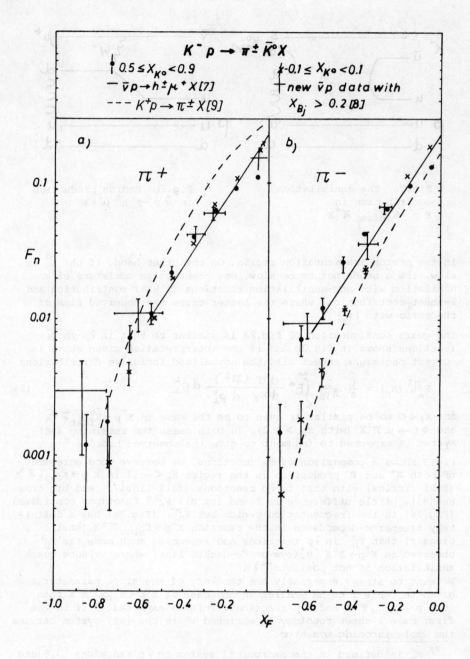

Fig.3 A comparison of F_n functions

TABLE I   n-values gained from a fit to $F_n^{\pi^\pm} = A(1-x)^n$

Reaction	x_F Region	n	Comments
$K^-p \to \pi^+ \bar{K}^0_{fast} X$	$-0.8 \leq x < -0.4$	2.77 ± 0.45	probably annih. reaction
$\pi^+ \bar{K}^0_{slow} X$	"	3.56 ± 0.42	mixed
$K^+p \to \pi^+ X [9,12]$ (32 GeV/c)	$-1.0 \leq x < -0.5$	3.0 ± 0.2	non annihil. reactions
$K^-p \to \pi^- \bar{K}^0_{fast} X$	$-0.8 \leq x < -0.4$	3.22 ± 0.48	probably annih. reaction
$\pi^- \bar{K}^0_{slow} X$	"	4.14 ± 0.52	mixed
$K^+p \to \pi^- X [9,12]$ (32 GeV/c)	$-1.0 \leq x < -0.5$	3.9 ± 0.7	non annih. reactions
$K^-p \to \pi^+ \Lambda_{fast} X$	$-0.8 \leq x < -0.4$	1.37 ± 0.44	probably annih. reaction
$\pi^+ \Lambda_{slow} X$	"	2.59 ± 0.36	
$\pi^- \Lambda_{fast} X$	"	2.12 ± 0.45	probably annih. reaction
$\pi^- \Lambda_{slow} X$	"	2.83 ± 0.5	

3. PION PRODUCTION IN THE REACTION $K^-p \to \Lambda \pi^\pm X$

In order to test further the annihilation assumption of the slow
$\bar{u}$-quark of the incoming K^-, we investigate another reaction in which
a fast s-quark can be identified: $K^-p \to \Lambda_{fast} \pi^\pm X$. We expect again
$R \approx 1$ in the proton fragmentation region.
Figs.4a and 4b show R for $K^-p \to \Lambda_{fast} \pi^\pm X$ and $K^-p \to \Lambda_{slow} \pi^\pm X$ respectively. The ratio is near 1 for fast Λ's, but it approaches
a value of 3 at $x_F \approx -0.6$, if the Λ's are slow.
As indicated in fig.5 the expected total spectator system $(ud - \bar{u}\bar{d})$
after the production of the Λ is different from that in $\bar{\nu} p \to$
$\mu^+ h^\pm X$ and $K^-p \to K^0_{fast} h^\pm X$ reactions (fig.2). It has baryon quantum number O. We could therefore expect a similarity with meson
production in $\bar{p}p$ annihilation reactions.

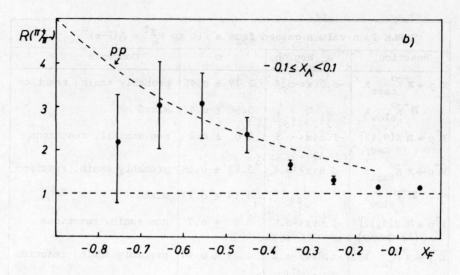

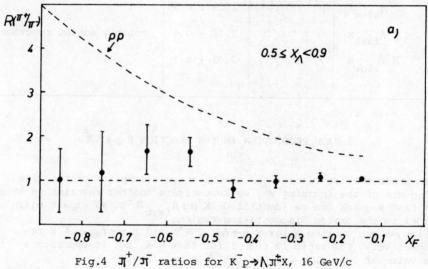

Fig.4 π^+/π^- ratios for $K^-p \to \Lambda \pi^\pm X$, 16 GeV/c

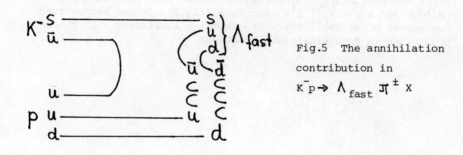

Fig.5 The annihilation contribution in

$K^-p \to \Lambda_{fast} \pi^\pm X$

In fig.6 a comparison is done of F_n functions in $K^-p \to \Lambda_{fast}\pi^{\pm} X$ reactions (full circles) with those in pp annihilation reactions at 12 GeV/c [11](eye-guide line = dashed line) both in the proton fragmentation region. Even more striking - the comparison with single quark fragmentation functions observed in $\bar{\nu}p$ reactions in the region $x_F > 0$ (dashed region) - indicates that only one of the two chains in fig.5 is developed. In order to underline the difference to (ud)-diquark fragmentation, found in $\bar{\nu}$ p reactions($x_F < 0$) and $K^-p \to \bar{K}^0_{fast}\pi^{\pm}X$, the latter are indicated as double-dashed region and full triangles on the same figure.

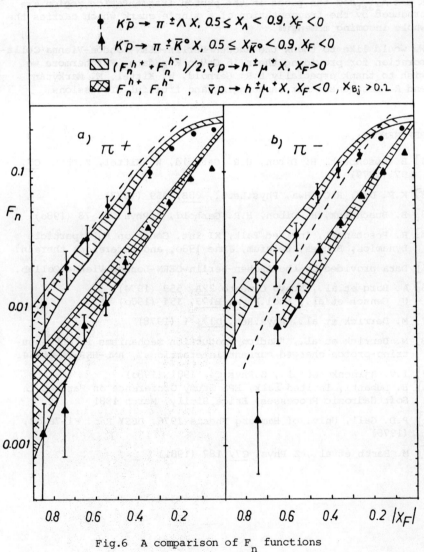

Fig.6 A comparison of F_n functions

4. CONCLUSIONS

The experimental features - a π^+/π^- ratio of 1 in the proton fragmentation region for the reactions $K^-p \to \bar{K}^o_{fast}\pi^{\pm}X$ and $K^-p \to \Lambda_{fast}\pi^{\pm}X$ and the similarity of their $F_n^{\pi^{\pm}}$ -functions to those in $\bar{\nu}p \to \mu^- h^{\pm}X$ reactions for $x_F < 0$ and $x_F > 0$ respectively, support the following conjectures:

It is possible to select the valence $u\bar{u}$ quark annihilation reaction by demanding fast $\bar{K}^o$'s or Λ's. (ud)-diquark fragmentation is observed in the reaction $K^-p \to \bar{K}^o_{fast}\pi^{\pm}X$, whereas in the reaction $K^-p \to \Lambda_{fast}\pi^{\pm}X$ the pions in the proton fragmentation region are produced by the fragmentation of a single quark which carries the whole incoming momentum.

We would like to thank the Aachen-Berlin-CERN-London-Vienna-Collaboration for providing their 16 GeV/c K^-p data. Furthermore we wish to thank especially H.R. Gerhold, W. Kittel, M. Markytan and R. Peschanski for encouraging and fruitful discussions.

REFERENCES:

[1] B. Buschbeck, H. Dibon, H.R. Gerhold, W. Kittel, Z.Phys. C3, 97 (1979)

[2] K.P. Das, R.C. Hwa, Phys.Lett., 68B, 459 (1977)

[3] B. Buschbeck, H. Dibon, H.R. Gerhold, Z.Phys.C7, 73 (1980)

[4] R. Peschanski, Invited Talk, XI Int. Conf. on Multiparticle Dynamics, Bruges, Belgium, June 1980, and references therein.

[5] Data provided by the Aachen-Berlin-CERN-London-Vienna Collab.

[6] A. Borg et al., Nuovo Cimento 22A, 559 (1974)
U. Gensch et al., Nucl.Phys. B173, 154 (1980)

[7] M. Derrick et al., Phys.Rev. D17, 1 (1978)

[8] M. Derrick et al., "Hadron production mechanisms in anti-neutrino-proton charged current interactions", ANL-HEP-PR-80-54.

[9] I.V. Ajinenko et al., Z.Phys.C4, 1981 (1980)

[10] H. Lubatti, Invited Talk, EPS Study Conference on Partons in Soft Hadronic Processes, Erice,Sicily, March 1981

[11] P.D. Gall, Univ. of Hamburg Thesis 1976, DESY Rep. F1-76/02, (1976)

[12] M. Barth et al., Z.Phys. C7, 187 (1981)

FRAGMENTATION OF QUARK AND DIQUARK JETS
INTO PIONS IN K^-p INTERACTIONS AT 70 GeV/c

Paris - Rutherford - Saclay Collaboration
presented by A.M. Touchard - LPNHE - Tour 32
4, place Jussieu - 75230 Paris Cedex 05 - France

ABSTRACT

The pion (π^+, π^-, π°) production in the K^-p interactions at 70 GeV/c (BEBC) are found to be very similar in both K^- and proton fragmentation regions to their production in $\nu(\bar{\nu})p$ interactions as expected from quark parton models. Quark and diquark fragmentation functions D_u^π, D_{uu}^π and D_{ud}^π are extracted from our data.

INTRODUCTION

In this paper we present the experimental determination of the quark and diquark fragmentation into pions, extracted from the data on K^-p interactions at 70 GeV/c since, in fragmentation models, the longitudinal cross-section distributions of the produced pions can be related simply to these fragmentation functions.

The analysis is based on a 0.87 event/μb experiment performed in the Big European Bubble Chamber, BEBC, filled with H_2 and exposed to an r.f. separated K^- beam. The final sample corresponds to about 12,000 primary interactions with 4,500 γ's, 1400 K_s°'s and 400 Λ's. The selection used to obtain these events and the corrections needed to calculate the $\pi^\pm$ and π° inclusive distributions have been exten- sively reviewed in our publications [1,2,3] (Note that estimation of charged kaon contributions has been made by consideration on neutral kaon production).

MESON FRAGMENTATION

According to the fragmentation model [4], pion production in the K^- fragmentation can be written in terms of the quark fragmentation functions as :

$$\frac{1}{\sigma} \frac{d\sigma^{K^- \to \pi^{\pm}_{\circ}}}{dx} \simeq \frac{1}{2} \left[D_s^{\pi^{\pm}_{\circ}}(x) + D_u^{\pi^{\pm}_{\circ}}(x) \right]$$

Isospin and charge invariance considerations reduce the number of independent D_q^{π} functions to three : $D_u^{\pi^+}$, $D_u^{\pi^-}$, D_s^{π}. Furthermore, following Field and Feynman [5], we have used the approximation $D_u^{\pi} \simeq D_s^{\pi}$ since the creation of a $\pi^-(\bar{u}d)$ from u-quark or the creation of π^- from s-quark requires the creation of the same two pairs of new quarks. Finally since the π° spectrum is also used in determining these fragmentation functions through the isospin relation $D_q^{\pi^{\circ}} = (D_q^{\pi^+} + D_q^{\pi^-})/2$ [6] we get :

$$\frac{1}{\sigma} \frac{d\sigma^{K^- \to \pi^+}}{dx} \simeq D_u^{\pi^-}(x)$$

$$\frac{1}{\sigma} \frac{d\sigma^{K^- \to \pi^-}}{dx} \simeq \frac{1}{2} \left[D_u^{\pi^-}(x) + D_u^{\pi^+}(x) \right]$$

$$\frac{1}{\sigma} \frac{d^{K^- \to \pi^{\circ}}}{dx} \simeq \frac{1}{4} \left[3 D_u^{\pi^-}(x) + D_u^{\pi^+}(x) \right]$$

The results obtained for $D_u^{\pi^+}$ and $D_u^{\pi^-}$ are shown on fig. 1, the region $0.7 < x < 0.9$ is covered only by the π^+ and π° data since in this region the π^- cross-section is dominated by the leading K^--effect.

We compare our results to some $\nu(\bar{\nu})p$ data [7] with W>4 GeV and to the Field and Feynman functions [5]. The agreement between these experiments and our data is rather good in the range x,z>0.30 provided we take into account the following remarks :

- In $\nu(\bar{\nu})p$ experiments, $D_{u(d)}^h$ represent the fragmentation of a pure u(d) quark only at high values of z, where the contribution of sea quarks is negligible. In addition, the dispersion of these data is due to the fact that they represent $D_u^{h^\pm}$ or $D_d^{h^\mp}$ and not $D_u^{\pi^\pm} = D_d^{\pi^\mp}$

- In our experiment the main difficulty is probably due to the production of diffractive resonances (Q^-, L^-, ..) which may contribute to the inclusive distribution of the $K^- \to \pi^-$ channel for values $x < 0.7$.

PROTON FRAGMENTATION

Pion production in the proton fragmentation region can be written in terms of the diquark fragmentation functions as :

$$\frac{1}{\sigma} \frac{d\sigma^{p \to \pi^\pm}}{dx} \simeq \frac{1}{3} D_{uu}^{\pi^\pm}(x) + \frac{2}{3} D_{ud}^{\pi^\pm}(x)$$

From the isospin invariance we have $D_{ud}^{\pi^-} = D_{ud}^{\pi^+}$ and thus three variables remain independent $H_1 \equiv D_{uu}^{\pi^+ud}$, $H_2 \equiv D_{ud}^{\pi^+} = D_{ud}^{\pi^-}$ and $H_3 \equiv D_{uu}^{\pi^-}$. The inequality $H_1 > H_2 > H_3$ is expected to hold since in H_1 each u-quark can contributes to the creation of a π^+ while in H_2 only one u-quark can contributes and in H_3 none of these u-quarks can contribute to the creation of a π^-. Furthermore, if baryon production is dominant in the diquark fragmentation Andersson et al[4] give also the approximate relation : $2H_2 \sim H_1 + H_3$. Then follows :

$$\frac{1}{\sigma} \frac{d\sigma^{p \to \pi^+}}{dx} \simeq \frac{2}{3} H_1(x) + \frac{1}{3} H_3(x)$$

$$\frac{1}{\sigma} \frac{d\sigma^{p \to \pi^-}}{dx} \simeq \frac{1}{3} H_1(x) + \frac{2}{3} H_3(x)$$

When we solve this system with our experimental data we obtain significant values for $H_1(x)$ and values compatible with zero for $H_3(x)$ over the range $-0.8 < x < -0.2$. We show our $H_1(x)$ distribution on fig. 2 with the data on $H_1(x)$ and $H_3(x)$ extracted from Bell et al[7] on νp interactions. These data on $H_1(x)$ fit our distribution very well and $H_3(x)$ is much smaller. Furthermore $2H_2(x)$ taken from data on $\bar{\nu}$p interactions [8] is shown also on fig. 2 and agree very well with the $H_1(x)$ distribution giving support to the above approximation.

CONCLUSION

- The quark fragmentation functions, $D_u^{\pi^\pm}$, extracted from data on K^-p interactions are compatible with those obtained with the recent lepton data and with the parametrization proposed by Field and Feynman in early 1978. This confirms that the current fragmentation region for lepto-production and the K^- fragmentation region can be described by the same quark fragmentation.

- In the proton fragmentation region, the diquark fragmentation functions have been determined and shown to be in a very good agreement with those obtained in $\nu(\bar{\nu})$ experiments and to obey the relation:

$$D_{uu}^{\pi^+} \simeq 2D_{ud}^{\pi^\pm} \gg D_{uu}^{\pi^-}$$

REFERENCES

1. J.M. Laffaille et al. Z. Physics C2, 95 (1979).

2. J.M. Laffaille et al. Nucl. Phys. To be published.

3. L. de Billy et al. Nucl. Phys. To be published.

4. B. Andersson et al. Phys. Let. 69B, 221 (1977).
   B. Andersson et al. Phys. Let. 71B, 337 (1977).
   A. Capella et al. Z. Physics C3, 329 (1980)

5. R.D. Field and R.P. Feynman, Nucl. Phys. B136, 1 (1978)

6. L.M. Sehgal - Proceedings of the 1977 International Symposium on Lepton and Photon interactons. Hamburg 1977, p. 837.

7. a) J. Bell et al. Phys. Rev. D19, 1 (1978)
   b) P.C. Bosetti et al. Nucl. Phys. B149, 13 (1979)
   c) J.P. Berge et al. Proceedings of the International Conference on Neutrinos. Bergen 1979, vol. 2, p. 328.
   d) H.J. Lubatti, Bergen 1979, vol. 2, p. 543 (data on $\bar{\nu}$Ne $\rightarrow$ h$^-$x and νNe $\rightarrow$ h$^-$x).
   e) M. Derrick et al. Phys. Let. 91B, 470 (1980)

8. M. Derrick et al. Phys. Rev. D17, 1 (1978).

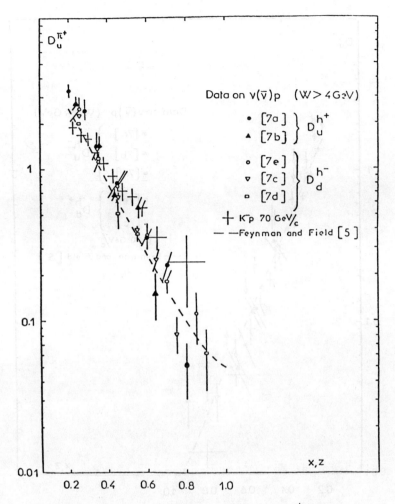

Fig. 1a. Quark fragmentation function $D_u^{\pi^+}$ obtained in K^-p interactions at 70 GeV/c. Data on $D_u^{h^+}$ and $D_d^{h^-}$ from $\nu(\bar{\nu})p$ experiments are also reported.

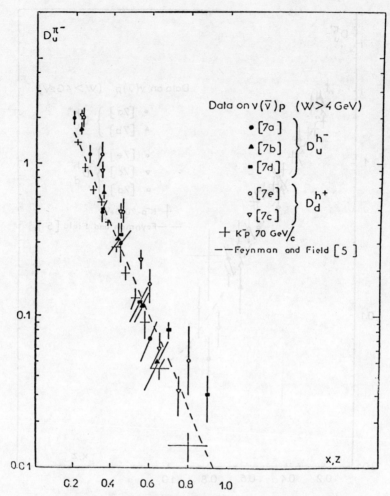

Fig. 1b. Quark fragmentation function $D_u^{\pi^-}$ obtained in K^-p interactions at 70 GeV/c. Data on $D_u^{h^-}$ and $D_d^{h^+}$ from $\nu(\bar{\nu})p$ experiments are also reported.

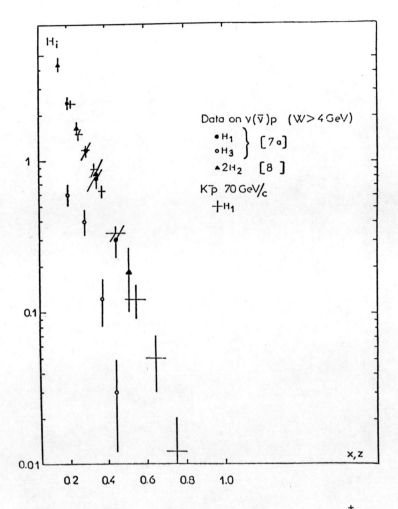

Fig. 2. - Diquark fragmentation function $H_1 = D_{uu}^{\pi^+}$ obtained in K^-p interactions at 70 GeV/c. Data on $H_1 = D_{uu}^{h^+}$, $H_2 = D_{ud}^{h^{\pm}}$, $H_3 = D_{uu}^{h^-}$ from $\nu(\bar{\nu})p$ experiments are also reported.

STUDY OF REACTIONS pp → (ω°, ρ°, η°) + ANYTHING AT 300 GeV/c

H. Wald, W.A. Mann, J. Schneps
Tufts University, Medford, Massachusetts 02155
W.D. Shephard, N.N. Biswas, J.M. Bishop, N.M. Cason
V.P. Kenney, R.C. Ruchti
University of Notre Dame, Notre Dame, Indiana 46556
R. Engelmann, J. Hanlon, T. Kafka, F. LoPinto, S. Sommars
State University of Stony Brook, Stony Brook, New York 11794
R. Ammar, R. Davis, N. Kwak, R. Riemer, R. Stump
University of Kansas, Lawrence, Kansas 66045

ABSTRACT

Primary charged prongs together with associated gamma conversions have been studied using 6900 inelastic pp interactions at 300 GeV/c recorded in the Fermilab 15-foot diameter hydrogen bubble chamber. Using 606 events in which gamma pairs have candidate $\pi°$ masses, a 3.5 standard deviation enhancement at the $\omega°$ is observed in $m(\pi^+\pi^-\pi°)$. For inclusive processes (pp → $\omega°$ + anything) we measure 36±14 mb. Decays $\eta° → \pi^+\pi^-\pi°$ are not seen, implying σ(pp → $\eta°$ + anything) < 37 mb at 90% C.L. Using $\pi^+\pi^-$ invariant mass distributions, $\rho°$ production in multiplicities 6, 8 to 10, and 12 to 14 prongs, is determined to be 1.1 ± 0.4, 2.7 ± 1.1, and 2.1 ± 1.9 mb respectively. Our measurements indicated that inclusive $\omega°$ production exceeds $\rho°$ and $K^*(890)$ rates at 300 GeV/c. We find that about half the $\omega°$ signal occurs with $|y_{cm}| < 1.0$, and that roughly 70% of $\omega°$'s arise from multiplicities greater than 10 prongs.

INTRODUCTION

The extent to which short-range correlations and clustering behavior in high energy pp collisions reflect abundant production of pseudoscalar, vector and tensor resonances and/or new dynamical structures, remains unclear.[1,2] At energies $\sqrt{s} \gtrsim 20$ GeV, inclusive $\rho°$ and $K^{*\pm}(890)$ production on the order of a few to tens of millibarns have been established,[3,6,7,8,9] and production rates of higher mass states have been given.[6,9,10] However, accurate inclusive rates for $\eta°$, $\omega°$, $f°$, A_1, A_2, g... at a single energy are not yet available. Lack of knowledge about the total meson resonance contribution is becoming more problematic as phenomenology of low-p_t-hadron-hadron reactions progresses; theoretical explanation of these most prevalent hadronic processes in terms of underlying parton dynamics may eventually be impeded.[11,12,13]

We report results from a new analysis of pp interactions at 300 GeV/c, recorded in a 23,000 picture exposure of the hydrogen-filled Fermilab 15-foot bubble chamber. About 9100 measurable inelastic pp interactions are observed with a primary fiducial volume extending 280 cm along the beam direction. Associated vees and gammas from conversion have been measured in the entire sample. Production prongs have been measured in about 80% of the events, and more measurements are in progress. Our experimental methods, including momentum and topology-dependent weighting for gammas and vees, have been described in detail previously, in studies of π° correlation moments, and measurements of inclusive γ, K_S°, Λ°, $\bar{\Lambda}^\circ$, $K^{*\pm}$, and $\Sigma^{*\pm}(1385)$ production at 300 GeV/c.[4]

EXPERIMENTAL PROCEDURE

For the present study, fiducial volume restrictions on gamma vertices were relaxed in order that an overall conversion efficiency of 11.5% could be achieved, which is higher than the 10.0% efficiency used in our earlier work. Previous measurements of gamma conversion pairs in this exposure yielded a π° signal in the invariant mass distribution of $\gamma\gamma$ pairs (see Fig. 4 of T. Kafka et al., Ref. 4). We then selected events having gamma pairs with masses $0.11 \leq m(\gamma\gamma) \leq 0.15$ GeV/c^2, and subjected them to a special measurement pass, in which both primary and associated gamma tracks in the events were measured on film-plane digitizing machines, taking typically ten points per track. Special care was taken to get quality measurements on as many tracks in these events as possible. In the 606 events which were processed in this way, a prong track reconstruction efficiency of 90% was achieved, which corresponds to a prong pair reconstruction efficiency for "$\pi^+\pi^-$" combinations of 84%. The $\gamma\gamma$ invariant mass spectrum obtained after these procedures is shown in Fig. 1. The "π°" signal receives roughly equal contributions from low and high charged multiplicities, as indicated by the shaded and hatched histogramming in the figure.

RESULTS

From the distribution in Fig. 1 we select π° decay candidates to be those $\gamma\gamma$ pairs whose masses fall between .12 and .14 GeV/c^2. Pi-zero four vectors are constructed using the 3-momenta of these $\gamma\gamma$ pairs together with the nominal π° mass. Invariant mass combinations $\pi^+\pi^-$"π°" are then constructed to yield the distribution displayed in Fig. 2. A clear enhancement of 125 combinations above a hand-drawn background is apparent in 3 bins around the $\omega^\circ(783)$ mass, which corresponds to a signal of about 4.0 standard deviations. Inclusive phase space for this plot, as indicated by the distribution of three-pion masses resulting from combining pions from different events having identical multiplicity, peaks well above the ω° region as Fig. 2 itself suggests. Possible contributions to Fig. 2 from Δ° or N^* reflections have been examined and may contribute significant misidentified combinations to masses at and above 850 MeV/c^2. It is possible that Δ° reflections may contribute a tail of up to 20 combinations in the ω° enhancement region. Consequently we subtracted this amount from the apparent signal. The stability of the ω° signal under variation of the $\gamma\gamma$ mass interval defining π° four-vectors was examined. As shown superimposed in Fig. 2, sharpening the $\gamma\gamma$ mass to .13 $< m(\gamma\gamma) <$ .14 GeV/c^2 results in a 40% reduction in the number of $\pi^+\pi^-\pi^\circ$ combinations; the ω° remains visible although

146

it is also reduced by about 40%. We take the weighted average of ω° cross sections obtained using these two different "π°" selections to represent the inclusive cross section at 300 GeV/c :

$$\sigma(pp \rightarrow \omega^\circ + anything) = 36 \pm 14 \text{ mb}.$$

The inclusive signal is not confined to the central region; only 50% of the ω°'s have $|y_{cm}| < 1.0$. Additionally, we find that 70% of inclusive ω°'s arise from multiplicities greater than ten prongs. Consequently, we do not regard our observed signal to be in qualitative disagreement with the null result reported by R.D. Kass et al., at 400 GeV/c[5], in which a similar 15-foot bubble chamber exposure was used to search for $\omega^\circ \rightarrow \pi^+\pi^-\pi^\circ$ production in $n_{ch} \leq 10$-prong events.

Using the "tight" π° definition with $0.13 < m(\gamma\gamma) < 0.14$ GeV/c^2 described above, we searched for the $\eta^\circ(549)$. Since the η° is near $\pi^+\pi^-\pi^\circ$ threshold, binning in 10 MeV was used, as shown in Fig. 3. No statistically significant enhancement in the η° region is found, and the corresponding 90% confidence level upper limit for inclusive η° production is

$$\sigma(pp \rightarrow \eta^\circ + anything) < 37 \text{ mb}.$$

The result is an improvement over the 45 mb upper limit previously obtained using the $\gamma\gamma$ invariant mass distribution of this experiment.

Production of ρ° mesons has been examined semi-inclusively in 6 through 14-prong interactions. As indicated by Fig. 4 for 6 and 8 prongs, a broad shoulder is discernable in each of the $\pi^+\pi^-$ invariant mass distributions which is centered at the $\rho^\circ(776)$. In each distribution, the mass interval 0.54 to 1.10 GeV/c was fitted to an incoherent sum of a P-wave Breit-Wigner Amplitude-squared plus a quadratic polynomial background (dotted curve in Fig. 4). Best results were obtained in fitting with the ρ° mass and width fixed at 760 MeV/c and 155 MeV/c respectively, with chi-squares per degree of freedom ranging between 0.7 and 1.9. The ρ° rates determined by fitting, for topologies 6, 8 to 10, and 12 to 14 prongs, are respectively 1.1 ± 0.4 mb, 2.7 ± 1.1 mb, and 2.1 ± 1.9 mb (see Table I). Summing over these values, the ρ° rate from 6 through 14 prongs is found to be 5.9 ± 2.2 mb.

A difficulty in using the standard fitting method to extract an inclusive ρ° rate from an overall $\pi^+\pi^-$ invariant mass distribution is indicated by Fig. 5, where the $\pi^+\pi^-$ distribution from 4 through 16 prong topologies is displayed. In our data the superposition of $\pi^+\pi^-$ mass distributions obscures ρ° enhancements which are apparent when individual topologies are separately examined. However, the semi-inclusive rates measured here are consistent with an overall inclusive rate of order 10 mb, which is the value reported from study of 205 GeV/c pp reactions in the FNAL 30-inch bubble chamber.

In summary, we list our inclusive and topological cross sections for ω°, η°, and ρ° mesons in Table I. A compilation of all the inclusive meson cross sections which have been determined using the 300 GeV/c reactions of this experiment, is given in Table II.

147

FOOTNOTES AND REFERENCES

[1]. R. Diebold, in Proceedings of the 19th International Conference on High Energy Physics, Tokyo, 1978, edited by S. Homma, M. Kawaguchi and H. Miyazawa (Physical Society of Japan, Tokyo 1979).

[2]. J. Whitmore, in Proceedings of the 19th International Conference on High Energy Physics, Tokyo, 1978, edited by S. Homma, M. Kawaguchi and H. Miyazawa (Physical Society of Japan, Tokyo 1979).

[3]. 205 GeV/c pp: R. Singer et al., Phys. Lett. 60B, 385 (1976); E.L. Berger, R. Singer, G.H. Thomas, T. Kafka, Phys. Rev. D15, 206 (1977); R. Singer et al., Nucl. Phys. B135, 265 (1978).

[4]. 300 GeV/c pp: T. Kafka et al., Phys. Rev. D19, 76 (1979); F. LoPinto et al., Phys. Rev. D22, 573 (1980).

[5]. 400 GeV/c pp: R.D. Kass et al., Phys. Rev. D20, 605 (1979).

[6]. 405 GeV/c pp: H. Kichimi et al., Phys. Rev. D20, 37 (1979); A. Suzuki et al., Nucl. Phys. B172, 327 (1980).

[7]. 100, 200, 360 GeV/c $\pi^- p$: P.D. Higgins et al., Phys. Rev. D19, 65 (1979).

[8]. 53 GeV/c pp: G. Jansco et al., Nucl. Phys. B124, 1 (1977).

[9]. 53 GeV/c pp: G. Van Dalen et al., Phys. Rev. Lett. 41, 1761 (1978).

[10]. 53 GeV/c pp: D. Drijard et al., CERN/EP 81-12, January 1981.

[11]. K. Kinoshita, in Proceedings of the 1979 KEK Summer School, Tsukuba, Japan, 143 (1979).

[12]. A. Capella, U. Sukhatme, and J. Tran Thanh Van, Zeitschrift für Physik C3, 329 (1980).

[13]. J.F. Gunion, "Quarks and Gluons in Low-p_t Physics", SLAC-PUB-2607, September 1980.

TABLE I: CROSS SECTIONS FOR $\omega°$, $\eta°$, AND $\rho°$ MESONS IN 300 GeV / c pp COLLISIONS

Resonance	Event Sample	Cross Section
$\omega°$(783)	Inclusive	36 ± 14 mb
	$\|y_{cm}\| < 1.0$	17 ± 11 mb
	$n_{ch} > 10$ prongs	28 ± 11 mb
	$n_{ch} \leq 10$ prongs	11 ± 5 mb
$\eta°$(549)	90% conf. level limit	< 37 mb
	best estimate	∿ 12 ± 12 mb
$\rho°$(776)	6 prongs	1.1 ± 0.4 mb
	8 prongs	1.3 ± 0.8 mb
	10 prongs	1.4 ± 1.3 mb
	8 + 10 prongs	2.7 ± 1.1 mb
	12 + 14 prongs	2.1 ± 1.9 mb

TABLE II: INCLUSIVE CROSS SECTIONS FOR MESONS FROM THIS EXPERIMENT

Resonance	Inclusive Cross Section	Mode Examined
$\omega^\circ(783)$	36 ± 14 mb	$\pi^+\pi^-\pi^\circ$, $\pi^\circ \to \gamma\gamma$
$\rho^\circ(776)$	6 thru 16 prongs: 5.9 ± 2.2 mb	$\pi^+\pi^-$
$\eta^\circ(549)$	< 37 mb at 90% c.l.	$\pi^+\pi^-\pi^\circ$
	$\sim 12 \pm 12$ mb, best estimate	$\pi^+\pi^-\pi^\circ$
	< 45 mb at 90% c.l.*	$\gamma\gamma$
	$\sim 9 \pm 6$ mb, best estimate*	$\gamma\gamma$
π°	134.9 ± 5.8 mb*	Inclusive γ's
	110 ± 20 mb*	$\gamma\gamma$
$K^\circ/\bar{K}^\circ$	13.8 ± 1.6 mb†	K°_S
$K^{*\pm}(890)$	4.4 ± 1.4 mb†	$K^\circ_S\pi^+ + K^\circ_S\pi^-$

* T. Kafka et al., Phys. Rev. D19, 76 (1979).
† F. LoPinto et al., Phys. Rev. D22, 573 (1980).

FIGURE CAPTIONS

Fig. 1: The invariant mass of $\gamma\gamma$ pairs used in the ω° and η° searches of this experiment. The distribution is the result of "dedicated remeasuring" (10 points per track) of events which upon initial measurement (6 points per track) yielded gamma pairs with $.11 \leq m(\gamma\gamma) \leq .15$ GeV/c^2. Masses of gamma pairs from multiplicities greater than 8 prongs and greater than 10 prongs are shown with diagonal and hatched shading respectively.

Fig. 2: The invariant mass distribution $m(\pi^+\pi^-\pi^\circ)$ from inclusive processes pp $\to \pi^+\pi^-\pi^\circ$ + anything. A π° is taken to be a $\gamma\gamma$ pair having invariant mass between .12 and .14 GeV/c^2. The distribution obtained by taking π°'s to have $m(\gamma\gamma)$ between .13 and .14 GeV/c^2, is displayed superimposed.

Fig. 3: Invariant mass of $\pi^+\pi^-\pi^\circ$ combinations near threshold, shown in 10 MeV/c^2 binning. The π° definition used in Fig. 2, that $.13 < m(\gamma\gamma) < .14$ GeV/c^2, is imposed.

Fig. 4: Mass distribution of $\pi^+\pi^-$ combinations from 8 and 10 prong events, and from 6-prong events (lower distribution). Dotted curve on 8-10 prong distributions displays quadratic background from fit which also includes a P-wave Breit-Wigner shape for the ρ° meson (see text).

Fig. 5: Invariant mass distribution of $\pi^+\pi^-$ from 4 through 16-prong topologies.

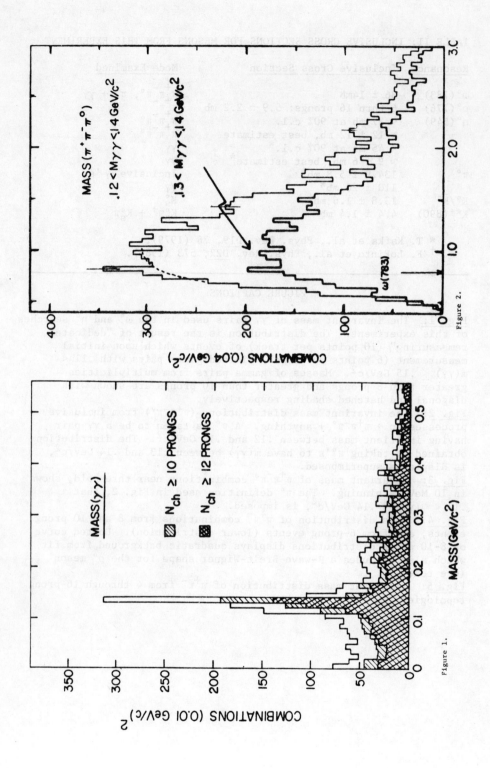

Figure 2.

Figure 1.

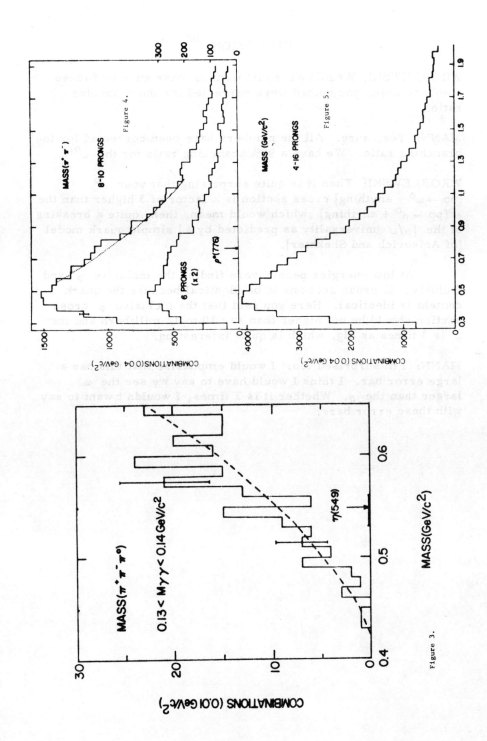

Figure 4.

Figure 5.

Figure 3.

DISCUSSION

WROBLEWSKI, WARSAW: I just want to make sure that these numbers which you quoted were corrected for the branching ratio.

MANN: Yes, sure. All the numbers have been corrected for the branching ratio. We have a 90% branching ratio for the ω^o's.

WROBLEWSKI: Then it is quite surprising that your ω^o [$pp \rightarrow \omega^o$ + anything] cross section is a factor of 3 higher than the $\rho^o[pp \rightarrow \rho^o$ + anything] which would mean, then, quite a breaking of the [ρ/ω universality as predicted by a] simple quark model [of Anisovich and Shekhter].

At low energies people were finding the inclusive ρ and inclusive ω cross sections to be identical because the quark content is identical. Here you said that the inclusive ρ cross section should be not higher than say 10 to 12 millibarn and the ω is 3 times as big, which is quite interesting.

MANN: I'm surprised too. I would emphasize that this has a large error bar. I think I would have to say we see the ω larger than the ρ. Whether it is 3 times, I wouldn't want to say with these error bars.

THE SEARCH FOR RESONANT STATES FROM 1 TO 5 GeV

John A. Poirier
Physics Department
University of Notre Dame

The reaction studied is pi- + CH $\rightarrow$ V^c + V^c + pi$^\pm$'s + $\underline{X}$. The $\underline{f}$inal states include two vees, e.g. $K^O K^O$, $K^O \Lambda$, $K^O \bar{\Lambda}$, or $\Lambda \bar{\Lambda}$ accompanied by additional pions or baryons. The Fermilab experiment (Ref. 1) involves a 200 GeV/c pi-beam (10% of the data was taken at 250 GeV/c) incident on an active target of 20 scintillation counters. This pion beam momentum was limited by the 350 GeV primary proton beam energy available from the accelerator. The experiment was performed in the M6W beamline utilizing the Multiparticle Spectrometer Facility (Ref. 2). The collaborators involved in this experiment are:

ARIZONA - Jenkins, Lai, Pifer, Chen,
        Lin, LeBritton, Johnson;

FERMILAB - Greene, Fenker;

FLORIDA STATE - Albright, Diamond, Hagopian,
        Lannutti, Goldman, Piper;

NOTRE DAME - Davis, Poirier

TUFTS - Napier, Mann, Schneps;

VANDERBILT - Waters, Webster, Williams, Roos;

VIRGINIA TECH - Ficenec, Collins, and Trower.

An effective technique to search for new narrow resonances has been to probe the definite quantum state J,P = 1-. For example this quantum state is probed by e+ + e- and by mu+ + mu- states, and has revealed the series of J/psi and upsilon resonance states. We wish to extend this quantum number filter technique to different quantum states. The trigger for our experiment selected two vee events which can be, for example: a) a pair of neutral K's whose J,P states are restricted to 0+, 2+, 4+ ..., and b) lambda + anti-lambda resonances which would have an I-spin state of I=0. Adding additional charged pions to these final states allows a more general search for additional J, P, and I-spin possibilities. The run was approved for 800 hours; we have used up about 450 hours of this running time. Our initial run obtained 1.7 million triggers on magnetic tape; of those triggers about 1.4 million were two vee triggers, and the remaining were diagnostic triggers like single beam tracks. We have so

far analyzed 24% of the data. This talk presents preliminary results based on this sample. About 8% of the data
triggers reconstruct to two vee events through the pattern recognition program.

The apparatus is shown in Figure 1.   It has an incident pi- beam that goes through two proportional wire
chamber stations. Next is the active scintillator target
followed by proportional wire chambers before the decay
region.   The decay region is filled with helium gas to
minimize neutral interactions that might fake a vee
event. Next are proportional wire chambers upstream of a
super-conducting spectrometer magnet which is followed by
proportional wire chambers downstream.   A hodoscope Čerenkov counter distinguishes pions from protons. Magnetostrictive spark chambers track the particles in the
most downstream area.

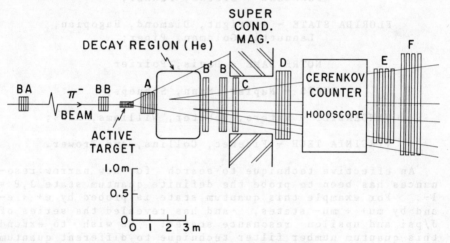

Figure 1.   Side view of the experimental apparatus (not
to scale). A rough scale is indicated.

The topology of a typical event is superposed on Figure 1.   A pi- beam interacts in one of the scintillation
targets and produces two charged primary tracks and two
neutral vees.   Only one of the charged tracks goes
through the spectrometer and the two neutral vees each
decay into two charged decay products which go through
the rest of the spectrometer. We calculate the effective
masses of various possible configurations of these daugh-
ter vees with or without additional pions to look for
parent resonant states.

The trigger for the experiment was in two parts: a pre-trigger and a post-trigger. The pre-trigger was a small scintillator counter after the spectrometer magnet that signaled a disappearing beam particle. In such a case the post-trigger counted the number of tracks in the A region (M) and the number of tracks that were in the B', B, C, and D regions (N). The post-trigger demanded that N = M + 4. This is characteristic of two neutral V particles that decay each into two charged tracks thus increasing the number of charged tracks by four. (When we count the tracks in the A region (M), we look only at those traversing the central portion of the A station since they are the only tracks which can make it through the spectrometer.) This is the basic data trigger of the E580 experiment. We have rationed the high multiplicity events so as to decrease the dead time; thus we were able to pick up essentially all of the events of low multiplicity. Figure 2 shows the distribution of multiplicities that we obtained in our event sample. The peak at three charged primary tracks is a result of this trigger bias.

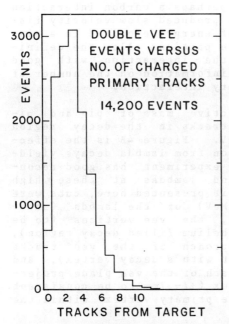

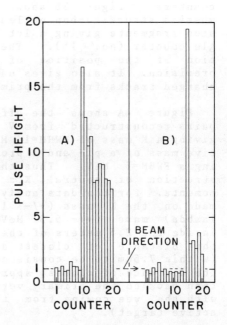

Figure 2. The distribution of charged primary tracks in our sample of the rationed triggers.

Figure 3. Pulse height distributions in the active target scintillation counters for two events.

The experimental resolution of the effective mass squared is a function of the resolution of the momenta of the neutral kaons and the opening angle between the two. For our experiment, the opening angle error dominates; this error in angle comes mainly from the uncertainty in the position of the primary vertex. To minimize this uncertainty, the target was constructed of twenty scintillation counters each one 6mm thick x 32mm x 32mm. The pulse height of each counter was digitized and recorded for each trigger. In addition these pulse heights were fast summed and used in the trigger to signal an interaction which took place in the target. The pulse height distribution for two events is shown in Figure 3. They were taken from the first seven events on a randomly chosen raw data tape. The most probable pulse height of a single particle traversing each counter at normal incidence is shown by a dashed line at "1" in the Figure. In Figure 3A one sees the Landau statistical fluctuations of a single track going through counters 1-9. The interaction is signaled by a big pulse height in counter 10 with the pulse heights progressively decreasing in counters 11-20 as the wide angle tracks leave these downstream counters. Figure 3B shows perhaps a carbon interaction where a struck carbon nucleus produced slow velocity fission fragments giving a lot of energy deposited in a single counter (no. 15). These patterns allow the definition of the position of the interaction with good precision. It also gives us information on the number of charged tracks from the primary interaction.

Figure 4A shows the effective mass of pi+ and pi- pairs reconstructed from V tracks in the decay region giving a K mass of FWHM=14 MeV. Figure 4B is the effective mass of a pi- and a proton from lambda decays yielding a FWHM=5 MeV. Thus the experiment has good reconstruction of neutral K's and lambdas at these high momenta. For the data analysis presented here, cuts were made on the K mass (+/- 17MeV) or the lambda (anti-lambda) mass (+/- 5.5 MeV), the vee vertices (to be inside the 2.5 meters of the helium filled decay region), the distance of closest approach of the vee tracks (within 7.5mm to be consistent with a decay vertex), and the distance of closest approach of the vee plane projection back to the primary vertex (+/- 7mm to be consistent with the vee coming from the primary vertex within the active target).

Figure 5 shows the K star region where we look at the invariant mass of either a pi+ or a pi- together with a neutral K. There appears to be an excess of events above the background in the region of the K-star (890 MeV). The bottom inset is the negative K-star(890) + pi+ mass spectrum. Figure 6 shows the corresponding plot for a lambda plus a charged pion, with some excess of events in

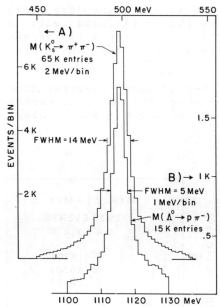

Figure 4. The mass distribution of vee decays in the 2.5m decay region for: A) pi+ and pi- in the K mass region; and B) p and pi- in the lambda mass region.

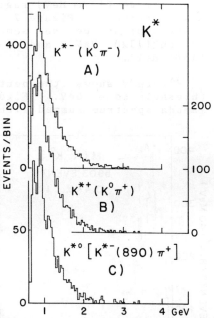

Figure 5. The mass distributions for: A) neutral K and pi-; B) neutral K and pi+; and C) K*(890) and pi+ from neutral K* decays.

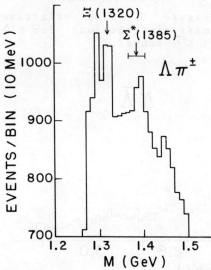

Figure 6. The effective mass distributions for a lambda plus a charged pion.

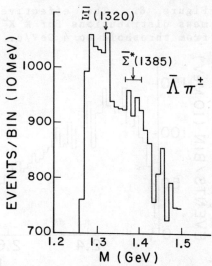

Figure 7. The mass distributions for an anti-lambda plus charged pion.

the cascade (1320 MeV) region and in the sigma star (1385 MeV) region. Figure 7 shows the anti-lambda with a charged pion; we see some activity in the anti-cascade region (1320 MeV) and perhaps at the anti-sigma star (1385 MeV).

Figure 8 shows the neutral K-pair mass spectrum from threshold to 4 GeV. Figure 9 shows the neutral kaon + lambda spectrum summed with that of the K + anti-lambda.

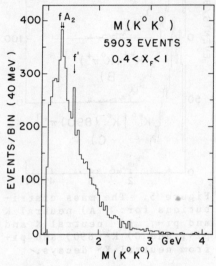

Figure 8. The effective mass distributions for $K^o K^o$ from threshold to 4 GeV/c.

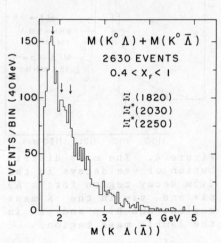

Figure 9. The effective mass distributions for $K^o \Lambda$ + $K^o \overline{\Lambda}$ summed.

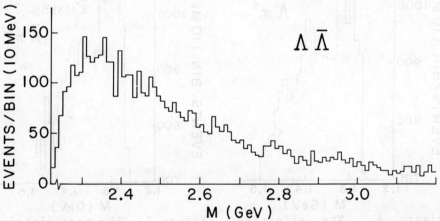

Figure 10. The effective mass distribution for lambda plus anti-lambda pairs from threshold to 3.2 GeV.

Figure 10 is the effective mass for the lambda + anti-
lambda.   Figure 11 shows the   combination of events that
involve a neutral K, pi+, pi-, and pi-, a possible decay
mode of the D- (1868 MeV)   or other bare charm object.
Figure 12 is the effective mass distribution plot of the
combination of particles: K, K, pi-, pi-, pi+ which
could be a signature of beauty, charm, or strange decays
as well as a non-strange object with I=1 or more.   The
Geneva-Lausanne group (Ref. 3) has reported evidence for
an A meson with J,P=6+, I=1, which decays into KK channels with a mass of 2.45 +/- 0.13 GeV.   We seem to see
indications of old as well as newer resonances.   With
further analysis of cuts and backgrounds,   these indications may be further strengthened.

The construction of drift chambers has been started at
at Notre Dame in order to replace the magnetostrictive
spark chambers.   When completed,   this change should
increase our data rate by a factor of five.   We think we
can improve our counting,   fast triggering,   and tracking
to gain perhaps another factor of two in data.   Thus we
can gain an order of magnitude more in event statistics
with the remaining 1/2 of our approved running time.   The
experiment will have a sensitivity of approximately 200
events/nanobarn,   or about a factor of 1000 improvement
over the present world's sample of two vee events in this
mass region (Ref. 4).   These vastly increased statistics

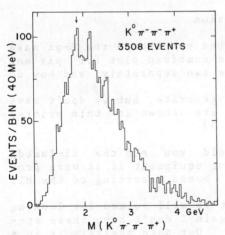

Figure 11.   The mass distribution for a neutral K,
pi+, pi-, and pi-.   The
arrow indicates the bare
charmed D- mass of 1868
MeV.

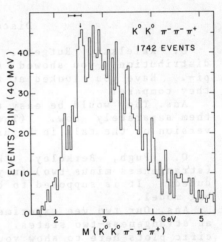

Figure 12.   The mass distributions for a K, K, pi-,
pi-, pi+.   The point (with
error bars) is the 2450 MeV
A (J=6)   K$\bar{\text{K}}$ resonance of
I=1 (Ref. 3).

159

in an experiment with good effective mass resolution looking at rare strange states could easily lead us to exciting new physics discoveries!

References:

1. E-580 proposal entitled: "A Search for Narrow and Broad Resonances Decaying into Double Vees from pi- p Interactions at 300 GeV/c Using the Fermilab MPS".

2. "Multiparticle Spectrometer at Fermilab", the Caltech-Indiana-Fermilab-UCLA-Illinois C. C. Collaboration; by A. Dzierba, E. Malamud, and D. McLeod; a Fermilab Report.

3. Pittsburgh-Geneva-Lausanne contribution to the XX International Conference on High Energy Physics, Madison, Wisconsin (July, 1980), Paper No. 1011, "Observation of a Spin 6 Isospin 1 Boson Resonance in the Charged KK-bar System".

4. See, for example: Phys. Rev. D16, 2098 (1977); and the E-110 proposal entitled "Proposal to Study Multiparticle Peripheral Physics at NAL".

Discussion:

Q. Kalelkar, Rutgers. When you showed the K-pi mass distribution, you showed the combined plot for pi+ and pi-. Have you looked at the two separately and how do they compare?
Ans. They would be easy to generate, but we don't have them separately now. (They are shown in this printed version of the talk in Figure 5).

Q. Pugh, Berkeley. Would you see the di-lambda (strangeness minus two) in your equipment if it were produced? It is supposed to be bound according to the MIT bag model.
Ans. Our two vee experiment is well suited for looking at strangeness-two states. Again I don't have these specific plots here to show you. Our data analysis is in a very preliminary stage; we welcome your suggestions of these ideas.

INCLUSIVE STRANGE RESONANCE PRODUCTION IN pp, π^+p and K$^+$p INTERACTIONS AT 147 GeV/c

(Presented by M. Kalelkar, Rutgers University)

D. Brick, H. Rudnicka, A. M. Shapiro and M. Widgoff
Brown University, Providence, Rhode Island 02912, U.S.A.

R. E. Ansorge, W. W. Neale, D. R. Ward and B. M. Whyman[1]
University of Cambridge, Cambridge, England

R. A. Burnstein and H. A. Rubin
Illinois Institute of Technology, Chicago, Illinois 60616, U.S.A.

E. D. Alyea, Jr.
Indiana University, Bloomington, Indiana 47401, U.S.A.

L. Bachman[2], C. Y. Chien, P. Lucas[3], and A. Pevsner
Johns Hopkins University, Baltimore, Maryland 21218, U.S.A.

J. T. Bober, T. A. J. Frank, E. S. Hafen, P. Haridas, D. Huang[4],
R. I. Hulsizer, V. Kistiakowsky, P. Lutz[5], S. H. Oh, I. A. Pless,
T. B. Stoughton, V. Suchorebrow, S. Tether,
P. C. Trepagnier[6], Y. Wu[4], and R. K. Yamamoto,
Laboratory for Nuclear Science and Department of Physics,
Massachusetts Institute of Technology,
Cambridge, Massachusetts 02139, U.S.A.

F. Grard, J. Hanton, V. Henri, P. Herquet, J. M. Lesceux,
P. Pilette, and R. Windmolders,
Universite de l'Etat, Mons, Belgium

H. de Bock, F. Crijns, W. Kittel, W. Metzger, C. Pols,
M. Schouten and R. Van de Walle,
University of Nijmegen and NIKHEF-H, Nijmegen, The Netherlands

H. O. Cohn
Oak Ridge National Laboratory, Oak Ridge, Tennessee 37830, U.S.A.

G. Bressi, E. Calligarich, F. Carminati, C. Castoldi,
R. Dolfini, and S. Ratti
Istituto di Fisica Nucleare and Sezione INFN, Pavia, Italy

R. DiMarco, P. F. Jacques, M. Kalelkar, R. J. Plano,
P. Stamer[7] and T. L. Watts
Rutgers University, New Brunswick, New Jersey 08903, U.S.A.

E. B. Brucker, E. L. Koller and S. Taylor
Stevens Institute of Technology, Hoboken, New Jersey 08903, U.S.A.

L. Berny, S. Dado, J. Goldberg and S. Toaff,
Technion, Haifa, Israel

G. Alexander, O. Benary, J. Grunhaus, R. Heifetz, and A. Levy,
Tel-Aviv University, Ramat-Aviv, Israel

W. M. Bugg, G. T. Condo, T. Handler, E. L. Hart and A. H. Rogers,
University of Tennessee, Knoxville, Tennessee 37916, U.S.A.

Y. Eisenberg, U. Karshon, E. E. Ronat, A. Shapira,
R. Yaari and G. Yekutieli
Weizmann Institute of Science, Rehovot, Israel

T. W. Ludlam[8], R. Steiner, and H. D. Taft,
Yale University, New Haven, Connecticut 06520, U.S.A.

---

[1] Present address: University of Liverpool, Liverpool, England.
[2] Present address: Université de Neuchâtel, Neuchâtel, Switzerland.
[3] Present address: Duke University, Durham, North Carolina 27706 U.S.A.
[4] On leave of absence from the Institute of High Energy Physics, Beijing, Peoples Republic of China.
[5] Present address: College de France, Paris, France.
[6] Present address: Automix, Burlington, Massachusetts 01803, U.S.A.
[7] Permanent address: Seton Hall University, South Orange, New Jersey 07079, U.S.A.
[8] Present address: Brookhaven National Laboratory, Upton, New York 11973, U.S.A.

ABSTRACT

We have studied the inclusive production of the $K^{*\pm}(890)$ and $Y^{*\pm}(1385)$ in pp, π^+p and K^+p interactions at 147 GeV/c. The experiment used the Fermilab 30-inch hydrogen bubble chamber with the hybrid spectrometer system. Results are based on a sample of 1916 observed K_S and 932 observed Λ. Inclusive cross sections are given for $K^{*\pm}$ and $Y^{*\pm}$ production from the three beams, and comparisons are made with experiments at other energies. Feynman x and transverse momentum-squared distributions are also calculated. The results suggest that the K^{*-} is entirely produced in the central region, while the K^{*+} includes a component from beam fragmentation. Comparisons are made with the additive quark model.

INTRODUCTION

In this paper we report final results on inclusive $K^{*\pm}(890)$ and $Y^{*\pm}(1385)$ production in pp, π^+p and K^+p interactions at 147 GeV/c, all in the same experiment. The experiment was performed in the Fermilab 30-inch hydrogen bubble chamber with proportional wire chambers (PWC) located both upstream and downstream. A total of 400,000 pictures was taken in two runs, with different beam ratios in each. The final sample had an overall composition of 51% π^+, 9% K^+, and 40% p.

A total of 6744 vees constituted the final data sample. After resolving ambiguities, there were 1916 K_S, 932 Λ, 164 $\bar{\Lambda}$ and 3732 γ's. Table I summarizes the event sample.

Table I. No. of K_S, Λ, $\bar{\Lambda}$ and γ for each beam type

Vee	p beam	π^+ beam	K^+ beam
K_S	882	912	122
Λ	464	414	54
$\bar{\Lambda}$	63	91	10
γ	1732	1755	245

INCLUSIVE $K^{*\pm}(890)$ CROSS SECTIONS

In order to calculate inclusive cross sections for $K^{*\pm}(890)$ production, we first weighted each observed K_S to correct for detection inefficiency due to the imposition of a minimum length cut and decay volume restriction. The average escape weight was 1.98 for K_S. We then made a weighted $K_S\pi^\pm$ effective mass distribution for K_S events from all three beams. The resulting distribution is shown in Figure 1. A clear peak is apparent in the vicinity of the $K^*(890)$.

163

Mass distributions for $K_s\pi^+$ and $K_s\pi^-$ were separately fitted for all three beam types to yield the numbers of resonant events in each channel. To determine the inclusive cross sections, the numbers of escape-weighted events were corrected to account for scan inefficiency, measurement and geometrical reconstruction losses, and neutral decay modes of the K_s. A further weight of 2.0 was applied to include the contribution from K_L, and finally the branching ratio of 2/3 for $K^{*\pm} \to K^o\pi^\pm$ was incorporated. Table II summarizes the inclusive cross sections.

Table II. Inclusive cross sections for K^{*+} and K^{*-} production.

Reaction	σ(mb)	$\sigma(K^*)/\sigma(K^o+\bar{K}^o)$
$pp \to K^{*+}$ + anything	1.5 ± 0.3	$.15 \pm .03$
$pp \to K^{*-}$ + anything	1.2 ± 0.2	$.12 \pm .02$
$\pi^+ p \to K^{*+}$ + anything	1.3 ± 0.2	$.16 \pm .03$
$\pi^+ p \to K^{*-}$ + anything	0.7 ± 0.2	$.08 \pm .02$
$K^+ p \to K^{*+}$ + anything	3.2 ± 1.3	$.27 \pm .11$
$K^+ p \to K^{*-}$ + anything	0.6 ± 0.5	$.05 \pm .04$

INCLUSIVE DISTRIBUTIONS FOR $K^{*\pm}(890)$

Figures 2 and 3 show the Feynman x dependence of K^* production by the proton and π^+ beams respectively. In both cases we note an approximate equality of the K^{*+} and K^{*-} cross sections in the central region, followed by an excess of K^{*+} over K^{*-} in the beam fragmentation region. In order to make a rough quantitative measure of this observation, we have chosen x = .2 as a dividing line between the central and beam fragmentation regions, and then calculated the cross section ratio $\sigma(K^{*-})/\sigma(K^{*+})$ in the two regions for both beams. This ratio is shown in Table III, and does exhibit a marked difference in the two regions. In particular, K^{*-} production is at least four standard deviations less than K^{*+} production in the beam fragmentation region.

Table III. Ratio of K^{*-} to K^{*+} cross sections

Beam	$|x| < .2$	$|x| > .2$
p	$1.1 \pm .4$	$.16 \pm .17$
π^+	$.7 \pm .3$	$.13 \pm .20$

We have also produced transverse momentum squared distributions for $K^{*\pm}$ from the proton and π^+ beams. Experiments at lower energies have found a universal slope of about 3.4 $(GeV/c)^{-2}$, independent of beam, for particles of mass about 1 GeV. Our values, shown in Table IV, are somewhat lower. However, there is some evidence[1] that all p_T^2 slopes drop slightly with incident beam energy. To this end we show in Figure 4 the variation of the slope with beam energy for K^{*+} and K^{*-} production by proton and π^+ beams. The trend of the data suggests a decrease.

Table IV. Fitted p_T^2 slope parameters in units of $(GeV/c)^{-2}$

Beam	K^{*+}	K^{*-}
p	2.7 ± .4	3.1 ± .5
π	2.9 ± .4	2.8 ± .6

INCLUSIVE $Y^{*\pm}$(1385) PRODUCTION

We used our sample of Λ events to study $Y^{*\pm}$ production. We followed a procedure analogous to that for the K^* study. The observed Λ's were weighted for geometric detection efficiency, providing an average escape weight of 1.91. We than examined weighted $\Lambda\pi^+$ and $\Lambda\pi^-$ effective mass distributions from each of the three beams. In general we observed enhancements in the vicinity of the Y^*(1385), but these were less pronounced than the peaks for the K^*.

Fits to these mass distributions were made using a smooth polynomial background and a simple Breit-Wigner shape for the Y^*, with a mass of 1385 MeV and a width of 40 MeV. To determine inclusive cross sections, the numbers of escape-weighted events were corrected to account for scan inefficiency, measurement and reconstruction losses, neutral decay modes of the Λ, and the branching ratio of .88 for the $Y^{*\pm} \rightarrow \Lambda\pi^{\pm}$ decay.

Table V summarizes the inclusive cross sections for Y^{*+} and Y^{*-} production. Where upper limits are given, they are at 95% confidence level. For the proton beam the cross sections for the two charge states of the Y^* are very close to each other, suggesting that the production is primarily in the central region. By contrast, the meson beams from this experiment do not provide any evidence for Y^{*-} production, while there is a non-negligible signal for Y^{*+}.

Table V. Inclusive cross sections for Y^{*+} and Y^{*-}

Reaction	$\sigma(mb)$	$\sigma(Y^*)/\sigma(\Lambda)$
$pp \rightarrow Y^{*+}$ + anything	$.38 \pm .13$	$.09 \pm .03$
$pp \rightarrow Y^{*-}$ + anything	$.40 \pm .13$	$.10 \pm .03$
$\pi^+ p \rightarrow Y^{*+}$ + anything	$.29 \pm .07$	$.16 \pm .04$
$\pi^+ p \rightarrow Y^{*-}$ + anything	$<.10$	$<.06$
$K^+ p \rightarrow Y^{*+}$ + anything	$.08 \pm .18$	$.06 \pm .13$
$K^+ p \rightarrow Y^{*-}$ + anything	$<.17$	$<.12$

Table V also gives the ratio of Y^* to Λ total cross sections. If we assume that Y^* production in pp reactions is primarily in the central region (as suggested by the equality of Y^{*+} and Y^{*-} cross sections) we can make a quantitative test of the additive quark model.[2] This model makes predictions for particle production ratios in the central region. Many of these predictions are dependent on a strange-quark suppression factor λ, whose numerical value is not predicted by the model. Moreover, experimental data have suggested a momentum dependence for λ. However, the Y^*/Λ ratio in the central region is independent of the suppression factor and is predicted to have the value 0.25, both for Y^{*+} and Y^{*-}. Using the fraction of our Λ cross section in the central region[3] and assuming that all the Y^* is central, we obtain $\sigma(Y^{*+})/\sigma(\Lambda) = .25 \pm .09$ and $\sigma(Y^{*-})/\sigma(\Lambda) = .27 \pm .09$, which are in excellent agreement with the additive quark model.

ACKNOWLEDGMENTS

This work was supported in part by the U.S. Department of Energy, the National Science Foundation, the U.S.-Israel Binational Science Foundation, and the Dutch F. O. M. We gratefully acknowledge the efforts of the 30-inch bubble chamber crew, and the scanning and measuring personnel at the participating institutions.

REFERENCES

1. For a recent review, see J. Whitmore, Proceedings of the XIX International Conference on High Energy Physics, edited by S. Homma, M. Kawaguchi and H. Miyazawa, Tokyo, 1978, p. 63.
2. V. V. Anisovich and V. M. Shekhter, Nucl. Phys. B55, 455 (1973).
3. D. Brick et al., Nucl. Phys. B164, 1 (1980).

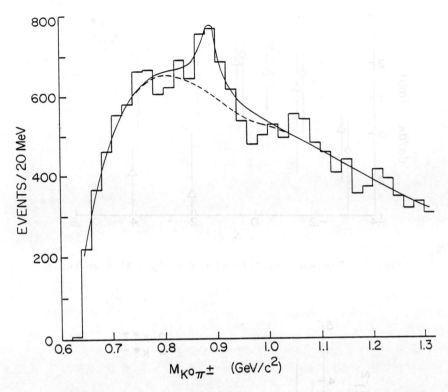

Fig. 1. Escape-weighted ($K^0\pi^+ + K^0\pi^-$) effective mass distribution from all three beams.

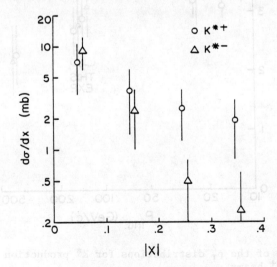

Fig. 2. Feynman x distribution for $pp \to K^* +$ anything.

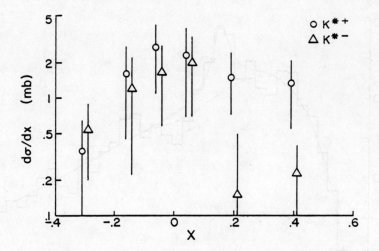

Fig. 3. Feynman x distribution for $\pi^+ p \rightarrow K^* +$ anything.

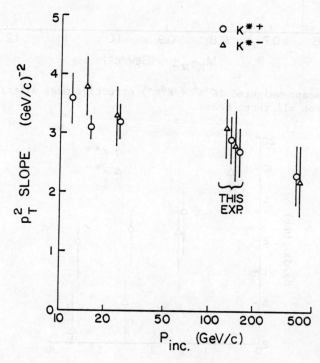

Fig. 4. Slope of the p_T^2 distributions for K^* production from proton
and π^+ beams.

DISCUSSION

BIAŁAS, KRAKOW: My question is about production of K^{*-} in K^{+}-induced reactions. Is it also energy independent?

KALELKAR: Oh, in the K^{+} reaction I can't give you an energy dependence. There were so few events we only have 125 K^{o}'s, so I can't really make a statement.

CERN/EP/0464R/RTR/ed
10 July 1981

INCLUSIVE STRANGE PARTICLE PRODUCTION AND VECTOR AND TENSOR MESON PRODUCTION IN K^+p INTERACTIONS AT 70 GeV/c

Brussels-CERN-Genova-Mons-Nijmegen-Serpukhov Collaboration

Presented by
R.T. Ross
CERN, Geneva, Switzerland

ABSTRACT

Results on K^n, $\bar{\Lambda}^0$, Λ, K^* and ρ^0 production from an exposure of BEBC filled with hydrogen, to a 70 GeV/c K^+ beam are presented. Inclusive channel and differential cross sections are presented. Scaling of the structure functions with data at 32 GeV/c is observed in the dominant x-region for each reaction and in some cases over the full x-region. Polarization and spin density matrix measurements are given where obtainable. In particular a significant polarisation for forward produced $\bar{\Lambda}$'s is observed.

INTRODUCTION

This paper summarises the main results presented in three papers submitted to this conference from the 70 GeV/c K^+p experiment in BEBC [1-3]. Due to lack of space some results and detailed descriptions of methods presented in the original papers will be omitted. The results on vector and tensor meson production should be considered as preliminary.

THE DATA

The data were obtained from an exposure of BEBC to an RF separated K^+ beam of 70 GeV/c nominal momentum. Details of data taking, scanning, measurement and reconstruction are given in ref. [4]. The results presented come from a sample of approximately

50 000 events, corresponding to about 3 events/microbarn. A
modified version of the HYDRA kinematics program optimized for
V^0/γ fitting was used to fit the usual K_s^0, Λ, $\bar{\Lambda}$ decay hypotheses and
e^+e^- pair production at each V^0/γ vertex. The K_s^0, Λ and $\bar{\Lambda}$
samples were fully corrected for losses due to decay or interaction
outside the fiducial volume or close to the production vertex,
losses due to kinematic ambiguities and losses in the scanning,
measurement or fitting procedures. For the K^n, Λ and $\bar{\Lambda}$ results
cross sections were obtained by normalising the corrected numbers of
V^0's in each topology to the primary total topological cross
sections as previously published [4].

INCLUSIVE NEUTRAL KAON PRODUCTION

The cross sections for the reactions

$$K^+p \rightarrow K^n + X \tag{1}$$

$$K^+p \rightarrow 2K^n + X \tag{2}$$

$$K^+p \rightarrow 3K^n + X \tag{3}$$

are shown in fig. 1 with the corresponding data at other energies
(here K^n stands for K^0 or $\bar{K}^0$). The cross sections for reaction (1)
continues to rise with increasing energy and this rise can be
attributed to the rapid rise in the multikaon cross sections. The
reactions (1), (2) and (3) can be interpreted in terms of final
states of exclusive strangeness

$$K^+p \rightarrow K^0 + X_{ns}$$

$$\rightarrow \bar{\Lambda} + X_{ns}$$

$$\rightarrow K K \bar{K} + X_{ns}$$

$$\rightarrow K K \Lambda + X_{ns}$$

$$\rightarrow K \bar{K} \bar{\Lambda} + X_{ns}$$

$$\rightarrow K \Lambda \bar{\Lambda} + X_{ns}$$

where X_{ns} denotes a system which does not contain any strange
particles. These reactions can be related to the observed inclusive
cross sections using the method and assumptions described in
refs [5] and [6] yielding the cross section values summarised in
table I. We find that at this energy the reaction $K^+p \rightarrow \bar{K}^0 + X$
corresponds to $\sim 16\%$ of reaction (1), compared with $\sim 8\%$ at 32 GeV/c.

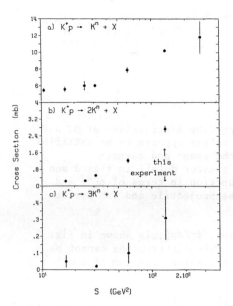

Fig. 1 – Energy dependence of cross sections for reactions (1)-(3)

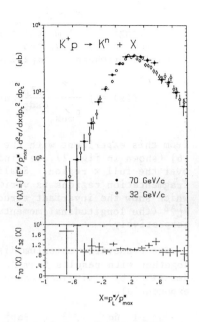

Fig. 2 – Invariant structure function for reaction (1)

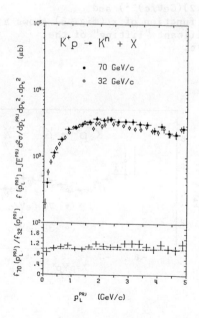

Fig. 3 – $f(P_L^{PRJ})$ distribution for reaction (1)

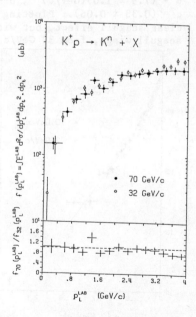

Fig. 4 – $f(P_L^{LAB})$ distribution for reaction (1)

The longitudinal momentum distribution $d\sigma/dx$, where $x = p_L^*/p_{max}^*$ for reaction (1) shows that the K^n's are produced mainly in the forward hemisphere. Comparing the invariant structure function

$$f(x) = \int \frac{E^*}{p_{max}^*} \frac{d^2\sigma}{dx dp_t^2} dp_t^2$$

from this experiment with the corresponding distribution at 32 GeV/c [6] (shown in fig. 2), we find that scaling appears to be satisfied over the full x region. Scaling in the beam- and target-fragmentation regions is verified in greater detail in figs 3 and 4 which show the invariant structure function in terms of p_L^{PRJ} and p_L^{LAB} (the longitudinal momentum in the projectile and lab system respectively).

The transverse momentum distribution, $d\sigma/dp_t^2$, is shown in fig. 5 together with results from 32 GeV/c. This distribution cannot be described by a single exponential, but requires the sum of two exponentials

$$\frac{1}{\sigma} \frac{d\sigma}{dp_t^2} = (1 - c)\, e^{-\alpha p_t^2} + c\, e^{-b p_t^2}$$

Fitting this expression up to $p_t^2 = 1.8$ (GeV/c)^2 yields $\alpha = (7.9 \pm 1.0)(\text{GeV/c})^{-2}$, $b = (2.8 \pm 0.2)(\text{GeV/c})^{-2}$) and $c = (0.33 \pm 0.06)$. Plotting $\langle p_t \rangle$ as a function of x (fig. 6) shows a clear seagull effect, but with no significant "lifting" of the seagull wings, from 32 GeV/c to 70 GeV/c.

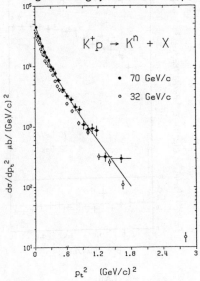

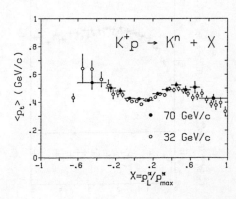

Fig. 5 - Transverse momentum distribution for reaction (1)

Fig. 6 - "Seagull" plot for reaction (1)

For certain model predictions it is desirable to compare the x
distribution of prompt K^n's, those not arising from the decay of
resonances. Fig. 7 shows the invariant structure function
distributions for all K^n, and for prompt K^n after correction for
those K^n's from the decay of $K^*(890)^+$ and $K^*(890)^0$ (see ref. [1]
for details). The prompt K^n distribution (x > 0.2) is flatter than
that for all K^n's (the latter sample giving a slope n = .78 ± .08
when fit to $(1 - x)^n$ in the region x > 0.2).

Fig. 8 shows the predictions of various models compared with the
prompt K^n distribution. Predictions of a recombination model [7,8]
using the parameters for the quark structure functions advocated in
refs [9,10] and the phase space and recombination function from
ref. [7], give a good description of the data. Similarly the Lund
fragmentation model [11], using the quark fragmentation functions
from Field-Feynman model [12] (but with $s\bar{s}:u\bar{u}:d\bar{d}$ = 1:3:3, and
considering only direct K^0 production) describe the data equally
well. On the other hand if the quark distribution functions used in
the QRM are included in the fragmentation model approach, as is well
known, the predictions fall much too steeply with increasing x.

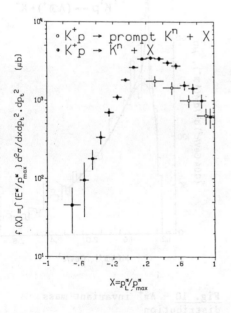

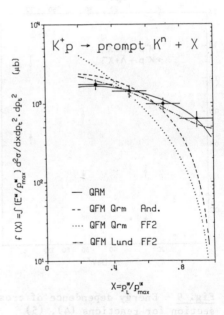

Fig. 7 - Invariant structure
function distribution for all
and prompt K^n

Fig. 8 - Invariant structure
function for prompt K^n
compared with model predictions

175

The cross sections for the reactions

$$K^+ p \to \Lambda + X^{++} \tag{4}$$

$$K^+ p \to \bar{\Lambda} + X^{++} \tag{5}$$

are given in table I and shown with data from lower energies in fig. 9. The cross sections increase approximately as $\ln(s)$. Fig. 10 shows evidence for $\Sigma(1385)^+$ production. Fitting the mass distribution $\Lambda\pi^+$ and $\Lambda\pi^-$ with relativistic Breit-Wigner function and five parameter background [2] gives cross sections for $\Sigma^+(1385)$ and $\Sigma^-(1385)$ of .20 ± .06 mb and .07 ± .05 mb respectively, indicating that some 25% of Λ's come from $\Sigma^\pm(1385)$ decays.

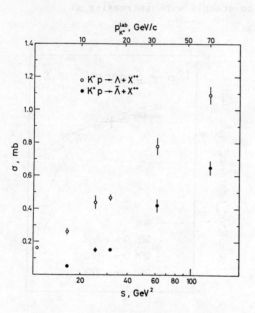

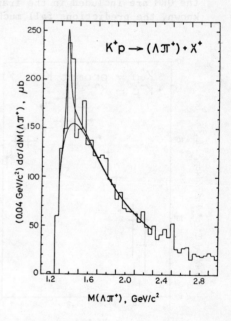

Fig. 9 - Energy dependence of cross section for reactions (4), (5)

Fig. 10 - $\Lambda\pi^+$ invariant mass distribution

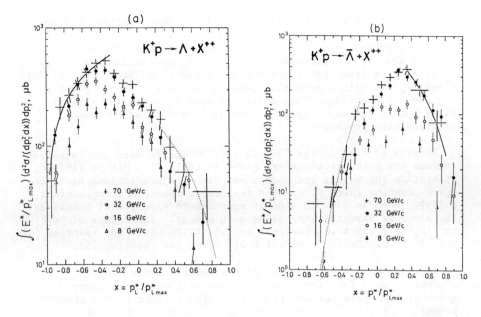

Fig. 11 – Invariant structure function distribution for (a) reaction (4) and (b) reaction (5)

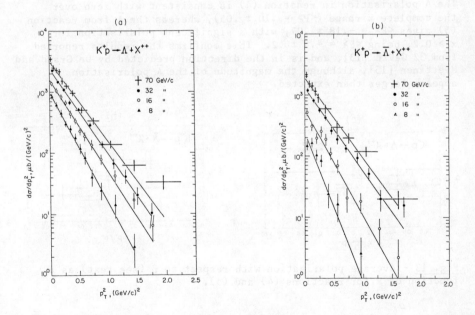

Fig. 12 – Transverse momentum distributions for reactions (4) and (5)

The invariant structure functions, $f(x)$, for reactions (4) and (5) are shown in fig. 11, with data from lower energies [13]. The Λ's are produced preferentially in the target-fragmentation region whereas the $\bar{\Lambda}$'s are produced in the beam-fragmentation region. The distributions from both reactions are close to scaling over the full x region, but with a possible increase in the region $-.5 \lesssim x \lesssim 0$ for the $\bar{\Lambda}$ production. This effect is more clear in the rapidity distribution and may suggest a rise in $\bar{\Lambda}$ production by beam fragmentation processes.

Results of fits to the form $(1 - |x|)^n$ for forward and backward regions for both reactions are given in table II. Within limited statistics the values are consistent with the predictions based on quark counting rules [14]. The transverse momentum distributions, $d\sigma/dp_t^2$, shown in fig. 12 show an approximate exponential behaviour, the slope as expected increasing with energy. The values of $\langle p_t^2 \rangle$ and $\langle p_t \rangle$ are $.35 \pm .01$ $(GeV/c)^2$ and $.51 \pm .10$ GeV/c for reaction [4] and $.35 \pm .02$ $(GeV/c)^2$ and $.51 \pm .10$ GeV/c for reaction (5).

Fig. 13 shows the polarisation distributions for Λ and $\bar{\Lambda}$ production as a function of x, together with data from lower momenta. The polarisation is defined with respect to the production normal

$$\hat{n} = (P_{\Lambda(\bar{\Lambda})} \times P_{p(K^+)})/|P_{\Lambda(\bar{\Lambda})} \times P_{p(K^+)}|$$

The Λ polarisation in reaction (4) is consistent with zero over the complete x range ($\langle P \rangle = .10 \pm .05$), whereas the $\bar{\Lambda}$ from reaction (5) gives $\langle P \rangle = -.18 \pm .07$, with a significant polarisation for $x > 0.2$, being $\langle P \rangle = -.5 \pm .2$. This confirms the results reported from 32 GeV/c [13], and is in the direction predicted by De Grand and Miettinen [15], although the magnitude of the $\bar{\Lambda}$ polarisation appears larger than expected.

(a)

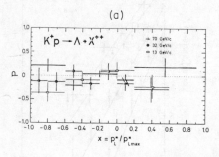

(b)

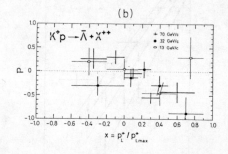

Fig. 13 - Average polarisation with respect to $\hat{n}$ (see text) as a function of x for reactions (4) and (5).

178

The results presented in this section concern the reactions

$$K^+ p \rightarrow K^{*+}(890) + X^+ \qquad (6)$$

$$K^+ p \rightarrow K^{*0}(890) + X^{++} \qquad (7)$$

$$K^+ p \rightarrow \rho^0(770) + X^{++} \qquad (8)$$

together with some results on $\bar{K}^{*0}(890)$, $K^{*-}(890)$, $K^{*0}(1430)$ and $K^{*0+}(1430)$ production.

The $K^n\pi^+$ mass spectrum (fig. 14) shows clear evidence for $K^{*+}(890)$ production. Fitting this distribution to the expression

$$\frac{d\sigma}{dm} = (BG) \times (1 + \alpha \; BW), \text{ where}$$

the background term $BG = a(M - M_{threshold})^b \exp(-cM + dM^2)$, where a, b, c, d and α are free parameters, and BW represents a relativistic p-wave Breit-Wigner function with resonance mass and width from Particle Data Group Tables (the width being increased to allow for the experimental resolution), yields the cross section values in table I. It should be noted that all these resonance cross section determinations are sensitive to the parametrisation assumed for the background. This is reflected in the systematic errors quoted.

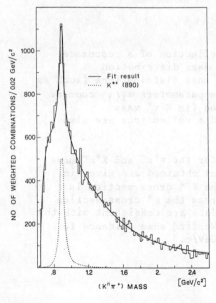

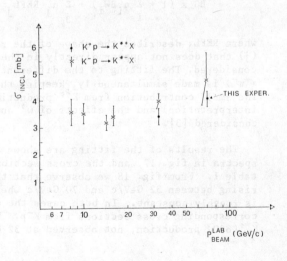

Fig. 14 $K^n\pi^+$ mass distribution

Fig. 15 $K^{*\pm}$ inclusive cross section from $K^\pm p$

In the region $x_F > .6$ a $K^{*+}(1430)$ signal is observed and the corresponding cross section (calculated as above, but with a d-wave Breit-Wigner) is also in table I, together with an estimate of a weak $K^{*-}(890)$ signal indicated in the $K^n\pi^-$ mass distribution.

The energy dependence of the $K^{*+}(890)$ cross section is shown in fig. 15, together with the corresponding K^-p data. A small increase in the cross section between 32 GeV/c and 70 GeV/c is observed, and this is associated with an increase in the cross section associated with higher charge multiplicities.

The methods used to calculate channel and differential cross sections for the neutral resonances are necessarily more complicated than for the K^{*+} states. The reason is that in addition to the large combinatorial background effects there is the problem of accounting for the reflection of a given resonance into the distribution of the invariant mass when a different track mass assignment is used. The latter effect arises since we have practically no mass identification, hence all opposite charged track combinations must be tried as $\pi^+\pi^-$, $K^+\pi^-$ and π^+K^- interpretations. The effect is very clearly illustrated in fig. 16 where the $K^+\pi^-$ and $\pi^+\pi^-$ mass interpretation for $X_{\pi\pi}(K\pi) > 0.8$ is shown. A clear $K^{*0}(890)$ found in the $K^+\pi^-$ interpretation, and the strong peak in the $\pi^+\pi^-$ mass is completely accounted for as a reflection of the $K^{*0}(890)$ signal

The fitting method is described in detail in ref. [3]. The mass distributions for a given interpretation (eg. $m_{\pi^+\pi^-}$) is fitted to an expression of the type

$$BG \times (1 + \sum_i \alpha_i BW_i) + \sum_j \alpha_j REFL_j$$

where $REFL_j$ describes the shape of the reflection of a resonance (j) that does not appear directly in the mass distribution considered. The fitting to the different mass distributions (such as $\pi^+\pi^-$) is made simultaneously, keeping the parameters $\alpha_{i(j)}$ common. The small contribution from K^{*0} production (in $K^-\pi^+$ mass interpretation) and the effects of ω^0 and ϕ reflections are also considered [3].

The results of the fitting are shown for the $\pi^+\pi^-$ and $K^+\pi^-$ mass spectra in fig. 17, and the cross sections obtained are given in table I. From fig. 18 we observe that the K^{*0} cross section is rising between 32 GeV/c and 70 GeV/c whereas the ρ^0 cross section is roughly constant. In both cases the data are consistent with the corresponding cross sections from K^-p. We find some evidence for $\bar{K}^{*0}(890)$ production, not observed at 32 GeV/c.

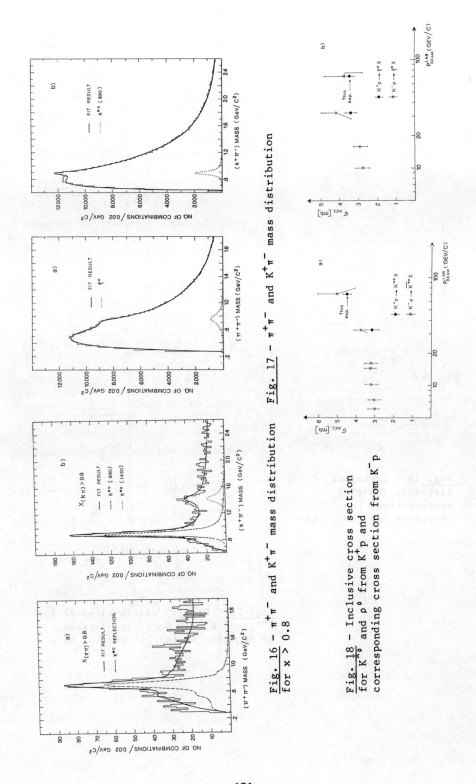

Fig. 16 - $\pi^+\pi^-$ and $K^+\pi^-$ mass distribution for x > 0.8

Fig. 17 - $\pi^+\pi^-$ and $K^+\pi^-$ mass distribution

Fig. 18 - Inclusive cross section for K^{*0} and ρ^0 from K^+p and corresponding cross section from K^-p

181

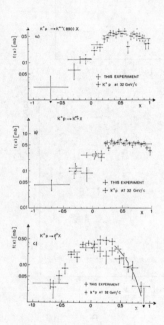

Fig. 19 - Invariant
structure function
distributions for
reactions (6), (7), (8)

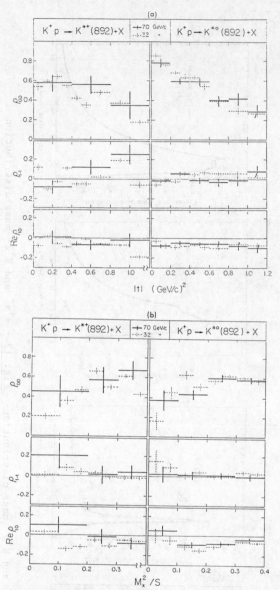

Fig. 20 - Density matrix elements for
K^{*+}, and K^{*0} in Gottfried-Jackson,
as function of (a) $|t|$ and (b) M_X^2/s

Differential distributions for reactions (6), (7) and (8) were
determined by repeating the above fit procedures for selected ranges
of the variables studied. Since the experimental resolution is a
strong function of the momentum the resonance widths used in the
fits were different for different x regions. Fig. 19 shows the
invariant structure function distributions, $f(x)$, for these
reactions, compared with data from 32 GeV/c [16]. For the K^{*+} the
results are consistent with scaling over the full x region, while for
the K^{*0} scaling is observed in the beam fragmentation region. Both
distributions are essentially flat in the region $x > 0.3$, possibly
related to diffractive production processes. For the ρ^0 the $f(x)$
distributions are compatible with the lower energy data in the beam
and target region but there is some difference in the region
$0. < x < 0.4$. The fitted slopes in $(1 - |x||)^n$ for these reactions
are included in table 2. The $d\sigma/dp_t^2$ distributions for K^{*+0} and
ρ^0 production (not shown) are very similar to the results at
32 GeV/c.

The spin density matrix elements for $K^{*+}(890)$ and $K^{*0}(890)$ have
been studied in the Gottfried-Jackson and transversity reference
frames. For the analysis the cuts $M_X^2/s < 0.4$ and $|t| < 1.2$ (GeV/c)^2
were applied to yield a system with relatively small background, and
the density matrix elements were
obtained by a maximum likelihood
fit to the usual expressions for
the decay distribution of a pure
spin 1^- state. Fig. 20 shows
the results as a function of $|t|$
and M_X^2/s using the Gottfried
Jackson frame, and we note the
results are in agreement with
32 GeV/c data.

Recent model calculations in
the framework of the recombination picture [17] predict
values of the spin density
matrix elements in the
transversity frame. Fig. 21
shows these data as a function
of $|p_T|$, for $0.4 < x_K^* < 0.7$,
with the predictions from this
model (valid over the x_K^*
range used).

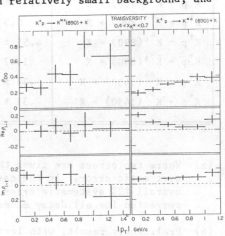

Fig. 21 - Density matrix
elements for K^{*+} and K^{*0} in
transversity frame, as a
function of $|p_T|$

SUMMARY

The inclusive cross sections for the reactions are summarised in
table I, together with a comment on the energy dependence observed.

TABLE I Cross section results

Reaction	Cross section mb[(a)]	Energy dependence comment
$K^+ p \to K^n + X$	$10.20 \pm .16$	Rising
$K^+ p \to K^0 + X$	$8.52 \pm .20$	-
$K^+ p \to \bar{K}^0 + X$	$1.68 \pm .13$	-
$K^+ p \to \Lambda + X$	$1.09 \pm .05$	} Rising
$K^+ p \to \bar{\Lambda} + X$	$0.65 \pm .04$	
$K^+ p \to K^0 + X_{ns}$	$3.70 \pm .41$	-
$K^+ p \to \bar{\Lambda} + X_{ns}$	$0.34 \pm .07$	-
$K^+ p \to \rho^0 + X$	$3.42 \pm .24 \pm .5$	Constant
$K^+ p \to K^{*+}(890) + X$	$4.07 \pm .3 \pm .3$	Slight increase
$K^+ p \to K^{*-}(890) + X$	$0.3 \pm .2^{(b)}$	Increase?
$K^+ p \to K^{*0}(890) + X$	$4.55 \pm .21 \pm .7$	Rising
$K^+ p \to \bar{K}^{*0}(890) + X$	$1.47 \pm .70^{(b)}$	Not seen at 32 GeV/c
$K^+ p \to K^{*+}(1430) + X$	$.33 \pm .08 \pm .10^{(c)}$	Compatible with
$K^+ p \to K^{*0}(1430) + X$	$.31 \pm .06 \pm .10^{(c)}$	32 GeV/c

(a) Where two errors are given the second represents the estimated systematic error. In the case of K^n, Λ and $\bar{\Lambda}$ there is an overall $\pm 2\%$ systematic error (not included). All values are corrected for all decay modes except $K^*(1420)$ results.

(b) Preliminary result, with large systematic error.

(c) For $x > 0.6$, only corrected for $K\pi$ decay modes.

For the main reactions studied differential distributions were presented. Fits of the invariant structure function to the form $(1 - |x|)^n$ are summarised in table 2 together with a comment on the scaling behaviour.

TABLE II Results of fits of $(1 - |x|)^n$ to the invariant
structure function, and comment on scaling behaviour

Reaction	Scaling behaviour	Fits of $(1 - |x|)^n$ x range Lower	Upper	n
$K^+p \rightarrow K^n + X$	Scales all x	0.2	1.0	$.78 \pm .08$
$K^+p \rightarrow \Lambda + X$	Close to scaling all x	.2	.8	$1.66 \pm .76$
		-1.0	-0.3	$0.77 \pm .09$
$K^+p \rightarrow \bar{\Lambda} + X$		.3	.8	$2.14 \pm .57$
		$-.8$	$-.2$	6.0 ± 1.6
$K^+p \rightarrow \rho^0 + X$	Scales beam and target region	.4	1.0	$2.24 ^{+1.2}_{-.8}$
$K^+p \rightarrow K^{*+}(890) + X$	Scales all x	.2	1.0	$.11 \pm .07$
$K^+p \rightarrow K^{*0}(890) + X$	Scales x > .2	.2	1.0	$.03 \pm .04$

Polarisation results for Λ show values consistent with zero over
the complete x region whereas the $\bar{\Lambda}$ production exhibits a
significant polarisation in the region $x > .2$. Spin density matrix
elements for K^{*0} and K^{*+} are in agreement with results at
32 GeV/c.

Acknowledgements

It is a pleasure to thank our scanning and measurement staff at our
respective laboratories and the operating crews of BEBC and the SPS
accelerator and members of the EF Division for their help with the RF
beam.

185

REFERENCES

[1] Inclusive neutral kaon production in 70 GeV/c K^+p interactions, this Collaboration, submitted to the conference, University of Nijmegen, preprint HEN-202.

[2] Inclusive Λ and $\bar{\Lambda}$ production in K^+p interactions at 70 GeV/c, this collaboration, submitted to the conference, CERN/EP 81-51.

[3] Production of vector and tensor mesons in K^+p interactions at 70 GeV/c, a preliminary analysis, this collaboration, submitted to the conference.

[4] M. Barth et al., Z. Physik C. Part. and Fields 2 (1979) 285.

[5] C. Cochet et al., Nucl. Phys. B124 (1977) 61.

[6] P. Chliapnikov et al., Nucl. Phys. B133 (1978) 93.

[7] R.C. Hwa and R.G. Roberts, Z. Physik C, Particles and Fields 1 (1979) 81.

[8] Z. Dziembowski et al., Int. Rep. Nijmegen HEN 126.

[9] P. Chliapnikov et al., Nucl. Phys. B148 (1979) 400.

[10] P. Chliapnikov, Proceedings of XIth Int. Symposium of Multiparticle Dynamics, Bruges (1980) 2.

[11] B. Andersson et al., Phys. Lett. 71B (1977) 337 and 69B (1977) 221.

[12] R.D. Field and R.P. Feynman, Nucl. Phys. B136 (1978) 1.

[13] (a) P.V. Chliapnikov et al., NUcl. Phys. B112 (1976) 1;
     (b) W. Barletta et al., Nucl. Phys. B51 (1973) 499;
     (c) P.V. Chliapnikov et al., Nucl. Phys. B131 (1977) 93.

[14] J.F. Gunion, Phys. Lett. 88B (1979) 150.

[15] T.A. de Grand and H.J. Miettinen, Univ. of California, Santa Barbara, preprint HCSB TH-27 (1981).

[16] (a) I.V. Ajinenko et al., Z. Phys. C5 (1980) 177;
     (b) P.V. Chliapnikov et al., Nucl. Phys. B176 (1980) 303.

[17] H.I. Miettinen, private communication and T.A. Grand and M.I. Miettinen, Phys. Rev. D23 (1981) 1227.

DISCUSSION

MORRISON, CERN: Richard, if you compare the number of $\bar{\Lambda}$'s you had with the previous talk you had many more, because the K^+ tends to give $\bar{\Lambda}$. And I think there was something about diffractive in this [a significant production of $\bar{\Lambda}$'s diffractively via $K^+ \to \bar{\Lambda}p$]. You said the $\bar{\Lambda}$ cross section was rising, perhaps something like log (s) or log (s) to some power. Can you in some way separate out the diffractive $\bar{\Lambda}$'s and see whether the cross section is rising?

ROSS: Well, if you look up the $\Lambda\bar{\Lambda}$ distributions you see, I think, a lot of the rise is coming from an increase in the $\Lambda\bar{\Lambda}$ cross section. So that does not directly answer your question but it's a hint in that direction. Do any of my colleagues have any more learned comments?

MORRISON: I think you want to perhaps separate in terms of the x. The $\Lambda\bar{\Lambda}$ tends to be produced near x = 0. Whereas if it's diffractive it would be $\bar{\Lambda}p$ and the combination would be very forward and it would give you some way of separating out the diffractive. Then you could look at variation of cross section if there are such events.

ROSS: We should continue on that line.

FIAŁKOWSKI, KRAKOW: You have shown there is a big asymmetry in the n value of forward and backward $\bar{\Lambda}$'s. So you have evidently much more forward $\bar{\Lambda}$'s.

ROSS: We have a lot more forward so its a beam fragmentation process, but that does not distinguish between diffractive and non-diffractive processes.

MULTIPARTICLE FRAGMENTATION IN K^+p
INTERACTIONS AT 32 GeV/c

CERN-SOVIET UNION COLLABORATION

E.A. DE WOLF[*]
Interuniversity Institute for High Energies,
ULB-VUB, Brussels

Universitaire Instelling Antwerpen, Wilrijk, Belgium

ABSTRACT

Results are presented on inclusive production of resonant and non-resonant particle systems produced in K^+p interactions at 32 GeV/c. We compare $K^{*+}(892)$, $\overline{\Sigma}^{*-}(1385)$, $\Sigma^{*\pm}(1385)$ and ρ° inclusive x-spectra with the ones of non-resonant $K^\circ_s\pi^\pm$, $\overline{\Lambda}\pi^-$, $\Lambda\pi^\pm$ and $\pi^+\pi^-$ pairs at the same effective mass. Resonance-particle pairs $K^{*+}\pi^\pm$, $\Sigma^{*+}\pi^-$ and $\overline{\Sigma}^{*-}\pi^-$ are also studied together with non-resonant triplets $K^\circ_s\pi^+\pi^\pm$, $\Lambda\pi^+\pi^\pm$, $\overline{\Lambda}\pi^-\pi^-$.
The invariant x-spectra of resonant particle pairs decrease less rapidly with x then the corresponding non-resonant pairs. Comparison with quark-recombination predictions indicates that the single particle or resonance and multiparticle systems are probably created off a single valence quark or diquark instead of carrying all possible valence-quarks.
These results are confirmed by an analysis of K^{*+} and ρ° production in association with same hemisphere trigger particles.

INTRODUCTION

Studies of low p_T hadron hadron collisions have provided strong evidence that the constituents of projectile and target may play an active and individual role in the production of small p_T, large momentum hadrons. Nevertheless no clear answer as yet exists as to the nature of the dynamical mechanisms involved in soft collisions. Popular models may be classified in two groups : quark-recombination[1] (QR) models and quark-fragmentation (QF) models[2]. In the first class it is assumed that incoming valence quarks remain essentially undisturbed during the collision and can subsequently recombine with each other or with sea-quarks to form hadrons of large longitudinal momentum. The QF-picture views non-diffractive scattering as a process whereby one of valence quarks is slowed down, the remaining valence quarks which carry most of the initial momentum then fragment into hadrons.

At a phenomenological level, both pictures are surprising successful and fit well a large number of single particle data. However, these successes can hardly be considered conclusive evidence for their validity. Single particle inclusive processes are sensitive to one valence quark or diquark only and as such do

[*] Bevoegdverklaard Navorser NFWO, Belgium.

not reflect the very different roles attributed to valence quarks
in QR or QF models. In QR, all valence quarks may participate ma-
ximally in producing a large momentum hadronic system; in QF only
one valence quark or diquark is active in producing such systems,
whereas the held-back ("wounded") quark contributes to the central
region. Significant tests of the models are therefore only possible in experiments which are sensitive to the joint momentum distribution of valence quarks. Analyses of two- and more particle
fragmentation processes seem well suited for this purpose e.g.
by comparing two-particle systems of rather low effective mass
(one of the particles may be a resonance) which in the QR-picture
may inherit varying numbers of valence quarks, depending on the
quantum numbers of the pair. A test of this type involving
resonant and non-resonant $\Lambda\pi^+$ pairs was recently proposed by
Fialkowski[3]. This kind of analyses is particularly interesting in
kaon beams since the strange valence quark is easily detected and
identified.

In this paper we review recent work[4-5] on multiparticle
fragmentation involving strange particles and resonances, performed by the Brussels-Mons-Serpukhov collaboration using K^+p data
at 32 GeV/c. Due to lack of space, we refer to the original
papers for a full description of the methods and other experimental details.

THE DATA

The data were obtained from several exposures of the H_2 bubble
chamber MIRABELLE to a r.f. separated K^+ beam of 32.1 ± 0.2 GeV/c
at the Serpukhov accelerator. Results on strange particles and
resonances are based on a sample of 750K pictures yielding 21207
K_s^0's, 2953 Λ's and 1683 $\overline{\Lambda}$'s, unambiguously identified. All cross
sections quoted are corrected for the usual losses and processing
efficiencies. Resonance cross sections, corrected for unseen
decay modes and branching ratios, were determined by fitting the
corresponding two-particle effective mass distributions to a
superposition of a relativistic Breit-Wigner function and an
exponential background. Charged particles are only identified
for laboratory momenta $p_{LAB} < 1.2$ GeV/c or in 4C-fit channels;
otherwise they were treated as pions.

EXPERIMENTAL RESULTS

In Fig. 1 are shown the densities $\rho(x) = \frac{1}{\sigma} d\sigma/dx$ of $K^n\pi^\pm$, $\Lambda\pi^\pm$
and $\overline{\Lambda}\pi^-$ pairs as a function of Feynman x for several effective mass
intervals of the pairs. In general $\rho(x)$ depends weakly on the
effective mass of the pairs, except for $K^n\pi^+$ and $\Lambda\pi^+$ in the kaon
and proton fragmentation region, respectively, the density being
largest for combinations in the $K^{*+}(892)$ and $\Sigma^{*+}(1385)$ mass bands.
Furthermore, $\rho(x)$ for $K^n\pi^-$ coincides with $\rho(x)$ for $\overline{\Lambda}\pi^-$-pairs at
all masses (solid line) and is also similar to $K^n\pi^+$ upto x ∿ 0.7
for $K^n\pi^+$ masses below the K^*-band.

For the mass-intervals shown in Fig. 1 we determined separately the resonance and background contribution in each x-inter-
val. The ratios of resonance to background for K^{*+} (and also K^{*0}),
$\Sigma^{*\pm}(1385)$, $\overline{\Sigma}^{*-}(1385)$ and ρ^0 are plotted in Fig. 2 as a function

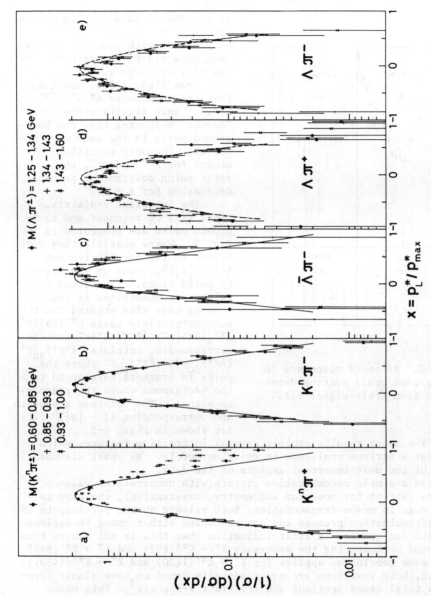

Fig. 1. Inclusive Feynman-x densities of two-particle systems in intervals of effective mass.

191

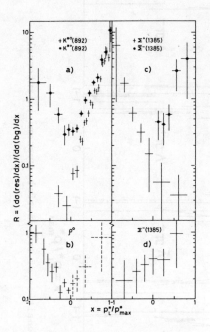

Fig. 2. Ratio of resonance to background cross section determined from Breit-Wigner fits.

of x. The ρ° data are from ref.6 for $x(\pi^+\pi^-) < 0$. For $x(\pi^+\pi^-) > 0$, only events with observed K_S° were used to avoid K^+ contamination of the π^+'s at large x.

From Figs. 1-2 we conclude that the resonances $K^{\ast +,\circ}$, $\overline{\Sigma}^{\ast -}$, $\Sigma^{\ast +}$, ρ° have significantly "harder" x-spectra than the background pairs in the same mass interval. The only exception occurs for $\Sigma^{\ast -}$ and $\Lambda\pi^-$, the ratio being constant or perhaps decreasing for x < 0.

The invariant x-distributions f(x) of resonant and background pairs are presented in Fig. 3. Where possible they were fitted to the parametrization $A(1 - |x|)^n$, shown on the figure by solid lines. The fitted n-values are summarized in Fig. 5.

We have also studied the resonance+particle pairs $K^{\ast +}(892)\pi^{\pm}$ $\Sigma^{\ast +}(1385)\pi^-$, $\overline{\Lambda}^{\ast -}(1385)\pi^-$ and the corresponding triplets $(K^n\pi^+)_{BG}\pi^+$ $(\Lambda\pi^+)_{BG}\pi^-$, $(\overline{\Lambda}\pi^-)_{BG}\pi^-$ where the pairs in brackets correspond to the background under the resonance signal. The obtained x-spectra and corresponding $(1 - |x|)^n$ fits are shown in Figs. 4-5.

The above results exhibit several interesting features which present a serious challenge to existing models. We shall discuss some of the most important aspects of the data.

In a simple recombination picture with uncorrelated valence quarks (except for momentum and energy conservation), one expects that e.g. in meson fragmentation, both valence quarks participate in the recombination process and can recombine either among themselves or with sea quarks. A first indication that this is not always true is found by comparing the processes $K^+ \rightarrow K^{\ast +}(892)$ and $K^+ \rightarrow K^{\ast \circ}(892)$ (the same conclusion applies for $K^+ \rightarrow K^{\ast +}(1430)$ and $K^+ \rightarrow K^{\ast \circ}(1430)$). Indeed, both reactions are experimentally found to have almost identical total cross sections and invariant x-spectra[7]. This means that the contribution of a diagram where the initial $\bar{s}$ and u valence quarks recombine into the same particle (possible for $K^{\ast +}$ but not for $K^{\ast \circ}$) is apparently small or altogether absent in non-diffractive production. The dominant mechanism should therefore be recombination of $\bar{s}_V$ with either a u or d sea-quark, yielding equal cross sections and x-spectra if the sea is SU(2) symmetric.

At the multiparticle level, the recombination model predicts that uncorrelated particle pairs, with each particle carrying a valence quark, should be "faster" than systems or particles in-

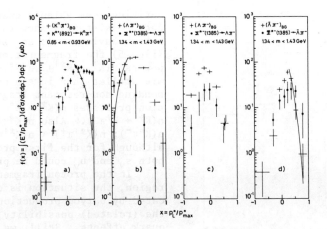

Fig. 3. Invariant x-spectra for resonances and corresponding background. Curves are fits to $(1 - |x|)^n$.

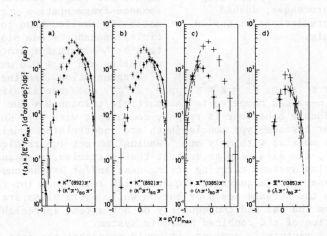

Fig. 4. Invariant x-spectra for resonance + pion pairs and "background + pion triplets. Curves are fits to $(1 - |x|)^n$.

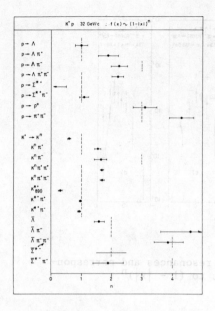

Fig. 5. Fragmentation powers
for various K^+ and proton fragmentation processes; dashed
lines are predictions of Gunion
counting rule.

heriting less valence quarks. Our
data consistently show the opposite behaviour. As an example
(fig. 5) the fragmentation
powers $n(K^+ \to K^{*+})$, $n(K^+ \to K^{*0})$
are smaller than in the corres-
ponding non-resonant fragmentation processes $n(K^* \to K^0\pi^+)$,
$n(K^+ \to K^+\pi^-)$. Also $n(K^0\pi^+) \sim$
$n(K^0\pi^-)$; $n(K^{*+}\pi^+) \sim n(K^{*+}\pi^-)$
although for the first process
both $\bar{s}_v$ and u_v could be present.

In the proton fragmentation
region, the situation is complicated by baryon production and
the (related) possibility of diquark effects. Still, we again
observe $n(p \to \Lambda) < n(p \to \Lambda\pi^+)$;
$n(p \to \Sigma^{*+}) < n(p \to \Sigma^{*+}\pi^-)$;
$n(p \to \Lambda\pi^+) \sim n(p \to \Lambda\pi^-)$;
$n(p \to \Sigma^{*+}) < n(p \to \Lambda\pi^+)$ contrary
to naive QR expectations. The
data show on the contrary a
rather different systematics.
From Fig. 5 we observe that n-
values for single particle/re-
sonance fragmentation are systematically smaller than for par-
ticle/resonance + one pion systems by ~ 1 unit for proton fragmentation and by > 0.6 units for
K^+ fragmentation. Adding an
extra π^+ or π^- (e.g. $K^+ \to K^0\pi^+\pi^{\pm}$, $p \to \Lambda\pi^+\pi^-$) does not significantly
change the n-values compared to the particle/resonance + one pion
systems. Such a behaviour is roughly compatible with e.g. Kuti-
Weisskopf or cascade-type models (both are uncorrelated longitudinal
phase space models) with only one "leading" object (quark/diquark).

In all, the data seem to suggest that particles, resonances and
multiparticles systems carrying strangeness and/or baryon number are
driven by one valence-quark or diquark which recombines (or fragments) into the strange meson or baryon while extra pions are
created from the sea and consequently do not affect appreciably the
x-distribution of the combined hadronic system.

If this picture is correct, it would mean that the data show
evidence for a strong dynamical anti-correlation effect among valence quarks of the type suggested in fragmentation or dual-sheet
models.

In a previous section we pointed out that the processes
$p \to \Sigma^-(1385)$ and $p \to \Lambda\pi^-$ behave differently and seem to contradict
the rule that n(resonance) < n(background). However, this diffe-
rence is only apparent. Indeed since $\Sigma^-(dds)$ can only inherit one
d-valence quark it should have a steeper x-distribution and bigger
n-value than the background $\Lambda\pi^-$ pairs which may pick-up a (ud)-
diquark or larger average momentum. Consequently, in QF or QR-

models, one expects the resonance to background ratio to decrease with increasing $|x|$.

In Fig. 5 we also indicated the predictions of the Gunion counting rule[8] first formulated for single particle processes. Although the agreement is generally very satisfactory for such reactions[9], several discrepancies are found for multiparticle processes. In particularly, the counting rule predicts equal n-values for resonances and uncorrelated pairs, contrary to the data.

In the previous sections we concentrated on x-spectra of single and multiparticle systems. A more differential way to study such processes consists in looking for production properties of particles and resonances in association with so-called "trigger" particles. This approach is particularly relevant in the QR picture where one may assume that a large $|x|$ trigger with suitably chosen quantum numbers contains one or several of the initial valence quarks. In this case, additionally produced hadrons should reflect the composition and momentum distribution of the remaining-spectator-valence quarks. Previously published results[10] on $\pi^{\pm}$'s produced with fast $\pi^{\pm}$, K_s°, Λ etc. in the same hemisphere were in good agreement with the QR model[11] but also with QF models[12]. To provide further tests of the models we have extended our previous analysis to a study of the resonances K^{*+}, ρ° associated with $\pi^{\pm}$, K^n, p, $\Delta^{++}(1232)$ and Λ triggers. As an example, we show in Fig. 6 the invariant x-spectra for the reactions $K^+ p \to K^{*+}(890) + \pi^{\pm} + X$ in two intervals of $x(\pi)$, the trigger momentum; $\tilde{x}$ is defined as $\tilde{x}(K^+) = x(K^*)/(1 - x(\pi))$. The curves are fits to $(1 - x)^n$ with the following results for $x(K^*) > 0.3$:

$$x_{\pi^+} > 0.1 \qquad\qquad n = 0.91 \pm 0.07$$

$$x_{\pi^-} > 0.1 \qquad\qquad n = 0.96 \pm 0.015$$

$$x_{\pi^+} > 0.2 \qquad\qquad n = 0.72 \pm 0.19$$

$$x_{\pi^-} > 0.2 \qquad\qquad n = 0.77 \pm 0.18$$

From Fig. 6 it is seen that the K^*-spectra are similar in shape and normalization for π^+ and π^- triggers, and decrease less rapidly for higher trigger momenta. The last feature contrasts with earlier observations[10] that x-spectra depend weakly on the trigger momentum for large x. The rather low values $x(\pi^{\pm}) > 0.2$ adopted for reasons of statistics may partly be responsible for the observed scaling violation. The n-values quoted above are significantly larger than $n(K^+ \to K^{*+}) = 0.30 \pm 0.06$ found for the total inclusive K^{*+} spectrum corrected for diffractive and two-body channels[7]. A similar difference is observed for the processes $K^+ \to K^{*+}$ and $K^+ \to K^{*+} \pi^{\pm}$ with $n(K^+ \to K^{*+} \pi^{\pm}) = 0.94 \pm .04$. In view of the observed flattening of the triggered K^*-spectrum at higher trigger $x(\pi^{\pm})$ one cannot exclude however that it approaches the untriggered spectrum for $x(\pi^{\pm}) \to 1$.

The insensitivity of the K^{*+} spectra to the trigger charge agrees with the observation that $K^{*+}\pi^+$ and $K^{*+}\pi^-$ have similar x-spectra but is somewhat surprising in the context of QR-models. The trigger π^+, but not π^- may be produced off the u-spectator valence quark, the s-valence quark recombining with a sea-quark to K^{*+}. The latter process should also lead to steeper x-spectra for π^- than for π^+ when associated with a K^{*+} trigger. However the data of ref. (10) show that the ratio π^+/π^- is compatible with unity for all $x(\pi^\pm)$ and moreover independent of the $(K^n\pi^+)$-effective mass. This seems to suggest that, once the s-valence quark recombines or fragments into K^{*+} or into a $(K^n\pi^+)$ system, the spectator u valence-quark no longer participates in any appreciable way to produce fast π^+'s.

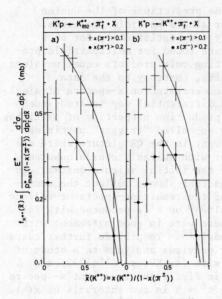

Fig. 6. Invariant x-distributions of K^{*+} with π^+ and π^- triggers. Curves are fits to $(1 - x)^n$.

SUMMARY AND CONCLUSIONS

Starting from the recombination picture which we adopted to analyse data on single- and multiparticle systems produced in the beam and target fragmentation regions we observed several interesting and partly surprising facts.

i) Resonances carry on the average more momentum than the non-resonant particle pair of comparable effective mass even when both particles in the pair could carry a valence quark. On the one hand, this perhaps proves that resonances are indeed more directly produced and are thus not merely final-state interactions among directly emitted stable hadrons. More important however, this observation is in striking disagreement with the usual QR-picture

ii) The x-distributions of single particle/resonance and of multiparticle systems also consistently suggest that they do not carry the maximum number of valence quarks allowed by quantum numbers, but are created off a single "leading" object : valence-quark or diquark

iii) The x-dependence of resonances (K^*, ρ°) produced in association with same hemisphere triggers also shows properties which are difficult to understand in the QR model.

Summarizing, our data perhaps point to the existence of a strong anti-correlation among valence quarks during the soft collision process such that one valence quark effectively slows down, and merely behaves like a sea-quark, whereas the fast quark/diquark recombines or fragments. Obviously, this picture contains elements of the "wounded-quark" concept, popular in models for hadron-nucleus collisions[13]. At the same time, it fits well into the ideas of QF and dual-sheet models.

Future high energy and high statistics experiments should be able to elucidate more clearly the questions raised by our medium energy data and hopefully solve the long-standing controversy concerning the nature of soft hadron production mechanisms.

It is a pleasure to thank Bill Shephard and collaborators for organizing such an enjoyable conference. I am also grateful to P.V. Chliapnikov and many of my colleagues and participants at the Symposium for interesting and enlighting discussions.

REFERENCES

1. R.C. Hwa, Phys. Rev. D22, 759 (1980); 1593 (1980)
2. B. Anderson et al., Phys. Lett. 69B, 221 (1977); 71B, 337 (1977); Nucl. Phys. B135, 273 (1978)
   A. Capella and J. Tran Thanh Van, Phys. Lett. 93B, 146 (1980)
   G. Cohen-Tannoudji et al., Phys. Rev. D19, 3397 (1979); D21, 2699 (1980)
3. K. Fialkowski, Acta Phys. Pol. B11, 659 (1980)
4. E.A. De Wolf et al., A study of multiparticle fragmentation in K^+p interactions at 32 GeV/c, submitted to this conference
5. I.V. Ajinenko et al., A study of K^{*+} and ρ° production with meson and baryon triggers, submitted to this conference
6. P.V. Chliapnikov et al., Nucl. Phys. B176, 303 (1980)
7. P.V. Chliapnikov et al., Comparison of strange antibaryon and strange meson production in K^+p interactions at 32 GeV/c, preprint IIHE-81.2 (Brussels), unpublished
8. J.F. Gunion, Phys. Lett. 88B, 150 (1979)
9. D. Denegri et al., Phys. Lett. 99B, 127 (1981)
10. E.A. De Wolf et al., Z. Phys. C - Particles and Fields 8, 189 (1981)
11. E. Takasugi and X. Tata, Particle correlations in the recombination model ..., DOE-ER-03992-377, University of Texas preprint, 1981
12. G. Gustafson, Europhysics study Conf. Partons in soft hadronic processes, Erice 1981, unpublished.

13. For a review see A. Bialas, Projectile fragmentation in hadron-nucleus interactions at high energies; Proc. XI Int. Symposium on Multiparticle Dynamics, Bruges, Belgium; E.A. De Wolf and F. Verbeure eds., 1980, 171.

DISCUSSION

HWA, OREGON: My feeling is that perhaps one is putting too much burden on all these models. You want to apply these ideas to experiments where the energy is [only] 32 GeV with many particles being produced and there are a lot of kinematical constraints and we demand that all the valence quarks play an important role, to contribute to data. I feel that we probably just cannot extract that much out of this data at that energy. If it were at much higher energy where I really can think of scaling distributions for the various valence quarks, and sea quarks, and similar things happen that would be very . . . . [Written comment by Hwa: The model assumes scaling x distributions for the quarks which is untrue at your energy. (Buschbeck et al. showed significant s-dependence.) While the simple recombination model may be applicable for one valence quark at large x, it is dangerous to consider all valence quarks. One of the valence quarks is likely to be affected by central interaction.]

DE WOLF: Rudy, I think there is a standard answer to that. First of all, the Mirabelle experiment has very high statistics and it's the only place were we can really do this trigger business with any kind of confidence. Secondly, whenever we have some successes which confirm the models, people always forget the arguments about the energy dependence. They say, okay, it's very fine. One has to decide what one really wants to believe.

FIAŁKOWSKI, KRAKOW: I have only one question. Is the background which was compared with the resonance states averaged to the same mass as the resonance?

DE WOLF: It's exactly the same mass. As a matter of fact, we have done several things but the idea is rather simple. You fit the mass with some x value.

FIAŁKOWSKI: So you are sure that it's not a kinematical effect.

DE WOLF: As a matter of fact we have studied at lower masses and it gives very similar or almost exactly the same distribution as what we got for our background. But of course we make that the background is physical.

FIAŁKOWSKI: So it seems to show a real contradiction to this naive recombination prediction.

DE WOLF: It's very different. I mean it's not just a little bit different, it's completely different.

BIAŁAS, KRAKOW: I would just like to repeat what I said this morning. I do not think that these beautiful data represent a decisive argument against the recombination model. Even in the recombination model one quark interacted and the other didn't. From the point of view of the recombination model, these data provide direct evidence that the interaction [wounded] constitutent quark fragments differently than the spectator quark [that wounded quarks do not contribute significantly to the projectile fragmentation region]. Let me repeat that people who work with scattering from nuclear targets knew for several years that these two quarks play very different roles and they behave very differently. So I am very happy that it was finally found also in elementary [hadron-hadron] inter-actions.

OBSERVATION OF ρ⁰-MESON SPIN ALIGNMENT IN p̄p INTERACTIONS AT 22.4 AND 5.7 GeV/c.

B.V.Batunya, I.V.Boguslavsky, N.B.Dashian[x],
I.M.Gramenitsky, R.Lednický, S.V.Levonian[xx],
V.L.Lyuboshitz, L.A.Tikhonova, V.Vrba, Z.Zlatanov.
Joint Institute for Nuclear Research, Dubna.

E.G.Boos, V.V.Samojlov, Zh.S.Takibaev, T.Temiraliev.
Institute of High Energy Physics, Alma-Ata, USSR.

S.Dumbrajs, J.Ervanne, E.Hannula, P.Villanen.
Department of High Energy Physics, University of
Helsinki,Helsinki, Finland.

R.K.Dementiev, I.A.Korzhavina, E.M.Leikin, V.I.Rud.
Institute of Nuclear Physics, Moscow State University,
Moscow, USSR.

J.Böhm, J.Cvach, I.Herynek, M.Jireš, M.Lokajíček,
P.Reimer, J.Řídký, J.Sedlák, V.Šimák.
Institute of Physics, Czechoslovak Academy of Sciences,
Prague, CSSR.

M.Suk, J.Valkárová, J.Žáček.
Nuclear Centre, Charles University, Prague, CSSR.

A.M.Khudzadze, G.O.Kuratashvili, T.P.Topuriya,
V.D.Tsintsadze.
Institute of High Energy Physics, Tbilisi State
University, Tbilisi, USSR.

[x]Yerevan Institute of Physics - Yerevan, USSR.
[xx]Institute of Physics - Moscow, USSR.

ABSTRACT

Studying the ρ^o-meson production in $\bar{p}p$ interactions at 22.4 GeV/c and in 4-prong annihilation
channels of $\bar{p}p$ interactions at 5.7 GeV/c, we have
observed an essential ρ^o-meson spin alignment. The
values of the ρ_{oo} element of the ρ^o-meson spin density matrix (the z-axis is directed along the normal
to the production plane) are equal to 0.61 ± 0.06 and
0.55 ± 0.03, respectively. The absence of such an
effect in pp interactions at 24 GeV/c and also the
essentially larger ρ^o production cross section in $\bar{p}p$
interactions at 22.4 GeV/c make it possible to connect
the observed ρ^o-meson spin alignment with the annihilation processes. This effect could be interpretted as
a spontaneous polarization of quarks and antiquarks
during the stage preceding their recombination to
mesons.

Study of the spin dependence of multiparticle
production is now of great interest due to observation
of unexpected strong spin effects in several experiments at high energies[1-6].

Below we present results of an experimental investigation of the spin effects in ρ^o production in $\bar{p}p$
interactions. In a preliminary study [7] of the reaction
$\bar{p}p \rightarrow \rho^o + X$ at 22.4 Gev/c we have obtained an indication of ρ^o spin alignment at the level of two standard deviations. In the present paper this result is
confirmed using essentially larger statistics and at
two values of primary momenta - 5.7 and 22.4 GeV/c.
In the analysis we use about 35000 events (1.1 weighted
ev. per μb) of $\bar{p}p$ interactions at 22.4 GeV/c and
about 35 000 4-prong annihilation events (3.3 ev. per

μb) of $\bar{p}p$ interactions at at 5.7 GeV/c. The angular
distributions of the pions from ρ° decays have been
obtained by fitting the $\pi^{+}\pi^{-}$ effective mass distributions over a range 0.32 – 1.1 GeV/c^{2} in different
angular intervals. The standard fitting formula has
been used: superposition of the usual exponential background multiplied by the two-pion phase space factor,
relativistic p-wave Breit-Wigner distribution multiplied by the background and ω°-meson reflection. Four
free parameters have been fitted: background slope,
fractions of ρ°- and ω°-mesons and ρ° mass; the ρ°
width has been fixed at a value of 150 MeV.

In fig. 1 we present the distributions of the cosine of polar angle θ_{t} in the t-channel helicity frame:

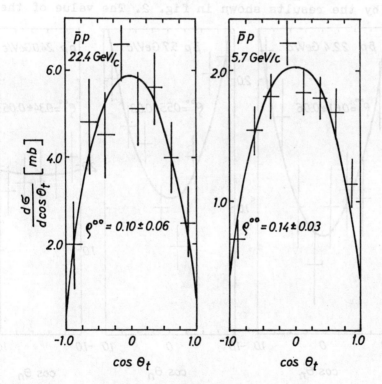

Fig. 1. cos θ_{t} distributions.

z-axis is directed along the beam (target) momentum in the rest frame of ρ^0 produced in the forward (backward) hemisphere in the c.m.s. Both distributions at 22.4 and 5.7 GeV/c are essentially different from the uniform one and agree with our previous result [7]. The character of these distributions (maximum at cos Θ = 0) remains unchanged for other ρ^0 production analyzers (z-axis) lying in the reaction plane (e.g., for the c.m.s. reaction axis). In then follows that the vector of the ρ^0 production amplitude is predominantly directed along the normal $\hat{n}$ to the reaction plane (the ρ^0 spin predominantly lying in the reaction plane). Further the ρ^0 decay angular distribution with respect to $\hat{z} \parallel \hat{n}$ should be close to $\cos^2\Theta_n$- distribution. Indeed, this is confirmed by the results shown in fig. 2. The value of the

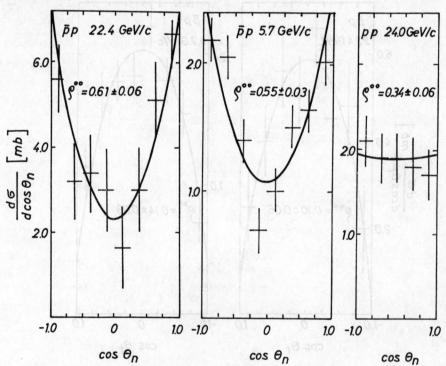

Fig. 2.   cos Θ_n distributions.

204

ρ_{oo}^n element of the ρ^o spin density matrix ($\hat{z} \parallel \hat{n}$) is found to be 0.61 ± 0.06 and 0.55 ± 0.03 [x] at 22.4 and 5.7 GeV/c, respectively. Thus we observe a deviation from the uniform distribution ($\rho_{oo} = 1/3$) at the level of 4.5 and 7 standard deviations.

It should be noted that the analogous distribution for the reaction pp $\to \rho^o$ + X at 24 GeV/c [8] practically coincide with the uniform one (our fit of these data yields $\rho_{oo}^n = 0.34 \pm 0.06$). This fact and also the essentially higher ρ^o production cross section in $\bar{p}p$ interactions at 22.4 GeV/c (7.8 ± 0.7 mb as compared with 3.5 ± 0.4 mb for pp interactions at 24 GeV/c) make it possible to connect the observed ρ^o spin alignment with annihilation processes. Assuming the nonannihilation ρ^o production is represented by pp data we obtain $\rho_{oo}^n = 0.83 \pm 0.13$ in annihilation reactions at 22.4 GeV/c.

It is worth noting that an essential ρ^o spin alignment may be expected in multiperipheral models. However, for $\rho^{o\prime}$s emitted by meson trajectories (nonannihilation reactions) a large ρ_{oo} value is expected when the z-axis is close to the reaction axis [9]. In the case of $\rho^{o\prime}$s coupled to baryons (annihilation reactions) the usual ρ-baryon couplings (no $\mathcal{E}_{\mu\nu\rho\sigma}$) do not yield a large ρ_{oo}^n value. Despite the Reggeization is important it is unlikely that it could lead to a large value of ρ_{oo}^n. E.g., although the spin density matrix of ρ^--meson produced near the backward direction in the reaction $\pi^- p \to p\rho^-$

[x] We have obtained the $\cos\theta_n$-distribution at 5.7 GeV/c also by the subtraction method using the angular distributions under the ρ^o peak (0.66–0.80) GeV/c^2 and in the nieghbouring regions (0.60–0.66) and (0.82–0.88) GeV/c^2. The resulting value of ρ_{oo}^n equals 0.50 ± 0.05.

has been found essentially dependent on the Δ_δ - trajectory parameters [10], all the solutions of ref.[10] give $\rho^n_{oo} \lesssim 1/3$. This is in contradiction with the experimental data on this reaction at 9 and 12 GeV/c [11] which yield $\rho^J_{11} = 0.40 \pm 0.02$ and $\rho^J_{1-1} = 0.20 \pm 0.06$ in the Jackson frame implying $\rho^n_{oo} = \rho^J_{11} + \rho^J_{1-1} = 0.60 \pm 0.06$. We point out that this experiment and our results suggest the same mechanism of ρ-meson production both in $\bar{N}N$ annihilation processes and in $\pi N \rightarrow X + \rho$ reactions with high momentum transfer.

We should like to recall here about the effect of spontaneous radiation polarization [x] of ultrarelativistic electrons (positrons) moving in transversal magnetic or electric fields [12]. The time of spontaneous polarization is proportional to the energy squared and the field cubed. In electron-positron accumulators the polarization reaches its maximal value of 92% during several minutes (positrons are polarized along and electrons oposite to the field direction). An analogous effect can take place for fast quarks excited in processes characterized by high momentum transfer or in annihilation reactions when the initial colorless hadrons are totally destroyed and the quarks move for a large enough time in the transversal color field. Simple estimates "by analogy" show that the polarization time here has a reasonable order of 10^{-23} sec.

It should be pointed out that a measurement of the spin density matrix for other vector meson resonances (ω^o, K^*, φ) in annihilation reactions or in high momentum transfer meson-nucleon interactions would be very useful.

[x] This effect is mostly determined by the particle magnetic moment. Thomas precession only effectively weakens the contribution of the Dirac magnetic moment.

We wish to thank the CERN-Prague Collaboration for their $\bar{p}p$ data at 5.7 GeV/c. We acknowledge useful discussion with A.M.Baldin, S.B.Gerasimov and H.I. Miettinen.

REFERENCES

1. G.Bunce et al. Phys.Rev.Letters 36, 1113 (1976).
   F.Lomanno et al. Phys.Rev.Letters 43, 1905 (1979).
   S.Erhan et al. Phys.Lett. 82B, 301 (1980).
   K.Raychaudhuri et al. Phys.Lett. 90B, 319 (1980).
2. K.Heller et al. Phys.Rev.Letters 41, 607 (1978).
3. R.D.Klem et al. Phys.Rev.Letters 36, 929 (1976).
   J.Antille et al. Phys.Lett. 94B, 523 (1980).
4. A.V.Efremov, Yad.Fiz. 28, 166 (1978).
5. B.Andersson et al. Phys.Lett. 85B, 417 (1979).
6. T.A. De Grand, H.I. Miettinen, HU-TFT-90-47, Helsinki (1980).
7. B.V.Batunya et al. Nucl. Phys. B137, 29 (1978).
8. V.Blobel et al. Phys. Lett. 48B, 73 (1974).
9. S.Fenster, J.Uretsky, Phys.Rev. D7, 2143 (1973).
   E.M.Levin, M.G.Ryskin, Yad.Fiz. 19, 904 (1974).
10. C.C.Shih, Phys.Rev.Letters 22, 105 (1969).
11. P. Benkheiri et al. Lett.Nuovo Cim. 20, 297 (1977).
12. A.A.Sokolov, I.M.Ternov, DAN SSSR 153, 1052 (1963).
    V.L.Lyuboshitz, Yad.Fiz. 4, 69 (1966).

CHARGED AND NEUTRAL STRANGE PARTICLE PRODUCTION IN 300 GEV/C PROTON-NEON INTERACTIONS

D. Minette, U. Camerini, A.R. Erwin, W.F. Fry, R.J. Loveless, D.D Reeder
The Department of Physics
University of Wisconsin
Madison, WI 53706 USA

S.A. Azimov Sh. V. Inogamov, E.A. Kosonowski, V.D. Lipin, S.L. Lutpullaev
K. Olimov, K.T. Turdaliev, T.M. Usmanov, A.A. Yuldashev and B.S. Yuldashev
The Physical Technical Institute
The Uzbec Academy of Sciences
700084 Tashkent, U.S.S.R.

Abstract

The data on the production of charged particles and neutral strange particles are presented. It is found that the total cross section, σ_{in}^{pNe} = 356$\pm$13 mb. The kaon and lambda cross sections in the pp backward hemisphere, rapidity < 3.23, are found to equal 68+8 mb. and 32$\pm$5 mb., respectively. Both pion production and kaon production are found to differ from pp interactions. Furthermore, the K^o_s/pi$^-$ ratio also varies from that seen in pp interactions.

I. INTRODUCTION

Within the last few years, there has been renewed interest, both theoretical and experimental, in hardron nucleus interactions at high energy. As a number of authors have pointed out[1], these interactions can provide important information on the space-time structure of strong interactions. In particular, they can provide information on the time period before hadronization takes place.

In this paper, we present data from inelastic 300 Gev/c proton-neon interactions. The multiplicity data has been presented before[2], so we will limit our discusion of the charged multiplicity data to the total cross section and the proton multiplicity distribution. In addition we will consider the inclusive production of pions and strange particle production.

II. EXPERIMENTAL PROCEDURE

The data come from an exposure of the 30 inch Fermilab bubble chamber to a 300 Gev/c diffractive proton beam. Pi$^+$, K$^+$, and u$^+$ contamination is estimated to be less than 0.2%.

The bubble chamber was filled with a light neon-hydrogen mixture (30.9$\pm$0.7% molar Ne). The radiation length and the density of this mixture are 128.1 cm and 0.249 g/cm^3, respectively. The average number of incoming primary protons is observed to be 2.66 $\pm$0.01 per frame.

The prong multiplicity data presented below were obtained from the double scanning of 29134 frames of which 5765 were rejected due to an interaction upstream from the fiducial volume. The double scan effeciency was greater than 99% for events with 4 or more prongs. The scanning effeciency for events with one to three prongs was lower, 87%. Further, there were systematic scanning losses for elastic one prong (proton-neon) and two prong (proton-proton) interactions. Approximately 30% of the elastic two prong events were lost due to a short ($L \leq$ cm.) recoil proton.[3] Almost all the one prong interactions were missed, since their detection required a scattering angle of > 2. Therefore, our sample consists of the inelastic proton-neon and proton- proton events plus 70% of the elastic proton-proton events. After all corrections, the number of analyzed events (sum of pNe and pp interactions) is found to be $N_{tot}(pNeH_2) = 9296 \pm 140$.

A subsample of 454 of these events have been measured and analyzed. These measurements were reconstructed by the reconstruction program TVGP. After one remeasurement pass, >96% of the tracks were sucessfully measured.

In a seperate vee scan, 41188 frames of film were scanned for events containing neutral strange particles. The initial vee scanning effeciency was 70%. A total rescan was initiated and is now in progress. The scanning efficiency for this scan is approximately 90%. This gives a total scanning effeciency of 96%.

In addition to the scanning losses, there is a loss of high energy vees due to our short fiducial length for neutral decay, 30 cm on the average. This means that we have poor statistics at high rapidity, as did the 300 Gev/c pp experiment.[4] We, however, cannot assume symmetric production about the pp center of mass rapidity 3.23. Therefore, we are limiting our discussions in this paper to the backward hemisphere in the pp center of mass, rapidity < 3.23.

The main vertex fiducial volume is between X=28 and X=-20. The vee fiducial volume extended further back, to X=-25. Further, the vees were subject to a neutral length cut L>2 cm.

III. CROSS SECTION

The interaction cross section for protons in our NeH_2 mixture can be written as:

$$\sigma(pNeH_2) = \delta \sigma_{in}(pNe) + 2(1-\delta)\sigma^*(pp), \qquad Eq\ 1$$

where δ is the molar fraction of neon; $\sigma_{in}(pNe)$ is the inelastic cross section of pNe interactions at 300 Gev (as noted above elastic pNe events were not seen); and $\sigma^*(pp)$ is the "visible" total cross section for pp:

$$\sigma^*(pp) = \sigma_{tot}(pp) - 0.3\,\sigma_{el}(pp) \qquad Eq\ 2$$

30% of the elastic cross section is subtracted because, as mentioned above, 30% of the proton-proton elastic scatters are missed due to short recoil protons.

The interaction cross section $(pNeH_2)$ measured in this experiment is (162.8 ± 2.5) mb. The quoted error is statistical only. The systematic error due to uncertainties in definition of the number of incoming protons and the length of fiducial volume is estimated to be less than 4%.

Using our measured values of $(pNeH_2)$ and the pp cross sections $(\sigma_{tot}(pp)$ and $\sigma_{el}(pp))$ at 300 Gev/c[4], we obtain the inelastic cross section for PNe interactions:

$$\sigma_{in}(pNe) = 356.0\pm13.0 \text{ mb.} \qquad \text{Eq 3}$$

Next, we will consider the strange particle cross section. Since we are only considering vees produced in the pp backward hemisphere, we will not quote the total cross section for strange particles. Rather, we will consider the cross section in the backward hemisphere. This cross section will be quoted for kaons and lambdas.

The production cross section is defined as the product of the average number of particles produced per event and the total cross section. In order to obtain the pNe cross section for K_S^0s and s, we will use give the $pNeH_2$ cross section and subtract the contribution from H_2 (see above). The average number of kaon and lambdas in the backward hemisphere per event is $0.160\pm.0014$, and 0.077 ± 0.009, respectively. Using our value for the $(pNeH_2)$ cross section, we have. $\sigma_{KS}(pNeH_2) = 26.0\pm2.3$ mb, and $\sigma_{\Lambda}(pNeH_2) = 12.6\pm1.6$ mb, Since pp interactions are symmetric with respect to the center of mass, the production cross section in the pp backward hemisphere is just half the total production cross section. Thus we have,

$$\sigma_{back}(pNeH_2) = \delta \, \sigma_{back}(pNe) + (1-\delta)\sigma(pp) \qquad Eq \text{ 4}$$

Using Eq. 4 and the pp kaon and lambda cross sections[4], we obtain the kaon and lambda total production cross section in the pp backward hemisphere: $\sigma_{K_S^0 \, back}(pNe) = 68\pm8$ mb., and $\sigma_{\Lambda back}(pNe) = 32\pm5$ mb.

IV. THE MULTIPLICITY OF SECONDARY IDENTIFIED PROTONS

Fig. 1 gives the multiplicity distribution for identified protons, that is protons with momenta $0.13 \leq p \leq 1.2$ Gev/c, in pNe interactions. The solid line is the prediction of a model which assumes that secondary protons hadron-nucleus interactions are independently emitted in subsequent intranuclear rescattering.[5,6] This model gives the distribution of the number of observed protons, $P(n_p)$, in terms of the probability of ν inelastic collisions, $P(\nu)$.[2,7] Using this model again, we obtain the distribution for $P(\nu)$ shown in Fig. 1 as a dashed curve (upper scale). The mean number of interactions is $\langle \nu \rangle_{pNe} = 2.00$. The mean number of interactions for a given number of observed protons is given in Table I. The model fits the data well, as can be seen in Fig. 1, except at $n_p = 0$ and $n_p \geq 9$. The disagreement at high and low proton multiplicity can be due to the limited momentum range for proton identification: $0.13 < p < 1.2$ Gev/c, and to the misidentification of heavy fragments, (d, t, He, etc.) as protons. The contamintion of these fragments in the identified proton sample is estimated to be less than 10%.[9]

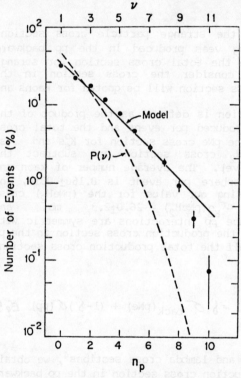

Fig 1: proton multiplicity dist.

Table I.

n_p	$\langle \nu \rangle$
0	1.33
1	1.72
2	2.09
3	2.34
4	2.82
5	3.17
6	3.54
7	3.87
8	4.17
9	4.48
10	4.80
all	1.99

The average number of inelastic intranuclear collisions $\langle \nu \rangle$ as a function of the number of identified protons: n_p

VI. INCLUSIVE CHARGED PION PRODUCTION

In our examination of inclusive charged pion production, we will consider only the negative pions. We are considering only the negative pions because of contamination of the positives by protons.

In order to investigate the effects of multiple scattering, two subsamples were considered. The first sample consists of events with no evidence of more than one nucleon being involved in the interaction. The second sample consists of events that contain strong evidence for more than one nucleon being involved in the interaction.

The first sample contains events that are consistent with proton-proton or proton-neutron scattering. That is, the total charge ≤ 2 and $N_p < 2$. These events consist of all of our pp interactions and the pNe interactions that show no evidence for more than one recoil proton. (Since we are defining this sample in terms of the excess charge, as well as the number of visible protons, this definition does not depend on this proton being slow enough to identify by ionization.)

The second sample consists of events that have more than one nucleon involved in the interaction. This sample contains events with total charge > 3 or $N_p > 1$. These events have more than one recoil proton.

The first sample, which we will label the "pN" sample, contains 57% of the events. 58% of the "pN" events are from hydrogen and 42% are from neon. The second sample, which we will label the "multiple proton" sample, contais 36% of the events. All of the events in this sample are from neon. The remaining 7% of the events have $N_p < 2$ and

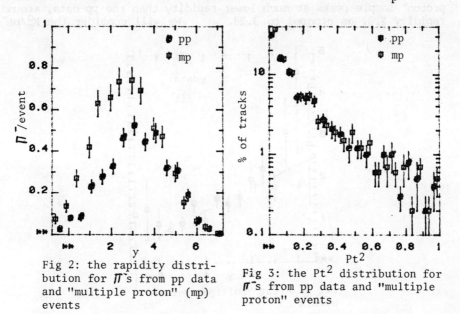

Fig 2: the rapidity distribution for Π^-s from pp data and "multiple proton" (mp) events

Fig 3: the Pt^2 distribution for Π^-s from pp data and "multiple proton" events

213

total charge = 3, and are not used in this analysis. The transverse momentum squared and the rapidity distributions for the "pN" sample are shown in Figs. 2 and 3. We see from the rapidity distribution, Fig. 2, that this sample is a fairly good representation of nucleon-nucleon scattering. The rapidity distribution is symmetric about the pp center of mass, y=3.23; and the mean multiplicty (3.59±0.10) is fairly close to the mean multiplicity for true pN data (3.35±0.05).

The rapidity and transverse momentum squared distributions for the "multiple proton" events are also shown in Figs. 2 and 3. We see from the rapidity distribution in Fig. 3 that the "multiple proton" events look very different from pN events. The mean multiplicity is much higher, 6.38±.20. Furthermore, this distribution is not symmetric about the center of mass; there is an excess of pions at lower rapidity.

We can see this best by considering the R distribution. R is the ratio of the pion multiplicity in the "multiple proton" sample vs the "pN" events. Looking at Fig. 4, we see that R is approximatetely one at high rapidity and rises steadily with lower rapidity.

Looking at the Pt^2 plots for the "pN" and the "multiple proton" samples, in Figs. 2 and 3, respectively, we see there is no major change in the Pt^2 distributions between these samples. The only real difference is in the total number of pi^-s in each sample.

VI. INCLUSIVE KAON PRODUCTION

The K_S^0 rapidity distributions for the pp data and the "multiple proton" events are shown in Fig. 5. We see that the "multiple proton" sample peaks at much lower rapidity than the pp data, around rapidity 2.0, as opposed to 3.23. We will consider the K_S^0/pi^-

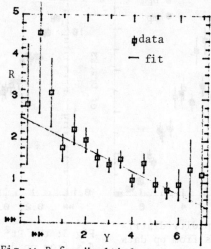

Fig 4: R for "multiple proton" events

214

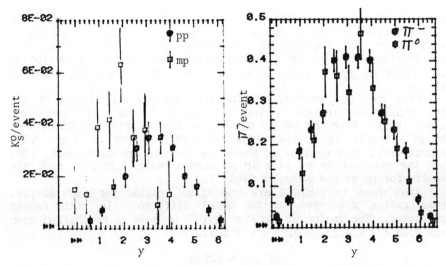

Fig 5: the rapidity distribution of K_S^0 for pp and "multiple proton" (mp) events

Fig 6: the rapidity distribution for pp π's as given by the pp π^0 and the pN η^- models

ratio to see if this peak is different from the shift in rapidity seen in the pi⁻ distribution. To maximize statistics, we will do this in two rapidity bins: 0.78 < y < 2.23 and 2.23 y< 3.23. In order to give the K_S^0/pi⁻ ratio, we need the number of pi⁻s per event in this range. For our data, this can be taken directly from Fig. 3, and is shown in Table 2. Obtaining this ratio for other pp data is more difficult because the p⁻ spectrum from pp interactions is not presently available. However, there are two fairly accurate methods of estimating this. The first method is to use the published pi⁰ rapidity spectrum. Since the production mechanisms for pi⁻s and pi⁰s are very similar, the shape of the rapidity distributions should be approximately the same. The second method is to use the pi⁻ rapidity distribution from our "pN" sample. Since it is close to

Table II

	pp data		multiple proton events	
	$0.73 \leq y \leq 2.23$	$2.23 \leq y \leq 3.23$	$0.73 \leq y \leq 2.23$	$2.23 \leq y \leq 3.23$
Pi⁻/event	0.70±0.04	0.81±0.04	1.78±0.11	1.60±0.10
K_S^0/event	.043±.006	.065±.006	.144±.021	.073±.018
K_S^0/Pi⁻ ratio	.062±.008	.080±.007	.081±.012	.046±.011

symmetric about the center of mass, it should be a fairly accurate representation of pi⁻s produced in proton-nucleon interactions. Further, we can improve on the accuracy by symmetrizing this distribution about the pp center of mass. We do this by averaging the values at $\pm y$ in the center of mass. The result is shown in Fig. 6.

From the first method, we find that $22.0\pm2\%$ and $22.3\pm2\%$ of the pi⁻s should be in the range $.78 < y < 2.23$ and $.23 < y < 3.23$, respectively. From the second method, we find $21.8\pm1\%$ and $25.0\pm1\%$ of the pi⁻s shoul⟨ ⟩e in the range $.78 < y < 2.23$ and $.23 < y < 3.23$, respectively. The results of these two methods are consistent. So, with the mean number of pi⁻s in pp events being 3.25^8, we have the results for pp events shown in Table 2.

Also shown in Table 2 are the $K_S^0/$pi⁻ ratios for each sample. These ratios are seen to be quite different, in both ranges considered. The region of maximum $K_S^0/$pi⁻ is seen to be shifted down from $2.23 < y < 4.23$ to $0.73 < y < 2.23$.

VII. CONCLUSION

In this work we have presented data on inlelastic pNe interactions at 300 Gev/c. The inelastic, K_S^0 inclusive production, and inclusive production cross sections are seen to be 356 ± 13 mb, 68 ± 8 mb. and 32 ± 5 mb respectively. The multiplicity distribution of secondary indentified protons is consisted with a model that considers these protons to be products of independent emmisions of nucleons involved in hA interactions.

We see throughout our data that pNe collsions look quite different from pp collisions. The average multiplicity of n⁻ is higher in pNe than it is in pp. We see that R, the ratio of the pi⁻ multiplicities in the "multiple proton" sample and the "pN" sample, is strongly dependent on rapidity. The K_S^0 rapidity for the "multiple proton" events is seen to peak at lower rapidity than the pp K_S^0s. Further, the $K_S^0/$pi⁻ ratio is also seen to change.

References

1. K. Gottfried, Ref/ TH/1735 — CERN (1973).
   L. Bertocci, Report IC/75/67, Triest (1975).
   K. Zalewski, Multiplarticle production on Nuclei at Very High Energies, ed G. Bellini et al., Trieste (1977), p. 145.
   N.N. Nikolaev et al., Ref. TH 2541–CERN (1978).
2. S.A. Azimov et al, Phys. Rev D23 p. 2512 (1981)
3. A Firestone et al., Phys. Rev. D10, 2080 (1974).
4. A Sheng, et al., Phys. Rev. D11 1733 (1975).
5. B. Andersson et al., Phys Lett. 73B, 343 (1978).
6. J. Cincheza et al., Nucl. Phys. B158, 280 (1979).
7. Y. Afek et al., Preprint Technion — PH 76 87 (1976).
8. Yu.M. Shabelsky, Nucl. Phys. B132, 491 (1978).
9. V.S. Barashenkov, V.D. Toneev, in the book "Interactions of High Energy Particles and Atomic Nuclei with Nuclei", Atomizadat, Moscow, 1972, (in Russian).

MULTIPARTICLE PRODUCTION ON HYDROGEN, ARGON AND XENON TARGETS IN A STREAMER CHAMBER

C. Favuzzi, G. Germinario, L. Guerriero, P. Lavopa, G. Maggi,
C. de Marzo, M. de Palma, F. Posa, A. Ranieri, G. Selvaggi,
P. Spinelli, F. Waldner
University of Bari, Bari, Italy

A. Bialas, W. Czyz, T. Coghen, A. Eskreys, K. Eskreys,
D. Kisielewska, P. Malecki, K. Olkiewicz, K. Sliwa, P. Stopa
University of Krakow, Poland

W. H. Evans, J. R. Fry, C. Grant, M. Houlden, A. Morton,
H. Muirhead, J. Shiers
University of Liverpool, Liverpool, U. K.

M. Antič, W. Baker, F. Dengler, I. Derado, V. Eckardt, J. Fent,
P. Freund, H. J. Gebauer, T. Kahl, R. Kalbach, A. Manz, P. Polakos,
K. P. Pretzl, N. Schmitz, P. Seyboth, J. Seyerlein, D. Vranič,
G. Wolf
Max-Planck-Institut für Physik und Astrophysik, München, Germany

F. Crijns, W. Metzger, C. Pols, T. Schouten, T. Spuijbroek
University of Nijmegen, Nijmegen-NIKHEF, Netherlands

N. Sarma, Visitor at CERN from Bhaba Institute, Bombay, India

presented by P. Seyboth

ABSTRACT

Interactions of 200 GeV protons and antiprotons on hydrogen,
argon and xenon are studied with a 2 m streamer chamber. Momenta
and charges of produced particles are mesured and for momenta below
600 MeV particle identification was possible. There are small
differences between proton and antiproton reactions. A saturation
of the multiplicity of fast secondaries is observed in events with
a large number of slow protons and nuclear fragments. Also the
rapidity plateau height of fast secondaries increases only slowly
from argon to xenon. Quark constituent models seem to be preferred
by our data.

INTRODUCTION

Nuclear targets are potentially a powerful tool for the study
of elementary interactions and the structure of hadrons. The
nucleus is believed to probe the time behaviour of hadron interactions. In recent years there has been a large number of papers,
both experimental and theoretical, on various aspects of high
energy collisions with nuclei[1]. The experimental data are not yet
detailed or precise enough to differentiate between the models.

Previous experiments have established that hadron formation time far exceeds transit time through the nucleus. In order to make a detailed study of hadron-nucleus interactions we need precise momentum measurement of all the final-state charged particles. To this end, we have performed an experiment with xenon and argon gas targets and with a liquid hydrogen target situated inside a streamer chamber, with a magnetic field, and a downstream spectrometer. The near 4π solid angle coverage of the streamer chamber and its excellent multi-track efficency make it a proper tool for this experiment. In this paper we present preliminary experimental results from the inter-action of 200 GeV/c protons and antiprotons with hydrogen, argon and xenon targets. The film measurements are still in progress and we will eventually obtain significantly increased statistics.

EXPERIMENTAL PROCEDURE

A schematic layout of the apparatus is shown in Fig. 1. The three gap streamer chamber was filled with a helium-neon gas mix-ture at atmospheric pressure and operated with ± 350 KV, 12 ns high voltage pulses. The memory time was reduced to ~2μs by adding 0.1 ppm of the electronegative gas SF_6 to the helium-neon mixture. The visible volume of $2 \times 1.4 \times 0.72$ m^3 was photographed by three stereo cameras fitted with single-stage electrostatic image intensifiers (type VARO8605) at a demagnification of 60. The setting error in space was 0.4 mm. The chamber was inside a superconducting magnet giving a 1.5 Tesla field. The downstream spectrometer consisted of seven magnetostrictive wire chambers and, in the beam region, three

NA 5 Layout

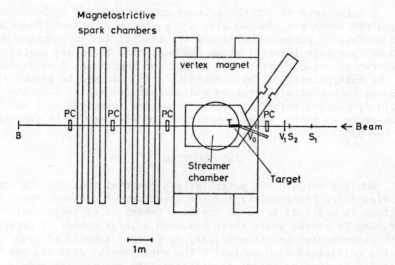

Fig. 1: Layout of the experiment

additional multiwire proportional chambers. The measuring accuracy was $\Delta p/p = 0.0025$ p (GeV) in the streamer chamber and $\Delta p/p = 0.0005$p (GeV) for tracks also passing through the spark chambers.

The nuclear target consisted of a mylar tube, 50 cm long, 3cm in diameter and ~50 microns thick. It was filled with xenon or argon gas at 9 atmospheres. Protons with momentum, p, less than 100 MeV/c, or pions with p < 35 MeV/c stop inside the target. The liquid hydrogen target was 35 cm long and 2 cm in diameter.

The incident beam was defined by a set of counters S_1, S_2, and veto counters V_0, V_1 and, 10 metres down-stream, there was a 2 cm diameter scintillation counter, B, onto which the beam was focused. The interaction trigger was a $S_1 \cdot S_2 \cdot \bar{V}_0 \cdot \bar{V}_1 \cdot B$ coincidence, which corresponded to $|t| \geq 0.042$ (GeV/c). From a Monte-Carlo calculation we estimate that this condition vetoes less than 2% of inelastic events. A disc Čerenkov counter in the beam was used to identify protons and antiprotons. The beam rate was $2\text{-}8 \cdot 10^4$ per second.

We made careful checks for biases, particularly multiple interactions. All tracks apparently coming from a single vertex (the vertex itself is not visible) were measured and reconstructed. Any event which had even a single track which came from the target but did not come from the vertex was rejected from the sample. Where possible we used range, ionisation and decay signature to identify particles. The chamber characteristics meant that we could only resolve pions from protons by their ionization over the momentum range 100 MeV/c <p < 600 MeV/c.

After the interaction the nucleus is left in a highly excited state which then evaporates nucleons. In order to eliminate the evaporated fragments we use the ionization density and range information to eliminate heavy particles with momenta $\leq$ 600 MeV/c, or we consider negatively charged particles only.

RESULTS

In table I we summarize the average and dispersion of the distributions of mesons ($<n_M>$, D^M) and negative particles ($<n_->$, D).

Table I: average multiplicities and dispersions

	$<n_M>$	$D^M = <n_M^2> - <n_M>^2$	$<n_->$	$D^- = <n_-^2> - <n_->^2$
p Xe	16.53±0.29	10.34±0.24	6.79±0.12	4.37±0.11
p Ar	13.10±0.34	7.98±0.31	5.50±0.15	3.64±0.15
$\bar{p}$ Xe	18.39±0.31	10.92±0.24	8.59±0.13	4.75±0.10
$\bar{p}$ Ar	13.63±0.28	8.14±0.21	6.70±0.13	3.74±0.10

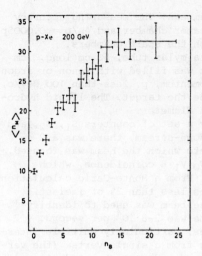 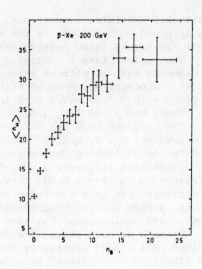

Fig. 2: number of produced mesons $\langle n_M \rangle$ versus the number of baryons
and nuclear fragments n_B in the event

The average negative multiplicity is more than one unit larger for
$\bar{p}$ induced than for p induced reactions. The distributions are broader than a Poissonian distribution, indicating the presence of
correlations between particles.

Fig. 2 shows the average multiplicity of produced mesons $\langle n_M \rangle$
in events versus the number of observed slow, heavy particles n_B.
After a fast linear rise of the number of mesons with n_B a saturation effect is observed. We may interpret the number of heavy particles as a measure of the number of collisions taking place in the
nucleus. Then the observation of a saturation is in qualitative
agreement with the additive quark model[2], where a maximum of three
quarks from the beam particles can interact.

Fig. 3 shows the ratio R of the average meson densities produced on nuclear and proton targets as a function of the rapidity
$y = 1/2 \ln (E+p_\parallel) / (E-p_\parallel)$ in the laboratory frame. There is a clear
indication of intranuclear cascading for $y \lesssim 1$, a slow decrease in
a plateau like central region, and a clear depletion of fast particles ($y \gtrsim 5$). There is only a small increase in plateau hight
from Ar to Xe. The data above the cascading region are qualitatively reproduced by both quark models[2,3] (CQM) and multiple scattering
models[4] (MSM). One may parameterize the dependence on the nucleon
number A by the parameter

$$\bar{\nu} = A \cdot \sigma_{hp} / \sigma_{hA} \approx 0.7 \cdot A^{0.31}$$

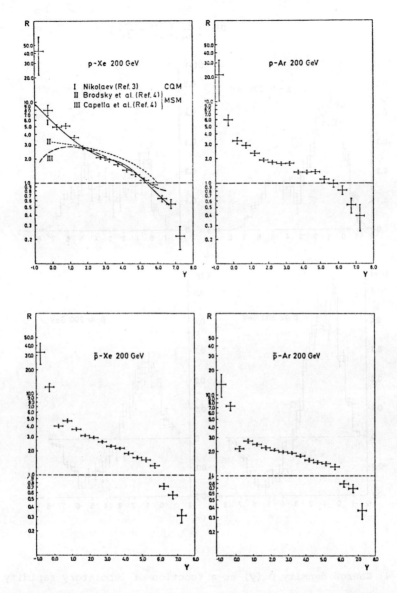

Fig. 3: Ratio R of the average charged meson density produced by
200 GeV p,p̄ on Ar, Xe targets to that produced on a hydrogen target
versus laboratory rapidity y.

$$D(y) = \frac{1}{N_{ev}} \left(\frac{dn+}{dy} - \frac{dn-}{dy} \right)$$

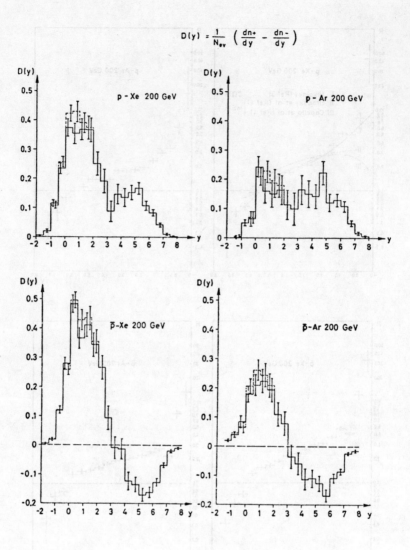

Fig. 4: Charge density D (y) as a function of laboratory rapidity y.

Table II shows a comparision of the predictions of the two classes of models with our data at $y_{CMS} = 0$ ($y_{LAB} = 3.03$ at 200 GeV).

Table II: Ratio R of charged meson densities produced at $y_{CMS} = 0$ compared to model predictions

	experiment	MSM[4]	CQM[3]	$\bar{\nu}$
$R_{p\ Ar}$ $(y_{CMS}=0)$	1.85±0.09	1.88	1.75	2.2
$R_{p\ Xe}$ $(y_{CMS}=0)$	2.00±0.08	2.43	2.12	3.1
$R_{p\ Xe}$ / R_{Ar}	1.15±0.10	1.30	1.21	

Fig. 4 shows the average charge density as a function of rapidity. One clearly sees a separation of the projectile fragmentation region and the target fragmentation region where cascading occurs.

CONCLUSION

We present data from an experiment of p,$\bar{p}$ interactions on heavy nuclear targets in which for the first time momenta and charges of all produced secondaries are mesured. The produced particle multiplicities are higher in $\bar{p}$ than in p induced reactions. The observation of saturation effects and the slow rise of the produced particle density in the central region between Ar and Xe targets favour quark constituent models. There is clear evidence for some cascading at small values of rapidity which indicates that some kind of formation length is important in the explanation of the data. At large rapidity our data confirms the depletion of the particle density observed previously.

REFERENCES

1. See reviews e.g.: W. Busza, Acta Phys. Pol. **8**, (1977) 333; D.S. Barton, Proc. XI Int. Symp. on Multiparticle Dynamics, Bruges (Belgium), 1980.
2. A. Bialas, W. Czyz, Acta Phys. Pol. **B10**, (1979) 831.
3. N.N. Nikolaev, S. Pokorski, Phys. Lett. **80B**, (1979) 290; N.N. Nikolaev, A.Ya. Ostapchuk, Reprint CERN-TH-2527 (Oct.1978)
4. S.J. Brodsky et al., Phys. Rev. Lett. **39**, (1977) 1120; J.H. Kühn, Phys. Rev. **18D**, (1978) 338; A. Capella, A. Krzywicki, Phys. Rev. **18D**, (1978) 3357.

Neutral and Charged Pion Production
in π⁻-p and π⁻-Ne Collisions

W. D. Walker

Physics Department, Duke University

Durham, North Carolina

ABSTRACT

In this paper we examine the correlation between neutral and charged pions mainly at 200 GeV/c in π⁻-proton collisions. The events are divided into diffractive and nondiffractive categories. With this separation, the correlation between neutral and charged pions in either category is smaller than without the separation. The residual correlation can probably be accounted for by resonance production. A simple model is proposed involving the production of $I = 0$ pairs of pions which can account for the data.

In the second part of the paper we examine π⁻-Ne collisions. A review of the systematics of π⁻-Ne collisions is given.

R, the ratio of the average multiplicity of π's produced in π⁻-Ne and π⁻-p collisions, was measured to be 1.22+0.05 for both π± and π⁰. Nearly identical to lower energy measurements, R is strongly dependent on the number of observed protons ($P<1.0$ GeV/c) in the event. KNO scaling is observed for the corrected multiplicity distributions.

Our study of the pion single particle spectra show the excess particle production of the neon target over that of hydrogen is largest in the target fragmentation region and nearly constant (1.28+0.02) in the central production region. The dependence of the pion rapidity distribution on the number of observed protons is consistent with the projectiles having collisions with one or two nucleons in the neon nucleus.

It is found that limiting fragmentation for π⁻-Ne collisions is reached for bombarding energies considerably higher than 200 GeV/c. In this respect π⁻-p and π⁻-Ne collisions are very similar. The proton momentum distribution is energy independent with the protons carrying proportionally less of the outgoing momentum as the bombarding energy increases.

The study of multipion production has been of active interest for nearly forty years[1]. In this paper we consider two aspects of the production of pions. First, we look at the production of π^o's and the correlation with the production of charged pions. Second, we examine the production of pions in π^--Neon collisions. This study produces insights not attainable in π^-nucleon collisions since there are sometimes multiple collisions involved.

In Figure 1 we show the multiplicity of π^o's as a function of the number of charged tracks produced for π^--p collisions at incident energies ranging from 10 to 350 GeV. We have combined several sets of observations for the data taken at 200 GeV. In particular we add the data taken in our π^--Ne-H_2 run at 200 GeV[2]. The chamber was filled with a 30% H_2-Ne mixture. This resulted in a radiation length of 1.2 meters. From the sample of events the hydrogenic events were separated out and used in the present study involving π^o production.

Figure 1 shows a wide variation of dependence of π^o multiplicity on charged multiplicity. At low bombarding energy there is very little dependence of π^o multiplicity on charged multiplicity whereas at high energy there is considerable. We also note that for 2, 4 prong events that the π^o multiplicity is independent of energy. We know that these topologies are dominated by diffraction dissociation. In Figure 2 we show the mean π^o multiplicity as a function of the number charged pions produced. There is a strong even-odd effect. This also would seem to indicate the importance of the diffractive processes in determining the value of $\bar{n}_o$, the mean number of π^o's.

We have attempted to separate the diffractive and the non-diffractive components of pion components in the various topological classes of events.

Separation of Diffractive and Nondiffractive Events

A method has been devised to separate the diffractive and non-diffractive components in low multiplicity events. Often, for the diffractive case, a low mass object is excited whose decay products lie entirely in either forward or backward c.m. hemisphere. Lamsa[3] et al have used this effect by noting the observed difference in the rates for production of even and odd numbers of tracks in each hemisphere and using a KNO scaling curve as a background estimate. We use a variation of this method in estimating the fractions of diffractive events in a given topology.

We first separate the events to have the correct charge distribution. The products of the dissociated π^- should have a net negative charge in the forward hemisphere, with proton dissociation giving a net positive charge in the backward hemisphere. Such events are the sample from which the number of diffractive events is determined. The cases with zero net charge in the forward hemisphere are probably dominated by the nondiffractive events. We find that a KNO type scaling law works fairly well in describing the distribution of multiplicities in the forward hemisphere. The scaling curves are shown in Figure 3. The triangles are the even-even configuration points, with the circle the odd-odd configuration. The abscissa in these plots is the scaling variable $n_f/2\bar{n}_f$ where n_f and $\bar{n}_f$ are the number and mean number of forward going tracks respectively. The ordinate is the ratio of the number of events at a given value of this variable to the total number of even-even events for the topology noted.

The distribution of even-even configurations is well represented by the solid curve 'triangle fit' shown in the figures, with a $x^2/D.F.$ of .5 for the entire data sample. The excess events in the odd-odd category above the scaling curve is taken to be the diffractive effect. Numerical results of the background sub - traction are given in Table I. The even-odd effect in the number of pions is again seen in this Table. The class of events with an identified proton shows a consistently larger fraction of diffractive events than those of the same multiplicity without a proton. This agrees with the results in Figure 2.

Neutral to Charge Ratio

We have attempted an overall fit of the various pieces of data shown in Figure 1. To do this it is assumed that the average number of π^0's associated with the diffractive events is 1.60, independent of multiplicity and energy. With this assumption and knowing the fraction of diffractive events we are then able to calculate the average number of π^0's associated with the various charged multiplicities of non-diffractive events. We proceed then to fit these data using a linear relationship $\bar{n}^0 = A \cdot n^{\pm} + n_0^0$. The values of these slopes A and intercepts n_0^0 are given in Table II. We note in passing that we have always tried to take the protons out of the multiplicity counts. The curves shown in Figure 1 that are labelled 125, 200, 360 were calculated on this basis. We have assumed the diffractive cross sections found at 147 GeV/c for calculating the results at 100, 147, 200, 360.[4] The diffractive cross sections presumably change slowly with energy. Unfortunately,

227

the slope is best estimated for high multiplicities where the frequencies are low and also the scanning efficiencies for finding π^0's may be dropping.

In Table II we show the Mueller Moments giving the correlation between charged and neutral pions for the total sample and for the nondiffractive events. In calculating the moments we have used our estimates of the fraction of diffractive component and then calculated the multiplicity of the nondiffractive component. This procedure should be accurate for 6 and 8 prong events.

Conclusion

The main points of this part of our paper is that it is important to separate diffractive and nondiffractive components of pion production. Figure 1 shows that the number of π^0's associated with 2 and 4-prong events is nearly energy independent which is characteristic of diffraction dissociation at high energies. In comparison, the number of π^0's associated with 10-prong events is strongly energy dependent. The even-odd alternation seen in Figure 2 can also be understood in terms of varying amounts of diffraction dissociation with topology. The data shown are consistent with the results of the separation described in this paper.

The correlations of charged and neutral pions are often characterized by moments whose values are quite dependent on whether a separation is done. We believe that we have demonstrated that the correlation between the number of neutral and charged pions is weaker than would be estimated without the separation. The value of $\frac{\partial \bar{n}_0}{\partial n_{\pm}}$ =A determined after the diffractive-non-diffractive separation at 200 GeV is between .1 and .2. Higgins [5] et al. have made a determination of the inclusive number of ρ^0's π^--p interactions. Their result is that about 10-15% of the $\pi^{\pm}$'s associated with $\rho^{\pm}$ production by assuming $\bar{n}_{\rho 0} = 1/2(\bar{n}_{\rho^+} + \bar{n}_{\rho^-})$. From the data of Higgins et al. [5] we estimate A =.1 from production. We also expect some correlation of π^0's with charged π's from ω^0 production. Our overall conclusion is that after the resonance production is taken into account the correlation between neutral and charged pions is relatively weak.

A simple model that accounts for many of the characteristics seen is to suppose that most of the pions are produced as low mass I=0 pairs. The correlation seen between neutral and charged pions is then mainly the result of ρ and ω^0 production. At 200 GeV we estimate that

50-60% of the pion production occurs through the low mass (500-600 MeV/C) mode.

II.   Introduction to the study of π-Ne Collisions

Since the very early days in the history of particle physics, it has been noted that the multiplicity of pions produced in π-nuclear and π-nucleon collisions were not terribly different.[6] With the advent of Fermilab and the CERN SPS, these results have been further quantified. The qualitative features of these results can largely be explained in terms of the Lorentz time dilation and the fact that the collision time must be short compared to the birth time of the particles.

Experiment

The results of this experiment have been derived from a small (50,000 picture) exposure of the 30" Fermilab bubble chamber to 200 GeV π⁻ mesons. The chamber was filled with the neon-hydrogen mixture (31 molar percent neon; density = 0.255 g/cm^3). The pictures were carefully scanned, and interactions in a fiducial volume were selected for measurement. The results from our initial scan have been reported earlier.   We report here the results of measurements of the outgoing tracks. The methods used for the weighting of event types and tracks are to be found in a thesis by Band.[7] We have also measured electron pairs which are the result of γ-conversions in the liquid.

We will now outline the major results from this experiment which come from the measurement of about 2000 events. The measurements were done in three successive passes through the film with a mangia-spago (SUNY-Albany) and then an automatic measuring machine at Duke. The events were classified as hydrogenic or neonic on the basis of whether there was an even or odd number of tracks, the number of identified protons. Figure 4 shows the expected (from previous experiments) numbers of events with various numbers of minimum ionizing tracks. The cross hatched part of the histogram is the expected number of hydrogen events. There seems to be a 20-30% contamination of the true hydrogen events by collisions that show hydrogen-like qualities. For the hydrogen-like events, the rapidity distribution for π⁺ were compared with the results of our experiments at 200 GeV with those from a hydrogen filled chamber. The agreement was excellent except for the highest energy tracks where our measurement accuracy was poor.

Results

A. Scaling

Figure 5 shows the neon topological cross section distribution. For low multiplicities, large cross sections are shown for $N_{min}=3$, 5, 7 which is expected for π diffraction dissociation. ($N_{min}=1$ is low because of poor scanning efficiency.) In our earlier publications, we showed that KNO multiplicity scaling seems to work for the comparison of π^--p and π^--Ne interactions at different energies. We display this comparison in Figure 6 and give a numerical comparison of multiplicities at 10.5 and 200 GeV in Table III. An examination of Figure 6 shows that the agreement between the scaled π^--p and π^--Ne distributions is good for modest multiplicities but is no better than ±20% at high multiplicities. The ratio of π^--Ne to π^--p multiplicites is 1.22±.05.

B. Number of Protons

One of the objective indications of the complexity of the collisions in neon seems to be the number of slow protons (i.e. identifiable by ionization) emitted from the collision. In Figure 7 the proton multiplicity distribution is shown from collisions at bombarding energies of 10.5 and 200 GeV. The two distributions are nearly identical. We show in Figure 8 the average number of $\pi^{\pm}$ mesons (minimum ionizing tracks plus identifiable pions) versus the number of slow protons. To compare the results at 10.5 and 200 GeV we have scaled the average number of π's (at a given N_p) by dividing by the average number of charged π mesons for all π^--Ne collisions. The agreement between the two sets of data is quite good. The progression from low to high multiplicity which is strongly correlated with the number of protons; which correlation may be associated with a progression from diffractive dissociation to one and then two or more nondiffractive collisions. Figure 9 shows a similar plot including both $\bar{n}_{\pm}$ and $\bar{n}_{\gamma}$ as a function of the number of protons (N_p). The similarity of the $\bar{n}_{\gamma}$ versus N_p and the $\bar{n}_{\pm}$ versus N_p suggests that most of the γ's are the result of π^0 decay and that $\bar{n}_{\pi^0}=(\bar{n}_{\pi^+}+\bar{n}_{\pi^-})/2$.

In summary, for this section we have found that the number of slow (greater than minimum ionizing) protons is an objective way of classifying interactions. A plausible explanation of the correlations found is that the number of slow protons is mainly correlated with the impact parameter of the incident projectile. The distribution of number of slow protons seems to be nearly independent of bombarding energy.

Single Particle Spectra

One of the more interesting comparisons to be made is that of the single particle spectra of π^+, π^-, π^0 etc. with the corresponding spectra from π^--p collisions. The distributions in $P_\perp^2$ for π^--Ne and π^--p are indistinguishable and are not shown. Figure 10 shows the longitudinal momentum spectrum for π^--Ne compared with π^--p. Figure 11 shows the π^- longitudinal momentum distributions for π^--Ne collisions at bombarding energies of 10.5 and 200 GeV. One of the interesting results is that for low momenta (P <200 MeV/c), the number of particles is less at 200 GeV than at 10.5 GeV. Figure 12 shows a comparison of the integrated longitudinal spectra ($-0.4 \leq P \leq 0.2$ GeV/c) for π^--Ne and π^--p collisions plotted against $s^{-1/2}$. There seems to be very similar behavior when comparing π^--p and π^--Ne collisions. The assymptotic values for the differential cross sections for hydrogen and neon seem to differ by about a factor of 2.

In Figures 13 and 14 we display the rapidity distributions obtained from π^--p and π^--Ne collisions. The distributions are qualitatively quite similar. There are, however, differences in that the median rapidity is lower by about 0.5 units for π^--Ne collisions and the plateau value is higher by a factor of about 1.3. In Figure 15 we plot the ratio of π^--Ne to π^--p cross sections as a function of the laboratory rapidity. The curve through the data points is from a parton theory due to Brodsky and Gunion.[8]

D. Proton Spectrum

In our work at 10.5 GeV on $\pi^\pm$-Ne interactions we could directly determine the proton spectrum by interpreting the data using charge symmetry arguments. In that experiment we found that the fast protons and neutrons carried away a sizeable fraction of the incoming momentum. We cannot do as precise a determination here since we do not have π^+ data available. We can, however, estimate the number of minimum ionizing protons and their spectra by comparing our positive and negative particle spectra. Figure 16 shows the results of the comparison. In the comparison we have used the fact that the ratio R = (no. of π^+)/(no. of π^-) for $-0.4 \leq P \leq 0.2$ GeV/c is 0.91±.08. In this range the protons and pions can be readily distinguished by ionization. Whitmore et al.[9] have found that the ratio R has a constant value of 0.91±.05 over a much wider range of momentum

$(-0.2 \leq P \leq 1.6$ GeV/c). By using the factor 0.91, the spectrum shown in Figure 16 has been deduced. The momentum spectra of protons for bombarding energies of 10.5 and 200 GeV seem to be remarkably similar. We also deduce the same number of minimum ionizing protons at both energies.

E. γ-Rays

For each interaction we have measured the directions of the γ-rays from the main vertex. The measurements of the momenta of the electron pairs was found not to be very useful because of bremsstrahlung energy loss. For particles of zero rest mass the pseudo rapidity, $\eta = -\log \tan \theta/2$, and the rapidity, y, are the same.

In Figure 17, we show a comparison of rapidity distributions for γ's coming from hydrogen and neon collisions. This rapidity comparison seems quite similar to our previous comparisons for charge tracks.

F. Breakdown by N_p

As mentioned earlier, the number of identified protons gives a simple means of dividing the collisions. In Figure 18 the rapidity distribution for charged pions is shown with the events classified by the number of slow protons. In our work at 10.5 GeV[12] we found that for a similar breakdown the mean rapidity decreased for the events with higher numbers of protons. The results here show a similar behavior. Figure 19 shows the mean rapidity for $\pi^{\pm}$ and γ as a function of the number of protons, N_p. The mean value of y varies nearly one unit of rapidity. As in our previous work the variation of <y> can be interpreted in terms of a progression from single to double or even triple collisions in the neon nucleus. It should be noted that a similar but smaller variation in <y> as a function of total multiplicity in π^--p collisions can be found.

III. Conclusions

We have made comparisons of various aspects of π^--Ne collisions at 200 GeV with corresponding quantities in π^--p collisions. It is found that π^- collisions with neon differ relatively little with those on hydrogen. Further, the differences seem smaller for 200 GeV collisions than for those at 10.5 GeV. The number of particles produced close to y=0 is less at 200 GeV than at 10.5 GeV. In this respect the π^--Ne and π^--p collisions are quite similar. In contrast to results at 10.5 GeV,

the nucleons coming form π^--Ne collisions at 200 GeV do
not carry a very large fraction of the product momentum.
The laboratory proton spectrum, although not well measured at 200 GeV, seems similar to that obtained from
10.5 GeV collisions. It is possible that for the high
multiplicity events that nucleons carry a sizeable fraction of the event momentum.

The multiplicity in π^--Ne collisions is greater than in
π^--p collisions by a factor of 1.21 ± 0.04. The value of
the plateau cross section in the rapidity spectrum is
greater by a factor of 1.3 than the corresponding
quantity in hydrogen.

Various models explain some features of the data,[10]
but none are completely satisfactory. The Energy Flux
model elucidates the small increase in multiplicity with
increasing A, the Coherent Tube[11] model is consistent
with the more massive target found Kinematically in π-Ne
collisions, and the Independent Parton model can explain
the higher plateau region in the rapidity distribution
which is observed at 200 GeV. The E.F. model fails with
respect to the observed increase of R with energy; the
C.T. model predicts an increase in the width of the
rapidity distribution which is not observed. The I.P.
model allows for the composite structure of the pion, and
Consequently that the height of the rapidity distribution
is greater than in the case of π^--p collisions without an
increase in width.

References

1. E. Fermi, Prog. Theo. Phys. 5, 5770 (1950).
2. J.R. Elliott et al., Phys. Rev. Lett. 34, 607 (1975).
3. J.W. Lamsa et al., Phys. Rev. D18, 3933 (1978).
4. 10.5 (a) J.R. Elliott et al., Nucl. Phys. B133, 1
             (1978).
    25   (b) J.W. Elbert et al., Nucl. Phys. B19, 85
             (1970).
    100 (c) N.N. Biswas et al., Nucl. Phys. B167, 41
             (1980).
    150 (d) D. Brick et al., Phys. Rev. D20, 2123 (1979)
    200 (e) D. Ljung et al., Phys. Rev. D15, 3163 (1977).
    360 (c)
5. P. Higgins et al., Phys. Rev. D19, 65 (1979).
6. U. Camerini, W. Lock and D. Perkins, in "Progress in
    Elementary Particle and Cosmic Ray Physics", Vol 1,
    Interscience Publishers (New York, 1952).
7. H.R. Band, Ph.D. Thesis, Duke University (1980).
8. S.J. Brodsky, J.F. Gunion and J.H. Kuhn, Phys. Rev.
    Lett. 39, 1120 (1977).
9. J. Whitmore et al., Phys. Rev. D16, 3137 (1977).

10. K. Gottfried, Phys. Rev. Lett. 32, 957 (1974).
11. (a) G. Berlad et al., Phys. Rev. D13, 161 (1976).
    (b) F. Takagi - Proc. of the 19th ICHEP - Tokyo.
12. W. Yeager et al, Phys. Rev. D 16, 1294 (1977)

TABLE I.

Diffractive Cross Sections

	Total σ in mb.	Diff. σ in mb.	Fraction Diffractive
π-p	.96 ± .08	.92 ± .11	.96 ± .08
(2π)^o	.84 ± .07	.61 ± .10	.73 ± .1
(3π)$^-$-p	2.14 ± 0.8	1.33 ± .13	.62 ± .06
(4π)o	2.02 ± .08	.55 ± .08	.27 ± .04
(5π)$^-$p	1.83 ± .1	.38 ± .05	.20 ± .026
(6π)^o	2.72 ± .1	.10 ± .06	.037 ± .020
(7π)$^-$p	1.58 ± .08	.27 ± .04	.17 ± .025
(8π)^o	2.72 ± .10	.18 ± .08	.066 ± .030

TABLE II.

Neutral-Charged Correlation

E_i	A	n_o^o	$f_2^{\pm o}$ N.D.	$f_2^{\pm o}$
100	.16 ± .05	2.00 ± .2	.91 ± .35	1.17 ± .4
150	.18 ± .077	2.03 ± .5	1.67 ± .4	3.51 ± .4
200*	.18 ± .05	2.6 ± .4	1.98 ± .4	3.67 ± .4
360	.25 ± .06	2.4 ± .4	2.3 ± .4	5.6 ± .4

$$\bar{n}_{+-}^o = A\, n_{+-} + n_o^o$$

$= $ Number of π^o's for n_{+-} charged pions

TABLE III.

Average Charged Track Multiplicity

Pi- neon	200.0 GeV/c	10.5 GeV/c
<minimum tracks>	9.41 ± 0.21	3.91 ± 0.03
<scan protons>	1.77 ± 0.10	1.50 ± 0.08
<fast protons>	0.44 ± 0.14	0.46 ± 0.12
<protons>	2.21 ± 0.17	1.95 ± 0.14
<pi+, pi->[†]	8.97 ± 0.25	3.45 ± 0.05
<pi+, pi->[*]	9.46 ± 0.27	3.74 ± 0.06
$R^{†} = \frac{<pi+,pi->}{<pi+,pi->}$	1.21 ± 0.04	1.11 ± 0.03
$R^{*} =$	1.22 ± 0.05	1.20 ± 0.04
<gammas>	9.10 ± .27	3.54 ± .3
$<Pi>^{0}$	4.55 ± .24	1.77 ± .15
$R' = \frac{<\pi^{0}>_{Ne}}{<\pi^{0}>_{H}}$	1.21 ± .05	1.09 ± .07
$D = <Pi^{2}> - <Pi>^{2}$	5.34 ± .67	2.04 ± .03
<Pi>/D	1.86 ± .24	1.83 ± .06

[†]All Neon Events

[*]All non-diffractive neon events

'Calculated for non-diffractive events using NMIN
uncorrectd for fast protons

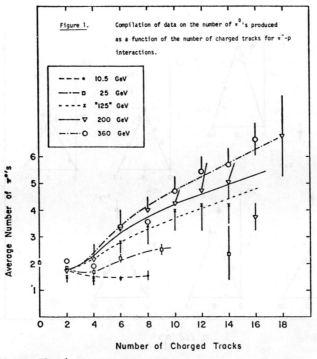

Figure 1. Compilation of data on the number of π^0's produced as a function of the number of charged tracks for $\pi^- \text{-} p$ interactions.

- - - - • 10.5 GeV
— · — □ 25 GeV
- - - - × "125" GeV
——— ▽ 200 GeV
— · · — ○ 360 GeV

Fig. 1

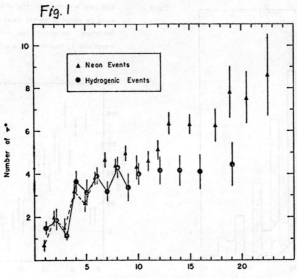

▲ Neon Events
● Hydrogenic Events

Fig. 2 Figure 2. The number of π^0's as a function of the number of minimum ionizing tracks (minimum positive and negative tracks plus any identified π^-).

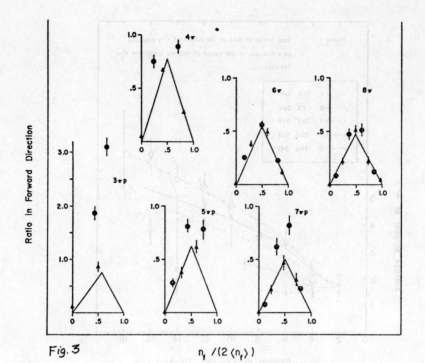

Fig. 3

$n_f / (2 \langle n_f \rangle)$

Ratio in Forward Direction

<u>Figure 3.</u>   Plot used to estimate fraction of diffractive events.
n_f is the number of charged tracks in the forward
hemisphere. The ratio is that of the number of
events with n_f forward to the total number of even-even events.

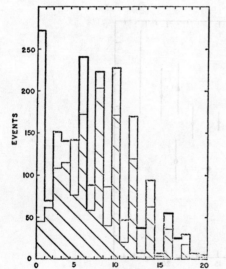

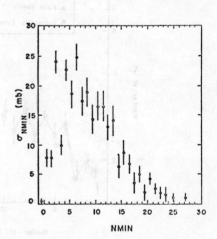

Figure 4   Hydrogen and hydrogenic multiplicity distributions.

Figure 5.   Neon topological cross sections.

238

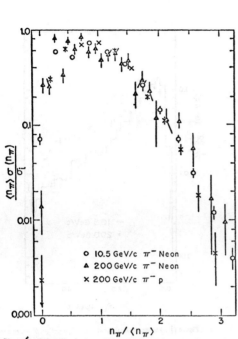

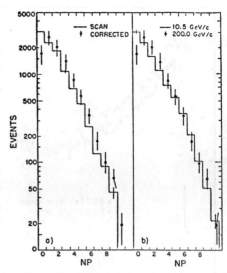

Figure 6 KNO scaling in pi-neon interactions. All inelastic events. Hydrogen multiplicities were modified by assuming 0.6 protons per event.

Figure 7. Proton multiplicity distributions.
a) Scan distributions were corrected for fast protons.
b) 10.5 and 200 GeV/c proton multiplicity distributions.
All distributions are normalized to 10000 events.

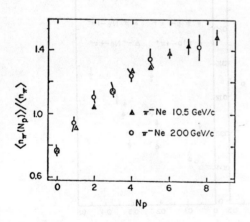

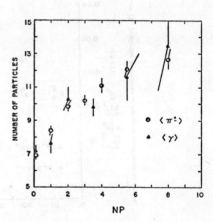

Figure 8. Scaled pion multiplicity vs NP. Data sample was all inelastic events.

Figure 9. Pion multiplicity vs NP. Data sample was all inelastic events.

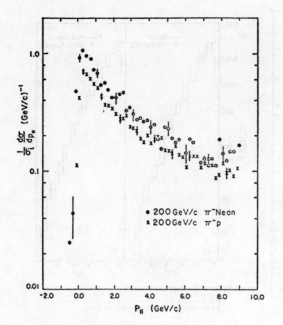

Figure 10 Pi- longitudinal momenta for neon and hydrogen.

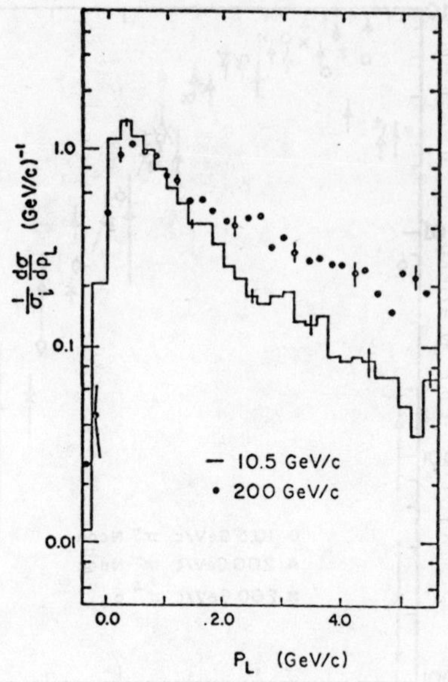

Figure contains data points:
- 200 GeV/c π^-Neon
- 200 GeV/c π^-p

Axes: $-\frac{1}{\sigma_i}\frac{d\sigma}{dp_{\parallel}}$ (GeV/c)$^{-1}$ vs $P_{\parallel}$ (GeV/c)

Figure 11 Pi- longitudinal momenta at 200 and 10.5 GeV/c neon.

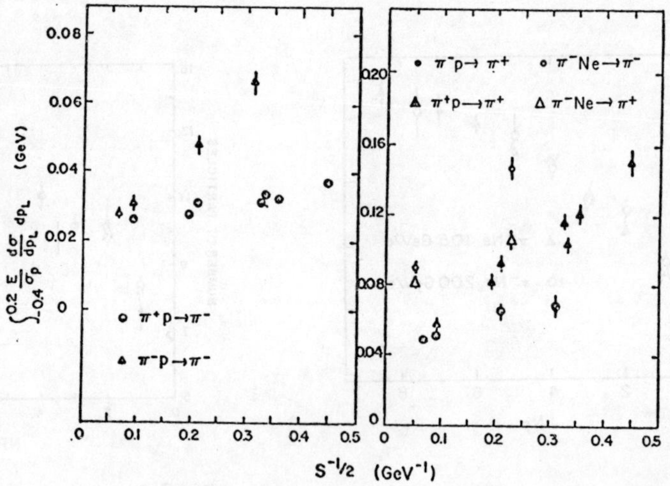

- — 10.5 GeV/c
- 200 GeV/c

Axes: $-\frac{1}{\sigma_i}\frac{d\sigma}{dp_L}$ (GeV/c)$^{-1}$ vs P_L (GeV/c)

Figure 12 Projectile dependence in the target fragmentation region.

$\int_{-0.4}^{0.2} E \frac{1}{\sigma_p}\frac{d\sigma}{dp_L} dp_L$ (GeV) vs $S^{-1/2}$ (GeV^{-1})

- $\pi^+ p \to \pi^-$
- $\pi^- p \to \pi^-$
- $\pi^- p \to \pi^+$
- $\pi^- Ne \to \pi^-$
- $\pi^+ p \to \pi^+$
- $\pi^- Ne \to \pi^+$

240

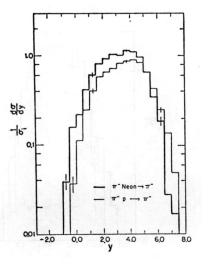

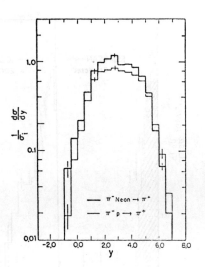

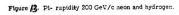

Figure 13. Pi- rapidity 200 GeV/c neon and hydrogen.

Figure 14. Pi+ rapidity at 200 GeV/c for hydrogen and neon.

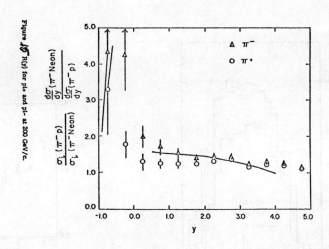

Figure 15. R(y) for pi+ and pi- at 200 GeV/c.

241

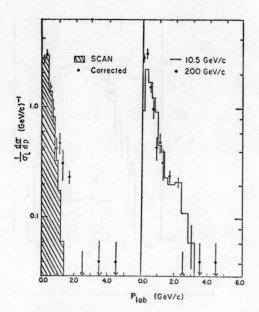

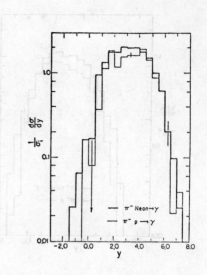

Figure 16. Proton momentum distribution.
a) Scan proton momentum distribution was corrected for minimum ionizing protons.
b) Proton distributions of 200 GeV/c and 10.5 GeV/c are compared.

Figure 17. Gamma rapidity 200 GeV/c neon and hydrogen.

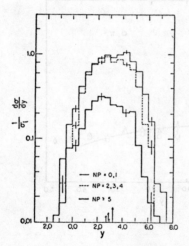

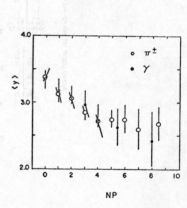

Figure 18. Pion rapidity vs NP. Pi± rapidity for NP = 0,1; NP = 2,3,4; and NP ≥ 5. The ⟨y⟩ of each distribution is represented by arrows with NP = 0,1 corresponding to the rightmost arrow.

Figure 19. Average rapidity vs NP.

242

DISCUSSION

WROBLEWSKI, WARSAW: Do you have an estimate of the [average] number of fast protons in your 200 GeV π-Ne data? The same estimate that you did at 10 GeV.

WALKER: The data are not as good. What I did was to estimate the momentum carried by pions out of the collision and then, just by subtraction, estimate the momentum carried by nucleons. That, for these complex collisions, is on the order of 30 to 40% of the incoming momentum.

WROBLEWSKI: You are talking about momentum, not about the number of protons?

WALKER: Right. You know, you can count and show there are anywhere from 1 to 3 fast protons, but it's a question of how fast they are. I mean, if they are 2 GeV it's not very interesting. Our current experiment at Fermilab, E597, should make direct identification of fast protons.

WROBLEWSKI: I would like to comment, we are measuring now π-neon at 30 and 64 GeV. This is the Seattle-Strasbourg-Warsaw collaboration. We have estimated, by essentially using your method, the average number of fast protons per event and it seems that this number does not change between 10 and 64 GeV. Also we have some preliminary measurements for the laboratory [momentum] spectrum of these protons and again it seems to be exactly the same as what you have found at 10 GeV. That's why I asked you the question of what happens at 200.

MINETTE, WISCONSIN: I have one comment on that. We compared our fast protons [at 300 GeV] to lower energies also. We found (although we are not quite through with the analysis) in essence the same thing you found. That the proton spectrum is essentially the same as it is at lower energies.

BIAŁAS, KRAKOW: Can you show the last transparency again? Can you comment a little more about the last point [of your conclusions]?

WALKER: Well, if what has been said is true, what we found at 10 GeV is that it appears that in a sizeable fraction of the interactions a large fraction of the momentum is carried out by nucleons, struck nucleons. It remains to be seen whether this is true at 100 or 200 GeV. If it is, then I think it is likely that there has to be some rather efficient mechanism for knocking these protons out.

SESSION B1

SOFT-HADRON PHYSICS: THEORY I

Tuesday, June 23, a.m.

Chairman: F. Verbeure
Secretary: N. M. Cason

CENTRAL PRODUCTION AND SMALL-ANGLE ELASTIC SCATTERING IN THE VALON MODEL

Rudolph C. Hwa

Institute of Theoretical Science, University of Oregon
Eugene, Oregon 97403 U.S.A.

ABSTRACT

The valon model is applied to the problem of particle production in the central region. Using the recombination mechanism for hadronization, a constraint is obtained on the plateau heights of produced pions and protons. The available data are in rough agreement with that constraint. The problem of elastic scattering at small angles is then investigated in the valon model. It is shown that various familiar properties follow as simple consequences of the model. They are: additive quark model for total cross sections, droplet model for $d\sigma/dt$ at small t, and Regge behavior with some new constraints on Regge residues.

I. INTRODUCTION

By regarding a valence quark as carrying its own cloud of sea quarks and gluons, collectively called a valon, the valon model has been successful in describing a variety of processes involving hadrons, either as initial particles, or final, or both. Such processes include deep inelastic scattering, quark and gluon fragmentation in jets, hadron production in soft processes in the fragmentation region, form factors at low Q^2, pion decay, and quark attenuation in nuclear matter.[7] But for the valon model to be universally relevant to all hadronic processes, it must also be able to describe particle production in the central region as well as elastic scattering. These are the problems that the Regge approach has a long-standing claim to success, although it has not led to any insight into the structure of hadrons. In this paper these problems are investigated in the framework of the valon model, thereby eliminating two glaring omissions in the applicability of the model.

As a brief summary of the characteristics of the valons, in contrast to the quarks, we list below their distributions in momentum fractions in a hadron denoted respectively by $G_v^h(y)$ and $q_i^h(x)$. The momentum sums are

$$\sum_v \int dy\, y\, G_v^h(y) = 1 \qquad \text{vs.} \qquad \sum_i \int dx\, x\, q_i^p(x) \sim 0.5$$

The valon distributions for proton can be determined either from deep inelastic scattering[2] (DIS) or from form factor[3] (FF). They are, respectively,

$$\text{DIS:} \quad \begin{aligned} G_U^p &= 8y^{0.65}(1-y)^2, \\ G_D^p &= 6y^{0.35}(1-y)^{2.3}, \end{aligned} \qquad \text{FF:} \quad \begin{aligned} G_U^p &= 10.8y^{0.9}(1-y)^2 \\ G_D^p &= 4.6y^{0.1}(1-y)^{2.8} \end{aligned}$$

The quark distribution in a proton, on the other hand, are Q^2 dependent; at $Q^2 \simeq 10 \text{GeV}^2$ they are

$$q_{u,\text{val}}^p = 5.5\ x^{-0.18}(1-x)^{3.2}, \quad q_{d,\text{val}}^p = 2\ x^{-0.35}(1-x)^{3.54}$$

$$q_{\text{sea}}^p \simeq 0.02\ x^{-1}(1-x)^{5.9} + 0.1\ x^{-1}(1-x)^{13.6}$$

Note that the large x behaviors of G_v^p and q_{val}^p differ by roughly one unit in the exponent of $(1-x)$. Moreover, G_v^p vanishes as $y \to 0$, while q_v^p diverges. For pion, the valon distribution inferred from lepton-pair production (LPP) and FF are, respectively,

$$\text{LPP:} \quad G_v^\pi = 1, \qquad\qquad \text{FF:} \quad G_v^\pi = 1.8[y(1-y)]^{0.3}$$

whereas the quark distribution at large x behaves roughly as[4] $(1-x)^1$. The reason why $G(y)$ is spread out more toward the high momentum side than $q(x)$ is clearly that a valon at y includes all the partons in the valon with momenta $x \leq y$.

II. CENTRAL PRODUCTION

Our interest here is in the application of the valon model to the production of particles in the central region. Let us take pp collision for definiteness. In the valon model we would have a 6-body scattering problem with 3 valons incident on 3. A way of approaching the 6-body problem is to order the scattering amplitude into a sum according to the number of initial valons participating in the collision. Thus the first set of amplitudes would consist of 2-valon interactions, one valon from each proton, permuted over all valons; they contribute to central production. Then there is a sub-set involving 3-valon interactions, and so on. The non-interacting ones are spectators with respect to central production but are active participants in the production of particles in the fragmentation region. Since the spatial extension of a valon is small compared to that of a hadron,[1] it is reasonable to assume that the 2-valon amplitudes are more important than the amplitudes involving three or more valons. In the following we shall for definiteness consider only two-valon interactions, i.e. when the impact parameter between valons, not between hadrons, is small.

The point of view adopted is very different from the DTU approach[5] which emphasizes the two-chain structure due to quark-

diquark separation. There is, however, no obvious conflict since
the DTU scheme is a consequence of unitarity and is therefore concerned with the physical particles in the final state of a multiparticle production process with no definite specification about
the time evolution of the hadronic system after impact. The
"quark" lines in the dual diagrams are mainly to keep track of the
quantum numbers, and have not been shown to correspond to the
quarks probed in hard processes, although such interpretations have
often been made. The valon model specifies what initially takes
place in a collision: it depends only on the impact parameter
between two valons with no statement made about their colors, or
the colors of the valence quarks contained therein. What happens
as time evolves is hard to describe precisely except that the
partons in the valons interact with short-range order. Thus it is
the wee partons that interact, but only those that belong to the
two valons that overlap in transverse plans. This picture bears no
obvious relationship to the mechanism of $3-\bar{3}$ separation in DTU since
it refers to the initial stage of the interaction. It is not
impossible that in the course of time the partons evolve in such a
way that locally (in rapidity) hadronization takes place through
recombination but globally unitarity is satisfied in the DTU way.

Having stated the framework for central production we now
investigate the implications of hadronization through the recombination mechanism. Unlike the fragmentation region where the x
distributions for various detected particles need to be determined,
in the central region there is just the flat distribution of a
rapidity plateau. One can only compare the plateau heights for
various final-state particles, short of investigating questions on
correlations. We shall consider the problem at asymptotic energies
so that the plateau width can be assumed wide. As it turns out, the
recombination model implies a strong constraint on the plateau
heights.

Two valons that overlap in transverse coordinates interact
through their wee partons and produce a central plateau (of partons).
For the purpose of applying the recombination model, we consider the
saturated sea by converting all the gluons to quark-antiquark
pairs.[1,6] Let h denote the height of the resultant plateau for
quarks or antiquarks having definite flavor and polarization. These
quarks and antiquarks recombine and give rise to a hadronic plateau
which we denote by H^h for hadron of the type h, summed over all
polarizations. Since plateau heights refer to the rapidity space,
we write the recombination formula in terms of the rapidity variable
Y (instead of the usual notation y to avoid confusion with the
momentum-fraction variable for valons used in Sec. I)

$$H^{\pi} = \int F(Y_1, Y_2) \; R^{\pi}(Y_1, Y_2, \; Y=0) \; dY_1 \; dY_2 \tag{1}$$

$$H^p = \int F(Y_1, Y_2, Y_3) \; R^p(Y_1, Y_2, Y_3, \; Y=0) \; dY_1 \; dY_2 \; dY_3 \tag{2}$$

Note that the hadron rapidity is evaluated at Y=0.

Since only quarks and antiquarks in the central region would recombine to form hadrons at Y=0, $F(Y_1,Y_2)$ and $F(Y_1,Y_2,Y_3)$ are multiquark distributions in the plateau. As a first approximation we ignore the effect of correlation on those F's. Then obviously $F(Y_1,Y_2) = h^2$. Of course, this is for c.m. energy high enough such that q and $\bar{q}$ have the same plateau heights. Moreover, it is for direct hadronization into a pion of a definite charge because h refers to q and $\bar{q}$ having definite flavor and polarization. Contributions to pions from resonance decays would have spin complications, which we discuss below. For the direct formation of proton, on the other hand, we would have $F(Y_1,Y_2,Y_3) = 2h^3/3$. The 2/3 factor appears because (a) the proton has two polarization states, and (b) the transverse-space wave function of a proton favors the D valon to be between the two U valons,[1,7] implying a suppression factor which we estimate to be about 1/3 on the basis of the simple counting that a random distribution in a one-dimension array would have three possibilities: DUU, UDU, and UUD.

Since these F functions are constant in the central region, they can be moved to the outside of the convolution integrals in (1) and (2). What remains for the integrals are just integrations over the recombination functions which are normalized such that those integrals are unity.[6] Hence we obtain

$$H^{\pi}_{dir} = h^2 \quad , \qquad H^{p}_{dir} = \frac{2}{3}h^3$$

Clearly, the recombination model in general and the recombination function R in particular play a crucial role in arriving at this result. Eliminating h, we have

$$(H^{\pi}_{dir})^{1/2} \quad = \quad (\frac{3}{2} H^{p}_{dir})^{1/3} \tag{3}$$

From the derivation it is also clear that the proton plateau height can be replaced by that for antiproton, again reflecting validity only at asymptotic energies.

To check (3) experimentally is not easy since the usual inclusive cross sections include resonance decays. It is therefore necessary to convert (3) to a relationship on H^{h}_{tot} where

$$H^{h}_{tot} = H^{h}_{dir} + H^{h}_{ind}$$

H^{h}_{ind} being the contribution from indirectly produced hadrons. Estimation of the relative proportion between H^{h}_{dir} and H^{h}_{ind} depends on the flavor symmetry of the hadronic system assumed and other considerations.[8] An example of such a study is the work of Anisovich and Shekhter.[8] We give here a very simple estimate based on spin-isospin symmetry.

For pion we consider the 16 states of $\pi, \rho, \omega,$ and η. From

their dominant decay modes we get a ratio of 3 to 30 for directly
to indirectly produced pions. Hence, $H^{\pi}_{dir} = H^{\pi}_{tot}/11$. For proton
there are 20 states for N and Δ, and the direct-to-indirect ratio
for proton production is 2 to 8. So $H^p_{dir} = H^p_{tot}/5$. Any further
refinement of these estimates would not make a great deal of difference, nor would the result be much more reliable. Using these estimates we therefore obtain from (3)

$$R(\frac{\pi^+}{p}) \equiv H^{\pi^+}_{tot}/H^p_{tot} = (120/H^{\pi^+}_{tot})^{1/2} \qquad (4)$$

Experimentally, the ratio is about 10, but for $\pi^+/\bar{p}$ it is about
20 in pp collisions at ISR. Evidently, at ISR energies the valence
quarks still make contributions to the central production of π and
p. We take the average of the two numbers as an asymptotic estimate
and adopt $R \simeq 15$. For the r.h.s. of (4) we find from Ref. 10 that
the plateau height for all charged perticles varies from 1.4 to 1.9
over the ISR range. We thus infer that H^{π}_{tot} for a definite charge
state is roughly half of that, yielding for the r.h.s. of (4) a
range from 11 to 13. In the logarithmic scale from which R is
extracted, the agreement is remarkably good. We should not expect
a more precise verification of (4) since (4) itself is not a rigorous consequence of (3). We conclude that the no-parameter prediction of the recombination model is adequately supported by the
existing data in the central region.

III. ELASTIC SCATTERING AT SMALL ANGLES

In considering small momentum-transfer reactions, whether it is
form factor at small Q^2,[1,3] or elastic scattering, it is necessary
to consider valon distributions in both the longitudinal and transverse directions. We use the notation $V^h_j(x,r)$ for valon j in hadron
h with x being the longitudinal momentum fraction of the valon and
r the transverse coordinate measured from the center-of-x. The
Fourier transform in r is denoted by

$$G^h_j(x, \vec{q}) = \int d^2r\, e^{-i\vec{q}\cdot\vec{r}} V^h_j(x, \vec{r})$$

and is related to the G-function discussed in Sec. I by

$$G^h_j(y) = G^h_j(x=y,\vec{q}=0) \qquad (5)$$

At low Q^2 $(Q^2=\vec{q}^2)$ where the structure of valons cannot be resolved,
the charge form factor for hadron h is[3]

$$F_h(Q^2) = \sum_j e_j \int_0^1 dx\, G^h_j(x,\vec{q}) \qquad (6)$$

In the following we shall show what bearing this has to elastic
scattering.

251

According to the valon model described in the beginning of Sec. II, the elastic scattering amplitude T can be represented mainly by the diagram shown in Fig. 1, where the circles signify wave functions of valons and the rectangle denotes valon-valon elastic scattering amplitude $\hat{T}$. There are other terms involving multi-valon interactions which are negligible at small angles. In impact-parameter space the elastic h-h' scattering amplitude may be written as

$$T_{hh'}(s,\vec{b}) = \sum_{jj'} \int \frac{dx}{x} \frac{dx'}{x'} d^2r d^2r' v_j^h(x,\vec{r}) v_{j'}^{h'}(x',\vec{r}') \hat{T}_{jj'}(\hat{s},\vec{b}+\vec{r}-\vec{r}')$$

which translates in the $\vec{q}$ space to

$$T_{hh'}(s,\vec{q}) = \sum_{jj'} \int \frac{dx}{x} \frac{dx'}{x'} G_j^h(x,-\vec{q}) G_{j'}^{h'}(x',\vec{q}) \hat{T}_{jj'}(\hat{s},\vec{q}) \tag{7}$$

where

$$T_{hh'}(s,\vec{q}) = \int d^2b\, e^{-i\vec{q}\cdot\vec{b}} T_{hh'}(s,\vec{b})$$

The pictorial representation of (7) by Fig. 1 becomes more transparent when it is recognized that v_j^h is the square of valon wave function in h so its Fourier transform $G_j^h(x,\vec{q})$ is a convolution of the wave functions in momentum space, exactly as depicted in Fig. 1. Eq. (7) is the central equation in the valon model for elastic scattering at small angles. It should be regarded as a two-body irreducible Born term that is to be unitarized. We give below a number of consequences that follow directly from it.

1. Total Cross Section

Let $t=\vec{q}^2 = 0$. Apply optical theorem to both T and $\hat{T}$ using $\text{Im}T_{hh'}(s,0) = s\sigma_{hh'}^{tot}$ and $\text{Im}\hat{T}_{jj'}(\hat{s},0) = \hat{s}\hat{\sigma}_{jj'}$. Since $\hat{s} = xx's$ one obtains from (5) and (7)

$$\sigma_{hh'}^{tot}(s) = \sum_{jj'} \int dx\, dx'\, G_j^h(x)\, G_{j'}^{h'}(x')\, \sigma_{jj'}(\hat{s}) \tag{8}$$

If $\sigma_{jj'}(\hat{s})=\hat{\sigma}$, i.e. independent of $\hat{s}$ and the flavors of j and j', then it can be taken outside of the summation and integral. The

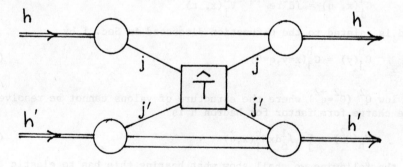

Figure 1

252

resultant double integral is then factorizable; in fact, it is equal to one because of the normalization condition on G_j^h. Hence, we have

$$\sigma_{hh'}^{tot} = \begin{matrix} 9 \hat\sigma & \text{for pp} \\ 6 \hat\sigma & \text{for } \pi p \end{matrix}$$

This is the result of additive quark model in that $\sigma_{pp}/\sigma_{\pi p} = 3/2$. The valon model goes further to identify $\hat\sigma$ as a meaningful cross section between valons. In reality $\sigma_{ij}(\hat s)$ is flavor dependent because we know that $\sigma_{Kp} < \sigma_{\pi p}$; moreover, it does depend on $\hat s$ since $\sigma_{hh'}^{tot}$ increases as $\ln s$ or $\ln^2 s$. These are detail properties that do not upset the general sensibleness of the model. In fact, (8) provides a clear avenue for refinement of the additive quark model, and places the focus more properly on the valon-valon interaction.

2. Elucidation of the Droplet Model

The droplet model as originally proposed by Wu-Yang[11] and later improved by Chou-Yang[12] asserts that the elementary interaction in the scattering of hadrons has to do with the attenuation of matter wave as hadrons pass through each other and that matter density is proportional to the charge density of a hadron. We show how these hypotheses emerge naturally from (7).

In a scaling model the scattering amplitude is proportional to s as $s \to \infty$. Let $T_{jj'}(\hat s,t) \to \hat s A(t)$. It then follows from (7) that

$$\frac{1}{s} T_{hh'}(s,t) = A(t)[\sum_j \int dx G_j^h(x,t)][\sum_{j'} \int dx' G_{j'}^{h'}(x',t)] \qquad (9)$$

In the droplet model $A(t)$ is taken to be a constant so that the t dependence arises entirely from the quantities in the square brackets. Let us denote the latter by $n_h \rho_h(t)$, where n_h is the number of valons in h. Then for a proton

$$\rho_p(t) = \int dx \frac{1}{3}[2G_U^P(x,t) + G_D^P(x,t)] \qquad (10)$$

This is precisely the Fourier transform of matter density in the valon model because the hadronic "stuff" (in the language of the droplet model) consists of only two U and one D valons. At t=0, ρ_p is normalized to one. In the droplet model the t dependence of $\rho^P(t)$ is unknown, so it is assumed that it is the same as that of the charge form factor. At low momentum transfer the charge form factor, according to (6), is

$$F_p(t) = \int dx [\frac{4}{3}G_U^P(x,t) - \frac{1}{3}G_D^P(x,t)] \qquad (11)$$

for t sufficiently small such that the internal structure of the valons may be neglected.[13] Clearly, the Wu-Yang hypothesis $\rho_p(t) = F_p(t)$ is correct, provided $G_U^P = G_D^P$.

Thus the droplet model can be derived from the valon model in the approximation of flavor independence. The dip structure

253

of $d\sigma/dt$ that follows from adding successive terms in an eikonal expansion is therefore also expected in the valon approach. Moreover, there are definite guidelines on how to improve the calculation. Flavor dependence of the valon distributions as exhibited in Sec. I can readily be applied to give a more reliable description of $\rho_p(t)$. For $t > 1\text{GeV}^2$ (11) is not valid unless it is supplemented by the valon form factor. Thus the Wu-Yang hypothesis, though sensible 15 years ago, should not be regarded as trustworthy for $t > 1\text{GeV}^2$ even in the flavor-independence approximation. The t dependences of G_j^h are discussed in Ref. 3.

At large t, (9) itself is unreliable. That is because for elastic scattering at finite angles multi-valon scattering terms of the type suggested by Landshoff are no longer negligible and may even be dominant.[13] Although large-angle scattering is not our concern here and will not be pursued further, it is another area of investigation where the valon model would have very specific predictions.

3. Regge Behavior

We define the cross-channel partial-wave amplitude by the Mellin transform

$$T_{hh'}(\ell,t) = \int_o^\infty ds \ s^{-\ell-1} T_{hh'}(s,t) \tag{12}$$

Then it follows from (7) that

$$T_{hh'}(\ell,t) = \sum_{jj'} G_j^h(\ell,t) G_{j'}^{h'}(\ell,t) \hat{T}_{jj'}(\ell,t) \tag{13}$$

where

$$G_j^h(\ell,t) = \int_o^1 dx \ x^{\ell-1} G_j^h(x,t) \tag{14}$$

and $\hat{T}(\ell,t)$ is related to $\hat{T}(\hat{s},t)$ as in (12) except for s being replaced by $\hat{s}$. Note that it is the moments of $G_j^h(x,t)$ that connect the hadron to valon partial-wave amplitudes.

Suppose that $\hat{T}_{jj'}$ is singular at $\ell = \alpha_k(t)$. Writing it in the Regge representation (ignoring spin complication)

$$\hat{T}_{jj'}(\ell,t) = \sum_k \frac{\gamma_k^j(t)\gamma_k^{j'*}(t)}{\ell - \alpha_k(t)} \tag{15}$$

we obtain

$$T_{hh'}(\ell,t) = \sum_k \frac{\Gamma_k^h(t)\Gamma_k^{h'*}(t)}{\ell - \alpha_k(t)} \tag{16}$$

where

$$\Gamma_k^h(t) = \sum_j \gamma_k^j(t) G_j^h(\alpha_k, t) \tag{17}$$

It is therefore a particular property of the valon model that the Regge singularities of the hadronic amplitude are due entirely to the singular structure of the valon amplitude. Thus the locations of the singularities depend on the nature of valons, but not of the hadrons. The Regge residues, on the other hand, involve the moments of the valon distribution in hadron, a dependence that very reasonably expresses the hierarchy of the hadronic substructure. In this respect the valon model makes a strong connection between the scattering and bound-state natures of the hadrons. Indeed, (17) appears to offer bountiful possibilities for interesting phenomenology.

From (5) and (14) one can derive a special property for the $\ell = 1$ moment, namely: at $t = 0$, $G_j^h(\ell=1, t=0) = 1$. It follows from the normalization condition on $G_j^h(x)$. Thus for vector exchange, i.e. Pomeron, $\Gamma_{\alpha=1}^h(0) = \sum \gamma_{\alpha=1}^j(0)$. This is yet another expression of the result of the additive quark model already discussed in Sec. III.1. One can consider other applications of (17) by using the G_j^h functions as given in Sec. I and in Ref. 3 for the t dependence. In Regge phenomenology j and k are constrained, so in many reactions the sum in j may be trivial, and $\gamma_k^j(t)$ may be cancelled in ratios of the Regge residues $\Gamma_k^h(t)$.

Phenomenology aside, it is clear that the valon model has brought to focus the central role that valon-valon interaction plays in hadronic reactions. Significant progress can be made only if we can understand $T_{jj'}$, better. We have already remarked that at high energy the dominant process in valon-valon scattering involves the interaction of the wee partons. Since perturbation theory is inapplicable, it is hard to make precise a description of the process. Qualitatively, one expects successive gluon-emission and pair-creation subprocesses (unfortunately in the language of perturbation theory) and multi-gluon exchange between wee partons arising from the two valons, as depicted schematically in Fig. 2(a). These are presumably the basic type of subprocess from which the Pomeron is built up. The wee-parton distribution in a valon specifies both the normalization and energy dependence of the corresponding contribution to the total cross section.

The most impostant sub-dominant term, when valon j' is the "antivalon" of j, is when the valence quark in j annihilates the valence antiquark in j' as shown in Fig. 2(b). This, of course,

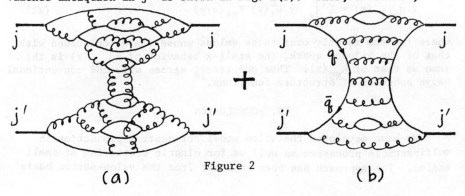

Figure 2

(a)   (b)

corresponds to the planar dual diagram. Calling this the annihilation contribution, we have

$$\sigma_{jj'}^{ann}(\hat{s}) = \int \frac{dx_1}{x_1} \frac{dx_2}{x_2} K_{NS}(x_1) K_{NS}(x_2) \sigma_{q\bar{q}\rightarrow X} \tag{18}$$

where $K_{NS}(x)$ is the non-singular quark distribution in a valon relevant for soft hadronic reactions.[1,6,14] From dimensional consideration, as in Drell-Yan process,[15] we take the $q+\bar{q}\rightarrow X$ annihilation cross section to be inversely proportional to its c.m. energy which is $x_1 x_2 \hat{s}$. Hence, we get

$$\sigma_{jj'}^{ann}(\hat{s}) \sim \hat{s}^{-1} \prod_{i=1,2} \int_{\bar{x}}^{1} dx_i \, x_i^{-2} K_{NS}(x_i) \tag{19}$$

where we take the lower limit of integration $\bar{x}$ to be proportional to $\hat{s}^{-1/2}$ because we are studying finite-energy effects which we represent by a rounding-off of the scaling distribution in the integrand in the small-x_α region.

If $K_{NS}(x) \sim x^{1-\alpha}$ at small x, then the integral in (19) behaves as $\hat{s}^{\alpha/2}$, so $\sigma_{jj'}^{ann}(\hat{s}) \sim \hat{s}^{\alpha-1}$. Using this result and the fact $\hat{s} = xx's$ in (8), we thus obtain for the leading term in the non-Pomeron contribution to $\sigma_{hh'}$

$$\sigma_{hh'}^{nonP}(s) \propto s^{\alpha-1} G_j^h(\alpha) G_{j'}^{h'}(\alpha) \tag{20}$$

where $G_j^h(\alpha) \equiv G_j^h(\alpha, t=0)$. The proportionality factor, left unspecified in (20) is, evidently, according to (17), $\gamma_k^j(0) \gamma_k^{j'}(0)$ where k specifies α. Sums over j and j' are implied. Clearly, it is the small-x behavior of $K_{NS}(x)$ that determines the Regge behavior of $\sigma_{hh'}$ with $\alpha = \alpha_k$. Note that we have derived this result in the valon model without any reference to deep inelastic scattering.[1,6] Since the valence quark distribution in hadron h is

$$x q_{val}(x) = \sum_j{}' \int_x^1 dy \, G_j^h(y) K_{NS}(x/y) \tag{21}$$

where the sum is only over those valons whose flavor coincides with that of the valence quark, the small-x behavior of $xq_{val}(x)$ is the same as that of $K_{NS}(x)$. Thus our result agrees with the conventional Regge behavior of structure functions.[16]

IV. CONCLUSION

We have developed the valon model for central production in multiparticle processes as well as for elastic scattering at small angles. The approach has been to start from the valon-parton basis and derive observable consequences at the hadronic level. The results have been highly supportive of the general relevance of the valon model, already found to be successful in other areas of hadronic physics.

It is significant to reduce the hadron-hadron scattering problem to one that involves only valon-valon interaction. Two areas of work are now in view. One is to cement that reduction by investigating all phenomenological consequences of that relationship. It would further elucidate the connection between bound-state and scattering problems of hadrons. The other area would be to explore more deeply the valon-valon interaction itself. Most notably is the question of the origin of the small-x behavior which gives rise to Regge behavior. Problems related to charge exchange can now also be initiated.

We have just begun the path to a more satisfactory understanding of two-body reactions at the constituent level. Many complications, such as those due to mass and spin, will present hurdles along the way. But obviously a model is in good health when it is ready to confront obstacles of that nature.

This work was supported in part by the U.S. Department of Energy under Contract No. EY-76-S-06-2230.

REFERENCES

1. For a review see R.C. Hwa, Erice lectures in Partons in Soft Hadronic Processes, ed. R.T. Van de Walle (to be published in 1981).
2. R.C. Hwa and M.S. Zahir, Phys. Rev. D23, 2539 (1981).
3. R.C. Hwa and C.S. Lam, OITS-158 (1980).
4. C.B. Newman et al., Phys. Rev. Lett. 42, 951 (1979).
5. A. Capella, Erice lectures, loc. cit.
6. R.C. Hwa, Phys. Rev. D22, 1593 (1980).
7. R. Carlitz, S.D. Ellis, and R. Savit, Phys. Lett. 64B, 85 (1977); N. Isgur, G. Karl and D.W.L. Sprung, Phys. Rev. D23, 163 (1981).
8. V.V. Anisovich and V.M. Shekhter, Nucl. Phys. B55, 455 (1973).
9. M. Antinucci et al., Lett. Nuovo Cimento 6, 121 (1973).
10. W. Thomé et al., Nucl. Phys. B129, 365 (1977).
11. T.T. Wu and C.N. Yang, Phys. Rev. 137, B709 (1965).
12. T.T. Chou and C.N. Yang, Phys. Rev. 170, 1591 (1968).
13. P.V. Landshoff, Phys. Rev. D10, 1024 (1974).
14. L. Gatignon, R.C. Hwa, and R.T. Van de Walle, OITS-170, to be published in the Proc. for EPS International Conference on High Energy Physics, Lisbon, 1981.
15. S.D. Drell and T.M. Yan, Phys. Rev. Lett. 25, 316 (1970).
16. H.D.I. Abarbanel, M.L. Goldberger and S.B. Treiman, Phys. Rev. Lett. 22, 500 (1969); H. Harari, ibid. 24, 286 (1970); J. Kuti and V.F. Weisskopf, Phys. Rev. D4, 3418 (1971).

DISCUSSION

De GRAND, SANTA BARBARA: I'd like to go back and talk about the central region for a minute. It looks as if, in the model which you are making, if I'm allowed to adjust the number of partons in the central region by diddling some parameter (and let's forget about fitting data), if I double the number of quarks and antiquarks in the central region then I don't double the number of mesons. I increase the number of mesons in the central region or the height of the plateau by a factor of 4. To me that seems a bit unphysical, but since you are the author of this model it must not to you and why not?

HWA: Why is it unphysical?

De GRAND: Well, if I have twice as many quarks around and twice as many antiquarks around I would think that since a meson is a quark and an antiquark I would have twice as many mesons around. But you seem to be getting 4 times as many mesons around because you are proportional to a density times a density.

HWA: Did you notice that I had to consider all the rhos and everything? That's how I got . . . .

De GRAND: Yes, but forget flavor. This is in a world in which there is one flavor and no resonances. You will still get this result. I would think that I could imagine writing down some theory of a world where there is one flavor. Resonances will give you cascading, but still at some level you are making four times as many of some sort of pre-particle if I double the number of quarks than you would have had I not done so.

HWA: Yes, that sounds right. I hadn't thought about that. I don't know if it's unreasonable or not but that seems to be what comes out.

De GRAND: That is the reason why Miettinen and I never wrote the long paper.

SUKHATME, ILLINOIS: I have two short questions. One is, I consider the energy dependence of the height [of the central plateau] to be one of the successes of the [Orsay] DTU model. Do you have any ideas on that? Secondly, your connection between charge and matter density seems rather sensitively dependent on the fact that you are using a proton. You were using

2/3 and 1/3. Does it work for a neutron? Same calculation.

HWA: For the first question, rising of the plateau has to do with the valon-valon interaction and I'm just beginning to look into this problem. So far, I certainly cannot give you an answer why the plateau is rising. I don't think it is out of the scope, but I just haven't looked at that. The second question about the neutron: I studied the neutron form factor and can see why the total charge is zero but the neutron charge radius is not zero. For the neutron proton scattering, whether you want to use a neutron matter density and how neutron matter density differs from a neutron -- well it's not a charge density . . . . That's a good question. I can't imagine something really basic will go wrong. Things turn out to be so physical. I just haven't thought about it. In fact, all this was done only last week or so.

MORRISON, CERN: The test of any theory is the predictions it makes. Maybe I missed it, but you are a little bit short on predictions. I have two questions for you. One was, you talked of elastic scattering. Where do you predict the first dip and the second dip to be in your elastic scattering?

HWA: These are all good questions. I only just began on the elastic part. The first place is to get the beginning of eikonal iteration, which Chou-Yang has done. It just pops out in my eyes that their hypothesis is right in front of me. It is not really needed to make such an assumption. It is right there. And if it is reasonably connected to the form factor and you have the first step correct then, after, you just use iteration. I would think the dips would come out as they would predict if I wanted to take over their whole machinery. On the other hand, Andrzej [Białas] may have a different view on that.

BIAŁAS, KARKOW: Some time ago we [A. Białas, K. Fialkowski, W. Slominski, and M. Zielinski, Acta Physica Polonica (1976)] did the calculation of elastic [p p] scattering in a model like that [the 3 quark (valon) picture]. Although the beginning of the curve can be really very well described by this model (I mean the forward peak), when we tried to look for the second of the scatterings and so on we couldn't really get any reasonable agreement with the data. The dip is coming almost invariably somewhere between 0.8 and 1.0 GeV, which is too close to the beginning, and also the secondary maximum becomes too high. These two things are actually related. Obviously, in such models you get too much double scattering and multiple scattering compared to the data. I cannot exclude that, by some very clever choice of the [proton] wave function, you can't actually finally fit the data, but with any simple way -- trying to, for example, get the wave function from form factors and things like - - it simply doesn't work. I mean, there has to be some new idea. There are two ways, as far as I can say. One is that there is extra glue inside which is not attached to valence quarks. The other thing is that perhaps these valons are actually more complicated than one thinks. That's my experience.

HWA: May I just ask, what did you do different from what Chou-Yang did?

BIAŁAS: The difference between Chou-Yang and our calculation is that "Chou-Yang" effectively means that they take an infinite number of constituents inside the proton. I mean that there is a continuum of constituents. Whereas if you take only 3 that makes a tremendous difference. That's just the problem.

HWA: They never mentioned constituents.

BIAŁAS: Sure, sure, of course. But effectively the calculation is as if you make an infinite number of constituents. If you translate the Chou-Yang picture.

HWA: But, the way I did it, I showed that the first term . . .

BIAŁAS: The first term is alright.

HWA: The first is only involving three valons.

BIAŁAS: With the first term everything is okay. This is a trivial term. Everybody will get it. Once you take some geometrical picture you get it.

MORRISON: The second question is, one of the recent interesting results is the frequent production of charm. Do you have any idea as to how you get such a large cross section for charm in the forward direction?

HWA: A very good question, and I wish I had a good answer.

MULTIPARTICLE CORRELATIONS [x)]

I.M.Dremin

P.N.Lebedev Physical Institute, Moscow, USSR

Abstract

Recent experimental and theoretical results dealing
with correlations in multiparticle low-p_T production processes
in high energy hadron collisions are reviewed. Some new possible
correlation effects at very high energies are discussed.

Contents

1. Introduction
2. Charge correlations
3. Rapidity correlations
4. Azimuthal correlations
5. Conclusions

[x)] Invited Talk submitted to XII International Symposium on
Multiparticle Dynamics, Notre Dame, USA, June 21-26, 1981

1. Introduction

The latest years hardly can be named as the years of furious activity in studying multiparticle correlations (in contrast to 1975-78). Now this field reminds of a poorly-fed child. But it is healthy and deserves being reviewed because, first of all, some new interesting data appeared in a wide energy range and, then, the prospects of learning new correlation effects become more realistic in the yet unexplored region of forthcoming accelerators (CERN, Brookhaven etc.).

From previous studies we have learned that the secondary particles produced in low-p_T hadron collisions are strongly correlated [1]. Phenomenologically this correlation is often attributed to the two-step nature of the process [2,3] : the massive hadronic systems (resonances, clusters, fireballs) are first created and then they decay into the observed secondary particles. The average number of pions per cluster according to different estimates [1-9] varies between 3 and 6. In my opinion, the notion of a cluster is surely much wider than the usual low-mass resonances, it includes some more massive formations.

Besides the pragmatic problem of the size of clusters there is an important question, why the Nature chooses such two-step mechanism with production of massive clusters.

The only palliative answer, we have, is provided by the multiperipheral interpretation [3,4,10] which shows that the massive intermediate states are necessary to generate the proper Pomeron trajectory ensuring the nearly constant total cross-section, proper multiplicity and slow contraction of the diffraction peak in elastic scattering.

The problems of the origin of clusters and of their nature have also been disputed. Many people believe in multiperipheral mechanism of their production even though other approaches (hydrodynamics, dual scheme, coherent states etc.) are advocated too. What concerns the nature of clusters, one can argue [3] that the high-mass nonresonant clusters are as usual quantum objects as low-mass resonances. Their decay used to be described by the statistical approach but one should remember that it is an additional assumption which can, in principle, be replaced by another one. For example, at the intermediate stage they can decay into usual resonances which subsequently produce final particles.

More generally, this is a problem of secondary interactions of particles produced. Clusters correspond to short-range ordering while rescattering (multiladder) mechanism reveals the long-range correlations.

Still we lack the treatment of clusters in more pretentions quark-gluon terms. We do not pretend to apply QCD at full strength to low-p_T processes because of problems with confinement Even in the framework of some simplified quark models (fragmentation, recombination etc.) calculations are done at the level of one-particle distributions in the fragmentation region.

The simplest correlations have been tried in [35]. A few attempts of calculation of single-particle distributions and of average multiplicity for the gluon radiation graphs in QCD have been done [11]. In my opinion, the radiation mechanism is not so important in the central rapidity region but one should consider an alternative picture of the break of color strings stretched by the sepaŗaͭing color charges [12] .It is in charge of multiparticle correlations there.

Thus, their study can reveal the deepest unknown features of the confinement mechanism.

Let me say a few words about the general structure of the talk. I shall review mostly the problems and conclusions without going very deep into the details of experiment and theoretical calculations which can be found in the papers referred to. Reminding briefly the well-known important results at the beginning of each section, I come then to the recent ones.

2. Charge correlations

The simplest correlations of final particles are related to their charge states. They used to provide the very general information about average strength of correlations and charges of clusters.

Let us remind briefly some facts. The average multiplicity $\langle n \rangle$ increases as $\ln s$ or $\ln^2 s$. The charged particle multiplicity distributions are wider than Poisson law at energies above 50 GeV. Their maxima shift to higher values at higher energies but not as fast as maxima of Poisson distributions with the same average multiplicity. The correlation functions (of charged particles) $f_2 = \langle n(n-1) \rangle - \langle n \rangle^2$,

$$f_3 = \langle n(n-1)(n-2) \rangle - 3 \langle n(n-1) \rangle \langle n \rangle + 2 \langle n \rangle^3, \ldots$$

are positive at energies above 50 GeV and fastly increase (see Fig.1) while they must be equal to zero for Poisson distributions. Let us note that at the highest accelerator energies they exceed even the KNO-predictions (the dashed curve in Fig.1). The ratio $\langle n \rangle / D \equiv \langle n \rangle / \sqrt{\langle n(n-1)\rangle}$ is weakly decreasing with energy and it is close to its value of 2, following from KNO-scaling.

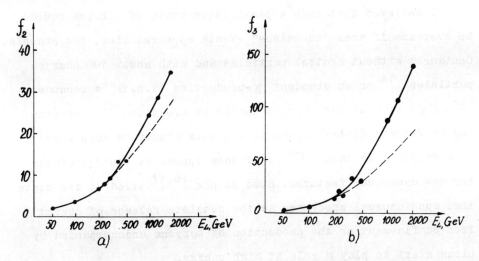

Fig.1. Energy dependence of correlation functions f_2 and f_3 (pp-interaction) .The full curve is drawn to guide the eye, the dashed curve interpolates the KNO-predictions.

This is an indication of strong correlations of particles produced. It was often used in many models, in particular, for treatment of clusterization of pions [3,10,13]. The additional information, say, about the linear increase of the number of backward produced pions with the number of pions in the forward hemisphere or about the linear increase of neutral versus charged particles, as well as about charge zone characteristics or about charge transfer correlations showed us that high-mass clusters can be created and there should be charged clusters besides the neutral ones.

However even these qualitative conclusions seemed to be model dependent. In general, due to the averaged nature of the measured distributions, their smooth features could be described in many ways (independent cluster emission model, multiperipheral cluster scheme, coherent states of clusters etc.).

I believed that such a pessimistic state of affairs could be overcome if some "anomalous" events appeared like, for example, Centauros without neutral particles and with about 100 charged particles [14] or an abundant γ-production (i.e. π^{o}'s because $\pi^{o} \rightarrow 2\gamma$) without the charged particles creation [15]. According to Poisson distributions such events should be suppressed by a factor less than 10^{-10}. If they appear it could indicate the new dynamical features. Some people [16,17] tried to speculate that such unusual processes as the complete release of quarks from confinement or the production of baryons unaccompanied by pions start to play a role at high energy.

It was a surprise for me to learn that such exotic events could be simply explained by proper treatment of isotopic spin conservation in hadronic processes [18]. It is proved for the oversimplified model but the general origin of the conclusion seems very plausible.

Let us imagine the process of nucleon-nucleon interaction as proceeding via creation of two leading baryons and some coherent state of pions $^{x)}$. Then isospin of the pionic system does not exceed two.

---

x)
Previously the coherent states of pions were considered, for example, in [19] to fit the multiplicity distributions and charge correlations with a general conclusion that clusters should play an important role. The assumption of the coherence of pions is very crucial.

The state with isospin I and its z-component I_z can be obtained from the usual coherent state of pions [20]

$$|f\rangle = e^{-\frac{c}{2}} \cdot \exp\left\{ \int d^3k \sum_i f_i(k) a_i^+(k) \right\} |0\rangle \tag{1}$$

(here $c = \sum_i c_i = \sum_i \int d^3k |f_i(k)|^2$ $(i = +,-,0)$, a_i^+ are production operators of pions, $|0\rangle$ is a pion vacuum) by averaging over the angles of the vector $\vec{f}$ in isospace with the corresponding spherical function $Y_{I,I_z}(\vartheta, \varphi)$ [20]:

$$|f; I, I_z\rangle = e^{-\frac{c}{2}} \int d\Omega_e \, Y_{I,I_z}^*(\vartheta, \varphi) \exp\left\{ \int d^3k \, \vec{f}(k) a^+(k) \right\} |0\rangle, \vec{f} = f\vec{e} \tag{2}$$

(here $\vec{e}$ is an unit vector). It is easy to derive from (2) the distributions of pions in different charge states.

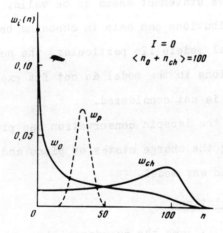

Fig.2. The charged (w_{ch}) and neutral (w_o) particle distributions for a coherent system of pions $(\langle n \rangle = 100)$ with the total isospin $I = 0$. The dahsed curve shows the would — be Poisson distribution of (neutral) pions.

Leaving aside the final formulas (see [18]) let me explain the situation by Fig.2 where the pion distributions for the zero total isospin of pion system at average number of pions equal to 100 are reproduced. The distributions of neutral and charged

pions are shown by solid lines. They are much wider than those
for arbitrary isospin state (from 0 to 100) which are narrow
Poisson laws (as it follows from (1)) described by the dashed
line. The requirement of the definite isospin corresponds to
averaging over many Poisson curves with different positions
of their maxima. Thus the events without neutral (or charged)
pions can amount to some percents of all events and therefore
be easily observed.

The model considered does not look very realistic. One
could take into account the different from (1) symmetries of
the pion system in k-space and the broader primary distributions
but the qualitative statement seems to be valid. The precise
form of the distributions can help in choosing between the
different dynamical models .In particular, the neutral -char-
ged pion correlations in the model do not fit experimental data
[19] if clustering is not considered.

To conclude, the isospin conservation can produce strong
correlations among the charge states of pions and should be taken into account in any model.

3). Rapidity correlations

The rapidity y_i and the pseudorapidity η_i of the i-th
particle in an event are defined as

$$y_i = \frac{1}{2} \ln \frac{E_i + p_{i\,\shortparallel}}{E_i - p_{i\,\shortparallel}} ,\qquad (3)$$

$$\eta_i = - \ln tg \frac{\theta_i}{2}\qquad (4)$$

where E_i , $p_{i\,\shortparallel}$, θ_i are the energy, the longitudinal momentum
and the angle of i-th particle.

In terms of inclusive cross-sections the two- and three-particle correlation functions are defined by the formulas.

$$R_2(y_1, y_2) = \frac{\rho(1,2) - \rho(1)\rho(2)}{\rho(1)\rho(2)} ,$$

(5)

$$R_3(y_1, y_2, y_3) = \frac{\rho(1,2,3) + 2\rho(1)\rho(2)\rho(3) - \rho(1,2)\rho(3) - \rho(1,3)\rho(2) - \rho(2,3)\rho(1)}{\rho(1)\rho(2)\rho(3)}$$

(6)

where

$$\rho(k) = \frac{1}{\sigma_{in}} \frac{d\sigma}{dy_k} , \quad \rho(k,\ell) = \frac{1}{\sigma_{in}} \frac{d^2\sigma}{dy_k dy_e} , \quad \rho(k,\ell,m) = \frac{1}{\sigma_{in}} \frac{d^3\sigma}{dy_k dy_e dy_m} \quad (7)$$

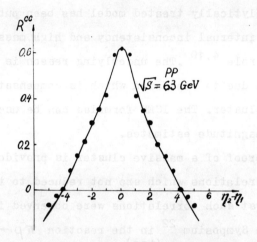

Fig.3. Rapidity correlation of two charged particles
in pp-interactions at $\sqrt{s}$=63 GeV.

It is well known (see Fig.3) that the correlation of two charged particles is short-ranged (the half-width of the inclusive
distribution R_2^{cc} is about 2 units of rapidity), rather large
($R_2^{cc}(0,0) \approx 0.65$), almost independent of the incident energy and
of the individual rapidities for a fixed rapidity difference.
All these features find a natural explanation within the cluster
models and are used for estimates of cluster parameters [4,5,7,10].
However the final results still differ appreciably. There is a
strong contradiction between the results obtained from inclusive

and semi-inclusive data [3-4] if both samples are treated according to the independent cluster emission model. While the former favours massive clusters (5-6 particles), the latter shows smaller number (3-4) of pions per each cluster .It means that there are correlations which are not taken into account by the model. Such a correlation (namely, a repulsion of clusters) is properly considered in multiperipheral cluster models (the simplified but analytically treated model has been published [21]). There is no internal inconsistency and high mass clusters play an important role [4,10] .The underlying reason is the negative correlation due to repulsion which is compensated by the larger size of a cluster. The ICEM formulas can be used only for the order of magnitude estimates.

The further proof of a massive cluster is provided by the multi-particle correlations which are not reduced to the two-particle ones. First such correlations were observed in 1978 by the hosts of the Symposium [22] in the reaction $\pi^- p \rightarrow \pi^+ \pi^- \pi^- X$ at 200 GeV/c with the result $R_3^{(+--)}(0,0,0) = 0.26 \pm 0.17$.

Now the three-particle pseudorapidity correlations have been studied [23] at 67, 200 and 400 GeV for pN-interactions and at 50 and 200 GeV for $\pi^- N$-interactions in emulsion (the total statistics is about 7000 events). The three-particle (charged) correlator R_3^{ccc} in the central region ($\eta_i \approx 0$) increases with the incident energy what strongly differs from the behaviour of two-particle correlators (see Table I). Experimental values have been compared with the Monte-Carlo results of the modified cylindrical phase space (MCPS) in which the experimental multiplicity and semi-inclusive pseudorapidity distributions for each sample are reproduced. It appears that the difference between the experimental and MCPS values of $R_3^{ccc}(0,0,0)$ increases also with the incident energy and the positive excess becomes statistically meaningful at 400 GeV(see Table I, column 2).

Table I

Interaction, energy		$R_3^{ccc}(0,0,0)$	$R_3^{ccc}(0,0,0) - \left[R_3^{ccc}(0,0,0)\right]_{MCPS}$
pN	67 GeV	0.15 ± 0.07	-0.07 ± 0.08
	200 GeV	0.21 ± 0.05	-0.05 ± 0.05
	400 GeV	0.39 ± 0.05	0.13 ± 0.06
$\pi^- N$	50 GeV	0.17 ± 0.09	-0.03 ± 0.10
	200 GeV	0.67 ± 0.07	-0.04 ± 0.08

Even in the independent cluster emission model such a correlation would require the 4-5-particle clusters.

The energy about 200 GeV seems to mark the level at which the three-particle correlations appear. The further check of the correlations rise is very desirable.

To get the quantitative results about cluster sizes one should fit the functions R_2 and R_3 by the theoretical curves obtained from Monte-Carlo calculations for different models with appropriate values of parameters. To simplify the job it was earlier proposed to analyze the rapidity gaps ($r = y_{i+1} - y_i$) between the adjacent particles [5] Their distribu-tion at large rapidity separations defines the cluster density ρ in the rapidity scale ($\sim e^{-\rho r}$) while at smaller gaps it is determined by the average number of pions K per cluster ($\sim e^{-\rho Kr}$).

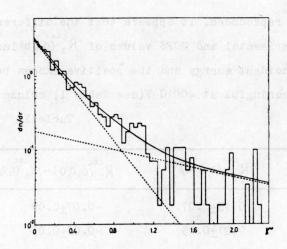

Fig.4. Rapidity gaps in pN-interactions at
400 GeV (Two exponents are shown by
dotted lines)

Again this method was recently used by the Indian group [9] for
pN-interactions at 67, 400 and $\sim$ 1000 GeV. The typical rapidity
gap distribution is shown in Fig.4. The authors [9] conclude that
one should ascribe 5-6 pions per cluster at all energies and
the cluster multiplicity increases with energy. However, this
conclusion also suffers from all the diseases of the independent
cluster emission model.

Still, the only typical feature of the rapidity gap distribution, i.e. the break at some value of r , requires larger
statistics than in Fig.4. More impressive ones can be the rapidity interval distributions where intervals $r_m^n = y_{i+m+1} - y_i$
contain $m > 0$ particles in n-particle events. They possess the
maximum and some width which can be used to fix the cluster
properties. They seem to provide quite sensible test of various

models.This conclusion is drawn from the comparison of the multiperipheral cluster model [7,10,24] with pp and $\bar{\pi}p$ -data in the energy range from 60 to 200 GeV. While a lot of other model distributions (including the rapidity gaps) have been shown to fit experiment nicely [10] the radipity intervals at values of m around $n/2$ reveal some substructure which is not as rich in experiment as in the model.

Nevertheless, there are some indications from experimental $\bar{\pi}p$-data at 40 GeV [25a] on the possible multimaximum structure of such distributions (see Fig.5). No statistically significant structure was observed in Kp at 32 GeV and pp at 69 GeV [25b].The careful study of rapidity intervals was stimulated by the idea that the clusters repulsion inherent in the multiperipheral approach could produce such a structure. Its appearence was

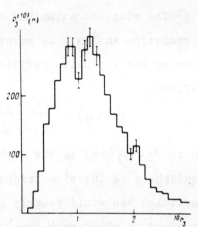

Fig.5. Double-maximum structure of the 3-particle rapidity interval distribution in 10-prong events for $\bar{\pi}p$ -interactions at 40 GeV/c.

demonstrated [21] within the framework of the generalized independent cluster emission model incorporating some multiperipheral features. Namely, it is supposed that the clusters may not be produced at rapidity separations smaller than a fixed value of Δ . Such a model gives rise to irregular rapidity interval distributions with maxima positions determined both by the cluster masses and by their (large enough) rapidity

separation Δ . The similar structure appears [26] in multiladder
graphs (corresponding to Regge cuts in elastic processes).
Thus, such an irregular behaviour would uniquely determine the
multiperipheral dynamics with its inherent repulsion. Therefore,
some other data analysis is very desirable.

4. Azimuthal correlations

The simplest azimuthal characteristic which is used for the
correlation analysis is an azimuthal asymmetry $B(\Delta y)$ as a function of the inclusive rapidity distance between two pions Δy
defined as

$$B(\Delta y) = \frac{N_>(\Delta y) - N_<(\Delta y)}{N_>(\Delta y) + N_<(\Delta y)}$$

where $N_{>(<)}(\Delta y)$ is the number of pions in the opposite (same)
hemisphere as the pion trigger. For completely independent
particles one would roughly get $B(0) \sim 1/\langle n \rangle$, while for particles
combined in clusters $B(0) \sim 1/K$ in the rapidity correlation range.
Fig. 6 reminds the earlier obtained results [27,10] for $\pi^- p$ at
40 GeV/c and for pp at 200 GeV/c and $\sqrt{s}$ = 52.5 GeV. The stepwise dependence reveals the clusterization of pions at short
rapidity distances. The decay multiplicity of a cluster K slightly
increases with energy what disagrees with the statement of the
Indian group [9] but does not contradict to expectations from the
multiperipheral models with the wide cluster mass spectrum [10].

Some traces of such cluster can be possibly guessed from
the statement of [28] about rather drastic qualitative change of
the azimuthal correlations in pN-interactions at energies
about 100 GeV. At lower energies they do not contradict to the
statistical type models (the modified cylindrical phase space-MCPS-

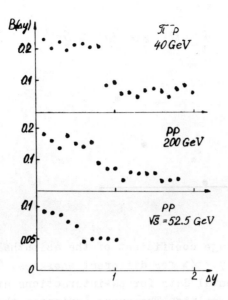

Fig.6. The azimuthal asymmetry $B(\Delta y)$ as a function
of the rapidity separation of two charged
particles ($\pi^-\bar{p}$ at 40 GeV and pp at 200 GeV
and $\sqrt{s}$ = 52.5 GeV)

with the experimental multiplicity and semi-inclusive rapidity

distributions). However, at higher energies there is a noti-

ceable difference from such models in any range of rapidities.

This is demonstrated by Fig.7 where the behaviour of the average

coefficient $\langle\alpha\rangle$ of the azimuthal asymmetry is shown. Here

$$\langle\alpha\rangle = \frac{1}{N}\sum_{\kappa=1}^{N}\alpha_\kappa \; ; \; \alpha_\kappa = \sum_{i\neq j}^{n_\kappa}\cos\varepsilon_{ij}/\sqrt{n_\kappa(n_\kappa-1)}$$

where N is the total number of events, $\cos\varepsilon_{ij} = \vec{P}_{\perp i}\vec{P}_{\perp j}/P_{\perp i}P_{\perp j}$,

n_κ is the charged particle multiplicity in the k-th event.

Experimental values of $\langle\alpha\rangle$ are closer to zero at energies

exceeding 100 GeV than the predictions of MCPS shown in

Fig.7 by the curve. Strong azimuthal correlations indicate the

tendency of particles to group into clusters or jets.

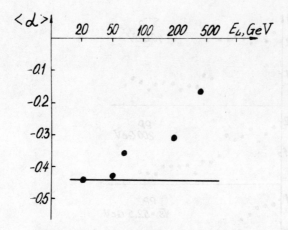

Fig.7. The average coefficient of the azimuthal
asymmetry $\langle \alpha \rangle$ for different energies. The
experimental data for pN-interactions are
shown by points. The curve indicates the
results of the modified cylindrical phase
space model.

The same conclusion was obtained from the more refined
analysis of groups of particles within the definite pseudorapidity interval $r_m = \eta_{i+m+1} - \eta_i$. The coefficients of the azimuthal
asymmetry α_m were determined for such groups. It is shown[28]
that there are multiparticle correlations increasing with energy
and having the short-range behaviour. The most remarkable effect
is pronounced at small r_m for large m as it is demonstrated
by Fig.8.

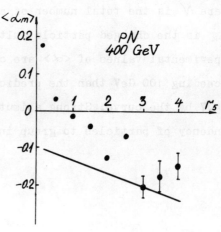

Fig.8. The coefficient of the
azimuthal asymmetry
as a function of the
rapidity intervals
containing 5 charged
particles in pN-inte-
ractions at 400 GeV.
The modified cylindri-
cal phase space model
does not fit experimen-
tal points.

I would like also to mention that some azimuthal correlations have been treated [29] by the similar method at extremely high energies ($\sim$ 100 TeV) in cosmic-ray studies of gamma-families. It is claimed that some new effects appear. Very high-p_T groups (clusters, jets?) of particles are emitted with large probability which exceeds the possible extrapolations of the well-known p_T^{-4} law. Even though there are some theoretical papers [30,31] with the predictions of a softer decrease law in some range of transverse momenta (like p_T^{-2}) the situation is still unclear both from experimental and theoretical viewpoints.

Some experimentalists [32] attempted to separate groups of particles correlated both in polar and azimuthal angles in individual events of pN-interactions at 200 GeV. It has been concluded that such events appear much more often than in the simple models.

Among them there are special events where the groups are concentrated (see Fig.9a) within the same narrow ring in the target diagram (i.e. in the plane perpendicular to incident direction). Possibly such a feature could be occasional due to usual fluctuations. One has to test for correlations on event by event basis. However if the further analysis reveals the larger portion of such events they would require some dynamical origin to be explained.

I would like to draw your attention to the speculation [30] that such events show the temporary release of quarks of a hadron from confinement when entering another hadron. The classical electrodynamic model for such a process would be the electron confined within a waveguide and passing from one of them to another one separated by a gap filled by a matter. The waveguides reproduce the confinement conditions while the matter should be hadronic one in the case of quarks. The ring-like structure of the emission of an electron passing across the gap is easily verified. Similarly, the gluons emitted by quarks appear within the rings in the target diagram producing the

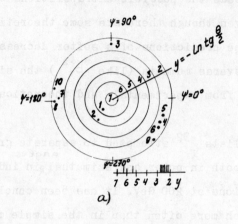

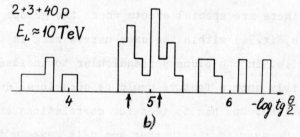

Fig.9. The ring-like events.
a) The target diagram and the pseudorapidity distribution of the ten-prong event in pp-interaction at 200 GeV.
b) The pseudorapidity histogram of the 2+3+40 p event at ~ 10 TeV. The rings at large c.m.s. angles are indicated by arrows.

final hadronic jets. The azimuthal asymmetry of particles within the ring should be observed in an individual event.

Even if by some reasons such a transition gluon radiation

278

is suppressed there could be observed the gluon radiation in the same ring with a lower intensity analogous to Cherenkov radiation of photons. One can show that at superhigh energies ($\gtrsim$ 10 TeV) the refractivity of the hadronic medium becomes positive because it is proportional to the real part of the forward. scattering hadronic amplitude which (according to experimental data) becomes positive above some threshold. It is the necessary condition for Cherenkov-like radiation of gluons by quarks.

Some events with the ring-like structure were observed also in cosmic-ray studies [33]. The pseudorapidity distribution for one of them at incident energy about 10 TeV is shown in Fig.9b. Besides the two fragmentation groups of particles there are two ring-like groups in the pionization region appearing at predicted [30] values of angles about 60-70° in c.m.s.

I shall not enter into the details of the calculations and their predictions (see [30]). I would like just to stress the importance of such extremely specific correlations for revealing the underlying dynamics of quark confinement and properties of the hadronic medium.

5. Conclusions

The recent studies of multiparticle correlations have confirmed the previous conclusions that besides the two-particle correlations there exist the multiparticle ones increasing with energy. It favours the notion of clusters irreducible to low-mass resonances.

Still there are many open questions which need to be solved. I mentioned just some of them:

1. To which extent do the conservation laws change our conclusions about dynamical correlations in multiparticle systems? Do the symmetry properties in internal degrees of freedom (isospin etc.) of a multiparticle system impose some severe restrictions at high energies?

2. If the particles are emitted in groups which repulse, is it reflected in multiparticle correlations, for example, in the form of the substructure of the rapidity interval distributions?

3. If such substructure exists, does it indicate the inhomogeneous distribution of quarks within hadrons?

4. Can quarks confined within a hadron react coherently as some medium and emit gluons according to classical laws? The ring-structured events are asked for.

5. Do the high transverse momenta appear at superhigh energies more often than according to p_T^{-4}-law?

I did not mention many important problems which were discussed during the last years (for example, the similarity of low-p_T hadron-hadron reactions with e^+e^-, ℓp and high-p_T processes etc.).. The experiment will surely produce new insights to all these problems at the new generation of accelerators. From the theoretical standpoint the situation does not seem to be as promising because the miraculous confinement is still a hard nut to crack.

Acknowledgements

I thank Prof.W.D.Shephard for the invitation to give this talk and for his enormous efforts.

References

1. L.Foa Phys. Reports 22, 1 (1975);
   J.Whitmore Phys. Reports 27, 1 (1976).

2. I.M.Dremin, A.M.Dunaevskii, Phys. Reports 18, 159 (1975)
   and references therein. See also Proceedings of IX International Symposium on Multiparticle Dynamics, Tabor, 1978;
   talks by I.Pless, W.Shephard, E.Feinberg, I.Dremin.

3. I.M.Dremin, E.L.Feinberg, Fizika Elementarnych Chastitz
   i Atomnogo Jadra 10, 996 (1979) (Engl. transl.: Elementary
   Particles and Atomic Nuclei); E.L.Feinberg, Invited talk at
   XVIII International Cosmic Ray Conference, Paris, 1981.

4. E.L.Berger, Nucl. Phys. B85, 61 (1975).

5. C.Quigg, F.Pirila, G.H.Thomas Phys. Rev. Lett. 34,290(1975).

6. T.Ludlam, R.Slansky Phys. Rev. D12, 59 (1975).

7. M.I.Adamovich et al. Nuovo Cim. 33A, 183 (1975).

8. Ts.Baatar et al Yad. Fiz. 26,1022 (1977) (Sov. J.Nucl. Phys.,
   26, 541 (1977)).

9. P.K.Sengupta et al. Phys. Rev. D20,601 (1979);
   S.Roy et al. Phys. Rev. D21,24 97 (1980);
   W.S.Arya et al. Phys. Rev. D21, 3060 (1980); Phys. Rev. D22,
   2652 (1980);
   I.K.Daftari et al. Phys. Rev. D23, 14 (1981).

10. I.M.Dremin, A.M.Orlov, E.I.Volkov Preprint P.N.Lebedev
    Physical Institute N 120 (1978).

11. J.F.Gunion Preprint SLAC-PUB-2607, 1980.

12. C.B.Chiu 14th Rencontre de Moriond, Les Arcs, 1979, v.2,
    p.587; A.Casher, H.Neuberger, S.Nussinov Phys. Rev. D20,
    179 (1979).

13. V.G.Grishin et al. Nuovo Cim. Lett. 8, 590 (1973);
    A.S.Kurilin et al. JINR-preprint D2-11833 (1978);
    P.J.Hays et al. Phys. Rev. D23, 20 (1981).

14. C.M.C.Lattes, Y.Fujimoto, S.Hasegawa, Phys. Rep. 65,
    151 (1980).

15. M.Schein, D.M.Haskin, R.G.Glasser, Phys. Rev. 95, 855 (1954).

16. J.D.Bjorken, L.D.McLerran Phys. Rev. D20, 2353 (1979).

17. Dias De Deus, W.A.Rodrigues, Jr. CERN preprint TH2676 (1979).

18. I.V.Andreev, Pisma v ZhETP 33, 384 (1981) (JETP Lett.)

19. L.J.Reinders Acta Phys. Pol. B7, N 2 (1976); Z.Golab-Meyer,T.
    W. Ruijgrok. Acta Phys. Pol. B8, 1105 (1977); W.Czyz,
    T.W.Ruijgrok, Acta Phys. Pol. B9, 433 (1978).

20. J.C.Botke, D.J.Scalapino, R.L.Sugar Phys. Rev. D9,813 (1973).

21. A.M.Orlov Yadernaja Fizika 22, 524 (1980) (Sov. J.Nucl.
    Phys.).

22. V.P.Kenney et al. Nucl. Phys. B144, 312 (1978).

23. S.A.Azimov et al. Doklady AN Uzbekh SSR 9, 31 (1980).

24. M.I.Adamovich et al. Trudy FIAN 108, 3 (1979)
    (Proceedings of P.N.Lebedev Physical Institute)

25. a) I.M.Dremin, T.I.Kanarek, A.M.Orlov Pisma v ZhETP 29,
       724 (1979) (JETP Letters).
    b) V.V.Babinzev et al. Preprint IHEP 80-181, Serpukhov,
       1980.

26. V.A.Zoller, private communication.

27. M.Basile et al Nuovo Cim. 39A, 441 (1977).

28. S.A.Azimov et al Preprint PTI AN Uzbekh SSR, 2-80,
    1980; Jadernaja Fizika (to be published).

29. S.A.Slavatinskii et al. Paper presented at XVIII International
    Cosmic Ray Conference, Paris, 1981; S.A.Azimov et al
    Paper presented at XVIII International Cosmic Ray
    Conference, Paris, 1981

30. I.M.Dremin Pisma v ZhETP 30, 152 (1979).
    (JETP Lett. 30, 140 (1979)); Jadernaya Fizika 33, N 5 (1981).
    (Sov. J. Nucl. Phys. to be published).

31. L.V.Gribov, E.M.Levin, M.G.Ryskin,Preprint LIYaF N 637,
    1981; Phys. Lett. to be published.

32. N.A.Marutjan et al. Jadernaya Fizika 29, 1566 (1979).

33. A.V.Apanasenko et al. Pisma v ZhETP 30, 157 (1979).
    (JETP Lett. 30, 145 (1979)); I.M.Dremin, M.I.Tretyakova,
    Paper presented at XVIII International Cosmic Ray Conference,
    Paris, 1981.

34. S.N.Ganguli, D.P.Roy Phys. Rep. 67, 201 (1980).

35. H.Fukuda et al. Progr. Theor. Phys. 65, 961 (1981).

DISCUSSION

FIAŁKOWSKI, KRAKOW: I think that the Andreev explanation of
Centauro events is beautifully simple but, unfortunately, it can't
be true because it neglects resonances completely. The same
isospin argument tells that most of the resonances should be $\rho^\pm$
and ω [as compared to ρ^0] which all decay into π^0. So, by the
same argument, it tells that the number of π^0's should be rather
high [and the probability of an event having no π^0's is consequently
very small].

DREMIN: Well, first of all, I said that this model is not a very
realistic one but it produces the effect about 10 or maybe even
larger effect. So if the distributions are wider then you cannot
change it by such an effect. But the coherent state of pions was
applied to many reactions actually and it was specially published in
Acta Physica Polonica. There were many papers! [laughter] (I
answer Fiałkowski.) Then Ruijgrok and some other people worked
on it and they said that they could explain some correlations between
π^0's and charged particles and so on.

FIAŁKOWSKI: Yes, but my point is only that to say that there is a
reasonable probability of having no π^0's in the event tells also
that you could have a reasonable probability to have no ρ^+, ρ^- and
no ω^0 in the event. That, altogether, is extremely unlikely at the
level of 10^{-5} or so if you believe in the inclusive estimates of
resonances.

DREMIN: Well, we should look for the data at higher energies, I
suppose, to answer your question. It can be only an experimental
answer.

KENNEY, NOTRE DAME: Just a reminder that when you look at
correlations, two particle correlations, it makes a great deal of
difference whether the two particles have the same charge or have
different charges. For unlike charges, the correlations increase
with energy and fall off with increasing multiplicity. Whereas this
is not true for like-charge pairs. When you look at correlations
for all charges, these differences get washed out. This is of course
a problem for experiments which don't have magnetic fields, some
emulsion experiments, but it is something which we all have to take
into account.

DREMIN: But, I think that the theoretical explanation depends on
what is the multiplicity within the cluster. If it is not very large,
then you don't look for large correlations between the particles with
the same charges.

SOFT GLUONS, BARYON-ANTIBARYON PRODUCTION AND LOW $p_\perp$-PHYSICS IN THE LUND MODEL

Bo Andersson
Department of Theoretical Physics, University of Lund, Sweden

(Work done in collaboration with G. Gustafson, T. Sjöstrand, G. Ingelman, O. Månsson, I. Holgersson)

ABSTRACT

I briefly review the Lund soft hadronisation model, in particular the gluon model and show how to apply it to the emission of soft gluons (transverse momentum effects) and baryon-antibaryon production. Finally I remark upon a semi-classical reaction mechanism for low $p_\perp$-interactions leading to the well-known fragmentation model predictions, and plead for some more precise data on strange particle- and baryon production in order to differ between the models on the market.

1. INTRODUCTION

I have at earlier conferences in this series [1] presented some details of the Lund program on the soft hadronisation of quark and gluon jets and on the similarities to low $p_\perp$ hadronic reactions. As the audience is to a large extent new I will briefly, in section 2, discuss some features of the model before I turn to the developments of the last year.

There are some interesting and rather subtle effects on the soft hadronisation from the emission of soft gluons which will be discussed below in section 3. During the last year the Lund model has also been amended with a mechanism for baryon-antibaryon production and I will discuss these features in section 4 below. Finally in section 5 I would like to say a few words on ordinary low $p_\perp$-interactions, in particular the way we figure the reaction mechanism leading to the so-called fragmentation model suggested by the Lund group in 1977.

Due to time-limitations I will not present detailed predictions in section 5, although I am happy to see that most of what our experimental friends have presented here is in very good agreement with our results, but instead be satisfied with a few general remarks.

2. The Lund soft hadronisation model and the stateconfiguration of the string system

The main features of the Lund model for the soft hadronisation of quark and gluon jets are as follows. If a quark and an antiquark go out with large momenta, a linear colour force field (or colour flux tube) is stretched between them. This field breaks into pieces (mesons) by the production of quark-antiquark pairs. The force field is assumed to have no excited transverse degrees of freedom. This property which can explain the observed polarization of inclusively produced Λ-particles [2], implies that transverse momentum is locally conserved. Thus the quark and antiquark in a produced pair have equally large and opposite transverse momenta $k_\perp$.

Massless quark pairs can be produced classically in one point and pulled apart by the field. The production of quarks with mass or transverse mass can be regarded as a tunnelling process, which gives a Gaussian distribution in $k_\perp$ and also a suppression of the heavier strange quarks.

In terms of the transverse mass $\mu_\perp = \sqrt{k_\perp^2 + \mu^2}$ of the produced quark one obtains the resulting probability distribution [3]:

$$dP = N \exp{-\left(\frac{\mu_\perp^2 \pi}{\kappa}\right)} d^2k_\perp \qquad (1)$$

where N is a normalisation constant and κ the constant strength of the force field (or the energy per unit length stored in the field). If κ is determined from charm spectroscopy one obtains a mean (meson) $p_\perp$ around 350 MeV/c.

For a detailed discussion of the space-time and energy momentum structure of the model I have to refer to the original papers [4]

There is a strict ordering of the mesons with regard to flavour (called rank) which on the average also agrees with the ordering in rapidity. However, ordering with regard to production time is Lorentz frame dependent; slow mesons are produced first in any Lorentz frame. In this sense it is an "inside-out cascade". Thus, e.g. with a large boost along the leading quark we come to a Lorentz frame in which the first rank meson is also produced first in time. We assume that the probability for a particular break up situation is determined by the available number of final states, and this implies an iterative structure as in cascade jet models [5]. This feature can be easily implemented in a Monte Carlo generation process [6].

A relativistically invariant generalization of a one-dimensional linear force field to three dimensions is provided by the dynamics of a massless relativistic string. On such a string it is possible to have a localized excitation, a kink, which carries energy and momentum and which moves with the velocity of light. Such a kink is pulled back by the string by a force which is twice as strong as the one acting on an endpoint quark. The kink thus acts much like a gluon. (In QCD with N colours the ratio between the forces on a gluon and a quark is expected to be $2/(1 - 1/N^2)$, which gives 9/4 for 3 colours and 2 for infinitely many colours.) We have used such a kink as a model for a gluon and calculated the gluon jet fragmentation with the same methods as used for the quark jets. The resulting pion distribution is shown in Fig. 1 together with the corresponding distributions for a quark jet.

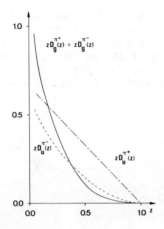

Fig. 1 Fragmentation function D_g^π for pions in a gluon jet.
For comparison the fragmentation of a u-quark into
π^+ and π^- is also shown.

The idea is thus that in the initial perturbative period, in,
e.g. an e^+e^--annihilation event, gluon emission corresponds in our
model to a kinklike excitation on the force field. In the following
soft hadronization the force field is stretched without any further
transverse excitations. How the stringlike field is stretched and
breaks into pieces (hadrons) in an $e^+e^- \to q\bar{q}g$ event is illustrated
in Fig. 2

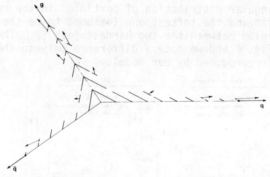

Fig. 2 The space-time development of a quark-anti-quark-gluon event.
The quark and antiquark move along the directions marked q and $\bar{q}$ and
are at the endpoints of a string field. The gluon is a pointlike
energy-momentum carrying piece of the string moving along the direction g, thereby causing a triangular shape of the outmoving string
field. The field breaks by the production of $q\bar{q}$-pairs and the directions of the final state mesons are marked by arrows when they become
independent entities. (Note that the slowest mesons in the cms are the
first ones to emerge and also take the largest pieces of the string.)

We note in particular that the relativistic invariance of the model implies that we avoid the selection of a fixed point in a particular Lorentz frame (e.g., the cms) to connect jets. A very essential feature is that the force field is stretched from the quark to the antiquark <u>via</u> the gluon. This implies that the mesons are produced around two hyperbolae in momentum space (cf. Fig. 3) , which leads

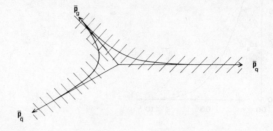

Fig. 3   The momentum space distribution of the final state
          particles which appear in the mean along two hyperbolae.
          The size of the hatched area indicates the size of the
          transverse momentum fluctuations in a string field
          without excited transverse degrees of freedom.

to correlations between multiplicity and angular distribution [7]. An asymmetry of the type predicted here is observed by the JADE-collaboration at PETRA. The QCD matrix element implies that usually the softest jet is the gluon jet. The JADE-group has studied the angular distribution of particles in the region between the hardest jet and the softest one (assumed to be the gluon jet) and in the region between the two hardest jets [8]. Their result is shown in Fig. 4 and we note a difference between the two regions which is well reproduced by our model.

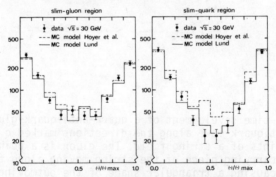

Fig. 4.   The angular distribution θ/θ_{max} of the particles in the
          region between the hardest and the softest jets,
          slim-gluon region, and between the two hardest jets, slim-
          quark region. The data are from the JADE-collaboration at
          PETRA [8].

288

The model also provides a natural cut off for the divergencies connected to soft and colinear gluons. For a colinear gluon the energy in the field between the gluon and the quark (or antiquark) is so small that it cannot break and produce a meson at the end. The first break will be on the other side of the kink, and the quark and the gluon will look just as a single quark-jet. In this way we obtain also a smooth transition between the 3-jet and 2-jet events. Similarly for soft gluons there will only be a small "bump" on the field configuration. For more detailed descriptions cf. ref. [9].

In order to give a quantitative meaning to these statements it is necessary to discuss in some detail the determination of the different possible states for the string systems. We will assume that the state configuration of the string system is determined from perturbative QCD during a brief space-time period (of the order of $1/\Lambda_{QCD} \sim 0.5$ fm). Gluon emission then corresponds to kink-like excitations on the resulting force field. From then on, during the soft hadronization, there are no more transverse excitations and the force field will stretch in a smooth way and break up as described before. Then we obtain the cross section for $e^+e^- \rightarrow q\bar{q}g$ in terms of scaled energy variables in the CM-frame $x_1 = 2E_q/W$, $x_2 = 2E_{\bar{q}}/W$, $x_3 = 2E_g/W$ (with the obvious normalization $\Sigma x_i = 2$) neglecting quark mass corrections [10]:

$$\frac{1}{\sigma_o} \frac{d^2\sigma}{dx_1 dx_2} = \frac{2}{3} \frac{\alpha_s}{\pi} \frac{x_1^2 + x_2^2}{(1-x_1)(1-x_2)} \tag{2}$$

In [9] it is shown that the situations when there are one or more mesons related to the emitted gluons (from now on true hard gluon emission) can be disentangled by the requirements:

$$2\gamma \le x_i \le 1 - 2\gamma \qquad i = 1, 2$$

$$(1-x_1)(1-x_2) > \gamma (x_1 + x_2 - 1) \tag{3}$$

with γ defined by

$$\gamma = \left(\frac{M_o}{W}\right)^2 \qquad M_o \approx 2.5 \text{ GeV} \tag{4}$$

289

The requirements in eq.(3a) are usually referred to as "thrust-cuts" on collinear gluons while eq.(3b) defines a situation in which a centrally (in phase-space) emitted gluon is hard enough to permit related particle emission

It should be understood that the requirements in eqs.(3) and (4) are "theoretical" conditions, based on the properties of the Lund model. Many of the events which in that way are classified as 3-jet events are experimentally indistinguishable from 2-jet events. The experimentally observed rate of 3-jet events (which correspond to a scale $M_o \gtrsim 5$ GeV/c^2) on the 10-20 % level at 30 GeV [8,11] is, however, the same. As both particle production and transverse momentum transfer in connection with gluon emission are localized phenomena in phase-space (connected to a scale of 1-2 units in pseudorapidity), the precise value of the M_o scale is a matter of convention.

3. THE SOFT GLUON EFFECTS ON THE TRANSVERSE MOMENTUM

In this section we will discuss the infrared divergences associated with the emission of soft and collinear gluons.

We note that while the infrared cut off in QED is determined by the adjustable experimental energy resolution, in QCD it is fixed by the masses of the available hadrons. Therefore in order that a description of the strong interaction in terms of quarks and gluons should be meaningful, a necessary requirement is infrared stability. By this we mean that the effects of a soft or collinear gluon emission on the observable hadron momenta should vanish when one approaches the singularity. A branching of a single gluon into several gluons should not be observable in case all the original gluon energy ends up in one final state hadron. In e.g. e^+e^--annihilation events these requirements imply a continuous transition between 3- and 2-jet events (and between 4- and 3-jet events, etc.). The emission of a soft or collinear gluon should result in an event still essentially of quark-antiquark jet character and the transition between $q\bar{q}gg$ and $q\bar{q}g$ events should be continuous when e.g. the invariant mass of the two gluons becomes small.

In the Lund model this requirement is trivially satisfied when the gluons are treated as transverse excitations on a string-like force field. (In contrast, a model in which jets are assumed to fragment independently and then are joined in the origin in the CM frame is not explicitly infrared stable (or Lorentz invariant).)

We will again for simplicity consider a situation relevant to e^+e^--annihilation and in order to determine the state configuration of the string system we will use the first order matrix element in perturbative QCD for $e^+e^- \rightarrow q\bar{q}g$ as discussed above. In order to obtain a better understanding of the effects of soft central and collinear gluon emission (a precise partition is given below) we consider the situation in the particular coordinate system where the gluon is moving transversely to the direction of the back-to-back motion of the $q\bar{q}$-pair (Fig. 5).

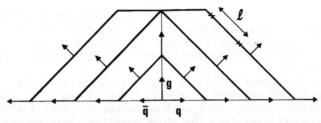

Fig. 5

The gluon transverse momentum is during the classical motion of the string system transferred to two string segments which after the gluon is stopped will move one in each direction. It is noteworthy that the transverse momentum carried by a string piece like the one marked out in the figure, is independent of the size of $k_\perp$ and only depends on the length of the string segment [4,12], i.e. in particular is proportional to the longitudinal (the projection on $q\bar{q}$-direction) size. In the example of Fig. 5 this proportionality factor is $\kappa/\sqrt{2}$.

As we have remarked upon above, the longitudinal state of a string piece corresponding to a meson m is Lorentz contracted, i.e. proportional to m/E with E the energy in the system where the gluon longitudinal momentum is zero. Therefore a gluonic disturbance will only affect mesons with rapidities y close to the gluon rapidity y_0, and we conclude that any such disturbance will fall of like $\exp(- |y-y_g|)$ for $|y-y_g| \gtrsim 1$. We note that a rapidity range of about 2 units is also the typical one in the Lund model for very hard gluon emission. If we e.g. would plot the multiplicity as a function of the angular variable pseudorapidity (defined in a coordinate system like Fig. 5), the multiplicity would increase above the general background (related to the q- and $\bar{q}$-jets) inside an angular range of typically 1 unit in pseudorapidity on each side of the gluon rapidity. The central value would increase and the half-width of the "bump" would decrease with increasing gluon energy but "the range of the disturbance" is independent of this energy. A similar structure will appear in e.g. an energy flow investigation.

For soft central gluons we obtain essentially that a gluon with transverse momentum $\vec{k}_{\perp g}$ and rapidity y_g will effect mesons in the rapidity range $(y,y+dy)$ like

$$\vec{k}_{\perp g} \frac{N'}{\cosh(y-y_g)} \, dy \qquad (5)$$

with N' a normalization constant.

The arguments given above and the fact (derived in ref. [4]) that the median production time at rest for a meson of rest-size ℓ is given by ℓ/c implies that gluons emitted at rapidities more than about a unit apart will not interfere. If there are several (soft) gluons emitted into a certain angular range, the effect will be very similar, although the straight string segments of Fig. 5 will be modified into several broken segments, i.e. for many soft gluons be "rounded off".

There is a further effect on the final state particle distributions which is not noticeable in Fig. 5 due to the particular coordinate system chosen. There will be an obvious recoil effect on the q- and q̄-particles from the gluon emission. This recoil contribution will be distributed among the fast particles along the q- and q̄-directions (cf the discussion above in connection with Fig. 2) in accordance with the fractional energy momentum [4], i.e. for the gluon emission discussed above as

$$ -\vec{k}_{\perp g} dz \approx -\vec{k}_{\perp g} \, dy \, exp(|y| - y_m) \qquad (6) $$

with y_m the maximum rapidity of a meson

$$ y_m = ln\left(\frac{W}{m}\right) \qquad (7) $$

Thus the particles in the fragmentation region $|y| \simeq y_m$ are affected by the soft central gluon emission. On the other hand it is obvious that the combined effects from eqs (5) and (6) will imply that the collinear gluon emission, i.e. $y_g \simeq y_m$ will give only minor contributions to the meson spectra.

Actually, a Monte Carlo study shows that the combined effects of eqs (5) and (6) imply for gluons collinear to the quark that in situations with $x_2 > 1 - 2\gamma$ (in terms of the variables defined in connection with eqs. (3)) the $p_\perp$-effects are negligible compared to the general zero-point Gaussian fluctuations in eq. (1). This implies that the only rapidity range over which we have to consider soft gluon radiation is approximately described in the cms by

$$ |y_g| \lesssim \frac{1}{2} ln\left(\frac{W^2}{6M_o^2}\right) \qquad (8) $$

(i.e. independently of the $k_\perp$ of the gluon). Further we will limit the $k_\perp$ of the soft gluons to $k_\perp < M_0$ (for larger $k_\perp$ we are back in situations corresponding to gluon radiation with associated particle production as described above).

In accordance with ref. [13] we will assume that the emission of such soft gluons can be treated as a Poissonian process. Then the probability to obtain e.g. a resulting quantity $\vec{p}_\perp$ from many increments $\vec{k}_\perp$ each determined by a Poissonian emission governed by dn̄ is given by [13,14]:

$$\frac{d^2P}{d^2p_\perp} = \frac{N}{(2\pi)^2} \int d^2b \, exp(i\,\vec{p_\perp}\vec{b}) \, exp\{-\int d\bar{n}(1 - exp\,i\vec{k_\perp}\vec{b})\} \qquad (9)$$

with N a normalization constant, depending upon the allowed $k_\perp$-range. The quantity $d\bar{n}$ in eq.(9) corresponds to the mean number of gluons emitted, and eq.(3) implies the following result for $d\bar{n}$

$$\frac{d\bar{n}}{dk_{\perp g} \, dy_g \, d\phi} = \frac{2\bar{\alpha}_s}{3\pi^2} \frac{1}{k_{\perp g}} \qquad (10)$$

with ϕ the azimuthal angle).

We have used the Lund Monte Carlo [6] to study the effects of soft gluon emission on the event structure in e^+e^--annihilation according to the principles outlined above. For central hadrons the main result is an increase in the effective Gaussian width in (1) from $\simeq 0.35$ to $\simeq 0.42$ GeV/c. The non-Gaussian nature of the soft gluon contribution is effectively masked by $p_\perp$ from the tunneling process, from particle decays and from true hard gluons.

The combined recoil effect on the fragmentation region hadrons corresponds to the two jets not being back-to-back. Since the original jet axis is not known in $e^+e^- \to q\bar{q}$, it is not meaningful to specify exactly how the recoil is shared between the two sides.

The total recoil from soft gluons is approximately 800 MeV/c. The shape of the recoil $p_\perp$ distribution will be of an almost Gaussian character, with significant deviations only at small $p_\perp$. In passing we note that the resulting change in event structure is not all that different from the one obtained with the Field-Feynman recipe [5] of giving also the primary quarks a $p_\perp$ with respect to the jet direction. The sizes of these effects depend on the value of the effective coupling constant $\bar{\alpha}_s$ in eq.(10). We have, remembering our assumption in section 2 that the state configuration is determined during a brief space-time period, used the value $\bar{\alpha} \approx 0.2$ which seems relevant to $Q^2 = W^2$ values in the high energy PETRA range. In case we used instead a value of $\bar{\alpha}_s$ relevant to $Q^2 = M_0^2$ we obtain effects of about double the size.

In leptoproduction the direction of the outgoing, struck quark is known. It is then meaningful to separate what is traditionally called primordial $k_\perp$ into two pieces [15]. One is the true primordial $k_\perp$ due to Fermi motion inside the proton. This should give a $k_\perp$ of the order of 500 MeV/c compensated in the target fragmentation region. The other contribution is due to soft gluon emission from the struck quark, which is compensated rather centrally. The size of the latter $k_\perp$ contribution depends on the cuts used for

293

hard gluons, but it will generally be larger than the true primordial $k_\perp$. In Fig. 6 we show the result of including the soft gluons [16] together with a corresponding calculation when the soft

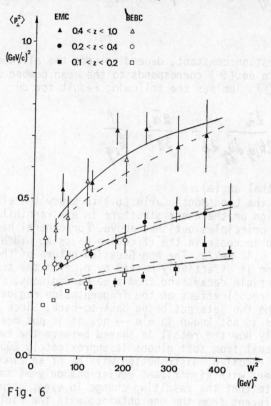

Fig. 6

gluons are neglected and a large value of the primordial $k_\perp$ is used
$(\langle k_\perp^2 \rangle = 0.6 (GeV/c)^2)$.

4. BARYON-ANTIBARYON PRODUCTION

In this section I will present a simple model for baryon-antibaryon production in quark and gluon jets. The basic assumption is that a diquark-antidiquark pair (a colour antitriplet-triplet) can be produced in a colour force field in a way similar to the production of a quark-antiquark pair. The diquark colour antitriplet is treated as one unit and the suppression of baryon-antibaryon production, as compared to the meson production, is determined by the larger diquark mass compared to the quark mass. It is a "minimal" model in the sense that while it takes due account of the symmetry of the baryonic quark states and other kinematical constraints, it has a minimum of new dynamical assumptions.

The production probability of heavy quark flavors in a colour field with no transverse excitation is determined by eq.(1). We note that eq.(1) implies a suppression of strange quarks and even more for charmed quarks as

$$u : d : s : c \sim 1 : 1 : 0.3 : 10^{-11} \tag{11}$$

We will assume that the production of a diquark and an antidiquark (which is also a colour antitriplet-triplet pair) can be treated as a similar tunneling process, implying that the production probability is determined by the same expression as for a quark-antiquark pair (eq. (1)).

This assumption is consistent with the observed baryon production rate and also gives the following observable properties, testable in future experiments:

1. A common Gaussian $p_\perp$ spectrum is obtained for all primary mesons and baryons. It should be remembered, however, that further $p_\perp$ contributions come from hard and soft gluons [9] and that decays of resonances modify the spectra differently for baryons and mesons.

2. A baryon and an antibaryon will be neighbours in rank and thus close in rapidity.

3. Equation (1) implies that for $m > \sqrt{\kappa}/\pi \approx 250$ MeV a small increase in m will give a large change in the probability. Thus strange diquarks will be much suppressed compared to nonstrange diquarks. This implies that e.g. a Λ is more often produced together with a K and an $\bar{N}$ than together with a $\bar{\Lambda}$. It also implies that production of Ξ and Ω is strongly suppressed.

All these properties would be modified if the diquarks were produced in a stepwise manner with one quark produced first and the other later (cf ref. [3]). In such a picture more dynamical assumptions are needed. Our attitude is to compare the available data to the prediction of a simple model in order to find out if there are experimental features which indicate the need for new dynamics.

For the values of diquark masses used in our approach I refer to ref. [17].

We also note the important fact that a baryon is a symmetric system of three quarks. A basic property of the Lund jet model is the assumption that the production of a certain $q\bar{q}$ pair is determined by the density of available final states. When a diquark joins a quark to form a baryon, we therefore weight the different flavour and spin states by the probability that they form a symmetric 3-quark system. This implies that all states in the 56-multiplet become equally probable. For the relative probabilities for a diquark and a quark to join into a baryon cf. ref. [17]. The resulting picture is in very good agreement with the data both from e^+e^--annihilation (4-30 GeV) and leptoproduction (EMC data) as can be seen from Figs 7 and 8. An interesting fact is that in the Lund gluon model described above there is a larger production of baryon-antibaryon pairs in a gluon jet than in a quark jet. We note that as the gluon drags along the force field, the first rank hadron

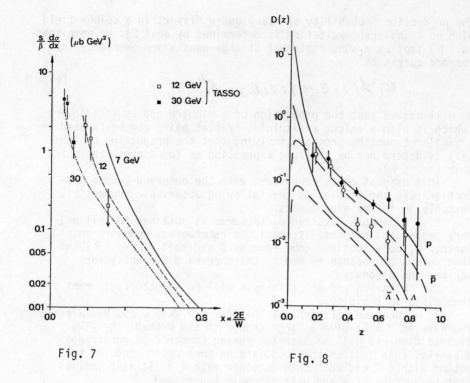

Fig. 7   Fig. 8

of a gluon jet will be a baryon with the same probability as anywhere else inside the colour field. In a quark jet, however, the first rank hadron always contains the original quark. Therefore baryons will be first rank hadrons in quark jets roughly half as often as higher rank hadrons or first rank hadrons in gluon jets. Hence, whereas the probability to have a p or a $\bar{p}$ as fastest charged particles varies between 8% (u) and 6.5% (d) for quark jets, it is 10% for gluon jets.

This means that we do expect a higher baryon fraction on "onia" resonances($J/\psi,\Upsilon,t\bar{t}$?) than off them, but that the difference should become less pronounced at higher energies. However, if the indications from the DASPII-group of a considerable difference in baryon production on and off the Υ-resonance is substantiated, there is a need for new physical mechanisms of baryon production in gluon jets (our results indicate that the fraction of p+$\bar{p}$ among the charged particles should rise from 6% off to 8% on the Υ-resonance). I would finally like to mention the implications for polarization from the model, although space does not permit an analysis [17].

5. LOW-$p_\perp$-PHYSICS

In 1977 the Lund group suggested, essentially on phenomenological grounds, the so-called fragmentation model i.e. that the final state particle distributions in low $p_\perp$ hadron physics should be similar to a weighted mean of the corresponding valence quark distributions as measured in hard processes.

The fact that the single particle pionic distributions (or more generally the charged particle distributions which are completely swamped by the pions) are similar, is of course nice. However, anybody who knows the mean multiplicity and the mean $p_\perp$ can, using the time-honoured longitudinal phase-space models, do a good fitting job on these distributions. In order to learn any dynamics from the final state particles it is unfortunately necessary to go to the less frequently produced strange particles or, even better, the baryons (antibaryons). For the iterative cascade jet models presently in fashion (the Lund model presents a semi-classical justification) it is necessary to investigate the rank neighbours in order to pin down any dynamics.

Due to the suppression of strangeness a strange and antistrange particle found in an event (or even more a baryon-antibaryon pair) are in general such neighbours. All kinds of information is interesting in this respect - I would like to challenge anybody to present an understanding of e.g. K^--spectra in proton fragmentation regions or to predict the correlations in $p_\perp$ azimuthal angle or rapidity for a baryon-antibaryon pair without extensive modifications of a longitudinal phase-space model or even an iterative diquark jet model. Further, one should also be able to understand the sizeable polarisation effects presented at this conference. I would like to suggest also investigations of relative polarisation effects in e.g. $\Lambda\bar{\Lambda}$-production for ΛK^*-states etc. as interesting future possibilities.

Nevertheless, in case fragmentation models should continue to be successful, it is necessary to understand the basic dynamical reason why they work. In particular it is necessary to pinpoint the fact that in this scheme there is no energy-sharing among the valence constituents in the final state - one of them is supposed to be carrying all the energy of the impining hadron. We note, however, that in the framework of the Lund model the final state particle energies and masses are taken from a linear confining force field. Therefore it is only necessary to arrange the transfer of energy to the force field in such a way that one of the valence constituent quantum numbers is at the endpoint of the force field when all the energy has been deposited in the field. We have in ref. [18] presented a semi-classical reaction mechanism, where the main ingredient is that the hadrons are extended objects and that the valence constituents, therefore, in general are apart in space-time. Therefore if two hadrons partly overlap only a piece of each is initially affected by e.g. exchange of gluons etc. like in the Low-Nussinov model. Causality implies that the remainder of the hadrons

will move under the influence of the surrounding confining vacuum
pressure during the time necessary for a disturbance (i.e. a message)
to reach them. For a relativistically covariant model this causality
requirement corresponds to a very long (time-dilated) development
time, during which the impinging hadronic energy is transferred to
the confining field. To be more precise, we note that the colour of
a quark is due to asymptotic freedom, not well localised but rather
distributed like a blob around the quark. A meson can be imagined
as two blobs confined in a bag as in Fig. 9. It is thus e.g. red in
one end, antired in the other, and "white" in the middle.

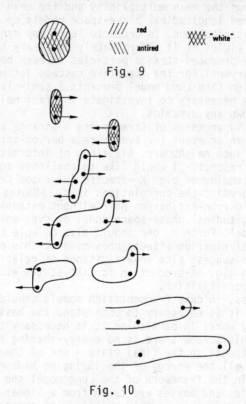

Fig. 9

Fig. 10

In Fig. 10 we show two meson bags approaching each other. (They are
Lorentz contracted.) If the coloured parts overlap, the two bags
can unite to form one larger bag. This bag is stretched to a colour
flux tube as in Fig. 10. The bag pressure, or alternatively the
force in the stringlike colour force field, retards the meson remnants.
As seen in Fig. 10 the colliding "blob" will lose its momentum first,
and when it is stopped the other "blob" will be retarded. The bag is
thus stretched to the same length as if all the energy were carried
by one of the valence quarks. Although the dynamics is very different,

298

when the energy in the colour field is divided into final state
hadrons we obtain essentially the same result as for a quark jet,
with one of the original valence quarks in the first rank meson.
   The model can be made sufficiently precise to predict cross
sections via an overlap function formalism, diffraction via
essentially the Good-Walker mechanism [19] but all these things would
lead us too far outside the time limits.

References

1. The Goa Conference Proceedings 1979
    The Bruges Conference Proceedings 1980
2. B. Andersson, G. Gustafson, G. Ingelman, Phys. Lett. 85B (1979)417
3. J. Schwinger, Phys. Rev. 82 (1951) 664
    A. Casher, H. Neuberger, S. Nussinov, Phys.Rev. D20 (1979) 179
    E. Brezin, C. Itzykson, Phys. Rev. D2 (1970) 1191
    B. Andersson, G. Gustafson, T. Sjöstrand, Zschr. f. Physik
    C6 (1980) 235
4. B. Andersson, G. Gustafson, C. Peterson, Zschr. f. Physik
    C1 (1979) 105
    B. Andersson, G. Gustafson, Zschr. f. Physik C3 (1980) 22
5. A. Krzyvicki, B. Petersson, Phys. Rev. D6 (1972) 924
    F. Niedermeyer, Nucl. Phys. B79 (1974) 355
    B. Andersson, G. Gustafson, C. Peterson, Nucl. Phys. B135 (1978) 273
    R.P. Feynman, R.D. Field Nucl. Phys. B136 (1978) 1
6. T. Sjöstrand, B. Söderberg, LU-TP 78-18
    T. Sjöstrand, LU-TP 80-3
    G. Ingelman, T. Sjöstrand, LU-TP 80-12
7. B. Andersson, G. Gustafson, T. Sjöstrand. Phys. Lett.
    94B (1980) 211
8. A. Petersen, contribution to XVth Rencontre de Moriond, JADE-
    Collaboration, Phys. Lett. 101B (1981) 129
9. B. Andersson, G. Gustafson, T. Sjöstrand, LU-TP 81-4
    T. Sjöstrand, LU-TP 80-3
10. J. Ellis, M.K. Gaillard, G.G. Ross, Nucl. Phys. B111 (1976) 253
11. P. Söding, Rapporteur talk at the EPS International Conference
    on high energy interactions, Geneva 1979
    B. Wiik, Rapporteur talk at the Madison Conference 1980
12. X. Artru, Orsay preprint LPTHE 78/25
13. G. Parisi, R. Petronzio, Nucl. Phys. B154 (1979) 427
14. Yu Dokshitzer, D. Dyakonov, S. Troyan, Proceedings of the
    13th Winter School, Leningrad (1978)
15. B. Andersson, G. Gustafson, G. Ingelman, T. Sjöstrand LU-TP 80-6
    (to be published in Zschr. f. Physik C)
16. B. Andersson, G. Gustafson, G. Ingelman (to be published)
17. B. Andersson, G. Gustafson, T. Sjöstrand LU-TP 81-3
18. B. Andersson, G. Gustafson, I. Holgersson, O. Månsson, Nucl.Phys.
    B178 (1981) 242
19. G. Gustafson, LU-TP 81-1, Contribution to IXth International
    Winter Meeting on Fundamental Physics, Sigüenza, Spain

DISCUSSION

McLERRAN: UNIV. OF WASHINGTON: Does your model have a prediction for the production of glueballs?

ANDERSSON: No.

PIETRZYK, CERN: Can you tell us some more differences for the case where you have on the end of your string $3\bar{3}$ or $8\bar{8}$ except for having more baryons?

ANDERSSON: No, actually I would say myself, you know, that evidently the spectrum would become different in principle. But due to, once again, what I said before (that energy is not very large in present day gluon jets) it's not so easy to see it. Let me show you in this spectrum. If you look upon this spectrum, what can you see? Well, you can see over here that this is a u quark going to a π^+, this is a d quark going to a π^-. This is a gluon going to either π^+ or π^-. However, this is a scaling variable. If you only have an energy of, let's say, 10 or something like that, you are only seeing down to .1, which is over here. You can see that you don't see very much difference. If you are able to get to 100 then you will see essentially more particles produced in a gluon jet than in a quark jet. But that is not very easy to see. Then I also have the prediction of how particles will be distributed in p_T. Finally, maybe I should say that one prediction which we think is nice is that the baryon spectrum in connection with lepto-production events will be influenced by the fact that baryons are produced also in the gluonic jets. That means that the p_T spectrum for baryons as a function of θ (if you take $< p_T >$ as a function of θ) will be essentially flat for baryons, while for pions it shows this typical going up and then going down. There is a difference which is a direct prediction of the model in that connection.

PIETRZYK: Do you expect to have any difference between a gluon jet and a hadronic system in an 8 color state composed of quarks, etc.?

ANDERSSON: What one should expect in a gluonic jet is, as I say, that in principle you should get more particles but you need very large energies in order to see it. However, in low p_T hadronic physics (and I'm sorry I never got time to talk about that) it seems to be working anyway. What I should say in that connection was rather that the particular model we have been working on in that case is different from the ones which people talked about as gluon models. They are also different from what is called the Nussinov-Low mechanism in which you have a gluon in the middle. We expect you to see essentially something very similar to triplet jets in each end.

BIALAS, KRAKOW: Is it the right question to ask about the transverse size of your jet or tube in fermis?

ANDERSSON: Well, yes. It's 1 GeV per fermi which essentially determined it and .3 is about the order of 2 fermis, isn't it?

BIAŁAS: Aha, so this will be just 1 GeV expressed in terms of fermis.

ANDERSSON: That's essentially it. You expect something of that order, you know. That's an ordinary constant of nature, it seems.

ALAM, VANDERBILT: I'd like to know if you have any specific predictions about the number of protons or Λ's per event on the Υ as compared to the continuum region.

ANDERSSON: That's a very good question. Thank you very much. I will have. At the present time the data which are around seem to indicate that there are very many baryons which are produced on the Υ as compared to off the Υ. We find more baryons but we do not find the number which you see in the literature. So if that number should be confirmed by further experiments this model is out.

ALAM: I would like to add that we have made such measurements at CLEO and we do see some evidence for an enhanced proton and Λ production at the Υ as compared to the continuum. But our statistics are poor and certainly it is not as large as the results from DORIS.

ANDERSSON: Ours was exactly the same. Our statistics was not as poor because we have done it by Monte Carlo.[laughter] But you know I think we should go together and we should have a clean look upon this. You know if I would say 16% now or if I say 20 that might depend upon the way you look upon it and the way you really do it. So let's talk about that. But I can say directly that if the present day experimental data which have been published would stand up (I don't know whether they have been published, by the way, but anyway they have been around.) then this model is out and nature is much more interesting.

ALAM: Well, I'll present some results from CLEO later.

MORRISON, CERN: I'm not quite clear about how you actually use charm in your model, and this question of causality comes in. But you say that in each coordinate system the slowest [i.e. local] particles combine first. Let me take two examples. Suppose in neutrino interactions, a quark, which goes by itself, is red, and suppose all the neighboring quarks are not anti-red. What happens?

ANDERSSON: You know, to be precise, neither you nor I can see a quark and its color. Okay? What we are claiming is only that inside this kind of force field, which is essentially a classical force field, you have lots and lots and lots and lots of very small gluons. Okay. Inside this, what we are saying is only that there are so many possible gluons which can produce quark-antiquark pairs that there is no difficulty for anybody to find a partner. The only difficulty is essentially that the particular break (if you make a particular break) would mean that you get so many final states, and we are weighting each particular break (that's the idea in the model) by the number of final states that would be available by making that break.

MORRISON: Do you mean the gluon knows this quark is a red and the gluon says "Therefore I must produce an anti-red."?

ANDERSSON: We could start a very philosophical discussion now about what is quantum mechanics and what is not quantum mechanics. But in principle I think myself that the Copenhagen interpretation is that nature knows.

MORRISON: Nature does know? I see.

ANDERSSON: But I should say, maybe, that charm is very strongly suppressed in this soft hadronization. It's down to a factor of 10^{-11} as compared to .3 for an $s\bar{s}$ pair. So you are never in our model producing charm anticharm by a soft process. The way which we can do it is either by a kind of Drell-Yan-like process, something like this gluon fusion model, or we can also do it by diffraction. I would be very willing to tell you more about that but I think my Chairman would be very, very sorry.

HWA, OREGON: I don't have a question but I have a comment. Hopefully, the experimentalists looking for gluon jets can help me. I don't know how easy it is (I'm sure it is not so easy) to look for

η' in e^+e^- experiments. If η' is to be interpreted as a glueball, then η' should be produced more profusely in the gluon jet than in the quark jet on account of the triple-gluon coupling.

ANDERSSON: I'm not an expert myself on glueballs, you know, Rudy. What I would say is essentially that we built this model in order not to have any glueballs around. It might very well be that there are lots of glueballs around. One of my friends over there does know a lot about it.

PETERSON, SLAC: I have investigated this problem by having an octet string and making a Monte Carlo generation allowing for η''s etc. It turns out that unless you can detect the η''s by themselves it is very hard to disentangle those jets from the standard quark jets. A lot of assumptions go into it as far as primordial functions and k_T are concerned.

DUAL TOPOLOGICAL UNITARIZATION APPROACH
TO
SINGLE PARTICLE SPECTRA

K. Hirose

Institute of Physics, Teikoku Women's University
Moriguchi, Osaka 570, Japan

T. Kanki

Institute of Physics, College of General Education
Osaka University, Toyonaka 560, Japan

ABSTRACT

We analyze the recent Serpukhov data on inclusive single particle spectra in the framework of the dual topological unitarization
models, the original one as well as the modified version. We calculate multiregge branching process accurately by taking account of
the longitudinal momentum conservation. It is found that the original model is unable to reproduce the large increase in π^+/π^- and
K^+/K^- ratio for $x \to 1$. In order to find a model working both in the
central and the fragmentation regions, we study the modified version
in which the first branching at the end-vertex of the dual chain is
treated differently from others. This modified model resolves the
difficulty on the above ratios and realizes Ochs' ansatz. The resonance productions play the important role for the single particle
spectra and are included here. Interesting informations on the
mechanism of multihadron productions are obtained.

1. INTRODUCTION

For small p_T multihadron productions, many features in the central region have been well understood in the framework of the dual
topological unitarization (DTU) model[1,2] and its modified version.[3]
On the other hand, recent experiments have shown that in the fragmentation region, the free parton picture, associated with the recombination model,[4] works very well.[5] Although we write quark lines
in DTU model, quarks are not free; the string assumed implicitly by
the duality carries momenta from one quark line to another. Experimental cross sections $x d\sigma/dx$, however, show no sizable change for
the transition from the central to the fragmentation region. It is
therefore important to find the bridge connecting these two models
smoothly.

We have recently shown[6] that the DTU model is in the marked
disagreement with the data of the particle spectra in the fragmentation region. In order to find a model working in both the central
and the fragmentation regions, we have studied[7] the modified version
of DTU. We introduced flexible functions to the end-vertices of the
DTU ladder keeping other parts of the ladder in the original multiregge form. This modified DTU model resolves the above disagreement
and also realizes Ochs' ansatz, i.e., the similarity between the

asymptotic behaviors for x→1 of the single particle spectra and of the structure function of the incident hadron. By fitting the existing data, we have further extracted informations on the mechanism of the fragmentations and the whole dynamics of the multiparticle processes. The role of the resonance productions is confirmed to be basically important. In this note, we would like to present the review of these investigation.[6,7] This would give some guide to more refined future theory being able to explain both the pionization and the fragmentation.

We do not deal with the socalled dual sheet model[3] nor the jet universality. Instead, we shall calculate multiregge branching process very accurately by taking account of the (longitudinal) momentum conservation. We would like to study how well our modified multiregge bootstrap (DTU) scheme works.

2. FORMALISM

Let us consider the DTU diagram shown in Fig.1. Here, the inside part between dotted lines at a and b is identified with the original DTU amplitude for the meson-meson collision. But we modify the end-vertex functions corresponding to the outsides of these lines. [The motivation for the modification will be seen in the first paragraph of Sec.4.]

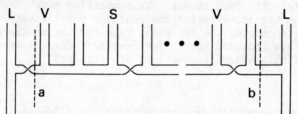

Fig.1. A DTU diagram

2.1 The original DTU model

In order to calculate the inclusive cross sections in the original DTU model, we employ the method of the bootstrap integral equations,[6,8]

$$n^i(x) = n_L^i(x) + \int_0^{1-x} \frac{dy}{y} n_L^i(y) n^i(\frac{x}{1-y}), \quad (i=T,R) \quad (1)$$

where

$$n^i(x) = \frac{1}{\sigma^i} \int dp_\perp^2 E \frac{d\sigma^i}{dp^3}, \quad (2)$$

and T, R indicate the contributions from whole diagrams of Fig.1 and from the reggeon diagrams respectively, and $\sigma^T = \sigma^P + \sigma^R = as^{\alpha_P-1} + bs^{\alpha_R-1}$.

Fig.2 is the schematic illustration of (1): $n_L^i(x)$ correspond to the first term of the r.h.s. of this schematic equation and represent the <u>leading particle</u> distributions and are parametrized as[6,8]

$$n_L^T(x) = 2g^2 x^\delta (1-x)^{\alpha_P - 2\alpha} \quad \text{and} \quad n_L^R(x) = g^2 x^\delta (1-x)^{\alpha_R - 2\alpha}, \tag{3}$$

where $\alpha_T \equiv \alpha_P$, α_R and α are the intercepts of output pomeron, output reggeon and input (internal) reggeons, respectively. $n_L^i(x)$ satisfies the normalization condition,

$$\int_0^1 \frac{dx}{x} n_L^i = 1, \tag{4}$$

which shows that there is one and only one leading particle.

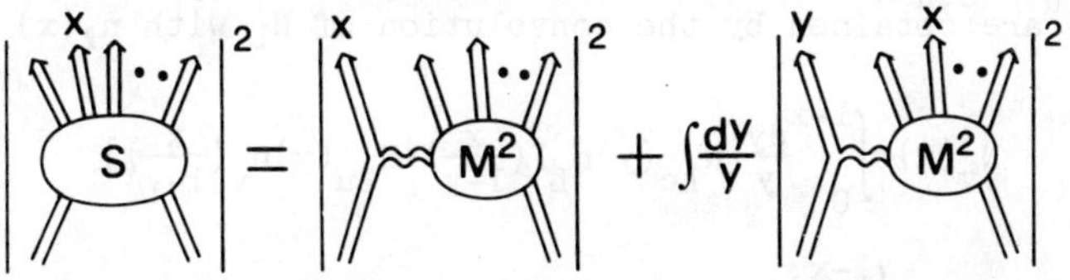

Fig.2. The schematic illustration of Eq.(1).

Excellent property of this scheme is that the solutions of (1) satisfy the (longitudinal) <u>momentum conservation</u> exactly. This is the difficult problem in the ordinary calculations in the rapidity space. When $\delta=1$, the insertion of (3) into (4) readily leads the famous <u>Lee-Veneziano bootstrap</u> condition[1] of the DTU chain. Thus the DTU chain is described by $\delta=1$. The (bare) pomeron intercept $\alpha_P=1$ is derived by familiar bootstrap condition $\alpha=\alpha_R(=1/2)$, which leads to $n_L^T(x)=2g^2 x=x$ ($\because g^2=1/2$). The leading particle is likely to have large momentum! Subsequent use of the iteration in (1) leads the strong ordering of rung-momenta on average.

In order to extract the role of the valence quarks, we consider the rung (v) as another valence-sea(v-s) combination. The distribution of this rung (meson), $n_v(x)$, satisfies

$$n_v(x) = \int_0^{1-x} \frac{dy}{y} n_{Lc}(x) n_L^T\left(\frac{x}{1-y}\right) + \int_0^{1-x} \frac{dy}{y} n_{Lu}(x) n_v\left(\frac{x}{1-y}\right), \tag{5}$$

where $n_L^T(x)=n_{Lu}(x)+n_{Lc}(x)$ [u=uncross and c=cross]. The contribution by sea-sea (s-s) combination is obtained by $n_s(x)=n^T(x)-n_L^T(x)-n_v(x)$. The solutions of (1) [for the T-sector] and (5) are easily obtained by using Mellin transformation technique:

$$n^T(x)=1, \quad n_v(x)=x^{1/2}(1-x^{1/2}), \quad n_L^T(x)=x, \quad n_s(x)=1-x^{1/2}, \tag{6}$$

for $\alpha=\alpha_R=g^2=1/2$. These expressions will be used for the inside part of the diagram in Fig.1 [the modified chain].

2.2 The modified DTU model

In order to modify the end-vertex, we provide the following flexible forms for the leading particle terms as an input [We denote, by $N_\ell(x)$, those quantities in the modified model which correspond to $n_\ell(x)$ in Sec.2.1.];

$$N_{Lu}(x)=\xi B(\delta_u,\eta_u+1)^{-1}x^{\delta_u}(1-x)^{\eta_u},$$

and

$$N_{Lc}(x)=(1-\xi)B(\delta_c,\eta_c+1)^{-1}x^{\delta_c}(1-x)^{\eta_c},$$

(7)

where $\delta_{c,u}$, $\eta_{c,u}$ and $\xi(0\leq\xi\leq1)$ are the adjustable parameters. $N_v(x)$ and $N_s(x)$ are obtained by the convolution of N_L with $n_\ell(x)$ of (6):

$$N_v(x)=\int_0^{1-x}\frac{dy}{y}[N_{Lc}(y)n_L^T(\frac{x}{1-y})+N_{Lu}(y)n_v(\frac{x}{1-y})]$$

(8)

$$N_s(x)=\int_0^{1-x}\frac{dy}{y}[N_{Lc}(y)(n_v(\frac{x}{1-y})+n_s(\frac{x}{1-y}))$$

$$+N_{Lu}(y)(n_L^T(\frac{x}{1-y})+n_s(\frac{x}{1-y}))].$$

(9)

2.3 Resonance productions and decays

A rung of the DTU ladder represents both stable and unstable hadrons. $N_\ell(x)$ relates to the direct productions of stable hadrons and the productions of resonances. For resonance decays, we shall take the following simple picture:

(1) Only the vector [nonet] and the decuplet baryon productions are taken into account. All other resonances are disregarded. The production rate of vector mesons from the L, s and v-rungs are denoted by $C_{L,s,v}$. We assume $C_s=C_v\equiv C\neq C_L$. For baryon, we follow the SU(6) scheme for the ratio of the octet and decuplet productions.

(2) For simplicity, all resonance decays are treated as two-body decays* with the isotropic angular distributions in their rest system. We neglect the $s\bar{s}$ production in the decay vertex. Thus, for the indirect process, $N_\ell(x)$ are replaced by $R_\ell(x)$:

$$R_\ell(x)=\int_x^1\frac{dz}{z}N_\ell(x)\gamma(\frac{x}{z}), \quad [\ell=s,v \text{ and } L].$$

(10)

* This assumption would not be so bad because three body decays are rare, except for ω. Since we will discuss only charged mesons, our treatment would not lead much errors for ω-decay.

For the isotropic decay, we have $\gamma(x)/x \sim \theta(\Lambda-x)/\Lambda$, where $\Lambda \sim 1$ for the vector meson and $\Lambda \sim (1-(m_N/m_\Delta)^2 \sim 0.42$ for the baryon decay.

2.4 Constituent model for initial hadrons

There are two ways to consider the connection between the DTU chain and the incident hadron. The first one, the one body model of the end-vertex, is the ordinary model where the incident hadron is treated like an "elementary". This model, however, gives the serious disagreement[6] with the experimental data. The second model, we take in this paper, is the <u>constituent</u> model; here, the momentum of only one of the constituent quarks (CQ) in the hadron is shared into the chain. And thus, y_{max} of the chain is limited by the rapidity of the CQ. Other CQ's are treated as spectators.

Now the single particle cross section can be written as

$$\left(\frac{x}{\sigma}\frac{d\sigma}{dx}\right)_{H \to h} = \frac{1}{n_i} \sum_{i=1}^{n} \int_x^1 dz F_{i/H}^{CQ}(z) G_i^h\left(\frac{x}{z}\right), \qquad (11)$$

where the sum is taken over all CQ's. For the CQ distributions,

$$F_{i/H}^{CQ}(z) = B(B_i+1, A_i+1)^{-1} z^{B_i} (1-z)^{A_i}, \qquad (12)$$

and use the values of A and B given in Table I. By introducing the quark flavor counting [the broken $SU(3)^*$ with the S-quark suppression factor F^2], we have the expressions for G_i^h [κ_i^ℓ and τ_i^ℓ are the flavor counting factors. Their explicit values are given in Ref.7.]:

Table I. The parameters of the structure functions

H	PROTON[10,11]		$K^-(K^+)$[9]	
i	u	d	$\bar{u}$ (u)	s ($\bar{s}$)
A	1.75	2.10	2.0	1.0
B	0.55	0.20	1.0	2.0

* We, however, assume the same intercept for all internal reggions.

$$\begin{pmatrix} G_i^{\pi,K} \\ G_i^{res} \end{pmatrix}_{dir} = \sum_{\ell=v,s,L} \begin{pmatrix} 1-C_\ell \\ C_\ell \end{pmatrix} \kappa_i^\ell N_\ell(x),$$ (13)

and

$$\left(G_i^{\pi,K} \right)_{indir} = \sum_{\ell=v,s,L} C_\ell \tau_i^\ell R_\ell(x).$$ (14)

3. THE BEHAVIORS OF THE CROSS SECTIONS FOR $x\to1$

Since Ochs has pointed out the similarity between the asymptotic behaviors for $x\to1$ of the cross section of $P+P\to\pi^\pm+X$ and of the proton structure function,[4] many people like to analyze the data by means of the recombination model which is expected to lead such similarities simply. Recent elavorate studies[10] of the recombination model, however, show that the derivation of the similarity contains much complications with regard to the evolution of the structure function. In our DTU model, we can easily derive such similarities, if we provide to give some property to the modified end-vertex functions. Let us discuss this very interesting fact.

3.1 The leading particle distribution

In our model, the leading particle distribution is given as

$$\left(\frac{x}{\sigma}\frac{d\sigma}{dx}\right)_{lead} \propto \sum_i \kappa_i^L \int_{1-x}^1 \frac{dz}{z} \int_0^z dy\, \delta(x-(1-z+y)) \times$$

$$\times F_i^{CQ}(z) N_L\left(\frac{y}{z}\right)\left(\frac{xz}{y}\right)$$ (15)

$$\sim x(1-x)^{\eta_c} \sum_i \kappa_i^L \int_{1-x}^1 dz\, z^{B_i-\eta_c-1}(1-z)^{A_i}\left(\frac{z+x-1}{z}\right)^{\delta_c-1}$$

$$+ \text{ (similar term for the "uncross" vertex)},$$ (16)

where we have used (7) and (12). When $\eta_{c,u}$ satisfy the condition

$$\eta_{c,u} > B_i,$$ (17)

the integrand of (16) becomes singular for $z\to0$ and then, for $x\to1$, the integral is dominated by the vicinity of the lower bound;

$$\left(\frac{x}{\sigma}\frac{d\sigma}{dx}\right)_{lead} \underset{x\to1}{\sim} (1-x)^\eta \int_{1-x}^1 dz\, z^{B_i-\eta-1} \sim (1-x)^{B_i}.$$ (18)

B_i is equal to the power of the spectator distribution, i.e.,

$F^{SP}(x)=F_i{}^{CQ}(1-x) \backsim x^{A_i}(1-x)^{B_i}$. This is nothing but the Ochs similarity. If $\eta<B_i$, the power in (18) depends also on δ and we have a complicated situation. Hence we assume (17) in what follows.

3.2 Valence term in the DTU chain

For the (v) rung in Fig.1, we insert (8) and (13) into (11), and obtain

$$\left(\frac{x}{\sigma}\frac{d\sigma}{dx}\right)_v \propto \sum_i \kappa_i{}^v \int_x^1 dz F_i{}^{CQ}(z) \int_0^1 \frac{du}{u} N_{Lc}(u(1-\frac{x}{z}))$$

$$\times n_L{}^T(\frac{x/z}{1-u(1-(x/z))})+(\text{the "uncross" term}), \qquad (19)$$

where $u=y/(z-x)$. For $x\to 1$, the arguments of N_{Lc} and $n_L{}^T$ go to, respectively, 0 and 1. Since $n_L{}^T(1)=1$, $N_{Lc}(x)\backsim x^{\delta_c}$ and $\delta_c<<\delta_u$ [See later.], (19) becomes

$$\left(\frac{x}{\sigma}\frac{d\sigma}{dx}\right)_v \backsim \int_x^1 dz F^{CQ}(z)(1-\frac{x}{z})^{\delta_c} \backsim (1-x)^{A_i+\delta_c+1} \; ; \text{ for } x\to 1. \qquad (20)$$

The (s) rung contribution has steeper decreasing for $x\to 1$. The vector meson spectra now has the power $\backsim A+1$ [$\because \delta_c<<1$, See Sec.4.], instead of $\backsim A$, suggested by Ochs. In the proton fragmentation, A=3 was used in the early recombination model as for the valence distribution, but recent elaborate models[10,11] gives $A\backsim 2$ for CQ (valon) distribution. Since we have taken the latter view point,[11] the power $A+1\backsim 3$ is quite pleasant.

4. NUMERICAL FITS OF THE DATA

When $\delta_u=\delta_c=1$, $\eta_c=\eta_u=0$ and $\xi=0.5$ in (7), the model reduces to the original DTU. The hatched regions in Fig.3a and 3b are the predictions for π^+/π^- and K^+/K^- ratios in the proton fragmentation region. [The region corresponds to $0\leq C$(the production rate of vector meson)≤ 1.] The large increase for $x\to 1$ could not be explained in this case (the original DTU). Here, the requirement of the strong ordering is satisfied over all branchings, including of course the each end-vertices. Thus, the (v) rung in Fig.1 does not have large momenta, and gives insufficient valence effects to produce the large increase of these ratios. Here, the strong ordering means that most of modmnta are being carried by the string but not quarks. This situation implies that at least the end-vertices should be modified. A small value of δ_c enhances the momentum of the (v) rung. From this consideration, we assume $\delta_c=0.1$ but $\delta_u=1$. [The small $\delta_c/\delta_u\backsim 0.1$ is effective to produce the large increase of the ratios π^+/π^- and K^+/K^-.] For $\eta_{c,u}$ we assume that (17) is satisfied for all B_i in Table I. Then we take $\eta_c=\eta_u=3$. Now, remarkable improvements are

311

obtained as seen in Figs.3. $\xi=0.1$ gives the best fit. Using these parameter values, let us further analyze the recent Serpkhov data on $K^{\pm}+P\rightarrow h+X$.[13]

4.1 Resonance productions in K-fragmentation

These productions are theoretically much simpler than the indirect (ps-meson) case. Since we are dealing with the normalized cross section $(xd\sigma/dx)/\sigma$, we have no free parameter which shifts all cross sections in Figs.4-5, simultaneously. [The coupling constant for the end=vertex is canceled out. Hence, (7) are common to the Kaon and proton fragmentation.] Nevertheless, one sees that our overall normalization is excellent. This is because our model satisfies the longitudinal momentum conservation. Indeed, the integration of $(xd\sigma/dx)\sigma$ over x gives the momentum sum rule.

For ρ^0, K^* and ϕ productions, normalizations of individual curves depend on the the choices of C, C_L in (13) and of the s-quark suppression factor F. F=0.2 and C=0.5 lead the best fit to the cross sections near $x\sim0$. But we need $C_L=0.3<C$ for the fits in the region $x>0.4$. This means that some suppression of the vector meson productions at the end=vertex is required. The curve for $C=C_L=0.5$ is also shown in Fig.4a. The slopes of the the-

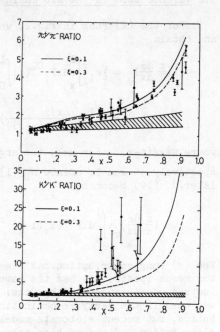

Fig.3. The particle ratios in P-P collision. The hatched regions are the predictions of the original DTU, while the curves are the ones of the modified model. The data point is from Ref.12.

oretical curves for $x\rightarrow1$ follow the asymptotic form (18). ρ^0 data shows steeper decreasing than K^* and ϕ data. If the $\bar{u}(u)$ and $s(\bar{s})$-valence CQ in $K^-(K^+)$ have the same distribution, the same slope is obtained. The values of parameters given in Table I suggested by other experiments show that $s(\bar{s})$ is faster than $\bar{u}(u)$ and lead the reasonable fits. The relative magnitude of the leading to the chain $[(v)+(s)]$ contributions is important to give the bending in $0\leq x\leq0.4$.

4.2 Resonance productions in the proton fragmentation

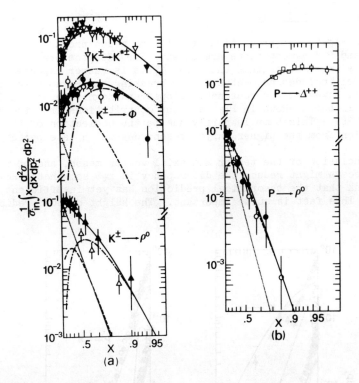

Fig.4. The spectra of the resonance productions. The
solid curves are the prediction of our modified DTU
model; dotted, dashed and dash-dotted curves are the
(s), (v) and leading contributions, respectively.
The dash-dot-dotted line is the prediction for the
case $C_L=C=0.5$. The open (closed) circles and triangles represent Serpkhov data[13] on $K^+(K^-)$ fragmentations.

If all produced (uuu) states are identified with $\Delta^{++}(1236)$, we
predict the cross section twice as large as the data in Fig.4b. We
are then forced to assume that a half of (uuu) is $\Delta^{++}(1236)$ and the
remainder should be higher baryon resonances. The predicted curve
for $P\to\Delta^{++}$ in Fig.4b corresponds to this case.

The $P\to\rho^0$ spectrum for $x>0.4$ is dominated by the (v) term whose
behavior is given in (20): Now the power in (20) is $\sim A+1$ [instead
of A by Ochs.]. As seen in Fig.4b, the CQ distributions given in
Table I gives the beautiful fit with the power $A+1$. It should be
noted that the power is different between the leading particle term
and (v) term. This is in contrast to the recombination model.

313

4.3 Indirect productions

In Fig.5, the predictions for the ps meson productions both in the K and P fragmentations are shown. For $x \sim 0$, the predicted cross sections are smaller than the data. This is rather expected because our model includes no mechanism for breaking the Feynman scaling. We have, however, the reasonable slopes and magnitudes for $x > 0.4$. The small bumps shown in $P \to \pi^{\pm}$ are the contributions from the decay of $\Delta(1236)$. This bump is partly due to the neglection of the contribution from the higher baryon resonances which is expected to be small.

Inclusion of the tensor and axial vector meson, and of three=body decays might reduce the discrepancy in $x \sim 0$ somewhat. However, we think that the theoretical prediction has yet insufficient magnitude. This fact is quite important. The height of our predicted

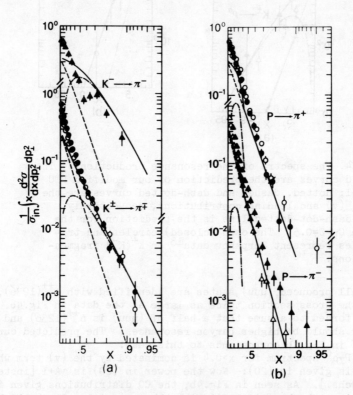

(a) (b)

Fig.5. The inclusive cross sections of the pion productions. The solid lines are our prediction: The dash=dotted curves are the contributions from the leading particles or the decays of them. The dashed ones represent the (s)+(v) contributions. The open (closed) circles and triangles correspond to the $K^+ + P (K^- + P)$ data.[13]

central plateau is saturated by the multi-regge bootstrap condition, $\alpha_P=1$ and $\alpha_R=\alpha=g^2=1/2$. As far as this condition is respected, the dual sheet model will not produce the correct height either.

5. SUMMARY AND CONCLUSIONS

We have shown in Sec.4 that the DTU model with the modification of the end-vertices can give beautiful fits for the vector meson and Δ production data. For ps mesons, it gives less conclusive results. Let us summarize our findings:

(a) Ochs' ansatz can be realized even in the DTU model, if the condition (17) is satisfied. Our derivation is rigorous and contains no unclear factor such as a recombination function.

(b) For $K^{\pm}\to\pi^{\pm}$ and $P\to\pi^{\pm}$, the decay of the leading particles is quite important. If we ignore this [In most previous analyses, we did so.], no reliable conclusion can be drawn.

(c) The vector meson productions at the end-vertex are suppressed somewhat as compared with those at an internal vertex.

(d) About one half of the produced (uuu) states correspond to $\Delta^{++}(1236)$. The remainder must be higher resonances.

(e) The central plateau produced by the DTU bootstrap, $\alpha=\alpha_R=1/2$ and $\alpha_P=1$, has insufficient height to explain the experimental cross sections for $x\stackrel{\sim}{\sim}0$.

(f) In the "cross" (uncross) end-vertex, 78(67)% of the initial CQ momentum is transfered into the chain, and the "cross" vertex dominates. [This should be compared with the corresponding value 33% for the internal vertices.] This property is somewhat similar to that of the dual sheet model.

It remains to study why the dual string has to behave differently at the end and the internal vertices.

We would like to thank Dr. E. Takasugi for careful reading of the manuscripts and critical discussions on the presentation.

REFERENCES

1. H. Lee, Phys. Rev. Lett. <u>30</u>, 719 (1973); G. Veneziano, Phys. Lett. <u>52B</u>, 220 (1974); Nucl. Phys. <u>B74</u>, 365 (1975).

2. H. M. Chan, J. E. Paton and S. T. Tsou, Nucl. Phys. <u>B86</u>, 13 (1975); N. Papadopoulos, C. Schmid, C. Sorensen and D. M. Webber, Nucl. Phys. <u>B101</u>, 189 (1975); M. Fukugita, T. Inami, N. Sakai and S. Yazaki, Phys. Rev. <u>D19</u>, 187 (1979), and further references therein.

3. A. Capella, U. Sukhatme, C. I. Tan and J. Tran Thanh Van, Phys. Lett. <u>81B</u>, 68 (1979); G. Cohen-Tannoudji, A. El Hassouni, J. Kalinowski and R. Peschanski, Phys. Rev. <u>D19</u>, 3397 (1979).

4. W. Ochs, Nucl. Phys. <u>B118</u>, 397 (1977); K. P. Das and R. C. Hwa, Phys. Lett. <u>68B</u>, 459 (1977); D. Wl Duke and F. E. Taylor, Phys. Rev. <u>D17</u>, 1788 (1978); E. Takasugi, Proceedings of XI Int. Symposium of Multiparticle Dynamics, Bruges (1980).

5. W. Kittel, Proceedings of X Int. Symposium on Multiparticle Dynamics, Goa (1979).

6. K. Hirose and T. Kanki, Osaka preprint, OS-GE 81-30 (1981).

7. K. Hirose and T. Kanki, Osaka preprint, OS-GE 81-31 (1981).

8. J. Finkelstein and R. D. Peccei, Phys. Rev. $\underline{D6}$, 2606 (1972); R. J. Yaes, Phys. Rev. $\underline{D10}$, 941 (1974), $\underline{D12}$, 805 (1975).

9. For the Kaon structure function, see, Kwan-Wu Lai and R. L. Thews, Phys. Rev. Lett. $\underline{44}$, 1729 (1980).

10. R. C. Hwa and M. S. Zahir, Oregon preprint, OITS-139 (1980).

11. K. Hirose and T. Kanki, OS-GE 80-26; Prog. Theor. Phys. $\underline{65}$, 1391 (1981).

12. J. R. Johnson et al., Phys. Rev. $\underline{D17}$, 1292 (1978).

13. D. Denegri et al., Phys. Lett. $\underline{98B}$, 127 (1981). C. Cochet et al., Nucl. Phys. $\underline{B155}$, 333 (1979).

DISCUSSION

HWA, OREGON: I'm just a little bit confused about just the essence. What have you done to DTU to modify it to get it right? I'm sorry. I just missed the key point.

KANKI: I just introduced a phenomenological factor for the end vertex function . . . two end vertex functions. And the internal part is still the original DTU chains.

HWA: So, as far as the fragmentation region goes you don't use the fragmentation [function] of a diquark. You put in the valon distribution in some way to give you the correct x distribution.

KANKI: I don't use the fragmentation of a diquark.

HWA: So that's modified. In place of that what did you do?

KANKI: So I took the valon picture here. Two quark lines here are the spectator. This line is not a baryon . . . maybe a parton quark. So this state is baryon-baryon plus some parton quark. And then this term [words cannot be understood] some resonant states. Actually this is just the ordinary idea of DTU.

INCLUSIVE PRODUCTION OF MESON RESONANCES
AND THE SEA-QUARK DISTRIBUTIONS IN THE PROTON

Qu-bing Xie *
Institute of Theoretical Science, University of Oregon
Eugene, Oregon, 97403, U.S.A.
and
Center for Particle Theory, Department of Physics
University of Texas, Austin, Texas 78712, U.S.A. **

ABSTRACT

We have extended the application of the recombination model
to the production of meson resonances. The x distributions of
mesons, produced in the pp reactions, including resonances, were
found to display certain simple relations, which are independent
of specific assumption about the sea-quark distributions. These
predictions can be used both to examine the model and to differentiate between the distributions of strange and non-strange sea-
quarks. We use the resonance production to determine directly
the sea-quark distributions. And, with the distribution of sea-
quark known, the inclusive distribution shape of produced mesons
can then be determined without any further adjustable parameters.

I. INTRODUCTION

After the discovery of the similarity between the π^+ x-distribution in the proton fragmentation region and that of valence
quark u in the proton,[1] Das and Hwa proposed the recombination
model.[2] In the model a pseudoscalar meson is formed by a valence
quark q_1 picking up a sea-quark $\bar{q}_2$ from the initial proton, and
the inclusive distribution of meson and the distribution of sea-
quark are connected intuitively. But neither deep-inelastic
lepton–nucleon scattering nor high-mass muon-pair experiments
can give the exact distribution of sea-quark for individual flavors.[3,4] In order to get reliable information on the sea quarks
some authors have used the original recombination model to determine the sea-quark distribution from inclusive cross section
of pseudo scalar mesons;[5,6] however there exist two important
shortcomings. First, there are some ambiguities and arbitrari-
ness in the original model which make it difficult to discover
the intrinsic relations among basic quantities and among the
reactions. Second, most of the pseudo-scalar mesons observed

---

* On leave of absence from Department of Physics, Shandong University, Jinan, Shandong, Peoples Republic of China.
** Address until February 1982.

in the experiments are from resonance decay,[7] so its x distribution has substantial contribution from the decay spectra of resonances, especially in smaller x region. The seq-quark distributions are concentrated in x < 0.5, and consequently can only affect the normalization of the distribution in the high x region, not its shape.

Aiming at the first shortcoming, Hwa has reformulated the earlier model.[8] In this paper, we propose a way which can overcome the second shortcoming by using the distribution shapes of meson resonances to determine directly the distributions of sea-quarks of every flavor. Conversely, if we know these distributions, we can predict the inclusive distribution shape for every kind of meson without any additional parameters.

II. THE IMPROVED RECOMBINATION MODEL AND THE EXTENTION OF ITS APPLICATION TO RESONANCE PRODUCTION

In the recombination model the inclusive distribution of meson M is:

$$\frac{x}{\sigma}\frac{d\sigma}{dx} = f^M(x) = \iint F^{P \to M}(x_1,x_2)R^M(x_1,x_2;x)\frac{dx_1}{x_1}\frac{dx_2}{x_2} \tag{1}$$

where, $F(x_1,x_2)$ is the $q_1 - \bar{q}_2$ joint momentum probability distribution for the initial proton, and $R^M(x_1,x_2;x)$ is the $q_1 - \bar{q}_2$ recombination function.

Hwa has explicitly introduced the concept of the valon.[9] Let $G_{v/h}(y)$ describe the valon distribution in a hadron. Its normalization is

$$\int_0^1 G_{v/h}(y)dy = 1 \tag{2}$$

and it satisfies the momentum sum rule:

$$\int_0^1 G_{v/h}(y)ydy = \frac{1}{3} \quad (h = \text{nucleon}) \tag{3}$$

$$= \frac{1}{2} \quad (h = \text{non-strange meson}) \tag{4}$$

Using the deep-inelastic neutrino scattering data Hwa determined

$$G_{v/N}(y) = 6 \cdot 56 y^{1/2}(1 - y)^2 , \tag{5}$$

and

$$G_{vv/N}(y_1,y_2) = 19 \cdot 9[y_1 \cdot y_2(1 - y_1 - y_2)]^{1/2} \tag{6}$$

The valon distribution in a non-strange meson, which satisfies eqs. (2) and (4) is

$$G_{v/M}(y) = [B(j,j)]^{-1}[y(1-y)]^{j-1} \tag{7}$$

and the two-valon distribution is

320

$$G_{vv/M}(y_1, y_2) = [B(j,j)]^{-1}(y_1, y_2)^{j-1}\delta(y_1 + y_2 - 1) \qquad (8)$$

For the pion j is already determined from the massive lepton-pair production:

$$j = 1 . \qquad (9)$$

Hence, the valon distribution in a pion has very simple form:

$$G_{v/\pi}(y) = 1 , \qquad (10)$$

$$G_{vv/\pi}(y_1, y_2) = \delta(y_1 + y_2 - 1) . \qquad (11)$$

Let $K(z)$ denote the invariant distribution of finding a quark with momentum fraction z in a valon of the same flavor, and $L(z)$ denote the unfavored distribution. K can be expressed as:

$$K(z) = K_{NS}(z) + L(z) \qquad (12)$$

where $K_{NS}(z)$ is determined from leptoproduction data at low Q^2:[9]

$$K_{NS}(z) = 1 \cdot 2(z)^{1 \cdot 1}(1 - z)^{0 \cdot 16} \qquad (13)$$

The unfavored distribution $L(z)$ of course comes from the sea-quark in the valon; we adopt the same canonical form:

$$L(z) = \alpha(1 - z)^\beta \qquad (14)$$

The parameters α, β will be discussed later.

In the new recombination model the absolute square of the wave function $<V_1(y_1)V_2(y_2)|\pi>$ describes not only the probability of finding the two valons of a pion at y_1 and y_2, but also the probability of forming a pion from two valons at the same y_i values. Thus the invariant recombination function for a pion is

$$R^\pi(x_1, x_2; x) = \frac{x_1 x_2}{x^2}\delta(\frac{x_1}{x} + \frac{x_2}{x} - 1) . \qquad (15)$$

This turns out to correspond to the case with $j = 1$ in eq. (7). But if this were true, we could determine the recombination function for any other meson only after we get its valon distribution from experiment. For some stable particles like p and π, we can extract its valon distribution from the deep-inelastic lepton scattering data, but we cannot do it for meson resonances.

We have first studied non-strange resonance mesons, including π, ρ, ω, f, A_1, A_2, ϕ, f', $\cdots$. Because all their $G_{v/M}$ must satisfy the conditions (3) and (4), the simplest symmetric forms can only be of the type shown in (7) and (8); hence only the j value may be different. We have determined $f^M(x)$ by numerical integration and have found both the normalization and shape are insensitive to the value of j. This important fact, that the

x distribution of the meson is determined only by the distributions of q_1 and $\bar{q}_2$ in the initial proton, is precisely what is
required by the recombination mechanism. Thus, in fact, we do
not need to consider the valon distribution of different mesons,
but can use (15) as the recombination function for all mesons including resonances.

In this work, we assume first that the shapes of the distributions of the three sea-quarks are independent of flavor but
the normalization of strange sea is lower by a factor λ. Because
the average momentum fraction carried by the valence quarks is

$$\bar{x} = 3\iint dxdy G_{v/N}(y)K_{NS}\left(\frac{x}{y}\right) = 0.45 \; , \tag{16}$$

on the basis of saturated sea we obtain as a constraint on L

$$1 - \bar{x} = 0.55 \tag{17}$$

$$= (12 + 6\lambda)\iint dxdy G_{v/N}(y)L\left(\frac{x}{y}\right)$$

i.e., we only need two parameters λ and β to determine the sea-
quark distributions:

$$x\bar{u}(x) = x\bar{d}(x) = 3\int_0^1 G_{v/N}(y)L\left(\frac{x}{y}\right)dy = \frac{32.67}{12 + 6\lambda}(\beta+1)\int_x^1 \sqrt{y}(1-y)^2\left(1-\frac{x}{y}\right)^\beta dy \tag{18}$$

$$x\bar{s}(x) = xs(x) = \lambda x\bar{u}(x) \tag{19}$$

III. THE SHAPE OF INCLUSIVE MESON DISTRIBUTION

For every meson, we can substitute eq. (15) for the recombination function into eq. (1) to determine the shape of $f(x)$, thus
the shape difference can be determined by $F^{p \to M}(x_1,x_2)$. We divide
$F^{p \to M}(x_1,x_2)$ into two terms $F^{(1)}(x_1,x_2)$ and $F^{(2)}(x_1,x_2)$:

$$F^{(1)}(x_1,x_2) = \int dy G_{v/N}(y)A(x_1,x_2;y) \tag{20}$$

with q_1 and $\bar{q}_2$ from one valon, and

$$F^{(2)}(x_1,x_2) = \iint dy_1 dy_2 G_{vv/N}(y_1,y_2)B\left(\frac{x_1}{y_1},\frac{x_2}{y_2}\right) \tag{21}$$

with q_1 and $\bar{q}_2$ from two different valons.

In general, for different mesons the functions A and B are
different. We list A and B in Table 1 for various mesons, where
$L'(z)$ denotes the invariant distribution of finding a strange
quark s (or $\bar{s}$) with momentum fraction z in a valon.

322

Table 1

M	$A(x_1,x_2,y)$	$B(x_1/y_1,x_2/y_2)$
1. π^+, ρ^+, A_2^+, A_1^+,...	$(2K_{NS} + 3L)L$	$(4K_{NS} + 6L)L$
2. π^0, ρ^0, ω, f, A_2^0, A_1^0,...	$(\frac{3}{2}K_{NS} + 3L)L$	$(3K_{NS} + 6L)L$
3. π^-, ρ^-, A_2^-, A_1^-,...	$(K_{NS} + 3L)L$	$(2K_{NS} + 6L)L$
4. K^+, $K^{*+}(890)$, $K^{*+}(1430)$,...	$(2K_{NS} + 3L)L'$	$(4K_{NS} + 6L)L'$
5. K^-, $K^{*-}(890)$, $K^{*-}(1430)$,... $\overline{K}^0$, $\overline{K}^0(890)$, $\overline{K}^0(1430)$,...	$3LL'$	$4LL'$
6. ϕ, f',...	$3L'L'$	$4L'L'$

From Table 1 we arrive at the following predictions:

(1) The shape of x distribution of each kind of meson included in a row in Table 1 should be the same, e.g., the shapes of ρ^0, ω, f, A_2^0, A_1^0, g^0, h, are to be the same.

(2) When x is comparatively small, L is the dominant contribution, i.e., A ~ 3LL, B ~ 6LL, and the shapes of mesons in rows 1, 2, and 3 are basically the same. When x is relatively large, K_{NS} dominates. Then, for example A ~ $2K_{NS}L$ and B ~ $4K_{NS}L$ for ρ^+, and A^+, A ~ $K_{NS}L$ and B ~ $2K_{NS}L$ for ρ^- and A^-, thus

$$f^{\rho^+}(x)/f^{\rho^-}(x) = f^{A^+}(x)/f^{A^-}(x) \approx 2 \qquad (22)$$

(3) If L' = λL, and the shapes of mesons in rows 1, 4 and 5, 6 are exactly the same,

$$f^{\rho^+}(x)/f^{K^{*+}(890)}(x) = f^{A_2^+}(x)/f^{K^{*+}(1430)}(x) = f^{K^{*-}(890)}(x)/f^{\phi}(x)$$
$$= f^{K^{*-}(1430)}(x)/f^{f'}(x) = \cdots = 1/\lambda .$$

If the above ratios depend on x, then L' $\neq$ λL and the strange sea-quark has a different shape.

(4) When L' = λL, even if x is small, the shapes of mesons of type 5, 6 fall off more rapidly in comparison with the previous 4 kinds.

(5) If L' = λL, then from the ratio of the pair of particles either of type 1, 4 or particles of type 5, 6, we can determine λ. If L' $\neq$ λL, we can directly determine L', in terms of the x distribution of particles of type 6, and then the distribution $xs(x) = x\overline{s}(x)$.

(6) In case of large x and L' = λL

$$f^{K^{*+}(890)}(x)/f^{\rho^-}(x) = f^{K^{*+}(1430)}(x)/f^{A_2^-}(x) = \cdots \cong 2\lambda \qquad (23)$$

323

The first two predictions, independent of any concrete properties of K_{NS} and L, can be used to check the present theory; and the last four can be used to determine the properties of sea-quark distribution.

IV. COMPARISON WITH DATA

We obtain explicit formulas of x distribution of the six kinds of meson which are included in Table 1:

$$f^M(x) = \frac{1}{x}\int_0^x dx_1 \{6.56 \int_x^1 dy y^{\frac{1}{2}}(1-y)^2 ([e \cdot K_{NS}(\frac{x_1}{y}) + f \cdot L(\frac{x_1}{y})] \cdot L(\frac{x-x_1}{y-x_1})$$
$$+ L(\frac{x-x_1}{y}) \cdot [9 \cdot K_{NS}(\frac{x_1}{y-x+x_1}) + h \cdot L(\frac{x_1}{y-x+x_1})])$$
$$+ 19.9 \int_x^{1-(x-x_1)} dy_1 \int_{x-x_1}^{1-y_1} dy_2 [y_1 y_2 (1-y_1-y_2)]^{\frac{1}{2}} \cdot [\ell \cdot K_{NS}(\frac{x_1}{y_1})$$
$$+ m \cdot L(\frac{x_1}{y_1})] L(\frac{x-x_1}{y_2})] \tag{24}$$

Here e, f, g, ··· are various definite constants for each kind of meson that is listed in Table 2.

Table 2

M	e	f	g	h	l	m
1. π^+, ρ^+, A_1^+, A_2^+...	1	2	1	1	4	6
2. π^0, ρ^0, ω, f, A_2^0, A_1^0...	0.75	2.25	0.75	0.75	3	6
3. π^-, ρ^-, A_2^-, A_1^-...	0.5	2.5	0.5	0.5	2	6
4. K^+, $K^{*+}(890)$, $K^{*+}(1430)$	λ	2λ	λ	λ	4λ	6λ
5. K^-, $K^{*-}(890)...$, $\bar{K}^0$, $\bar{K}^{*0}(890)$,...	0	1.5λ	0	1.5λ	0	4λ
6. ϕ, f'	0	$3\lambda^2$	0	0	0	$4\lambda^2$

Now we first determine the parameters β, λ from the vector meson data, and then compare the meson inclusive distribution data with theory.

Figure 1 shows the ρ^0 x-distributions in 147 GeV/c pp reactions, and ρ^0 x-distribution of target proton fragmentation in 147 Gev/c π^+p reactions.[12] To fit these data, we derive: $\beta = 1.5$. The solid line is a theoretical curve by taking $\beta = 1.5$. Figure 2 gives the 24 Gev/c data. The normalization of ρ^0 points has been reduced by a factor of 0.18. The solid lines in the figures are theoretical curves of ρ^+ and ρ^0. Note the difference between these two curves is very small. Thus we take $f^{\rho^+}(x)/f^{K^{*+}(890)}(x) \approx f^{\rho^0}(x)/f^{K^{*+}(890)}(x) \approx 1/0.18$, or $\lambda = 0.18$.

324

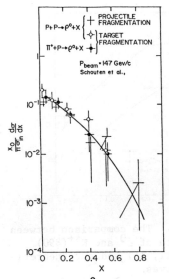

Fig. 1. The ρ^0 x-distributions
and our theoretical curve by
taking $\beta = 1.5$.

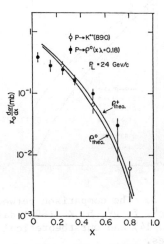

Fig. 2. The x-distributions of
$K^{*+}(890)$ and $\rho^0(770)$. The solid
lines are our theoretical curves
of ρ^+ and ρ^0.

Figure 3 shows the comparisons between the ρ^0, f, g^0, h,
$K^{*-}(890)$ data[14] and our theoretical curves. Here, the normalizations are arbitrary as in the data. The curve for $K^{*-}(890)$ are
normalized to the same value at the x = 0 point of ρ^0. The data
agree well with our predictions, i.e. the shapes of the x distribution of ρ^0, f, g^0, h are the same, while that of $K^{*-}(890)$ falls off
more rapidly with x.

Figure 4 shows the comparisons between the data[11,15] of ρ^0 and
$K^{*+}(890)$ produced in the fragmentation region of proton in K^+p interactions at 32 Gev/c and the theoretical curves (solid lines).
Note even though the energy is small, the agreement is still very
good.

The $f^{\pi^+}(x)/f^{K^+}(x)$ data is given in Figure 6.[17,18] This data
should be considered as the consequence of resonance decay. In
fact, if the ratio $f^{\rho^+}(x)/f^{K^{*+}(890)}(x)$ is constant (i.e. L' = λL),
the ratio of π^+ to K^+, which are mostly from ρ and K^{*+}, respectively, should not be constant. In the decay $K^{*+} \to K^+\pi^0$, the larger
momentum is carried by K^+; but in $\rho^+ \to \pi^+\pi^0$ the moments of π^+ and π^0
are the same. As a result the ratio $f^{\pi^+}(x)/f^{K^+}$ should increase for
small x, and decrease for large x, just like what is shown in Figure
6. Hence the data are not inconsistent with L' = λL.

Figure 7 shows the comparison between the data[17,18] of
$f^{\pi^-}(x)/f^{K^-}(x)$ and the theoretical curve. In the larger x region,
the lower experimental values may result from reasons similar to
those mentioned above. From Table 1 we can see that the yield of
K^{*-} is much less than that of K^{*+} in pp reactions. Only a comparatively small fraction of the K^-'s is from K* decay and the large

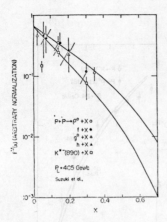

Fig. 3. The comparison between the ρ^0, f, g^0, h, $K^{*-}(890)$ data (Ref. 14) and our theoretical curves.

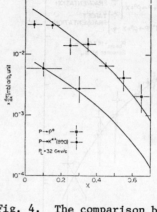

Fig. 4. The comparison between the data of ρ^0 and $K^{*+}(890)$ (Ref. 11, 15) and our theoretical curves.

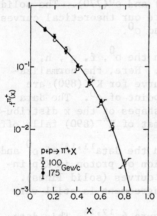

Fig. 5. The comparison of the π^+ data (Ref. 16) and the theoretical curve (solid line). The agreement is excellent in both shape and normalization over the whole x range.

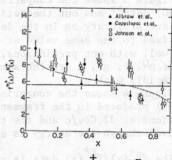

Fig. 6. The $f^{\pi^+}(x)/f^{K^-}(x)$ data versus x.

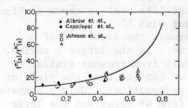

Fig. 7. The comparison between the data (Refs. 17, 18) of $f^{\pi^-}(x)/f^{K^-}(x)$ and the theoretical curve.

fraction is from the K^+K^- decay mode of non-strange resonances (like A_2, f', etc.), as compared to the case of K^+. Consequently, the deviation between data and theory is obviously less than the ratio $f^{\pi^+}(x)/f^{K^+}(x)$.

The quantity $f^{K^+}(x)/f^{K^-}(x)$ is independent of any concrete assumption about strange sea. In Figure 8 one can see that the theory agrees well with data.[17,19] In Figure 9 the comparison $f^{\pi^+}(x)/f^{\pi^-}(x)$ is given.

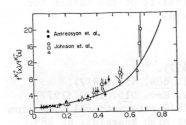

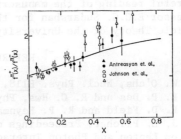

Fig. 8. The comparison between the data of $f^{K^+}(x)/f^{K^-}(x)$ and the theoretical curve.

Fig. 9. The comparison between the data of $f^{\pi^+}(x)/f^{\pi^-}(x)$ and the theoretical curve.

By the same reasoning as the (22), when x is large, the ratio $f^{\pi^+}(x)/f^{\pi^-}(x)$ should be around 2. The dispersion of available data[17,19] is compatible.

V. THE SEA-QUARK DISTRIBUTIONS

Using the $\lambda = 0.18$ and $\beta = 1.5$, we immediately obtain the non-strange and strange sea-quark distributions:

$$\bar{xu}(x) = \bar{xd}(x) = 6.24 \int_x^1 y^{\frac{1}{2}}(1-y)^2 (1-\frac{x}{y})^{1.5} \, dy \,, \qquad (25)$$

$$xs(x) = x\bar{s}(x) = 1.24 \int_x^1 y^{\frac{1}{2}}(1-y)^2 (1-\frac{x}{y})^{1.5} \, dy \,. \qquad (26)$$

These sea-quark distributions lead to the following fractions of momentum carried by the various quarks:

valence quarks: $\qquad 3\int dxdy G_{v/N}(y)K_{NS}(\frac{x}{y}) = 0.450$

up and down sea-quarks: $\qquad 12\int dxdy G_{v/N}(y)L(\frac{x}{y}) = 0.503$

strange sea-quarks: $\qquad 6\lambda\int dxdy G_{v/N}(y)L(\frac{x}{y}) = 0.045$

Our results are somewhat different from those previously obtained by fitting pseudo-scalar meson data.[5] For example, the amounts of momentum carried by the strange sea are twice that of their value 0.020 of Ref. 5. However, the basic behavior is similar, that is, the sea-quark distributions rapidly fall off as the x increases and are concentrated in the small x region of $x < 0.5$.

Most of the work reported here was done in the Institute of Theoretical Science, University of Oregon. I would like to thank Professor Rudolph C. Hwa for his guidance and many invaluable discussions. I am also grateful to Professor W. D. Shephard for his consideration and his efficient arrangement, which made this conference particularly enjoyable. Thanks are also due to Professor Charles Chiu and Professor Duan Dicus for discussions and a careful reading of the manuscript. Finally, special thanks to Professor E.C.G. Sudarshan for the hospitality of the Center for Particle Theory of the University of Texas.

REFERENCES

1. W. O'chs, Nucl. Phys. B118, 397 (1977).
2. K. P. Das and R. C. Hwa, Phys. Lett. 68B, 459 (1977).
3. R. D. Field and R. P. Feynman, Phys. Rev. D15, 2590 (1977).
4. J. E. Pilcher, Proceedings of the 1979 International Symposium on Lepton and Photon Interactions at High Energies, p. 185.
5. D. W. Duke and F. E. Taylor, Phys. Rev. D17, 1788 (1978).
6. N. N. Biswas et al., Phys. Rev. D19, 1960 (1979).
7. Xie Qu-bing, Phys. Energiar Fortis et Phys. Nucl. 4, 439 (1980).
8. R. C. Hwa, Phys. Rev. D22, 1593 (1980).
9. R. C. Hwa, Phys. Rev. D22, 759 (1980).
10. J. Bartke et al., Nucl. Phys. B107, 93 (1976).
    G. Jansco et al., Nucl. Phys. B124, 1 (1977).
    C. Cochet et al., Nucl. Phys. B155, 333 (1979).
11. Ajinenko et al., Z. Physik C5, 177 (1980).
12. M. Schouten et al., HEN-187 (1981).
13. K. Bockmann et al., Nucl. Phys. B166, 284 (1980).
14. A. Suzuki et al., Nucl. Phys. B172, 327 (1980).
    A. Suzuki et al., Nuovo Cim. Lett. 24, 449 (1979).
15. P. V. Chlapnikov et al., Nucl. Phys. B176, 303 (1980).
16. G. W. Brandenburg and V. A. Polychronakos, The Preliminary Data of Fermilab Experiment No. E118 (see Ref. 8).
17. J. R. Jonson et al., Phys. Rev. D17, 1292 (1978).
18. P. Cappiluppi et al., Nucl. Phys. B79, 189 (1974).
    M. G. Albrow et al., Nucl. Phys. B56, 333 (1973); B73, 40 (1974).
19. D. Antreasyan et al., Phys. Rev. Lett. 38, 112 (1977); 38, 115 (1977).

SESSION B2

SOFT-HADRON PHYSICS: THEORY II

Tuesday, June 23, p.m.

Chairman: W. Kittel
Secretary: N. M. Cason

PERTURBATIVE QCD FOR HADRONIC PROCESSES AT LOW-p_T

J.F. Gunion
Department of Physics
University of California, Davis, CA 95616

ABSTRACT

The perturbative QCD approach to low-p_T physics is reviewed. In this approach the diagrams of lowest order in α_s that can result in a given low-p_T process are computed. For instance, the one gluon exchange model of Low and Nussinov is used for the total cross section; one gluon radiation diagrams for particle production in the central plateau region of the final state; and exclusive low order diagrams for fast particle production in the final state. The overall consistency with experiment is remarkable and important insights emerge. In particular, a number of experimental signatures for the QCD approach are focused on which contrast sharply with expectations in traditional Regge/multiperipheral approaches. As one example, no low order QCD analogue of the "triple-Pomeron" behavior for $d\sigma/dm^2$ is found. A second example is the low order QCD prediction that target jet fragmentation in deep inelastic scattering on a proton target and hadronic jet fragmentation in a proton-proton collision can only be the same for fast particles (as observed experimentally) if the underlying jets are different. This difference is natural in the Low-Nussinov picture of hadronic scattering.

INTRODUCTION

It is useful to divide the discussion into three sections related to the type of QCD diagrams which dominate in low order.

I. Cross sections:

a) Low-Nussinov model for hadron-hadron collisions;
b) diffractive excitation in hadron-nucleus collisions.

II. Distribution of Produced Particles:

a) gluon radiation model for the central plateau;
b) gluon radiation predictions in the "triple-Pomeron" region.

III. Fragmentation Spectra:

a) quark fragmentation;
b) deep inelastic target jet fragmentation;
c) fragmentation in hadronic collisions.

Equations are minimized and physical insight emphasized especially where standard approaches yield different results. A review of the basic features of this approach and many details appear in Reference 1. Here we focus on the physical structure of the calculations and on a number of very recent results.

I. CROSS SECTIONS

a) Low-Nussinov Model[2,3]

The Low-Nussinov model for π-π scattering is illustrated in Fig. 1. Scattering occurs in lowest order via one-gluon exchange.

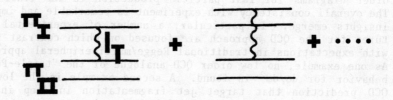

Fig. 1. Diagrams for π-π scattering via one-gluon exchange.

There are four amplitude diagrams which interfere with one another when the total cross section is computed by squaring the sum of amplitudes.[3] A diagram in which the gluon attaches to a pion's quark is opposite in sign relative to the related diagram in which the gluon attaches to the antiquark. The interference term between these two diagrams is proportional to $-f_{q\bar{q}}(\ell_T^2)$ where ℓ_T is the transverse momentum carried by the exchange gluon and $f_{q\bar{q}}$ is a form factor arising from the routing of ℓ_T through a pion bound state wave function. The square of any one diagram is proportional to 1 in the same normalization. The resulting cross section takes the form[3]

$$\sigma_T^{\pi\pi} = \frac{8}{9} \int \frac{d^2\ell_T}{\ell_T^4} \alpha_s^2 \ (2) \ [1-f_-^{I}(\ell_T^2)] \ (2) \ [1-f_-^{II}(\ell_T^2)]. \quad (1)$$

Important features include the following:

a) The cross section is s-independent; the conventional "Pomeron" is replaced by two-gluon exchange. (Vector gluons, as in

QCD, are necessary for σ to be s-independent.)

b) The color-singlet-bound-state interference structure discussed above (1) regulates the $1/\ell_T^4$ infrared divergence from the gluon propagators. (The relevance of calculations which sum infrared divergences[4] at the quark level is not clear.)

c) There is a hadron size effect.[3] Roughly the form factor which appears in (1) is given by

$$f_{q\bar{q}} (\ell_T^2) \sim \frac{1}{1+4 \ \ell_T^2/m_v^2} \tag{2}$$

where m_v is a "vector dominance" type mass scale. A hadron composed of massive quarks has a correspondingly large m_v value and small "size". The $[1 - f_{q\bar{q}} (\ell_T^2)]$ factor for such a hadron is small on average and the total cross section correspondingly small.

d) Quark counting rules apply.[3] The (2) $[1 - f_{q\bar{q}} (\ell_T^2)]$ factor for a pion is replaced by (3) $[1 - f_{qq} (\ell_T^2)]$ in the case of a proton, with $f_{q\bar{q}} \cong f_{qq}$ for quarks of similar mass.

e) The normalization of σ_T is reasonable. For πp collisions we take $m_v \sim m_\rho$ and find that $\alpha_s \cong 2$ yields the observed inelastic cross section.

In this approach exclusive diffractive production channels such as πp → A₁p or πp → A₂p occur in lowest order via diagrams involving 2-gluon exchange as in Fig. 2. The amplitude is not automatically s-independent. Indeed the s-independent part of the amplitude survives only if the helicity wave function of the quarks in the final diffractive hadron is the same as that of the initial hadron; this is because the exchange gluon conserves helicity at high s. Thus the only s-independent

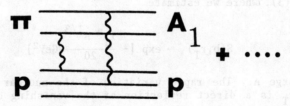

Fig. 2. The 2-gluon exchange diagrams for diffractive πp→A₁p production.

exclusive channel diffractive reactions should be those obeying a "Morrison rule"[5] which states that changes in the orbital state are allowed but not changes in the helicity state.

b) Diffractive Excitation

In a more general context diffractive excitation of an incoming hadron on a target arises due to the variation of the S-matrix seen by the incoming hadron.[8] For diffractive excitation of a meson state we can write[7]

$$\psi'(b,\vec{r}_T,x) = S(b,\vec{r}_T)\ \psi(\vec{r}_T,x) \qquad (3)$$

where $\psi(\vec{r}_T,x)$ is the wave function of the incoming meson and $\vec{r}_T$ and x are the internal meson coordinates; $\vec{r}_T$ = transverse $q\bar{q}$ separation, and x = quark light cone "+" - momentum fraction. The impact parameter of the collision is b and we have implicitly assumed that the target is unaltered by the collision. The outgoing diffractive state wave function is denoted by ψ'. Clearly if S were independent of b and $\vec{r}_T$ the outgoing state would be proportional to the incoming and no excitation would occur. All models predict, however, that S varies with $\vec{r}_T$ and b. The most dramatic variation occurs in a QCD model of the Pomeron. This is because the coupling of a colored gluon to a color singlet bound state hadron must vanish in the limit of zero hadron size, $r_T \equiv |\vec{r}_T| \to 0$, whereas the coupling strength is substantial at large r_T.

A dramatic example of the effects of this rapid variation is produced by the case of a pion incident on a heavy nuclear target.[7] We require that no net color be transferred to any of the target nucleons. The process will then be diffractive since color neutralizing radiation of colored quanta to fill the rapidity gap between the pion and any given struck nucleon will not be necessary. (In lowest order the incoming π must exchange two gluons with each nucleon that it interacts with.) The excited pion state is written as in (3), where we estimate

$$S(b,r_T) \sim \exp\left[-\frac{r_T^2\ A^{1/3}}{20}\ \text{GeV}^2\right] \qquad (4)$$

for large A. The rapid variation of the nuclear target S-matrix with r_T is a direct reflection of the vanishing of the π-nucleon interaction in the Low-Nussinov model as $r_T \to 0$. Much as a plane wave incident on a pinhole spreads out on the opposite side, the enhancement of small r_T in (4) causes an enhancement of large k_T

(k_T = relative q-$\bar{\text{q}}$ transverse momentum) and large masses, m, of the diffractively-produced-excited q-$\bar{\text{q}}$ state. The enhancement is much greater for a QCD-Pomeron model than in other approaches. To estimate the cross section for producing a q-$\bar{\text{q}}$ state orthogonal to the original pion we rewrite (3) as

$$\psi'(b',\vec{r}_T,x) = \bar{S}(b)\,\psi(\vec{r}_T,x) + [S(b,\vec{r}_T) - \bar{S}(b)]\,\psi(\vec{r}_T,x) \qquad (5)$$

where $\bar{S}(b) \equiv \int dx d^2 r_T |\psi(\vec{r}_T,x)|^2\, S(b,\vec{r}_T)$. The first term is simply the survival amplitude of the incident state, while the second term is a superposition of excited hadron states orthogonal to the incident state. For small r_T where $S(b,r_T) \cong 1$ we obtain the excited state cross section

$$\frac{d\sigma}{dx d^2 r_T} = \int d^2 b\,(1 - \bar{S}(b))^2\,|\psi(\vec{r}_T,x)|^2 = \sigma_{el}|\psi(\vec{r}_T,x)|^2 \qquad (6)$$

where σ_{el} is the elastic scattering cross section. Since $S(b,\vec{r}_T)$ is so rapidly varying in r_T at large A, we need only know $\psi(0,x)$ to calculate the cross section as a function of k_T or integrated over r_T. QCD predicts a definite form for $\psi(0,x)$ that is normalized by the $\pi \to \mu\nu$ decay rate:[8]

$$\psi(0,x) = \sqrt{12}\,x(1-x)\,f_\pi \qquad (7)$$

with $f_\pi \sim .1$ GeV. Combining (6) and (7) we find the k_T dependent cross section

$$\frac{d\sigma}{d^2 k_T}\,(\pi \to \text{orthogonal diffractive state}) \qquad (8)$$

$$\sim 50\text{ mb GeV}^{-2}\,\exp[-10\,k_T^2/A^{1/3}\text{ GeV}^2].$$

The cross section integrated over r_T values ($r_T < r_T^{max}$) such that $S \cong 1$ is, from (6), the product of the elastic scattering cross section of the projectile ($\propto A^{2/3}$) with the probability that a state of size r_T is present and transmitted. This latter probability is proportional to $(r_T^2)_{max}$ which in turn behaves as $A^{-1/3}$ from (4). Thus

$$\sigma\,(\pi \to \text{orthogonal diffractive state}) \cong 16\text{ mb }A^{1/3}. \qquad (9)$$

Both the $\langle k_T \rangle$ value and total integrated cross section size are far bigger than would be typical of other approaches.

An interesting application of these ideas[7] is to note that, if the Fock states containing charm-anticharm pairs of the proton are intrinsically small[9] (so that the transmission factor S is approximately unity over the entire range in r_T of the Fock state), then

$$\sigma(\text{diffractive charm production}) = \sigma_{el} P_{charm} \qquad (10)$$

where P_{charm} is the probability that the projectile contains a small intrinsic charm state. A value[10] of $P_{charm} \cong 2\%$ yields charm production cross sections of the same size as those seen at FNAL;[11] in addition, the large x_F values observed would be natural in diffractive production.

Thus we see that a color based mechanism for hadronic collisions agrees with basic total cross section phenomenology, but has a number of testable features of an unusual and striking nature. In the following sections we will presume that the Low-Nussinov gluon exchange diagram is the dominant interaction in hadronic collisions and derive consequences for the central multiplicity region, the triple-Pomeron region, and the fragmentation region of a typical hadronic collision. QCD will continue to play a special role, yielding novel and unexpected effects, often remaining consistent with experimental facts that standard theories find "uncomfortable".

II. DISTRIBUTION OF PRODUCED PARTICLES

a) The Central Plateau

The prototype QCD calculation of particle production is that envisioned in the jet calculus approach to the e^+e^- final state.[12] The primordial final state contains separating quark and antiquark color triplet states. Final state particle production is initiated by perturbatively produced radiation which begins immediately after the creation of the q and $\bar{q}$ jets. Eventually sufficient numbers of perturbatively produced quanta are present that on-shell hadrons can be formed (non-perturbatively) from overlapping color quanta. The non-perturbative stage is presumed to occur too "late" to influence over all distributions at asymptotic energies - the underlying perturbatively produced color quanta largely determine the distribution and multiplicity of final state particles.

In the Low-Nussinov (L-N) gluon exchange model of a hadronic collision a similar sequence of events occurs.[13] The gluon exchange leads to a primordial final state consisting of separating color octets[13] which in turn results in color radiation/bremsstrahlung followed by non-perturbative hadron formation. From a jet-calculus point of view the asymptotic relationship between the final states of e^+e^- annihilation and of a hadron collision will depend upon the comparative distribution of radiation products in the two cases.

We outline the results[14] of this comparison at the simplest level - a calculation of the gluon radiation multiplicity from one gluon emission diagrams. If the radiation is Poisson distributed this calculation is sufficient. We already know from higher order e^+e^- calculations[15] that this is not precisely the case but hope that the low order comparison will nonetheless be representative.

In e^+e^- annihilation there is only the one-gluon-radiation diagram of Fig. 3, in an appropriate gauge. Defining the momentum of the radiated gluon as $(k = (k^+, k^-, k_T)$ with $k^{\pm} = k^0 \pm k^3)$

$$q = (x\sqrt{s}, \frac{q_T^2}{x\sqrt{s}}, q_T, 0) \tag{11}$$

in a frame where $p_T' = 0$, we obtain[14] in the central region

$$\frac{dn}{d^2 q_T d\eta} = \frac{\dfrac{d\sigma^{1-gluon}}{d^2 q_T \, d\eta}}{\sigma^{0-gluon}} = \frac{C_F \alpha_s}{\pi^2} \frac{1}{q_T^2}, \tag{12}$$

where $\eta = \frac{1}{2} \ln q^+/q^- = \ln(x\sqrt{s}/q_T)$ is the rapidity of the radiated gluon. Integrating over q_T^2 from a lower limit, Λ^2, determined by the non-perturbative confinement scale, and an upper limit, s, determined by kinematics we find a flat rapidity distribution,

$$\frac{dn}{d\eta} = \frac{C_F \alpha_s}{\pi} \ln(s/\Lambda^2). \tag{13}$$

Integrating over the allowed rapidity range we obtain

$$\langle n \rangle = \frac{C_F \alpha_s}{\pi} \ln^2 s + C_1 \ln s + C_2 . \tag{14}$$

The generalization to include the non-Poissonian effects of higher order calculations leads to the form[15]

$$\langle n \rangle \cong a + b \exp[c(\ln s/\Lambda^2)^{\frac{1}{2}}] . \tag{15}$$

Both calculations suggest a rapid rise with s of the gluon plateau height at $\eta \sim 0$ and a faster- than- log rise of the total gluon multiplicity. However, the connection between these gluon multiplicity calculations and the observed hadron multiplicity is tenuous. It is conventional to assume that dn(hadron) ≅ dn(gluon). Thus forms (14) and (15) have both been used to fit the observed e^+e^- charged particle multiplicity. Using (14), for instance, and an estimate of $\frac{1}{2}$ charged hadron per gluon (as in a simple gluon → quark pair → hadron model) one obtains an excellent fit with $\alpha_s \cong 1$ and $\Lambda \sim .2$ GeV, to both <n> and $\left.\frac{dn}{d\eta}\right|_{\eta=0}$. This procedure is almost certainly too extreme. At current energies Monte Carlo studies suggest[16] that much of the rise of the e^+e^- plateau height is kinematic in origin due to the opening up of resonance channels in the Feynman Field fragmentation model[17] only a small perturbative component from the $q\bar{q}g$ final state is required in these fits. Asymptotically, however, the perturbative component should dominate.

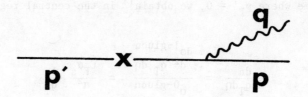

Fig. 3. One gluon radiation in e^+e^- annihilation.

What is the corresponding result in the Low-Nussinov model for a hadronic collision? The dominant diagrams[14] for the central region are illustrated in Fig. 4.

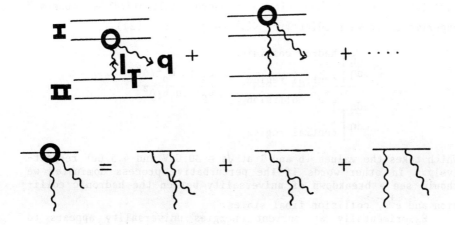

Fig. 4. Dominant diagrams for one-gluon emission in the Low-Nussinov model. In the qg → qg emission subprocess diagrams a) and b) correspond to "pre-emission" of the gluon while in (c) the gluon is radiated after the L-N interaction ("post-emission").

The result[14] for the multiplicity in the central "plateau" region is

$$\frac{dn}{d\eta d^2 q_T} = \frac{C_A}{\pi^2 q_T^2} \alpha_s \frac{\int \frac{d^2 \ell_T}{\ell_T^4} \alpha_s^2 \frac{\ell_T^2}{(\ell_T - q_T)^2} (2) [1 - f_{q\bar{q}}^{I}((\ell_T - q_T)^2)] \quad (2) [1 - f_{q\bar{q}}^{II}(\ell_T^2)]}{\int \frac{d^2 \ell_T}{\ell_T^4} \alpha_s^2 (2) [1 - f_{q\bar{q}}^{I}(\ell_T^2)] (2) [1 - f_{q\bar{q}}^{II}(\ell_T^2)]} \quad (16)$$

which has the following limits:

$$\frac{dn}{d\eta d^2 q_T} \sim \begin{cases} \frac{C_A}{\pi^2} \frac{\alpha_s}{q_T^2} & \text{for } q_T^2 < \langle \ell_T^2 \rangle \\[2ex] \propto \frac{1}{q_T^4} & \text{for } q_T^2 > \langle \ell_T^2 \rangle \; . \end{cases} \quad (17)$$

Thus at low $q_T^2 < \langle \ell_T^2 \rangle$ we obtain C_A/C_F (=9/4) times the result (12) for $e^+ e^-$ annihilation,[18] but at high q_T^2 the $1/q_T^2$ "tail" is lost and the spectrum is cut off. Consequently if we again identify

339

(12)-(13) and (16)-(17) with the <u>hadron</u> multiplicities observed experimentally we predict (for $\langle \ell_T^2 \rangle \propto m_\rho^2$ - Eq. (2))

$$\frac{\left.\dfrac{dn}{d\eta}\right|_{\substack{\text{hadron collision}\\ \text{central region}}}}{\left.\dfrac{dn}{d\eta}\right|_{\substack{e^+e^- \text{ collision}\\ \text{central region}}}} = \frac{C_A}{C_F} \frac{\ln m_\rho^2/\Lambda^2}{\ln s/\Lambda^2} \qquad (18)$$

which takes the values .6 and 1 at $\sqrt{s}$ = 30 GeV and 4.5 GeV respectively. In other words if the perturbative process dominates we should see a breakdown of universality betwen the hadronic collision and e^+e^- collision final states.

Experimentally at current energies universality appears to hold.[19] This is to be expected if the dominant component in the final multiplicity is non-perturbative in origin as in the e^+e^- Monte Carlo studies mentioned earlier; kinematic and phase space effects upon the development of jets in the two processes could be sufficiently similar to yield the observed universality of the gross multiplicity. Asymptotically in s, however, we predict an increasing lack of universality. A more sensitive test at current energies is to compare the distribution in p_T^2 of the final state hadrons in the e^+e^- and hadronic collision cases. The q_T^2 spectra of Eqs. (12) and (17) predict that $\langle p_T^2 \rangle_{e^+e^-}$ has a perturbative component which rises logarithmically with s whereas $\langle p_T^2 \rangle_{\text{hadron-hadron}}$ does not. This predicted difference appears to be substantiated by recent analyses presented at this conference.[20] We anticipate that higher order corrections will not substantially alter the hadronic-collision prediction.

The origin of the different q_T^2 spectra is physically transparent. In e^+e^- the off-shell virtual photon can directly supply sufficient off-shellness to the radiating quark of Fig. 3 that any $q_T^2 < s$ is possible. In contrast the quarks of the incoming hadrons of Fig. 4 are only off-shell on a scale of order $\langle \ell_T^2 \rangle$; this limit applies to the L-N exchange gluon as well. Since it is this L-N gluon which must supply the off-shellness to the radiating quark, only $q_T^2 \lesssim \langle \ell_T^2 \rangle$ is possible.

Other theoretical features of this calculation are discussed at length in Refs. 1 and 14. The most important feature is the necessity of including all three subprocess radiation diagrams of Fig. 4 ;[19] in two of them the radiation gluon is first emitted and then the L-N gluon exchange occurs while in the remaining diagram the L-N interaction occurs first and then the radiated gluon is emitted - "pre-emission" and "post-emission", respectively. These diagrams form a gauge invariant set and interfere destructively with one another. Indeed it is only the non-abelian nature of the QCD couplings that prevents their cancelling completely in the central region. It would be incorrect to include only pre-emission diagrams (as done in multiperipheral model approaches) or post-emission diagrams (as done in color separation models) or to ignore interference phenomena. Another point is that is is impossible for a soft-gluon in the incoming hadron Fock-state to materialize as an on-shell final state gluon unless a L-N gluon exchange occurs and is such that the two perturbative vertices are contiguous in a Feynman diagram sense as in Fig. 4; the off-shellness of the quark which gives rise to the on-shell soft gluon must be supplied by a nearby L-N exchange gluon.

A final phenomenological point is to note that the variation on an event by event basis of the multiplicity with the ℓ_T of the exchanged L-N gluon leads to a long range correlation between forward and backward multiplicities event by event. Collisions with large forward multiplicity ($\eta > 0$) will on the average be expected to have large backward multiplicity as well ($\eta < 0$). Such a correlation is observed in the data.[21]

b) The "Triple-Pomeron" Region

This region is conventionally defined to be the limit $s \to \infty$ followed by $m^2/s \to 0$, where one considers a reaction of the type

$$H_1 H_2 \to X H_2 \qquad (19)$$

with $m_X \equiv m$. In this limit one of the colliding hadrons is left unexcited on one end of the rapidity axis while at the other end of the rapidity axis there is an excited state of mass m. In between the two final state systems is a large rapidity gap. In conventional Regge theory there is a natural mechanism for producing such events based on the so-called "triple-Pomeron" coupling.[22] In QCD such events, in the lowest order L-N exchange approach, must arise from Feynman diagrams like the one illustrated in Fig. 5 for $H_1 = H_2 = \pi$.

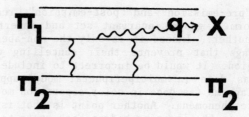

Fig. 5. A sample QCD amplitude for $\pi\pi \to X\pi$ with $m_X \equiv m$.

First note that two-gluon exchange to the unexcited π_2 is required. One gluon exchange would turn the incoming π_2 into a color octet state which would then be required to radiate soft-gluons (as just discussed) filling the rapidity plateau with hadrons; only two gluon exchange allows π_2 to remain in a color singlet state and thus allows a large rapidity gap (in the $s \to \infty$ limit) between π_2 and X. The lowest order mechanism for forming X is the addition of a single gluon (momentum q) to the incoming π_1 Fock state. This gluon can arise perturbatively from many locations (10 for each basic 2-gluon exchange amplitude, only one of which is illustrated in Fig. 5). With q defined as in Eq. (11) (in a frame where $(p_{\pi_1})_T = (p_{\pi_2})_T = 0$ in the initial state) the m^2 spectrum is determined by the x-spectrum of this extra intrinsic gluon. In order to reproduce the dm^2/m^2 spectrum of a triple-Pomeron coupling a dx/x spectrum is required. Explicit calculation[23] reveals that all such terms cancel for diagrams of the QCD type. For instance, for Fig. 5 there are two "Glauber" cuts (one before and one after the perturbative attachment of q); each alone has a dx/x spectrum but they have opposite signs and cancel. Intuitively we anticipate this result since two-gluon exchange leaves the $\pi_{(2)}$ state in a color singlet; no net color is exchanged and there is no need for soft gluon radiation. Thus one expects the same over all features as for an abelian (colorless) elementary exchange attached to $\pi_{(2)}$.

As we learned in the previous section, a), the dx/x spectrum cancels in the elementary abelian gluon exchange case.

The result is that in the context of the QCD, L-N gluon exchange approach there is no low order mechanism for a dm^2/m^2 spectrum. The natural result of a perturbative QCD calculation is[23]

$$\frac{d\sigma}{dm^2} \propto \frac{1}{(m^2)^2} + \text{background} \qquad (20)$$

where the background terms will fall with s at fixed m^2. Surprisingly available data appear to agree with this QCD prediction. A typical fit[24] at FNAL energies is

$$\frac{d\sigma}{dm^2} \sim \frac{1}{(m^2)^{1.85}} + A/s; \tag{21}$$

even when allowed for, the triple-Pomeron, $1/m^2$ term is set to zero by the computer fitting programs; currently there is no evidence for such a term and certainly it is not dominant at FNAL energies. We await comparable analyses at ISR energies.

III. FRAGMENTATION SPECTRA

The third distinctive region discussed in traditional low-p_T phenomenology is the so-called "fragmentation region" in which one observes, inclusively, a single fast hadron carrying a large fraction of the maximum allowed energy. The classic example is

$$pp \to \pi + X \tag{22}$$

where asymptotically, in the center of mass, $x_F \equiv 2p_\pi/\sqrt{s}$ is larger than $\sim .5$. Before discussing this situation it is useful to briefly review quark fragmentation in e^+e^- annihilation.

a) Quark Fragmentation

In Fig. 6 we illustrate the simplest QCD diagram for fragmentation of a quark in the e^+e^- final state to a fast meson.

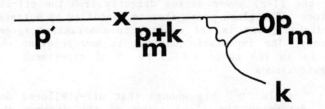

Fig. 6. One diagram for Meson Fragmentation in the e^+e^- final state.

In a frame with $p_T' = 0$ we define

$$p_M = (z \sqrt{s}, \frac{p_T^2}{z\sqrt{s}}, p_T, 0) \qquad (23)$$

so that z is the "+" momentum of the fragmenting meson as a fraction of the "+" momentum of its source quark. Asymptotically z becomes equal to the usual fragmentation variable

$$z \cong \frac{p_M \cdot q}{q^2} . \qquad (24)$$

As $z \to 1$, it is easy to calculate from Fig. 6 that the fragmenting quark is forced far off-shell

$$(p_M + k)^2 \overset{z \to 1}{\sim} \frac{p_T^2 + m^2}{z(1-z)} \equiv "Q^2" \to \infty . \qquad (25)$$

Also the gluon is forced far off-shell. The contribution of this amplitude (calculated in Feynman gauge) to the distribution of the meson is[25]

$$\frac{dN}{dz d^2 p_T} \equiv D_{M/q}(z,p) \overset{z \to 1}{\sim} \frac{1}{6\pi} \frac{(1-z)^2}{p_T^4} C_F^2 \alpha_s^2("Q^2") I_M^2("Q^2") \qquad (26)$$

where I_M can be directly related to the meson form factor in the Brodsky-Lepage formalism[8]

$$F_M(Q^2) = \frac{4\pi C_F \alpha_s(Q^2)}{Q^2} I_M^2(Q^2) . \qquad (27)$$

In (26) the (1-z) power arises directly from the off-shell quark propagator. (The gluon propagator is cancelled by a trace factor.) The predicted power is, of course, in substantial agreement with experiment. The important question is how relevant this simple diagram is in the range z = .5 to .9 of experiment. There are three basic issues.

i) Is "Q^2" big enough that off-shellness does indeed determine the (1-z) power of the diagram shown? The answer is yes. For $\langle p_T^2 \rangle \sim .3$ GeV^2 and $m^2 \sim .1$ GeV^2, "Q^2" $\gtrsim 1$ GeV^2 and propagators are sufficiently off-shell that the suppression obtained asymptotically as $z \to 1$ is already taking place. One should compare to the early days of deep inelastic scattering where $x_{Bj} = -q^2/2p \cdot q$ is theoretically relevent on the basis of a quark diagram calculated as $q^2 \to \infty$ and was experimentally relevent even at $1 < q^2 < 5$ GeV^2.

ii) Is "Q^2" large enough that perturbative calculations at the lowest order are sufficient? The answer is probably not. Just as in comparing deep inelastic data at a detailed level to theoretical calculations at low q^2, one must at low "Q^2" consider more complicated diagrams than Fig. 6 involving real gluon emission and virtual gluon vertex corrections. However, also as in deep inelastic scattering, one anticipates that the gross prediction, i.e. the power law in the present case, of the lowest-order calculation is not greatly modified.

iii) Is the normalization of the result (26) plus other diagrams contibuting at the same order large enough to contribute substantially to the data? This question is, of course, related to ii) since a large normalization of the low-order calculation relative to data suggests that higher-order corrections are not overwhelming. The answer depends on the definition of α_s and how large α_s at "Q^2" is. It has been suggested that higher order corrections are minimized in the momentum subtraction scheme of defining α_s.[26] In this scheme $\alpha_s($"Q^2"$)$ is substantial at low "Q^2" and the normalization of our low-order contribution is significant. A comparison at fixed high $p_T^2 > 2$ GeV^2, where "Q^2" is > 5 GeV^2, would help minimize scheme dependence in the comparison, but data for this has not been published.

In any case, encouraged by the success of this lowest-order off-shell approach to quark fragmentation, we extend it to other situations. We will search for the lowest-order perturbative graph with the leading $(1-z)$ power law as $z \to 1$ that can be drawn for a given fragmentation process. We claim that higher order corrections may yield logarithmic modifications for moderate z but will not significantly alter the basic power law obtained for a given fragmentation process from the leading low-order diagrams(s).

b) Deep Inelastic Target Jet Fragmentation[27]

In deep inelastic scattering we believe that the fragmenting target jet has an underlying two quark content and tend to view fragmentation of this two-quark jet independently of the process which established the jet. The main point of this section is to show that this is incorrect in the $z \to 1$ limit and more generally whenever the off-shell calculations (of the type done for e^+e^- annihilation in a)) are relevant. We only quote results. Fragmentation of a simple two quark system to a meson in lowest order

345

of QCD perturbation theory is sketched in Fig. 7. As s of the
meson approaches 1 the quark labelled q_2 <u>must</u> by pure kinematics,

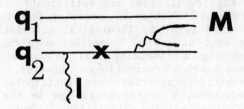

Fig. 7. Diquark fragmentation at lowest order in QCD. X=off-
         shell quark line.

be far off-shell, just as in the e^+e^- case considered earlier.
This off-shellness must come from somewhere. In Fig. 7 the gluon
labelled ℓ provides this off-shellness. Provided there are no
hidden additional suppressions coming from the source supplying the
ℓ-gluon one obtains from the single off-shell quark propagator

$$D_{M/\text{two-quark jet}}(z) \sim (1-z) . \qquad (28)$$

The deep inelastic situation, is however, completely different! Typical diagrams for proton target jet fragmentation to a π^+
are illustrated in Fig. 8.

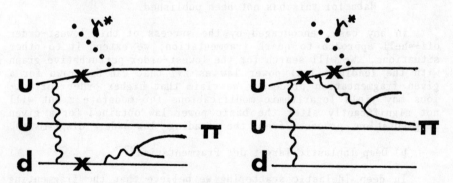

Fig. 8. Typical diagrams for target jet fragmentation to a
         π^+ in deep inelastic scattering.

In both amplitudes two quark propagators are forced far off-shell
(as always gluon off-shell propagators are compensated by trace

346

factors) and one obtains

$$D_{M/target\ jet}(z) \sim (1-z)^3 . \qquad (29)$$

(The rule for the power is 2x(# of uncompensated off-shell propagators)-1.) The virtual photon γ^* is the only source of off-shellness in this situation and properly routing this off-shellness so that the π^+ can have z→1 is not as straightforward as in Fig. 7. The power law of Eq. (29), which results from the need for an "extra" off-shell quark propagator relative to Fig. 7, is in excellent agreement with experiment.[28] As far as I know this is the only calculation of this power law currently in the literature.

c) Hadronic Collision Fragmentation[29]

It has often been claimed that the experimental observation of the power law (29) in deep inelastic scattering is what one expects by comparison to the observation of

$$D_{\pi/p}(z) \sim (1-z)^3 \qquad (30)$$

in pp collisions. (Here $z \equiv x_F$.) The idea is that in pp collisions one of the quarks in the fragmenting proton is "slowed down" and that the fragmenting system has the same two quark nature as for the deep inelastic target jet. In the z→1 off-shell approach this reasoning is completely erroneous. In fact, the power laws observed, (29) and (30), are predicted to be the same <u>only because the fragmenting jets are different</u>.

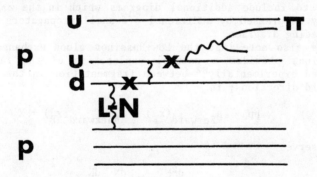

Fig. 9. One of several low order leading diagrams for pp→π^+X. X's indicate off-shell quarks as z→1.

A typical diagram in the Low-Nussinov model of pp collisions that is responsible for a leading contribution to $pp \to \pi^+ X$ is given in Fig. 9. There are two off-shell propagators in this case because the off-shellness provided by the Low-Nussinov gluon must be routed through an additional quark compared to the elementary two-quark fragmentation depicted in Fig. 7. The prediction of this model is precisely the observed power law Eq. (30) and relies crucially on the presence of three quarks in the fragmenting proton jet. Thus neither $D_{M/p}$ nor $D_{M/target\ jet}$ corresponds to "naive" two-quark jet fragmentation, in this off-shell approach.

As an aside we mention that higher Fock states for the fragmenting proton in the deep inelastic target jet and proton-proton collision cases do not alter either the power laws (29) and (30) or the comparison to "two-quark" fragmentation when the two-quark system is in a higher Fock state. In all cases the extra $q\bar{q}$ pairs in the higher Fock states are transmitted without off-shell suppression into a higher Fock state of the meson fragment.

Thus it is incorrect to assume that the experimental similarity of $D_{\pi^+/target\ jet}$ to $D_{\pi^+/p}$ implies an underlying similarity of the fragmenting jets. In the off-shell approach the similarity in the D's requires a dissimilarity in the fragmenting jets.

The results of the off-shell propagator procedure for all types of single particle fragmentations.

$$H_1 p \to H_2 X \qquad (31)$$

have been thoroughly covered in Ref. 1 and Ref. 29. The predicted power laws compare well with those observed. In Ref. 1 it was also shown that only a restricted class of diagrams contributes to the leading power law of a given fragmentation. Other approaches such as the valon/recombination[30] model or dual topological model[31] appear to include additional diagrams which in the $z \to 1$ limit are clearly suppressed by additional off-shell propagators relative to the leading diagrams.

We also note that the Low-Nussinov gluon exchange model for collisions provides a natural explanation of the factorization observed experimentally[32] between fragmentation in the forward and backward directions; in

$$pp \to \pi_{Forward}(x_F) + \pi_{Backward}(x_B) + X \qquad (32)$$

one observes

$$\frac{d^2 N}{dx_F\ dx_B} \cong \frac{dN}{dx_F} \frac{dN}{dx_B} . \qquad (33)$$

In the Low-Nussinov gluon exchange model no longitudinal momentum is transferred between beam and target constituents; no constituents are annihilated or slowed down. Thus the beam and target will fragment simultaneously according to their separate single particle fragmentation spectra as in (33).

IV. SUMMARY

In conclusion we first emphasize that the current universality of e^+e^-/deep inelastic multiplicities compared to hadron-hadron multiplicities should not survive asymptotically according to the perturbative QCD calculation based on the Low-Nussinov one-gluon-exchange approach to hadron-hadron collisions. At current energies, however, multiplicity appears to be primarily controlled by phase space considerations which could have a universal character. A more sensitive discriminator is provided by $\langle p_T^2 \rangle_{final\ hadrons}$ which is predicted to show little growth in hadronic collisions compared to that seen in e^+e^- annihilation. Some evidence in favor of this prediction was presented at this conference.

In the fragmentation region the observed single particle power laws are easily understood using the Low-Nussinov gluon exchange interaction mechanism coupled with minimal low-order QCD diagrams for the fragmentation. These diagrams dominate rigorously in the $z \to 1$ limit where internal propagators are kinematically forced far off-shell. At the moderate z values of experimental measurement higher-order corrections may be important but appear not to greatly alter the leading asymptotic power law predictions. Work is in progress to consider two particle correlation spectra in fragmentation. There is a tendency for other models to yield harder spectra for pairs of particles than for a single particle; e.g. in pp collisions $n(\pi^+\pi^-) < n(\rho)$. This is contrary to experiment. In the off-shell QCD approach the diagrams for two-particle emission are more complicated than those for one-particle emission, leading to more off-shell propagators; thus we anticipate better agreement with experiment.

Finally we have seen that the perturbative QCD model of hadronic interactions has several unique features that distinguish it from more conventional approaches. Most important are the large enhancement of diffractive excitation expected in, for example, π-Nucleus collisions and the predicted absence of the "triple-Pomeron" dm^2/m^2 spectrum in diffractive single nucleon collisions.

Thus a simple, successful and predictive approach to low-p_T interactions can be formulated on the basis of low-order QCD perturbative diagrams. Certain additional definitive tests have been suggested. If the model agrees with these additional experimental tests the inevitable question of how to understand its success within the context of non-perturbative confinement will take on increased urgency.

349

REFERENCES

1. J.F. Gunion, "QCD and Low-p_T Physics", to be published in "Partons in Soft Hadronic Processes", Ettore Majorana Centre for Scientific Culture, Erice, Italy.
2. F.E. Low, Phys. Rev. D12, 163 (1975); S. Nussinov, Phys. Rev. Lett. 34, 1286 (1975).
3. J.F. Gunion and D.E. Soper, Phys. Rev. D15, 2617 (1977). The original model of Ref. 2 differs substantially from that presented in this paper.
4. Alan White, CERN-TH-Preprint '81.
5. See D. Morrison's talk at this conference.
6. J. Pumplin and E. Lehman, Zeit. f. Phys. C, to be published.
7. G. Bertsch, S.J. Brodsky, F. Goldhaber, and J.F. Gunion, Institute for Theoretical Physics Preprint, May 1980, Phys. Rev. Lett. to be published.
8. G.P. Lepage and S.J. Brodsky, Phys. Rev. D22, 2157 (1980).
9. S.J. Brodsky, P. Hoyer, C. Peterson, and N. Sakai, Phys. Lett. 93B, 451 (1980); and S.J. Brodsky, C. Peterson, and N. Sakai, SLAC-PUB-2660 (1980).
10. J.F. Donoghue and E. Golowich, Phys. Rev. D15, 3425 (1977).
11. H. Wachsmuth, Proceedings of the Photon-Lepton Conference, Fermilab, 1979.
12. See the review of K. Konishi, Proceedings of the XI International Symposium on Multi-Particle Dynamics, Bruges, Belgium (1980).
13. Early discussion of such comparisons appears in: S.J. Brodsky and J.F. Gunion, Phys. Rev. Lett. 37, 402 (1976) and Proceedings of 7th International Colloquium on Multiparticle Reactions, Munich 1976; J.F. Gunion, Proceedings of the 13th Rencontre de Moriond, Vol. I (1978).
14. G. Bertsch and J.F. Gunion, UCD-81-4.
15. A. Bassetto, M. Ciafaloni, and G. Marchesini, Nuclear Phys. B163, 477 (1980), and Phys. Lett. 83B, 207 (1979). W. Furmanski, R. Petronzio, and S. Pokorski, Nuclear Phys. B155, 253 (1979).
16. See H.V. Martyn, Proceedings of the 1981 Moriond Conference, Les Arcs, France.
17. R. Feynman and R. Field, Nucl. Phys. B136, 1 (1978). The calculations discussed by Ref. 16 include all modifications as incorporated by the experimental groups at DESY.
18. S.J. Brodsky and J.F. Gunion, Phys. Rev. Lett. 37, 402 (1976).
19. See the summary talk by W. Kittel at this conference as well as the many relevant shorter contributions. See also M. Basile et al., Phys. Lett. 92B, 367 (1980) and Ref. 20.
20. W.T. Myer et al., paper #6 presented at this conference.
21. S. Uhlig et al., Nucl. Phys. B132, 15 (1978); T. Kafka et al., Phys. Rev. Lett. 34, 687 (1975).
22. R.D. Field and G.C. Fox, Nucl. Phys. B80, 367 (1964).
23. G. Bertsch and J.F. Gunion in preparation.
24. R.D. Schamberger et al., Phys. Rev. D17, 1268 (1978).

25. J.F. Gunion and W. Iley, in preparation; the other contributing graph (in Feynman gauge) is more complex in nature and will be dealt with in this paper.
26. W. Celmaster and D. Sivers, Phys. Rev. D23, 227 (1981).
27. J.F. Gunion, in preparation.
28. See for instance: N. Schmitz, Proceedings of 1978 Lepton Photon Symposium (Fermilab), for the experimental distribution. See also H. Lubatti, Proceedings of "Partons in Soft Hadronic Processes", Erice ('81).
29. J.F. Gunion, Phys. Lett. 88B, 150 (1979). The spectator counting rules given in this reference are valid only for single particle spectra.
30. See the lectures by R. Hwa in "Partons in Soft Hadronic Processes", Erice ('81).
31. See the lectures by A. Capella, ibid.
32. G.J. Bobink et al., Phys. Rev. Lett. 44, 118 (1980).

DISCUSSION

ANDERSSON, LUND: In connection with the multiplicity growth, I seem to understand that you have your logarithmic growth in the center [of phase space] due essentially to the fact that you do have a $1/k_T^2$ [tail] in your gluon spectrum.

GUNION: Yes, there is a component [to the logarithmic growth] in e^+e^- annihilation which comes from the $1/Q^2$ tail. Whether that is the only component is this question of how important is the phase space.

ANDERSSON: Let me just state my question before you answer. You know it is an interesting fact which I have been aware of for a few years (and I think that a lot of other people also are aware of even if we don t discuss it) and that is essentially that the multiplicity growth in the center seems to be centered at very small p_T's. In our own experiment (as a matter of fact, a very beautiful experiment done at the ISR despite the fact that it is very difficult to get very close to the beam pipes) these people [the British-Scandinavian group at the SFM with G. Jarlskog as spokesman] were able to derive the pion spectra etc. down to about 85 MeV in p_T for somewhere in the center. The interesting fact is that between 85 MeV and up to about 150, let's say 170 MeV, there is a strong growth with energy over the ISR range in the multiplicity. Outside of 200 MeV there is no growth.

GUNION: That seems like it could fit in with this idea that the growth in the small p_T region could be kinematic, because I think that is where kinematic effects would occur . . at the small p_T domain. Whereas, if it was the perturbative growth, then one would see it at large p_T^2.

VAN HOVE, CERN AND SANTA BARBARA: I have a couple of questions. The first one is an experimental one concerning the existence or otherwise of the triple Pomeron and, more completely, the $1/M^2$ part of the high-mass diffraction dissociation spectrum. Obviously the data on this to look at are the ISR data [at the highest available energy] where, if there is any evidence [on the M^2 dependence], it is to be found most clearly. And this has to be done before you can make any statement on whether one needs it or not. Then another question I would like to ask is concerning your central multiplicities. Maybe you could show the diagram, the basic diagram, by which you calculate the multiplicity in the central region in the hadron case.

GUNION: That's, of course, a set of diagrams but there they are.

VAN HOVE: Now obviously these diagrams are missing certain elements in order for color to be neutralized between the forward and the backward hemisphere because there is a color octet being exchanged.

GUNION: Yes, of course. We are just doing two comparable comparisons between e^+e^- annihilation and hadron collisions. . . that is we are computing, at the moment, just the lowest-order gluon radiation. One has to worry about the higher-order corrections which are related.

VAN HOVE: Now another reason why you expect that, in reality, the diagrams are much more complicated is that the multiplicities that you want to obtain at high energy are very high. You never can get the high multiplicity from a diagram with a small number of legs unless you interpret as a Regge-Mueller diagram.

GUNION: No, wait a minute. Let me give you an example. If you had an Abelian theory then you can compute the multiplicity of radiated photons just from the one-photon emission diagram because they are Poisson-distributed.

VAN HOVE: Exactly.

GUNION: So if QCD were Poisson-distributed then this calculation would be exactly right. Now, of course as we know, QCD is not exactly Poisson-distributed.

VAN HOVE: No, it is not Abelian group. So things are much more complicated.

GUNION: It is not an Abelian group. But, nonetheless, the jet calculus experience in the e^+e^- case shows us that the one-gluon calculation of the multiplicity assuming Poisson distribution is not too misleading.

VAN HOVE: I see that point, but I believe that, in the jet calculus, by the fact that you only use tree diagrams you are well off. Whereas, in your case it is much more complicated. Once you have several of these emissions there are going to be overlaps between your lines which will induce, perhaps, further cancellations.

GUNION: No, I would claim that, in the leading-log approximation, you would have to use tree diagrams again in this case because of this contiguity requirement. You would discover that you couldn't populate the central region if you tried to separate things.

VAN HOVE: I see. And you believe that that is commuting then? I don't think it will commute.

GUNION: Commute with what?

VAN HOVE: I mean the SU(3) algebra will still give you complications.

GUNION: Yes, very much like the e^+e^- case. If we had had an Abelian group, then we would not have gotten that extra exponential enhancement of the one-gluon $\log^2$s. I expect such an effect, of course, in the hadron-hadron case . . . that, instead of getting a constant, I'll probably get a little bit of increase from the non-Abelian structure in the higher orders. But I still expect this to be a representative calculation in that I expect there to be a difference between the e^+e^- and hadron. But of course until we have done the higher-order calculations there is no guarantee of that. That's obviously something that needs to be done.

HWA, OREGON: I have three comments to make. First is about your comment on the e^+e^- fragmentation function calculation in the recombination model. You said that we do not take the off-shell-ness into account. That is incorrect. We certainly do take the off-shell into account.

GUNION: Not the "z goes near 1" off-shell-ness.

HWA: When we calculate the fragmentation function, we use the perturbative QCD to calculate the quark distribution and then recombine. When we recombine, we only recombine quarks or valons at low momentum, low off-shell.

GUNION: I understand. But none of your quark lines are off-shell.

HWA: They are all off-shell. But they are not off-shell by an amount controlled by z.

HWA: Okay, they are controlled by . . . .

GUNION: By another Q^2, a phenomenological decreasing Q^2.

That's the difference I wanted to bring in.

HWA: The second point is that your emphasis on the off-shell-ness comes in because you consider x or z going very close to 1. Therefore, it is only off-shell when x is very close to 1. But yet you compare with data where x is not very close to 1.

GUNION: That's the point that I make. How far back can you extrapolate?

HWA: This is always the question.

GUNION: That involves higher-order corrections.

HWA: Higher-order corrections. And I never feel convinced that, for the data we have at hand, it forces us to worry that much about off-shell-ness since x is not that near 1 for low-p_T reactions. Thirdly, you talk about the gauge invariance, but that seems to be a burden of those who do perturbative QCD calculations. Because when you do perturbative calculations you do have to worry about that. But, if you want perturbative calculations, you do have to worry about the size of the diagrams and the many other terms which you don't consider. For others, who would just give up from the very beginning because they don't think that you can take into account order diagrams, gauge invariance is not a burden any longer.

GUNION: Yes, but then why do you trust your normalization? If you haven't accounted for the interaction mechanism . . . .

HWA: We don't trust it on the basis of perturbative QCD calculations.

GUNION: But, if you ignore the interaction mechanism and that interaction mechanism affects normalization, how can you . . . ?

HWA: I calculate the normalization from looking at nuclear attenuation, for example. I know how much momentum is lost. I know a fast quark loses a small amount of momentum. Otherwise we would not have factorization. Therefore I know the normalization.

GUNION: I don't see how that tells you the normalization, I guess. Just one other comment on the relevance of this diagram. As I say, one indication of relevance is the normalization of the diagram that I compute. It's certainly a diagram that appears. It is the dominant one as z goes near 1. If it were very tiny, then I would have to agree with you that probably it is irrelevant. But, as I say, if you use a momentum subtraction scheme which is some way of controlling the higher-order corrections then actually normalization is quite reasonable. But certainly there is still room for disagreement until the higher-order corrections have been done.

RAJA, FERMILAB: Does the Low-Nussinov Pomeron Reggeise? And if so what is the first realization?

GUNION: Well, nobody has attempted to compute the higher order. (This is again a question of higher order correction as to how the Reggeization might take place or not take place.) Now Alan White, of course, has done a computation involving quark-quark scattering where you have this whole infrared mess and there you get a Reggeization. But I don't think it is relevant because there is no infrared problem if you use color-singlet scattering hadrons. So one has to repeat what has been done in the higher-order calculations to try to see if one can systematize things in order to understand whether there is Reggeization occurring or what.

RAJA: But, if it does Reggeise, would the first realization not be gluonium rather than a quarkonium state?

GUNION: Yes, it would be a gluonium state or something like that.

EDITORS NOTE: The paper by J. Gunion contains material from two separate talks given at the Symposium. The material in Section III was presented separately in a talk following that of U. Sukhatme in Session D. The rest of the discussion occurred after that talk.

BIAŁAS, KRAKOW: I would like to point out that in the analysis of A-dependence in the scattering from nuclei you can separate the fragmentation of quarks from the fragmentation of, let's say, [diquarks] or something because they give different A dependence. In fact such analysis was done and it's quite consistent with what Sukhatme said . . . . that at high x a diquark essentially fragments independently. At low x it becomes mixed, and so on. So anyway you really can have a handle on that by studying A-dependence. A-dependence separates different contributions.

GUNION: I understand what you are saying but, even in your situation, as it goes through the nucleus there has to be something that is supplying the off-shell-ness to the diquark in order to allow the fragmentation. And so, what you are seeing there is analogous to the target jet fragmentation in deep-inelastic scattering. It is not the simple response of the diquark to some internal probe. It is analogous, though, to the deep inelastic fragmentation.

ANDERSSON, LUND: Can you explain to me the reason why your Nussinov gluon must necessarily attach to your d quark when you want to make a π^+?

GUNION: Oh, I only drew one of many diagrams for the proton.

ANDERSSON: Why couldn't it directly apply to that quark which you are going to use in connection with your π^+?

GUNION: Well, it can but you get the same number of off-shell propagators.

ANDERSSON: Why?

GUNION: Well, I'd just have to draw it for you. I have, though, considered all the diagrams of course. They often have the same behavior.

VAN HOVE, CERN AND SANTA BARBARA: Do I understand correctly that now the latest version of the counting rules is $(1 - x)$ with an exponent which is twice the number of off-shell propagators minus 1. Is that now the general rule? These rules have changed very often, of course, as the facts demanded. I would like to have a clear statement of the rule today.

GUNION: The correct statement is that you must calculate the underlying QCD diagram. Now I wrote down some rules that are applicable only to the hadronic fragmentation situation in the Physics Letters paper. Those rules are only good for the hadronic fragmentation situation. The general statement is that you have to calculate the diagram and see what you get. And the correct thing is what you say. As you look for the number of uncompensated off-shell propagators . . . .

VAN HOVE: That's what you mean by hadron fragmentation. Obviously you looked at hadrons.

GUNION: But I mean in the situation where you have a hadron like a proton colliding with some other hadron producing a fragment in the final state. In hadron-hadron collisions.

VAN HOVE: Then the accepted rule does not apply here?

GUNION: Certainly not to the target jet case. No.

PIETRZYK, CERN: You mentioned that in the case of proton fragmentation you have three fragmenting quarks. In the case of di-quark fragmentation you have two fragmenting quarks. So naively one could think that the difference is 33%. One can use different language. One can say that the proton is composed of three valence quarks and plenty of gluons and plenty of sea quarks and, in case of deep inelastic scattering, one removes a very little piece of it . . . just a pointlike quark. And in both cases a hadronic system composed of plenty, plenty, plenty partons . . . 10, 20 . . . with dimensions of one fermi is fragmenting. So one can just naively think that there should be very little difference [in the fragmentation]. It's just a problem of the language one is using.

GUNION: Well, as we discussed, the rules that I write down are strictly valid only as this z variable goes to 1 and things are forced off-shell. And then there is this question of continuity back to the experimental region. Regardless of how many extra quark-anti-quarks or gluons you have, the rules that I just gave are correct as z goes near 1.

SCHMITZ, MUNICH: You discussed the fragmentation into π^+. What about the [fragmentation into] π^-? Is it the same?

GUNION: Well, neglecting the isospin breaking inside the proton it would be the same. But of course there is some suppression of the d quark inside the proton. So you have to take that into consideration.

SCHMITZ: But for the diquark fragmentation there is a difference between π^+ and π^- experimentally, as I will show.

GUNION: That's right. And in a naive approach like this you don't quite get that because you don't take account of whatever dynamics it is that is causing the d quark to be different from the u quark inside the proton. You have to have a more complete Bethe-Salpeter treatment to do that.

HWA, OREGON: I agree with you that, as the saying goes, "any similarity is purely coincidental" with very different physics involved. I feel that we ought to ask the experimentalists to look not for similarity but for differences. It's kind of like the early

days of deep inelastic scattering where people say "Its scaling!"
So all the experimentalists say its scaling until someone says
"It should not be scaling!" so everybody begins to see that there is
scaling violation. So, also here, if we stay away from the $x = 1$
region, then clearly in the e^+e^- case, or in the deep inelastic
scattering case, there is a large Q^2. That is not present in the
low-p_T hadronic reactions and there ought to be differences. It
would be nicer for experimenters to find the difference rather than
find the similarity.

GUNION: Well, I agree.

DeGRAND, SANTA BARBARA: Have you looked at target
fragmentation? (Notice I did not say the word that everyone else
here is saying.) Are there differences between target fragmentation and soft-hadron fragmentation other than the trivial ones of
[flavor] quantum number conservation? For instance, are power
laws different for other things . . . say K^- or . . . ?

GUNION: Obviously one interesting question is exotic things like
K^-, for instance [which] was mentioned. There are a lot of
diagrams that you get involved with and you start having to
compute them all. So far I think there are [not many differences]
despite the difference between the two jets.

PREJUDICES AND SCENARIOS IN SMALL TRANSVERSE MOMENTUM PHYSICS

Thomas A. DeGrand

Department of Physics, University of California, Santa Barbara, CA
93106

ABSTRACT

I review contemporary theoretical ideas about the dynamics
of hadronic interactions and hadronization at large energy and small
transverse momentum.

INTRODUCTION

In this talk I will summarize recent work on the theory of
small transverse momentum reactions. This subject has always been
difficult to model from any first-principle point of view. That
statement remains true these days, even though we have a candidate
theory for the strong interactions-quantum chromodynamics. In the
other subjects discussed at this conference we are able to think of-
and in some cases to carry out - comparisons of the data with
predictions of QCD. Such is not the case in the subject of small
P_t reactions - the relevant momentum scales are small so that the
effective couplings of quarks and gluons are strong, and therefore
(to give my conclusions in the first paragraph) there are no
reliable QCD predictions for small transverse momentum reactions.
Instead of a theory of small P_t reactions, we have a
metatheory - a set of general rules which serve as a framework
for discussion. That metatheory is the parton model as formulated
by Feynman over ten years ago. Most contemporary models of small
$-P_t$ reactions are a combination of elements of the parton model
and of what the inventor of the model imagines to be relevant
ingredients of QCD, prejudices about dynamics leading to scenarios
for the dynamics of the reaction.
I would like to discuss first the metatheory (as a theory
itself is lacking) and various current models in the context of the
metatheory. I will probably be a bit severe in my judgements - the
problem is, no one can really calculate anything and so the discussion is often reduced to the level of esthetics: is one model
which emphasizes certain features of QCD and the parton model better
than one which emphasizes different features, when (as usually is
the case) both models fit data equally well? "De gustibus non
disputandum est," is unfortunately the rule rather than the
exception.
The outline of the talk will proceed as follows: first
I will talk about the metatheory, the parton model, in the light of
various QCD-based prejudices. Then I will discuss hadronization in
the central region, and at large x_F. As a sample problem I will
briefly discuss polarization in inclusive hadron production since

361

it is the one subject in this field which I have been thinking about.
Then some conclusions.

THE METATHEORY PLUS COMPLICATIONS

Feynman's book[1] tells us that the assumptions underlying
the parton model are
1) The wave function of a hadron in the infinite momentum
frame is made of partons whose
2) transverse momentum is limited,
3) wave function depends only on x_F and
4) behaves like dx/x at small x so that some partons, the
"wees", always have finite momentum as $p \to \infty$.
5) The behavior of the "wees" is the same for all hadrons.
6) The probability of a rapidity gap y in the parton
distribution is proportional to $e^{-\Delta y}$.
7) Finally, parton interactions are short range in rapidity.
This implies short range correlations in rapidity, the
independence of the central plateau with respect to
particle type, and factorization.
It is still difficult to realize the parton model quantitatively using QCD, although many qualitative features are present.
For example, we assume that the proton "at rest" is a very simple
object - three valence quarks. When we boost to an infinite
momentum frame, time dilitation generates a more complicated wave
function by "stretching out" or enhancing the quantum fluctuations
into higher quark number states (which are always present in a
quantum field theory) into a finite fraction of the wave function.[2]
We cannot calculate the actual mechanics of this process, and two
classes of guesses have grown up to describe it. They are inspired
by the two Feynman graphs (1a) and (1b).

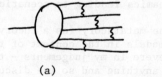   

     (a)   (b)
Fig. 1

Graph (a) is supposed to represent a Bethe-Salpeter (BS)
wave function. The rationale for looking at it is that the wave
function, including the many parton component, was prepared far in
the past and has held together ever since; therefore the partons
must all be bound to and exchanging momentum with one another. From
this picture follows all the counting rule phenomenology of
Blankenbecler, Brodsky, Gunion, and collaborators.[3]
Graph (b) is inspired by structure function evolution and
will lead to additive quark model predictions for reactions. It
also gives a "natural" explanation of the dx/x wee spectrum (which
in the BS language only arises for infinite-component Foch states).
Rudy Hwa likes this picture.[4]

In principle (and in practice, too) both sets of graphs
contribute to any given reaction, and except for isolated regions
of phase space where honest perturbation theory calculations can be
performed, favoring one set over another is so far pretty much a
question of taste.

One rule which has become very complicated with the advent
of OCD is the requirement of short range correlations. Massless
vector bosons (gluons) are designed to give long range correlations
(that's how flashlights work). It is commonly thought that short
range correlations arise because hadrons are gauge singlets - bound
states characterized by some correlation length Λ - which decouple
from all gluon excitations of wavelength $\lambda \gtrsim \Lambda$. This solution to the
problem will not be seen in any calculation which is of finite
order in perturbation theory since bound states are an intrinsically
non-perturbative phenomenon. The same color screening is also
thought to account for limiting transverse momentum effects -
hadrons are soft, extended objects.[5]

Semiclassical models in which the proton remains three
valence quarks (in a bag?) until the moment of collision do not
seem to me to be in accord with parton model tenets or (more
importantly) with the little time dilitation argument just alluded
to.[5,6,7] Some versions of semiclassical models predict exponentially damped production of heavy flavors, $n^\sim e^{-m}$, which appear to
be ruled out by charm production data (see below): they predict
a ratio nonstrange: strange: charm $\sim 1: 1/3 : 10^{-10}$. In my opinion
semiclassical descriptions are of only limited applicability for
the hadronization problem which is a strong coupling problem
totally dominated by the effects of quantum fluctuations.

THE LOW-X_F REGION

Now that we have prepared an initial state we can collide
it with a target. What happens next is completely unknown. In the
parton model with its short range correlations the left-moving wees
of one particle interact with the right-moving wees of the other
particle. The coherence of the incident wave functions is destroyed
and the partons fall apart/decay/recombine into a many-hadron final
state. This transformation first affects the wees and then
propagates outward in rapidity until it reaches the leading partons,
who are the last to hadronize. This "inside-outside" cascade is
the antesis of fragmentation models in which fast partons decay into
slower hadrons and partons which themselves decay, etc. It has been
shown to be the correct picture of hadronization in solvable
confining models, such as the Schwinger model.[8]

Indeed, it was also shown years ago that in reactions such
as e+e- annihilation sequential fragmentation generally could not
result in color neutralization. If the original quark and antiquark
don't communicate someone will have fractional charge, and confinement is lost.[9]

Incidentally, fragmentation functions are also found in two
dimensional models. In two dimensional QCD when one calculates
inclusive production a+b→c+X (a,b,c hadrons) one finds the same

fragmentation functions as in e+e- annihilation

$$\frac{d\sigma}{dx}(a+b\to c+X)=G(x)=\frac{d\sigma}{dx}(e^+e^-\to c+X)$$

This universality arises nonperturbatively and occurs because regardless of the origin of the initial quark, it is rescattered through intermediate bound state resonances, and it loses knowledge of its origin.[10] In four dimensions the G(x)'s are thought to be similar but not necessarily identical.

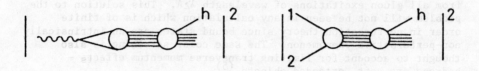

Fig. (2): H H and e^+e^- interactions in two dimensional QCD.

The two salient features of the small x region are the constancy of cross sections and the height of the plateau in the central region.

QCD provides no totally convincing explanation of the constancy of total cross sections. Most of us presume that it is related to the presence of vector gluons in the perturbative sector of the theory, since vector exchange leads to constant cross sections and (in lowest nonvanishing order, exchange of two gluons), an imaginary elastic scattering amplitude.[5] The main reason for skepticism is that the small momentum carried by the gluons in these models makes perturbation theory untrustworthy. In Low's model, for example, to get σ_{pp} = 40 mb requires α_s = 2.

Ratios of cross sections are much easier to understand. They are controlled by the transverse size of the hadrons. Thus $\sigma(pp) > \sigma(\Sigma p) > \sigma(\Xi p)$ and $\sigma(\pi p) > \sigma(Kp) > \sigma(\phi p) > \sigma(\psi p)$. In this model the $\pi p/pp$ ratio reflects the number of constituents only so long as the wave functions are dominated by minimum-constituent configurations of one $q\bar{q}$ pair for the meson and one qqq triplet for the baryon.

The height of the central plateau is another thorny problem. Here the hot question is whether or not dN/dy or its integral, the multiplicity <n> is the same in pp and e^+e^- collisions. Theory differs in its predictions:

a) Color separation models[3] give $n(pp) = c\; n(e^+e^-)$ where c = 9/4, as long as color interference effects are neglected in the former case. When they are included the s-dependence of $n(pp)$ and $n(e^+e^-)$ are quite different, with $n(e^+e^-)>n(pp)$ at high energies. Practitioners of these models have made an implicit assumption that soft-gluon multiplicity, calculated perturbatively, is directly proportional to hadron multiplicity in the central region.

b) Simple two-chain models[11] also favor c>1 since e^+e^- collisions produce only one chain. One could check these models by comparing correlations among hadrons in the central region; they should be different in pp and e^+e^-.

c) Leading particles are an important part of <n>. Should one remove the leading protons in pp[12]? What about the leading quarks in e^+e^-?

And there is an experimental problem: a much larger fraction of e^+e^- cross section involves charm than pp reactions do. The multiplicity from the decay of hadrons containing these particles is high and at present machine energies they populate the whole rapidity region. This enhances e^+e^- multiplicity over what it would be if only u and d quarks were being produced.

The best experiment is to compare dN/dy's near y=0, rather than multiplicity at large enough energies and with a small enough y bin that all leading particle effects can be excluded.

LARGE X_F

It is well known that the valence quantum numbers of a fragmenting hadron are found in its fast fragments in a hadron reaction or at large x_{Bj} in deep inelastic scattering, and one cannot help but wonder if there is a connection. Depending on its quantum numbers, one can imagine that a produced hadron can carry off 0, 1, 2, or 3 valence quarks from the proton. It is an important first question to ask whether this picture is self consistent by considering proton fragmentation into many final states with probabilities a_i for i valence quarks to combine into a baryon. This subject has been analyzed by Van Hove[13] who finds that a consistent parameterization can be made, with

$$a_0 = 0.05\pm0.03 \qquad\qquad a_2 = 0.30\pm0.16$$
$$a_1 = 0.20\pm0.04 \qquad\qquad a_3 = 0.45\pm0.17$$

so that 95% of the time baryon number conservation involves at least one valence quark.

We are also beginning to see experiments on fragmentation of hadrons which have lost a quark in a hard scattering, such as target fragmentation in deep inelastic scattering. The reactions may serve to distinguish among models designed to explain triquark fragmentation. As a first question, do the two leading quarks stick together and form a baryon or fragment independently? Sukhatme, Lassila, and Orava have recently analyzed this question and find that the answer is somewhere inbetween.[14] I would like to see model-builders test their models on these systems.

(A parenthetical polemic: in this subject the words "diquark fragmentation" often appear. I do not like them. Are proton initiated reactions "triquark fragmentation"? The words also imply a connection between pointlike probe-induced reactions and correlations in soft reactions ($pp\to\pi\pi X$ for example) where one may not exist, and in fact does not exist below a rather shallow level in all models with which I am familiar).

Models for the large X behavior of hadron distributions may be classified according to their response to Ochs' observation,[15] which I will express as a question: Is there any direct relation between parton distributions measured in low-Q^2 deep inelastic scattering and hadron distributions in hadron-hadron collisions?

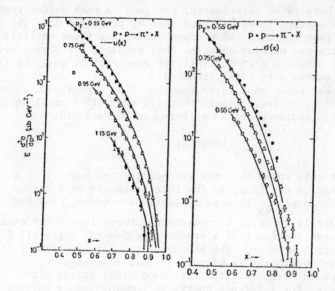

Fig. 3: Ochs' observation: lines are quark distributions from deep inelastic scattering; data, $Ed\sigma/d^3p(pp \rightarrow \pi X)$.

People who like Bethe-Salpeter equations[3] answer No, shapes of spectra are governed by the minimum number of spectators, both as members of bound states or radiated as final state particles. Members of this school like to quote spectra in terms of counting rules

$$\frac{dN}{dx} = \sum_i c_i (1-x)^{p_i}$$

where p= 2x(number of "bound state" spectators)+(number of "radiated" spectators)-1 and the sum on i runs over all possible contributions. The values of the constants c_i lie outside the model. B.S. model predictions have generally been quite successful as the accompanying figures show[16, 17].

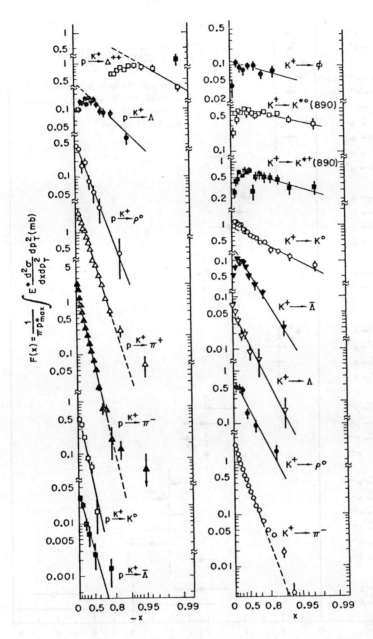

Fig. (4) Inclusive data[16] and counting rule-type power fits for a variety of reactions.

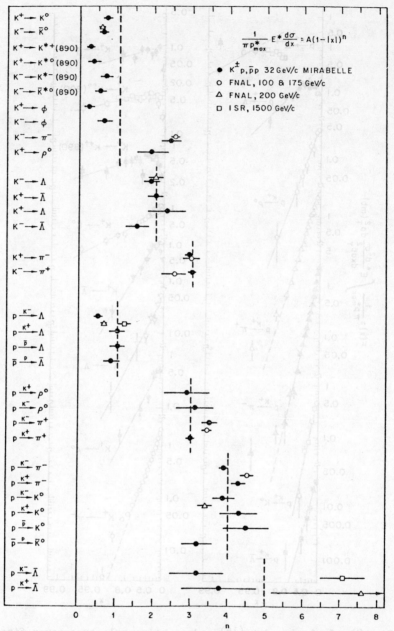

Fig. (5): How do the counting rule powers[17] agree with power law fits to data[16]? The predictions are the dotted lines.

These models are based firmly on perturbative QCD, involving as they do the calculation of bound-state acattering via the evaluation of Feynman graphs for the exchange of massless quarks and gluons. In "extreme" regions of phase space, when all particles not inside bound states are far off shell, the predictions of these models are the predictions of QCD - for form factors, quark fragmentation at very large x, elastic scattering at large angle, etc. But when one computes hadronization spectra with these models, one is making an additional assumption: that there is a smooth continuation from large Q^2 to small Q^2, and that the ghosts of perturbation theory (the p_i's) survive into the nonperturbative regime.

People who like recombination models[4] say Yes. In these models, meson production is given by

$$\frac{dN}{dx_F} \propto \int \frac{dx_v}{x_v} \frac{dx_s}{x_s} F_{vs}(x_v, x_s) R(x_v, x_s, x_F) \delta(x_v + x_s - x_F)$$

where F_{vs} is a combined probability function for a valence quark to be found at x_v and a sea quark at x_s and R is a "recombination function" which measures the relative probability that the quarks at x_v and x_s will recombine into the meson. Neither F nor R may be measured directly by experiment, so a complete model must assume (or fit them to) some functional form. In particular, one makes contact with Ochs' conjecture by the following chain of argument:

F_{vs} at large $x_v + x_s$ is biggest when x_s is very small, where $F_{vs} \sim F_s(o) F_v(x_v + x_x)$ and F_v looks like the structure function for a quark as measured in low-Q^2 deep inelastic scattering. Generally R is rather smooth so that

$$\frac{dN}{dx_F} \propto F_v(x_F) \cdot \int \frac{dxs}{x_s} R(x_F - x_s, x_s) \text{ where the integral is}$$

slowly varying. In the limit of no variation, Ochs' conjecture is exact.

Model building here centers around the F_{vs}'s and people have tended to be very elaborate in their construction. The Kuti Weisskopf model has been popular[18] (as much for its computational ease as for anything else). Rudi Hwa[4] has an interesting model based heavily on a perturbative QCD scenario with the following assumptions:

1) The quarks and gluons seen in deep inelastic scattering at moderate Q^2 evolve from "clumps" of partons, the "valons", whose dynamics dominate the low Q^2 hadronic wave function: three valons bind together to make a proton, for example.

2) Recombination occurs via the sequence

i→valons in i→quarks→valons in f→f

(actually the quarks are the valons in f). Valon distributions are determined using the QCD evolution equations and SLAC data; the valons-to-quarks transition by a perturbatively-inspired ansatz. The fit to data is good. As before, there is an additional assumption of smoothness from high to low Q^2, although no explicit calculations of inclusive production are performed at large Q^2.

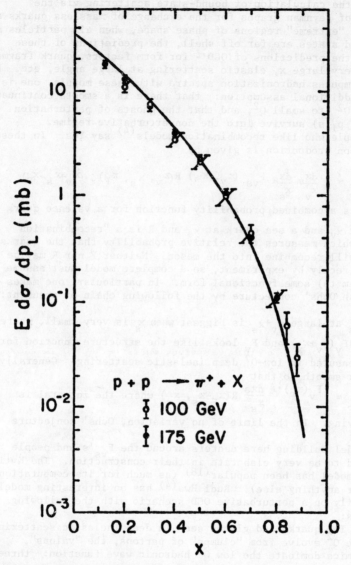

Fig.(6): Hwa's fit of $pp \rightarrow \pi^+ X$.

It is very hard to extend these models to the central region or to study disfavored fragmentation (such as $p \rightarrow K^-$). At low X most of the proton's momentum is carried by glue: one must parameterize both the glue's x distribution and its relative propensity to fragment into different kinds of matter. Answers are much more model dependent than for the favored case.

Fragmentation models [7, 11, 19] also say No. In these models

$$\frac{dN}{dx} \sim \int_x^1 \frac{dz}{z} F_{q/p}\left(\frac{x}{z}\right) D_{h/q}(z)$$

with F a quark distribution in the proton and D a fragmentation function for the hadron from the quark. If D is the same fragmentation function measured in e^+e^- annihilation then F has <u>no</u> relation to the $F_{q/p}(x)$ measured in deep inelastic scattering - <u>if</u> it were, $\frac{dN}{dx}$ ($p \rightarrow \pi$) would be too steep. (As a parenthetical comment, one often sees $F_{q/p}(x) \sim \delta(1-x)$. This choice of F is not allowed in any realistic bound-state model for a hadron, where no constituent is allowed to carry all the momentum of the bound state with non-vanishing probability). To my eye these models can fit data as well (or poorly) as the other models and to my unprejudiced mind there doesn't have to be any relation between $F_{q/p}$ and vW_2. And indeed, nowadays all models use some combination of recombination and fragmentation so that differentiation between models is not found on the conceptual level, but at the level of details.

Recent measurements of sizable inclusive charm production at large X_F have their best explanation in B.S. models, where $D°$ or Λ_c production arises from recombination of the uudc$\bar{c}$ Fock state of the incedent proton's wave function.[20] All five quarks have the same velocity, so $\langle x_c \rangle = \langle x_{\bar{c}} \rangle$ is large, about 0.3, because the charmed quark's mass is so big. This distribution is reflected in the flat or peaked shape of the $D°$ or Λ_c spectrum. It is hard for me to imagine how the data can be reproduced with a valon-type picture, since the charmed quark distribution required to fit the data is more sharply peaked than the light valence quarks' distributions, from which it is fragmented. It is also hard for me to imagine how to reproduce the data with a semiclassical model in which the c$\bar{c}$ pair is produced by string popping with an e^{-cm^2} spectrum - the exponential suppression is enormous.

The production of "hidden charm" - Ψ, Ψ', etc. - in the model must be greatly suppressed compared to "open charm" production in order to avoid conflict with Ψ production data. A 10^{-4} suppression is required, which seems to me to be hard to obtain. The basic idea is that if anything happens to a c$\bar{c}$ state after it is produced, it will decay into D's or Λ_c's, while an excited c$\bar{u}$ or cud state can only decay into open charm. So Ψ production is suppressed compared to Λ_c or D.

A SAMPLE PROBLEM: POLARIZATION

A problem which has shown some interesting developments in the past year involves the study of polarization of inclusively produced baryons in pp and pN collisions. The polarization is measured with respect to the normal to the scattering plane, $\vec{n}$, defined for $B \to B' + X$ as $\vec{n} = \vec{P}_B \times \vec{P}_{B'}$; the polarization asymmetry is

$$P = \frac{\sigma(B'\uparrow) - \sigma(B'\downarrow)}{\sigma(B'\uparrow) + \sigma(B'\downarrow)}$$

Very good data now exists, and shows that (for $B'=\Lambda$) P is negative, depends nearly linearly on the Λ's transverse momentum, weakly on x_F and (nearly) not at all on S. Other B' 's have been measured:

$$P(\Lambda) = -P(\Sigma^+) = P(\Xi^\circ) = P(\Xi^-). \quad (21)$$

Trying to explain these remarkable results illustrates all the difficulties inherent in the field. I know of two models: neither is particularly compelling.

If we first assume a recombination-type model and use the quark model to look for regularities we learn that the Λ data require the strange quark to recombine slightly more spin down than up (the ud diquark is spin zero) but the Σ/Λ ratio (= -1) requires that the (spin one this time) ud diquark also recombine, this time slightly more up than down (If they do nothing, $\Sigma^+/\Lambda = -1/3$). As the s quark is a sea parton and the u and d quarks are valence partons, we see that the data are explained by the simple rule: slow quarks recombine with their spins down, fast quarks with their spins up. This rule also naturally accounts for the Ξ data and makes lots of predictions ($P(K^- \to \Lambda)/P(p \to \Lambda) = -1$, for example). (22) Now the question is, what dynamics give this rule?

$B \to B'$ transition	Polarization	
$p \rightleftharpoons n,\ \Sigma^- \rightleftharpoons \Xi^-\ \Sigma^+ \rightleftharpoons \Xi^0$	$-\frac{20}{21}\epsilon + \frac{1}{42}\delta$	
$p \rightleftharpoons \Sigma^+,\ n \rightleftharpoons \Sigma^-,\ \Xi^- \rightleftharpoons \Xi^0$	$\frac{1}{3}\epsilon + \frac{2}{3}\delta$	Fig. (7): Predictions of
$p, n \rightleftharpoons \Lambda^0$	$-\epsilon$	the "fast-up, slow-down"
$\Sigma^+, \Sigma^-, \Xi^-, \Xi^0 \rightleftharpoons \Lambda^0$	$-\frac{2}{3}\epsilon + \frac{1}{6}\delta$	model(24): ϵ is the quark asymmetry, δ the diquark
$p, n \rightleftharpoons \Sigma^0$	$\frac{1}{3}\epsilon + \frac{2}{3}\delta$	asymmetry. The data
$\Sigma^+, \Sigma^-, \Xi^-, \Xi^0 \rightleftharpoons \Sigma^0$	$-\frac{20}{21}\epsilon + \frac{1}{42}\delta$	favor $\epsilon = \delta$.
$p \rightleftharpoons \Xi^0, \Xi^-, \Sigma^-$ $n \rightleftharpoons \Xi^0, \Xi^-, \Sigma^+$	$-(\frac{1}{3}\epsilon + \frac{2}{3}\delta)$	
$\pi, K^+ \to \Lambda$	$-\frac{1}{2}\delta$	
$K^- \to \Lambda$		

A semiclassical model due to Andersson, Gustaffson, and Ingelman (23) assumes that during hadronization the S-quark is produced by the popping of a string of color flux attached to the fast valence quarks. The string is assumed to be in the ground state with respect to its transverse degrees of freedom. If the $s\bar{s}$ pair is produced at some position x_o, the quarks must tunnel some distance $\sim M/K$ from x_o before they can go on shell (K is the string tension). If they have some $k_\perp$ with respect to the string, they will also have an orbital angular momentum $\ell = \vec{r} \kappa_\perp$ perpendicular to the string's axis. A $k_\perp$-S correlation is introduced by the assumption that the spin of the s quark tends to oppose its ℓ. This model has a phenomenological problem in my opinion in that it seems to account only for the "slow quark down" part of the quark model rule, giving

$$P(\Sigma^+)/P(\Lambda) = -1/3, \text{ not } -1.$$

Another semi-explanation has been proposed by Miettinen and me.[24] We noticed that the confinement force increases the x_F of the strange quark from a small value x_s (~ 0.1) in the proton where the s is part of the sea to $\frac{1}{3} x_F$ in the Λ when the s is a valence quark.

In addition, this force is not parallel to the quarks' velocity when the Λ and hence the quarks which form the Λ have transverse momentum. This is a natural situation for a spin-transverse correlation to arise which will be able to account for all the qualitative features of the data: opposite polarizations for fast and slow quarks since confinement slows down the fast quarks, no asymmetry for $p \rightarrow \bar{\Lambda}$ since all the antiquarks in the proton have identical x distributions, and a bigger effect as x_F increases since the change of the sea quarks' momentum also increases. If the long distance confining potential is a world scalar, then Thomas procession is the origin of the asymmetry, and we predict the right sign for the asymmetry. However, at this point we lose control of the calculation and are once again confronted with a strong coupling problem. Our crude fits to the data hint that we may be close to understanding the kinematics of polarization, but the dynamics are still mysterious.

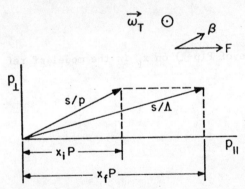

Fig. (8): Kinematics of slow quark recombination from the proton (S/P) into the Λ(S/Λ). $\vec{\omega}_T$ is the Thomas frequency, proportional to $\vec{F} \times \vec{\beta}$.

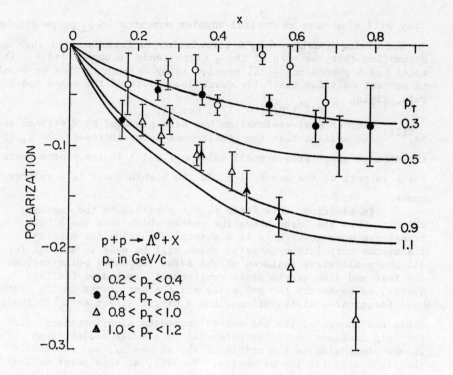

Fig. (9): Dependence of $P(p \to \Lambda)$ on x_F in the model of ref[24]. Data from ref.(21).

CONCLUSIONS

The phenomenology of small transverse momentum physics is in excellent shape. Using simple theoretical ideas it is possible to fit nearly all available data. Future tests of correlations will serve to discriminate between the various models which I have discussed, as will improved charm and polarization experiments. Hopefully surprises await!

And yet the theory of small transverse momentum physics is in rather poor shape. The connection of the phenomenology with some underlying Lagrangian which describes the short-distance behavior of hadronic interactions and with some renormalization group scheme which allows a continuation to the long-distance regime is not understood at all. For example, why does perturbation theory-based phenomenology work so well? No one knows. These are the problems we theorists must understand before we can say that we have small transverse momentum reactions under control, and before we can attain that understanding we have a long way to go.

ACKNOWLEDGEMENTS

I would like to thank Bill Shephard and his colleagues at Notre Dame for organizing such an enjoyable conference. I have benefited from useful discussions with Jack Gunion, Carsten Peterson, Bob Sugar, and L. Van Hove. I thank the Institute for Theoretical Physics for its hospitality. This work was supported by the National Science Foundation.

REFERENCES

(1) R.P. Feynman, Photon-Hadron Interactions, W.A. Benjamin, Inc., 1972.

(2) Cf. the discussion by J. Bjorken, Lectures given at the International Summer Institute in Theoretical Physics, DESY, Hamburg, 1975; SLAC-PUB-1756.

(3) For complete references, see J. Gunion's talk, these proceedings.

(4) For complete references, see the talks of R. Hwa and Xie Qiu-bing, these proceedings.

(5) F.E. Low, Phys. Rev. D12, 163(1975), S. Nussinov, Phys. Rev. Lett. 34, 1286(1975).

(6) A. Casher, H. Neuberger, and S. Nussinov, Phys. Rev. D20, 179 (1979).

(7) B. Andersson, G. Gustafson, and C. Peterson, Z. Physik C1, 105 (1979).

(8) A. Casher, J. Kogut, and L. Susskind, Phys. Rev. D10, 732(1974).

(9) J. Kogut, D. Sinclair, and L. Susskind, Phys. Rev. D7, 3637(1973).

(10)R. Brower, J. Ellis, M. Schmidt, and J. Weis, Nucl. Phys. B128, 125(1977).

(11)For complete references, see A. Capella, U. Sukhatme, and J. Tran Tranh Van, Z. Physik C3, 329(1980).

(12)M. Basile, et al., Phys. Lett. 95B, 311(1980).

(13)L. Van Hove, CERN preprint TH.2997(1980).

(14)U. Sukhatme, K. Lassila, and R. Orava, preprint FERMILAB-Pub-81/20-THY (1981) and U. Sukhatme, these proceedings.

(15)W. Ochs, Nucl. Phys. B118, 397 (1977).

(16)D. Denegri, et al. Phys. Lett. 98B, 127(1981).

(17)J.F. Gunion, in the Proceedings of the XI International Symposium on Multi-Particle Dynamics, Bruges, Belgium, 1980, E. DeWolf and F. Verbeure, eds.

(18)T. DeGrand and H.I. Miettinen, Phys. Rev. Lett. 40, 612(1978)
T. DeGrand, Phys. Rev. D19, 1398(1979).
E. Takasugi and X. Tata, Texas preprint (1981) plus references therein.

(19)B. Andersson, G. Gustafson, and C. Peterson, Phys. Lett. 69B, 221(1977) and 71B, 337(1977).

(20)S.J. Brodsky, C. Peterson, and N. Sakai, Phys. Rev. D23, 2745 (1981) and C. Peterson's talk, these proceedings.

(21)K. Heller, in the Proceedings of the XX International Conference on High Energy Physics, Madison, Wisconsin, 1980.

(22)At this conference R. Ross presented data for $K^+ \to \bar{\Lambda}$ which are in agreement with the fast-up, slow-down model (his preprint quotes the opposite convention for $\vec{n}$ (i.e. $\vec{n} = \vec{P}_\Lambda \times \vec{P}_K$). The magnitude of the asymmetry is much larger than for $p \to \Lambda$. See his talk in the Proceedings.

(23)B. Andersson, G. Gustafson, and G. Ingelman, Phys. Lett. 85B, 417(1979).

(24)T. DeGrand and H.I. Miettinen, Phys. Rev. D23, 1227(1981) and preprint UCSB-TH-27(1981).

DISCUSSION

RATTI, PAVIA: I apologize. I'm an experimentalist so I will ask
a very naive question. I think that, apart from the very simple kind
of counting rules such as average multiplicity etc., the models that
you have been talking about will always end up with Monte Carlo
calculations. I mean, it is very difficult to do explicit calculations.
If I remember correctly, almost everybody picks up some
structure functions in x, which is one variable, and tries to adopt
p_T in one way or another. Now we know that if, empirically, we
make a longitudinal phase-space calculation with the proper x
distribution and p_T distribution we explain everything. Is energy
and momentum conserved in all the evolution of the process that
takes place in dressing up the quarks with physical particles?
Otherwise, I cannot get the difference between the physical meaning
and getting the x distribution and p_T distribution which are
difficult to distinguish. We don't know how to reject any theory if
the input is close to the output.

DeGRAND: Well, I think that the naivest answer I can give, which
may not be what you want, is that in any field theory, as you know,
energy and momentum need only be conserved between the very
initial state and the very final state. If the description was anything semiclassical, then energy and momentum would be very
nearly conserved step by step as you go through things. But in a
relativistic field theory things fluctuate off shell and then back on.

MORRISON, CERN: When you were talking about Bo Andersson's
work you raised the same question as I tried to raise this
morning, which is essentially a question of causality. As you
produce a certain number of quarks and antiquarks in various ways,
they have different colors. How do they know to link up with one
another? Is it right in terms of time? I mean, if you have anti-
red here and red down there, is there time for them to get
together? Is the interaction slowed because there wasn't enough
time? What happens?

DeGRAND: It is generally very complicated. The quark here who
has some color or is in some color state, if I can imagine that,
will be interacting with every other color around him. Presumably what happens is that, if you can imagine taking a red quark
and knocking him a long way away in phase space from all of his
neighbors, he will interact with all his neighbors as if they were
one colored object. You can think of him being knocked out and
trailing a string behind him or something like that. But I can't

really give a very good calculation. I know if I do various sorts of strong-coupling calculations that there is a lot of twinkling around, but that the whole wave function overall must be a color singlet. In that sense, since the total wave function (I bring two protons in and that's a color singlet) is a color singlet, the intermediate state is a color singlet although part of the singlet may be locally colored. He does know about the color of his neighbors. I think the best way to think about this thing is maybe in perturbation theory. Perturbation theory may try to fool us but it is at least a self-consistent way of looking at these things. If you go back and recall what Jack [Gunion] was talking about, where you have verious sorts of emission, beforehand and afterhand everything is real. The colored particles realize that they are part of a color singlet and they sometime do things and then get thrust on-shell by whatever happens to them. In sort of a wrong time-ordering. And sometimes the right time ordering, or the time ordering we think is the right time ordering, is the imporatnt thing for what is going on. In a quantum field theory you often absorb stuff before you have to, if you know what I mean.

MORRISON: But do you see any problems with time?

DeGRAND: If you do it right, no. But you ought to take all time orderings into account. It may be that you can choose some gauge or choose some method of calculation in which some time ordering is much less important than another one. But we are dealing with a gauge field theory, and I can very easily do a gauge transformation or some sort of transformation in which time is very important. You know, the calculations Jack is doing that have 9 Feynman diagrams which are important are in some particular gauge. If he does another particular gauge he might have 35 diagrams whose absolute square, when all added up, will give him the same as these 9 diagrams, but everything has a different time ordering. It's like the question over here earlier about energy momentum conservation. In a field theory all possible time orderings are possible and you get them!

GUNION: Just a quick comment on the quark counting. You said that you were worried about the higher Fock states preserving quark counting for the total cross section. It turns out that, if you just simply use a bremsstrahlung model for creating the extra quark-antiquark states, then the number of extra quark-antiquarks is proportional to the original number of valence quarks.

DeGRAND: Oh sure, but what if I want to take a Fock state of, say, a proton with four quarks and one antiquark?

GUNION: That's what I'm saying. The extra number of quark-antiquark pairs is proportional, in fact, to the original number of valence quarks. A proton will have more extra $q\bar{q}$ pairs than a pion by exactly the ratio of 3 to 2.

DeGRAND: Oh yeh, you're in fine shape as long as, at some level, I can say a proton is three quarks and there is no admixture of . . . .

GUNION: That's right. You have to imagine that the extra $q\bar{q}$ pairs somehow knew about the original number of valence quarks.

DeGRAND: I was trying to be hard nosed and say you should worry about it. Has anyone looked at the three-gluon exchange with the Low Pomeron?

GUNION: Well, I've started to. It's a terrible mess! But that's part of the question that is the analog that we have to do to what Alan White had done, because we now have to include the infrared regularization. It's a very terrible business, let me tell you.

ALAM, VANDERBILT: Wroblewski has shown some very nice comparisons of multiplicities from e^+e^- and lepton-hadron and hadron-hadron. Did I understand you correctly to say that such comparison is not really valid and justified?

DeGRAND: No, I wouldn't say that. I would say that you are better off looking at dN/dy than the absolute multiplicity.

ALAM: But his fits were pretty nice.

DeGRAND: Fine. No one has presented fits at this conference which are not nice. Maybe I'm sawing off my limb after me, but I think that truth is more important than being able to fit the data. Or calculations from a Lagrangian at some level, which we can't do yet, are more important than fitting the data. I'm just trying to be nasty and picky about that. I would prefer to see the dN/dy distributions. I'm worried about how much charm there is in the e^+e^- end where it is influencing things. I can't say anything else.

ALAM: Basically, you are saying that the fit in terms of the particular variables that Wroblewski used may be purely coincidental, that it's an accident that they come out on top of each other from three different processes.

DeGRAND: Well, I did it to quote the bible here. You have said it.

KITTEL, NIJMEGEN: [A fit has been found at $y = 0$.]

DeGRAND: Yes, that's nice. And now you have to go off and measure it at 15 different energies and then Jack [Gunion] will show you how they are different. Jack's predictions are for dN/dy where $y = 0$ and not for N. That is the thing which it is best to try to compare.

SUKHATME, ILLINOIS, CHICAGO CIRCLE: Well, Tom clearly has a lot of worries. In order to make him sleep better tonight, let's set one worry to rest. In the fragmentation approach, the key element is that there is an asymmetric distribution . . . that the faster colored system carries most of the momentum. There are two ways of understanding that. First, if you use a classical DTU approach, then this [fast system] distribution comes out as $(1 - x)$ to some negative power obtained from Regge intercepts. This has been shown a long time ago by Chew, Matsuda, and other people. A more heuristic way of saying it is that the thing which remains in the middle is a wee quark. Those are the only ones which have time to interact. It is the remaining colored system which is carrying away the bulk of the momentum. It is not that the system must have that momentum. It's that if you have a proton carrying all the momentum and you hold back something with very little momentum, the remaining system naturally has all the rest. But it's a complicated system.

DeGRAND: The only comment I can make is that I'll be much happier when you can start it with a Lagrangian and a cutoff prescription.

SUKHATME: That has to come later.

SESSION C

HEAVY NUCLEUS INTERACTIONS

AND

QUARK-GLUON PLASMA

Tuesday, June 23, p. m.

Chairman: D. R. O. Morrison
Secretary: N. M. Cason

THE QUARK–GLUON PLASMA AND THE LITTLE BANG*

L. McLerran
University of Washington, Seattle, WA 98195

ABSTRACT

A space-time picture of the fragmentation and central regions is presented for extremely high energy head-on heavy nucleus colli-sions. The energy densities of the matter produced in such colli-sions are estimated. Speculations concerning the possible formation of a quark-gluon plasma are discussed, as are possible experimental signals for analyzing such a plasma.

The potential experimental study of ultra-relativistic colli-sions between very heavy nuclei presents an exciting challenge to both theorists and experimentalists. The large multiplicities, 10^3 - 10^4, arising from head-on collisions at ISR energies will pro-vide a fresh problem for detector technology. High energy densities and large multiplicities generated in such collisions offer a new panorama in which novel features of QCD may be probed.

The studies of Willis and colleagues have elucidated the possi-bilities for employing state-of-the-art detection systems to analyze heavy nucleus-nucleus collisions at ISR energies.[1] Such systems are capable of providing sophisticated calorimetry, tracking, momentum resolution, and particle identification. Equally sophisticated systems capable of carrying out similar analysis might be employed at the SPS, or the proposed Venus collider at LBL.

Recent theoretical studies provide a simple space-time descrip-tion of ultra-relativistic nucleus-nucleus collisions.[2] These studies concentrate on head-on collisions — that is, collisions with impact parameters b $\lesssim$ 1 fm. Geometrical considerations show that head-on collisions comprise $(1/2R)^2 \sim 1/2\%$ of all uranium-uranium collisions.

Head-on collisions are easily distinguished from peripheral collisions by the multiplicities of collision products. Assuming such multiplicities in nucleus-nucleus collision grow as the baryon number of the nuclei, A, the multiplicity in a head-on collision at ISR energies is $n \sim 10^3$ - 10^4. Fluctuations in peripheral colli-sions could rarely simulate such a large multiplicity.

We shall now proceed to present the results of an analysis of nucleus-nucleus collision, using the quark-parton model.[3-5] Such a collision may be analyzed in the target rest frame. The projectile nucleus appears in this frame as a thin, Lorentz-contracted pancake. This projectile pancake passes through the target during the colli-sion. The target is heated and compressed during this passage. The central region between the two nuclei is heated when the projectile emerges from the target. The projectile is heated and compressed in

*Work supported in part by United States Department of Energy.

the final stage of the collision.

A hot, dense, baryon-rich configuration of hadronic matter is formed in the nucleus fragmentation region. The energy density of this hadronic matter may be $\varepsilon \sim 2$ GeV/fm^3 for heavy nuclei such as uranium. This energy density is larger than that of the energy density of matter inside a proton,

$$\varepsilon_p \sim \frac{1}{\frac{4}{3}\pi r_p^3} \sim \frac{1}{2} \text{ GeV/fm}^3 (r_p \sim .8 \text{ fm}),$$

and may be large enough for formation of a quark-gluon plasma.[6] The decay products of the nuclear fragmentation regions are identified by their rapidities,

$$y = \ln \frac{E + p''}{m}, \tag{1}$$

where E is the particle's energy, p'' its longitudinal momentum, and m its mass. For heavy nuclei such as uranium, the width of each of the nucleus fragmentation regions is $y \sim 3$-4 units. The total accessible rapidity interval at ISR energies is $y \sim 6$-8 units. A study of the central region at ISR energies is severely limited by this kinematic restriction. The central region might, however, be studied at Isabelle energies.

If the multiplicity in the central region grows as the nuclear baryon number, A, and if the multiplicity continues to grow as it does at ISR energies, the energy density of hot, hadronic matter in the central region may be as high as $\varepsilon \sim 2$ GeV/fm^3 for heavy nuclei such as uranium. This energy density is as large as that of the fragmentation region, and a quark-gluon plasma may form. If the multiplicity grows as $A^{2/3}$, the energy density is probably too small ($\varepsilon \sim 300$ MeV/fm^3) to produce a plasma.

There is recent cosmic ray data for nucleus-nucleus collisions which favor a multiplicity growth proportional to A.[7] In Fig. 1, a

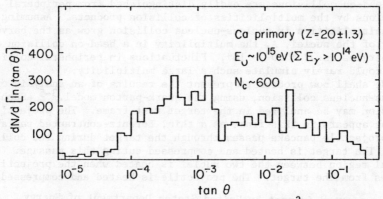

Fig. 1. A pseudo-rapidity distribution for a 10^3 TeV calcium nucleus-emulsion event.

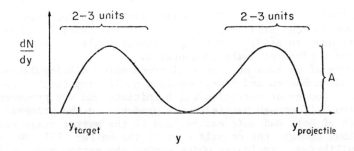

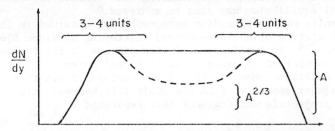

Fig. 2. Rapidity distributions for head-on nucleus-nucleus collisions: (a) The baryon number distribution (nucleon minus anti-
nucleon, (b) The meson distribution assuming heights in the central
region proportional to A(---) and $A^{2/3}$(---).

pseudo-rapidity plot for the interaction of a 10 TeV calcium nucleus
with, presumably, a carbon nucleus is shown. About 600 charged particles are produced. The density of particles in the central region
is consistent with a $\log^2$ multiplicity growth determined at ISR
energies and an A dependence, but is inconsistent with an $A^{2/3}$ dependence and a $\log^2$ s multiplicity growth.

A distinction between matter produced in the fragmentation
regions and the central region is found in the baryon number density.
This density is the difference between the number of baryons and
antibaryons per unit volume. There should be only a small baryon
number density in the central region. If a quark-gluon plasma forms
in the central region, the study of the transition region between
the fragmentation regions and central region would allow a study of
the dependence of characteristics of the quark-gluon plasma on baryon number density.

A hypothetical rapidity distribution for baryon numbers (baryons minus antibaryons) for a head-on collision between nuclei of
baryon number A is shown in Fig. 2(a). The baryon number is concentrated in a nucleus fragmentation region of the width $\Delta y \sim 2\text{-}3$. The
heights of these fragmentation regions are proportional to A.

A hypothetical rapidity distribution for mesons for this collision is shown in Fig. 2(b). If the height of the central region
were proportional to $A^{2/3}$, it would be cleanly isolated from the
fragmentation regions at Isabelle energies.

A feature of nucleus-nucleus collisions which distinguishes
them from hadron-hadron and hadron-nucleus collisions is the extremely large number of particles which initiate and are produced in
the collision. If enough particles are produced in the primeval
distribution of hot, hadronic matter, and if the matter stays hot
and dense long enough, the constituents of the matter will come into
thermal equilibrium. Estimates which employ the parton model indicate that if the height grows as A, kinetic equilibrium is established in the fragmentation region and in the central region. This
conclusion is verified within perturbative QCD for the fragmentation
region.[8] These perturbative computations also tentatively conclude
that chemical equilibrium may also be achieved.[8]

The results of the preceding paragraphs are obtained in the
parton model picture of nucleus-nucleus scattering provided Bjorken
and Gottfried.[3-4] An implication of this picture is the inside-
outside development of cascades in hadronic collisions.

In the collision shown in Fig. 3, a massless, inelastically
produced fragment separates from the projectile hadron. After a
time t, the projectile and fragment have separated

$$\Delta r_\perp \sim \frac{p_\perp}{p''} t \qquad (2)$$

and

$$\Delta r'' \sim (\frac{p_\perp}{p''})^2 t. \qquad (3)$$

In the local $p'' = 0$ frame of the fragment, this distance is

$$\Delta r_\perp \sim \Delta r'' \sim \frac{p_\perp}{p''} t. \qquad (4)$$

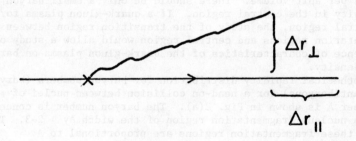

Fig. 3. Inelastic production of a particle.

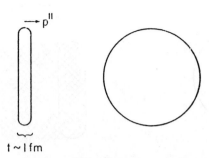

$\xrightarrow{\quad} p^{\parallel}$

$t \sim 1 \, \text{fm}$

Fig. 4. Nucleus-nucleus scattering before a head-on collision.

The fragment may be properly included as part of the projectile hadron's wave function if $\Delta r_{\perp}$, $\Delta r''\,\stackrel{<}{\scriptstyle\sim}\,1$ fm. The time needed for a fragment to materialize in a cascade is, therefore, $t \sim p''$, which is to say slow fragments materialize before the fast fragments.

Before a nucleus-nucleus collision takes place, the projectile nucleus appears as a Lorentz contracted pancake (Fig. 4). This pancake has a limiting thickness of $t \sim 1$ fm as a consequence of the low momentum wee parton component of the nuclear wave function.[3]

Not much intranuclear cascading takes place as the projectile nucleus traverses the target. Since the cascade is inside-outside, the high momentum projectile fragments are produced after the projectile has traversed the target nucleus. The projectile nucleus does not slow significantly as it passes through the target nucleus. This phenomenon of nuclear transparency leads to the conclusion that approximate Feynman scaling should be appropriate for nucleus-nucleus interactions at energies high enough so that the projectile and target nuclei begin to pass through one another.

Low momentum pions, $p''\,\stackrel{<}{\scriptstyle\sim}\,R/R_0$ where $R_0 \sim 1$ fm and R is the nuclear radius, are produced within the nucleus and may become trapped in the nucleus fragmentation region. These pions are probably initially produced with a distribution typical of proton-proton interactions, and are generated by each of the A target nucleons.

The number of pions trapped in the nucleus fragmentation region may be estimated using pp inclusive scattering data. At ISR energies, this number is 3-4 pions/nucleon for a uranium-uranium collision. This number should be a slowly varying function of energy, $n \sim \ln E$, and nucleon number $n \sim A^{1/3}$ at the highest ISR energies. The energy per nucleon which is trapped in the $p'' = 0$ frame of the struck target is $E/N \sim 3-4$ GeV/nucleon. The longitudinal γ of the struck nucleus is $\gamma \sim 2$.[2]

As the projectile nucleus traverses the target, the target becomes compressed. A simple kinetic theory estimate of the compression is

$$C \sim 2\gamma \sim 3-4. \tag{5}$$

The energy density in the fragmentation region may be estimated from the compression C, the trapped energy per baryon E/N and the

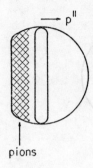 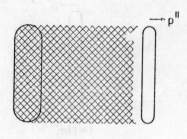

pions

Fig. 5. The projectile nucleus   Fig. 6. Matter forming in the
sweeping through the target.   central region.

energy density of nuclear matter, ρ_0, as

$$\varepsilon \sim \rho_0 \ C \ E/N \sim 2 \ \text{GeV/fm}^3. \tag{6}$$

A picture of the projectile nucleus sweeping through the target
is shown in Fig. 5. The nucleus is shown being compressed, and
pions are shown being produced in the wake of the projectile.

After the projectile emerges from the target, hot matter begins
to form in the central region between the nuclei. This matter forms
a hot fire tube which eventually joins the two nuclei.[8] The radiation from this fire tube forms the central rapidity region. The
matter forming in the central region is shown in Fig. 6.

The energy density in the central region is estimated to be $\varepsilon \sim$
2 GeV/fm^3 for uranium–uranium collisions at ISR energies and $\varepsilon \sim$
4 GeV/fm^3 at Isabelle energies. This estimate assumes the height of
the central plateau in nucleus–nucleus collisions varies as A and
with a lns growth as measured at ISR energies. The resulting value
of ε varies as $A^{1/3}$ and ln E_{cm}.

At some large time, $t \sim p''$, the hot matter produced in the central region catches up with the projectile nucleus and the projectile begins to fragment. By this time, the target and central
rapidity region particles have fragmented.

Estimates using the parton model and perturbative QCD agree
that there is sufficient time to establish kinetic equilibrium in
the fragmentation region.[2,9] During this time, the constituents of
the hot, hadronic matter scatter many times. Perturbative QCD estimates also tentatively indicate that chemical equilibrium may be
established in the fragmentation region.[9] Parton model estimates
also suggest that kinetic equilibrium is attained in the central
region.

Perhaps the most compelling reason for performing high energy

388

nucleus-nucleus collisions is to
produce a new phase of matter.
This new phase is an unconfined
quark-gluon plasma. If this new
phase of matter is produced, our
current theoretical understanding of confinement will be tested.
Since our current understanding
relies on models of the hadronic
vacuum, nucleus-nucleus collisions probe the vacuum. Nucleus-
nucleus collisions might provide
the first confrontation between
experiment and a theoretical
conception of the vacuum since
the classic Michelson-Morly experiments disproved the aether
hypothesis.

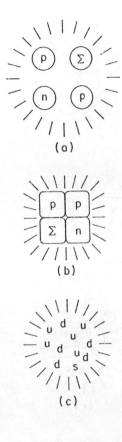

(a)

(b)

(c)

The formation of a quark-
gluon plasma at high temperatures
and densities is easily understood
as a consequence of the composite-
ness of hadrons. Consider first
the zero temperature finite baryon
number density (baryon minus antibaryon) limit. At zero temperature
and small baryon number density,
matter is a degenerate baryon gas.
(Fig. 7(a)). As the baryon density
is increased, a critical density is
reached when the baryons overlap.
(Fig. 7(b)). Above this density,
quarks and gluons within individual
baryons are no longer confined and
may travel throughout the entire
hadronic system (Fig. 7(c)).

Fig. 7. Confinement and deconfinement at zero temperature and finite baryon density: (a) Low density, (b)
Critical density, and (c)
High density.

This transition should occur
at an energy density of approximately that of a proton, $\varepsilon_p \sim$
500 MeV/fm^3. This energy density
is several times larger than that
of nuclear matter, $\varepsilon_{NM} \sim$ 150 MeV/fm^3.

At zero baryon number density and low temperatures, hadronic
matter is a gas of mesons and baryon antibaryon pairs (Fig. 8(a)).
At a critical temperature, the hadrons overlap, and quarks and gluons
may be exchanged among the hadrons (Fig. 8(b)). Above this temperature, quarks and gluons may travel throughout the hadronic system
(Fig. 8(c)). This transition occurs at an energy density $\varepsilon \sim \varepsilon_p$.

The crucial feature of these analyses is that increasing
temperature or baryon number densities results in overlapping
hadrons, and when hadrons overlap, de-confinement of quarks and
gluons may occur. These considerations suggest the phase diagram of

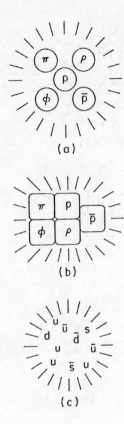

(a)

(b)

(c)

Fig. 8. Confinement and deconfinement at finite temperature and zero baryon density:
(a) Low temperature, (b)
Critical temperature, and
(c) High temperature.

Fig. 9. A hypothetical phase
diagram for hadronic matter.

Fig. 9. Reasonable estimates of T_c
and $_c$ are $T_c \sim$ 150–300 MeV and $\eta_c \sim$
3–10 η_0 where η_0 is nuclear matter
density.[10]

The breakdown and realization of
chiral symmetry may be studied in
finite temperature and density systems.[11] If chiral symmetry was realized at zero temperature and density,
the proton and neutron would have
zero mass, and the pion mass would be
of order of the rho mesons mass. At
zero temperature, chiral symmetry is
spontaneously broken and the nucleons
acquire finite mass. The pion is a
Goldstone boson of this symmetry
breakdown, and would have a zero mass
in the limit of zero up and down
quark masses.

The breakdown of chiral symmetry
may be the result of condensation of
σ-mesons in the vacuum. These σ-
mesons are composites of quark-antiquark pairs. Near the critical temperature and density at which confinement disappears, the σ-mesons may become ionized. The σ-meson condensate
might disappear, and chiral symmetry
could be restored. This restoration
might have interesting consequences
for the rates of pion and baryon production. Near the critical temperature, there might also be strong

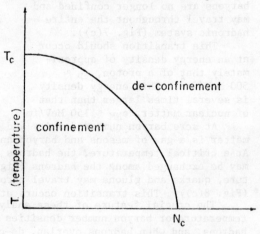

390

correlations in pion and baryon production amplitudes.

At extremely high temperatures and baryon densities, the dynamics of a quark-gluon plasma is well approximated by that of an ideal relativistic gas of quarks and gluons.[12] For high energy densities, the average momentum transfers for quark and gluon collisions are large. The asymptotic freedom of quark and gluon interactions makes effects of collisions small.

If extremely high energy densities were produced in ultra-relativistic nucleus-nucleus collisions, perturbative QCD computations could be tested. At densities which are more likely to be found, nonperturbative effects are probably important, at least for bulk properties of the system. The high energy tails of distributions nevertheless may be perturbatively calculated, and their effects determined.

The recent QCD Monte-Carlo revolution has made non-perturbative computations a practical reality. Computations of thermodynamic parameters for finite temperature systems are particularly simple.[13-15] Such computations are much simpler than corresponding computations of hadronic-hadronic spectra and wave functions.

Finite temperature SU(2) theories of gluons in the absence of fermions are being exhaustively studied.[13-15] The deconfinement transition has been found in these Monte-Carlo simulations,[13-14] and the equation of state has been studied.[15] In Fig. 10, the results of a Monte-Carlo computation of $e^{-F/T}$ is shown, where T is the temperature and F is the free energy of an isolated static quark. Below a critical temperature T_c the free energy is infinite, but for $T > T_c$ the free energy becomes finite and quarks are no longer confined. At this temperature, there is also a sharp peak in the specific heat of the gluon gas, a peak which also signals deconfinement. The deconfinement transition is generated by a breakdown in a discrete dynamical symmetry of QCD and the full implications of this symmetry breakdown are not yet understood. Monte-Carlo computations suggest, however, that the confinement-deconfinement transition in QCD is somewhat analogous to a paramagnetic-ferromagnetic transition in an Ising spin system, and the QCD system may have the rich structure characteristic of such a system.

More recent work has also established the deconfinement transition in SU(3) gluon theories.[16] Realistic computations involving fermions may even become practical in the not too far distant future.[17-18] Nucleus-nucleus collisions may provide the first quantitative test of non-perturbative QCD.

Electromagnetic signals may provide a clean probe of the plasma dynamics in its hot, early stages.[1,9,19] Electromagnetically interacting particles can penetrate through the hot, hadronic matter produced in the collision, and are radiated from the system as a whole. These particles are produced most copiously in the early stages, since the interactions between constituents of the plasma are then most energetic.

Lepton pairs and photons may provide clean probes of the plasma dynamics. These particles are directly emitted from the charged constituents of the plasma. These electromagnetic probes are very penetrating and are radiated from the entire volume of the plasma.

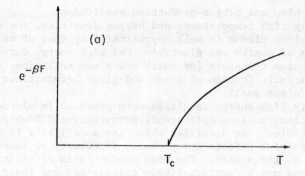

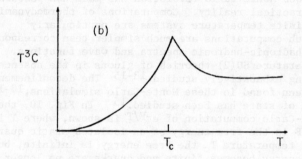

Fig. 10. Some results from Monte-Carlo simulations of the thermo-
dynamics of a gluon gas at finite temperature: (a) The statis
quark free energy F, (b) The specific heat C.

For a head-on collision, the multiplicity is $n \sim A^{4/3}$ arising from
the volume of the plasma $\sim A$ and the confinement time $\sim A^{1/3}$ (in the
fragmentation region). These electromagnetic particles are most
copiously produced in the hot, dense early stages of the plasma
evolution when the most scattering between the most energetic plasma
constituents takes place.

A large background for electromagnetic particles produced in
the plasma is direct particle production, such as Drell-Yan pair
production, generated by hard quark scattering in the initial high
energy nucleus-nucleus collision. These directly produced particles
dominate plasma produced particles at large $p_\perp$, and large masses for
lepton pairs. Gluon bremsstrahlung corrections to direct production
mechanisms generate high $p_\perp$ leptons and photons, while the plasma
production falls off as an exponential of $p_\perp$ scaled by the temperature. High mass Drell-Yan pairs are suppressed by inverse powers
of the mass compared to exponentials of mass scaled by the temperature for plasma pairs.

The plasma lepton and photon yield may be enhanced relative to
direct leptons and photons in the $X_f > 1$ kinematic region. This
kinematic region is forbidden for pp collisions. Rescattering and
thermalization of nucleons and pions generates such a component.

Fermi motion corrections to direct processes also generate particles in this region, but may be reduced at large enough X_f. The obvious problem for measurements in this kinematic region is the very low rate of lepton and photon production.

Another severe background for photons is from π^0 decays. The total number of such photons is $n \sim 10^2 - 10^3$ in the fragmentation region. The total number of plasma photons may only be $n \sim 1 - 10$.[9] The π^0 background might however be reduced at $X_f \gg 1$ since the π^0's arise from the nuclear surface which is cooler than the hot interior from which energetic photons arise.

As has been emphasized by Willis, low $p_\perp$ photons and low mass, $M < M_\rho$, and low $p_\perp$ lepton pairs may provide a signal for the quark-gluon plasma dynamics. In this kinematic region, backgrounds from resonance decays and Drell-Yan pairs may be small. This kinematic region deserves careful theoretical study to determine if signals from the quark-gluon plasma may be disentangled from contributions arising from conventional processes in hadron-hadron scattering.[20-21]

Energetic hadrons also provide information about the plasma dynamics. These energetic hadrons are very penetrating in nuclear matter. An energetic quark or gluon might directly penetrate from the hot plasma interior to the plasma surface. There is a large background, however, from high $p_\perp$ hard scattering of quarks and gluons in the initial hadron collision. This background might not be so severe for $X_f > 1$. The flavor composition of the plasma might be studied using energetic hadrons arising from the quark-gluon plasma.

Low $p_\perp$ hadrons are copiously produced at the plasma surface, and when the plasma has cooled enough for hadrons to coalesce. There are nevertheless signatures of the early stages of the plasma evolution in these plasma debris. For example, consider strange particle production.[22] The production of $s\bar{s}$ quark pairs rapidly takes place in the early stages of plasma formation. This production is not Zweig rule suppressed since colored $s\bar{s}$ quark pairs may be produced by one gluon annihilation. If the plasma lives for a long enough time, the strange quark component comes into chemical equilibrium with the up and down quark components.

Many strange anti-quarks are radiated away in kaons. The $K^+(u\bar{s})$ and $K^0(d\bar{s})$ mesons are emitted preferentially to the $K^-(\bar{u}s)$ and $\bar{K}^0(\bar{d}s)$, since the u and d quarks have large Fermi energies due to the finite baryon number of the plasma. These kaon emissions increase the plasma strangeness.

As the plasma expands and evaporates, baryons are formed. These baryons might be formed in a sea of up and down quarks which is greatly enriched with strange quarks. In the SU(3) symmetric limit where $n_u \sim n_d \sim n_s \gg n_u - n_{\bar{u}} \sim n_d - n_{\bar{d}}$, there may be copious strange baryon production. Assuming an SU(3) symmetric coalescence of quarks to form baryons, the fraction of strange baryons is

$$f \overset{\sim}{=} 1 - \left(\frac{2}{3}\right)^3 = \frac{19}{27} \sim \frac{2}{3}. \qquad (7)$$

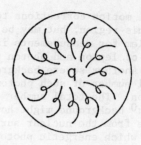

Various collective excitations of the Yang-Mills plasma might appear in nucleus-nucleus collisions. The Prasad-Sommerfield chromomagnetic monopole might be produced.[23] The effects of finite temperature and density instantons might be studied.[24] Domains of wrong Z_N field configurations might be produced.[25]

Fig. 11. An isolated quark surrounded by an ensemble of color neutral quarks and gluons.

Truly exotic phenomena might appear in such collisions. For example, if free quarks were unconfined at zero temperature,[26-27] they might be produced in a nucleus-nucleus collision more easily than in a hadron-hadron collision.[28-29]

In a finite temperature plasma, free quarks and gluons are present when $T > T_c$. This is true even for theories in which there are no light mass fermions. The octet color charge gluons may shield the triplet charge of heavy quark sources.

Suppose a quark or gluon surrounds itself with a color neutral ensemble of quarks and gluons as shown in Fig. 11. This ensemble might be regarded as a quark-gluon plasma with a finite color charge. The quark or gluon surrounded by hadronic matter is reminiscent of the picture of a free quark proposed by DeRujula, Jaffe, and Giles.

In the plasma, the color field of the excess quark or gluon might generate a Coulombic strength field at the surface of the plasma ball, $E \sim Q/4\pi R^2$. If the QCD vacuum does not confine color, this field strength might be insufficient to produce quark-antiquark or gluon-antigluon pairs as the plasma-ball separates from another plasma-ball of opposing color charge.

REFERENCES

1. W.J. Willis, CERN-EP/81-21, March (1981).
2. R. Anishetty, P. Koehler, and L. McLerran, Phys. Rev. $\underline{D22}$, 2793 (1980).
3. J. Bjorken, Lectures at DESY Summer Institute (1975); SLAC Preprint Pub.-1756 (1976)(unpublished).
4. K. Gottfried, Proceedings of the Fifth International Conference on High Energy Physics and Nuclear Structure, Uppsala (1973).
5. A. Bialas, Proceedings of the First Workshop on Ultra-Relativistic Nuclear Collisions (1979), p. 63.
6. For a review and further references see E.V. Shuryak, Phys. Rev. $\underline{61}$, 71 (1980).
7. R.W. Huggett et al., Paper submitted to the 17th International Cosmic Ray Conference, Paris, France.
8. This picture of nucleus-nucleus scattering is reminiscent of the Flying Fire-sausage picture of hadron-hadron scattering advocated by G. Preperatta and G. Valenti, CERN Preprint CERN-EP 80-47 (1980).
9. K. Kajantie and H.I. Miettinen, Helsinki Preprint HU-TFT-81-7 (1981), Z. Phys. C (in press).
10. H. Satz, Proceedings of the Fifth High Energy Heavy Ion Study, Berkeley, CA (1981).
11. N. Snyderman, SLAC Preprint, SLAC-PUB-2654 (1980).
12. J.C. Collins and M.J. Perry, Phys. Rev. Lett. $\underline{34}$, 1353 (1975).
13. L.D. McLerran and B. Svetitsky, Phys. Lett. $\underline{98B}$, 195 (1981); Santa Barbara ITP Preprint NSF-ITP-81-08 (1981).
14. J. Kuti, J. Polonyi, and K. Szlachanyi, Phys. Lett. $\underline{95B}$, 75 (1981).
15. J. Engels, F. Karsch, H. Satz, and I. Montvay, Phys. Lett. $\underline{101B}$, 89 (1981), Bielefeld U. Preprint, BI-TP 80/34 (1980).
16. K. Kajantie, C. Montonen and E. Pietarinen, Helsinki U. Preprint HU-TFT-81-8 (1981).
17. F. Fucito, E. Marinari, G. Parisi, and C. Rebbi, CERN Preprint, CERN-TH-2960 (1980).
18. D.H. Weingarten and D.N. Petcher, Phys. Lett. $\underline{99B}$, 333 (1981).
19. G. Domokos and J.I. Goldman, Phys. Rev. $\underline{D23}$, 203 (1981).
20. T. Goldman, M. Duong-van, and R. Blankenbecler, Phys. Rev. $\underline{D20}$, 619 (1979).
21. J.D. Bjorken and H. Weisberg, Phys. Rev. $\underline{D13}$, 1405 (1976).
22. R. Hagedorn and J. Rafelski, Phys. Lett. $\underline{97B}$, 180 (1980).
23. M.K. Prasad and C.M. Sommerfield, Phys. Rev. Lett. $\underline{35}$, 760 (1975).
24. B.J. Harrington and H.K. Shepard, Nucl. Phys. $\underline{B124}$, 409 (1977).
25. N. Weiss, Univ. of British Columbia Preprint, UBC-81 (1981); ibid., UBC-80 (1980).
26. A. DeRujula, R. Giles, and R. Jaffe, Phys. Rev. $\underline{D17}$, 285 (1978).
27. R. Slansky, T. Goldman, and G. Shaw, LA-UR-81-1378 (1981).
28. R.V. Wagoner and G. Steigman, Phys. Rev. $\underline{D20}$, 825 (1979).
29. L. McLerran, Proceedings of the Fifth High Energy Heavy Ion Study, Berkeley, CA (1981).

DISCUSSION

MINETTE, WISCONSIN: Have you thought about astrophysical applications of quark-gluon plasmas, such as describing the interior of neutron stars?

McLERRAN: Yes, about five years ago. It is very probable that there is a core inside a neutron star which is composed of a quark . . . not really a quark-gluon plasma, it's more a degenerate quark gas. The problem is that it is very hard to see any effects of that core. If you use a quark gas in the center of a star and you calculate its radius, it turns out to be what a radius of a neutron star is . . . 10 kilometers plus or minus a kilometer or two. Its moment of inertia turns out to be the moment of inertia of a neutron star. Just about everything turns out to be about what it would be for a neutron star. And they are (I don't know how many parsecs away they are. I'm not an astrophysicist.) a long way away. Even if you were lucky enough to get on the surface of one of these neutron stars you would still have to dig an oil well about three or four kilometers deep through nuclear matter before you would get to the quark core and learn about the physics which is going on there. That nuclear matter is very dense so it's going to be very hard to get any signal from there. It's a very tough business. There have been some speculations, based on observations of cooling rates of neutron stars, that there may be either a quark-gluon plasma core to these stars or, if there is pion condensation, the existence of this pion condensation will also allow for rapid cooling of the system, but . . . .

VAN HOVE, CERN AND SANTA BARBARA: On this same question, I think that it is more important that a physical manifestation of the quark-gluon plasma is in the early universe. When the age of the universe was about 10 microseconds, the temperature was 300 MeV and that is just above these possible temperatures for transition to the hadron gas phase. At one microsecond it was 1 GeV temperature, so hadronic matter was presumably a very hot plasma. Now, what are the consequences of this transition? First of all, the dynamics of the matter in the universe at that time was being strongly influenced by the transition. Of course, that does not necessarily leave relics right now except possibly for what you already mentioned in the case of nuclei . . . that there might be a slight mismatch of quarks and antiquarks and triplets of quarks and that possibly, when the ordinary QCD vacuum set in and began to separate the hadronic domains, there might have been some few quarks lost here and there and unmatched [i.e. unconfined in nucleons] which may then still exist in the universe.

I don't think there are reliable calculations possible until we know a little more about the phase transition for this, but this might be interesting as for physical manifestations all the same.

MESTAYER, CHICAGO: What would you see after the phase transition? Would the baryon/meson ratio suddenly change? How could it be seen experimentally?

McLERRAN: I think probably, if you really wanted something solid to look at, it would be a specific heat. If you really could put this system in a box (which you can't because you don't have a box with walls that good), I think probably the best way to study the system would be to make a plot of the specific heat of the system versus the temperature of the system. Then you would find a peak in it, presumably, if adding fermions to the system doesn't change the conclusions (which I expect it may not).

MESTAYER: How do you measure the specific heat?

McLERRAN: What you have to do with the specific heat is just measure the change in the energy density and change in temperature. You know what the size of your box is and you know what the energy inside the box is. Then you measure the change in temperature and then you can measure the corresponding change in average energy of the particles.

GARBINCIUS, FERMILAB: Is there any reason, other than the surface area to volume ratio, that you couldn't do all this stuff with just pp collisions?

McLERRAN: I think that, if this inside/outside cascade picture is correct for nucleus-nucleus collisions it would be very unlikely that there would be thermalization in a pp collision. The reason is that, when the particles are first produced in longitudinal phase space (think of a proton passing through another proton and scattering and then producing a shower) those produced particles, pions, are initially produced about a fermi away from any other pion in the inside/outside cascade picture. If they are initially produced a fermi away from any other pion in the cascade and if there is a large momentum gradient in the system (which there is because, after all, these pions are trying like mad to catch up with the projectile) then there will be no time for those pions to rescatter off one another. If they don't rescatter off one another then the system will not come into thermal equilibrium. There are other models, hydrodynamic models, which could give thermalization in a pp collision.

LBL-13029

RELATIVISTIC HEAVY ION EXPERIMENTS

Howel G. Pugh

Nuclear Science Division
Lawrence Berkeley Laboratory
University of California
Berkeley, CA 94720, U.S.A.

Invited paper presented at the XII International
Symposium on Multiparticle Dynamics, Notre Dame,
June 21-26, 1981

This work was supported by the Director, Office of Energy
Research, Division of Nuclear Physics of the Office of High Energy
and Nuclear Physics of the U.S. Department of Energy under Contract
W-7405-ENG-48.

RELATIVISTIC HEAVY ION EXPERIMENTS

Howel G. Pugh
Nuclear Science Division
Lawrence Berkeley Laboratory
University of California
Berkeley, CA 94720, U.S.A.

ABSTRACT

Objectives of high energy nucleus-nucleus studies are
outlined. Bevalac experiments on the formation of hot high-density
equilibrated nuclear matter are discussed. Future programs are
outlined, including research at the CERN ISR.

INTRODUCTION: DENSITIES AND TEMPERATURES

In this paper I will give a summary of the principal objectives
for studies of high energy nucleus-nucleus collisions. I will describe some experiments done at the Bevalac with emphasis on aspects
of most interest to higher energy studies, namely, whether an equilibrium state is produced and how its properties can be diagnosed.
Finally, I will discuss the next stages of experiment and the extension to very high energies to study quark-gluon plasmas, with the
$\alpha-\alpha$ experiments at the ISR as an important milestone along the way.
References 1-5 provide extensive material for further study.

1. Objectives

The primary objective for the study of very high energy
nucleus-nucleus collisions is to observe the properties of nuclear
matter under conditions of high density and temperature. Our
understanding of astrophysics requires a knowledge of the equation
of state of nuclear matter at densities ranging from a fraction of
normal to many times normal. Significant phase changes of nuclear
matter are suspected to occur in neutron stars; transitions to
quark or hadronic phases are possible at the centers of very dense
stars. Our understanding of cosmogenesis likewise requires a
knowledge of this subject: in the first seconds, if the "big bang"
theory is correct, the universe expanded and cooled rapidly through
a quark-gluon phase into a hot, high-density nuclear matter phase.
However, we do not know enough about the hadronic interaction to
predict the behavior of nuclear matter as the density is increased
or to predict much about the quark-gluon phase except its
existence. Experiment is therefore essential and high energy
collisions between heavy nuclei seem to be the only avenue. The
results of such studies may not only give us the empirical
information required for astrophysical and cosmological studies but
may also cast light on the hadronic interaction, especially on the
origin of confinement. If we can understand how nuclear matter
behaves when the nucleons merge into each other producing the
quark-gluon phase, we will understand the confinement mechanism.

400

2. Phase Diagram and Equilibrium Paths

Figure 1 shows a phase diagram for nuclear matter with some predictions of transitions from normal nuclear matter to hadronic

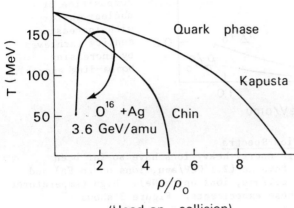

Fig. 1. Phase diagram for nuclear matter and trajectory followed during a heavy ion collision.

and quark phases. The presentation of such a diagram presupposes an equilibrium condition at each point, such as can be maintained in stars under the influence of "external", i.e., gravitational, forces. The region of the diagram that is accessible to low energy nuclear physics is confined to a small region near normal density. If, as in stars, we could increase the pressure while keeping the temperature low, we might find interesting new bound states such as pion condensates or density isomers as we proceed towards the quark-gluon phase. Increasing the temperature at constant density would also lead us to the quark-gluon phase, but we would be exploring a different aspect of its properties.

Heavy ion collisions produce at best a series of quasi-equilibrium states. The reaction path indicated by the arrow in Figure 1 is typical of a collision at Bevalac (or Synchrophasotron) energies. It was obtained from an intranuclear cascade calculation[6] using nucleon-nucleon data as input. The calculation shows that the transition to quark-gluon matter is approached even at these low energies. Such calculations are reliable as long as we do not stray too far from ordinary nuclear matter conditions. They provide a very useful basis of expectation with which to confront observations and provide information on a variety of features. They tell us that the high density period of the collision lasts only 10^{-23} or 10^{-22} seconds. We therefore have to disentangle the equation of state from the reaction dynamics, which are only to a limited extent under our control. However, there are many properties we can measure. For example, Figure 2 shows the prediction of another cascade calculation[7] how the composition of the nucleus must change as the density is increased by increasing the bombarding energy. At 1 GeV/amu, 20% of the nucleons should be converted to isobars.

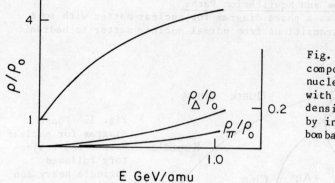

Fig. 2. Changing composition of nuclear matter with increasing density (achieved by increasing bombarding energy).

3. Temperatures from Pion Spectra

Experiments with high energy heavy ions have so far been carried out only at the Bevalac (2.1 GeV/amu, ions up to Fe) and the Synchrophasotron (4 GeV/amu, ions up to Ne). High temperatures are indeed reached in these experiments. Figure 3 shows π^-

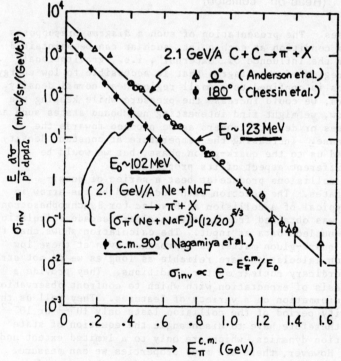

Fig. 3. Pion inclusive spectra measured in heavy ion collisions. Data from reference 8.

energy spectra from C + C collisions and Ne + NaF collisions at 2.1 GeV/amu.[8] (NaF is used to simulate a Ne target for the study of equal mass collisions.) The spectra are seen to be exponential-- this seems to be true at all angles and all energies; if the inverse slope is interpreted in terms of a temperature, the temperature is very high. The slope does not vary greatly with angle--for C + C at 0° it is $(123 \text{ MeV})^{-1}$ while for Ne + NaF at 90° it is $(102 \text{ MeV})^{-1}$.

If we plot the inverse slope (temperature) against bombarding energy, we find the systematic dependence shown in Figure 4.[9]

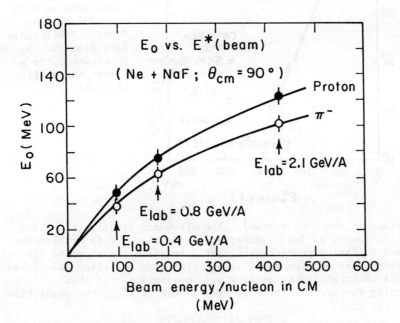

Fig. 4. Energy dependence of the "temperature" E_0 in Ne + NaF collisions. Data from reference 9.

For this, only data at 90° c.m. have been used because these may be the least dependent on assumptions made about the reaction mechanism. The temperatures observed reflect the final state of the colliding system and initial temperatures may be higher. We have little proof of the densities reached except insofar as the calculations that predict them seem to be in general accord with other observations. It should also be remarked that the inclusive spectra from which these temperatures are deduced reflect an average over impact parameters. Conditions for selected impact parameters are of greater interest, as will be discussed below.

Before turning from inclusive pion spectra to central collisions, it is of interest to show Figure 5. This presents data on pion production[8] at the laboratory energy of 200 MeV/amu (only

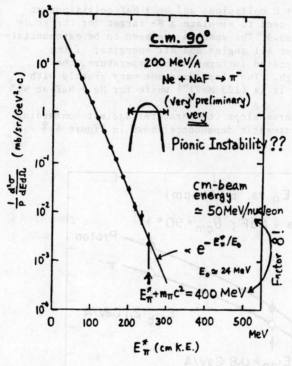

Fig. 5. Pion production at 90° with 200 MeV/amu beams. Preliminary data from ref. 8. No evidence for a pionic instability is found.

50 MeV/amu in the c.m. system). The microbarn cross sections are among the lowest so far measured at the Bevalac. This measurement was made to test a prediction[10] that a break in the slope of the invariant cross section would signal transient radiation associated with the onset of formation of a pion condensate. At this bombarding energy, no effect is seen as big as 1% of the prediction.

CENTRAL COLLISIONS

The question of whether and under what conditions equilibrium systems can be produced in heavy ion collisions is of capital importance to the whole field of research, and many studies are focused on this question. Most calculations predict that the optimum conditions for producing and studying high density states would be in head-on or "central" collisions.

1. Selection of Central Collisions and Pion Production

Central collisions are usually selected by means of a type of trigger developed by the Riverside streamer-chamber group. The principle is that in a grazing collision many nucleons in the projectile will not interact and will proceed in the forward direction ("target fragmentation"). At the highest energy of the Bevalac these fragments fall in a forward cone of half angle about 6°. Their momentum distribution seems to be consistent with the

Fermi distribution in the projectile nucleus combined with the beam velocity. The central trigger selects events in which few or no fragments of the projectile appear in the forward direction.

Figure 6 shows the schematic layout for such a trigger. The detectors respond to Z_i^2 where Z_i is the charge of each particle. Since Z_i^2 is greatest when all the charge is concentrated on one particle, the upstream detector has the maximum pulse height (corresponding to the beam particle) and the downstream detector has a continuous distribution of pulse height down to zero. Cascade calculations[11] indicate that for equal mass target and projectile this should give a unique measure of impact parameter. For unequal masses additional information would be necessary.

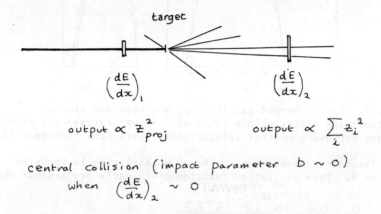

Fig. 6. Schematic diagram of the central trigger.

Figure 7 shows the effect of such a trigger selection on streamer chamber data[12] for ^{40}Ar + KCl collisions at 1.8 GeV/amu. The total number of charged particles per collision n_{ch} and the number of negative pions n_{π^-} are shown for unbiased selection of inelastic interactions (solid points) and for central collisions selected by the trigger (open points). The trigger selects 10% of the total inelastic cross section and corresponds to impact parameters less than about 2.4 fm. The effect of the trigger on both n_{ch} and n_- is striking.

The particle multiplicities shown in Fig. 7 extend up to very high values. The average number of charged particles in a central collision is $\langle n_{ch} \rangle \sim 42$, and the average number of pions is $\langle n_\pi \rangle = 3\langle n_{\pi^-} \rangle \sim 18$. It is clear that in a substantial fraction of events both nuclei must be completely disintegrated. The general picture is one of about 100 secondary particles (including neutrons) including about 20 pions.

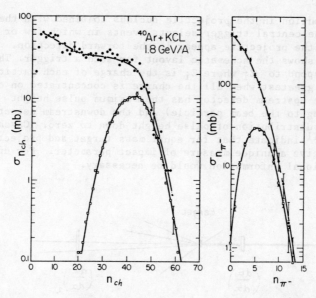

Fig. 7. Total charged particle n_{ch} and negative pion n
multiplicity distributions taken with (lower curves) and without
(upper curves) a central trigger bias. Data from reference 12.

The energy dependence of the π^- multiplicity is shown in
Figure 8. This excitation function tests early predictions that a

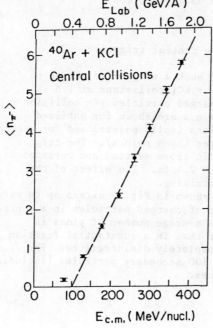

Fig. 8. Energy dependence of
multiplicity in central
^{40}Ar + KCl collisions. Data
from reference 12.

signature for pion condensation would be a step in the pion
multiplicity as a function of beam energy. There is no such step
at a level of more than a few per cent in the Bevalac energy range.

Isotropy of the particle emission and energy spectra is not
necessary for thermodynamical descriptions to be useful. For a
detailed discussion of this see Hagedorn[13]. However, it is
interesting to explore to what extent a global equilibrium is
reached in central collisions and to try to deduce the geometric
shape of the interaction region.

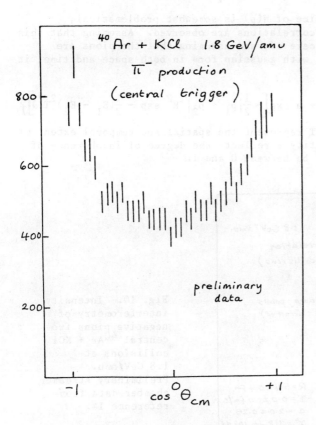

Fig. 9. Angular
distribution of
pions produced in
central ^{40}Ar + KCl
collisions at 1.8
GeV/amu. Prelimi-
nary data from
reference 14.

Figure 9 shows the angular distribution of π^- for central
collisions[14]. It is forward-backward peaked but not markedly
so. The energy spectra for central collisions are still being
analyzed. It will be interesting to compare them with the
inclusive data.

2. Intensity Interferometry

Intensity interferometry has become a much-used method for studying the source properties when multiple particle emission is probable[15]. Heavy ion collisions provide a new opportunity to use this technique. Basically, pairs of identical particles are observed (in this case negative pions) and the correlation function

$$C_2(\vec{p}_1, \vec{p}_2) = \frac{N(\vec{p}_1, \vec{p}_2)}{N(\vec{p}_1)N(\vec{p}_2)}$$

is measured. The choice of $N(\vec{p})$ is somewhat problematical, especially if strong correlations are observed. Assuming that this problem can be taken care of, and assuming that the pions are emitted from a source with gaussian form in both space and time, it can be shown that

$$C(\vec{p}_1, \vec{p}_2) = N\left\{1 + a \exp\left[-\frac{1}{2}|p_1 + p_2|^2 R^2 \exp{-\frac{1}{2}(E_1 - E_2)^2 T^2}\right]\right\}$$

The quantities R and T represent the spatial and temporal extent of the source. The quantity a reflects the degree of incoherence of the source and should be between 0 and 1.

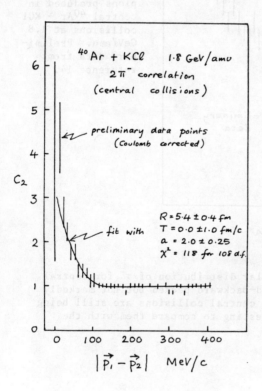

Fig. 10. Intensity interferometry of negative pions from central ^{40}Ar + KCl collisions at 1.8 GeV/amu. Preliminary streamer chamber data from reference 14.

In our experiments[14] the data were taken from the streamer chamber measurements. They were corrected for the small Coulomb repulsion between the pions. $N(\vec{p}_1,\vec{p}_2)$ was extracted using all possible pairs of pions and the background was calculated taking pairs of pions from different events of the same multiplicity.

Figure 10 shows preliminary results for the correlation and the extracted source parameters. The time parameter T is found to be zero with a large uncertainty. The source radius $R = 5.4 \pm 0.4$ fm, rather larger than the nuclear radius of about 4 fm.

The correlation parameter a seems to lie outside the range of values permitted by the simple theory. The preliminary value of 2.0 ± 0.25 is larger than any previously observed in such systems as πp, pp, etc., where it rarely approaches the value of unity. Recent theoretical work by Gyulassy[16] has shown that once the simple assumptions made about the source are relaxed, the parameter a may take on a wide range of values. Clearly this is a hot topic to pursue.

Even though it is not strictly valid, we have tried to separate out parts of the data that might reflect a nonspherical shape for the interaction region. We divided the data into forward, backward, and side cones of half angle 45° as shown in Figure 11. The extracted values for R, T, and a are also shown in the figure. The analysis suggests a nonisotropic source, but a more elaborate analysis should be performed, especially if the assumption of incoherence has to be abandoned as implied by the large value of the parameter a.

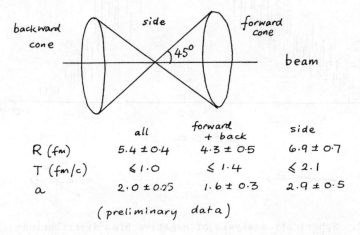

	all	forward + back	side
R (fm)	5.4 ± 0.4	4.3 ± 0.5	6.9 ± 0.7
T (fm/c)	$\leqslant 1.0$	$\leqslant 1.4$	$\leqslant 2.1$
a	2.0 ± 0.25	1.6 ± 0.3	2.9 ± 0.5

(preliminary data)

Fig. 11. Effect of selecting longitudinally or transversely emitted pion pairs on the parameters extracted by intensity interferometry. Preliminary data from reference 14.

3. Cluster Analyses

In order to extract dynamic information from cluster analyses, completely reconstructed events are desirable. So far, we have only a few dozen fully measured streamer chamber events, and these do not yet have particle identification. However, we have many events in which all the π^- have been measured, so it seemed interesting to try out the various methods on the negative pions by themselves. We have made thrust, sphericity, and minimal spanning tree analyses of these data. As an illustration a sphericity distribution is shown in Figure 12. This result appears interesting until we

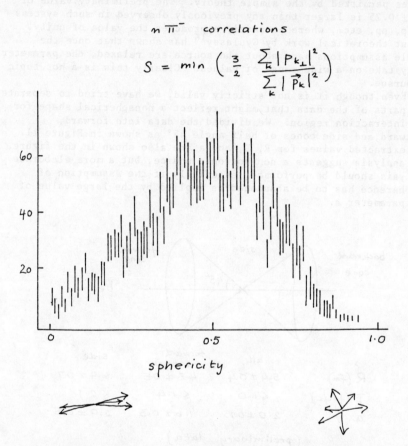

$$n\,\overline{\pi}^-\ \ correlations$$

$$S = \min\left(\frac{3}{2}\ \frac{\sum_k |P_{k\perp}|^2}{\sum_k |\vec{P_k}|^2}\right)$$

spbericity

Fig. 12. Sphericity analysis of negative pion distributions produced in ^{40}Ar + KCl central collisions at 1.8 GeV/amu. Preliminary data from reference 14.

compare with Figure 13 where the same results are shown divided by an artificial set generated by creating events of the same multiplicity by taking each pion from a different event, also of the same multiplicity. It is thus seen that the sphericity

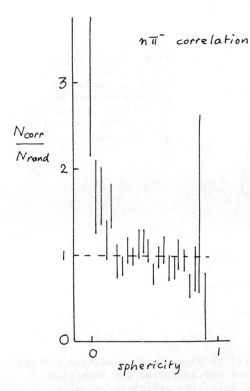

$n\,\overline{\pi}^-$ correlation

$\dfrac{N_{corr}}{N_{rand}}$

sphericity

Fig. 13. Ratio between the sphericity distribution of Fig. 12 and a distribution generated by choosing pions from different events. Preliminary data from reference 14.

distribution contains no new information beyond what was present in the multiplicity distribution and the pion spectra taken separately.

Such analyses are still in the earliest stages and we have not yet been able to apply them to data expected to contain important dynamic information.

4. Strange Particle Production

The Bevalac is close to the production threshold for K, Λ, Σ in nucleon-nucleon collisions. Figure 14 shows the thresholds and the effective nucleon-nucleon energy in ^{40}Ar + KCl collisions at 1.8 GeV/amu including the Fermi motion in both target and projectile. The majority of the particle pairs are above threshold for associated production but below threshold for pair production of kaons.

The Λ spectra have been measured at the streamer chamber for central collisions of ^{40}Ar + KCl at 1.8 GeV/amu.[17] The cross section for Λ production is 7.6 ± 2.2 mb for central collisions having a cross section of 180 mb (impact parameter <2.4 fm). This gives 0.04 Λ per central collision on the average.

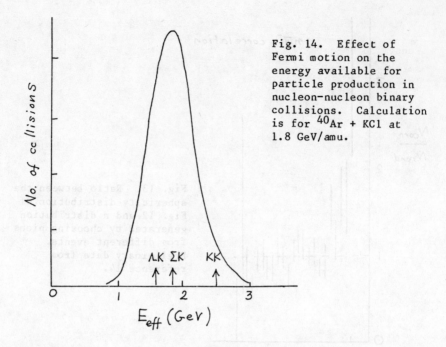

Fig. 14. Effect of Fermi motion on the energy available for particle production in nucleon-nucleon binary collisions. Calculation is for ^{40}Ar + KCl at 1.8 GeV/amu.

About 50 Λs have been observed as shown in the momentum scatter plot of Figure 15. These data are not corrected for detection efficiency. The shaded area has detection efficiency equal to zero. The circle shows the phase space limit for Λ production in free nucleon-nucleon collisions at the beam energy.

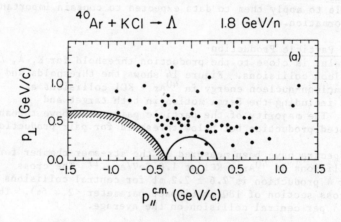

Fig. 15. Scatter plot of Λ production in the streamer chamber, as a function of $p_{\parallel}$ and $p_{\perp}$ (c.m.). Data from reference 17.

Table I shows the mean transverse and longitudinal Λ momenta (efficiency corrected). It is interesting that the average transverse momentum is about the same as the average longitudinal momentum. For an isotropic distribution it would be greater by a factor of $\pi/2$.

Table I

Average transverse and longitudinal momenta in the cm system for Λ production in ^{40}Ar + KCl at 1.8 GeV/amu, compared with various calculations

	$p_\perp$(GeV/c)	$p_\parallel$(GeV/c)
Data	0.49	0.43
AA	0.21	0.21
RS(fireball)	0.22	0.23
RS(initial)	0.28	0.64

The results of three simple attempts to explain the data are also shown in the table. The AA calculation includes only Fermi motion of the interacting nuclei. It clearly fails to introduce sufficient high momentum components. The RS (fireball) calculation shows the effect of introducing one rescattering of the (approximately the expected number) from an equilibrated system with a temperature determined from the pion spectrum. This also fails. The RS (initial) calculation uses one rescattering from a system in which all the nucleons still have their initial momenta. This boosts the longitudinal momenta by more than the required amount but still does not adequately explain the transverse momenta. The partial success for the third model is consistent with the idea that production occurs in the very first collisions, before the nucleon energies have fallen below the production threshold and a fortiori before an equilibrated system has been produced. However, this model has not explained the transverse momentum distribution. Additional data have been accumulated to study this further.

Since Λs are self-analyzing for polarization, it was easy to extract a measure of the polarization:

$$P = -0.10 \pm 0.05.$$

This result does not yet have the accuracy to complement high energy p-p data. Again we look to improved statistics using recent streamer chamber exposures.

The above measurements for central collisions are complemented by inclusive K^+ data.[8] These are shown in Fig. 16 where data from various angles have been combined. Once again it is found that rescattering of the outgoing particle is necessary to explain

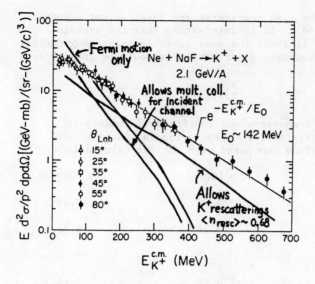

Fig. 16. The K⁺ spectrum in 2.1 GeV/amu Ne + NaF collisions. Preliminary data from reference 8.

the shape of the spectrum. Unfortunately the absolute cross section for production in nucleon-nucleon collisions is only known within a factor of two at these energies, so it is not known whether the discrepancy in absolute yield is significant.

Recent unpublished measurements at Berkeley[18] have shown that K⁻ production also occurs even though the threshold for K⁺K⁻ production (2.5 GeV) is well above the beam energy per nucleon. An interesting possibility is that the K⁻ is produced by secondary interaction of a Λ or Σ in the hot nuclear system. This would permit some interesting tests of chemical equilibrium and of the constitution of the nuclear system during the collision.

COMPOSITE PARTICLE PRODUCTION

Initial results at the Bevalac supported qualitatively the idea that deuterons and tritons are produced by coalescence of nucleons produced in some primordial fireball. Recent precise data shown some remarkable systematic behavior.[8]

It is found that a power law relation enables d, t, etc., inclusive spectra to be predicted from the proton spectra from the same reaction, i.e.,

$$E_A \frac{d^3 \sigma_A}{dp_A^3} = C_A \left(E_p \frac{d^3 \sigma_p}{dp_p^3} \right)^A$$

where the left-hand side refers to the production of fragments of mass A at momentum p_A and the right-hand side refers to the production of protons of momentum p_p where $p_p = (1/A)p_A$. The quantity C_A is a constant.

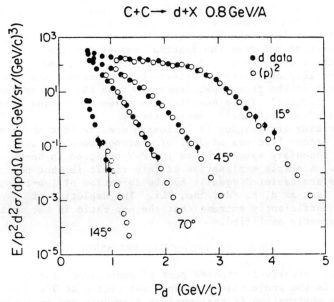

Fig. 17. Deuteron inclusive spectra from heavy ion collisions at 0.8 GeV/amu showing also the square of the proton inclusive spectra at the same angles. Data from reference 9.

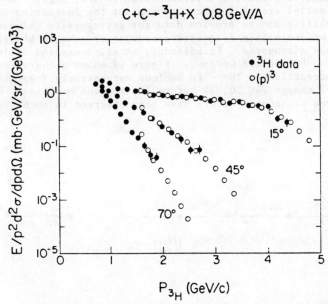

Fig. 18. Triton inclusive spectra from heavy ion collisions at 0.8 GeV/amu showing also the cube of the proton inclusive spectra at the same angles. Data from reference 9.

Figures 17,18 show how well this formula works.

While this result seems to imply a coalescence model, it would also result from local chemical equilibrium. Further information must be extracted not from the spectral shapes but from the values of C_A and their dependence on $A_{projectile}$, A_{target}, and E_{beam}. A variety of questions arise: Can we extract the source radius? What is the freeze-out density? Does the entropy change during the reaction? These questions have been addressed recently by Nagamiya[8,9] and Stöcker[19], among others.

One striking observation in inclusive spectra (not shown here) is a large excess (factors of 3-4) of neutrons compared with protons in secondary spectra below 100 MeV produced in Ne-Pb collisions. A simple explanation of this result is that neutrons and protons are depleted equally by the formation of low-isospin composites such as d, t, ^3He, ^4He, ^6Li. The depletion of protons is sufficiently extreme that the n-p ratio in the initial system is greatly amplified.

ANOMALOUS PROJECTILE FRAGMENTS

The most extensively studied part of phase space for heavy ion collisions is the projectile fragmentation region at 0°. Here the qualitative observation is that nuclear fragments are produced with velocities near the projectile velocity. The data are typically used to extract nuclear Fermi momenta. Some of the projectile fragments have very unusual neutron-proton ratios, e.g., ^{22}N or ^{44}S. Such nuclei are of interest to map out the boundaries of nuclear stability and to provide data for astrophysics calculations.

Among the projectile fragments some very remarkable objects have recently been discovered. Friedlander, et al.[20] exposed nuclear emulsions to ^{56}Fe at 1.8 GeV/amu. Figure 19 shows a characteristic chain of interactions. The ^{56}Fe nucleus successively fragments into particles of charge 24, 20, 11 before leaving the emulsion. As many as seven consecutive stars have been observed in such events.

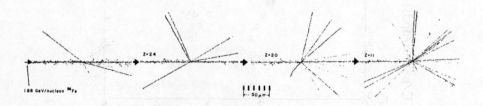

Fig. 19. A characteristic chain of interactions in emulsion following entry of a 1.88 GeV/amu ^{56}Fe (from the left). Data from reference 20.

416

For incident beam particles the distance before interaction can be used to extract a mean free path. Figure 20 shows such data for ^{16}O primaries and how a mean free path of 11.9 ± 0.3 cm is extracted. From similar data an empirical rule is derived:

$$\lambda_z = \Lambda_o \lambda^{-b} .$$

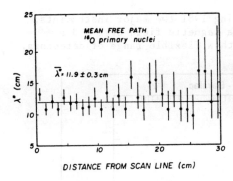

Fig. 20. Measured values of mean free path $\lambda*$ for 2.1 GeV/amu ^{16}O as a function of distance from entry into the emulsion. Data from reference 20.

If we try a similar analysis not on beam particles but on particles emerging from nuclear collisions (in the forward direction) we can use the above empirical formula to combine data with different z and accumulate good statistics. This yields the data of Figure 21, which do not follow a simple exponential absorption.

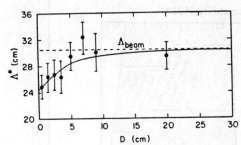

Fig. 21. Mean free path parameter $\Lambda*$ as a function of distance from the point of emission of the projectile fragments. The dashed line is the expected value. The solid line assumes a 6% admixture of "anomalons" with mean free path 2.5 cm. Data from ref. 20.

The deviation at small path lengths in Figure 21 can be explained if there is a 6% component of all fragments with a greatly enhanced interaction probability and a mean free path of 2.5 cm, less than that expected for any known nucleus, even uranium. Many speculations have focused on nuclear excitations involving quark degrees of freedom, but no theory has gained acceptance. We also await further experiments and other signatures beyond an enhanced interaction cross section.

417

1. Upgraded Bevalac

In 1982 the Bevalac will have beams of all ions. This will permit equal mass collisions to be extended up to the heaviest masses. In addition, enhanced intensities of such beams as [56]Fe will permit counter experiments whereas only emulsion experiments have been possible in the past. Figure 22 shows the expanded capability.

In addition, we completed during 1981 two major instruments: --the HISS spectrometer, a 3 Tesla magnetic field over a 3 m diameter, 1 m gap instrumented with a flexible range of detectors.

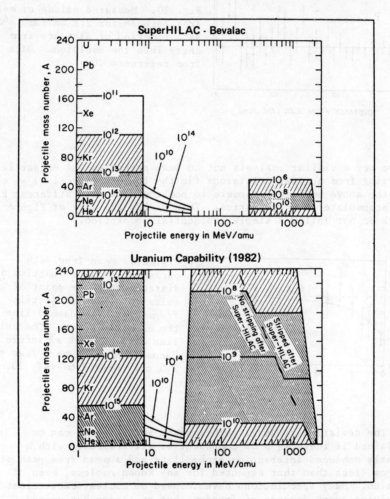

Fig. 22. Performance of the SuperHILAC-Bevalac now and after installation of a vacuum liner, presently in progress.

This will initially be concentrated on multiparticle measurements in the projectile fragmentation region, with missing mass resolution of about 1 MeV, and
--the GSI/LBL Plastic Ball/Wall, with over 1000 detector telescopes covering 96% of 4π. This will permit particle identification and energy measurement over a useful range of parameters and it will make investigations of many-particle correlations much easier.

For the longer term future, LBL plans to construct a much more powerful accelerator--VENUS--which will be described below.

2. Extension to Much Higher Energies

In order to probe the transition to a quark-gluon plasma much higher energies are predicted to be necessary[21].

Table II shows existing heavy ion accelerators and proposals for new ones. The beam momentum and range of rapidity ($y_{projectile} - y_{target}$) for each are given.

Table II

Existing and proposed accelerators for heavy ion studies, arranged in order of increasing c.m. energy. The momentum pc/A is indicated for ions with $Z/A = 1/2$, as is the rapidity range Δy between target and projectile.

	pc/A (GeV/amu)	Δy	$Z \lesssim 2$	$Z \lesssim 10$	$Z \lesssim 100$
Saturne	1.8	1.4	now	1981	prop
Numatron	2.6	1.7			prop
Bevalac	2.9	1.8	now	now	1982
Synchrophasotron	4.5	2.3	now	now	prop
CERN PS	13.5	3.4	now	prop	
SIS 100	15.0	3.5			prop
VENUS	25.0	4.0			prop
CERN SPS	200.0	6.1	now	prop	
CERN ISR	16.2	7.1	now	prop	
VENUS	25.0	8.0			prop

The Bevalac and the Synchrophasotron are the two presently operating heavy ion facilities. Saturne and the Numatron are expected to enter this energy range in the next several years. Saturne requires only successful operation of the CRYEBIS source. The Numatron, in Japan, is expected to be approved for construction this year.

At higher energies two major accelerators have been proposed: SIS 100 at GSI, Darmstadt, and VENUS at LBL. VENUS comprises both fixed target and colliding beam facilities, the latter being about 60% higher in energy than the ISR. In addition to these, the CERN

facilities that have already accelerated alpha particles are
obvious candidates for extension into the $Z \leqslant 10$ region, which
could be done with investment of about $10 M.

Figure 23 is a graphic representation of Table II, constructed
so as to explore the capabilities of each accelerator in terms of
parton concepts of the hadronic interaction. The target and
projectile rapidities are shown as a function of γ_{cm}. It is
assumed that target and projectile fragments (i.e., fragments of
the nucleons) will occur in a region within ±2 units of rapidity of
the target and projectile rapidities, respectively.

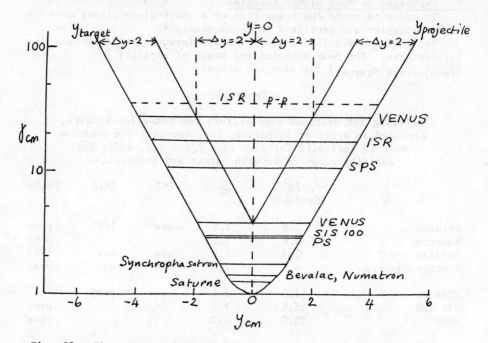

Fig. 23. The center of mass γ plotted versus c.m. rapidity for
target and projectile. Lines at $y_{target} + 2$ and $y_{projectile} - 2$
are intended to suggest the range of short-range rapidity
correlations. A clearly separated central rapidity region exists
at ISR and VENUS energies.

We thus see that at the four low energy accelerators the partons
from projectile and target may be expected to overlap. New states
involving all the quarks in both target and projectile might be
possible. At the highest energy accelerators, the projectile and
target fragmentation regions are well separated and there is a large
central region of created particles as well.

The energy of the ISR was well chosen to elucidate the rapidity
structure of the p-p collision. It falls by a factor of two for
ions ($Z/A = 1/2$). The VENUS design energy was increased above that
of the ISR to compensate for this factor. Note also that the VENUS

fixed target capability has a reasonably well-separated target and projectile fragmentation region.

The high energy accelerators would permit study of the central region of created quarks and gluons and of high density states involving the quarks of either target or projectile but not both.

In passing, note the values of γ. With $\gamma_{cm} = 25$, two colliding uranium nuclei would both be contracted to less than the thickness of a proton, providing the ultimate possibility of coherent multiquark interactions. With $\gamma_{lab} = 25$ (fixed target capability of a collider with $\gamma_{cm} = 25$) the projectile uraniun nucleus, viewed in the laboratory frame, is contracted to the thickness of a proton. In this case also, interesting coherent effects must occur and the entire collision must be considered at a parton level.

3. ISR Experiments

Recent α-α and p-α experiments at the ISR give our first look beyond p-p collisions. Some preliminary results are available[22], while much further data are being analyzed.

Starting from the most predictable quantity, R418 reports a preliminary uncorrected value of 255 ± 20 mb for the total inelastic cross section. Since their detector gave an uncorrected value for the p-p total inelastic cross section that was about 7% low, the α-α result should presumably be increased by about 7%, i.e., to 290 ± 20 mb. This may be compared with a Bevalac measurement of 276 ± 15 mb and a Dubna measurement of 304 ± 20 mb. Clearly there is no surprise.

R418 also report a measurement of the rapidity distribution of secondary particles. This is shown in Figure 24. The positive and negative distributions agree quite well near $y_{cm} = 0$ indicating a

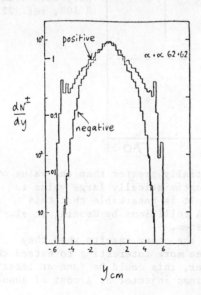

Fig. 24. Rapidity distributions for positive and negative particles produced in α-α collisions at 15.5 GeV/amu c.m. Preliminary data from R 418, ref. 22.

clear separation of the central region from the target
fragmentation region. The central value $(dn/dy)_{y=0}$ is difficult
to compare with the p-p value since the p-p data were taken at
twice the energy. However, after correcting for the known energy
dependence in p-p collisions it is found that

$$\left(\frac{dn}{dy}\right)_{\alpha\alpha,\,y=0} \Bigg/ \left(\frac{dn}{dy}\right)_{pp,\,y=0} = 1.8 \pm 0.1$$

This value is consistent with the constituent quark model
prediction of Bialas and Czyz[23], in which the central region
production results from the breaking of colored strings.

Another early result is on the p_T dependence of π°
production, which demonstrates the existence of coherent effects.
Figure 25 shows the ratio of the cross section to that for p-p

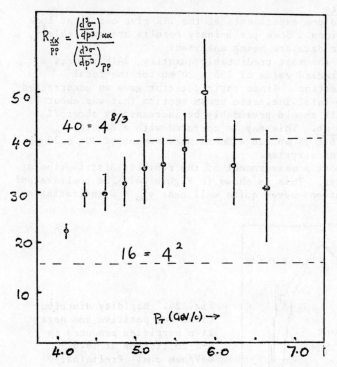

Fig. 25. Ratio
between π° pro-
duction in -
collisions and
p-p collisions
at 15.5 GeV/amu
c.m. Prelimi-
nary data from
R 108, ref. 22.

collisions. The yield is substantially greater than the value of
$A^2 = 16$, which would be the most optimistically large value in
the absence of coherent effects. It is remarkable that this
effect, previously observed in p-A collisions by Cronin, et al.[24]
should show up in such a small system.

It will be interesting to see the other results when they
become available. It would be even more interesting to extend the
value of A. As I indicated earlier, this could be done at least up
to $Z \sim 10$ by constructing a new linac injector at a cost of about

$10 M. Some of us are presently exploring the possibility of an interregional consortium to extend the life of the ISR for a program of light ion research after its scheduled closure for particle physics at the end of 1983.

4. Very Heavy Beams and Very High Energies

For a full program a dedicated accelerator is necessary, with beams of the heaviest ions and comprehensive facilities for both fixed target and colliding beam research. Figure 26 shows the layout for the VENUS facility at LBL[25], which is injected by beams from the existing SuperHILAC. It could be operating by the end of 1988 but has not yet been approved for construction.

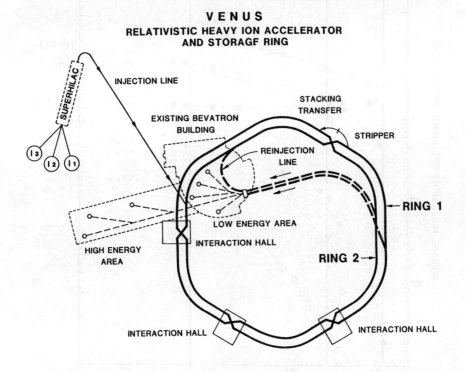

VENUS
RELATIVISTIC HEAVY ION ACCELERATOR
AND STORAGE RING

Fig. 26. One of the proposed layouts for the VENUS accelerator at LBL. The facility will include fixed target and colliding beam capabilities at 25 GeV/amu for Z/A = 1/2 (50 GeV protons, 20 GeV/amu uranium).

What could one expect in a U-U collision at these energies?
1) If the coherent enhancement of high p_T yields continues, we
may expect $1-2.10^6$ times the yield at high p_T compared with p-p
collisions.
2) If we use Landau theory to scale from p-p to A-A collisions, we
obtain the results shown in Figure 27. The left-hand scale shows
the multiplicity observed in p-p collisions, while the right-hand
scale shows the projected multiplicities for U-U collisions. The
latter are enormous. Note in particular the large yields of kaons,
which might permit production of multistrange objects.

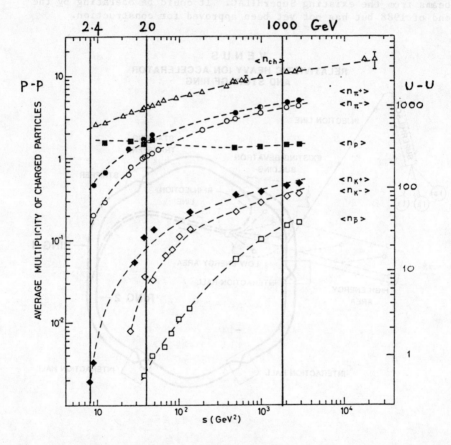

Fig. 27. Measured charged particle multiplicity in p-p collisions
and an extrapolation to U-U collisions at the same energy per
nucleon.

3) Finally, an important reason to go to large A is to create a system in which equilibrium has a chance to become established. In this context, Kajantie and Miettinen[26] have calculated the transition from quark-gluon plasma to hadron gas for a U-U collision. They find that there would be 50,000 gluon-gluon collisions and 4000 quark-gluon collisions in the cooling-down stage, surely enough to make statistical considerations not only valid but inescapable.

ACKNOWLEDGMENTS

I am grateful to my colleagues, especially Andres Sandoval and Rainer Renfordt, in the LBL/GSI streamer chamber collaboration for figures and unpublished data and to Shoji Nagamiya for providing figures and unpublished data from the LBL/INS collaboration.

This work was supported by the Director, Office of Energy Research, Division of Nuclear Physics of the Office of High Energy and Nuclear Physics of the U.S. Department of Energy under Contract W-7405-ENG-48.

REFERENCES

1. 1st Workshop on Ultrarelativistic Nuclear Collisions, Berkeley, May 21-24, 1979, LBL-8947
2. International Conference on Extreme States in Nuclear Systems, Dresden, February 4-9, 1980; Proceedings ed. H. Prade, S. Tesch, ZfK Rossendorf, ZfK-430 (1980)
3. Hakone Seminar (Japan-U.S. Joint Seminar) on High-Energy Nuclear Interactions and Properties of Dense Nuclear Matter, July, 1980; Proceedings eds. K. Nakai, A.S. Goldhaber, Hayashi-Kobo, Tokyo (1980)
4. Workshop on Future Relativistic Heavy Ion Experiments, GSI, Darmstadt, October 7-10, 1980; Proceedings eds. R. Bock, R. Stock, GSI 81-6 (1981)
5. 5th High Energy Heavy Ion Study, Berkeley, May 18-22, 1981; Proceedings ed. L.S. Schroeder, to be published early 1982
6. K.K. Gudima, V.D. Toneev, Dubna preprint E2-12624 (1979)
7. V.M. Galitzky, I.N. Mishustin, ref. 2, vol. 2, 131
8. S. Nagamiya, ref. 5 and LBL-12950 (1981)
9. S. Nagamiya, et al., LBL-12123 (1981)
10. M. Gyulassy, International Conference on Nuclear Physics, Berkeley, August 24-30, 1980; Proceedings eds. R.M. Diamond, J.O. Rasmussen, North Holland Publishing Company, Amsterdam, p. 395c.
11. Y. Yariv, Z. Frankel, Phys. Rev. $\underline{C20}$ (1979) 2227
12. A. Sandoval, et al., Phys. Rev. Lett. $\underline{45}$ (1980) 874
13. R. Hagedorn, ref. 4, p. 236
14. Preliminary data, LBL/GSI Streamer Chamber Collaboration
15. M. Gyulassy, S.K. Kauffman, L.W. Wilson, Phys. Rev. $\underline{C20}$ (1979) 2267 and references therein
16. Private communication, M. Gyulassy
17. J.W. Harris, et al., Phys. Rev. Lett. $\underline{47}$ (1981) 229

18. A. Shor, et al., contributed paper, ref. 5
19. Horst Stöcker, LBL-12302 (1981)
20. E.M. Friedlander, et al., Phys. Rev. Lett. <u>45</u> (1980) 1084;
    see also H.H. Heckman, LBL-12656 (1981)
21. L. McLerran, invited paper at this meeting; R. Anishetty, P.
    Koehler and L. McLerran, Phys. Rev. <u>D22</u> (1980) 2793
22. M.G. Albrow and M. Jacob, eds., Discussion Meetings between
    Experimentalists and Theorists on ISR and Collider Physics,
    Ser. 2, Nos. 2,3, CERN EP and TH preprints (1981);
    M. Faessler and M. Jacob, invited papers, ref. 5
23. A. Bialas, W. Czyz and L. Lesniak, Report INP-1141/PH and
    TPJU-10/81, Cracow (1981)
24. J.W. Cronin, et al., Phys. Rev. <u>D11</u> (1975) 3105
25. The VENUS Project, LBL PUB-5025 (1979)
26. K. Kajantie and H.I. Miettinen, University of Helsinki Research
    Institute for Theoretical Physics, Preprint HU-TFDT-81-7 (1981)

DISCUSSION

WALKER, DUKE: I just wondered, in the case of your anomalously short mean free path, whether it's been considered possible that you had a few neutrons tagging along [very closely correlated] with your charged tracks that apparently shorten your mean free path as a result of the fact that you can't resolve the interaction of the close-by neutron from the charged track interactions.

PUGH: They have looked downstream from the interactions to look, given the expected neutrons, for secondary interactions in the emulsion. They claim that's not a problem. I should mention maybe a couple of other models. Chaplin claimed that perhaps they were producing a toroidal nucleus, a nucleus with a hole in it, but I don't think that's realistic.

RAJA, FERMILAB: The correlation plot you showed between two negative pions. Do you have it between a π^- and a π^+?

PUGH: No, we haven't measured the π^+'s yet. [The correlation could be interpreted as a Bose-Einstein effect.] There is a similar effect for fermions in which [the correlation is negative], but we don't have the luxury at this point of calculations or looking at the background from unlike charges.

427

SESSION D

LEPTON - HADRON INTERACTIONS

Wednesday, June 24, a.m.

Chairman: J. A. Poirier
Secretary: N. N. Biswas

STRUCTURE OF QUARK JETS

Risto Orava

ABSTRACT

The evidence for quarks in the final states of deeply inelastic lepton scattering is reviewed and the detailed structure of the observed quark jets is described. Tests of quark fragmentation models and perturbative Quantum Chromodynamics (QCD) are analyzed.

1. Introduction

According to the quark-parton picture[1] distinct quark jets should result from the following "hard scattering" processes: (a) $e^+e^- \rightarrow$ hadrons, (b) $\ell N \rightarrow \ell$ +hadrons, (c) hh→ (high transverse momentum hadrons) + hadrons (Figure 1). This talk is built around the following three questions: (1) What is the evidence for the quark origin of the observed hadrons in hard scattering processes? (2) What is the detailed structure of the quark jets and how does it compare with the phenomenological models? (3) What is the relevance of the so called "tests of perturbative QCD" performed for the lepton produced hadrons?

Recently a few other hard scattering processes: $\gamma q \rightarrow G \gamma$ (QCD Compton effect), and $Gq \rightarrow \gamma q$ (Inverse QCD Compton effect) have been getting attention and as a last topic of this presentation I will talk about the QCD Compton effect (Figure 2).

2. Confined Quarks

2.1 Quarks in the Nucleon

The experimental evidence for spin-1/2 colored and fractionally charged quark-partons inside the nucleon comes from processes (b) and (c) above and is contained in the following four observations:

1. The scaling property of the nucleon structure functions implies pointlike partons.

2. The Callan-Gross relation for the nucleon structure functions $F_2(x)$ and $xF_1(x)$

$$(F_2(x)-2xF_1))/2xF_1(x) = 1$$

implies that the pointlike partons have spin 1/2.

3. The flavor relation for the nucleon structure function $F_2(x)$ measured in eN and $\nu(\bar{\nu})N$ interactions

$$F_2^{eN} = 5/18 \; F_2^{\nu(\bar{\nu})N}$$

implies that the spin-1/2 pointlike partons are fractionally charged.

4. The cross section measurements for the Drell-Yan process used to imply that the quarks involved in this hard scattering process were colored.

The evidence for partons other than these fractionally charged pointlike fermions comes from the observation that $\int dx F_2(x) \sim 0.5$, i.e., that about one half of the nucleon momentum is carried by constituents that are not counted in $F_2(x)$.

Another noteworthy observation is that the quark-antiquark ocean in the nucleon is not SU(3) symmetric, but there is an apparent suppression of strange quarks relative to up-and down quark flavors, i.e., $\bar{s}/\bar{u} = \bar{s}/\bar{d} < 1$.

This evidence tends to establish that there are quarks in the nucleon. In the following I shall argue that we have actually also found the quark hit in a hard scattering process (b). I will then discuss experimental features of these confined quarks.

2.2 Quark Jets

Which hard scattering process should we choose for our studies on confined quarks?

The simple quark-parton picture predicts universal absolute multiplicities of hadrons in the final states of all the hard scattering processes producing quark jets, i.e.,

$$\frac{dN^h}{dz} \equiv \frac{1}{\sigma tot}\frac{d\sigma^h}{dz} = \sum_i P_i D_i^{\;h}(z)$$

Here, z is the fraction of quark momentum carried by the final state hadron h ($z = E_h/E_{beam}$ for e^+e^- annihilation, $z = E_h/\nu$, with $\nu = E_\ell - E_\ell'$, for leptoproduction and $z = 2P_{||}/\sqrt{s}$ for muons produced in the Drell-Yan process, P_i represents the probability that the fragmenting quark is of flavor i

432

($\Sigma P_i = 1$) and $D_i^h(z)$ the probability of finding hadron h with energy fraction z among the fragmentation products of quark i; $D_i^h(z)$ is called "fragmentation function" for quark i to fragment into hadron h.

2.3 Theoretical Prejudice

(a) In e^+e^- annihilation to hadrons there are no complications due to the fragmenting "spectator" quarks. The probabilities P_i are given simply as

$$P_i = Q_i^2 / \sum_i Q_i^2$$

i.e., all quark flavors contribute according to the quark charge, Q_i, squared. We do not know what to expect for the flavor of an outgoing quark. We do not know the quark direction, a priori, either.

(b) In leptoproduction experiments we have to make a distinction between the electromagnetic and weak processes:

(b.1) In eN(μN) interactions the probabilities P_i are given as

$$P_i = Q_i^2 f_i(x) / \sum_i Q_i^2 f_i(x)$$

where $f_i(x)$ represents the momentum density distribution of quark i in the nucleon. All quark flavors thus contribute in the inclusive cross section. The outgoing quark direction - modulo the intrinsic transverse motion - is known, however, as the direction of the intermediate electromagnetic current $\vec{q}$.

(b.2) In $\nu(\bar{\nu})$N charged current interactions the probabilities P_i are given as

$$\nu N \begin{cases} P_u = f_d(x)/(f_d(x)+f_{\bar{u}}(x)\,(1-y)^2) \\[2ex] P_{\bar{d}} = f_{\bar{u}}(x)\,(1-y)^2/(f_d(x)+f_{\bar{u}}(x)\,(1-y)^2) \end{cases}$$

$$\bar{\nu} N \begin{cases} P_d = f_u(x)\,(1-y)^2/(f_{\bar{d}}(x)+f_u(x)\,(1-y)^2) \\[2ex] P_{\bar{u}} = f_{\bar{d}}(x)/(f_{\bar{d}}(x)+f_u(x)\,(1-y)^2) \end{cases}$$

433

By changing the variable $x = -q^2/2M\nu$ we can, in theory, tune the initial state quark flavor. We also know the jet axis - modulo the intrinsic quark motion and experimental uncertainties - as the intermediate (weak) current direction.

(c) In the Drell-Yan processes the probabilities P_i (for a final state muon) are given as

$$P_i = \frac{4\pi\alpha^2 s}{9M^4} \; \Sigma Q_i^{\;2} \{ x_A f_i^{\;A}(x_A) x_B f_i^{\;B}(x_B)$$

$$+ x_B f_i^{\;B}(x_B) x_A f_i^{\;A}(x_A) \}$$

where A and B refer to the initial state hadrons. Again we get contributions from all quark flavors into our inclusive cross section.

The obvious conclusion from the discussion above is that we should choose $\nu(\bar{\nu})N$ charged current interactions for our quark jet laboratory.

2.4 Experimental Prejudice

With electronic detectors we achieve high statistics or large numbers of events to be studied, but the events are not suited for the detailed "exclusive" type of analysis we require in our studies of confined quarks. In these studies each individual hadron should be accounted for. With a bubble chamber we are able to study hadronic final states in detail with 4π-acceptance. The statistical power of these studies is limited merely by the number of tracks instead of the number of events.

As the obvious conclusion from the discussion above we select the Fermilab 15-foot bubble chamber to detect the $\nu(\bar{\nu})N$ charged current events.

3. A Scenario for Jet Studies

More than two years ago we initiated a program of systematic studies on quark jets using the world's largest $\bar{\nu}_\mu N$ charged current event sample obtained in a bubble chamber exposure. For experimental details concerning this data sample photographed in the Fermilab 15-foot bubble chamber I Refer to Ref. 2.

434

Our scenario for the jet studies was the following:

3.1 Quark Origin; establish the quark origin of hadrons observed in deeply inelastic antineutrino (neutrino) scattering, calling the hadrons following the initial quark direction the "quark jet".[3]

3.2 Structure of Quark Jets; Analyze the structure of the quark jets in detail and compare with the quark fragmentation models.[4,5]

3.3 Tests of Perturbative QCD; Try to find relevant tests of perturbative Quantum Chromodynamics.

Here one should especially note that we did not wish to start from point 3.3, i.e., before we had understood the underlying experimental and phenomenological problems.

In the following I shall proceed according to the scenario given above but omit most of our results discussed elsewhere.[2-6]

3.1 <u>Quark Origin</u>

To check consistency of the simple quark-parton picture for deeply inelastic $\nu(\bar{\nu})$ scattering we first test the factorization hypothesis,

$$\frac{d\sigma^h}{dxdy} = \frac{G^2ME_{\nu(\bar{\nu})}}{\pi} \sum_i F_i(x)D_i^{h}(z)$$

when $-q^2 \rightarrow \infty$. The test is shown in Fig. 3 in which $(1/N_{ev})$ (dN^h/dz) is plotted in different x-intervals keeping $-q^2$ fixed. Within the experimental uncertainties - that are not too small - $d\sigma^h/dxdz$ factorizes.[3]

To check the scaling property of the $D_i^{h}(z)$ distributions we first note that these functions are universal (Fig. 4). Figure 5 then shows $D_i^{h}(z)$ functions measured in e^+e^- annihilation in different $\sqrt{s}$ regions. The scaling hypothesis is seen to be valid within ~10%.

Isospin conservation implies for any spin-1/2 quark jet

$$2D_i^{\pi^0} = D_i^{\pi^+} + D_i^{\pi^-}$$

and is tested in Fig. 4, where $D_i^{\pi^0}$ measured in an eN experiment is compared with $D_i^{\pi^+} + D_i^{\pi^-}$ measured in $\nu(\bar{\nu})N$ interactions. When the $\nu(\bar{\nu})N$ results are corrected for proton contamination good agreement is found.[4]

Isospin symmetry, on the other hand, implies

$$D_d^{\pi^+}(z) = D_u^{\pi^-}(z)$$
$$D_d^{\pi^-}(z) = D_u^{\pi^+}(z)$$

and is seen to be fulfilled (Fig. 6) iff proton contamination is accounted for.[4]

Scattered quark flavor should be identifiable if the quark "hadronizes" as described by the cascade models. There space-time structure of the hadronization process restricts possible realistic models and one expects that additive quark quantum numbers would be retained in the quark fragmentation region-modulo a "leakage" factor-, i.e., for any additive quark quantum number, N_q, one should find

$$<N>_{jet} = N_q - L_N$$

Where $<N>_{jet}$ is the corresponding net quantum number of the hadrons in the quark fragmentation region, averaged over many events. The leakage factor L_N is given as a weighted average over all quarks created in the quark jet cascade, i.e.,

$$L_N = \sum_i \gamma_i N_i \, ,$$

where γ_i is the relative probability of creating quark i in the quark jet cascade.

When applied to electric charge, equation (1) fails to be sensitive to the initial quark charge as originally suggested by Feynman. Electric charge measurement still identifies the scattered quark, however, and the prediction is

$$\begin{aligned}<Q>_{jet} &= Q_q - L_Q \\ &= 1-\gamma \text{ for u-quark jet} \\ &= \gamma \text{ for d- or s-quark jet}\end{aligned} \qquad (1)$$

To test this prediction we had to devise a method to account for the overlap between the quark and the two

spectator quark fragmentation regions. This overlap is
characteristic of relatively low W values accessible to the
present day experiments and causes inherent difficulty in
interpreting the experimental results on properties of quark
jets. To get rid of the overlap we used the empirical
observation that the average net charge of the hadrons
travelling forward in the hadronic center-of-mass system (in
the current direction) depended linearly on 1/W and provided
therefore means for an extrapolation. This extrapolation
(1/W → 0) gives the "pure", overlap free, net charge of the
hadronic shower and identifies the hadrons produced in
$\bar{\nu}N(\nu N)$ charged current interactions as coming from d-(u-)
quark fragmentation (Fig. 7).

Another method for struck quark flavor identification
was suggested by Field and Feynman and constitutes the
so-called "weighted charge" method.[7] One tries to suppress
the unwanted mixture of spectator quark fragments by a
weighting procedure, i.e., one defines the weighted charge
as

$$Q_w = \sum_i e_i z_i^r ,$$

where r is a small number between 0 and 1 and the sum is
over all hadrons in the jet.

The weighted charge distributions measured in our
$\bar{\nu}N(\nu N)$ charged current interactions are shown in Fig. 8.
They clearly identify the $\bar{\nu}$ jets (ν jets) as coming from
d-quark (u-quark) fragmentation.[3]

One may attempt further testing of the quark origin of
the hadrons by measuring the π^-/π^+ ratios in $\bar{\nu}N$ charged
current interactions, for example, but full account of the
experimental uncertainties will readily smear possible
interpretations.

We believe that we have established that there are
confined quarks in the final states of deeply inelastic
lepton scattering In the following we shall look into
detailed structure of jets that result from quark
fragmentation.

3.2 Structure of Quark Jets

3.2.1 Charge-Energy Correlations

The recursive scheme for the meson distribution in a
quark jet cascade is expressed by the integral equation

$$D(z) = f(1-z) + \int_z^1 \frac{d\eta}{\eta} f(\eta) D\left(\frac{z}{\eta}\right) \qquad (2)$$

where $f(1-z)$ represents the probability of emitting the "rank one" meson from the scattered quark, with the rest of the expression (2) containing the contribution from higher rank mesons.[7,8]

The expression for the net electric charge distribution in the jet is only sensitive to the distribution of rank one mesons, i.e.,

$$Q_{jet}(z) = (Q_q - L_Q) f (1-z)$$

with the average net charge of the jet given as (page 6)

$$\langle Q \rangle_{jet} = Q_q - L_Q$$

As a consequence of the recursive equation (2), on the other hand, we obtain

$$z D(z) = f(1-z)$$

which is exact for a power law $f(\eta) = (d+1)\eta^d$ and is also a good approximation for a more general $f(\eta)$ such as used in the fit by Field and Feynman.[7] Ochs and Shimada then note a basic result for primary meson production in a quark jet

$$z D(z) = Q_{jet}(z)/\langle Q \rangle_{jet} , \qquad (3)$$

i.e., that the energy and charge distributions in the jet are proportional to each others.[9] Here one should remember our previous results for $\langle Q \rangle_{jet}$ (page 6). $\langle Q \rangle_{jet} = 1 - \gamma$ for u-quark jets and $\langle Q \rangle_{jet} = -\gamma$ for d- and s-quark jets, i.e., that $\langle Q \rangle_{jet}$ is not a measure of the electric charge of the scattered quark. Resonance contributions might smear the result (3) and therefore, Ochs and Shimada propose a "clustering invariant" quantity as a ratio between the charge and energy flows $\Delta Q/\Delta \varepsilon$ in a given angular interval $\Delta \lambda$ with respect to the jet axis. Summing up all energies and charges in the angular interval $\Delta \lambda$ (and assuming that the meson p_T-distribution is independent of the meson rank) they obtain

$$\frac{\Delta Q}{\Delta \epsilon} = <Q>_{jet} = Q_q - L_Q .$$ (4)

In deeply inelastic antineutrino scattering we then expect to find the result

$$\frac{\Delta Q}{\Delta \epsilon} = -\gamma .$$ (5)

In neutrino scattering we should find

$$\frac{\Delta Q}{\Delta \epsilon} = 1-\gamma .$$

In Fig. 9 we test prediction (5) with our sample of $\bar{\nu}_\mu N$ charged current events. At small angles relative to the quark jet direction we obtain the same result as from our charge extrapolation (page 7).

At larger angles the mixture with the spectator quark fragments becomes important and eventually neutralizes the effect.

3.2.2 Charge Compensation

Another way to study the intrinsic structure of the quark jet cascades is the so called "rapidity gap" analysis invented for hadronic multiparticle production.[10] In fact interpretation of the rapidity gap distributions in the case of a single quark jet should be more straightforward than in the case of inclusive hadronic interactions. The rapidity zone distribution is defined as (Fig. 10)

$$Z(y^*) = \sum_i e_i \theta (y^*-y_i^*)$$

and can be thought as a measure of the length that the quark-antiquark lines of the quark jet cascade extend in rapidity space.

In Fig. 11 we show the first measurements of the average rapidity zone lengths for lepton produced hadrons. For comparison, results for hadronic intractions are also shown in the figure. We learn two things from our results: (1) At higher center-of-mass energies the charge in a quark jet seems to be locally compensated, i.e., average rapidity zones are short, $<\lambda_o>^{\nu N}=0.57\pm0.01$, compared to hadronic results or to the statistical model prediction of $<\lambda_o>^{stat}$=1.15, and (2) The overlap between the quark and spectator quark fragments is important at smaller c.m.s. energies.

3.2.3 Jet Multiplicities

A more conventional way of looking at the global characteristics of inclusive particle production is to look at the integrals over the quark fragmentation functions $D_i^h(z)$, i.e., particle multiplicities

$$<n>^h = \int dz D^h(z)$$

Figure 12 shows a comparison between the average charged multiplicities in e^+e^- annihilation and in our $\bar{\nu}$ induced hadrons in the quark fragmentation region.

The average jet multiplicities seem to be similar in these two hard scattering processes. As indicated by the logarithmic and power-low parameterizations of $<n>$ in Fig. 12 this type of a comparison is insensitive to fine structure of the quark jets.

A way to combine general characteristics of the charged particle multiplicity distributions at different c.m.s. energies is to use the so-called Koba-Nielsen-Olesen (KNO) scaling representation[11] where $<n>P_n(W)$ is plotted against the "scaled" multiplicity $n/<n>$ (Fig. 13). An interesting point to note here is that the recursive scheme incorporated in the cascade model predicts a Poisson distribution for the meson multiplicities whereas the experimental KNO - distribution is narrower than a Poisson distribution. In Fig. 13 also a comparison with hadronic forward multiplicities is presented. It is seen that the hadronic collisions result in a wider KNO-distribution than a Poisson or lepton induced distribution.

3.2.4 Quark Polarization [12]

In analogy with the use of hadron decay products to determine the initial hadron polarization it has been suggested by Nachtmann that the initial polarization might be reflected in the fragmentation products of the quark.[13] The quark hadronization process should not - because of parity conservation - induce any net polarization into the final state hadrons. In the following I shall present a test of the Nachtmann's hypothesis using our $\nu(\bar{\nu})N$ charged current interactions. Here, a crucial point is to select the proper quark fragments.

440

Following Ref. 13 I first define a parity - odd pseudoscalar quantity by the triple product

$$P = \frac{1}{W/2} (\vec{P}_1 \times \vec{P}_2) \cdot \vec{P}_3 \qquad (6)$$

where W/2 is the quark momentum in the hadronic c.m.s and $\vec{P}_1$, $\vec{P}_2$ and $\vec{P}_3$ are the three-momenta of the final state hadrons belonging to the quark jet and fulfill the conditions $|\vec{P}_1| > |\vec{P}_2| > |\vec{P}_3|$. Here, $\vec{P}_1$ is the momentum of the fastest particle in the jet, $\vec{P}_2$ is the momentum of the second fastest particle and $\vec{P}_3$ is the momentum of the third fastest one. The probability that the fragments of a left-handed quark have P>0 is given by $\langle\Theta(P)\rangle = R$ and the fragments have P<0 by $\langle\Theta(-P)\rangle = L$. Here $\Theta(P)$ is the step function: $\Theta(P) = 0$ for P<0 and $\Theta(P) = 1$ for P>0. I then define the right-left asymmetry parameter as

$$A_{RL} = (R-L)/(R+L).$$

We have observed earlier that the quark fragmentation products have universal distributions in transverse and longitudinal fractional momenta. This observation implies that R and L should be independent of the initial quark momentum. Therefore R and L can be considered as fundamental quark asymmetry parameters to be used for quark polarimetry.

To study possible parity-even correlations arising from the transverse polarization of the fragmenting quark, we define the quantity

$$S = \frac{1}{W/2} (\vec{P}_1 \times \vec{P}_2) \cdot n, \qquad (7)$$

where $\vec{P}_1$ and $\vec{P}_2$ are the three-momenta of the fastest and the second fastest particle in the quark jet, respectively, and n is the unit vector perpendicular to the $\bar{\nu}\mu$ plane. We then define an up-down asymmetry parameter

$$A_{UD} = (U-D)/U+D)$$

where U is the probability of having S>0 and D is the probability of having S<0.

In Fig. 14a the distribution of the parity-odd quantity P defined by Eq. 6 is shown for our $\bar{\nu}_\mu$ N charged current jets. The results for the quantities R and L are: R=0.495±0.032, L=0.505±0.033 giving for the asymmetry parameter $A_{RL} = (R-L)/(R+L)$ the vaue of $A_{RL} = -0.010±0.038$.

The distribution of the parity-even quantity S defined by Eq. 7 is shown in Fig. 14b. The corresponding asymmetry parameter $A_{UD}=(U-D)/(U+D)=0.011\pm0.027$.

To see possible effects from the overlap between the quark and spectator quark fragments we have plotted $A_{RL}(A_{UD})$ as a function of W^2. No diluting effects from this overlap are observed. No dependence on $Q^2= -q^2$ or $x=Q^2/2M\nu$ is seen either.

Among the charged final state hadrons in the current fragmentation region the fastest negatively charged particle is the most probable carrier of the original d-quark. Following the cascade picture of the quark jet I next select the special charge configurations where the fastest particle is negatively charged. This selection results in the following values of the asymmetry parameters: $A_{RL}=0.052\pm0.054$, $A_{UD}=0.059\pm0.038$. If we, on the contrary, require the fastest particle in the jet to be positively charged we should get $A_{RL}= -0.064\pm0.055$, $A_{UD}= -0.014\pm0.040$.

To conclude, our measurements indicate that one cannot use the asymmetry parameters measured in deeply inelastic leptoproduction for quark polarimetry or for the determination of the Weinberg angle as suggested in Ref. 13.

3.3 Tests of Perturbative QCD

A popular way to look for effects predicted by perturbative QCD is to evaluate moments of inclusive distributions in lepton production. These moment analyses frequently result in plots of the logarithm of one moment versus the logarithm of another moment.[14] The slope measured from the plots is then found to be in agreement with that predicted from ratios of anomalous dimensions γ_n in QCD; The basic QCD prediction reads for the moments $M_i(n,Q^2)= \int dx\ x^{n-1}F_i(x,Q^2)$,

$$M_i(n,Q^2) \simeq -\gamma_n \ln(\ln(Q^2/\Lambda^2))+\ln\tilde{M}_i^n ,$$

a plot of $\ln M_i(n,Q^2)$ versus $\ln M(n',Q^2)$ should result in a straight line.

In the following I shall first study sensitivity of the log-log plot as a test of perturbative QCD in general, and secondly analyze consequences of the incomplete separation of the quark fragmentation region from the spectator quark fragmentation regions.

442

No experiment measures fragmentation functions (or structure functions) over the full range of fractional momentum $z=p.p_h/p.q$. The experimenter, therefore, has to use some specific mathematical form for the fragmentation function in order to calculate the desired moments. A commonly adopted form by Monte Carlo model builders for the z-dependence of D^{NS} (NS=non-singlet) is

$$D^{NS}(z)=C\ z^{1/2}(1-z)^3 \qquad (8)$$

where C is a constant. Note that a similar assumption can be made indirectly by demanding a plateau in the rapidity distribution of the final state hadrons, i.e., $1/z$ behavior at small z. We generalize form (6) to

$$D^{NS}(z;Q^2)=C\ z^{f(Q^2)}(1-z)^{g(Q^2)}, \qquad (9)$$

where, e.g., we could demand $D^{NS}(z;Q_0^2)=D^{NS}(x)$ of eq. (8) for some $Q^2=Q_0^2$. In eq. (9), $f(Q^2)$ and $g(Q^2)$ are in general smoothly varying, gentle functions of Q^2 with acceptable forms being those eventually found reasonable by experimenters in parameterizing data. At $Q^2=Q_0^2$, the power $g(Q_0^2)$ could be reasonably specified by the quark counting rules.

Integration of eq. (9) determines the constant C in terms of the average charge of the jet $\langle Q\rangle_{jet}$ which might be expected to be constant at very large center-of-mass energy and very large Q^2,

$$\int dz D^{NS}(z,Q^2)=C\ B(1+f(Q^2);g(Q^2)+1)=\langle Q\rangle_{jet} ,$$

where $B(x,y)$ is the beta function. The average charge in a jet with W>3 GeV is experimentally more or less constant over the range of Q^2 available in present day data. If $\langle Q\rangle_{jet} \simeq$ constant, the Q^2 dependence of the functions f and g must tend to compensate as

$$C = \langle Q\rangle_{jet}/B(1+f(Q^2),g(Q^2)+1).$$

The nth moment of the distribution, e.g (9), is

$$M^{NS}(n,Q^2)=\int dz z^{n-1}D(z;Q^2)=\langle Q\rangle_{jet}\frac{B(n+f(Q^2);g(Q^2)+1)}{B(1+f(Q^2);g(Q^2)+1)}$$

We expect from examination of the data that $f(Q^2)$ and $g(Q^2)$ vary gently with Q^2 and that there exists some reference $Q^2=Q_o^2$ where $f(Q_o^2)$ and $g(Q_o^2)$ are simple numbers, perhaps 1/2 and 3, respectively, according to eg. (8).

Within experimental errors, the large z behavior of D in the data of Ref. 14 suggests that $|dg/dQ^2|<|df/dQ^2|$, implying that initially we try $g(Q^2)=g(Q_o^2)=\beta$=integer. We emphasize that kinematical effects inducing an apparent Q^2 dependence from improper elimination of small W events are such as to make the small w-region strongly varying with Q^2. Then, with this assumption that $dg/dQ^2=0$, we apply d/dQ^2 to the natural logarithm of the nth moment to get

$$\gamma(n) \equiv \frac{d}{dQ^2}(\ln M) = \langle Q \rangle_{jet} \frac{df(Q^2)}{dQ^2} \psi(n+f(Q^2)) - $$

$$\psi(2+f(Q^2)+\beta)-\psi(n+1+f(Q^2)+\beta)$$

where ψ is the derivative of the natural logarithm of the gamma function. In Fig. 16 we show a plot of the ratios of this function for n=7 and 3 and n=6 and 4, $\gamma(7)/\gamma(3)$ and $\gamma(6)/\gamma(4)$, versus $f(Q_o^2)$. The dashed lines indicate the experimental limits for these ratios, and the sets of four curves give these ratios for β=1,2,3 and 5. Clearly, for $1/4 \leq f(Q_o^2) \leq 2$, reasonable integer powers between 1 to 7 of $(1-z)$ yield slope ratios in good agreement with experiment. The general form $Cz^{f(Q^2)}(1-z)^{1-8}$ with $f(Q_o^2)$ ~ 1.0±0.8 thus leads to the observed ratios $\gamma(7)/\gamma(3)$ or $\gamma(6)/\gamma(4)$ which were attributed to QCD anomalous dimension ratios.

Based on the parameterization of the overlap between the quark and target fragmentation region (page 7) one can make a more quantitative prediction for the Q^2 evolution of the non-singlet fragmentation function. There the observed 1/W dependence of the overlap leads to a Q^2-dependence that coincides with the observed "scaling violations" of the quark fragmentation function.[15]

3.4 QCD Compton Effect

In the QCD Compton effect $\gamma q \rightarrow Gq$ (Fig. 2) the incoming (real) photon transfers its entire energy to the outgoing quark gluon system.[16] These events - in which the photon acts like an elementary field - constitute about 1 o/oo of the whole vector meson dominated photoproduction. The signature for the QCD Compton events is unique: two high transverse momentum jets - a quark and a gluon jet - plus a soft diquark jet traveling backward in the overall c.m.s. Thus there is no jet moving forward in the c.m.s.

The double differential cross section for the process can be written as

$$\frac{d^2\sigma}{dtd\nu} \simeq \frac{16\pi\alpha\alpha_s}{3s^2t}\left(\frac{u}{s} + \frac{s}{u}\right)\frac{F_2^{eN}(x)}{x}$$

where $\nu = -t/2x$ and s,t,u are the usual Mandelstam variables for the process. There one is interested in the limit s, $-t \to \infty$ with $-t/s$ fixed.

In Fig. 16 I present preliminary results from a Monte Carlo simulation of complete QCD Compton events with the incident photon energy of 130 GeV.[17] The diquark-being a color antitriplet - is taken to fragment as an antiquark. As one might expect the average particle multiplicities in the final states should be significantly higher for these "three-jet" events than for the usual vector meson dominated events (Fig. 16).

3.5 Conclusions

We have seen that the fractionally charged spin-1/2 quarks observed inside the target nucleons are also found as confined in the final state hadrons in deeply inelastic lepton scattering. We described structure of these quark fragmentation products - jets - using different methods of analysis. A good overall agreement with the inside-outside cascade picture that assumes only short range correlations between the produced hadrons was seen with the only exception of KNO-scaling for the jet multiplicities.

Tests of perturbative QCD were critically discussed and it was found that the tests performed for the quark fragmentation functions were not conclusive.

Acknowledgements

I am grateful to Kenneth Lassil a for reading the manuscript.

References

1. J.D. Bjorken, Proc. of 3rd International Symposium on Electron and Photon Interactions, Stanford, California (1967).

    R.P. Feynman, Phys. Rev. Lett. $\underline{23}$ (1969) 1415 and Photon-Hadron Interactions (W.A. Benjamin, New York, 1972).

    J.D. Bjorken and E.A. Paschos, Phys. Rev. $\underline{185}$ (1969) 1975.

    S.D. Drell and T.M. Yan, Phys. Rev. Lett. $\underline{25}$ (1970) 316.

2. R. Orava, Quark Jets, Fermilb Report FN-335 (1981).

3. J.P. Berge et al., Nucl. Phys. $\underline{B184}$ (1981)13.

4. J.P. Berge et al., Fermilab-Pub-80/96-Exp (1980).

5. J.P. Berge et al., Fermilab-Pub-81/30-Exp (1981).

6. R. Orava, Quark jets from deeply inelastic lepton scattering, Fermilb-Conf-81/21-EXP (to be published in Physica Scripta), Invited talk presented at the Arctic School of Physics, Lapland (1980).

7. R.D. Field and R.P. Feynman, Nucl. Phys. $\underline{B136}$ (1978)1.

8. V. Sukhatme, Phys. Lett. $\underline{73B}$ (1978) 478
    B. Anderson, G. Gustafson and C. Peterson, Nucl. Phys. $\underline{B135}$ (1978) 273.

9. W. Ochs and T. Shimada, Z. Physik C $\underline{4}$ (1980) 141.

10. A. Krzywicki and D. Weingarten, Phys. Lett. $\underline{50B}$ (1974) 265 and references cited therein.

11. F. Koba, H.B. Nielsen and P. Olesen, Nucl. Phys. $\underline{B40}$ (1972) 317.

12. V.V. Ammosov et al., Quark Polarization in Deep Inelastic Antineutrino Scattering, A paper to be published.

446

13. O. Nachtmann, Nucl. Phys. B127 (1977) 314.

14. See for example: J. Blietschau et al., Phys. Lett. 87B (1979) 281.

15. K. Lassila, F. Nezrick and R. Orava, Magic of Moments: QCD Tests in Fragmentation Functions?, A paper to be published.

16. H. Fritzsch and P. Minkowski, Phys. Lett. 69B (1977) 316.

17. R. Orava, A Monte Carlo program for generating QCD Compton events, Fermilab (1981), unpublished.

Figure Captions

Fig. 1: Illustration of the "hard" scattering processes and their description in the simple Quark-Parton Model: (a) e^+e^--annihilation to hadrons and the creation of a parton-antiparton pair ($p\bar{p}$) in e^+e^--annihilation, (b) deeply inelastic lepton-nucleon scattering and absorption of the intermediate current by a free parton p, and (c) muon pair creation in nucleon-nucleon ($N_1 + N_2$) collision and the corresponding process in QPM.

Fig. 2: The QCD Compton effect: the real photon is absorbed by one of the quarks in the target nucleon resulting in a large transverse momentum quark and gluon jet. In the center-of-mass system nothing moves in the incoming photon direction.

Fig. 3: $R = D^h(z, Q_0^2, x_i) / D^h(z, Q^2, x_i)$ with (a) $0.02 < x_1 < 0.20$, $0.1 < x_2 < 0.3$, (b) $0.01 < x_1 < 0.20, x > 0.2$, and (c) $0.1 < x_1 < 0.3, x_2 > 0.2$. The selection in Q^2 is for a, b and c: $3 < Q^2 < 10$ GeV^2/c^2.

Fig. 4: Fragmentation function $D^{h^+} + D^{h^-}$ for charged hadrons traveling forward in the hadronic c.m.s. in this experiment, in an ep-experiment, in a proton-proton experiment, and fragmentation function $D^h(z)$ for neutral pions in the ep-experiment. The solid line represents a parametrization by Field and Feynman with 0.27 as the magnitude of SU(3) symmetry violation in the quark jet cascade.

Fig. 5: PETRA results for $s\, d\sigma/dz$, where $z = p/p_{beam}$, with different s.

Fig. 6: Fragmentation functions D^h measured in this experiment compared to other measurements in $\nu(\bar{\nu})N$ interactions. Note that our D^h is corrected for proton contamination (see Ref. 4). The solid line is a parameterization by Field and Feynman.

Fig. 7: Average net charge of the hadrons traveling forward in the hadronic center-of-mass system as a function of 1/W. The dashed line represents a linear fit to the data points above W=3 GeV. The shaded area is a prediction from a Monte Carlo model that does not include the hypothesis of quark fragmentation.

Fig. 8: Weighted charge $Q_W^{\bar{\nu}(\nu)} = \Sigma z_i e_i r$, where e_i is electric charge of ith hadron i for antineutrino (neutrino) charged current induced hadrons traveling forward in the hadronic center-of-mass system (a,c) for r=0.2, and (b,d) for r=0.5. The solid curves represent the Field and Feynman predictions for the hadrons arising from the fragmentation of a u-quark with 10 GeV/c incident momentum and the dashed lines the corresponding predictions for the 10 GeV/c d-quark jets.

Fig. 9: Ratio $\Delta Q / \Delta \epsilon$ between the net charge flow and energy flow of the hadrons in an angular interval $\Delta \lambda$ relative to the jet axis in the forward c.m.s. hemisphere in our $\bar{\nu}$ charged current events.

Fig. 10: Illustration of the rapidity zone graph in a single event.

Fig. 11: Average rapidity range required for charge compensation in the quark jets, $\langle \lambda_o \rangle$, as a function of the c.m.s. energy for two different definitions of the "forward" rapidity zones: (I) excluding the zones which overlap with the target fragmentation region, and (II) allowing for one rapidity zone to extend into the target fragmentation region.

Fig. 12: Average charged particle multiplicity, $\langle n_{ch} \rangle$, of the hadrons traveling forward in the hadronic c.m.s. as a function of the c.m.s. energy squared, W^2. The solid line represents a power-law parameterization of the average multiplicity and the dashed line a logarithmic parameterization of the average multiplicity. The dashed-dotted line represents a fit to e^+e^- data with the vertical lines indicating typical uncertainties in the measurement.

Fig. 13: KNO-scaling representation of the charged multiplicities for hadrons traveling forward in the hadronic c.m.s. $\langle n_{ch} \rangle P_n$ is plotted against $n_{ch}/\langle n_{ch} \rangle$ where $P_n = \sigma_n/\sigma_{tot}$. The solid line represents a fit to $\pi^+ p$ - data at 8-16 GeV/c and to $\bar{p}p$ data at 22.4 GeV/c.

Fig. 14: Distribution of (a) the parity-odd quantity P and (b) the parity-even quantity S in our $\bar{\nu}$ charged current event sample.

Fig. 15: Ratios $\gamma(6)/\gamma(4)$ and $\gamma(7)/\gamma(3)$ for different values of the exponents $f(Q_o^2)$ and β.

Fig. 16: Results from a simulation of the QCD Compton events for (a) charged particle multiplicities and (b) effective masses in the final states. The smooth curves represent the simulation results. Expectations for vector meson dominated photoproduction are indicated by the stepped histogram.

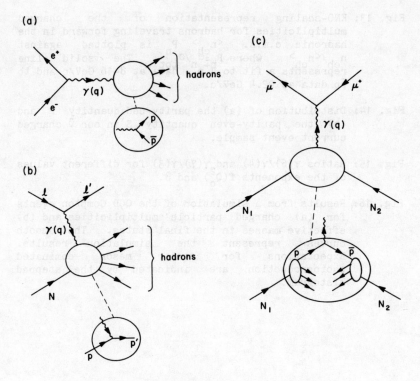

Fig. 1

QCD COMPTON EFFECT

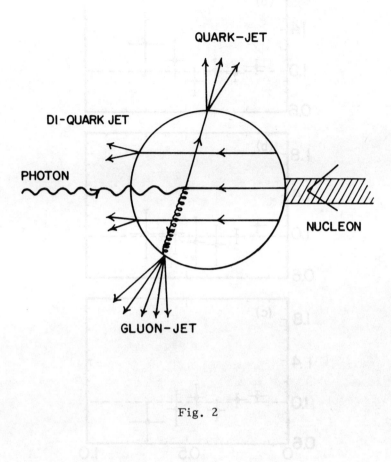

Fig. 2

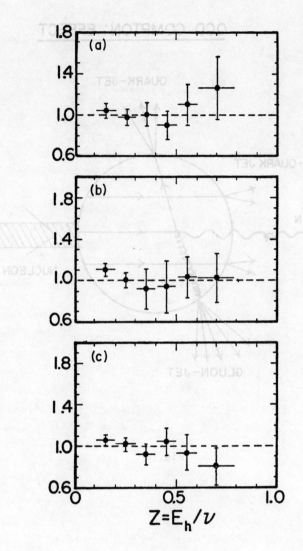

Fig. 3

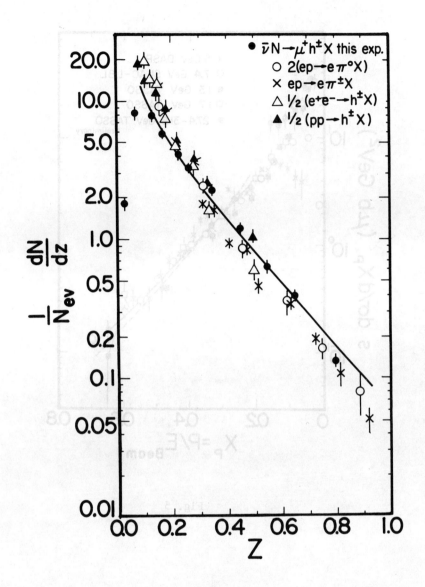

Fig. 4

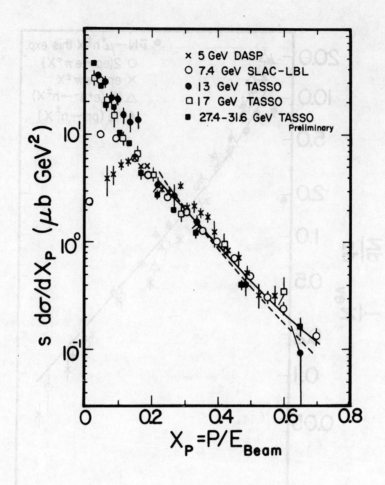

Fig. 5

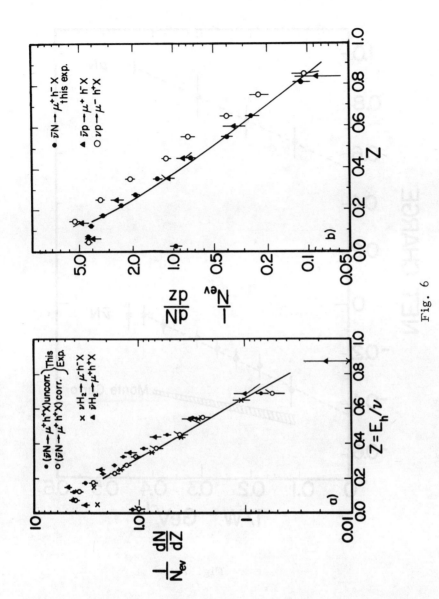

Fig. 6

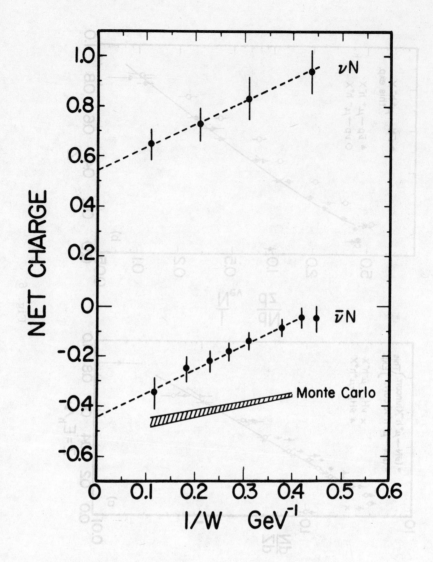

Fig. 7

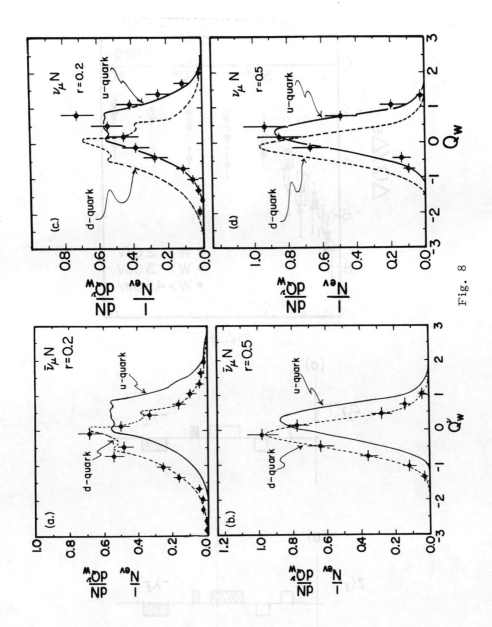

Fig. 8

457

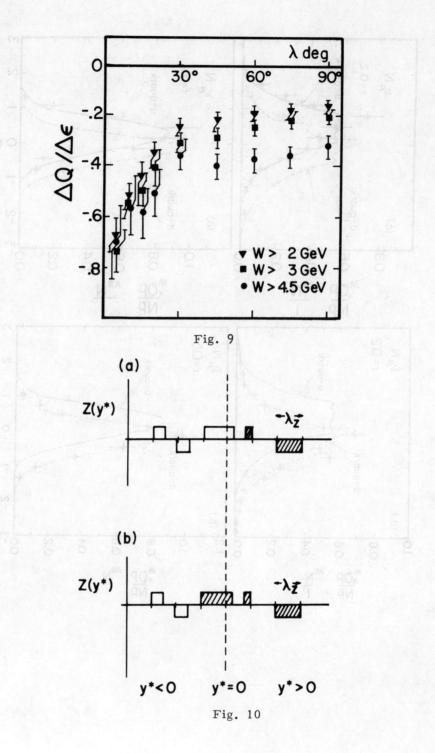

Fig. 9

(a)

$Z(y^*)$

$-\lambda_{\vec{z}}$

(b)

$Z(y^*)$

$-\lambda_{\vec{z}}$

$y^*<0$ $y^*=0$ $y^*>0$

Fig. 10

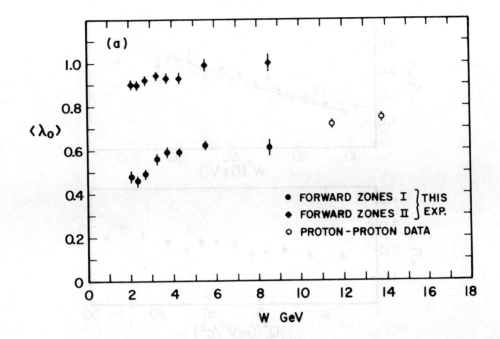

Fig. 11

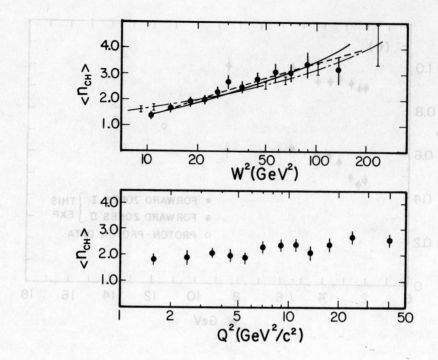

Fig. 12

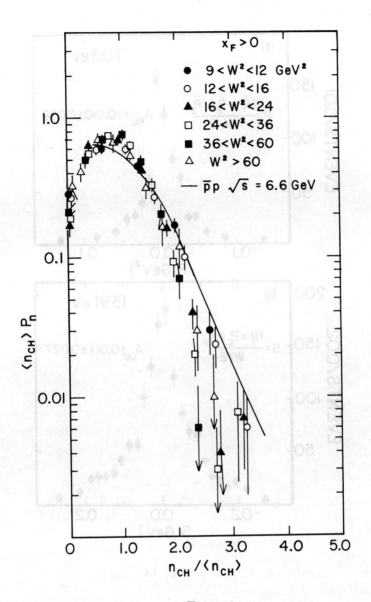

Fig. 13

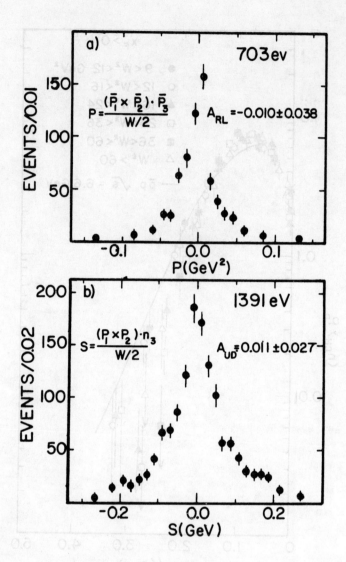

Fig. 14

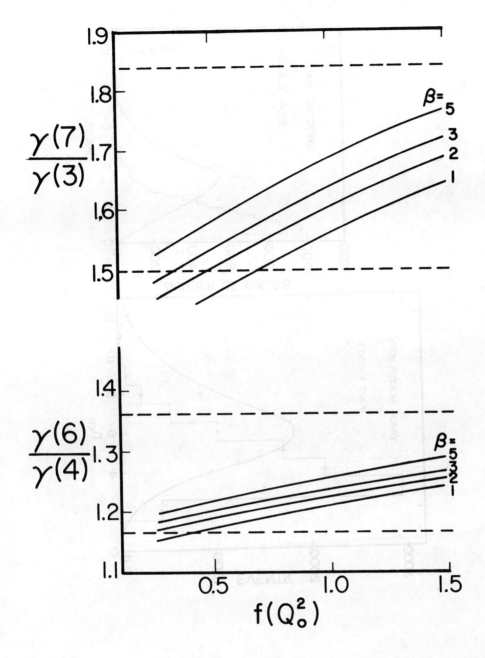

Fig. 15

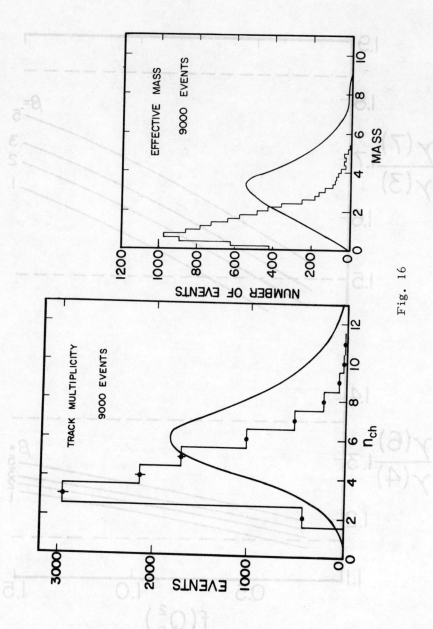

Fig. 16

464

DIQUARK JETS

Uday P. Sukhatme
University of Illinois - Chicago Circle, Chicago, Illinois 60680

K. Lassila and R. Orava
Fermi National Accelerator Laboratory, Batavia, Illinois 60510

[Talk presented by Uday P. Sukhatme]

ABSTRACT

Hadron jets formed from the fragmentation of diquarks are of special interest, since a diquark is a manifestly composite, spatially extended color source. This offers the opportunity of understanding the behavior and eventual fate of the initial quarks in a diquark during hadronization. Various experiments where diquark jets can be studied are critically discussed and a review of currently available data is presented. A recursive cascade model for diquark fragmentation is formulated in terms of two basic breakup vertices diquark $\rightarrow$ baryon + antiquark and diquark $\rightarrow$ meson + diquark. With reasonable assumptions, the model agrees well with existing data and yields useful quantitative parameterizations for diquark fragmentation functions into mesons and baryons. Present data suggest that the composite nature of a diquark is important during fragmentation, and the diquark cannot naively be treated as a single, point-like entity which becomes part of the baryon in a diquark jet.

I. INTRODUCTION

From the numerous talks[1] and papers presented at this conference, it should be obvious that the topic of diquark fragmentation is relatively new and fashionable. In this article, I would like to give an overview of the subject. In Sec. II, I will define precisely what I mean by a diquark and discuss why its fragmentation into a jet of hadrons is of interest. In Sec. III, I will survey a variety of experiments where diquark jets can be studied. In particular, I will discuss the virtues and drawbacks of various experiments, placing special emphasis on the numerous assumptions and corrections involved in extracting diquark fragmentation functions from data. A brief review of currently available diquark fragmentation data is given in Sec. IV. Finally, in Sec. V, I will discuss some phenomenological attempts to describe diquark jets. Since the recursive cascade model provides a useful way of looking at quark jets[2], I will describe an analogous recursive model for diquark fragmentation into mesons and baryons[3]. The model is based on two types of breakup vertices: diquark $\rightarrow$ baryon + antiquark and diquark $\rightarrow$ meson + leftover diquark. A model parameter 'a' measures the extent to which the composite nature of a diquark plays a role in its fragmentation. Currently available data are not very good, but they suggest that the diquark cannot simply be treated as a point source of color during fragmentation.

II. WHY ARE DIQUARKS INTERESTING?

The word "diquark" means different things to different people. For clarity, I define a diquark to be a baryon from which one current quark has been knocked out, say in a hard process via an electromagnetic or weak current interaction. Note that the above definition of a diquark does not imply any tendency for two quarks in a quiescent nucleon to form a loosely bound system.

Clearly, a diquark is a unique object, since it is both manifestly composite (essentially having a spatially extended structure similar to a hadron) as well as colored. A careful study of diquark fragmentation can offer new insight into how composite colored objects evolve into hadron jets. For example, some questions of obvious interest are:

(a) How is the baryon formed in a diquark jet and what is its momentum distribution?

(b) What is the fate of the two quarks in the initial diquark? Do they eventually become part of the same baryon? If this is not always the case, how often does is happen?

(c) Is the composite character of a diquark important, or can the diquark be simply treated as a color $\bar{3}$ point source?

(d) What are the diquark fragmentation functions into mesons, and what is the flavor dependence?

The phenomenological recursive model[3] described in Sec. V provides plausible answers to these questions.

III. EXPERIMENTS FOR STUDYING DIQUARK FRAGMENTATION

There are many feasible experiments for studying diquark fragmentation. Some of these experiments are listed in Table I and discussed below.

(a) Charged current interactions with $\nu, \bar{\nu}$ beams provide a relatively clean place for diquark studies, since the flavor of the fragmenting diquark is known with reasonable certainty. For example, the reaction $\nu p \to \mu^- X$ leaves a (uu) system [see Fig. 1a]. This is the dominant process at the naive parton model level. Even at this level, there is some contamination of the data from charged current interactions with sea quarks and antiquarks. Furthermore, there are calculable corrections due to QCD diagrams like the one shown in Fig. 1b, where a gluon from the target proton takes part in the interaction. Such a process leaves behind a colored triquark system. Note that Fig. 1a gives a 2-jet final state [u and (uu)] whereas Fig. 1b results in a 3-jet final state [q, $\bar{q}$ and triquark]. A similar discussion holds for electroproduction experiments. At the naive parton model level, the fragmenting diquark in deep inelastic ep or μp reactions is essentially of flavor content (ud), whereas for en and μn reactions, it is a mixture of (ud) and (dd) [see Table I].

(b) Hadronic collisons with a $\mu^+\mu^-$ trigger. This is a good process for studying diquarks if the naive parton model Drell-Yan diagram

Type of Reaction	Examples	Diquark System
ν reactions	$\nu p \to \mu^- x$	uu
	$\nu n \to \mu^- x$	ud
$\bar{\nu}$ reactions	$\bar{\nu} p \to \mu^+ x$	ud
	$\bar{\nu} n \to \mu^+ x$	dd
Electroproduction	$ep \to ex$ $\mu p \to \mu x$	$\frac{8}{9}$ ud + $\frac{1}{9}$ uu
	$en \to ex$ $\mu n \to \mu x$	$\frac{2}{3}$ dd + $\frac{1}{3}$ ud
Hadronic collisions with $\mu^+\mu^-$ trigger	$\pi^+ p \to (\mu^+\mu^-)x$	uu
	$\pi^- p \to (\mu^+\mu^-)x$	ud
	$\bar{p} p \to (\mu^+\mu^-)x$	$\frac{2}{3}$ ud + $\frac{1}{3}$ uu
Hadronic collisions with large-p_T hadronic trigger	$pp \to \pi^+ x$	$\approx$ud
	$pp \to \pi^- x$	$\approx$uu
Hadronic collisions with low-p_T multiparticle production	p fragmentation	$\frac{2}{3}$ ud + $\frac{1}{3}$ uu
	$\bar{p}$ fragmentation	$\frac{2}{3}$ $\overline{ud}$ + $\frac{1}{3}$ $\overline{uu}$
	n fragmentation	$\frac{2}{3}$ ud + $\frac{1}{3}$ dd

Table I: Diquark Systems occurring in various reactions. The assumptions required to establish diquark flavor content are described in the text.

467

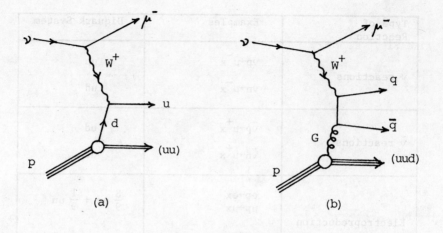

Fig. 1. Charged current interactions using a neutrino beam. Diquarks (uu) are produced by the dominant diagram shown in (a), but contamination of the type shown in diagram (b) is also present.

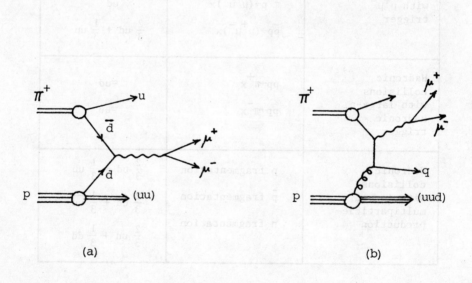

Fig. 2. π^+p Drell-Yan process at the naive parton model level gives rise to a (uu) diquark as shown in (a). However, a (uu) diquark is not left behind when a gluon from the proton interacts, as for example in diagram (b).

dominates. [The diagram for π^+p is shown in Fig. 2a]. Again, there is calculable contamination from triquarks coming from QCD diagrams like Fig. 2b, which must be taken into account.[4]

(c) <u>Hadronic collisions with a large-p_T pion trigger</u> offer yet another way of studying the fragmentation of diquarks. This scheme, which has been used in ISR experiments,[5] is based on the assumption that hard quark-quark scattering is the main mechanism underlying large p_T particle production. The charge of the large-p_T trigger pion is used to deduce the flavor of the hard-scattered quark and hence the flavor content of the forward-going diquark. In my opinion, the above method has numerous uncertainties and is a relatively unreliable way of getting diquark fragmentation functions. Firstly, the large-p_T trigger pion does not identify the flavor of the jet-initiating hard-scattered quark with high probability[6]; hence the flavor content of the forward diquark is not well-determined. Secondly, lowest order QCD calculations show that in addition to hard quark-quark scattering there is also substantial quark-gluon and gluon-gluon scattering at the <u>same Born diagram level of computation.[7]</u> This implies that there are a large number of large-p_T gluon jets, and that the forward going system is frequently a colored triquark, and not just a small contamination. Thirdly, there are indications that lowest order QCD calculations may acquire big higher order corrections at present energies.[8]

(d) <u>Low-p_T (soft) multiparticle production in hadronic collisions like pp, $\bar{p}p$, πp, etc.</u> In such processes, many successful phenomenological models (especially the fragmentation models based on dual topological unitarization DTU) suggest that the proton fragmentation region is mainly due to hadronization of a "diquark".[9] This is a model-dependent statement, but if true, then abundant low-p_T data could provide a detailed extraction of diquark fragmentation functions. However, it should be kept in mind, that no hard process is involved and interaction time scales are large - so it is not clear if the fragmenting "diquark" in soft multiparticle production agrees with the diquark definition of Sec. II.

IV. DIQUARK DATA: UNIVERSALITY

Figs. 3-5 essentially summarize currently available diquark fragmentation data[10] extracted from "hard" processes. Some of the new additional data presented at this conference has not been plotted.[1] It should be pointed out that the data are not very precise and correspond to relatively small values for the energy $\sqrt{W^2}$ of the fragmenting diquark. Nevertheless, there is reasonable agreement between data from various sources, supporting the idea of universality of diquark fragmentation functions.[10] Clearly the quality of the data can be substantially improved and this can be done most profitably at larger values of $\sqrt{W^2}$ which will eventually become available, for example at the Fermilab Tevatron. Note that in Figs. 3d and 4 the sum of (uu) fragmentation into p and π^+ is plotted. This is done because of the experimental difficulty in distinguishing between protons and π^+ in bubble chambers.[11] The ratio $r' \equiv D(dd \rightarrow \Lambda)/D(ud \rightarrow \Lambda)$ of unfavored

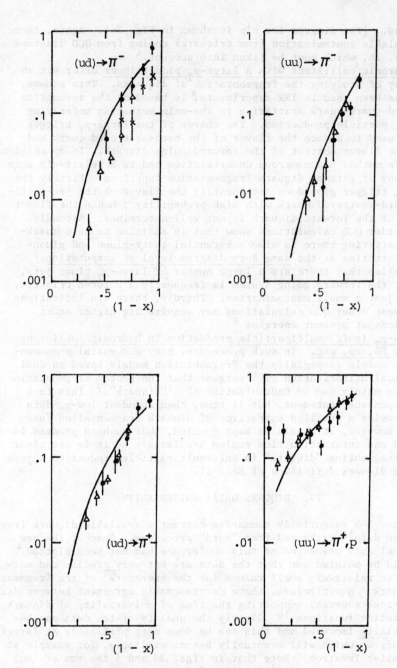

Fig. 3. A compilation of currently available data on diquark fragmentation. The quantity plotted is xD(x). References to experimental sources and methods used can be found in Refs. 3,10. Very recent data presented at this conference is in Ref. 1. The solid lines are calculated from the recursive cascade model when parameter 'a' equals 4.

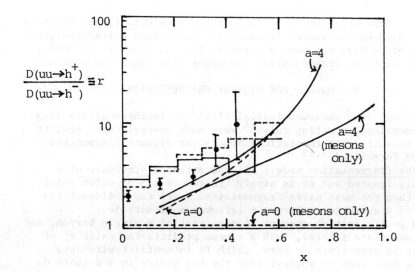

$$\frac{D(uu \rightarrow h^+)}{D(uu \rightarrow h^-)} \equiv r$$

a=4

a=4
(mesons
only)

a=0 (mesons only)

x

Fig. 4. The ratio r of (uu) diquark fragmentation functions into positive and negative hadrons. The predictions of the recursive cascade model for a = 0 and a = 4 are also shown. Data from Ref. 11.

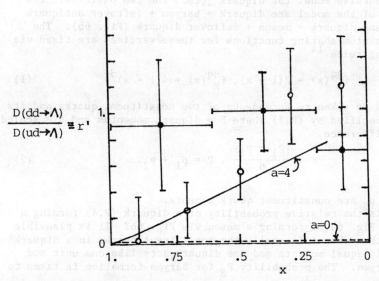

$$\frac{D(dd \rightarrow \Lambda)}{D(ud \rightarrow \Lambda)} \equiv r'$$

a=4

a=0

x

Fig. 5. The ratio r' of unfavored to favored diquark fragmentation functions into Λ's, compared with recursive cascade model calculations when the parameter 'a' has values 0 and 4. Data from Ref. 12.

to favored fragmentation into Λ's is taken from Ref. 12 and shown in Fig. 5. Although "diquark" fragmentation functions extracted from soft processes have not been plotted in Fig. 3, they are in good agreement with the fragmentation functions from "hard" processes.[1]

V. MODELS FOR DIQUARK FRAGMENTATION

Although some phenomenological fits[13] to leptoproduction data using dimensional counting rules[14] were made several years ago, it is only recently that theoretical models for diquark fragmentation have been formulated.[3,15]

(a) <u>Naive fragmentation model</u>: If the extended structure of a diquark is ignored and it is simply treated as a $\bar{3}$ of color point source, then the most naive fragmentation model for a diquark is:

diquark jet = baryon + leftover antiquark jet.

Here the entire initial diquark becomes part of a single baryon, and if resonances are ignored, such a scheme predicts the ratio r' of Fig. 5 to be approximately zero, which is in conflict with data. Thus, the data seem to suggest that the two quarks in a diquark do not always act coherently like one unit and become part of the same baryon.

(b) <u>Recursive cascade model</u>: Currently, this is the most complete model for diquark fragmentation.[3] If the two constituent quarks in a diquark do not always act like one unit, then it follows that sometimes they act like two independent quarks. This immediately suggests a recursive model for diquark jets. The two basic vertices (breakups) of the model are diquark $\rightarrow$ baryon + leftover antiquark (Fig. 6a) and diquark $\rightarrow$ meson + leftover diquark (Fig. 6b). The unknown momentum sharing functions for these vertices are fixed via the counting rules.[14]

$$f_B^{qq}(x) = 2(1 - x), \; f_M^{qq}(x) = 4(1 - x)^3. \tag{1}$$

The diquark is taken to be made up of two constituent quarks and its state is specified by (P,δ) where P = diquark momentum and δ = scaled momentum difference.

$$\delta \equiv \frac{|p_1 - p_2|}{P} \; , \; P = p_1 + p_2, \tag{2}$$

where, p_1, p_2 are constituent quark momenta.

What is the relative probability of a diquark (P,δ) forming a baryon via Fig. 6a or forming a meson via Fig. 6b? It is plausible to expect that when δ is small, the constituent quarks in a diquark have roughly equal momenta and the diquark acts like one unit and forms a baryon. The probability P_B for baryon formation is taken to be

$$P_B(\delta) = e^{-a\delta^2} \tag{3}$$

472

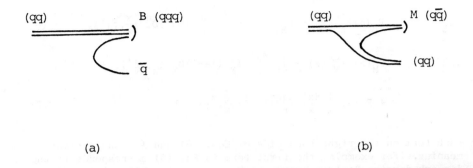

(a)   (b)

Fig. 6. The two basic vertices of a recursive model for diquark
fragmentation. (a) $(qq) \rightarrow B + \bar{q}$; (b) $(qq) \rightarrow M + (qq)$.

Then, it follows that the probability P_M for the diquark to act like
two separate quarks and form a meson is

$$P_M(\delta) = 1 - P_B(\delta) = 1 - e^{-a\delta^2}. \tag{4}$$

The quantity 'a' is a numerical parameter of the model. It should be
noted that if the parameter 'a' is taken to be zero, then $P_B(\delta) = 1$
which corresponds to the diquark always acting like a single coherent
unit and always forming a baryon. This is in fact the naive model of
the previous subsection (a).

On the other hand, a plausible numerical determination of the
parameter 'a' can be made by requiring short range correlation. Our
experience with low p_T multihadron production suggests that two par-
ticles "know" about each other if their rapidity separation $|y_1 - y_2|$
is less than one unit of rapidity. Since

$$\frac{x_1}{x_2} = \exp{(y_1 - y_2)}, \tag{5}$$

the criterion $|y_1 - y_2| \cong 1$ corresponds to $\delta_0 = (x_1 - x_2)/(x_1 + x_2)$
$\cong 1/2$. Therefore, the value $a = 1/\delta_0^2 = 4$ for the parameter 'a'
seems reasonable on physical grounds. For purposes of comparison,
we shall examine $a = 0$ and $a = 4$ results and compare both of them
with available diquark fragmentation data.

At this stage, one can write down the recursive cascade model
equations for the fragmentation functions of a diquark specified by
(P, δ). Assuming these functions scale, we obtain

$$D_B^{qq}(x;\delta) = P_B(\delta)f_B^{qq}(x) + P_M(\delta)\int_x^1 \frac{dx'}{x'} f_M^{qq}(1-x')D_B^{qq}(\frac{x}{x'};\delta'), \qquad (6)$$

$$D_M^{qq}(x;\delta) = P_M(\delta)\{f_M^{qq}(x) + \int_x^1 \frac{dx'}{x'} f_M^{qq}(1-x')D_M^{qq}(\frac{x}{x'};\delta')\} +$$

$$+ P_B(\delta)\int_x^1 \frac{dx'}{x'} f_B^{qq}(1-x')D_M^{\bar{q}}(\frac{x}{x'}) . \qquad (7)$$

Each term on the right hand side of Eqs. (6) and (7) has a clear meaning. For example, the first term in Eq. (6) corresponds to the diquark forming B(x) (a baryon with momentum fraction x) immediately in the first breakup. The second term corresponds to the situation where the initial diquark first forms M(1-x') (a meson with momentum fraction (1-x')) and a leftover diquark with momentum fraction x', and, subsequently, the diquark fragments to a baryon B(x). The scaled momentum difference δ' of the leftover diquark is taken to be[3]

$$\delta' = \left|\frac{\delta - (1 - x')}{x'}\right|. \qquad (8)$$

To make the model realistic, quark flavors must be introduced in Eqs. (6) and (7). This is done in the same manner as for quark jets[2] and as a result the momentum sharing functions acquire flavor indices. The probabilities P_u, P_d, P_s for $u\bar{u}$, $d\bar{d}$ and $s\bar{s}$ pair creation during a breakup are taken from Ref. 6.

$$P_u = P_d = \frac{4}{9}, P_s = \frac{1}{9}. \qquad (9)$$

Furthermore, integral eqs. (6) and (7) are written for an initial diquark specified by (P,δ). However, a physical diquark of momentum P has many possible allowed values of δ ranging for 0 to 1. Let the normalized probability distribution be $d\mathcal{P}/d\delta$. A plausible way to fix this probability distribution is to assume that the structure function of a quark in a diquark is analogous to the structure function of a quark in a pion. This yields[3]

$$\frac{d\mathcal{P}}{d\delta} = 0.3(1 + \delta)(1 - \delta)^{-\frac{1}{2}} \qquad (10)$$

which is peaked near $\delta = 1$. The average value of δ is

$$\bar{\delta} = \int_0^1 \delta \frac{d\mathcal{P}}{d\delta} d\delta = .72. \qquad (11)$$

Since $\bar{\delta} > \delta_o = .5$, clearly the diquark often acts like two separate quarks, if the parameter 'a' is = 4.

The fragmentation functions for physical diquarks are given by the convolution of $D_h^{qq}(x,\delta)$ with the probability distribution $d\mathcal{P}/d\delta$.

$$D_h^{qq}(x) = \int_0^1 D_h^{qq}(x;\delta)\left(\frac{d\mathcal{P}}{d\delta}\right) d\delta.\tag{12}$$

Note that the recursive model is now fully specified. Its predictions can be calculated from eq. (12), when flavor is included in eqs. (6) and (7) and they are solved by repeated iteration. The results are shown in Figs. 3-5. In general, the agreement with available data is satisfactory and the parameter value a = 4 seems to be preferred. In Fig. 4, the rise in the ratio r as x increases is largely due to the dominance of baryon production over meson production at large x. If the baryons in diquark jets (particularly protons) can be well-identified and removed, then the ratio $D(uu \to M^+)/D(uu \to M^-)$ for mesons only depends sensitively on the parameter 'a' as can be seen from the curves (marked mesons only) in Fig. 4. The curves in Figs. 4 and 5 show that the difference between a = 0 and a = 4 is pronounced whenever unfavored fragmentation is involved. Additional graphs showing (uu) and (ud) fragmentation functions into a variety of mesons and baryons can be found in Ref. 3. The meson multiplicity grows logarithmically with energy whereas each diquark jet (by construction) has a single baryon. The rapidity plateau height in a diquark jet is the same as in a quark jet – this is a consequence of the well-known fact[2] that knowledge of the initial flavor is rapidly lost with each successive breakup.

The above comparison with data suggests that the parameter 'a' is not zero. This implies that the diquark cannot simply be treated as a color $\bar{3}$ point source and in particluar the diquark does not always directly become part of a baryon. So, given a diquark, how often does it directly go into a baryon? This probability is given by $\int_0^1 e^{-a\delta^2}\left(\frac{d\mathcal{P}}{d\delta}\right)d\delta \cong .25$ when a = 4, whereas it is of course equal to one in the naive model when a = 0. Clearly, the precise value depends sensitively on the choice of the parameter 'a', and currently available data is not accurate enough to make a good determination. However it should be obvious that the field of diquark fragmentation has the potential to yield new insights into the behavior of partons during the jet formation process, especially as new and better data become available.

It is a pleasure to thank Professor W. Shephard and his Organizing Committee at Notre Dame for a very smoothly run and enjoyable conference.

REFERENCES

1. See the contributions to these Proceedings by A. Touchard, N Schmitz, B. Buschbeck, D. Zieminska, W. Kittel, T. DeGrand, L. Van Hove.
2. U. Sukhatme, Phys. Lett. 73B, 478 (1978) and Proc. of the XIII Rencontre de Moriond (1978), ed. J. Tran Thanh Van; R. Field and R. Feynman, Nucl. Phys. B136, 1(1978); F. Niedermayer, Nucl. Phys. B79, 355 (1974).

3. U. Sukhatme, K. Lassila, and R. Orava, Fermilab preprint 81/20 THY (1981).
4. G. Altarelli, R. K. Ellis and G. Martinelli, Nucl. Phys. B143, 521 (1978); J. Kubar-André and F. Paige, Phys. Rev. D19, 221 (1979).
5. D. Hanna, et al., Phys. Rev. Lett. 46, 398 (1981).
6. U. Sukhatme, Proc. Europhysics Conf. on "Partons in Soft Hadronic Processes", Erice, Sicily, ed. R. Van de Walle and S. Ratti (1981).
7. R. Feynman, R. Field and G. Fox, Phys. Rev. D18, 3320 (1978); J. Owens, Phys. Rev. D19, 3279 (1979); A. Krzywicki, J. Engels, B. Petersson and U. Sukhatme, Phys. Lett. 85B, 407 (1979).
8. R. K. Ellis M. Furman, H. Haber and I. Hinchliffe, Nucl. Phys. B173, 397 (1980).
9. A. Capella, U. Sukhatme, and J. Tran Thanh Van, Zeit. Phys. C3, 329 (1980) and references contained therein.
10. A compilation of diquark data was made by D. Hanna, et al., CERN preprint (1981). This paper and also Ref. 3 contain a list of references for diquark fragmentation data.
11. J. Bell, et al., Phys. Rev. D19, 1 (1979); D. R. O. Morrison, Proceedings of the XIX International Conference on High Energy Physics, Editors S. Hamma, M. Kawaguchi and H. Miyazawa (Phys. Soc. Japan, Tokyo, 1979) p. 354.
12. C.C. Chang, et al., Bull. Am. Phys. Soc. 25, 40 (1980). V. V. Ammosov et al., preprint Fermilab-Pub-80/44-Exp., submitted for publication (1980).
13. M. Fontannaz, B. Pire and D. Schiff, Phys. Lett. 77B, 315 (1978).
14. S. Brodsky and R. Blankenbecler, Phys. Rev. D10, 2973 (1974); S. Brodsky and J. Gunion, Phys. Rev. D17, 848 (1978).
15. B. Andersson et al., Lund preprint (1980), to appear in Nucl. Phys. B.; D. Beavis and B. Desai, Rutherford preprint RL-80-057 (1980) to appear in Phys. Rev. D.

DISCUSSION

FRIDMAN, VANDERBILT/SACLAY: I have a very experimental question. Why is your probability proportional to $e^{-a\delta^2}$ while the two-particle correlation function is always, for small rapidity differences, proportional to $e^{-a'\delta}$ [$1/a'$ being the correlation length], not δ^2.

SUKHATME: We simply chose one simple phenomenological form peaked near $\delta = 0$. That is what I meant. The thrust of the work was to get a first quantitative model which contains most of the physical features. Now if you would like $e^{-a\delta}$ that is clearly a reasonable thing to try.

FRIDMAN: Yes, because there are two-particle correlation results in this form.

SUKHATME: Yes, that in fact might be more reasonable.

ANDERSSON, LUND: If I understand you correctly, you have in your diquark fragmentation essentially only two steps. The baryon is either produced in the first step or in the second step. Is that correct?

SUKHATME: No.

ANDERSSON: How could you then explain the fact that your δ parameter is the same after the second step?

SUKHATME: It is not the same after the second step. I can show you. It changes because initially the separation is large. As you fragment further and further down the chain the separation decreases, δ changes, (and we have that) and eventually it forms a baryon.

ANDERSSON: Can you fit the K^- spectrum in the proton fragmentation region? Because that is critical in your approach, you know.

SUKHATME: We have basically concentrated only on hard reactions and we have not looked at this.

ANDERSSON: No, I'm just saying a proton fragmenting into a K^- is very critical in your approach, you know. Because what you are doing is essentially to take a diquark and add a quark which leaves you with an anti-quark. These anti-quarks can very often

477

produce a K^-. And it produces too many K^-'s in all approaches I have seen before.

SUKHATME: Absolutely. So if you have a diquark going directly into a baryon clearly you will produce too many K^-'s. That is why we have the option that the diquark does not always go into a baryon. It also has the chance of going into a meson. This decreases the number of K^- which you will produce so it goes in the right direction.

MANN, TUFTS: I agree with your accurate portrayal of the data on r' which is the [ratio of the] fragmentation of $dd \rightarrow \Lambda$ to $uu \rightarrow \Lambda$. However, I would like to point out that an experimental problem with measuring that quantity is that one very often has Y^*'s produced in the event and the Λ's come from Y^*'s rather than being directly produced. So when one takes an r' parameter one has to ask how the Y^*'s were treated.

SUKHATME: The model can incorporate resonances. It's just an additional complication. It's done for quark jets; it has to be done here.

GUNION, DAVIS: Don't you worry about the fact that, when you say you have a W^+ scattering on a proton, the system that's left behind is not just two quarks? That there is no reason in general to suppose that it doesn't have many extra quark anti-quark pairs? So, simply as a result of that, your r $[uu \rightarrow \pi^-/uu \rightarrow \pi^+]$ ratio wouldn't be 0, and various other types of things could happen starting from the very first stage . . . various different types of fragmentation.

SUKHATME: Yes, usually these things would happen at lowish x. They would dominate at lowish x.

GUNION: That's not necessarily true, I don't believe, because the extra stuff in the Fock space could go directly into the fragments.

SUKHATME: Yes, I understand. If you have an intrinsic component with strange or charm quantum numbers extending out to large x you would see corrections from that. I agree with you. If one takes that point of view, there will be effects.

BUSCHBECK, VIENNA: You have a large suppression factor for strange quarks. Is this given by new neutrino [νp and $\bar{\nu} p$] data? Because I think in e^+ and e^- it was different.

SUKHATME: Are you talking about this p_s/p_u?

BUSCHBECK: Yes.

SUKHATME: As I mentioned, we have taken it from the Nezrick group data. When you look at fast K's and π's the ratio of those two in a $\bar{\nu}$p event gives you this ratio and they end up with a value of about .25. The values which Kittel also quoted go anywhere from there to about .3. Something like that. So again this is a range of values which come from various experiments. Any of these is plausible.

BIALAS, KRAKOW: My question is about the fragmentation to mesons. At what value of the momentum is there a dominance of the second term? That means the independent fragmentation. I presume at high x it will be dominant.

SUKHATME: At high x baryon production is always dominant.

BIALAS: No, no, I don't consider baryons. I'm asking meson. Suppose you have mesons at high momentum? This will be the second graph, right?

SUKHATME: Yes, that is right.

BIALAS: My question is where is the limit where this graph really dominates in the meson production? I understand that at high x the second one is dominant; at low x the first one.

SUKHATME: Now, I'm with you. Meson production can come either by producing the baryon first and the meson afterwards or it can come from producing the meson first. This [crossover] is around, I think, .5, but I have to check to make sure.

RECENT RESULTS ON HADRON PRODUCTION IN A NEUTRINO-HYDROGEN EXPERIMENT WITH BEBC

Aachen-Bonn-CERN-München(MPI)-Oxford Collaboration

presented by N. Schmitz

ABSTRACT

The production of hadrons in charged current neutrino and anti-neutrino interactions with protons has been studied in BEBC filled with hydrogen and exposed to a wide-band horn-focussed neutrino beam generated by 350 GeV protons from the CERN SPS. Recent results on charged multiplicities, Feynman-x distributions, transverse momentum and jet properties of the events are presented.

A.) INTRODUCTION

In this paper we present some recent results on hadron production in neutrino-proton charged-current (CC) reactions. The data were obtained from an exposure of the bubble chamber BEBC, filled with hydrogen, to a wide-band neutrino beam generated by 350 GeV protons from the CERN SPS. The data sample consists of $\sim$8300 CC νp events with a visible energy $E_{vis} > 5$ GeV and a muon momentum $p_\mu > 3$ GeV/c. The secondary muon was identified by a two-plane external muon identifier EMI. The unmeasured neutrino energy E_ν was estimated in each event by a method [1,2] based on transverse-momentum balance. From E_ν the other relevant event and particle variables (effective mass W of all secondary hadrons, momentum transfer squared Q^2, Bjorken-x_B, Feynman-x_F, current direction $\hat{q}$) are then derived.

Charged hadrons were identified on the basis of bubble density, change of curvature and range in hydrogen, track residuals, break point probability and kinematic fits. The distinction between a proton and a π^+ for instance was in general possible by ionisation for momenta up to $\sim$1 GeV/c. All unidentified particles were assumed to be pions.

Further details of the experiment and of the data analysis can be found in previous publications of this collaboration [2-4].

In part of the present experiment an antineutrino beam was used to study also antineutrino-proton interactions. Some preliminary results on $\bar{\nu}$p scattering are included in this paper.

Most features of hadron production in neutrino and antineutrino proton scattering

$$\nu p \rightarrow \mu^- + \text{hadrons, i.e. } W^+ p \rightarrow \text{hadrons}$$
$$\bar{\nu} p \rightarrow \mu^+ + \text{hadrons, i.e. } W^- p \rightarrow \text{hadrons} \tag{1}$$

can be successfully described in the frame work of the quark-parton model (QPM) shown in Fig. 1. Neglecting the sea quarks, in νp scat-

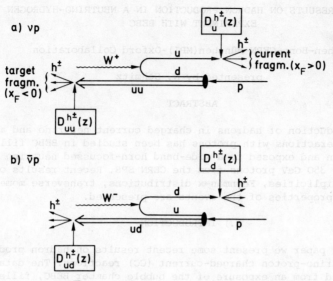

Fig. 1. νp and $\bar{\nu} p$ scattering in the quark-parton model.

tering the d-quark of the proton absorbs the positive current (W^+)
and becomes a forward going (i.e. in current direction) u-quark
which then hadronizes into the current fragments. The remaining uu-
diquark travels backward and fragments into the target fragments.

Thus in νp scattering the fragmentation functions $D_u^{h^{\pm}}(z)$ and $D_{uu}^{h^{\pm}}(z)$
of the u-quark and uu-diquark, respectively, into positive and nega-
tive hadrons $h^{\pm}$ can be studied. Here z is the energy or momentum
fraction carried by the hadron. In $\bar{\nu} p$ scattering on the other hand,
the negative current (W^-) is absorbed by a u-quark of the proton
and the partonic final state consists of a forward going d-quark and
a backward going ud-diquark. Here the fragmentation functions
$D_d^{h^{\pm}}(z)$ and $D_{ud}^{h^{\pm}}(z)$ of the d-quark and ud-diquark, respectively, can
be investigated.

A separation of the current and target fragments is possible, at
least approximately, at higher values of W (W $\gtrsim$ 4 GeV) whereas at
low W current and target fragments overlap. It turns out [5] that the
center-of-mass system of all final state hadrons is most appropriate
to carry out this separation by selecting as current (target) fragments those hadrons, which go forward (i.e. $x_F > 0$) (backward, $x_F < 0$)
in the hadronic cms.

B.) MULTIPLICITIES

The main results on the multiplicity n of charged hadrons in
νp scattering can be summarized as follows[4]:

1.) The average charged multiplicity <n> rises linearly with $\ln W^2$
above W$\approx$2 GeV, see Fig. 2. A linear fit gives

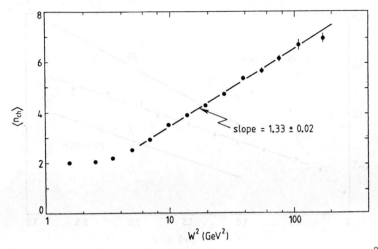

Fig. 2. Average charged multiplicity in νp scattering vs. W^2. The straight line is a fit of the form $\langle n \rangle = a + b \ln W^2$ for $W > 2$ GeV (see text).

$$\langle n \rangle = (0.37 \pm 0.02) + (1.33 \pm 0.02) \, \ln W^2 \quad (W \text{ in GeV}). \quad (2)$$

2.) At fixed W, $\langle n \rangle$ is almost independent of Q^2 (not shown).

3.) The dispersion

$$D = \sqrt{\langle n^2 \rangle - \langle n \rangle^2} \quad (3)$$

of the charged multiplicity distribution rises linearly with $\langle n \rangle$, i.e. $D = a + b\langle n \rangle$ (Wroblewski-relation[6]). Fig. 3 shows the dispersion D_- of the negative multiplicity (n_-) distribution as a function of $\langle n_- \rangle$. A linear fit above $\langle n_- \rangle \approx 0.5$ yields:

$$D_- = (0.36 \pm 0.03) + (0.36 \pm 0.03) \langle n_- \rangle . \quad (4)$$

Translating this fit into D vs. $\langle n \rangle$ ($n = 2 + 2n_-$ for νp) gives

$$D = 0.36 \langle n \rangle . \quad (5)$$

4.) The charged multiplicity distribution obeys KNO scaling[7]. In Fig. 4 the scaled multiplicity distribution $\langle n \rangle P(n,W)$ is plotted vs. $n/\langle n \rangle$ for various W intervals where $P(n,W)$ is the normalized charged multiplicity distribution ($\sum_n P(n,W) = 1$). Independence of W, i.e. KNO scaling is observed.

A similarity between νp scattering and $\overline{p}$p annihilation is observed both in Fig. 3 (same slope) and Fig. 4, whereas νp and pp scattering are different. This probably indicates that diffractive scattering, which is absent in $\overline{p}$p annihilation, is negligible in νp scattering whereas it contributes to pp scattering.

5.) In the average, more charged particles are produced in the forward

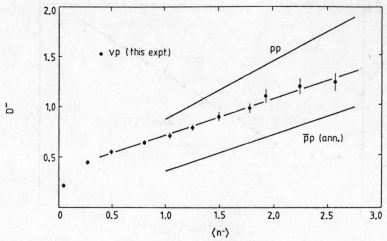

Fig. 3. Dispersion D_- as a function of $\langle n_- \rangle$ in νp scattering. The straight line is a fit of the form $D_- = a + b\langle n_- \rangle$ (see text). Also shown are straight lines representing data on pp scattering and $\bar{p}p$ annihilation.

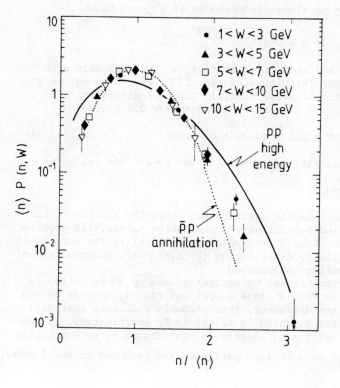

Fig. 4.

KNO-scaling distribution for five W intervals in νp scattering. The curves represent fits to pp scattering and $\bar{p}p$ annihilation data.

than in the backward cms hemisphere. This is seen in Fig. 5 which shows the average charged forward and backward multiplicities vs. W^2. A possible explanation for this difference is the fact that the baryon (mainly proton or neutron) in the reaction is emitted predominantly in the backward direction as the leading particle carrying a large fraction of the available backward energy and thus leaving less energy for the production of additional backward going hadrons. In the QPM the difference in forward and backward multiplicity implies a difference in the quark and diquark fragmentation.

6.) At larger fixed W, there seems to be no strong long–range correlation between the charged multiplicities in the forward and backward cms hemispheres. Fig. 6 shows $\langle n_F \rangle$ vs. n_B and vice versa for three W intervals. $\langle n_F \rangle$ is seen to be rather independent of n_B at higher W (and vice versa). This is expected from the QPM which predicts the current quark and the diquark to fragment independently from one another.

C.) QUARK AND DIQUARK FRAGMENTATION INTO CHARGED PIONS

In this section we study the fragmentation of the forward going quark and the backward going diquark into $\pi^{\pm}$ in terms of Feynman-x_F

$$x_F = p_L^* / p_{L\,max}^* \qquad (6)$$

where p_L^* is the longitudinal momentum of $\pi^{\pm}$ in the hadronic cms with

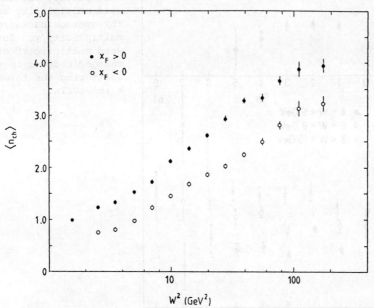

Fig. 5. Average charged multiplicity in the forward ($x_F > 0$) and backward ($x_F < 0$) hemisphere in the hadronic cms vs. W^2 in νp scattering.

respect to the current direction and p^*_{Lmax} its maximum kinematical value (calculated from a final state with a nucleon and a pion).

For this investigation only νp events with W > 3 GeV were retained leaving a data sample of 5360 CC events. This W cut is applied to remove the quasi-elastic resonance region, to work in a regime where the QPM is expected to be valid, and to reduce the overlap of current and target fragments. Furthermore all identified hadrons other than π± were removed and all unidentified hadrons were taken as pions. Finally, in order to remove the unidentified protons from the π+ sample (∼9%), an x_F dependent correction function has been applied to the raw π+ x_F-distribution. This function was determined from Monte-Carlo simulations. The correction turned out to be largest (∼20%) around $x_F ≈ 0.15$. Hadrons going backward in the cms hemisphere tend to be slow in the lab system and thus have a sizeable chance to be identified.

The normalized invariant x_F-distribution of π± is given by

$$F^{\pi^\pm}(x_F) = \frac{1}{N_{ev}} \frac{1}{\pi} \frac{E^*}{p^*_{Lmax}} \frac{dN^{\pi^\pm}}{dx_F} \approx \frac{1}{\pi} x_F D^{\pi^\pm}(x_F) \, . \qquad (7)$$

for larger $|x_F|$

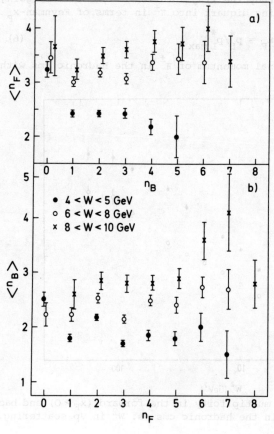

Fig. 6.

(a) average forward multiplicity vs. backward multiplicity and (b) average backward multiplicity vs. forward multiplicity of charged hadrons in νp scattering for three W intervals.

$\langle n_F \rangle$

n_B

a)

$\langle n_B \rangle$

n_F

b)

• 4 < W < 5 GeV
○ 6 < W < 8 GeV
× 8 < W < 10 GeV

N_{ev} is the number of events and $N^{\pi^\pm}$ the number of $\pi^\pm$ in these events. E^* is the $\pi^\pm$ energy in the hadronic cms. $D^{\pi^\pm}(x_F)$ are the fragmentation functions of the forward-going quark ($x_F > 0$) or the backward-going diquark ($x_F < 0$), see Fig. 1.

1.) Feynman scaling

In Fig. 7 $F^{\pi^\pm}(x_F)$ is shown (before correcting the π^+ distribution) for the two W intervals. Apart from the region of small $|x_F|$ the distributions are independent of W, i.e. Feynman scaling is fulfilled. In the subsequent figures of this section all events with W > 3 GeV are therefore taken together without any further subdivision in W.

2.) Comparison with the dimensional counting rule

The full points in Fig. 8 show the normalized invariant x_F distributions of $\pi^\pm$ for νp events with W > 3 GeV, after applying the correction function to the observed π^+ distribution. The errors are statistical only and do not include e.g. the uncertainty in the correction function.

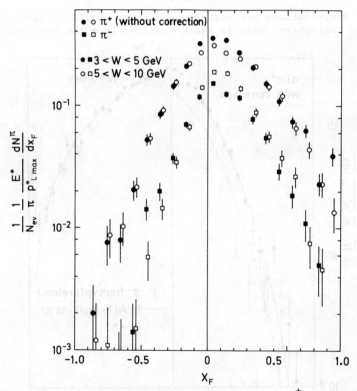

Fig. 7. Normalized invariant x_F distributions of π^+ (without correction for unidentified protons) and π^- in νp scattering for two W intervals.

Least squares fits of the form $A \cdot (1 - |x_F|)^n$ with A and n as free parameters have been carried out, separately in each hemisphere, to the full data points of Fig. 8 for $|x_F| > 0.2$. This cut was applied in order to avoid the region of small $|x_F|$ where the current and target fragments overlap. In the π^+ case the experimental point in the highest x_F interval has been omitted from the fit, since it lies outside the general trend by several standard deviations. This deviation may be due to diffractive scattering ($W^+ \to \pi^+$ via ρ^+ production) contributing at $x_F \approx 1$.

The fitted values for A and n are listed in Table I. The fits are shown by the full curves in Fig. 8; they describe the data points rather well. Applying a cut $x_B > 0.1$ (to reduce the sea contribution) leaves the A and n values practically unchanged. The data points and the n values in Table I show the following two main features:

- The π^- distribution falls off more steeply than the π^+ distribution in both hemispheres. This is expected from the QPM, since the π^+ can be the leading particle containing for $x_F > 0$ the original u-quark and for $x_F < 0$ a u-quark from the original uu-diquark. The π^- on the other hand contains only quarks created later on in the fragmentation process from the sea.

- The $\pi^\pm$ distributions are both steeper in the backward than in the forward hemisphere. This is again expected from the QPM, since a

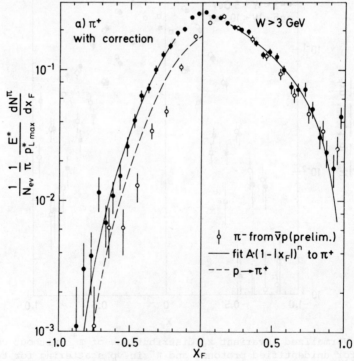

Fig. 8a. Caption on next page.

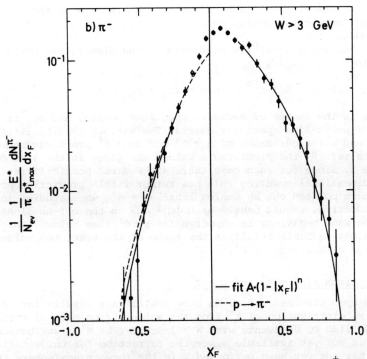

Fig. 8. Normalized invariant x_F distributions of (a) π^+ (after correcting for unidentified protons) and (b) π^- from νp events with $W > 3$ GeV (full points). The full curves show fits of the form $A(1-|x_F|)^n$ for $|x_F| > 0.2$ in each hemisphere. The open circles in Fig. 8a show the invariant x_F distribution for π^- from $\bar{\nu}p$ events with $W > 3$ GeV (preliminary). The dashed curves represent the invariant x_F distributions from a pp experiment [10].

Table I. Results of fitting $A \cdot (1-|x_F|)^n$ to the invariant x_F distributions $F^{\pi^{\pm}}(x_F)$ in the range $|x_F| > 0.2$ for νp events with $W > 3$ GeV (Fig. 8)

Pion	A	n	Predicted n values
π^+ $(u \to \pi^+)$	0.291 ± 0.010	1.23 ± 0.05	1, 0 and 2
π^- $(u \to \pi^-)$	0.235 ± 0.014	2.34 ± 0.12	2,3,5
π^+ $(uu \to \pi^+)$	0.378 ± 0.025	3.45 ± 0.15	1,2,3
π^- $(uu \to \pi^-)$	0.146 ± 0.024	4.84 ± 0.42	2,3,7

pion from the fragmentation of a single quark gets in the average a higher momentum fraction than a pion from the fragmentation of a diquark.

The behaviour $(1-x)^n$ is predicted by the dimensional counting rule [8] for large $|x_F|$:

$$xD(x) = A(1-x)^n \text{ with } n = 2n_H + n_{PL} - 1 \tag{8}$$

where n_H is the number of hadronic spectator quarks, and n_{PL} is the number of point-like spectator quarks. The various possibilities for n_H, n_{PL} and n for our cases of $u \rightarrow \pi^\pm$ and $uu \rightarrow \pi^\pm$ are discussed in detail in ref. 8; the predicted n values are given in the last column of Table I. Since for each case there are several possibilities for n, the dimensional counting rule has rather little predictive power. It has been pointed out by Gunion[9] that $uu \rightarrow \pi^+$, when studied in leptoproduction, should behave as $(1-|x|)^3$ due to the off-shellness as $|x| \rightarrow 1$. The same behaviour is expected for $p \rightarrow \pi^\pm$ (see below). The fitted n values in Table I fall in the range of the predicted values except for backward-going π^+.

3.) Comparison of $u \rightarrow \pi^+$ and $d \rightarrow \pi^-$

The open circles in Fig. 8a show preliminary results for the invariant x_F distribution of π^- from the antineutrino part of this experiment (2120 CC $\overline{\nu}p$ events with W > 3 GeV). (The π^+ distribution from $\overline{\nu}p$ is not yet available since the correction for unidentified protons has not yet been performed). In the forward hemisphere the π^- points from $\overline{\nu}p$ are in good agreement with the π^+ points from νp as expected in the QPM from isospin symmetry, which predicts $u \rightarrow \pi^+$ and $d \rightarrow \pi^-$ to be equal (see Fig. 1). In the backward hemisphere the two distributions are different.

4.) Comparison of $uu \rightarrow \pi^\pm$ and $p \rightarrow \pi^\pm$

We now compare the fragmentation of the uu-diquark into $\pi^\pm$ with that of the proton into $\pi^\pm$. Accurate data on $p \rightarrow \pi^\pm$ have been obtained by the Bonn-Hamburg-München collaboration[10] in a proton-proton experiment at 12 and 24 GeV/c, corresponding to cms energies of 4.93 and 6.84 GeV respectively. The average W in the present experiment is <W> = 5.57 GeV for νp events with W > 3 GeV. The normalized invariant x_F distributions at 12 and 24 GeV/c of ref. 10, which are very close to each other anyway, have therefore been averaged. The averaged pp data are represented by the dashed curves in Fig. 8.

From Fig. 8a it is seen that the π^+ distributions in νp scattering (backward) and in pp scattering have nearly the same slope but differ in absolute magnitude. As for the π^- distributions in Fig. 8b there is surprising agreement in shape as well as in absolute magnitude.

A quantitative prediction for proton fragmentation in proton-hadron collisions is given by the Lund model[11]. In this model the interaction takes place on a slow valence quark inside the proton which in the extreme case is stopped by the interaction so that the total proton momentum is carried by the remaining fragmenting diquark. Assuming equal interaction probability for each valence quark one thus obtains in terms of the invariant distributions (7):

$$F_p^{\pi^\pm}(x_F) = \frac{1}{3} F_{uu}^{\pi^\pm}(x_F) + \frac{2}{3} F_{ud}^{\pi^\pm}(x_F) . \tag{9}$$

This prediction is tested in Fig. 9 for π^- using the data from νp ($F_{uu}^{\pi^-}$), $\overline{\nu} p$ ($F_{ud}^{\pi^-}$) and pp ($F_p^{\pi^-}$) scattering. It is approximately fulfilled for $|x_F| > 0.2$. If one does not attribute the total proton momentum to the diquark, the right hand side of eq. (9) has to be replaced by a convolution integral; the points in Fig. 9 would then move to the right and the agreement can be improved.

5.) x_F distributions

In Fig. 10 we show the (unweighted) normalized x_F distributions

$$\frac{1}{N_{ev}} \frac{dN^{\pi^\pm}}{dx_F} = D^{\pi^\pm}(x_F) \tag{10}$$

for $\pi^\pm$ in νp events with W > 3 GeV, after applying the correction

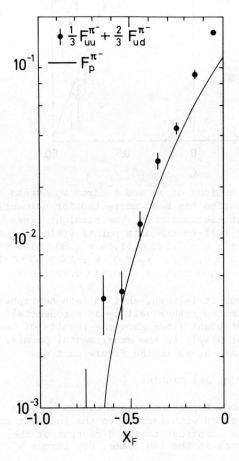

Fig. 9.

The points show $\frac{1}{3} F_{uu}^{\pi^-}(x_F)$ + $\frac{2}{3} F_{ud}^{\pi^-}(x_F)$ where $F_{uu}^{\pi^-}$ ($F_{ud}^{\pi^-}$) is the invariant x_F distribution of π^- with x_F < 0 from νp ($\overline{\nu} p$) events with W > 3 GeV. The curve shows the invariant x_F distribution of π^- in pp scattering [10].

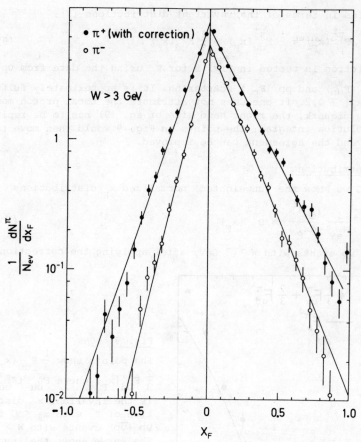

Fig. 10. Normalized x_F distributions of π^+ and π^- from νp events with $W > 3$ GeV. The π^+ distribution has been corrected for unidentified protons. The errors are statistical only. The straight lines show fits of the form $A \cdot \exp(-B|x_F|)$ to the data points yielding the following values for A and B: π^+: $A = 7.14 \pm 0.11$, $B = 4.80 \pm 0.06$ for $x_F > 0$; $A = 8.04 \pm 0.14$, $B = 7.43 \pm 0.09$ for $x_F < 0$. π^-: $A = 5.60 \pm 0.11$, $B = 6.23 \pm 0.09$ for $x_F > 0$; $A = 5.09 \pm 0.15$, $B = 10.64 \pm 0.22$ for $x_F < 0$.

function to the π^+ distribution. It is seen, that in each hemisphere the data points can be approximated rather well by an exponential over the whole x_F range. The straight lines show the results of least squares fits of the form $A \cdot \exp(-B|x_F|)$ to the experimental points. The fitted values for A and B are given in the figure caption.

D.) TRANSVERSE MOMENTUM, JET STUDIES

As usual in leptoproduction, the transverse momentum $\vec{p}_T$ of a single secondary hadron is measured with respect to the incident current ($W^\pm$) direction $\hat{q}$, which is identical to the direction of the system of all final state hadrons in the lab frame. For larger W

(W $\gtrsim$ 4 GeV) this direction can well be approximated by the direction
of the system of all charged hadrons.

There are three sources which can contribute to the transverse
momentum of a secondary hadron:

- Fragmentation of the partons into hadrons. In the normal case of
  the simple QPM (Fig. 1), where the final state partons (quark, diquark) have no transverse momentum, this is the only source of p_T.

- At higher W, hard QCD processes should become noticeable in some
  events, for instance the emission of a hard gluon by the current
  quark (see sketch). As the sketch indicates, already the forward-

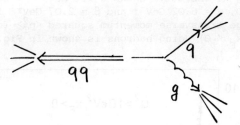

  going final state partons (quark, gluon) have a transverse momentum
  which contributes to the p_T of their hadronic fragments. The diquark
  on the other hand has no p_T. One thus expects a broadening of the p_T
  distribution in forward direction with increasing W whereas the p_T
  distribution of the backward going diquark fragments should not
  change with W. In other words, at higher W the jet of forward hadrons is expected to be wider than the backward jet. This expectation is not changed qualitatively if the other QCD processes are
  also taken into account [12]. Furthermore, at sufficiently high W,
  some events should show a planar structure (quark-gluon-diquark
  plane, see sketch) and, at even higher W, three separate hadronic
  jets (quark, gluon, diquark) should become discernable in the event
  plane.

- Primordial transverse momentum k_T of the partons inside the incident nucleon. This k_T is transferred to the final-state quark and
  diquark such that the event axis (e.g. sphericity axis) has an
  angle with respect to the current direction (see sketch). Thus the

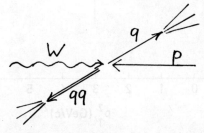

hadronic p_T distributions in the forward <u>and</u> backward direction
are broadened, independently of W.

We now confront some results from the νp experiment with these expectations.

1.) Transverse momentum

Fig. 11 shows the p_T^2 distribution of forward going ($x_F > 0$) hadrons for two intervals of W. At low p_T ($p_T \lesssim 1$ GeV/c) the p_T^2 distribution is well described by an exponential $e^{-\alpha m_T}$ where $m_T = \sqrt{p_T^2 + m_\pi^2}$ is the transverse mass and $\alpha \approx 6$ GeV^{-1}. At higher p_T however there is a tail which becomes more pronounced, thus broadening the p_T^2 distribution, with increasing W. The curve shows a fit $Ae^{-\alpha m_T} + Be^{-\beta m_T}$ with A = 22.13, B = 0.10, α = 6.02 GeV^{-1} and β = 2.07 GeV^{-1} for $W^2 > 50$ GeV^2.

The average transverse momentum squared $\langle p_T^2 \rangle$ for forward ($x_F > 0$) and backward ($x_F < 0$) going hadrons is shown in Fig. 12 (open circles)

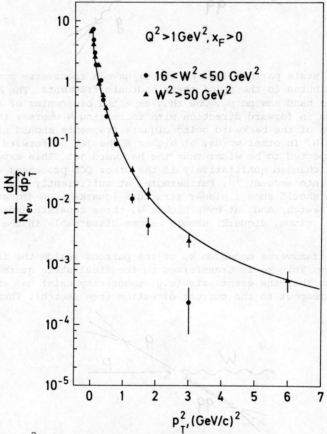

Fig. 11. p_T^2 distribution of forward going hadrons for two W regions in νp scattering. The curve is a fit of the form $Ae^{-\alpha m_T} + Be^{-\beta m_T}$ to the data points for $W^2 > 50$ GeV^2 (see text).

494

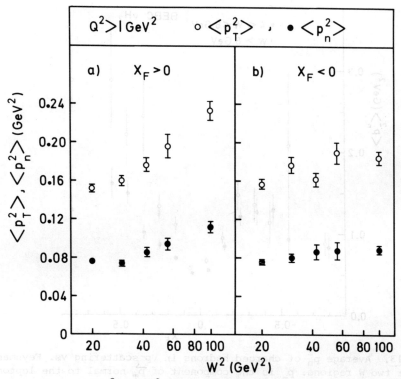

Fig. 12. Average p_T^2 and p_n^2 of charged hadrons going (a) forward and (b) backward in the hadronic cms vs. W^2 in νp scattering. p_n is the component of $\vec{p}_T$ normal to the lepton plane.

as a function of W^2. In forward direction $\langle p_T^2 \rangle$ increases with W thus indicating a broadening of the p_T distribution whereas in the backward direction $\langle p_T^2 \rangle$ is almost independent of W. The same figure also shows $\langle p_n^2 \rangle$ vs. W^2 where p_n is the component of $\vec{p}_T$ normal to the lepton (neutrino-muon) plane. This component is not affected by the uncertainty in estimating the neutrino energy E_ν and can thus be well measured. For an isotropic distribution around the current direction one expects

$$\langle p_n^2 \rangle = \frac{1}{2} \langle p_T^2 \rangle . \tag{11}$$

Fig. 12 shows that this relation is well fulfilled.

2.) Seagull effect

In Fig. 13 $\langle p_n^2 \rangle$ is plotted vs. x_F for two ranges of W. The observed increase of $\langle p_n^2 \rangle$ and thus of $\langle p_T^2 \rangle$ with $|x_F|$ is the well-known seagull effect. Furthermore at small W (4 < W < 6 GeV) a forward-backward symmetry is observed, whereas at large W (> 8 GeV) the seagull effect becomes asymmetric with $\langle p_n^2 \rangle$ being larger in forward than in backward direction. Thus for $x_F > 0$ $\langle p_n^2 \rangle$ and therefore $\langle p_T^2 \rangle$ increases with x_F and W.

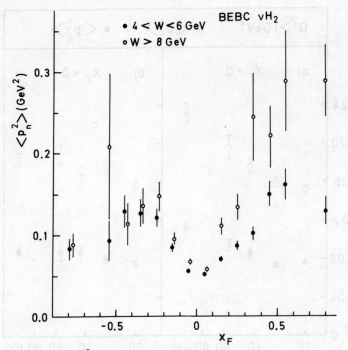

Fig. 13. Average p_n^2 of charged hadrons in νp scattering vs. Feynman-x_F for two W regions. p_n is the component of $\vec{p_T}$ normal to the lepton plane.

3.) Search for planar events

A search for planar events has been carried out according to the analysis of ref. 13. For each event an event plane was determined such that Σp_{Tin}^2 is a maximum and Σp_{Tout}^2 is a minimum. Here p_{Tin} and p_{Tout} are the components of $\vec{p_T}$ in and perpendicular to the event plane, respectively. The sum Σ extends over all charged final state hadrons in the event. A planarity P is then calculated for the event defined by

$$P = \frac{\Sigma p_{Tin}^2 - \Sigma p_{Tout}^2}{\Sigma p_{Tin}^2 + \Sigma p_{Tout}^2} \quad . \tag{12}$$

Furthermore, a transverse momentum dispersion

$$D = \frac{1}{\sqrt{n_F}} \widetilde{\Sigma} \, (p_T - \langle p_T \rangle) \tag{13}$$

is calculated for each event, where n_F is the number of forward going charged hadrons and $\widetilde{\Sigma}$ is taken over those hadrons. The average $\overline{D}$ of D over many events is of course zero and an event with a large value of D contains one or more forward hadrons with large p_T.

Fig. 14 shows the planarity distribution for events with $n_F \geqslant 3$, W > 4 GeV and $Q^2 > 1$ GeV^2. Included are events from a neutrino

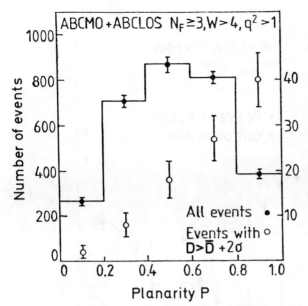

Fig. 14

Planarity distribution for events with W > 4 GeV, $Q^2 > 1$ GeV^2 and $n_F \geqslant 3$. The histogram (full points) is for all events (scale on left hand side), the open circles are for events with $D > \overline{D} + 2\sigma$ (see text) (scale on right hand side). Events of the ABCLOS collaboration[14] are included.

experiment with BEBC, filled with a Ne-H$_2$ mixture, by the ABCLOS collaboration[14]. The distribution of P is shown (a) without selection on D (full circles) and (b) for events with a D exceeding $\overline{D}$ by more than two standard deviations σ (≈ 0.2) of the D distribution, i.e. $D > \overline{D} + 2\sigma$ (open circles). The event sample (b) constitutes $\sim 3\%$ of sample (a) and consists of events with large W ($\geqslant 8$ GeV). It is seen that the events with large D have a rather strong tendency to be planar as compared to normal events. This occurrence of a small fraction of planar events may be due to hard gluon emission as discussed above. However it could also be a purely kinematical effect: The high p_T in an event with large D has to be balanced and since only a limited amount of energy is available, the balancing particles tend to lie opposite to the high p_T particle, thus making the event flat.

In Fig. 15 the angular energy flow in the event plane and in the plane perpendicular to the event plane (and containing the current direction) is shown for events with W > 4 GeV, $Q^2 > 1$ GeV^2 and with at least one particle with $p_T > 1$ GeV/c (168 events). The energy flow is defined by

$$\frac{d\varepsilon}{d\Theta} = \frac{1}{N_{ev}} \frac{\Sigma z_i}{\Delta\Theta} \tag{14}$$

where z_i is the cms energy fraction of a hadron and Θ the angle between the current direction and the projection of the hadron momentum in the event plane or in the perpendicular plane, respectively. The sum extends over all hadrons in the angular interval between Θ and $\Theta + \Delta\Theta$.

In the event plane the angular energy flow shows a dip in the forward direction which together with the maximum in the backward

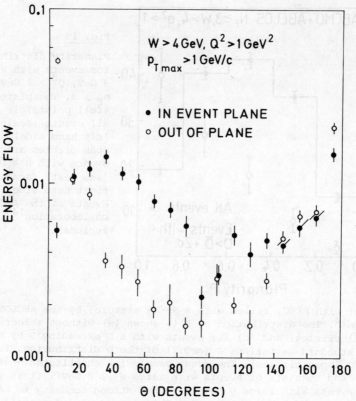

Fig. 15. Angular energy flow (with respect to current direction) in the event plane (full points) and in a plane perpendicular to the event plane (open circles) for events with $W > 4$ GeV, $Q^2 > 1$ GeV^2 and $p_{Tmax} > 1$ GeV/c. p_{Tmax} is the highest transverse momentum of a forward going particle in the event.

direction implies a three jet structure (two forward jets, one backward jet) in this plane. This structure could be due to hard QCD processes, but may also be a consequence of p_T balance. In the perpendicular plane the energy is accumulated, as expected, in the forward and backward direction.

Finally Fig. 16 shows for events with $W > 10$ GeV and $Q^2 > 1$ GeV^2 the distribution of the azimuthal angle ϕ around the direction of the charged hadron system (i.e. in the p_T plane) for forward going charged hadrons. The $\phi = 0$ axis is defined by the hadron with the largest p_T, p_{Tmax}. The ϕ distribution is shown (a) without selection on p_{Tmax} and (b) for events with a $p_{Tmax} > 1$ GeV/c. The distribution for sample (a) is rather isotropic with a slight excess, due to p_T balance, on the side opposite ($\phi \approx 180°$) to the hadron with the highest p_T. A much stronger excess of hadrons with ϕ near $180°$ is seen for sample (b). This sample shows in addition an accumulation of hadrons around $\phi \approx 0°$, i.e. of hadrons travelling on the same

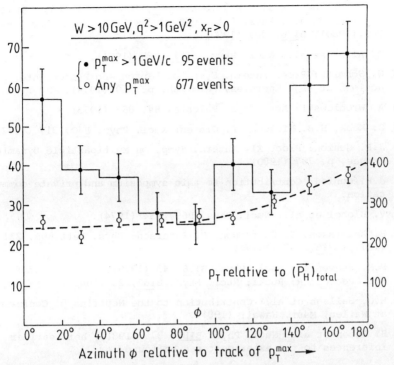

Fig. 16. Azimuthal angular distribution of forward going charged hadrons in events with $W > 4$ GeV, $Q^2 > 1$ GeV^2. $\phi = 0$ is defined by the hadron with the highest p_T. The open circles are for all events (scale on right hand side), the histogram (full points) is for events with $p_{Tmax} > 1$ GeV/c (scale on left hand side).

side with the p_{Tmax} particle and thus contributing to the flatness of the events. This accumulation can not be explained by p_T balance. It could be due to resonance production or indicate a two-jet structure (quark-gluon) in forward direction as predicted by QCD at sufficiently large W.

In conclusion one may say that the behaviour of the transverse momentum and the jet properties of some events with at least one high-p_T particle are in qualitative agreement with the expectations from QCD, although some of the features may be due to transverse momentum conservation.

It is a pleasure to acknowledge the work of the scanning and measuring teams at the several laboratories, and the staff at CERN for the operation of the SPS accelerator, BEBC chamber and associated equipment.

REFERENCES

1.) H.G. Heilmann: Bonn Internal Report WA21-int-1 (1978).

2.) J. Blietschau et al.: Phys. Lett. 87B, 281 (1979).

3.) J. Blietschau et al.: Phys.Lett. 86B, 108 (1979); 88B, 381 (1979); P. Allen et al.: Phys. Lett. 96B, 209 (1980); Nucl. Phys. B176, 269 (1980); B176, 333 (1980).

4.) P. Allen et al.: Nucl. Phys. B181, 385 (1981).

5.) N. Schmitz: Proc. Intern. Symp. on Lepton and Photon Interactions at High Energies, Fermilab, p. 359 (1979).

6.) A. Wroblewski: Acta Phys. Polonica B4, 857 (1973).

7.) Z. Koba, H.B. Nielsen, P. Olesen: Nucl. Phys. B40, 317 (1972).

8.) J.F. Gunion: Proc. XI. Intern. Symp. on Multiparticle Dynamics, Bruges, p. 767 (1980).

9.) J.F. Gunion: Contribution to this symposium and private communication.

10.) V. Blobel et al.: Nucl. Phys. B69, 454 (1974).

11.) B. Andersson, G. Gustavson, C. Peterson: Phys. Lett.69B, 221 (1977); 71B, 337 (1977).

12.) P.M. Stevenson: Nucl. Phys. B156, 43 (1979).
R.D. Peccei, R. Rückl: Nucl. Phys. B182, 21 (1981).

13.) H.C. Ballagh et al.: Contribution to the Neutrino 81 Conference at Wailea, Maui, Hawaii (1981).

14.) H. Deden et al.: Nucl. Phys. B181, 375 (1981); here earlier references to the experiment can be found.

DISCUSSION

SUKHATME, ILLINOIS CHICAGO CIRCLE: Could you comment
a little about how reliable is your correction for unidentified
protons? What you do? [How large is the correction and how
does it vary with x?] And could you show the curve before the
subtraction? Because I think it is a big effect in the diquark
region when x is beyond .5.

SCHMITZ: Unfortunately, I don't have all the plots with me but
let me say the following. We produce tape with Monte Carlo
simulated events, and the distributions according to which the
Monte Carlo events are produced are chosen such that all
quantities that you can measure directly are well reproduced. So
what goes essentially into the Monte Carlo are the lab momentum
distributions of the various particles and the Feynman x distribution. Also the transverse momentum distribution. So one first
makes sure that all distributions which can be measured are in
agreement with the distribution of the simulated Monte Carlo
events. Then one takes those Monte Carlo events and treats them
as genuine events. For instance, one would transform a proton
which has momentum larger than 1 GeV into a pion because at
about 1 GeV our possibility to identify a proton ends. Above 1
GeV we cannot tell a proton from a pion. Then I took the ratio
as a function of Feynman x between the original number of pions
and the number of pions after treating the Monte Carlo events as
our data. This ratio, then, is an x-dependent correction
function with which I must multiply my raw π^+ distribution. The
correction function looks about like this. [The correction function
was sketched.] In the backward hemisphere, of course, the
protons are slow and we can identify them. Then one starts mis-
identifying protons here. In the forward hemisphere there are
very few protons so the proton correction function tends to go to
1 again. At Feynman x about .15, the correction is largest and
amounts to about 20%.

FRIDMAN, VANDERBILT/SACLAY: If I remember well, the
PETRA results have shown that the planarity quantity, the thing
you use for defining a plane, is not very sensitive, not very
effective, to define a planar configuration.

SCHMITZ: That is true. If one plots, for instance, the average
transverse momentum squared in the plane and out of the plane as
a function of W, one sees a rise for both cases. Out of the plane,
one would expect constancy. That means that for most of the

events the plane is determined accidentally. It has nothing to do with the free partonic process. But our claim is that, if one selects events where at least one track has a high transverse momentum, then one gets a small subsample of 3% of all events where this plane is really meaningful . . . where it is not just a result of diagonalizing a momentum tensor but where it is really an indication that in that plane one has a free partonic process.

FRIDMAN: Yes, because you see in the continuum e^+e^-, which gives a $q\bar{q}$ which fragments, there is no plane in principle.

SCHMITZ: What I am saying is that for most of the events this plane has nothing to do with the free parton process but for some events where the transverse momentum is high we think that we have an effect.

FIAŁKOWSKI, KRAKOW: I think that your data on the lack of forward-backward correlations are not really so conclusive. Because I think I see some energy dependence in this 2/3 of the data. Maybe it's just fluctuations, but it looks like the red [low energy] data have negative correlations and the green [higher energy data] don't have it any more. Now obviously at low energy there is a negative energy-momentum-conservation correlation which seems to be overcome gradually. I would like to see it at still higher energy of course, if possible, where one should expect that the possible dynamic correlation, if existent, would show up.

SCHMITZ: You see I could go to higher energies. We have data up to 50 [GeV], but there are so few events. You see our energy distribution goes like this. It drops above 8 GeV.

FIAŁKOWSKI: But [if you would show the figure] I think there is rather clear evidence that there is a change with energy of the slope. I'm suggesting that the next bin would be already positive.

BIAŁAS, KRAKOW: Well, first of all, I would like to support what Fiałkowski says because he is also from Krakow. [laughter] And secondly a small comment is that in fact the slope [of this curve] which you expect can be calculated if you know the KNO [scaling] function. So in fact you can have a precise prediction for what is the slope. Probably you can do it and then we would know. The whole point of where we are at is whether this is just one chain or two chains. If this is one chain the correlations should build up with energy like in pp scattering. No, actually weaker, because the KNO function is not so broad in this case. So it will have less effect.

MANN, TUFTS UNIVERSITY: I think an interesting check on your 3-jet structure search would be to take your event Monte Carlo and fold in a [plausible] ρ/ω resonance contribution such as you find in the Field-Feynman model.

SCHMITZ: You mean, to explain the forward [region].

MANN: Yes. For example, the ρ's behave a bit differently from the typical pion. They are flatter in p_T and they grow with hadronic energy W. That may give you correlations which mimic jet effects.

BARDADIN-OTWINOWSKA, WARSAW: I have a comment related to the forward-backward multiplicities. We have measured these correlations in the K^-p interactions at the total center of mass energy of 14.4 GeV. And there indeed the slope is positive. The average forward multiplicity increases with the number of backward prongs and vice versa. It is perhaps interesting to point out that this correlation comes essentially from the central-region particles [$|x|$ less than 0.1]. It means there is no correlation between fast forward and fast backward particles. However, if the central region is included, then we get this positive slope.

EUROPEAN ORGANIZATION FOR NUCLEAR RESEARCH

CERN/EP 81-Draft
27 August 1981

CHARGE DISTRIBUTIONS IN νp AND ν̄p INTERACTIONS

Aachen-Bonn-CERN-Munich-Oxford Collaboration

Presented by: Douglas R.O. Morrison

1. INTRODUCTION

In this work, we present and discuss results on the following three questions:

- Can the charge on a quark be determined?

- Is it possible to separate target and beam fragments at present accelerator energies, e.g. by cutting at the centre of mass, or in the Breit frame?

- Where do the fragments of the current quark or of the target diquark system go? One would like to determine the distribution of characteristic fragments of these initial systems.

The results presented here were obtained by the ABCMO collaboration using a wideband neutrino beam in the BEBC bubble chamber filled with hydrogen. The data are based on 8000 neutrino-proton and 4000 antineutrino-proton charged current interactions.

2. DETERMINATION OF QUARK CHARGE

In fig. 1(a) and (b) the basic graphs for νp and ν̄p interactions are drawn giving the charge on the current quark and on the target

Presented at the XIIth International Symposium
on Multiparticle Dynamics, 21-26 June 1981, Notre Dame, USA

EP/0425P/DROM/ef

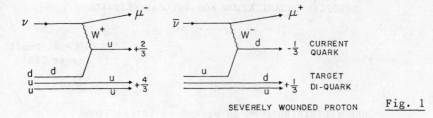

CURRENT QUARK

TARGET DI-QUARK

SEVERELY WOUNDED PROTON

Fig. 1

diquark system(*). We now consider what is called "quark leakage". As quarks cannot be emitted by themselves but only as hadrons, we have the typical situation as shown in fig. 2.

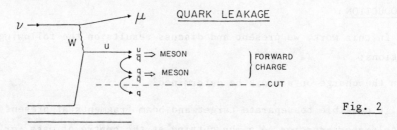

QUARK LEAKAGE

Fig. 2

If one tries to define the forward charge by making a cut, e.g. at the hadron centre of mass, then there is one extra quark left over. Thus, the "forward charge" is not simply that of the current quark but there is the leakage of another quark.

An indication of the size of this effect can be obtained as shown in fig. 3 for νp interactions. We consider three different energy regions – at low energy where only u and d quarks operate, the forward charge is $[u + (\text{average of } \bar{u} + \bar{d})] = [\frac{2}{3} + \frac{1}{2}(\frac{-2}{3} + \frac{1}{3})] = [\frac{2}{3} - 0.16] = +0.50$. At higher energies one must consider also production of $s\bar{s}$ pairs and at still higher energies also $c\bar{c}$ pairs. The production of these heavier $q\bar{q}$ pairs

---

(*) The target diquark system has been described as a "wounded proton" by A. Bialas, in the sense that it is essentially a proton from which a valence quark has been ejected. This seems an over-simplification, as the struck quark is not just emitted by itself, but leaves with an antiquark of the sea to give a hadron leaving another quark behind and in addition interacts with gluons. Thus the residual system is best described as a "severely wounded proton".

is restrained by their greater mass – the appropriate weighting factors are not well determined, but present indications give values of $\frac{1}{2}$ and $\frac{1}{20}$ for $s\bar{s}$ and $c\bar{c}$ respectively.

	Low energy	Higher energy	Still higher energy
	u,d quark	u,d,s quarks	u,d,s,c quarks

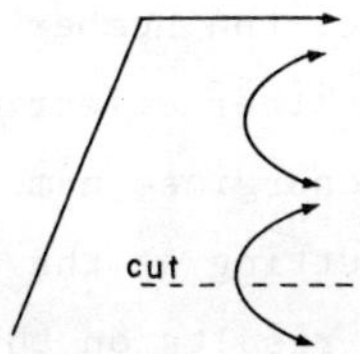

u
$\bar{q}$

q
$\bar{q} = \bar{u}, \bar{d}$ $\bar{q} = \bar{u}, \bar{d}, \frac{1}{2}\bar{s}$ $\bar{q} = \bar{u}, \bar{d}, \frac{1}{2}\bar{s}, \frac{1}{20}\bar{c}$

cut

q

Fig. 3

Forward charge $= \frac{2}{3} +$ (average correction).

$$= \frac{2}{3} + \frac{\frac{-2}{3} + \frac{1}{3}}{2}$$

$$= \frac{2}{3} + \frac{\frac{-2}{3} + \frac{1}{3} + \frac{1}{2} \cdot \frac{1}{3}}{1 + 1 + \frac{1}{2}}$$

$$= \frac{2}{3} + \frac{\frac{-2}{3} + \frac{1}{3} + \frac{1}{2} \cdot \frac{1}{3} + \frac{-1}{20} \cdot \frac{2}{3}}{1 + 1 + \frac{1}{2} + \frac{1}{20}}$$

$$= \frac{2}{3} - 0.16 \qquad = \frac{2}{3} - 0.07 \qquad = \frac{2}{3} - 0.08$$

$$= + 0.50 \qquad\qquad = + 0.60 \qquad\qquad = + 0.59$$

The three corresponding values for $\bar{v}p$ interactions would be -0.50, -0.40 and -0.41 (see also previous work given by Field and Feynman [1]). Thus the forward charge varies with the hadronic mass W. There are other forms of leakage e.g. cross-over leakage where the current quark is emitted backwards in the c.m. frame, or where the diquark is emitted forwards (forward baryon production).

Results for neutrino-proton [2] and antineutrino-proton interactions for the forward charge as a function of W^{-1} are shown in fig. 4. It may be concluded that while the results are not inconsistent with $\frac{1}{3}$ and $\frac{2}{3}$ charges on quarks, this cannot be considered a determination of the quark charge due to the various forms of leakage. It can be seen in fig. 4 that the difference between the forward net charges for vp and $\bar{v}p$, approaches the value of one as $W \to \infty$.

3. SEPARATION OF BEAM AND TARGET FRAGMENTS

It has been shown [3,4] that to separate beam and target fragments in neutrino- and hadron-proton interactions, it is best to cut at the centre of mass and this has been done to obtain the net forward charge as in fig. 4. It is conventional to discuss the variation of the net forward charge as a function of the energy W in terms of the number of quarks available and the leakage. However, there is another effect which is probably more important at the presently available energies, namely the overlap of beam and target fragments so that the cutting at the c.m. does <u>not</u> give a clean separation. We will now present results on this, firstly for hadron-hadron reactions and later for neutrino interactions.

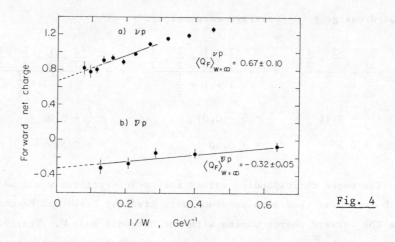

Fig. 4

Consider $\pi^- p$ interactions. The target proton gives positive fragments f^+ and negative fragments f^-. The charge distribution of these is sketched in fig. 5 as a function of rapidity.

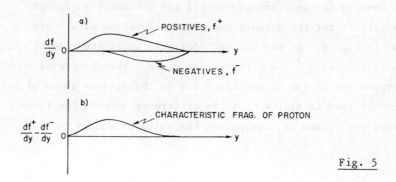

Fig. 5

508

On subtracting these distributions to obtain $df^+/dy - df^-/dy$ as in fig. 5(b), the resultant distribution may be considered the characteristic fragments of the proton and the area under the curve will be the proton charge of +1. Similarly, for the characteristic fragments of the incident π^-. These two curves of characteristic fragments, as shown in fig. 6 are theoretical and on subtracting them we obtain the experimental distribution dQ/dy, shown dotted. This experimental distribution of net charge is positive at low y and negative at high y with a cross-over value y_c, which has been found [4] in high statistics $\pi^- p$ and $K^- p$ collisions to be close to y = 0.

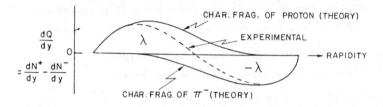

Fig. 6

For the experimental distribution we will define the total positive area as λ (the total negative area is then $-\lambda$).

The energy dependence of these effects is of importance. At very low energy the characteristic fragments of both beam and target are isotropic and hence overlap so that λ = 0. As the energy increases one expects the two sets of characteristic fragments to separate in rapidity and at very high energy to be completely separated giving λ = 1.0. Thus λ is the separation parameter.

The values of the net forward charge, and hence λ, have been collected [4] for $\pi^- p$ and $K^- p$ interactions between 10 and 360 GeV/c and are presented in fig. 7. It is not clear what variable λ should be plotted against - whether it be $s^{-1/2}$ or $s^{-1/4}$ depends on the reason (or reasons) for the overlap of the characteristic fragments (see discussion by Teper [5]). In fig. 7 the results are plotted against both $s^{-1/2}$ and $s^{-1/4}$. It may be seen that the results are consistent with the expectation that at infinite energy, the separation is complete, i.e.

$\lambda = 1$. However, it can be seen that at present accelerator energies λ is ~ 0.5 to 0.8, that is, there is considerable overlap of the characteristic fragments of beam and target fragments and hence no simple cut of the data will give a clean separation.

4. DISTRIBUTION OF CHARACTERISTIC FRAGMENTS OF BEAM AND TARGET IN Kp INTERACTIONS

We next consider the distributions of net charge in $K^{\pm}p$ interactions. This work has been done experimentally for 70 GeV/c $K^{\pm}$ beams by the Brussels-CERN-Genova-Mons-Nijmegen-Serpukhov Collaboration [6] who plotted $1/N \, (dN^+/dy - dN^-/dy)$ when N^+ and N^- are the number of positive and negative tracks, $N = N^+ + N^-$ and these are functions of the rapidity y.

We assume   T = characteristic fragments of target,
               B = characteristic fragments of beam.

The results expected are illustrated in fig. 8(a,b,c). The K^+p distribution is T+B and the K^-p distribution is T-B. Then, by adding and subtracting we obtain

characteristic fragments of proton = T = $1/2 \, (K^+p + K^-p)$,   (1)
characteristic fragments of kaon = B = $1/2 \, (K^+p - K^-p)$.   (2)

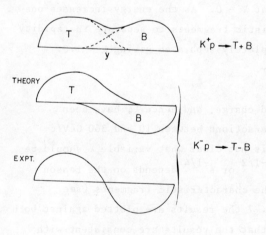

Fig. 8

It should be noted that there is no leakage here, no problems of quarks or cross-over effects. The experimental results obtained [6] are shown in fig. 9. The characteristic fragments of the kaons give a broad peak with FWHM of ~ 3 units of rapidity centred at $y \sim +2 = 1/2 \, y(max)$, plus a tail extending into the backward direction. The distribution of characteristic

fragments of the proton, fig. 9(b), through basically similar, has a
sharper peak as it contains diffractively scattered protons.

5. DISTRIBUTION OF CHARACTERISTIC FRAGMENTS OF CURRENT QUARK AND DIQUARK SYSTEM IN NEUTRINO INTERACTIONS

The above analysis is now repeated for neutrino and antineutrino
charged current interactions in hydrogen. The previous case dealt with
hadrons which have <u>integral</u> charges whereas in neutrino reactions we consider <u>fractional</u> charges on the current quark and diquark system as shown
in the Feynman diagrams of fig. 1. Thus, the current quark is considered
as the "beam" particle and the diquark system as the "target". Since it
is the <u>distribution</u> rather than the absolute value of the characteristic
fragments that we wish to determine, the distribution functions $B(y)$ and
$T(y)$ are normalized to a total value of unity. Hence, the distribution
of the characteristic fragments of the current quark in neutrino interactions is $2/3\ B(y)$. The expected charge distributions are sketched in
fig. 10. Then

$$\nu p \text{ charge distribution} = 4/3\ T + 2/3\ B \qquad (3)$$
$$\bar{\nu} p \text{ charge distribution} = 1/3\ T - 1/3\ B \qquad (4)$$

Solving these relations for T and B we obtain

$$B = \text{characteristics fragments of current quark} = 1/2\ (\nu p + 2\ \bar{\nu} p) \qquad (5)$$
$$T = \text{characteristics fragments of diquark system} = 1/2\ /\nu p - 4\ \bar{\nu} p) \qquad (6)$$

Here it has been assumed that u and
d quarks have the same shape of characteristic fragment distributions and also
the (uu) and (ud) diquark systems are
similar. Further it is assumed that
interactions on valence quarks dominate
(where the lepton interaction is with a
sea quark, in about half the cases the
struck quark will have the same charge
as the valence quark that interacts).

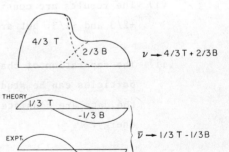

Fig. 10

The distributions of net charge in νp and $\bar{\nu} p$ interactions is given in the first column of fig. 11 for different ranges of W. In the second and third columns of fig. 11 the distributions of current quark and the target diquark systems are given being derived by using eqs (3) and (4). To guide the eye, solid lines have been drawn through the data. Considering the first columns of fig. 11, the shape of the net charge distribution for neutrinos gradually evolves as W increases until for $8 < W < 10$ GeV the shape resembles that predicted in fig. 10(a) with the height of the backward peak about twice the level in the forward direction. For antineutrinos the net charge forwards goes to higher negative values as W increases as shown in fig. 4.

The shape of the current quark and target diquark distributions in columns 2 and 3 respectively of fig. 11, are somewhat similar, both have a broad enhancement plus a tail extending into the opposite hemisphere (note the absence of diffractively produced protons). At higher W values the peak of this enhancement is about two units of rapidity from the maximum possible value of y. The width FWHM, of the enhancement is about two units of rapidity. There has been much discussion of what is the correlation length in strong interaction processes. These results suggest that this correlation length is about two units of rapidity. More data on this would be welcome.

6. CONCLUSION

(i) The results are consistent with fractional charges of quarks of +2/3 and -1/3, but are not conclusive proofs.

(ii) The separation of characteristic fragments of beam and target particles can be studied as a function of W^{-1} or $W^{-1/2}$ and the data are consistent with complete separation at infinite energy.

(iii) The best separation of fragments is a cut at $y = 0$. However, no simple cut in rapidity gives a clean separation at present

512

accelerator energies. For W = 10 to 20 GeV a cut at y = 0 gives 85 to 90% of the desired characteristic fragments plus 15 to 10% of the characteristic fragments from the other incident particle.

(iv) The distribution of characteristic fragments of the current quark and of the diquark system are basically similar and are as sketched below in Fig. 12 for W ∿ 10 GeV

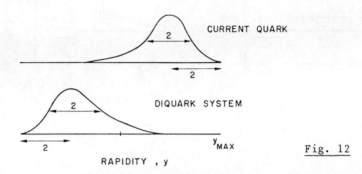

Fig. 12

Thus, there is a peak at ∿ 2 units of rapidity from y_{max} and with a FWHM of two units, which gives an estimate of the correlation length in strong interactions.

REFERENCES

[1] R.D. Field and R.P. Feynman, Nucl. Phys. B136 (1977) 1.

[2] P. Albin et al., Aachen-Bonn-Munich-Oxford Collaboration, Nucl. Phys. B181 (1981) 385.

[3] N. Schmitz, Aachen-Bonn-CERN-Munich-Oxford Collaboration, XXth Cracow School of Theoretical Physics, Zakopane, May-June 1980, Max Planck Institute, Munich, Report MPI-PAE/Exp.E1. 88.

[4] R. Göttgens et al., Aachen-Berlin-CERN-Cracow-Innsbuck-London-Vienna Collaboration, Zeitschr. für Phys. C9 (1980) 17.

[5] M. Teper, DESY preprint 81-007, Jan. 1981.

[6] M. Spyropoulou-Stassinaki, Brussels-CERN-Genova-Mons-Nijmegen-Serpukhov Collaboration, XIIth Int. Symposium on Multiparticle Dynamics, Notre Dame, June 1981.

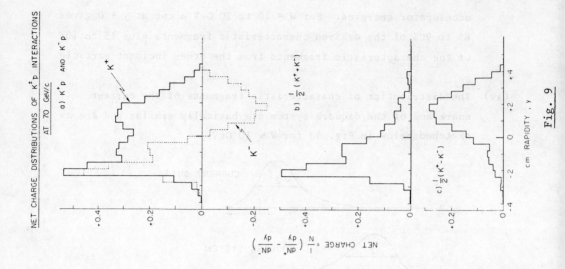

NET CHARGE DISTRIBUTIONS OF $K^{\pm}p$ INTERACTIONS AT 70 GeV/c

a) K^+p AND K^-p

b) $\frac{1}{2}(K^+ + K^-)$

c) $\frac{1}{2}(K^+ - K^-)$

NET CHARGE $= \frac{1}{N}\left(\frac{dN^+}{dy} - \frac{dN^-}{dy}\right)$

cm RAPIDITY , y

Fig. 9

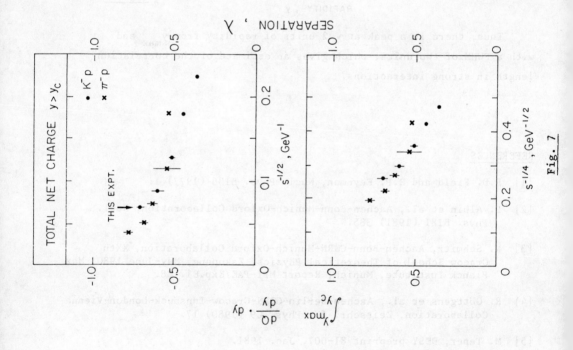

TOTAL NET CHARGE $y > y_c$

- K^-p
- π^-p

THIS EXPT.

$s^{-1/2}$, GeV^{-1}

$\int_{y_c}^{y_{max}} \frac{dQ}{dy} \cdot dy$

$s^{-1/4}$, GeV$^{-1/2}$

SEPARATION , γ

Fig. 7

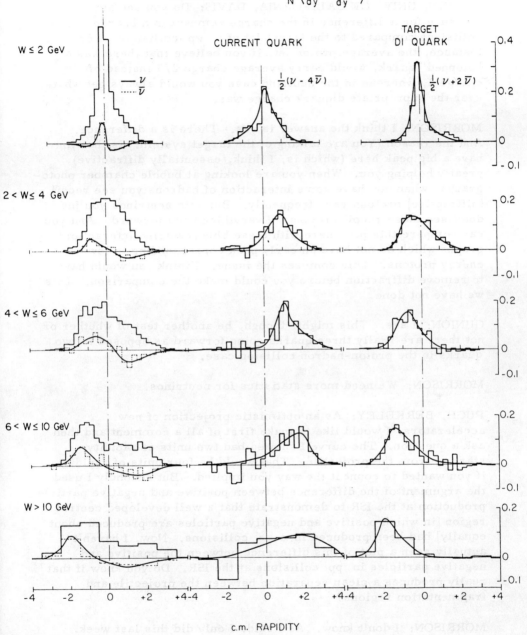

SEPARATION OF CHARGE FRAGMENTS OF CURRENT AND OF TARGET

DISTRIBUTIONS OF $\frac{1}{N}\left(\frac{dN^+}{dy} - \frac{dN^-}{dy}\right)$

W ≤ 2 GeV

2 < W ≤ 4 GeV

4 < W ≤ 6 GeV

6 < W ≤ 10 GeV

W > 10 GeV

CURRENT QUARK

$\frac{1}{2}(\nu - 4\bar{\nu})$

TARGET
DI QUARK

$\frac{1}{2}(\nu + 2\bar{\nu})$

— ν
···· $\bar{\nu}$

c.m. RAPIDITY

Fig. 11

DISCUSSION

GUNION, UNIV. OF CALIFORNIA, DAVIS: Do you see any evidence for a difference in the charge remnant in a hadron collisions compared to the target jet of a νp collision? For instance, the average proton jet, if you believe that there was a "stopped" quark, would carry average charge 2/3 instead of charge 1, whereas in the neutrino case you would always get whatever the appropriate diquark charge was.

MORRISON: I think the answer is yes. There is a difference. And the reason (You are talking of the target system.) is that you have a big peak here (which is, I think, essentially diffractive) greatly helping you. When you are looking at bubble chamber photographs, when you have some interaction of hadrons you see recoil [diffractive] protons very frequently. But with neutrinos you just don't see these recoil protons. Everything goes forward. And you can compare this peak here with these blue [neutrino interaction] graphs which do not have an early peak corresponding to low-energy protons. This confuses the issue. I think you would have to remove diffraction before you could make the comparison. This we have not done.

GUNION: I see. This might, though, be another test of whether or not there are really three quarks going forward as opposed to two quarks in the proton-hadron collision case.

MORRISON: We need more statistics for neutrinos.

PUGH, BERKELEY: As an optimistic projection of new accelerators, I would like to make first of all a comment and then ask a question. The curve I showed had two units of rapidity starting from the projectile. That would be four units of rapidity if you wanted to count it the way you implied. But secondly I used the argument of the difference between positive and negative particle production at the ISR to demonstrate that a well developed central region [in which positive and negative particles are produced about equally] had been produced in α - α collisions. Now, I haven't actually seen a plot of this difference between the positive and negative particles in pp collisions at the ISR. Do you know if that really produces a clean separation between the projectile and fragmentation regions?

MORRISON: I don't know. As I said, I only did this last week.

PUGH: The figure I have from the ISR discussion meeting last February implies that it would give quite a clean separation.

WROBLEWSKI, WARSAW: A comment on the π^-p and K^-p net charge variation with s. You pointed out that it seems to go to -1 [in the forward hemisphere] for a total net charge. Now since there is a diffractive component which contains complete charge separation (it's -1 in the forward hemisphere) you have to subtract it. Once you do this you see that it is not at all obvious that at infinite energy the net forward charge will be -1 for the non-diffractive collisions.

MORRISON: Well, that's the point of this graph. What you can do is you take the ISR energies which you see are somewhere around about here. And I don't think you will get to 1.

PRODUCTION OF HADRONS IN HIGH ENERGY νD INTERACTIONS

D. Zieminska,* C. Y. Chang, G. A. Snow, D. Son, P. H. Steinberg
University of Maryland
College Park, MD 20742 USA

R. A. Burnstein, J. Hanlon, H. A. Rubin
Illinois Institute of Technology
Chicago, Illinois 60616 USA

R. Engelmann, T. Kafka, S. Sommars
State University of New York
Stony Brook, New York 11794 USA

T. Kitagaki, S. Tanaka, H. Yuta, K. Abe, K. Hasegawa, A. Yamaguchi
K. Tamai, T. Hayashino, S. Kunori, Y. Ohtani, H. Hayano
Tohoku University
Sendai 980, Japan

C. C. Chang, W. A. Mann, A. Napier, J. Schneps
Tufts University
Medford, MA 02155 USA

(Presented by D. Zieminska)

ABSTRACT

Production of hadrons in the charged current neutrino interactions at Fermilab energies has been studied in the 15 foot
deuterium filled bubble chamber. Comparison of the number of
negative tracks emitted in the target fragmentation region by (uu)
diquarks from νp collisions and by (du) diquarks from νn collisions
reveals only small differences. If each quark separately fragments
it is predicted that there will be more negative particles produced by the (ud) system than by the (uu) system. We conclude
therefore that the (ud) system predominantly remains bound in the
subsequent fragmentation.

No scaling violation has been observed in the current fragmentation region.

INTRODUCTION

In this talk we discuss the fragmentation of diquarks produced
in high energy charged current ν-deuterium interactions. By isolating neutrino interactions on neutrons and protons separately, we
can study the hadronization of (ud) and (uu) diquark systems, respectively. We also study fragmentation properties of the u quark.

*On leave from University of Warsaw, Poland.

The data sample comes from an exposure of the Fermilab 15 foot deuterium filled bubble chamber to a single-horn focused wide-band neutrino beam produced by 350 GeV/c protons. The present analysis is based on measurements of 60% of the total exposure, corresponding to a flux of 3×10^{18} protons on target. We require that the sum of the longitudinal momenta of all secondaries is greater than 5 GeV/c. The charged current events are extracted by applying the kinematic method[1]. We require that P_{TR}, the μ^- transverse momentum relative to the total momentum of the other charged particles be greater than 1 GeV/c. The neutrino energy E_ν was calculated by the formula:

$$E_\nu = p_L^\mu + P_L^c + P_L^c \cdot |\vec{p}_T^\mu + \vec{p}_T^c|/P_T^c, \tag{1}$$

where $p_L^\mu(p_T^\mu)$ and $P_L^c(P_T^c)$ are the longitudinal (transverse) momenta of the μ^- and of the charged hadron system relative to the ν direction, respectively. We require $E_\nu > 10$ GeV and the Bjorken variable[2] $y < 0.9$. If not stated otherwise, the Bjorken x cut, $x > 0.2$, is made in order to select events in which a valence d quark is struck. We further require that the invariant mass of all final state hadrons W be greater than 4 GeV. This selection results in a sample of 2025 events in a fiducial volume of $15.6 m^3$. The separation into νn and νp categories is done in the following way:[3] events with an even number of prongs and those with an odd number of prongs including a visible spectator proton fall in the νn category; events with an odd number of prongs with no visible spectator belong to the νp category. By spectator, we mean an identified proton which is emitted backward in the laboratory frame, or which goes forwards with momentum $p < p_o = 340$ MeV/c. [This choice of p_o yields the ratio of the number of backward spectators to the number of forward spectators predicted using neutrino flux and cross section ratios.] We find the numbers of events in each category to be N = 1291 and $N_{\nu p}$ = 734.

DIQUARK FRAGMENTATION

The process of fragmentation of a diquark into hadrons has been studied theoretically by several authors[4,5,6]. Generally, the dominant question is whether the diquark acts as a single object, in which case the two valence quarks recombine with a quark to give a baryon, or whether the two quarks fragment independently. These processes are illustrated in Fig. 1. In the former case mesons

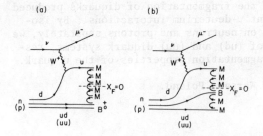

Fig. 1. Diagrams for νn and νp interactions: (a) coherent diquark fragmentation, (b) independent fragmentation of two valence quarks in the diquark.

520

(such as π^- or K^o) which contain a valence quark should be produced more frequently from a neutron target, particularly in the far backward region in the hadronic c.m. system. Consider the invariant function

$$F^h_{\nu p, \nu n}(x_F) = (2/\pi)(E^*/W)(dN^h/dx_F)/N_e$$

(E^* and x_F denote energy and Feynman variable, $x_F \equiv p_L/p_L^{max}$, of the hadron in the hadronic c.m. system and N_e denotes the number of events. We compare $F^-_{\nu n}(x_F)$ and $F^-_{\nu p}(x_F)$ for negative hadrons treated as π^-. Our results are shown in Fig. 2 and the ratio $F^-_{\nu n}(x_F)/F^-_{\nu p}(x_F)$ is shown in Fig. 3. We make the identifications $F_{\nu p} = F^h_{uu}$ and $F^h_{\nu n} = F^h_{ud}$

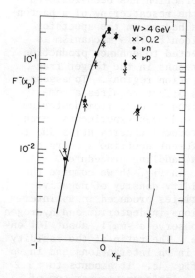

Fig. 2. Normalized Lorentz-invariant x_F distributions, $F^-(x_F)$, as defined in eq. (2), for negative hadrons in $\nu n \rightarrow \mu^- h^- X$ and in $\nu p \rightarrow \mu^- h^- X$, for $W > 4$ GeV and $x > 0.2$. For $x_F < 0$ they are called F_{ud} and F_{uu}, respectively. Also shown is the curve $A(1-x)^4$ found to fit the data of ref. [10].

Fig. 3. The ratio of $F^-(x_F)_{\nu n}/F^-(x_F)_{\nu p}$ for $W > 4$ GeV and for 3 regions of Bjorken x. The data $x > 0.2$ are compared with the model calculation of Sukhatme et al. [6] assuming independent fragmentation of quarks in a diquark. The νp data are not corrected for double scattering.

for $x_F < 0$ where F^h_{uu} and F^h_{ud} refer to the corresponding diquark systems. We observe no difference between F^-_{ud} and F^-_{uu} at $x_F < 0.2$. Similarly, the ratio of the rapidity distributions in the laboratory system is fairly constant and equal to 1.1-1.2 (see Fig. 4a).

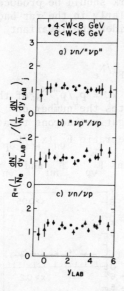

a) $\nu n / "\nu p"$

b) $"\nu p" / \nu p$

c) $\nu n / \nu p$

$R = (\frac{1}{N_e} \frac{dN^-}{dy_{LAB}})_i / (\frac{1}{N_e} \frac{dN^-}{dy_{LAB}})_j$

y_{LAB}

Fig. 4. The ratios of laboratory rapidity distributions of negative particles produced in νn and νp interactions. νn, "νp" and νp stand for the data from this experiment and νp stands for the data of ref. 9.

Our results suggest that the (ud) diquark predominantly remains bound to form a baryon, contrary to the prediction of reference 6. However, we cannot make a quantitative statement on diquark fragmentation because of an uncertainty in our data due to possible nuclear effects. In particular, the νp sample may contain up to 25% admixture of "double scattering" events where the original neutrino-nucleon interaction has been followed by the rescattering of a parton or a hadron on the spectator nucleon[7]. This mechanism is expected to enhance production of mesons in the target fragmentation region[8]. To assess this effect requires a comparison of data from deuterium and hydrogen experiments which introduces uncertainties due to different neutrino spectra and data handling procedures.

In fig. 4b we compare the rapidity density of negative particles produced in νp interactions in deuterium and hydrogen[9]. We observe a small, about 10% enhancement in the entire rapidity range. The ratio of the rapidity density of negative hadrons produced in νn interactions and in νp interactions in hydrogen is shown in fig. 4c. It amounts to ~ 1.2 and is rather independent of rapidity. Analogous comparison of x_F distributions for deuterium and hydrogen data is less reliable because of their sensitivity to experimental details.

Independent information on the fragmentation of (ud) and (uu) diquarks into mesons has been deduced by Hanna et al.[7] from pp collisions involving high p_T pion triggers. Assuming that large p_T quark-quark collisions dominate their high p_T triggered reactions, the authors have found that $F^-_{ud} = F^-_{uu} = (1 - x_F)^4$. Hence they conclude that the valence diquarks do not fragment independently. Furthermore, as shown in Fig. 2, our π^- distribution in the backward hemisphere has a similar dependence on x_F.

Below we discuss briefly some other possible tests and their feasibility in this experiment. One of them is the study of the ratio of (uu) fragmentation functions into π^+ and π^-. It would be high at large negative x_F according to the independent diquark fragmentation model, and equal to one in the case of coherent diquark fragmentation. However, the sample of "positive mesons" is heavily contaminated by unidentified protons which makes this test experimentally unavailable to us. The average number of positive particles (excluding identified protons, whose average number per νp event is 0.12 ± 0.01) and negative particles per event in the domain $x_F < -0.2$ are 0.45 ± 0.03 and 0.09 ± 0.01. Since most of these positive particles could be unidentified protons, our observed positive to negative ratio does not contradict the coherent diquark fragmentation model.

The ratio of the numbers of K^O mesons produced in νn and νp interactions is expected to increase as $-x_F$ increases according to the independent diquark fragmentation model, while it is expected to be equal to unity by the model of coherent diquark fragmentation. However, the power of this test is weakened by the fact that only the combined data on K^O and $\bar{K}^O$ production are accessible experimentally. For $x_F < -0.25$ we find $n_{ud}(K^O + \bar{K}^O)/n_{uu}(K^O + \bar{K}^O)$ = 0.9 ± 0.3 which is compatible with the coherent diquark fragmentation model, but does not discriminate against the independent fragmentation model[6] which predicts this ratio to be 1.2.

We finally comment on the comparison of Λ production in νn and νp interactions. This is of special interest since the Λ particle is the most clearly detected baryon, experimentally. The quark content of Λ is (uds). It can be readily produced in νn deep inelastic interactions by recombinations of the initial (ud) pair with an s quark from the sea. By contrast direct Λ production would be suppressed in νp interactions according to the model of coherent fragmentation of diquarks. This simple picture is complicated by the possibility of indirect production of Λ particles by the decay of Y^* resonances. Attempts have been made to determine the resonance contribution to the Λ particle production rates[11]. It was estimated that in a fraction $(28 \pm 13)\%$ of νp events the (uu) diquark behaves as independently fragmenting quarks.

QUARK FRAGMENTATION

We select current fragments as particles going forwards in hadronic c.m. system. Quark-Parton Model predicts the fragmentation of a quark into hadrons to be a universal process, independent of the reaction in which the quark has been created and independent of the four momentum transfer in the primary reaction[12]. In QCD however, a scaling-violating Q^2 dependence of quark fragmentation functions is expected[13]. We test these predictions by studying the z distributions of positive and negative hadrons:

$$D_u^{\pm}(z,Q^2) = \frac{1}{N_e(Q^2)} \frac{dN^{\pm}}{dz}(z,Q^2)$$

and their moments

$$D_m^{\pm}(Q^2) = \int_0^1 z^{m-1} D^{\pm}(z,Q^2)dz,$$

where $z = \dfrac{E_h}{E_H}$ is the fraction of the total energy of hadrons in laboratory system E_H, carried by a single hadron. Fig. 5 shows the u quark fragmentation functions obtained from the νn and νp samples. They agree with each other very well. Fig. 6 shows the dependence of the nonsinglet moments of the fragmentation functions, $D_m^{NS} = D_m^+ - D_m^-$, on log Q^2 for the combined νn and νp data. We observe no Q^2 dependence of the moments. When compared to the leading-order QCD prediction our data are consistent with the scale parameter Λ less than 200 MeV.

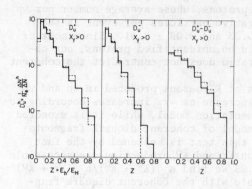

Fig. 5. z distributions of positive and negative hadrons per event in νn and νp interactions.

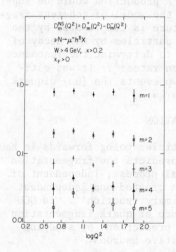

Fig. 6. Non-singlet moments of quark fragmentation functions for W > 4 GeV, x > 0.2 and x_F > 0 as a function of log Q^2 for m = 1,2,...5.

CONCLUSIONS

We conclude that our results are in agreement with the idea that the hadronization of a diquark occurs predominantly through recombination of the diquark with another quark to give a baryon. Our results for the u-quark fragmentation are consistent with predictions of Quark-Parton Model.

We thank the members of the Accelerator Division and Neutrino Department at Fermilab for their assistance in conducting this experiment, with special appreciation to the 15-foot bubble chamber staff. We also thank the scanning and measuring personnel at our laboratories for their careful work. This research is supported in part by the U. S. National Science Foundation and the U. S. Department of Energy.

REFERENCES

1. J. Bell et al., Phys. Rev. D19, 1 (1979).
2. We use the conventional deep inelastic scattering variables $x = Q^2/2M\nu$ and $y = \nu/E_\nu$ where Q is the lepton four momentum transfer, ν is the energy transfer between leptons and M is the proton mass.
3. For more detailed discussion see J. Hanlon, et al., Phys. Rev. Lett. 45, 1817 (1980), and J. Hanlon et al., in Proc. of Neutrino '79, edited by A. Haatuft and C. Jarlskog (Bergen, 1979) Vol. 2, p. 286.
4. M. Fontannaz, B. Pine and D. Schiff, Phys. Lett. 77B, 315 (1978).
5. T. A. deGrand, Phys. Rev. D19, 1398 (1979).
6. V. P. Sukhatme, K. E. Lassila and R. Orava, Fermilab-Pub-81/120-THY.
7. J. Hanlon et al., "Ratio of νn to νp Cross Section in High Energy Charged Current Interactions in Deuterium", paper submitted to the Conference of Neutrino Physics, Bergen, (1979).
8. H. Abramowicz et al., Nucl. Phys. B181, 365 (1981), A. Eskreys, et al., Nucl. Phys. B173, 93 (1980).
9. We adopted the method by H. G. Heilmann, Bonn Internal Report WA21-int-1(78) to calculate the neutrino energy and we relaxed the cut on Bjorken x. The νH_2 data was taken from N. Schmitz, Proc. of the 1979 International Symposium on Lepton and Photon Interactions at High Energies, edited by T.B.W. Kirk, and H.D.I. Abarbanel, Fermilab (1979), p. 359.
10. D. Hanna et al., Phys. Rev. Lett. 46, 398 (1981); see also the discussion of D. Deavis and B. R. Desai, Phys. Rev. D23, 1967 (1981).
11. C. C. Chang et al., Fragmentation of uu and ud di-quarks into Baryons in High Energy νD interaction. Abstract submitted to Notre Dame Conf.
12. R. P. Feyman, Photon-Hadron Interactons, Benjamin, Reading/Mass (1972).
13. J. F. Owens, Phys. Lett. 76B, 85 (1978); T. Uematsu, Phys. Lett. 79B, 97 (1978).

DISCUSSION

FRAGMENTATION OF U-QUARK JETS IN
HIGH ENERGY νD INTERACTIONS

C.C. Chang, W.A. Mann, A. Napier, J. Schneps
Tufts University, Medford, Massachusetts 02155
R.A. Burnstein, J. Hanlon, and H.A. Rubin
Illinois Institute of Technology, Chicago, Illinois 60616
C.Y. Chang, G.A. Snow, D. Son, P.H. Steinberg, D. Zieminska
University of Maryland, College Park, Maryland 20742
R. Engelmann, T. Kafka, and S. Sommars
State University of Stony Brook, Stony Brook, New York 11794
T. Kitagaki, S. Tanaka, H. Yuta, K. Abe, K. Hasegawa, A. Yamaguchi,
K. Tamai, T. Hayashino, S. Kunori, Y. Otani, H. Hayano
Tohoku University, Sendai 980,Japan

ABSTRACT

Inclusive rates for $\rho°$, $\omega°$, $\eta°$, and $K^{*\pm}(890)$ production in hadronic systems excited via high energy νD charged current interactions
have been determined using 7599 events recorded in the FNAL 15-foot
bubble chamber. In interactions from valence quarks ($X_{BJ} > 0.2$)
having hadronic energy $W > 4$ GeV, tracks forward in the hadronic CM
are taken to be remnants of fragmenting u-quark jets. For mesons
$(K° + \bar{K}°)$, $\rho°$, and $(\pi^\pm + K^\pm)$ in the jets, fragmentation functions
$D_u^h(z*)$ are measured in terms of hadronic CM energy fraction $z^* = 2$
E_h^*/W. Comparison of $\pi^+\pi^-$ invariant mass in jets to the inclusive
dipion spectrum is used to illuminate the role of $\omega°$, $\eta°$, and possible higher mass mesons in u-quark fragmentation. Differences between
these results and expectations of "standard" Field-Feynman fragmentation with jet energies $\langle W/2 \rangle = 3.24$ GeV are found, which however can
be minimized by tuning parameters γ, α_{ps}, and α_v in the model.

INTRODUCTION

In nearly all jet analyses to date, modeling of quark and/or
gluon hadronization into detected particles has been required; the
model most frequently invoked is the quark jet fragmentation scheme
of Field and Feynman (FFII)[1]. In this model there are several parameters which govern how mesons materialize from fragmentation of the
quark's initial energy, parameters which the authors initially set,
but which they indicate are to be "tuned" using data. The parameters
include:

γ = the probability of materializing a u$\bar{u}$ or d$\bar{d}$
     pair. Neglecting c$\bar{c}$ and b$\bar{b}$ production,

γ_s = $(1-2\gamma)$ is the probability for s$\bar{s}$ creation.

α_{ps}, α_v, α_T = relative probabilities for producing, pseudoscalar,
     vector, or tensor mesons.

a = constant in the function $f(\eta) = 1-a+3\eta^2$ which governs
     jet dynamics; $f(1-z)dz$ is the probability that the meson
     of first "rank" takes fraction z of the quark's momentum,
     leaving fraction $(1-z)$ for mesons following in rank.

In "standard" (as published) Field-Feynman quark jet, $\gamma = 0.4$ (so
that $\gamma_s = 0.2$), $\alpha_{ps} = \alpha_v = 0.5$, and a = 0.77.

That the quark parton model together with jet fragmentation of struck valence quarks provides an appropriate picture of high energy neutrino-nucleon interactions has been demonstrated using $\bar{\nu}_\mu \cdot N$ and $\nu_\mu \cdot N$ data, notably in the distribution of electric charge along the hadronic CM axis and in the energy-sharing ratio $Z_L = E^h/\nu$ for final-state hadrons. [2,3,4] In these studies, selections on the W^+. nucleon center-of-mass energy $W > 4$ GeV and $X_{BJ} = Q^2/2M > 0.2$ are required in order to separate current from target fragments[5] and to ensure predominance of interactions involving valence quarks. Recently, study of K_S° production in current fragmentation of d quarks induced via $\nu_\mu \cdot N$ reactions reported γ_s/γ to be 0.27 ± 0.46[6], rather than the value 0.5 assumed in the Field-Feynman standard jet. This result is in disagreement with an earlier measurement in an $e^- p$ experiment which reported $\gamma_s/\gamma = 0.13 \pm 0.03$.[7]

The goal of this work is to measure effects which characterize fragmentation of u-quark jets into mesons. Our results for u-quark jet fragmentation are used to confront the Field-Feynman model for jet energies induced in our experiment (W/2 ranges from 2.0 to 8.3 GeV/c^2 with mean value 3.24 GeV/c^2).

CHARGED CURRENT EVENT SAMPLES

Neutrino interactions were recorded in 328,000 picture exposure of the Fermilab 15-foot deuterium-filled bubble chamber exposed to a wide-band, single horn focussed neutrino beam; the neutrino beam was produced using a primary beam of 350 GeV/c protons, with a total flux of 4.9×10^{18} incident protons. Selection of final-state μ^- tracks in candidate charged current events was carried out using a kinematic method, described elsewhere.[8] An event's visible energy was required to be greater than 5 GeV; the incident neutrino's energy E_ν was estimated using the Heilmann method and was required to be greater than 10 GeV. Three different charged current samples were isolated: i) In studies involving inclusive primary production prongs (e.g. ρ° meson measurements), we used 7599 charged current events, corresponding to 60% of the exposure. ii) For studies involving K_S° decays we used events containing final-state V°'s from 70% of the exposure. iii) For studies involving $\gamma\gamma$ pairs we used events from 95% of the exposure.

INCLUSIVE MESON RESONANCES

By isolating enhancements in inclusive invariant mass distributions for particle combinations $\pi^+\pi^-$, $K_S^\circ\pi^\pm$, K^+K^-, $\gamma\gamma$, $\pi^+\pi^-\gamma$, and $\pi^+\pi^-\pi^\circ$, rates and/or upper limits have been determined for meson resonances ρ°, $K^{*\pm}(890)$, ϕ, η°, ω° and η'. In addition, the mass spectra for $\pi^+\pi^-$, $\pi^-\pi^-$, and $\pi^+\pi^+$ dipions have been examined for resonance structure with the aid of combinatorial background curves, estimated by pairing measured prongs from different events having identical multiplicities and similar W values. An incoherent super-position of ρ°, ω°, and η° contributions together with combinatorial background accounts for the bulk of correlated dipion mass observed in this experiment. The numbers of inclusive meson resonances per charged current event are presented in Table I. Details concerning methods of resonance extraction are given in Ref. 9.

MESON RATES IN U-QUARK JETS

With the $W^+ \cdot$ nucleon CM energy W required to be greater than 4 GeV/c^2 in order to disentangle current and target fragmentation processes, and with $X_{BJ} > 0.2$ required to ensure predominance of valence processes $W^+ + $ d-quark $\rightarrow$ u-quark, the charged current sample is reduced from 7599 to 2127 events. In studying this restricted sample, forward CM hemisphere particles are assumed to be produced in u-quark jet fragmentation, while particles backward in the CM are assumed to be remnants of fragmenting ud (neutron-target) or uu (proton target) diquark systems. In comparing data with Field-Feynman fragmentation, we use the variable $Z^* = E_h^*/E_{JET}^*$, which is the fraction of the quark jet energy carried by a final-state particle in the $W^+ \cdot$ nucleon center-of-mass. For W > 4 GeV events we assume E_{JET}^* to equal W/2, hence the distribution of (W/2) from data determines the u-quark jet energies to be used in a Field-Feynman Monte Carlo. For various settings of Field-Feynman parameters, samples of 24,000 u-jet processes were simulated. In bubble chamber neutrino experiments the transformation to the $W^+ \cdot$ nucleon system is well-defined to the extent that E_ν can be estimated. Consequently the Z^* distributions from Monte Carlo can be compared with $Z^* = E_h^*/(W/2)$ distributions for jet fragments ($Y_R > 0$ tracks) in the data. It should be noted that the region $Z^* > 0.2$ contains some particles from target fragmentation processes which have spilled over into the forward hemisphere; moreover, the Field-Feynman model is not expected to be accurate at low Z^*.

Using jet fragmentation particles only, the invariant masses for $K^{*\pm}$ have been examined. Best estimates for rates of $K^{*\pm}$ mesons in both u-quark jets and in diquark fragments are presented in Table I. The relative rate of $(K^\circ + \bar{K}^\circ)$ production in u-quark fragments is also given, from which we conclude that roughly 20% of K_s° in u-jets are the result of $K^{*\pm}(890)$ production.

The fragmentation function

$$D_u^{K^\circ}(Z^*) = \frac{1}{N_{ev}} \cdot \frac{dN^{K^\circ}}{dZ^*}$$

for K_s° mesons, which are particles of second or higher rank in the FFII hierarchy (neither K° nor $\bar{K}^\circ$ contain u quarks), is shown in Fig. 1. The dashed curve in the figure is obtained from the standard Field-Feynman u-quark jet for which $\gamma_s/\gamma = 0.5$. Although the shape of the distribution from the model is in rough agreement with the data, the predicted rate in the model is too high. Improved agreement with the model is obtained with $\gamma_s/\gamma = 0.3$ (dotted curve). Our $D_u^{K^\circ}$ distribution is also in accord with the ratio $\gamma_s/\gamma = 0.27$ extracted using $\bar{\nu}N$ data[6] for which K_s° at high Z^* are likely to be mesons of first rank in d-quark jets.

The $\pi^+\pi^-$ invariant mass from the jet can be fitted using a Breit$\cdot$Wigner form for the ρ° plus polynomial background, as was done in the overall inclusive distributions. For $Z^*(\pi^+\pi^-) > 0.2$ we find

$$(\rho^\circ/u\text{-jet}) = 0.17 \pm 0.05,$$

which is equal within errors to the inclusive charged current rate. The fragmentation function for ρ° mesons $D_u^{\rho^\circ}(Z^*)$ is shown in Fig. 2 together with the prediction of the FFII standard jet, for which pseudoscalar and vector meson rates are set according to $\alpha_{ps} = \alpha_v = 0.5$ with $\alpha_T = 0$. Once again, the model yields the appropriate shape for the fragmentation function but the predicted rate exceeds the observed one. Better agreement is obtained with the model when $\alpha_v = 0.4$ and $\alpha_{ps} = 0.6$ are used together with $\gamma_s/\gamma = 0.3$, as indicated by the solid curve in Fig. 2 (and Fig. 1). As shown in Tables I and II, estimated rates for $K^{*\pm}(890)$ resonances are in better agreement with the model when the latter parameters are used instead of the standard settings. Table I also shows that ρ° and $(K^\circ + \bar{K}^\circ)$ rates are quite different in u-quark versus diquark fragmentation, being more than twice as abundantly produced in u-quark jets.

Since nearly equal relative rates for ρ°, ω° and η° mesons together with an incoherent combinatorial background accounts for the entire inclusive $m(\pi^+\pi^-)$ spectrum in this experiment (Table I, and Ref. 9), it is of interest to plot the $\pi^+\pi^-$ mass distribution from the jet ($Z^* > 0$) together with the overall inclusive spectrum normalized to it as shown in Fig. 3. From the threshold of the ω° region around 480 MeV up to 2.0 GeV in $\pi^+\pi^-$ mass, the shapes of the two distributions are practically identical. This observation constitutes evidence, albeit indirect, that the ω° rate in u-quark jets must be roughly equal to the ρ° rate in the jet, as was found to be the case for ρ° and ω° inclusive charged current rates. On the other hand, the η° mass region in the jet distribution is depleted in comparison with the inclusive shape in the mass interval 360-440 MeV. This observation suggests that in u-quark jets of our energies, η° production is distinctly less than ρ° or ω° rates. As indicated by Table II which summarizes Field-Feynman fragmentation with reasonable variations in the γ and α parameters, this observation cannot be readily accomodated by the model. We note that in FFII fragmentation, neutral pseudoscalar particles such as the η° and η' are allowed to materialize nearly as easily as do π°'s -- very little penalty is imposed on them for their substantial mass. This arrangement would not seem suitable for jets with the modest energies induced in this experiment.

Similar comparisons of dipion masses in our data reveal that the η° is also nearly absent in diquark fragments. The bulk of the overall inclusive production is confined to interactions having $X_{BJ} < 0.2$ indicating a strong correlation with reactions involving sea quarks.

In Fig. 4 the FFII fragmentation functions for positive charged prongs $D_u^{h^+}(Z^*)$, where $h^+ \equiv (\pi^+ + K^+)$ are compared with the data. For either the standard setting or for any of the tuned settings shown previously having various γ and α values, the model yields very similar fragmentation curves which everywhere exceed the data beyond $Z^* = 0.4$. This behavior in $D_u^{h^+}(Z^*)$ can be partially remedied in the model by altering the constant "a" which appears in the functional form governing energy-partitioning among mesons of differing rank in the fragmentation chain. The FFII value a = 0.77 was assumed in the generation of curves from the model previously described (dotted, dashed and solid curves in Fig. 4, see Table II). If a is increased to 1.0, so that the dynamical function simplifies to the form $f(\eta) = 3\eta^2$, the model yields the dot-dashed distributions shown in Fig. 4. Although better correspondence with the data is achieved, the FFII prediction continues to exceed the data for Z^* values above 0.5 suggesting that more drastic changes in the energy-partitioning algorithm are needed.

FOOTNOTES AND REFERENCES

1. R.D. Field and R.P. Feynman, Nucl. Phys. B136, 1 (1978).
2. M. Derrick et al., Phys. Lett. 91B, 470 (1980); M. Derrick et al., ANL-HEP-PR-80-54, March 1981.
3. J.P. Berge et al., FERMILAB-PUB-80/62-EXP 7420.180, June 1980 (submitted to Nucl. Phys. B); V.V. Ammosov et al., FERMILAB-PUB-80/96-EXP 7420.180 (submitted to Nucl. Phys. B).
4. R. Orava, FERMILAB-CONF-81/21-EXP 7420.180, February 1981 (submitted to Physica Scripta).
5. N. Schmitz, in International Symposium on Lepton and Photon Interactions at High Energies, p. 379, FERMILAB, ed. T.B.W. Kirk and H.D.I. Abarbanel (1979).
6. V.V. Ammosov et al, Phys. Lett. 93B, 210 (1980).
7. I. Cohen et al., Phys. Rev. Lett. 40, 1614 (1978).
8. J. Hanlon et al., Phys. Rev. Lett. 45, 1817 (1980).
9. C.C. Chang et al., to be published in Proceedings of the International Conference on Neutrino Physics and Astrophysics ν'81, Wailea, Hawaii, July 1981.

TABLE I: PRODUCTION RATES FOR MESON
RESONANCES IN νD CHARGED CURRENT REACTIONS

Meson (Ref. 1)	(Meson/cc)[a]	(Meson/Diquark)[c]	(Meson/u-jet)[b] All z	(Meson/u-jet) $z^* > 0.2$
$\rho^\circ = (u\bar{u}-d\bar{d})/\sqrt{2}$	0.18±.03 (0.15±.03)[d]	0.07±.02	0.17±.05	0.17±.05
$\omega^\circ = (u\bar{u}+d\bar{d})/\sqrt{2}$	0.39±.19 (0.17±.02)[d]	–	($\wedge\rho^\circ$ rate)	($\wedge\rho^\circ$ rate)
$\eta^\circ = 1/2(u\bar{u}+d\bar{d}-\sqrt{2}\,s\bar{s})$	<1.0 at 90%CL (0.17±.04)[d]	–	$(.02^{+.06}_{-.02})$	–
$\eta' = 1/2(u\bar{u}+d\bar{d}+\sqrt{2}\,s\bar{s})$	<0.30 at 90%CL	–	–	–
$\phi = s\bar{s}$	0.009 .005	–	–	–
$K^{*+}(890) = u\bar{s}$	0.043±.010	0.008±.008	0.014±.014	0.012±.012
$K^{*-}(890) = \bar{s}u$	0.009±.005	0.006±.006	0.014±.008	0.012±.012
$K^\circ + \bar{K}^\circ = d\bar{s} + s\bar{d}$	0.164±.037	0.071±.016	0.146±0.14	0.116±.013

[a]Meson rate per charged current (CC) event, no W or X_{BJ} selections imposed.
[b]Meson system forward in the W^+ nucleon CM ($Y_R>0$), per CC event having W>4 GeV and $X_{BJ}>0.2$.
[c]Meson system backward in the W^+ nucleon CM($Y_R<0$), per CC event having W>4 GeV and $X_{BJ}>0.2$.
[d]Resonance rates in parentheses determined using combinatorial background.

TABLE II: MESON RESONANCES ($Z^* > 0.2$) PER
U-QUARK JET IN FFII FRAGMENTATION[†]

Meson	Standard Jet $\gamma=0.4$ $\alpha_{ps}=0.5$ $\alpha_v=0.5$ $a=0.77$ (dotted in Fig.'s)	$\gamma=0.45$ $\alpha_{ps}=0.75$ $\alpha_v=0.25$ $a=0.77$ (dashed in Fig's)	$\gamma=.435$ $\alpha_{ps}=0.6$ $\alpha_v=0.4$ $a=0.77$ (solid in Fig.'s)	$\gamma=.435$ $\alpha_{ps}=0.6$ $\alpha_v=0.4$ $a=1.0$ (dash-dotted in Fig.'s)
ρ°	0.24	0.14	0.21	0.21
ω°	0.34	0.21	0.33	0.28
η°	0.19	0.34	0.25	0.24
η'	0.28	0.45	0.35	0.31
ϕ	0.036	0.007	0.015	0.011
$K^{*+}(890)$	0.18	0.04	0.098	0.087
$K^{*-}(890)$	0.07	0.02	0.044	0.037
$K^\circ+\bar{K}^\circ$	0.21	0.11	0.146	0.145

As in the published "standard" model, we take $\alpha_T = 0$ and neglect
$c\bar{c}$ and $b\bar{b}$ production. The u-quark jet energies in the model have
been assigned according to the distribution of W/2 for CC events
in this experiment having W > 4 GeV and $X_{BJ} > 0.2$.

FIGURE CAPTIONS

Fig. 1: Data points (open circles) show the fragmentation function
for u-quarks into K° and $\bar{K}^\circ$ mesons $D_u^{K^\circ+\bar{K}^\circ}(Z^*)$ as a function
of the energy fraction $Z^* = E_{K_S^\circ}^* / (W/2)$ in the hadronic center
of mass. Curves show FFII predictions corresponding to different values of the $u\bar{u}$ production probability γ: $\gamma = 0.4$
(dotted curve), $\gamma = 0.45$ (dashed curve), and $\gamma = 0.435$
solid curve).

Fig. 2: Fragmentation function for u-quarks into ρ° mesons $D_u^{\rho^\circ}(Z^*)$
versus energy fraction Z^* in the hadronic center of mass.
Curves show FFII predictions corresponding to different relative probabilities for producing pseudoscalar or vector
mesons: $\alpha_{ps} = \alpha_v = 0.5$ (dotted curve), $\alpha_{ps} = 0.75$ and
$\alpha_v = 0.25$ (dashed curve), and $\alpha_{ps} = 0.6$, $\alpha_v = 0.4$ (solid
curve).

Fig. 3: Invariant mass $\pi^+\pi^-$ from u-quark jet fragmentation (W >
4 GeV, $X_{BJ} > 0.2$, and $Y_R^{\pi\pi} > 0$), shown with the inclusive
$\pi^+\pi^-$ distribution shown area-normalized and
superposed. Shaded histogramming shows regions in the jet
distribution which are depleted relative to the overall
inclusive distribution.

Fig. 4: Fragmentation function $D_u^{h^+}(Z^*)$ for u-quarks into positive
hadrons $h^+ = (\pi^+, K^+, p)$. Dotted, dashed and solid curves
correspond to FFII predictions with various γ, α_{ps}, α_v
settings (see Table II) but with parameter $a = 0.77$.
Dot-dashed curve shows prediction using $a = 1.0$, with
$\gamma = 0.435$, $\alpha_{ps} = 0.6$, and $\alpha_v = 0.4$.

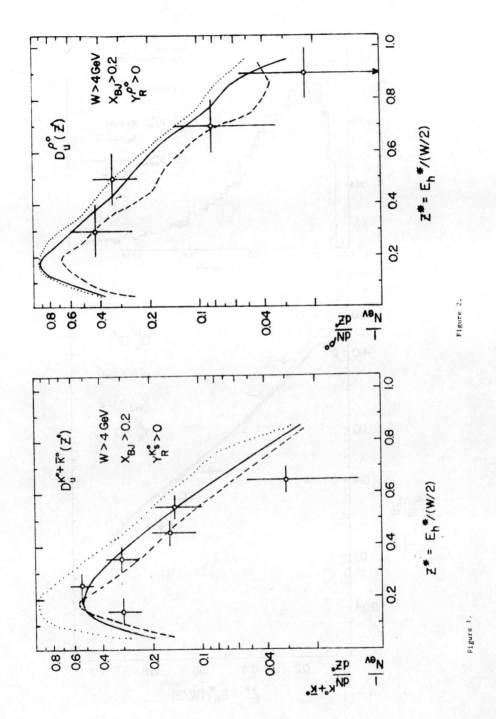

Figure 1.

Figure 2.

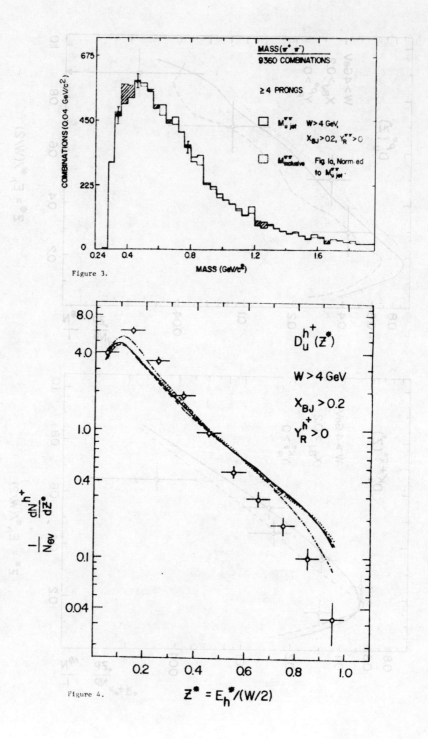

Figure 3.

Figure 4.

DISCUSSION

EDITORS' NOTE: This includes discussion for the previous talk by D. Zieminska as well as for the talk by Mann.

FRIDMAN, VANDERBILT/SACLAY: I have two technical questions. One [to Zieminska] is about the proton spectator. You defined the events which have a proton spectator forward [as events] where the momentum is smaller than 300 something [MeV/c] which is rather high for a spectator proton because in general it's [about 80] MeV/c. On the other hand you showed in some other article (We showed it also, but you too.) that you have among these events a lot of rescattering events. Have you made some correction for this? Or not?

ZIEMINSKA: No. Indeed our proton sample contains some double-scattering events. It would be better to compare our νn with νp from a hydrogen experiment. We do not have data with exactly the same cuts so it's impossible for us to do this.

FRIDMAN: No, you can take the proton spectrum.

ZIEMINSKA: However it is possible [to compare] with our proton sample. [Our] νp sample agrees with νp in hydrogen so double-scattering doesn't change very much.

FRIDMAN: The second technical question [to Mann] which I have for you is about the way you calculated E_z. I would like to know if I am [correct] that you should have some events with z bigger than 1.

MANN: I don't see any.

MORRISON, CERN (to Mann): In your results, as far as I could see, the ρ^o and the ω^o had very different cross sections. One was .18, the other was .39. Normally these are about equal.

MANN: There are two measurements there for the ω^o. The most direct one has a huge error because it comes from a signal which is something like [a few events]. Let me show you that. This [signal] is seen with direct bump hunting. This one is extracted with what I would characterize as a tuned background subtraction.

SUKHATME, ILLINOIS (to Zieminska): I'm a little confused about some of your conclusions. By looking at your data from ud and uu to Λ's one can say that the diquark does not usually stick together. And the same data (except now it is being used for π^-'s) says that the diquark does stick together. Now can you comment on why this is the case.

ZIEMINSKA: There is no contradiction in our data. We looked at all possible tests. Λ cannot be formed from a uu diquark, but Σ^{*+} can, and it can decay into Λ's. We looked at the $(\Lambda\pi^+)$ mass spectrum and we found that at least half of the Λ's come from Σ^* resonances. So there is no contradiction.

OWENS, FLORIDA STATE (to Zieminska): I just want to make a comment concerning your evidence or lack of evidence for scale violation in the fragmentation functions. Over the kinematic range that your data cover the logarithmic scale violations just don't amount to a large effect, so it's a little misleading to show curves which are quite consistent with the data with Λ = 300 MeV and then claim that there is no evidence for scale violations. The second thing is that it doesn't make any sense to quote a value of Λ without specifying what the higher order corrections are and what renormalization prescription is used. Since the higher order corrections have been calculated for fragmentation functions one should be careful to include them and to also specify the scheme which has been used to define your Λ.

ZIEMINSKA: I didn't like to stress this scaling violation problem, but last year it was discussed by people representing the BEBC collaboration. In the same kinematic range they found scaling violations when no cut on W was made. So I just repeated this.

OWENS: I presume though that the results you showed were done with a leading-log type of calculation.

ZIEMINSKA: Yes.

SAARIKKO, BONN (to Mann): In CERN WA21 experiment which Professor Schmitz described today we find some evidence for tensor meson production. We find the f to ρ^o ratio to be roughly 1 to 4. Do you have some upper limit for f production?

MANN: No. My statement comes from our study of the $\pi^+\pi^-$ mass spectrum. Essentially we see nothing at the f^o in our data. And our combinatorial background seems to describe everything above that [in mass], so that was the basis for our conclusion. Additionally, in the correlated $\pi^+\pi^-$ invariant mass [spectrum], high-mass stuff decaying into 3-body objects tends to fill up the region between the ρ and the ω. We don't see that. Our ρ stands cleanly away from the ω pile-up.

MEASUREMENT OF THE FRAGMENTATION OF THE HADRONIC SYSTEM
IN DIFFERENT FLAVOUR AND COLOUR STATES

B. Pietrzyk
CERN, Geneva, Switzerland

ABSTRACT

The lepton-pair trigger was used to select the hadronic system in different flavour and colour states; properties of its transformation into hadrons were measured in terms of charge flow, multiplicity distribution, and fragmentation function.

1. INTRODUCTION

Properties of low p_T hadrons produced in the hadron-hadron interactions have been studied since the beginning of investigations in high-energy physics. The proceedings of this series of conferences are witness to the experimental progress made in this field and also to the development of ideas to explain the information obtained.

In general it is very difficult to investigate the overall properties of hadron-hadron interactions, since the outgoing particles originate from a mixture of many different parton-parton interactions. The lepton-pair experiment, described here, allows the selection of different subprocesses, and hence the definition of the quantum numbers of the fragmenting system.

What do we find interesting in the physics of hadrons associated with lepton pairs? Figure 1 presents a graph of the dimuon pair production through the Drell-Yan mechanism in $\pi^- N$ interactions. Since the negative pion is composed of a down (d) and an anti-up ($\bar{u}$) quark and since the latter annihilates in the production of a lepton pair, the d quark continues its trajectory in the forward direction and fragments into hadrons. In a similar way a diquark from the nucleon fragments into hadrons in the backward direction. It is important to realize that knowing x and the mass of the lepton pair one can obtain the fraction of the momentum carried by the forward d quark x_d for every event[*].

Fig. 1 Graphs describing the parameters of a hadronic system accompanying the Drell-Yan and the J/ψ production.

[*] According to the formulae $x_d = 1 - x_{\bar{u}}$, $x_{\mu\mu} = x_1 - x_2$, $x_1 x_2 = mass_{\mu\mu}^2/s$ (where $x_1 = x_{\bar{u}}$).

For the J/ψ production we expect a different situation (Fig. 1). Since the J/ψ is produced mostly through the gluon-gluon fusion (at the energies of this experiment), both the d and $\bar{u}$ quarks are fragmenting in the forward direction. Again, here one knows the momentum of the fragmenting system event by event.

Therefore there are basic differences for these two cases. First of all, in the case of the Drell-Yan production, it is a system with a charge of $\frac{1}{3}$ which is fragmenting, while in the case of the J/ψ production it is the whole -1 charge. Do we see this difference experimentally? Secondly, for the Drell-Yan production (quark fusion) the fragmenting system is in the three-colour state, while for the J/ψ production (gluon fusion) the fragmenting system is in the eight-colour state. According to the QCD predictions one could expect a higher multiplicity by a factor of $\frac{9}{4}$ for the fragmenting of the eight-colour state[1]. Finally, what are the fragmentation functions? Particularly, is the d quark (accompanying Drell-Yan production) fragmenting as a "well-established jet" (WEJ).

In order to investigate these questions, properties of hadrons associated with the Drell-Yan and the J/ψ production were investigated and compared with the properties of hadrons produced in the normal hadronic interactions as measured in the same experiment. All these were compared with the properties of WEJs. Since the best examples of WEJs are those seen in the e^+e^- annihilation and since the Field-Feynman fragmentation describes them very well, for obvious technical reasons the Field-Feynman Monte Carlo (MC) simulation is taken as a model for WEJs[2].

The experiment was performed by the Saclay-Imperial College-Southampton-Indiana-CERN Collaboration (WA 11 - Goliath)[3] in the CERN SPS West Hall. It was a large open spectrometer, where in addition to the lepton pairs the momenta of charged hadrons were also measured. Space does not allow me to describe here the details of the experimental set-up and data analysis. More details can be found elsewhere[4].

2. EVENT SELECTION

Figure 2 shows the $\mu^+\mu^-$ invariant mass spectrum. It is worth noting an excellent mass resolution for the J/ψ (σ = 31 MeV). A total of 42,456 events with the lepton-pair mass between 2.95 and 3.25 were defined as those accompanying J/ψ production. As can be seen in Fig. 2, the background contamination below the J/ψ peak is very small. On the other hand, the 2126 events with the $\mu^+\mu^-$ mass between 3.85 and 9.3 were selected to represent the Drell-Yan production. Finally, these two sets of data were compared with those from normal π^-N interactions at 192.5 GeV/c in a beryllium target measured in the same experiment (13,500 events).

It is convenient to define the energy of the forward hadronic system as $x_{tot} = 1 - x_1$ (Fig. 1). In this way the longitudinal

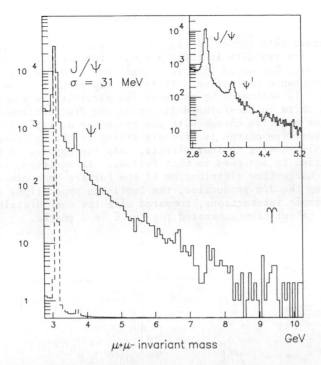

Fig. 2   μ⁺μ⁻ invariant mass spectrum measured in this experiment.
The dashed line shows the same spectrum in a linear scale.

momentum of the hadronic system is described in the units $\sqrt{s}/2$.
Owing to the different mass of the lepton-pair system the distribution of x_{tot} is different for the hadrons accompanying the J/ψ and
the Drell-Yan production, being slightly higher for the former.

One of the major problems of the analysis of the properties of
hadrons in this experiment was the fact that the interactions took
place not with the proton but with the beryllium. The results of the
hadron-nucleus[5] experiments show, however, that, while in the backward direction the nuclear effects dominate, in the forward direction
one can measure properties of hadrons produced in the elementary
interaction which are not very distorted by the nuclear effects.

Bearing in mind all these considerations, we will investigate
the properties of hadrons in terms of the following variables:
charge flow, multiplicity, and fragmentation functions. It is essential to note that all the experimental and MC curves and numbers presented in this paper are normalized to one (experimental or MC)
event.

3. CHARGE FLOW

The upper left corner of Fig. 3 shows the distribution of charge in the function rapidity in the π^-N c.m.s. for the J/ψ production (dashed line). The solid line represents the same distribution without the acceptance correction. It is clearly seen that the corrected distribution is dominated by the positive particles in the backward direction, while in the forward direction the distribution is dominated by the negative charge with the crossover at rapidity zero. Since the positive charge is strongly rising in the backward direction, clearly showing nuclear effects, only the physics in the forward direction is analysed in what follows. In addition, Fig. 3 shows the charge-flow distribution of the forward hadronic system accompanying the J/ψ production, the Drell-Yan production, and in the normal hadronic interactions, compared with the same distribution of the jet MC calculation generated for the $\bar{u}$ or d quarks.

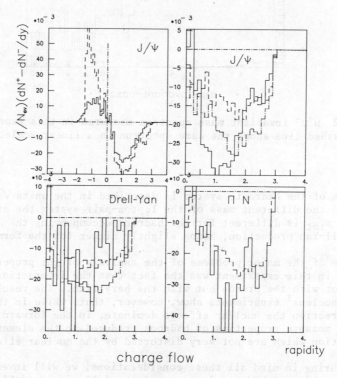

Fig. 3 Disbribution of charge versus rapidity. Upper left corner: distribution corrected for acceptance (dashed line) and not corrected for acceptance (solid line).
The other pictures show data corrected for acceptance (solid line), d quark MC (dashed line), and $\bar{u}$ quark MC (dash-dotted line).

Naively, the integration of these histograms over rapidity in the forward direction gives a very interesting number, the charge of a fragmenting system, i.e. $-\frac{1}{3}$ for the fragmentation of the d quark accompanying the Drell-Yan production, or -1 for the π^-N interactions or J/ψ production through the gluon-gluon fusion. Unfortunately, the reality is much more complicated; for example, the fragmentations in the forward and backward directions may overlap, the reinteractions in the beryllium target may add some positive charge also in the forward direction; finally, even in the fragmentation of the ideal d quark jet one does not expect to find a charge of $-\frac{1}{3}$. One should realize that for the balance of charge only the first (d) quark and the last antiquark in the fragmentation chain are important (the others cancel). If the last one is the $\bar{d}$, the $\bar{u}$, or the $\bar{s}$ with equal probability, then the net charge is really $-\frac{1}{3}$. On the other hand, if only the first two are produced, then the integrated charge is -0.5, while in the case of their production with the probability of 0.4, 0.4, and 0.2 (as in the case of Field-Feynman fragmentation), one can expect to observe a net charge of 0.4.

The experimentally measured charge in the forward direction after the acceptance corrections is -0.32 ± 0.05, -0.46 ± 0.07, and -0.64 ± 0.1 respectively, for the Drell-Yan production, the J/ψ production and for the normal hadronic interactions.

4. FRAGMENTATION FUNCTION

The simplest variable describing the fragmentation process in the z distribution, where z is the ratio of the x of every hadron to the x_{tot} of the hadronic system, as defined in Section 2. In this way fragmentation in the longitudinal momentum is investigated. The z distribution both in the linear and the logarithmic scale, for the hadrons accompanying the J/ψ and the Drell-Yan production and for the normal hadronic interactions in the forward direction, is presented in Fig. 4 and compared with the jet MC distribution. For the normal hadronic interactions there seems to be no difference in the fragmentation above $z = 0.1$, and an increase of particles is observed below. This increase at small z is also observed for the hadrons produced together with the J/ψ and the Drell-Yan production, but there it is accompanied by a little decrease in the number of particles at moderate z for the J/ψ production and with much smaller statistical significance for the Drell-Yan production.

However, one should be very surprised that these different fragmentations are so similar and so close to the jet MC calculation. There are plenty of reasons why they could be different:

a) The quark contents of the hadronic systems are different. In the case of the Drell-Yan production it is the d quark which is fragmenting, while in the case of the J/ψ or normal hadronic interactions, both of them are fragmenting.

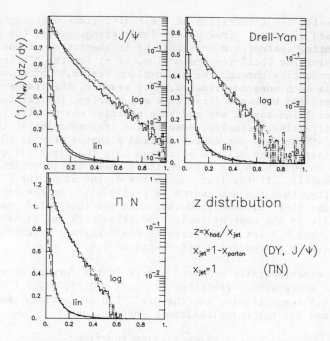

Fig. 4 z distribution describing the fragmentation both in the
linear and the logarithmic scale. Linear scale: data (solid line),
q MC (dashed line). Logarithmic scale: data (dash-dotted line);
q MC (dotted line). All curves have absolute normalization to one
experimental (or MC) event.

b) Colour charges are different. The forward hadronic system is in
the three-colour state in the case of Drell-Yan production,
mostly in the eight-colour state for the J/ψ production, and in
a mixture of different colour states in the case of normal hadronic interactions.

c) Valence quarks are not the only partons in the hadrons. There
are plenty of gluons and sea partons in the hadrons fragmenting
together with the valence quarks. In the case of e^+e^- annihilation we start with the point-like quarks, while in the soft fragmentation described here we have a system of fragmenting partons
(or quarks with a form factor). This is a difference between
current and constituent quarks.

5.   MULTIPLICITY

Since the mean energy of the forward hadronic system is lowest
for the Drell-Yan production, higher for the J/ψ production and
finally highest for the normal hadronic interactions, it is important

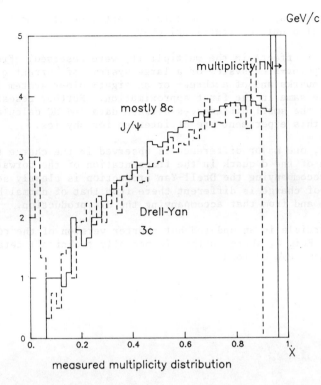

GeV/c

multiplicity ∏N→

mostly 8c
J/ψ

Drell-Yan
3c

measured multiplicity distribution

Fig. 5 The measured multiplicity distribution versus longitudinal
momentum of the forward-going hadronic system x_{tot}.

to take into account this trivial energy-dependence property. Figure 5
shows the mean measured multiplicity of the forward hadronic system
as a function of its x (x_{tot}). It is seen that the multiplicity is
very similar for both the Drell-Yan and the J/ψ production and it is
approaching the multiplicity of the normal hadronic interactions for
x = 1. One can therefore conclude that the QCD colour effects discussed above are not observed (of course in the energy region of this
experiment).

6. CONCLUSIONS

The hadronic system may be described by many variables: its
flavour (and correlated electric charge), its colour charge, its
geometrical size, the description of its "construction" (set of
structure functions).

In this work different hadronic systems have been investigated:
three- and eight-colour objects, fragmentation starting from a point-
like object (e^+e^- annihilation seen through the jet MC calculation)

and from a large system (all the data presented here), and fragmenting systems of a unit or fractional charge.

No colour effects in the multiplicity were observed. Fragmentation of the point-like system or a large system, of current quarks or
constituent quarks and of a three- or an eight-colour system is surprisingly the same in the first approximation. Further research will
soon show if the small differences between data and MC calculation
observed in this experiment are of interest for physics.

However, one major difference is observed in the charge flow.
The presence of the d quark in the fragmentation of the forward hadronic system accompanying the Drell-Yan production is clearly seen.
The balance of charge is different there from that of normal hadronic
interactions and from that accompanying the J/ψ production.

This article is an updated but shorter version of the report
presented in Ref. 4, where further (especially technical) details of
this experiment can be found.

REFERENCES

1. S.J. Brodsky and J.F. Gunion, Phys. Rev. Lett. <u>37</u>, 402 (1976).
2. The so-called Lund Monte Carlo program was used for the jet MC
   calculation: T. Slöstrand and B. Söderberg, Lund preprint LU TP
   78-18 (1978). It is equivalent to the Field-Feynman Monte Carlo
   calculation.
3. The members of the collaboration are: R. Barate, P. Bareyre,
   P. Bonamy, P. Borgeaud, M. David, F.X. Gentit, G. Laurens,
   Y. Lemoigne, G. Villet and S. Zaninotti (CEN-Saclay, Gif-sur-
   Yvette, France); P. Astbury, A. Duane, G.J. King, R. Namjoshi,
   B.C. Nandi, D. Pittuck, D.M. Websdale and J. Wiejak (Imperial
   College, London, England); J.G. McEwen (Southampton University,
   England); B. Brabson, R. Crittenden, R. Heinz, J. Krider and
   T. Marshall (Indiana University, Bloomington, Indiana, USA);
   B. Pietrzyk and R. Tripp (CERN, Geneva, Switzerland).
4. B. Pietrzyk, Properties of hadrons associated with lepton-pair
   production, to be published in Proc. Moriond Workshop on Lepton-
   Pair Production, Les Arcs, France, 25-31 January 1981 (Ed.
   J. Tran Thanh Van).
5. K. Braune et al., to be published.
   M. Faessler, New experimental results for particle production
   from nuclei, to be published in Ann. of Phys. (USA), 1981.

DISCUSSION

RAJA, FERMILAB: I have two points to make. One is that we have a similar experiment running at Fermilab and we would not dream of claiming that the dimuon spectrum is due to Drell-Yan because of the large π decay background outside the Ψ peak. Most of our dimuons come from decay of pions. And that is a large percentage of the effective mass outside the Ψ. So you have to subtract that out before you claim that they are prompt dimuons. That is the first point. The other point is that, of the Ψ production, about 30 to 40% comes from χ decay. That means that you are not really looking at a pure color octet system [if the χ's are formed by gluon fusion in the color singlet state (Carlson-Suaya model)]. So you have to correct for that too.

PIETRZYK: Your comment is quite right, but, of course, below the Ψ peak you can see that the correction from continuum background is anyhow very small. In principle, with our resolution we could still take Drell-Yan even between the Ψ and the Ψ' and there we have still some 10% of background. This background is absolutely not important, maybe one or two percent, above 4 GeV where the cut for $\mu^+\mu^-$ pairs was made. So, in fact, the only reason the data presented here were taken above the Ψ' was background. One could double statistics taking even between Ψ and Ψ'. Now the comment about original J/Ψ production. This is very difficult, a very important comment you made. I had no time to discuss this but all that I presented here was first-order analysis. Most of the J/Ψ produced is through 8-color gluons.

OWENS, FLORIDA STATE: I think the point about this octet of color is a little difficult to understand, because in any fusion type of calculation that you would use for J/Ψ production you'd have a large quark anti-quark component as well. [One must include the $q\bar{q} \rightarrow c\bar{c}$ subprocess as well as the $gg \rightarrow c\bar{c}$.] So the [remaining beam fragments] are going to be a mixture of a color triplet and octet depending on the relative ratios of quark anti-quark to Ψ or gluon gluon to Ψ. That depends on the structure functions that you use.

PIETRZYK: Experimental results show that if you use hadrons of about 40 GeV, which was the case in another experiment, then most J/Ψ are really produced through quark-quark fusion. If you go to higher and higher energies, production of J/Ψ through gluon-gluon annihilation is rising, and (again this refers to the next comment) at an energy of 185 GeV, where another experiment

(NA3 at CERN) measured J/Ψ production from $p\bar{p}\pi^{+}\pi^{-}$ etc. (One can work on it. We are working also on it.) it turns out that about 20 - 25% of J/Ψ may be produced through quark-quark fusion. So again this is a second-order effect. I have had no time to come to all details of this.

OWENS: I have one other comment concerning Raja's comment on the other mechanisms for producing the Ψ. If you use a fusion mechanism such as you've invoked here, then the Ψ, the Ψ' and the χ's are all going to be produced by the same type of mechanism because that fusion mechanism ignores color constraints at the Born term level. So therefore, aside from some slight smearing out of the kinematics due to the mass differences between the Ψ's and the χ's, there really shouldn't be any other effect on your analysis.

PIETRZYK: I agree with you.

WALKER, DUKE: This is a very naive question, but we have worked for several years on internal bremsstrahlung and there is lots of it. And as you go up in energy of the photon you continue to find this effect. Can there be appreciable production of μ pairs from internal conversion of high-energy photons . . . that is, bremsstrahlung photons from the hadronic collisions rather than the Drell-Yan process?

PIETRZYK: You mean in the continuum or J/Ψ?

WALKER: In the continuum.

PIETRZYK: There were calculations of this type and they showed that this type of process may be important at ISR energies and in the very, very forward direction of J/Ψ. There one might observe it, and, in fact, in a dimuon experiment at CERN by "Ting and company", they have probably observed such an effect. But at our energies it is just not important at all.

RESVANIS, ATHENS: I would like to take Raja's point one step further. (a) Could you show us this spectrum of the same-sign dimuons above 4 GeV? And (b): Your group has published two different numbers as far as χ-state production goes. But two years ago it was 8%, a year ago it was 40%. Do you have a final number now?

PIETRZYK: Not absolutely final, but we are still working on it.

The major problem was that the first number you mentioned was published when we had only some 5 or 9,000 Ψ's. We have now 42,000 Ψ's and we have an incredibly big statistical error on χ production which now is much more decreasing. In addition, a lot of work was done to understand acceptance. So this difference comes from this fact.

RESVANIS: Could you give us a number?

PIETRZYK: No, we still have about 30% of J/Ψ produced from χ. A final number should appear one or two months from now.

RESVANIS: Do you have a spectrum for the same-sign dimuons?

PIETRZYK: I don't have it with me, but, okay, this comes to a question of background. I don't know if I should come into details. It's quite tricky to get the background below this spectrum. You get this background shape from π decay etc., etc. You normalize the spectrum above the Ψ and then it goes down, down like this. So really you still have about 15% of background between Ψ and Ψ' but only a few percent above 4 GeV, and then it dies to 0.

RESVANIS: Last question, if I may. Do you resolve the χ states?

PIETRZYK: Yes.

RESVANIS: Which ones do you see?

PIETRZYK: What are masses? 3.51 and 3.55. In the same experiment we could observe photons through photon conversion in the spectrometer, in the target, or in foils or chambers. This way we get a very good resolution for photons but a very small conversion probability. It was typically 1%. So you have very good resolution for photons and one can look for invariant mass of Ψ and photon and, in this way, observe production of χ states and determine what is the source of production of J/Ψ.

SESSION E

HIGH TRANSVERSE MOMENTUM
HADRONIC INTERACTIONS

Thursday, June 25, a.m.

Chairman: A. Wroblewski
Secretary: J. M. Bishop

THEORETICAL DEVELOPMENTS IN LARGE TRANSVERSE MOMENTUM HADRONIC REACTIONS

J.F. Owens[†]
Physics Department, Florida State University
Tallahassee, Florida 32306

ABSTRACT

Calculations for high-p_T hadron and photon production using perturbative QCD are reviewed. The ambiguities inherent in both leading logarithm and next-to-leading order calculations are discussed. Techniques for improving the apparent rate of convergence of the perturbation series are also reviewed.

INTRODUCTION

It has now been about four years since calculations of high-p_T reactions based on perturbative QCD were first presented.[1] In addition to the basic two-body quark and gluon scattering subprocesses considered initially, subsequent calculations have included the effects of scaling violations in the parton distribution and fragmentation functions and of the running coupling, $\alpha_s(Q^2)$. Most of the high-p_T calculations performed to date have made use of the leading logarithm approximation.[2] Combined with the use of only the lowest order subprocesses, this limits the region of applicability to rather large values of $x_T = 2p_T/\sqrt{s}$. In this region at least it was hoped that the next-to-leading order corrections would be small.

Calculations going beyond the leading logarithm approximation have now been performed for a variety of processes.[3] From these results it is clear that rapid convergence of the perturbation series is by no means assured, even in kinematic regions where the relevant expansion parameter is relatively small. The truncation of the perturbation series introduces a number of well known ambiguities and some care must be exercized in order to avoid excessively large corrections to the lowest order predictions.

[†]Work supported in part by the U.S. Department of Energy.

In this talk I will first review the status of the leading logarithm calculations and discuss their inherent ambiguities. Next, I will discuss the calculation of next-to-leading order corrections and mention some of the theoretical uncertainties which remain after the inclusion of such terms. A brief review of the next-to-leading order calculations performed to date for high-p_T reactions will then be given.

LEADING LOGARITHM CALCULATIONS

The inclusive single particle invariant cross section is given by the expression in Eq.(1).

$$E\frac{d^3\sigma}{dp^3}(A+B\rightarrow h+X) = \sum_{abc} \int dx_a \, dx_b \, \frac{dz_c}{z_c^2} \, G_{a/A}(x_a,Q^2) G_{b/B}(x_b,Q^2)$$

$$D_{h/c}(z_c,Q^2) \, \frac{\hat{s}}{\pi} \frac{d\sigma}{d\hat{t}} \, (a+b \rightarrow c+x) \, \delta(\hat{s}+\hat{t}+\hat{u}) \qquad (1)$$

For a calculation using the leading logarithm approximation (LLA) the parton-parton scattering cross section, $d\sigma/d\hat{t}$, is calculated to lowest order in the running coupling $\alpha_s(Q^2)$ which, in turn, is calculated at the one-loop level. The parton distribution and fragmentation functions have a logarithmic dependence on the large momentum variable, Q^2, the definition of which will be discussed below. This Q^2 dependence is, of course, calculated using the LLA. Finally, in deriving Eq. (1) one assumes collinear kinematics for the colliding and fragmenting partons, i.e., no parton transverse momentum smearing effects are included.

Eq. (1) is expected to be valid in the kinematic region where there is only one large momentum scale. That is, the various Mandelstam variables for the hard scattering subprocess (denoted by a $\wedge$) should all be large. The region where $\hat{s}\sim-\hat{t}\sim-\hat{u}$ is also the region where the transverse momentum of the observed hadron is near its maximum value, $2p_T\sim\sqrt{s}$ or $x_T\sim1$. The variable Q^2 is chosen to be some combination of the large momentum variables which characterizes the momentum transfer in the hard scattering subprocess and which has the appropriate dimensions. In a kinematic region where Q^2 is large the running coupling should be small with the result that the lowest order hard scattering subprocesses should be dominant.

There are at least two well known ambiguities in

the leading logarithm calculation described above. The first of these concerns the definition of Q^2. At fixed values for the external kinematic variables, say s, p_T, and the center-of-mass scattering angle θ, increasing Q^2 by choosing an alternate definition causes a decrease in the predicted cross section. For example, suppose that two alternative definitions for Q^2 are related by a dimensionless function of x_T: $Q_2^2 = Q_1^2 f(x_T)$. Then,

$$\ln Q_2^2 = \ln Q_1^2 [1 + \ln f(x_T)/\ln Q_1^2] \qquad (2)$$

and the second term is dropped in the LLA--it contributes in the next-to-leading order. For very large values of $\ln Q^2$ such terms will be negligible, but for the kinematic regions where we presently have data they are not.

The second major source of uncertainty is related to the gluon distribution. At moderate values of x_T ($\lesssim 0.4$) the gluon-gluon and gluon-quark subprocesses give important contributions to the cross section and the choice of the gluon distribution can significantly affect the predictions. Whereas the quark distribution functions can be measured in deep inelastic scattering, we have no direct model independent measure of the gluon distribution. Although deep inelastic data can yield some constraints on the form of the distribution,[4] one cannot distinguish at present between a gluon of the form $(1-x)^4$ or $(1-x)^5$ in the region of Q^2 near 4 GeV^2, for example.

In order to obtain a quantitative estimate of the effect of these uncertainties I have calculated two sets of predictions for both high-p_T π^0 and γ production. The π^0 data were chosen since at present they cover the widest range in both energy and transverse momentum. In order to correctly calculate the π^0 cross section it is important to realize that at large values of p_T the two γ's from the π^0 decay are unresolved. Therefore, the π^0 trigger also accepts single γ triggers. This can be a significant effect in the large p_T region (above 10 GeV/c say) where the γ and π^0 yields become comparable. Specifically, the calculations presented here include three components:

1) the usual two-body QCD subprocesses with the π^0 appearing as a fragment of a parent quark or gluon,

2) direct γ production via the subprocesses $q\bar{q} \to \gamma G$ and $qG \to \gamma q$,

3) bremsstrahlung γ production from an outgoing quark. The fragmentation function $D_{\gamma/q}$ has been calculated in the LLA as discussed in refs. (5 and 6). The function $D_{\gamma/G}$ has been neglected here as it gives a negligible contribution in the high-p_T region.

Two separate sets of predictions will be presented corresponding to different commonly used gluon distributions and Q^2 definitions.

$$Q^2 = 2\hat{s}\hat{t}\hat{u}/[\hat{s}^2 + \hat{t}^2 + \hat{u}^2] \tag{3a}$$

Set A

$$xG(x, Q_0^2 = 4\,\text{GeV}^2) = 0.892\ (1+9x)\ (1-x)^4 \tag{3b}$$

$$Q^2 = -\hat{t} \tag{4a}$$

Set B

$$xG(x, Q_0^2 = 4\,\text{GeV}^2) = 2.676\ (1-x)^5 \tag{4b}$$

The Q^2 definition and gluon distribution in set A are those of ref. (7) while the gluon distribution in set B has the form suggested by the counting rules.[8] The Q^2 definition in set B represents an intuitive choice which would be relevant for quark-quark scattering of different flavors. In the dominant region of integration for the predictions shown here, this definition of Q^2 gives somewhat larger values for Q^2 than does the expression in Eq.(3a). This together with the harder gluon in Eq.(3b) results in the predictions from set A exceeding those from set B over the range in p_T considered here.

The resulting predictions are compared with the data[9-11] in Fig.(1). There is roughly a factor of two difference between the two sets of predictions over most of the kinematic region shown. In the region of p_T above about 8 GeV the upper curve, corresponding to set A, intersects the lower edge of the band covered by the data. In the intermediate region of p_T the theoretical predictions significantly underestimate the data. In most analyses parton transverse momentum smearing[7,12] or higher twist subprocesses[13] (constituent interchange model-CIM) have been invoked to explain this discrepancy. This region will be discussed further in a later section. For now, I will restrict my attention to the region above about 8 GeV.

556

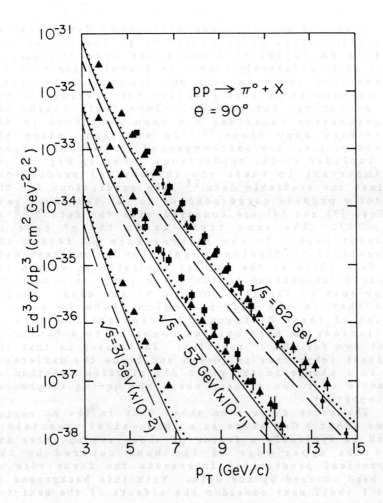

Fig. 1. Comparison between the π^0 predictions and the data. The solid and dashed curves correspond to the input sets A and B given by Eqs. (3) and (4), respectively. The dotted curves were obtained from the solid ones by replacing Q^2 with $Q^2(1-x_T)$ in the parton distribution and fragmentation functions. The data are from ref. (9) (triangles), ref. (10) (circles), and ref. (11) (squares). For clarity, the predictions and the data have been scaled by the factors shown in parentheses.

A second reaction for which some data exist at very high values of p_T is direct photon production. This is an important process for several reasons. First it is relatively simple in lowest order—for the pp reaction there is only one dominant subprocess. The usefulness of this reaction for testing QCD and for measuring, for example, gluon distribution and fragmentation functions has been mentioned in the literature many times.[14] In addition, since the direct γ, i.e., a γ unaccompanied by hadrons, signal was included in the predictions shown in Fig. (1) it is important to check the theoretical predictions against the available data.[15] The predictions for the direct γ process corresponding to the two input sets in Eqs. (3) and (4) are compared with the data[16-18] in Fig. (2). The same trend as in the π^0 case is apparent here. In the intermediate p_T region the theoretical predictions significantly underestimate the data while at the highest available p_T values the curves corresponding to the set A input intersect the lower edge of the band covered by the data. Notice also that in the high-p_T region there is about a factor of three difference between the two sets of predictions in this case as compared to a factor of about two for the π^0 reaction. The reason is that the dominant subprocess is $qG \rightarrow \gamma q$ and hence the difference due to a change in the gluon distribution function is enhanced over the π^0 case where the $qq \rightarrow qq$ subprocess is important.

The above discussion shows that in the p_T region above about 8 GeV there is a theoretical uncertainty which is typically a factor of about two or three and that the upper edge of the band covered by the theoretical predictions intersects the lower edge of the band covered by the data. With this background in mind I shall next consider the effects of the next-to-leading-order corrections.

NEXT-TO-LEADING-ORDER CALCULATIONS

It will be useful to first consider a simple example of a next-to-leading-order (NTLO) calculation. Inclusive compton scattering, $\gamma p \rightarrow \gamma + X$, has only one subprocess in lowest order while there are two in the next order:

Order α^2: $\gamma q \rightarrow \gamma q$

Order $\alpha^2 \alpha_s$: $\gamma q \rightarrow \gamma q G$ and $\gamma G \rightarrow \gamma q \bar{q}$.

These three subprocesses, convoluted with the appropriate distribution functions, can be used to calculate the invariant cross section. At this point,

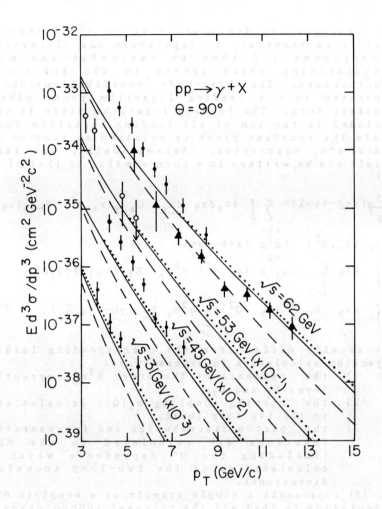

Fig. 2. Comparison between the direct γ predictions and the data. The notation is the same as in Fig. 1. The data are from ref. (16) (triangles), ref. (17) (closed circles), and ref. (18) (open circles).

the distribution functions have no Q^2 dependence.
Next it is necessary to regularize the ultraviolet
divergences and then to factorize the mass
singularities which appear in the three-body
subprocesses. The three-body contributions can be
separated into a leading logarithm piece plus a
constant term. The leading logarithm piece is then
included in the sum of all leading logarithm terms
while the constant piece is retained as part of the
order $\alpha^2 \alpha_s$ correction. Schematically, the final
result can be written in a form similar to that of Eq.
(1):

$$E\frac{d^3\sigma}{dp^3}(\gamma+p \rightarrow \gamma+X) = \sum_{abc} \int dx_a dx_b \frac{dz_c}{z_c^2} \; G_{a/\gamma}(x_a,Q^2) \; G_{b/p}(x_b,Q^2)$$

$$D_{\gamma/c} \; (z_c,Q^2) \; E\frac{d^3\sigma}{dp^3} \; (a+b \rightarrow c+x)$$

$$+ \int dx_b \sum_{i=1}^{2f} Gq_i/p \; (x_b,Q^2) \; \alpha^2 \alpha_s \; K_1 \; (\gamma q_i \rightarrow \gamma q_i G)$$

$$+ \int dx_b \; G_{G/p} \; (x_b, \; Q^2) \sum_{j=1}^{f} \alpha^2 \alpha_s \; K_2 \; (\gamma G \rightarrow \gamma q_j \bar{q}_j) . \qquad (5)$$

The results differ from the corresponding leading
logarithm calculation in several ways:
  1)   the   presence of the order $\alpha^2 \alpha_s$ correction
       terms K_1 and K_2,
  2)   the   running coupling $\alpha_s(Q^2)$ is calculated
       to the two loop level,
  3)   the   parton distribution and fragmentation
       functions are calculated to the NTLO
       (including the Q^2 dependence which is
       calculated using the two-loop anomalous
       dimensions).
Eq. (5) represents a simple example of a complete NTLO
calculation in that all the relevant subprocesses up
to order $\alpha^2 \alpha_s$ are included.
    Unfortunately, NTLO calculations are not free of
ambiguities and, therefore, some care is necessary
when discussing questions such as the rate of
convergence of the perturbation series. For example,
changing the definition of Q^2 alters the relative size
of the leading and next-to-leading terms as discussed
previously. It is possible to enhance the leading
contribution while decreasing the correction terms or
vice versa. Note, however, that the sum of the two
terms is not as sensitive to such changes as is either
term separately. In addition, one must choose a
particular renormalization prescription. The exact

result for a given calculation is independent of the particular renormalization convention chosen. However, when the series is truncated the answer can in practice depend on the convention used. Altering the convention changes the relative size of successive terms in the series and therefore affects the rate of convergence. Finally, there is a problem of mixing. One must be sure to include all of the relevant subprocesses up to the order being considered for a given reaction. For example, consider the gluon corrections to the subprocess $qq \rightarrow qq$ coming from $qq \rightarrow qqG$. The latter subprocess also has terms in it which appear as corrections (at the leading log level) to the $qG \rightarrow qG$ subprocess. Therefore, it can be misleading to associate the resulting constant correction term solely with the $qq \rightarrow qq$ subprocess.

Before considering specific NTLO calculations I would like to examine a related topic-phase space limitations on gluon emission. The Q^2 dependence encountered, for example, in parton distribution functions results in part from gluon bremsstrahlung. The emitted gluon transverse momenta are integrated over with an upper limit given by Q^2. In the leading logarithm approximation the nested integrals decouple and one obtains a series with terms of the form $(\alpha_s(Q^2)\ln Q^2)^m$. If one goes beyond the LLA the correct phase space limit which controls the gluon emission is put into the integration. For example, in deep inelastic scattering it is $xW = Q^2(1-x)$ which is the upper limit for the gluon transverse momentum integration.[19] In this case $G(x, Q^2)$ is replaced by $G(x, Q^2(1-x))$. This replacement sums up a portion of the non-leading contributions and thereby reduces the size of the corrections with respect to the modified leading term. In this particular example there are large corrections to the moments of the deep inelastic structure functions which behave as $\ln^2 n$. The above replacement is a way of including these in the leading term.

In order to see how such a replacement might affect the LLA high-p_T predictions, I repeated the calculations based on the input set A, Eq.(3), and replaced Q^2 by $Q^2(1-x_T)$ in the parton distribution and fragmentation functions. The variable x_T rather than x was used in order to avoid problems with small Q^2 values as x or z goes to one. This replacement gives a conservative estimate of the effect since the average value of x in the various integrations is approximately x_T. The results of this modification of the leading term are shown in Figs. (1) and (2). There is a modest enhancement of the predictions which is as large as 30% in the high-p_T region. The effect increases with increasing p_T since there is a progressively larger modification to the Q^2 definition. According to the above discussion the NTLO corrections should be decreased at the same time as the LLA predictions are increased, thereby improving the apparent rate of convergence of the series. This example will be discussed further below.

Next, I would like to discuss several calculations of next-to-leading-order corrections. The only published calculation related directly to high-p_T particle production is that of Ellis, Furman, Haber, and Hinchliffe.[20] They considered the corrections to the scattering of different flavor quarks, $qq' \rightarrow qq'$, coming from the subprocess $qq' \rightarrow qq'G$. They concluded that in the $\overline{MS}$ scheme the corrections could be as large as 100-300% depending on the kinematic region and also on the choice of the Q^2 definition. This conclusion must be viewed with some caution, however, as a result of the mixing problem discussed above. The claim of Ellis et al. that the perturbation series is out of control must be balanced against the contrary claim of Celmaster and Sivers.[21] They argue that by changing from the $\overline{MS}$ to the momentum subtraction scheme and by choosing $Q^2 = -\hat{t}(1-x_T)/2$ for the parton distribution and fragmentation functions that the corrections can be made smaller than about 40%. Their results are compared with those of Ellis et al. in Fig. (3).

Recently Furmanski and Slominski[22] have confirmed the calculation of Ellis et al. They also noted that it was possible to substantially reduce the size of the order α_s^3 corrections by choosing $Q^2 = \eta \hat{s}$ with $\eta = e^{-2} \sim 0.135$. Note that at $90°$ in the parton center-of-mass system (the dominant region of integration) and for $x_T = 1/2$ one has $-\hat{t}/2(1-x_T) = \hat{s}/8$. The results of these authors therefore appear to be consistent with the results of Celmaster and Sivers.

Now, if indeed the NTLO corrections can be made small by the above procedure, then the lowest order term itself should provide a reasonable estimate of the answer. I have repeated the calculations for both π^0 and γ production using the choices of Celmaster and Sivers. The results are virtually identical to the curves shown in Figs. (1) and (2) by the dotted lines. It is encouraging to note that in the p_T region above about 8 GeV the dotted curves are in good agreement with the data.

Several NTLO calculations have been made for high-p_T photon production. Aurenche and Lindfors[5]

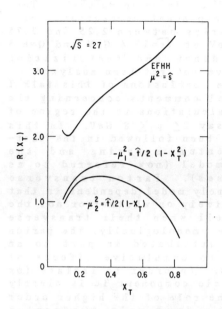

Fig. 3 Predictions for the ratio of the leading plus next-to-leading terms to the leading term alone. The EFHH curve is from ref. (20). This figure is from ref. (21).

calculated the subprocess $qq' \longrightarrow qq'\gamma$. This subprocess contributes a large leading logarithm bremsstrahlung term which was separated off, i.e., the direct photon contribution was considered separately. The residual correction term was found to be small, although the precise size relative to the lowest order contribution depended on the choice of the gluon distribution. This result is useful, but it does not represent a complete calculation in the sense that there are many more subprocesses of the order $\alpha\alpha_s^2$ which were not calculated.

Recently Contogouris, Papadopoulos, and Ralston have calculated the corrections due to the subprocess $qG \rightarrow qG\gamma$ in the soft gluon approximation. They find corrections which are typically 50-100% depending on the value of the transverse momentum. The terms which survive in the soft gluon limit do not depend on the renormalization prescription employed. However, there is no way to estimate the size of the remaining (uncalculated) correction terms. Furthermore, there is no way to judge the effect of alternative Q^2 definitions, etc. as discussed above. Although this technique does not constitute a complete calculation, it is nevertheless an interesting technique for isolating the large "π^2" contributions which, at least in the case of the Drell-Yan mechanism, make up a sizeable portion of the non-leading correction term.

Recently, Ellis, Martinelli, and Petronzio[24] have calculated the nonsinglet contribution to μ-pair production from the subprocesses $q\bar{q} \rightarrow \gamma^* GG$ and $qq \rightarrow qq\gamma^*$. The nonsinglet piece is that which is relevant for $(\bar{p}-p)p \to \mu^+\mu^- + X$ or $(\pi^- - \pi^+)p \to \mu^+\mu^- + X$. They find a multiplicative p_T dependent correction factor which in the $\overline{MS}$ scheme varies between 2.25 and 1.75 for p_T between 2 and 5 GeV at $\sqrt{s}=19.4$ GeV and $Q=6.5$ GeV. The scheme dependence and factorization prescription dependence have not yet been analyzed.

Before summarizing the conclusions of this talk I would like to make several comments concerning the extension of the high-p_T calculations to the region of intermediate p_T values, say $2 \leq p_T \leq 8$ GeV. In this region two approaches have been followed in the past—parton transverse momentum smearing and the constituent interchange model (now referred to as higher twist subprocesses). Parton transverse momentum effects are extremely model dependent in that the results depend sensitively on whether or not the partons are taken off shell when their transverse momentum is non-zero. Phenomenologically, the parton transverse momentum is attributed in part to an intrinsic component and to cumulative effects of higher order subprocesses. In order to isolate (or even to define) the intrinsic component it is clearly necessary to understand the role of the higher order terms. As an example, in μ-pair production estimates of the intrinsic average parton k_T have decreased from around 1 GeV to less than 0.5 GeV as more sophisticated treatments of higher order soft gluon effects[25] were included. A proper treatment of the intermediate p_T region must await a better understanding of the NTLO contributions which are best studied in the region of higher p_T.

The CIM gives contributions which are suppressed by powers of p_T with respect to the leading QCD (or lowest twist) subprocesses. Even so, it is possible that such terms can be important in the intermediate p_T region. These terms have, however, proven to be notoriously difficult to normalize dependably. Recently, Bagger and Gunion[26] have obtained new results for meson photoproduction using the subprocess $\gamma q \to Mq$. The subprocess normalization is determined by the meson form factor. Their results indicate that, in the case of π^- photoproduction, the CIM contribution can exceed the leading QCD predictions, at least in the intermediate p_T region. Meson photoproduction would therefore make a rather interesting testing ground for determining the interplay between the CIM and leading QCD terms.

564

CONCLUSIONS

Studies of large-p_T hadron reactions carried out in the last year have tended to reinforce the previous conclusion that by judicious choices for the renormalization prescription and Q^2 definition one could significantly reduce the size of the NTLO corrections with respect to the leading predictions. This flexibility, when combined with an appropriately chosen gluon distribution, yields predictions for high-p_T π^0 and direct γ production which are in agreement with the data in the p_T region above about 8 GeV.

The single largest source of uncertainty remaining in the large-p_T region is due to the uncertainty in the gluon distribution. Fortunately, we may soon be able to put new constraints on the gluon distribution. To do so we need two things:

1) a complete next - to - leading - order calculation for direct γ production
2) improved high energy high-p_T data for p and $\bar{p}$ beams.

It is likely that both will be available in the next year or so. This should help to reduce the uncertainty in the gluon distribution and, therefore, allow us to further refine the predictions for high-p_T hadronic reactions.

REFERENCES

1. B.L. Combridge, J. Kripfganz, and J. Ranft, Phys. Lett. 70B, 234 (1977).
2. For a review, see R.D. Field, AIP Conf. Proc. 55, "Quantum Chromodynamics", ed. by W. Frazer and F. Henyey (AIP, New York, 1979), p. 97.
3. For a review, see C.H. Llewellyn-Smith, AIP Conf. Proc. 68, "High Energy Physics - 1980", ed. by L. Durand and L.G. Pondrom (AIP, New York, 1981), p. 1345.
4. D.W. Duke, J.F. Owens, and R.G. Roberts, CERN preprint, Ref. TH.3095 (1981).
5. P. Aurenche and J. Lindfors, Nucl. Phys. B168, 296 (1980).
6. W. Frazer and J.F. Gunion, Phys. Rev. D20, 147 (1979).
7. R.P. Feynman, R.D. Field, and G.C. Fox, Phys. Rev. D18, 3320 (1978).

8. S.J. Brodsky and G.R. Farrar, Phys. Rev. Lett. 31, 1153 (1973) V.A. Matveev, R.M. Muradyn, and A.V. Tavkhelidze, Lett. Nuovo Cimento 1, 719 (1973).

9. A.L.S. Angelis et al., Phys. Lett. 79B, 505 (1978).

10. C. Kourkoumelis et al., Z. Physik C 5, 95 (1980).

11. A.G. Clark et al., Phys. Lett. 74B, 267 (1978).

12. J.F. Owens and J.D. Kimel, Phys. Rev. D18, 3313 (1978).

13. R. Blankenbecler, S.J. Brodsky, and J.F. Gunion, Phys. Rev. D18, 900 (1978); D. Jones and J.F. Gunion, Phys. Rev. D20, 232 (1979).

14. See, for example, L. Cormell and J.F. Owens, Phys. Rev. D22, 1609 (1980) and references therein.

15. For a review, see W.J. Willis, CERN preprint, CERN-EP/81-45 (1981).

16. A.L.S. Angelis et al., Phys. Lett. 94B, 106 (1980).

17. M. Diakonou et al., Phys. Lett. 91B, 296 (1980).

18. E. Amaldi et al., Nucl. Phys. B150, 326 (1979).

19. S.J. Brodsky and G.P. Lepage, SLAC-PUB-2447 (1979) and Proceedings of the 1979 SLAC Summer Institute, p. 133.

20. R.K. Ellis, M.A. Furman, H.E. Haber, and I. Hinchliffe, Nucl. Phys. B173, 397 (1980).

21. W. Celmaster and D. Sivers, Argonne preprint, ANL-HEP-PR-80-61.

22. W. Furmanski and W. Slominski, Jagellonian University preprint, May 1980.

23. A.P. Contogouris, S. Papadopoulos, and J. Ralston, McGill University preprint (1981).

24. R.K. Ellis, G. Martinelli, and R. Petronzio, CERN preprint, Ref. TH.3079 (1981).

25. G. Parisi and R. Petronzio, Nucl. Phys. B154, 427 (1979).

26. J.A. Bagger and J.F. Gunion, University of California, Davis, preprint (1981).

DISCUSSION

De GRAND, SANTA BARBARA: A crazy theoretical question. When people study e^+e^- annihilation they try to look at things which are insensitive to quark fragmentation functions. Has any one thought of any way of studying anything to do with large p_T physics where you're as insensitive as you can be, say, to the parton distributions of the gluons or the quarks in the proton? By integrating things appropriately?

OWENS: There have been some efforts in that direction. Normally you don't gain much by playing around with different regions of integration because you already have convolution to begin with. The results I have seen tend to concentrate on particle-antiparticle differences and that helps cancel out certain gluon effects. It's analogous to the calculation I referred to by Ellis, et al. where they looked at the non-singlet contribution to Drell-Yan production. This hasn't been possible thus far because, in the high-p_T region where we think this leading log prediction is working, we only have proton data. Now with the advent of antiprotons in the ISR certain comparisons will be possible. So that's an avenue which perhaps can be explored when the data become available.

BIAŁAS, KRAKOW: I'd like to refer to the problem of changing of scale which you were talking about. There was work done in Krakow recently by Furmanski and Slominsky (Krakow preprints TPJU 11/81, 12/81) just essentially considering the same thing. Their conclusion is slightly different from the one you had from the paper by Sivers. Namely, they claim that, to get these secondary corrections minimized, you have to use a scale which is about 1/10 of the total center of mass energy in the quark-quark scattering. So they say essentially 0.1 $\bar{s}$. You see. And that gives you the tremendous difference in scale. The phenomenology was not done. It was just theoretical work to minimize the second order. But it is clear that it will have fantastic effect on the scaling violation that you see. I'd like to emphasize that it is essentially one order of magnitude smaller scale than one would expect in the naive way. Like, let's say, Field, Feynman and Fox.

OWENS: Well, actually, depending on the kinematic region you are in, let's see. If you're at $90°$, $\hat{s}$ is twice $\hat{t}$, but the scale that I showed was $(\hat{t}/2)(1-x_T)$ where x_T is $1/2$. So that's a factor of 4, and another factor of 2, gives you a factor of 8. So it would be like $\hat{s}/8$ that I was using. So in fact that is the same order of magnitude as what they found. So I'd say they were quite compatible. I haven't seen that work. I'd like to see it.

CONTOGOURIS, MCGILL: I would like to point out that in a number of cases a large correction was found. Like, for example, in Drell-Yan, but in other cases also. That is mainly due to a π^2 term which comes entirely from soft gluons. Now that part of the correction does not depend on the normalization prescription, etc., apart from the choice of $d_s(Q^2)$. Therefore, if one chooses a reasonable d_s, that part of the correction is not ambiguous at all.

OWENS: Yes, I agree with you that that particular part doesn't depend on the prescription. But there is no guarantee. In fact, it is not true that the remaining interactions have to be positive-definite. There can be cancellations between the π^2 and the remaining piece of the calculations.

TAVERNIER, BRUSSELS: If you move on to collider energies, what is the p_T domain where you expect your leading log calculation to become valid? Is it still 10 GeV/c, or would it be a certain x_T value, or what?

OWENS: Strictly speaking, the leading log calculation would apply in the same region of x_T that we used here. So it would be x_T large. And the reason is that you have to avoid having too large momentum scales in the problem. You can pick up corrections which are proportional to say log x_T. Now if x_T is of order one than the correction is essentially zero. But if you were to stay at fixed p_T and go up in energy so that x_T got small these corrections which are going as log x_T could get large. So, being very conservative, I would have to say that the simple leading log calculations should be strictly valid in the large x_T region.

TAVERNIER: And you won't have experimental results soon.

OWENS: I should add one thing. The k_T effect, and possibly also the effects coming from higher twist terms, will still fall off very rapidly in p_T. So I may be ultraconservative here. These background terms to the leading log may in fact die off at a fixed p_T value, so that the leading log terms may in fact be accurate. But without having done the calculations and the next-to-leading-order corrections I just can't say.

MORRISON, CERN: You emphasized that the choice of the gluon

distribution was important, and you finished up by saying that you needed better proton and antiproton data to determine the gluon distributions. But is there any reason why you cannot take the gluon distribution determined from the neutrino data, for example the CDHS data we saw yesterday?

OWENS: If one looks at the entirety of deep inelastic data with an unbiased point of view, what you find is that you cannot discriminate reliably between the two gluon distributions which I used, a $(1-x)^5$ and a $(1-x)^4$. The data are just not sufficiently sensitive to determine this difference. The difference is not large until you get way out on the tails of these gluons and this is the region that's relevant for the high p_T calculations. There you run out of deep inelastic data and it's precisely the region where we need the distributions. So we've been looking at this problem and we just can't pin it down precisely.

DREMIN, MOSCOW: Can you say anything about non-perturbative contributions to the jets?

OWENS: Not really, no. That's something that I haven't looked at. Non-perturbative in what sense? I mean the k_T contributions, for example the smearing effects, would be one type of non-perturbative contribution.

DREMIN: I mean just in the field-theoretical sense that there is an average over the vacuum which is not mentioned.

OWENS: No that is something I haven't studied.

CONTOGOURIS, MCGILL: I would like to say that in connection with a gluon distribution, what counts, in present day calculations, is not so much the behavior near $x = 1$ but the behavior and the magnitude for x between, let's say, .1 and .4 or .5? And in this respect I think the data from deep inelastic and neutrino are quite right. They give exactly the same kind of gluon distribution as one gets by fitting large-p_T pion or direct photon production and so on.

OWENS: First of all you are absolutely right. I pointed out earlier that the average value of x which is relevant to the convolution integral is on the order of x_T. So that's right in the range that you mentioned. You are also right in that a gluon that goes as $(1-x)^5$ at some modest Q^2 like 5 GeV^2 is quite compatible with deep inelastic data. But so is one that goes as $(1-x)^4$. And that is exactly the comparison I was making here. And it gives rise to a factor of 2. Well, the choice of Q^2 scale combined with the choice of the gluon [distribution], gives something like a factor of 2 uncertainty in the theoretical predictions. At that level the deep inelastic data do confirm the phenomenology from the high-p_T calculations but if you want to go further than that the deep inelastic data are not going to sufficiently pin down the gluon distribution.

DEEP INELASTIC HADRON SCATTERING WITH A 2π CALORIMETER TRIGGER

C. Favuzzi, G. Germinario, L. Guerriero, P. Lavopa, G. Maggi,
C. de Marzo, M. de Palma, F. Posa, A. Ranieri, G. Selvaggi, P. Spinelli,
F. Waldner
University of Bari, Bari, Italy.

A. Bialas, W. Czyz, T. Coghen, A. Eskreys, K. Eskreys, D. Kisielewska,
P. Malecki, K. Olkiewicz, K. Sliwa, P. Stopa
University of Krakow, Krakow, Poland.

W.H. Evans, J.R. Fry, C. Grant, M. Houlden, A. Morton, H. Muirhead,
J. Shiers
University of Liverpool, Liverpool, U.K.

M. Antič, W. Baker, H. Bechteler, I. Derado, V. Eckardt, J. Fent,
P. Freund, H.J. Gebauer, T. Kahl, R. Kalbach, A. Manz, P. Polakos,
K.P. Pretzl, N. Schmitz, P. Seyboth, J. Seyerlein, D. Vranič, G. Wolf
Max-Planck-Institut für Physik und Astrophysik, München, Germany.

F. Crijns, W. Metzger, C. Pols, T. Schouten, T. Spuijbroek
University of Nijmegen, Nijmegen-NIKHEF, Netherlands.

N. Sarma, Visitor at CERN from Bhaba Institute, Bombay, India.

(presented by P. Seyboth)

ABSTRACT

Preliminary large p_T and large transverse energy cross sections
of 150, 300 GeV pions and protons on hydrogen measured with a large
acceptance, segmented calorimeter are presented. Little evidence is
seen for narrow jet production. Processes other than the scattering
of two constituents appear to dominate this deep inelastic hadron
scattering process.

INTRODUCTION

Previous experiments at the ISR (CERN) and Fermilab[1] have investigated deep inelastic hadron hadron collisions using a single particle
trigger at large transverse momenta p_T. It was found that the invariant cross sections do not decrease exponentially with p_T but rather
like a power law $E \cdot d^3\sigma/dp^3 = p_T^{-n} f(x_T)$ with $n \approx 8$. This was taken as

evidence for a hard scattering process of the parton constitutents, which subsequently fragment into hadron jets. Parton-parton scattering models in their original simplest version[2] with complete scaling would have predicted n = 4. It was assumed that the detected large p_T particle results from a rare configuration of the parton fragmentation in which it acquires most of the parton momentum. It was discussed that further insight into hard sacttering processes could be gained by triggering on particle jets using segmented calorimeters.

Three experiments at Fermilab E236[3], E260[4] and E395[5] have used calorimeters to trigger on particle jets at angles around 90° in the cms either on one side or on two opposite sides. The solid angle acceptance of each calorimeter arm was about 1-2 sr corresponding to the expected jet size. As expected the cross sections with this so called "jet trigger" were found to be 2 orders of magnitude larger than for the single particle trigger at large p_T. However it may be difficult to verify the existence of jets because the small acceptance of the calorimeter trigger could lead to a trigger bias selecting events which simulate jets due to statistical fluctuations.

In order to circumvent some of these problems we studied deep inelastic hadron-hadron collisions with an "unbiased jet trigger" using a segmented calorimeter with a solid angle acceptance of 2π in the azimuth and 50° to 130° in the cms polar angle. With this trigger we selected events in which a large transverse energy $\Sigma|p_T|$ was deposited in the calorimeter. We examined these events for constituent scattering processes leading to two jets of large p_T on opposite sides of the beam and two forward/backward spectator jets. Large p_T jets should manifest themselves as energy clusters observed in the segmented calorimeter.

APPARATUS

The layout of the experiment is shown in Fig. 1. The data presented here were obtained with 150 GeV/c π^- and 300 GeV/c p, π beams of $2 \cdot 10^6$ particles/sec incident on a 30 cm liquid hydrogen target using the calorimeters and spark chambers only (magnet off!). The spark chambers were used to verify that the event occurred in the target. The 2 m streamer chamber was employed for part of the data taking to determine the multiplicity of charged particles.

The ring calorimeter, a barrel of 3 m diameter with a 56 cm central hole, has a lead scintillator sandwich photon section (16 x 0.55 cm Pb-sheets) followed by an iron-scintillator sandwich hadronic section (20 x 5 cm Fe-sheets). Both sections are subdivided into 240 independent cells, each subtending about 9° in the c.m.s. polar angle and 15° in azimuthal angle. Combined wave-length shifting acrylic rods (doped with Yellow 323 and BBQ) were used to draw separated signals from the photon and hadron part of the calorimeter onto 240 pairs of photo-

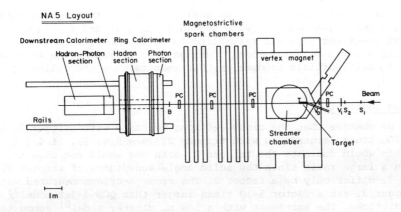

NA 5 Layout

Downstream Calorimeter   Ring Calorimeter

Hadron–Photon   Hadron   Photon
section   section   section

Magnetostrictive
spark chambers

vertex magnet

Rails

PC   PC   PC

B

Streamer
chamber

PC   Beam

T
V₀
V₁ S₂   S₁

Target

Im

Fig. 1 Layout of the experiment

multiplier tubes[6]. The obtained energy resolutions were $\sigma/E = 0.16/\sqrt{E}$
+0.01 for electrons and $\sigma/E = 0.86/\sqrt{E} + 0.02$ for hadrons. The downstream calorimeter, which covered the central hole of the ring calorimeter, measured the energy flow at small angles. The combined energy
measurements from both calorimeters was used to reject possible background. The distance of the calorimeters to the target was varied
with the incident beam energy such that the ring calorimeter always
covered $50°$ to $130°$ in the c.m.s. polar angle. The $\Sigma|p_T|$ trigger was
derived from the sum of the analog signals of all or part of the ring
calorimeter cells weighted by their radial distance from the beam
axis. The shape of the trigger pulses was recorded. Occasional background triggers due to Cerenkov light produced in the acrylic rods
was eliminated off-line by pulse shape analysis.

Data were taken with 3 types of trigger:

1. one arm trigger: quadrant of $\pi/2$ in azimuth;
2. two arm trigger: two opposite quadrants each of $\pi/2$ in azimuth;
3. full calorimeter trigger: all sectors of 2π in azimuth.

Triggers similar to 1. and 2. have been used at Fermilab by E236[3],
E260[4] and E395[5]. The use of trigger 3. allowed us to search for
evidence of jets in an unbiased manner.

PRELIMINARY RESULTS

The data are still <u>preliminary</u> since no unfolding of the calorimeter
resolution has been done and the energy calibration is not final. Be-
cause of this the $\Sigma|p_T|$ scale uncertainty is estimated to be approxi-
mately 10%, but this is not expected to affect our conclusions. All
quoted errors are of statistical nature only.

When discussing our results, we shall compare to two representative models, a QCD 4-jet model[7] with two large p_T jets and two forward backward spectator jets all resembling the jets observed in e^+e^- collisions and a low p_T cluster model[8] designed to reproduce the features of low p_T multiparticle production. Our results show the following features:

1. Cross sections

(a)   We observe that the cross sections measured with trigger 3. are 10-100 times larger than with trigger 2. (see Fig. 2). If 4-jet events would dominate the trigger 2. data one would not expect such a large ratio since the solid angle acceptance of trigger 2. and 3. differ only by a factor 2. The cross sections measured with trigger 3. are a factor 5-10 times larger than QCD 4-jet model[7] predictions. The agreement with a low p_T cluster model[8] seems to be better (Fig. 2).

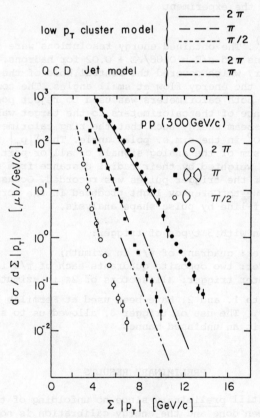

Fig. 2   Cross sections versus transverse energy $\Sigma |p_T|$ measured by the full calorimeter or some sectors of the calorimeter. Low p_T cluster- and QCD 4-jet model predictions are shown.

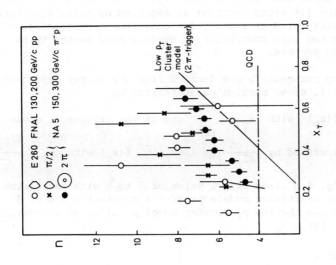
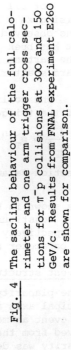

Fig. 4   The scaling behaviour of the full calorimeter and one arm trigger cross sections for π^-p collisions at 300 and 150 GeV/c. Results from FNAL experiment E260 are shown for comparison.

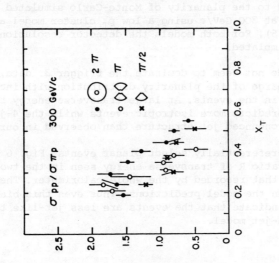

Fig. 3   Ratio of pp to π^-p cross sections as a function of x_T.

(b)   For all triggers the cross sections for pp decrease faster with rising $x_T = 2\,p_T/\sqrt{s}$ than for π^-p (see Fig. 3). A similar dependence was observed for jet cross sections as reported by refs. 4,5 where it has been interpreted as support for constituent scattering, since the partons in the pion contribute on average more energy to the hard scattering process.

(c)   The scaling parameter n was derived from the energy dependence of the trigger 1. cross section parameterized by

$$E\,\frac{d^3\sigma}{dp^3} \sim p_T^{-n}\,f(x_T) \quad \text{with} \quad x_T = \frac{2p_T}{\sqrt{s}}$$

and of the trigger 3. cross section parameterized by

$$\frac{d\sigma}{d\Sigma|p_T|} \sim (\Sigma|p_T|)^{-n}\,f(x_T) \quad \text{with} \quad x_T = \frac{\Sigma|p_T|}{\sqrt{s}} \; .$$

As shown in Fig. 4 n rises from a value of 5 to 8 with increasing x_T. Constituent scattering models predict a constant n = 4 or 3 respectively while the low p_T cluster model predicts an increasing n parameter at large x_T.

2. Event structure

The planarity P of the events has been calculated by performing a principal axis analysis of the transverse momentum distribution for each event as measured by the calorimeter. (Particle momenta were derived from the energy clusters observed in the calorimeter.) The planarity was defined as P = (a-b)/(a+b) with a (b) being the sum of the squares of the projected transverse momenta to the maximum (minimum) principal p_T-axis. For an isotropic event structure we would expect P = 0 and for pencil like jets P = 1. Fig. 5 shows the planarity distributions of events selected by the calorimeter trigger 3. from π^-p and pp collisions at 300 GeV/c for different trigger thresholds. These are compared to the planarity of Monte-Carlo simulated events for pp collisions at 300 GeV/c using a low p_T cluster model and a QCD 4-jet model (Fig. 5). For both models the detector resolution and acceptance were simulated.

Planar events do not seem to dominate the trigger 3. data, indeed there is little charge of the planarity distribution with increasing transverse energy in the events. At large transverse energy the low p_T cluster model predicts more isotropic events while the 4-jet model predicts a more pronounced jet structure than observed in our data.

Trigger 2 may preferentially select planar events. Fig. 6 shows for this trigger the ratio R of transverse energy seen in the two triggering quadrants to that recorded by the whole calorimeter. The data points fall between the model predictions. Thus even this biased trigger seems to indicate that the events are less jet-like than predicted by the 4-jet model.

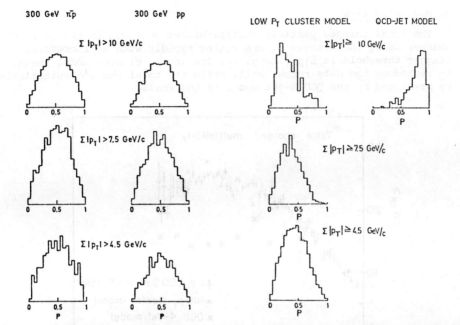

Fig. 5 Planarity distributions of events selected by the full calorimeter trigger from π^-p and pp collisions at 300 GeV/c for
different trigger thresholds. Results from a low p_T cluster
model and a QCD 4-jet model are shown for comparison.

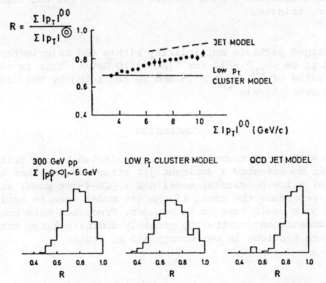

Fig. 6 Ratio of transverse energies measured in the triggering
quadrants to that recorded by the whole calorimeter for
events selected with trigger 2.

3. Multiplicities

The total charged particle multiplicities observed in the streamer chamber using the trigger 3. are rising rapidly with an increasing trigger threshold in $\Sigma|p_T|$ (Fig. 7). The low p_T cluster model seems to reproduce the data rather well, while the total charged multiplicity predicted by the QCD 4-jet model is too small.

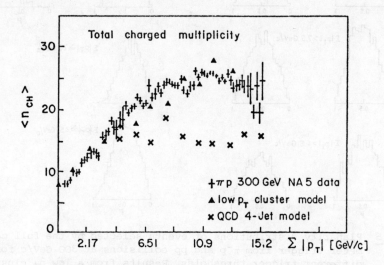

Fig. 7 The total charged multiplicity for πp collisions at 300 GeV/c as a function of the trigger threshold for the full calorimeter trigger.

The charged particle multiplicity within the calorimeter acceptance was found to be $\langle n_{CH} \rangle = 12$ for $\Sigma|p_T| > 10$ GeV/c. This is to be compared to a value of $\langle n_{CH} \rangle = 15$ and 10 predicted by the cluster- and QCD-model respectively.

CONCLUSION

Events with large transverse energy selected by the full calorimeter trigger do not show a dominant jet structure. Neither an extrapolation of a low p_T cluster model nor a QCD-4-jet model with e^+e^- type jets reproduce the data. If the jet model were to explain the data, the jets would have to be broader. Processes more complicated than two-constituent scattering probably dominate large transverse energy cross sections in our energy and p_T region.

REFERENCES

1) A summary of large transverse momentum hadronic processes is given by P. Darriulat in Ann. Rev. Nucl. Part. Sci. (1980) $\underline{30}$, 159-210.

2) S.M. Berman, J.D. Bjorken, J. Kogut: Phys. Rev. $\underline{D4}$ (1971), 3388.

3) V. Cook et al.: Fermilab-Pub-80/91 EXP 7180-236.

4) C. Bromberg et al.: Preprint CALT-68-738 (1979); Phys. Rev. Lett. $\underline{43}$, 565 (1979).

5) W. Selove: High p_T jet studies at Fermilab, Proceedings of the 14th Rencontre de Moriond 1979, Vol. I, 401.

6) V. Eckardt et al.: Nucl. Instr.& Meth. $\underline{155}$, 389 (1978).

7) In the QCD-model calculations of the cross sections the matrix elements for 2 constituent scattering were used. The quark and gluon distributions were obtained from νN and Drell-Yan experiments. Scale breaking effects were taken into account. A coupling constant $\alpha_S = 12\pi/25 \ln (1 + \frac{Q^2}{\Lambda^2})$ with $\Lambda = 0.5$ GeV and $Q^2 = 4 p_T^2$ was assumed. The model had 2 large p_T and two forward/backward spectator jets. Each of the two jet systems was made to resemble jets as observed in e^+e^- collisions, using the Feynman-Field fragmentation scheme. With these assumptions we find that 25% of the transverse energy in the calorimeter comes from the spectator jets while 95% of the energy of the scattered jets is seen by the calorimeter when selecting events with more than 10 GeV transverse energy. The predictions of the model are sensitive to the jet parameters, wider jets would tend to improve the agreement with our data.

8) The low p_T cluster model produces clusters with an average mass of $\langle m_c \rangle = 2$ GeV in a cylindrical phase space. The clusters decay isotropically with an average charged multiplicity of $\langle n_{CH} \rangle \sim 2.5$. the resulting p_T-distributions of the final particles have $\langle p_T^2 \rangle^{1/2} = 0.36$ GeV/c. Leading particle effects were taken into account and an overall KNO multiplicity distribution was enforced. The cross section were normalized to the total inelastic cross section. The model reproduced well the features of low p_T events observed in bubble chamber experiments.

DISCUSSION

OWENS, FLORIDA STATE: In the best possible situation, from the theoretical stand point, we would be able to calculate not only the high-p_T jet cross sections but we would also be able to calculate from the first principles the low-p_T interactions as well. In that case, one could take a QCD calculation and calculate the response of your 2π calorimeter trigger, and in that case we would be able to calculate and predict the kinds of things you are seeing. But we can't do that. At least not yet. And so, in fact, what we think we can predict would be just a rather small component of your calorimeter response, namely the jet cross section. And it appears that what you've shown us is that your calorimeter trigger is in fact not a good way of measuring jet cross sections . . . . that in fact you pick up a rather large response from events which are not associated with jet production at all. One way of phenomenologically accounting for that would be to consider an admixture of the two models you have shown. The QCD calculations seem to indicate that something like 10% of your 2π calorimeter trigger is accounted for by jets and the remainder is left over with tails from this low-p_T clustering model. So in fact you just have to consider a superposition of the two in order to understand your large cross sections. So in that sense I don't think your data are going to be shooting holes in QCD. I just think you have found that that is not the best way to look for jets.

SEYBOTH: I did not say that I'm shooting down or trying to shoot down QCD.

OWENS: I know you didn't but others have.

SEYBOTH: On the other hand, I think it's kind of worrying that this phase space model fits so well even the two-arm trigger cross section which should have, and the one-arm trigger cross section, which should be much less contaminated by this high-multiplicity type of event. That, I think, is still a worrying feature of the data.

FREHSE, CERN: Frankly, I don't find it very astonishing because, I think, if one simply does calculations for the cross section one would expect from the multiplicity in this region, one winds up with rather high cross section. I mean the graph with the highest E_T I think you have been showing, has been 10 GeV. So this means on the order of 20 particles if you say every particle has half a GeV, or 25 if you say it is below. So my precise question is: Did you calculate what cross section you would expect in the 2π

calorimeter simply from the normal multiplicity distribution? And I would bet it is exactly what you see.

SEYBOTH: No, that is what I did. I mean this is my cluster model. This is normalized to the known multiplicity.

FREHSE: So, in this respect, I don't understand your comment that it is worrying that this cannot be described by hard scattering models.

SEYBOTH: No, but I also predict the observed cross section for the situations where I restrict my transverse energy to smaller solid angles. This is what is worrying me.

FREHSE: But did you do the same calculation there for what you would expect from the observed multiplicity? I mean, it simply would wind up with the same conclusion. Even there the normal multiplicity of an event is simply too high, so that with this big solid angle you can't separate cross sections. I mean hard scatters.

SEYBOTH: Okay, but where are the jets?

MEYER, AMES LAB: As a useful cross check, have you tried looking for the class of events where on one side there is a single high-momentum particle and you can play the same games that people play at the ISR looking at the trigger-side jet and the away-side, and try to reproduce that data just as a check.

SEYBOTH: Unfortunately our statistics for this trigger are not such that we can do this very well. I mean the single-particle cross section is something like 4 orders of magnitude below our cross section, so we unfortunately don't have very much statistics to do this.

SCHMITZ, MUNICH: I nevertheless think what this experiment shows is the following: previous experiments have triggered only in one direction (single arm experiments) or in two directions (double arm spectrometers). They are more or less bound, if they look for high transverse momentum, to find jets because they see only part of the phase space. This experiment has a ring-shape calorimeter which covers the whole azimuthal range, and what the data show is that, in the previous experiments, which were interpreted as evidence for jets, they have probably seen only part of much more general events which can, at least to some extent, be explained just by a cluster model. So the interpretation of previous experiments in terms of jets, to my opinion, has become doubtful by this result.

VAN HOVE, CERN-SANTA BARBARA: My comment really is very close to the one of Norbert. I believe people who deal with jets, either experimentally or theoretically, should worry. To put it in perhaps too extreme terms, I doubt that there is any evidence for two jets . . . two-lateral-jet events . . . at Fermilab or the SPS at CERN as of now. Theorists may calculate them, but there is no evidence. The question at the ISR is somewhat different. There there is a better chance to have them, but even there, I think, to my knowledge the matter of establishing beyond doubt two-jet events as a well-separated class of collisions is much more difficult than has been thought throughout the many years of hard labor where people have attempted to do that. These are the warnings which I believe should be taken very seriously also by the QCD calculations.

CERN/EP/0476R/RG/ef
31 July 1981

STUDY OF EVENTS WITH AN IDENTIFIED HIGH TRANSVERSE MOMENTUM PARTICLE

AT THE ISR AT $\sqrt{s}$ = 63 GeV

D. Drijard, H.G. Fischer, H. Frehse, R. Gokieli[1], P. Hanke,
P.G. Innocenti, J.W. Lamsa[2], W.T. Meyer[2], G. Mornacchi,
A. Norton, O. Ullaland and H. Wahl[3]
CERN, European Organization for Nucl. Res., Geneva, Switzerland

W. Hofmann, M. Panter, K. Rauschnabel, J. Spengler and D. Wegener
Institut für Physik der Universität Dortmund, Dortmund, Germany

W. Geist, M. Heiden, W. Herr, E.E. Kluge, T. Nakada and A. Putzer
Inst. für Hochenergiephys. der Univ. Heidelberg, Germany

K. Doroba and R. Sosnowski
Institute for Nuclear Research, Warsaw, Poland

ABSTRACT

Complete events with an identified particle of p_T above
4 GeV/c have been studied using the Split Field Magnet (SFM)
facility at the CERN ISR. Results on charge correlations within
the away jet are presented. The x_E distributions and their
asymmetries are discussed. The results are compared with hard
scattering model predictions.

1. INTRODUCTION

Results presented here come from the R416 Experiment
performed at the CERN ISR using the Split Field Magnet facility.
The schematic view of the apparatus is shown in fig. 1.

---

[1]   On leave of absence from the Inst. for Nucl. Res., Warsaw, Poland.
[2]   Visitor from the Ames Lab., Iowa State Univ., Ames, Iowa, USA.
[3]   Now at Institut für Hochenergiephysik, Vienna, Austria.

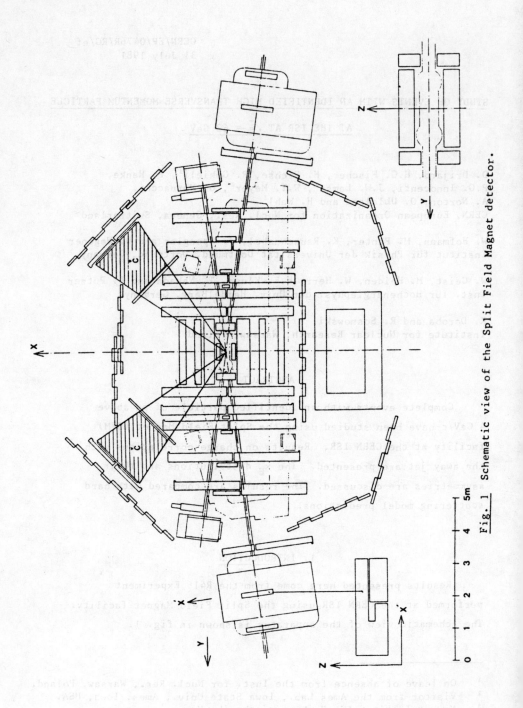

Fig. 1 Schematic view of the Split Field Magnet detector.

It consists of the set of multiwire proportional chambers, which provide an almost 4π geometrical acceptance. They are situated in the magnetic field, so that the momentum and charge of all charged secondaries are measured. Marked are two trigger regions, situated ∿ 50° with respect to the incoming beam directions. The transverse momentum range of triggers discussed here was between 4 and 6 GeV/c. Within each region there is a Cerenkov counter, allowing for the distinction between pions and heavier particles. The Cerenkov counter efficiency in this momentum range was better than 99.9%. The contamination of the non-pion sample by protons (antiprotons) was estimated to be ∿ 37% (12%), for the positive and negative particles respectively. Since the $\bar{p}$ admixture is rather small, in the following discussion we will assume all negative heavy particles to be K^-. Note also that all parton model arguments used here for K^- particles are perfectly valid for $\bar{p}$ as well.

As we know, triggering on a single particle with a large transverse momentum is a good way to study a hard scattering of proton constituents. Such events display a distinct four-jet structure, as is schematically shown in fig. 2. We distinguish there a so-called "trigger jet" around a high transverse momentum particle and on the opposite side in azimuth the "away jet". These two jets originate most probably from the hard scattering of proton constituents. We see also two "spectator jets", which stem from the left-over fragments of two protons.

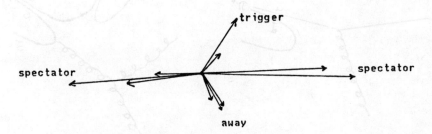

Fig. 2   Schematic picture of a 4-jet structure in a high p_T event.

When selecting events with a single high p_T-particles, we get trigger jet which fragments in a rather rare way - the trigger takes most of the jet energy. This effect, known as a "trigger bias", is usually disturbing, but in our case it is quite advantageous. It not only provides us with a good approximation of the jet direction, but also ensures that in most of the cases the fastest particle (normally the trigger) carries the flavour of the struck parton.

According to QCD, to the production of high transverse momentum events contribute mainly the following hard-scattering processes: quark-quark, quark-gluon and gluon-gluon scattering (this latter one we can neglect, as it usually does not bring enough energy to produce a high-p_T particle). This is therefore a place, where a first-order gluon effects would be involved. Now, the questions we aks ourselves from the experimental point of view are: what can we possibly learn about these processes? Can we distinguish different contributions? Is there some way to disentangle the quark-gluon and quark-quark scattering?

To try to answer these questions, let us first realize the important difference which exists between different kinds of triggering particles. Let us concentrate on two extreme cases from the point of view of parton ideas: on π^+ and K^- triggers only. They are shown in fig. 3.

(a)   (b)

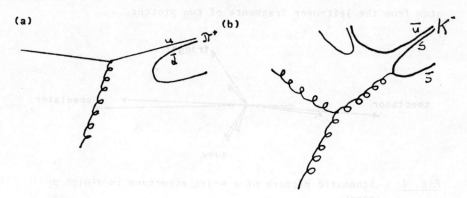

Fig. 3 Possible mechanisms of producing: (a) π^+ and (b) K^- triggers.

A π^+ particle consists of an u and $\bar{d}$ quark and so it most probably is a result of a fragmentation of an u valence quark from a proton, which has picked-up somewhere a $\bar{d}$ quark. On the contrary, the K^- particle has no common valence quarks with a proton, so it is a good candidate for originating from a scattered gluon. Note that there is no difference in experimental procedure for the two cases.

We will also use a few facts which follow from the parton kinematics. In fig. 4 we see the difference in the relative configuration of the trigger and away-jet, depending whether we had a quark-quark or quark-gluon scattering.

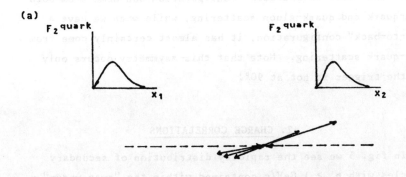

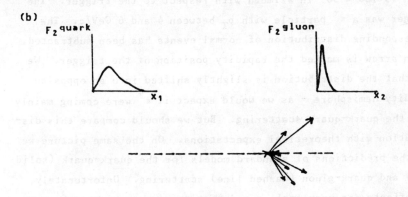

Fig. 4 Most probable configurations of the trigger and away jet for the (a) quark-quark and (b) quark-gluon scattering.

For a quark-quark case (fig. 4(a)) both partons have similar structure functions within a proton and so they will have, in the average, more or less equal momenta. In the result, after the scattering, in the total proton-proton centre of mass system, they will be opposite both in rapidity as well as in an azimuthal angle. This we will call a "back-to-back" configuration. The gluon structure function is, however, completely different from that of a quark - in the average quarks are much more energetic than gluons (fig. 4(b)). After the quark-gluon scattering then, due to the Lorentz boost, in the total p/p c.m.s. we will normally observe both jets to be within the same rapidity hemisphere. This configuration we will call "back-to-antiback". In different words: the "back-to-antiback" configuration can come from both quark-quark and quark-gluon scattering, while when we have a "back-to-back" configuration, it has almost certainly come from a quark-quark scattering. Note that this asymmetry occurs only when the trigger is not at 90°!

2. CHARGE CORRELATIONS

In fig. 5 we see the rapidity distribution of secondary particles with $p_T > 1$ GeV/c contained within the "away wedge" - which is $180 \pm 30°$ in azimuth with respect to the trigger. The trigger was a π^+ particle with p_T between 4 and 6 GeV/c. The corresponding distribution of normal events has been subtracted. By an arrow is marked the rapidity position of the trigger. We see that the distribution is slightly shifted into an opposite rapidity hemisphere - as we would expect if π^+ were coming mainly from the quark-quark scattering. But we should compare this distribution with theoretical expectations. On the same picture we see the predictions of standard models for the quark-quark (solid line) and quark-gluon (dashed line) scattering. Unfortunately, the effects are very small - and that means that we have to use all our available information and full precision to be able to answer our questions.

588

Let us look then at fig. 6. where are shown the same distributions as on the previous picture, but now plotted separately for positive and negative secondaries. They are also shown for both π^+ as well as for K^- trigger.

There are few important observations to be made. First, the distribution for a K^- trigger is much more symmetric than that for a π^+. Secondly, for both cases we have more positive than negative secondaries - what means that we always have valence quarks from the second proton scattered into the opposite ϕ hemisphere. And third - this difference is much bigger in the case of a K^- trigger than for π^+.

This last observation seemed to us so interesting that we decided to follow it up in a more quantitative way. In fig. 7 we see the ratio of the number of positive to negative secondaries within the above mentioned ϕ wedge, integrated over y and plotted as a function of a fragmentation variable x_E.

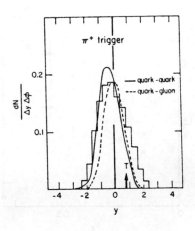

Fig. 5 Rapidity distribution of secondaries with p_T > 1 GeV/c within the "away wedge" for the π^+ trigger. Normal events background has been subtracted. The prediction for such distribution from standard quark-quark and quark-gluon scattering models are shown.

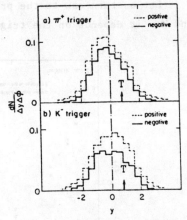

Fig. 6 Rapidity distributions of positive and negative secondaries of p_T > 1 GeV/c within the "away wedge" for (a) and (b) K^- trigger.

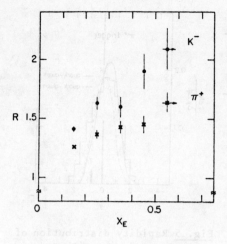

This variable is defined here as ratio of a p_T of a secondary to the p_T of a trigger. We see that for the K^- trigger the ratio of positive to negative secondaries in the "away wedge" very soon approaches the value of 2, which is the ratio of the number of u to d valence quarks within the proton. The data for a π^+ trigger are rising much slower - what means again that in this case the u quarks scatter also on something else than particles to the one for valence quarks, and this object is rather neutral.

<u>Fig. 7</u> Ratio of the rapidity distribution for positive partciles to the one fornegative particles, plotted for π^+ (crosses) and K^- (dots) trigger.

3. <u>FRAGMENTATION FUNCTION ASYMMETRIES</u>

Let us turn now to the problem of a fragmentation function and how it depends on the triggering particle. In fig. 8 are

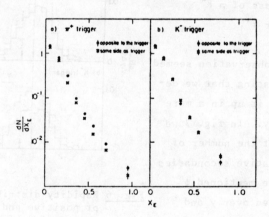

<u>Fig. 8</u> Fragmentation functions for particles on the opposite (dots) and same (crosses) side as the trigger for (a) π^+ and (b) K^- trigger.

590

presented distributions of a fragmentation variable x_E for both π^+ (fig. 8(a)) and K^- (fig. 8(b)) trigger in two cases - when the secondaries are within the same (crosses) or opposite (dots) rapidity hemisphere.

We still consider only particles being within the introduced above "away wedge". One can see that for the π^+ trigger the number of "back-to-back" particles is bigger than the number of "back-to-antiback" particles and that it is not so in the case of a K^- trigger.

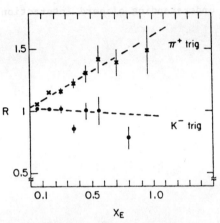

To be more quantitative, we see from fig. 9, where for each trigger the ratio of the wo distributions is plotted. The difference there between the two triggers is striking - the ratio is rising rapidly with x_E for the π^+ trigger and is constant (or even may be slightly decreasing) for the K^- trigger.

Fig. 9 Ratio at $1/N\ dN/dx_E$ dis-
tribution for particles being opposite to the trigger to the one for particles on the same side as trigger plotted for π^+ (crosses) and K^- (dots) trigger.

This agrees with our simple parton kinematics considera-
tions, if π^+ triggers were coming mainly from the quark-quark scattering and K^- triggers from quark-gluon scattering.

4. CONCLUSIONS AND OUTLOOK

We observe big differences in the event structure, depending on the nature of the high transverse momentum triggering particle. Since the same kinematics is involved in all cases, then any differences between events with different triggers must be

really of the dynamical nature. If we use the framework of parton models, then there is an evidence for a quark-gluon scattering.

We see also that data carry the necessary information to perform a detailed comparison with predictions of first order QCD, like gluon fragmentation function or relative importance of quark and gluon scattering, which is of primary importance in our understanding of hard interaction processes.

HIGH P_T HADRON PRODUCTION

M.D. Mestayer
University of Chicago
5640 South Ellis Avenue
Chicago, Illinois 60637

ABSTRACT

Recent results on the inclusive production of charged hadrons with transverse momentum up to 6 GeV/c are presented. The energy and angular dependence of the cross sections is examined as well as the P_T dependence. Particular attention is paid to comparing production by π^- and P beams. Particle ratios are consistent with simple quark counting schemes which assume that hard scattering dominates the production; they are also consistent with QCD calculations[8] but rule out some CIM predictions.[11]

In this talk I will discuss measurements of the production of single charged hadrons at values of P_T up to 6 GeV/c in $\pi^- p$ and pp collisions. The experiment was the latest in a series of Chicago-Princeton experiments[1] which use a small acceptance single-arm spectrometer to detect and identify hadrons. A specific aim of this experiment was to measure the dependence on P_T, x_T ($\equiv 2\, P_T/\sqrt{s}$), and angle of the production of different particle types for π^- beams, and to compare the cross sections for pion and proton beams. Measurements of these kind explore the underlying mechanisms responsible for hadron production. If valence quark scattering plays a large role in high P_T particle production, then the valence antiquark in the π^- should produce marked differences between pion and proton interactions, especially for the production of particles such as π^-, K^-, and $\bar{p}$ which contain the same valence antiquark as the π^-. We do see large differences between pion and proton collisions. The data are compared to current theoretical calculations as well as to the results of other experiments.[2,3,4] The calculations match the trends of our data; simple quark counting schemes also roughly describe our results.

The experiment has been described in previous publications.[5] The experiment was located in the experimental hall of the high intensity pion beam in the Proton West Area at Fermilab. Typical beam intensities during the data taking were $2-4 \times 10^9$ π^- pulse at an energy of 200 GeV, with a maximum intensity of over 10^{10} π^-/pulse. The detector consisted of a magnetic spectrometer with an acceptance $\frac{\Delta p}{p}\, \Delta\Omega$ of 420 μsr%, instrumented with two Cerenkov counters for identification of pions, kaons, and protons, a segmented shower counter for muon identification, and drift chambers for track reconstruction. The spectrometer viewed the target at an angle of 80 mrad from the beam, which corresponds to 90° in the c.m. frame for 300 GeV beams and 79° for 200 GeV incident beams. Additional 200 GeV pion beam runs were taken with four dipoles in the beam line to change the targeting angle by ±16 mrad, allowing data collection at c.m. angles of 90° and 67°. The transverse momenta studied ranged from .8 to 6.0 GeV/c. Details of the analysis will appear in a future publication.

The two sets of data taken at 90° at 200 and 300 GeV allow us to test the validity of the form predicted by naive dimensional counting,[6] $Ed^3\sigma/dp^3 = \frac{A}{P_T^n} (1-x_T)^b$, where $x_T \equiv P_T/(2\sqrt{s})$. A transverse momentum cut at 2.6 GeV/c was made to exclude the low P_T region where soft scattering dominates the cross sections. The results of the fit for each particle type are given in Table 1. For meson production the fitted values for n (the power of $1/P_T$) are consistent with a value of 8, which is the value observed for these reactions in p-p collisions.[7] The values of n for p and $\bar{p}$ production are both about 10, and are more alike than in the pp case. The values of b tend to be lower (especially for K^- and $\bar{p}$'s) than in the p-p case, presumably due to fewer quarks sharing the momentum in the initial state.

Figure 1 shows the ratios π^-/π^+, and $\bar{p}/p$ where, for example, π^-/π^+ is the ratio of produced π^- to produced π^+, all with the same incident π^- beam. Also represented on the figure are the same ratios measured in p-p collisions.[7] All of the pion-induced ratios are more constant in P_T than in the proton-induced cases. This is natural if hard scattering dominates, as the beam π^- can contribute a valence $\bar{u}$ quark to a π^-, K^-, or $\bar{p}$, whereas with a proton beam the $\bar{u}$ quark comes from the sea. Simple counting of the valence quarks gives $\pi^-/\pi^+=2.1/2$, and $K^-/K^+=3/4$ for a π^- beam. $\bar{p}$ production requires picking up two antiquarks, but the same simple counting of single valence quarks gives $\bar{p}/p=0.33$.

Figure 1 also shows predictions from a CIM calculation of Gunion and Jones[11] and a QCD calculation of Field.[8] The CIM calculations which had fit the production of each particle type in the p-p case very well (a special success being the prediction of a large proton flux) disagree with the pion-induced data by a large factor. The calculation of Field is in better agreement.

The existence of an initial valence antiquark would suggest that at large x_T incident π^-'s should produce more π^-, K^-, and $\bar{p}$'s than incident protons at the same energy. Figure 2 shows this to be so, with an enhancement factor of about 4 at a P_T of ~5 GeV/c. At this same P_T, moreover, π^- are as effective as incident protons at producing π^+ and K^+, in spite of starting with a neutral rather than a charge-two state. This is a reflection of the fact that the pion has a stiffer parton momentum distribution than the proton. The production of protons in π^-p collisions is approximately half that in pp collisions, consistent with simple quark counting in a quark quark scattering reaction.

In Figure 3 I present a comparison of the relative production of charged pions by the 200 GeV π^- beam and the 200 GeV proton beam. We have plotted the average of the π^+ and π^- cross sections, so that one can compare with previous π° measurements. The increase of the ratio agrees with a $1/(1-x_T)^2$ dependence, as expected from dimensional counting; and the two sets of data are consistent with one another.

In Figure 4, I show results for the ratios of the π^- to π^+ cross sections versus c.m. angle at a value of transverse momentum of 4 GeV/c. The solid line is the prediction of a QCD calculation from Reference 8. The dashed line, which comes closer to the data, is a similar recent calculation with a gluon fragmentation function which produces more hard particles[9] (i.e., more particles with a larger fraction of the gluon momentum). On the same plot are shown our measured point for the p-p case at 90° and the p-p prediction of Reference 8. The difference in valence quark content of the target and projectile in the π^-p case produces an asymmetry in the π^-/π^+ ratio with angle which is quite unlike the flat behavior in p-p collisions.[10] Our data is roughly consistent with QCD calculations; the two calculations shown illustrate how sensitive the data are to, for example, the gluon fragmentation function.

In conclusion, we have measured the cross section in $\pi^- p$ and pp collisions for the production of identified charged hadrons over the P_T range $0.8 < P_T < 6.0$ GeV/c. The data scale with P_T and X_T; the fitted parameters are tabulated and compared to the pp case. The comparisons agree well both with hard scattering calculations and with simple quark counting rules.

REFERENCES

1. D. Antreasyan et al., Phys. Rev. D 19, 764 (1979).
2. G. Donaldson et al., Phys. Rev. Lett. 36, 1110 (1976), Phys. Lett. 73B, 375 (1978), Phys. Rev. Lett. 40, 917 (1978).
3. M.D. Corcoran et al., Phys. Rev. Lett. 41, 9 (1978), Phys. Rev. D 21, 641 (1980).
4. C. Bromberg et al., Phys. Rev. Lett. 43, 561 (1979), Nucl. Phys. B171, 38 (1980).
5. H.J. Frisch et al., Phys. Rev. Lett. 44, 511 (1980). A more complete description is given in N.D. Giokaris, Ph.D. thesis, The University of Chicago (1981), and J.M. Green, Ph.D. thesis, The University of Chicago (1981).
6. S. Brodsky and G. Farrar, Phys. Rev. Lett. 31, 1153 (1973), V.A. Matveev, R.M. Muradyan and A.M. Tavkhalidze, Lett. Nuovo Cim. 7, 719 (1973).
7. The data presented in Table 1 on the fitted powers for proton-proton interactions are from Reference 1. The comparisons shown in Figures 4 and 5 are new proton-proton data taken with our present spectrometer. The two sets of data agree well.
8. R.P. Feynman, R.D. Field and G.C. Fox, Phys. Rev. D 18, 3320 (1978).
9. R.D. Field, private communication.
10. D. Lloyd Owen et al., Phys. Rev. Lett. 45, 89 (1980).
11. For a clear exposition of some of the problems of comparing predictions of QCD (quantum chromodynamics) with the present data, see R.D. Field, preprint CALT-68-696, and Phys. Rev. Lett. 40, 997 (1978). For an alternative approach using both hard scattering and CIM (constituent interchange model) terms, see D. Jones and J. Gunion, Phys. Rev. D19, 867 (1979).

FIGURE CAPTIONS

Figure 1   The ratio of (a) π^-/π^+, (b) K^-/K^+ and (c) $\bar{p}/p$ produced
            by 200 GeV and 300 GeV π^--p collisions. Also shown are
            the corresponding ratios in p-p collisions, and the
            theoretical predictions of Field (Reference 8 and
            Gunion and Jones (Reference 11).

Figure 2   The relative particle production versus P_T for a π^-
            beam versus a proton beam. The data are for 200 GeV
            beams, at 90° in the c.m. The lines are predictions
            of Reference 8.

Figure 3   A comparison of the relative production of pions for
            the π^- beam and the proton beam, at 90° in the c.m.
            The filled circles are the average of π^+ and π^- for
            the 200 GeV beams. Also shown are the data of Donaldson
            et al (Reference 2).

Figure 4   The angular dependence of the trigger ratio π^-/π^+ at
            a P_T = 4 GeV/c for the π^- beam. The solid curve is the
            prediction from Reference 8: the dashed curve is the
            same model with a harder gluon fragmentation. The
            proton datum at 90° and the corresponding prediction of
            Reference 8 are shown for comparison.

TABLE 1

	π^-p			pp	
	n	b	χ^2/DOF	n	b
π^+	8.6±0.5	7.0±0.6	6.6/4	8.2±0.5	9.0±0.5
π^-	7.5±0.5	8.9±0.7	2.2/5	8.5±0.5	9.9±0.5
K^+	7.3±0.6	8.6±1.1	1.1/4	8.4±0.7	8.3±0.6
K^-	8.4±0.5	8.2±0.5	9.3/4	10.1±1.5	11.5±1.3
p	9.9±0.8	6.4±1.5	1.8/4	11.8±1.6	7.3±1.1
$\bar{p}$	9.8±1.2	8.1±0.8	.96/1	8.8±1.8	14.2±2.0

The results of fits of the 200 and 300 GeV data to a form $Ed^3\sigma/dp^3 = (A/P_T^n)(1-x_T)^b$. The values for a proton beam from Reference 1 are also given for comparison.

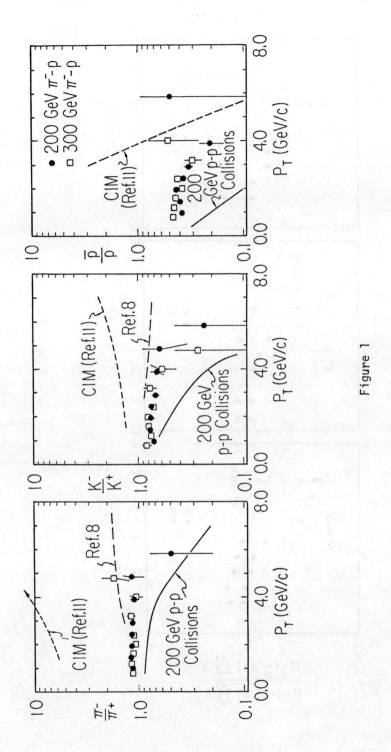

Figure 1

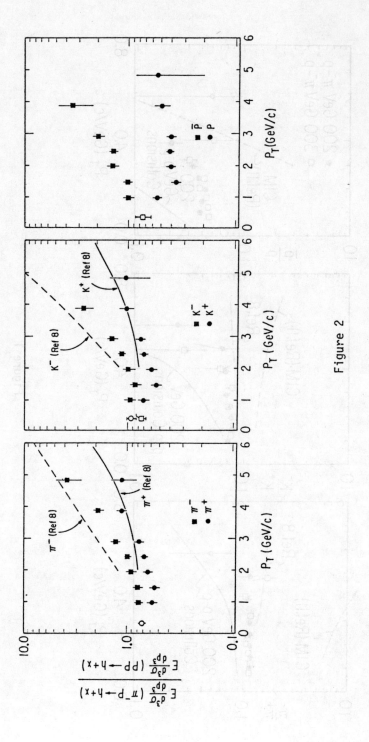

Figure 2

Figure 3

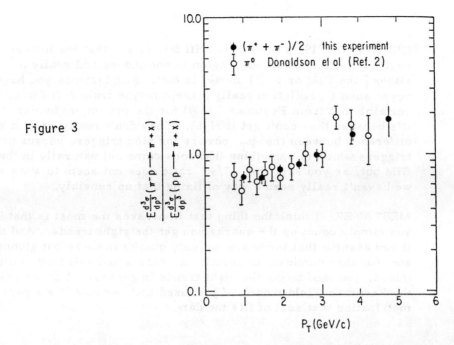

Figure 4

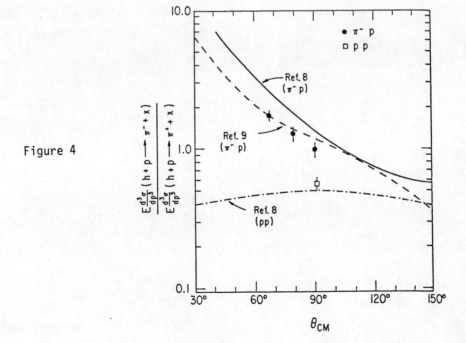

DISCUSSION

GUNION, DAVIS: I think it is still fair to say that the low-p_T region or intermediate-p_T region is not understood really in either [the CIM or F-F] model in fact. For instance you have never seen a prediction really (except maybe from Jeff Owens but certainly not from Feynman-Field) for the proton production. And it's because they can't get it right. They don't really predict much difference between the p_T powers for pion triggers versus proton triggers which is something that does come out naturally in the CIM but, as you say, the π^-/π^+ ratio does not seem to work and we haven't really seen a way of fixing that up sensibly.

MESTAYER: I think the thing that impresses me most is that if you simply count up the quarks you get the right trends. And then, if you assume that there are not only quarks in there but gluons and that they dominate in certain x values and will tend to dilute things, you tend to get the right trends in general. I'm not pushing the Feynman-Field model. I just used that because it's a parametrization that sort of fits the data.

REVIEW OF DIRECT PHOTONS AND ASSOCIATED PARTICLE PRODUCTION AT THE ISR*

L.K.RESVANIS

Physics Laboratory

University of Athens,

Athens, Greece.

ABSTRACT

We review the measurements of the hadronically produced high transverse momentum single photons.The correlations of the single photons with other associated particles have been measured by two expirements at the CERN ISR and we present their recent results on the study of the single-photon event structure.

We call DIRECT or SINGLE or PROMPT PHOTONS,photons that are NOT the decay products of mesons like π^0, η, η', ω... but are the direct product of the hard scattering of the costituents of hadrons.It is generally accepted now that direct photon production in hadronic interactions is an important test of Q C D. The importance of direct photon production comes from the "point like" nature of the photon and its well known electromagnetic coupling to charged particles in particular to the charged constituents of the proton,the quarks.Any hard scattering Feynman diagram that produces a gluon in the final state can also produce a final state photon instead, but with a cross-section down by a factor of a few per cent.The photon can escape as a free particle with all the P_T imparted to it in the hard scatter,while the gluon must fragment into hadrons (each one) of reduced P_T. This different behavior enhances the ratio of production of direct photons to single pions with the same P_T. Experiments to measure direct photons are extremely difficult,because of the copious production of mesons that decay into photons.There have been three definitive measurements of the signal by the following groups:Experiment R-806, Athens - Brookhaven - CERN (1) at the ISR, Experiment E-95, Fermilab-Johns Hopkins at Fermilab(2), and Experiment R-108, CERN-Columbia-Oxford-Rockefeller (3) at the ISR Q C D calculations were made by Halzen and Scott (4), Contogouris and Papadopoulos (5),Ruckl-Brodsky and Gunion (6), and Owens (7).
The dominant diagrams are:

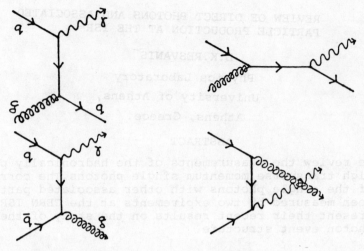

The qg→qγ is called the Compton diagram for obvious historical reasons. It gives an excellent probe of the gluon distribution in the photon. We can obtain useful information on the gluon distribution in the proton Fg/p, in fact this is a good method to determine Fg/p(x,Q²). Furthermore the P_T of the proton is balanced by a quark jet which in the case of p-p collisions should be mostly a u-quark jet. In fact since there are twice as many valence u-quarks as d-quarks in a proton and in the cross section we have the square of the charges we expect a ratio u:d∼8:1. This enhancement of u quark scattering will lead to asymmetries in the sign of the charge of the pions resulting from the fragmentation of these quarks. The leading pions tend to carry the charge of the quark. One would also expect that the Compton diagram gives rise to direct photons which are UNACCOMPANIED, in contrast to π⁰'s which are members of a jet.

The other diagram that I showed is the annihilation qq̄→γg and it is important in pp̄ and πp collisions. We can therefore use direct photons for tagging i.e. we can use the direct photons as a trigger to study quark jets (in pp collisions) and gluon jets (in p̄p collisions).

I should also mention the quark bremsstrahlung diagram:

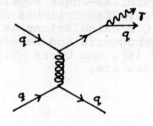

in this diagram the photon is accompanied by a quark
jet (again mostly u)

E X P E R I M E N T S

i) Measurement of direct photon signal: There are three
   experiments

   a) ISR-806 (Athens-BNL-CERN) [1]

   b) FNAL E-95 (Fermilab-John Hopkins) [2]

   c) ISR-108 (CERN-COLUMBIA-OXFORD-ROCKEFELLER) [3]

ii) DIRECT PHOTONS AND ASSOCIATED PARTICLES

   a) ISR-806 (Athens-BNL-CERN)

   b) ISR-806*807 (Athens-BNL-CERN)*

   *(BNL-CERN-COPENHAGEN-LUND-RUTHERFORD-TEL-AVIV)
                                                [7],[8].

   c) ISR-108 (CERN-Columbia-Oxford-Rockefeller)[9]

We review the Athens-BNL-CERN single photon measurement.
In order to avoid systematic problems introduced by coale-
scing γ's from π^0 decay the experiment was repeated at
three different distances from the Interaction Region, in
this way the two photons of π^0's were resolved up to a
$P_T \sim 10$ GeV. Various fiducial region cuts were made. The
longitudinal energy development of the showers and their
lateral spread were required to be consistent with that of
electromagnetic showers. It was further required that
there are no more than two showers in the calorimeter and
that the calorimeters be clean i.e. the difference between the total energy contained in the calorimeter minus
the sum of the energy assigned to the showers be less
than 35 MeV. This requirement excludes events with a
charged track in the calorimeter. We substracted π^0, η, η'
and ω induced backgrounds using the production cross
section measured by the experiment (10)
($\eta/\pi^0 = .55 \pm .05$, $\omega/\pi^0 = .9 \pm .2$, $\eta'/\pi^0 = .9 \pm .2$)
The shower reconstruction efficiency was calculated using
the EGS [11] Monte Carlo program coupled with our Reconstruction program . Cosmic ray background was estimated to
be < 5% in special cosmic ray runs. Background due to
hadrons e.g. neutrons, was calculated from tests at the
CERN proton synchrotron to be < 2% . Our final γ / π^0
numbers should be multiplied by $.85 \pm .15$ in order to
correct for the fact that events with a charged track
in the calorimeter were not included. In Figure 1, we
present our measurement for values of $\sqrt{s} = 31$ GeV, 45 GeV,
53 GeV and 63 GeV. The lines are from a calculation of
Contogouris et al [12] , the solid lines include soft

gluon corrections $O(a^2_s)$, the broken lines include first order QCD calculations only.

We conclude that we have measured direct photon production at 90° for $P_T > 4$ GeV/c. The signal rises with P_T and has little $\sqrt{s}$ dependence.

The Athens-BNL-CERN group made the first attempt to measure the difference in the structure of events with a high P_T π^0 or single photon (8). For π^0's where both energy and direction are known the rapidity in the c.m. is defined by $y = 0.5 \ln [(E-P_L) / (E+P_L)]$ while for charged tracks the pseudorapidity in the ISR frame $y = -\ln (\tan 1/2 \theta)$ is used. In the small y region covered by this experiment the two definitions are equivalent. The azimuthal angle Φ is defined in the transverse plane perpendicular to the incident proton beams in the overall c.m. Fig.2. shows the pseudorapidity difference between charged tracks (or π^0's) and trigger particles.The solid lines represent the expected Δy distribution for uncorrelated tracks,calculated assuming a flat y distribution. The curve is normalized to the number of events with $|\Delta y| > 0.6$. A clear excess is observed for the π^0 events over that expected from the uncorrelated background. A much smaller excess is observed for the single photon candidate sample and no excess is observed if one subtracts from the single photon candidates the expected background.

Similarly if one plots the azimuthal difference (Fig. 3) between the trigger particle and a π^0 one observes that on the same side there is a significant difference between the two samples, with the magnitude of the difference increasing with decreasing $\Delta\Phi$,while on the away side ($\Delta\Phi > 90°$) there is no significant difference between events with a π^0 or single photon trigger. So we con - clude that the single photon events have considerably lower numbers of associated charged tracks as well as π^0's on the trigger side. Most of the difference is concentra- ted in the small $\Delta\Phi$ and Δy region where the accompanying particles of the same "jet" are expected to be found and the relative magnitude of the effect is increased with increasing transverse momentum of the second particle. One would expect the above if direct photons were not part of a jet as the π^0's but were produced unaccompanied.

Further studies of the correlations were done using the combination of the Liquid Argon calorimeters and the Axial Field spectrometer at the ISR (8),(13).Fig.4.shows the combined apparatus.I shall present preliminary results. The trigger and e.m. energy measurement is again provided by the 806 Liquid Argon calorimeter and the 807 drift chamber information is recorded.Fig. 5a shows the 807 drift chamber information of a typical π^0 event and Fig.5b is similar,but,for a γ event; one can see the difference in the structure of these "typical" events.In Fig.6 we

define the variables to be used in this analysis ($h^{\pm}$
stands for a charged hadron,charged tracks are now measured in the 807 cylindrical drift chamber system).
Fig.7 shows the azimuthal difference $\Delta\Phi$ between a γ or π^0
and the nearest charged hadron.It is obvious that there is
no difference between the γ and π^0 events in the away
region but close to the trigger particle there are significantly fewer associated tracks in the γ-sample.
In this figure we also show the signal near $\Delta\Phi = 0°$,for γ's
after we subtract the expected background due to π^0's
and η's. If we assume that the associated multiplicity
due to processes other than hard scattering is given by
that observed at $\Delta\Phi \approx 90°$ we can see a small but significant excess of tracks correlated with single photons at
$\Delta\Phi = 0°$;we interpret this as evidence of quark bremsstrahlung production.
Fig.8 shows the X_E distribution ($X_E \equiv P$ (hadron)$\cdot P_\gamma / P_\gamma^2$)
of charged tracks opposite a γ or a π^0; there is no
obvious difference.Let us now measure the charge of the
associated particles. (We have checked for possible
charge bias in the detector and we set a limit of
δ (N^+ / N^-) < 0.1).In fig.9a we have plotted the ratio of
the mean number of positive to the mean number of negative
tracks, for the away side tracks vs X_E of the charged
tracks; there is no significant difference for the γ and
π^0 events. If we make the same plot but for events with X_E
(same side) $< .2$ this requirement selects events
that have only low momentum tracks in the direction of the
γ or π^0, we see in Fig.9b that for π^0 triggers the distribution is the same as in Fig.9a,but we note a rise in the
γ triggers above $X_E > 0.5$.
We would also expect a charge asymmetry in the associated
tracks in same side as the trigger γ if the bremsstrahlung
diagram is a significant source of photons,again we expect
more positive tracks since there are more u-quarks in the
proton. Fig. 10 shows some weak evidence that the charge
asymmetry may be larger for photon than for π^0 events.
Clearly we need more statistics(we have four times the
present sample)
So we conclude that we observe significantly fewer tracks
closely correlated indirection with γ's than with π^0's.
There is very little difference in the distribution of
charged particles opposite γ's and π^0's,with the exception
of their charge structure.Photon triggered events,especial-
ly those with little momentum flow in the direction of
the trigger, exhibit an excess of positive particles in
the recoiling system. This is what one would expect if the
direct photons were produced by the QCD Compton process.
We also observe an enhancement of charged tracks near the
direction of the γ's, in excess of what would be expected
from tracks that are associated with soft scattering
process.We interpret this as evidence of the fact that

direct photons are predominantly produced alone as one
would expect from the QCD Compton diagram but there is
a small contribution from bremsstrahlungproduction.
The apparatus of the ISR experiment 108 is shown on Fig.11.
The Pb-glass blocks are too coarse to resolve the two γ's
of a π^o so they use the conversion technique i.e. they take
advantage of the fact that the coil plus the cryostat
amount to about 1 radiation length.The probability for non-
coversionin one radiation length for single γ's is $\sim$ 0.47
and for π^o's is $\sim$0.25.Then they measure the fraction of
e.m. clusters that do not covert in the coil.From the above
they deduce the fraction of single γ's in the sample.Note
that with this method they cannot distinguish single γ's
from π^o's on an event by event basis.A statistical analysis
is only possible.Note also that an e.m.cluster can be due
to $\eta,\omega,\eta'....$so they really measure γ/ALL not γ/π^o.
They vetoe events with either a second energy cluster or a
charged track going through the same B-counter.This cut
introduces a similar bias to the no unassigned energy
requirement on the 806 calorimeters.Then they normalize
their data(sepaparately for the inside and the outside
Pb-glass arrays)to the γ/π^o = -.0 21$\pm$.012 for 3.5$<$P$_T$$<$5.0GeV/c
of an early experiment (Rome-BNL-Adelphi) (14).
There is a direct γ signal which is significant above
P$_T$ $\sim$6-8 GeV/c and rising with P$_T$ as shown in Fig.12.
It appears to be $\sqrt{s}$ depentet.When plotted against X$_T$,Fig.13,
it appears to scale.They define the $\Delta\Phi$ as the azimuthal
difference between the candidate γ or π^o and the closest
charged or neutral particle to it and they divide events
into two classes:
(i) Accompanied with $|\Delta\Phi| < 90^0$
(ii) Unaccompanied with $|\Delta\Phi| > 90^0$i.e.no particle within 90^0
    in Φ.
They recalculate γ/ALL for these two classes.The result is
shown on Fig.14 i.e. the γ/ALL ratio is much larger for the
unaccompanied sample as one would expect from the QCD
Compton diagram.
They plot then the ratio R$^{\pm}$ = (N$^+$ / N$^-$) versus X$_E$ (Fig.15)for
the unconverted plus unaccompanied sample(i.e.rich in single
γ's) and for the converted and accompanied sample(i.e. rich
in π^o's).Note that R rises with X$_E$ for all samples,but it
rises more steeply at higher P$_T$ and more steeply for the
unconverted and unaccompanied samples.
From the relative ambudances of γ and mesons in the sample
they calculate (Fig.16) the Ratio R$^{\pm}$ for pure γ's and
mesons.

C O N C L U S I O N S

Both 806/807 and 108-C.C.O.R. agree that the opposite
charge ratio is :
(a) bigger than 1.0
(b) higher for photon events than for π^0 events which
    means relatively more u-quark jets than d-quark jets.
These conclusions favour the process $gq \rightarrow \gamma q$.

*The members of the reviewed experiments are listed below:

Members of the ISR-806 group are:
E.Anassontzis,M.Diakonou,A.Karabarbounis,C.Kourkoumelis,
L.K.Resvanis,T.A.Filippas,E.Fokitis,C.Trakkas,A.Cnops,
E.C.Fowler,D.M.Hood,R.Palmer,D.Rahm,P.Rehak,I.Stumer,
C.W.Fabjan,T.Fields,D.Lissauer,I.Mannelli,W.Molzon,
P.Mouzourakis,A.Nappi and W.J.Willis.

Members of the FNAL E-95 group are:
R.M.Baltrusaitis,M.Binkley,B.Cox,T.Kondo,C.T.Murphy,
W.Yang,L.Ettliger,M.S.Goodman,J.A.J.Matthews and J.Nagy.

Members of the ISR-108 group are:
A.L.S.Angelis,H.J.Besch,B.J.Blumenfeld,L.Camilleri,
T.J.Chapin,R.L.Cool,C.del Papa,L.Di Lella,Z.Dinčovski,
R.J.Hollebeck,L.M.Lederman,D.A.Levintal,J.T.Linnemann,
C.B.Newman,N.Phinney,B.G.Pope,S.H.Pordes,A.F.Rothenberg,
R.W.Rusack,A.M.Segar,J.Singh-Sindhu,A.M.Smith,
M.J.Tannenbaum,R.A.Vidal,J.S.Wallece-Hadrill,J.M.Yelton,
and K.K.Young.

Members of the ISR-807 group are:
H.Gordon,R.Hogue,T.Killian,T.Ludlam,R.Palmer,D.Rahm,
P.Rehak,I.Stumer,M.Winik,C.Woody,O.Botner,V.Burkert,
D.Cockerill,C.W.Fabjan,T.Ferbel,A.HallgrenB.Heck,H.J.Hilke,
P.Jeffreys,G.Kesseler,H.J.Lubatti,I.Mannelli,W.Molzon,
B.S.Nielsen,L.Rosselet,E.Rosso,R.H.Schindler,D.W.Wang,
Ch.J.Wang,W.J.Willis,W.Witzeling,H.Bøggild:E.Dahl-Jensen,
I.Dahl-Jensen,G.Damgaard,K.H.Hansen,J.Hooper,R.Møller,
S.Ø.Nielsen,L.H.Olsen,B.Schistad,I.Akesson,S.Almehed,
S.Henning,G.Jarlskog,B.Lörstand,A.Melin,U.Mjörnmark,
A.Nilsson,M.G.Albrow,N.A.McCubbin,O.Benary,S.Dagan,
D.Lissauer and Y.Oren.

References

(1) M.Diakonou et al, Phys.Lett 87B,292 (1979) and
                Phys.Lett 91B,296 (1980).
(2) R.Baltrusaitis et al, Phys.Lett 88B, 372 (1979)
(3) A.Angelis et al, Phys.Lett 94B,106 (1980) and
                Phys.Lett 98B,115 (1981).
(4) F.Halzen et al,Phys.Rev.Lett 40,1117 (1978) and
     Proceedings of the XXth International Conference
     on High Energy Physics,Madison,Wisc.,pg.172
(5) A.Contogouris et al,Phys.Rev. D19,2607 (1980)
(6) R.Ruckl et al,Phys.Rev. D18,2469 (1978)
(7) L.Cormell et al,Florida State preprint,
                FSU-HEP-030780
(8)-M.Diakonou et al,Phys.Lett 87B,292 (1979)
    -W.Molzon,Moriond Workshop on Lepton Pair
                Production,Jan.81

    -Discussion meeting between experimentalists and
    theorists on ISR and Collider Physics,Series 2,
    Number 1, M.Jacob editor.
    -Direct photon production in Hadron Collisions,
    W.J.Willis,Talk presented at 4th Intern.Coloquium
    on Photon-Photon Interactions,Paris,8/4/81,
    CERN-EP / 81-45.

(9) J.T.Linnemann,Moriond Workshop on Lepton Pair
                Production,Jan.81
(10)C.Kourkoumelis et al,Phys.Lett 84B,277 (1979)
             and Phys.Lett 89B,432 (1980)
(11)R.L.Ford et al,SLAC-210,June 1978
(12)A.P.Contogouris et al,Mc Gill preprint 81-0723
(13)D.Cockerill et al,Physica Scripta 23,649 (1981)
(14)E.Amaldi et al,Phys.Lett 77B,240 (1978)

<u>FIGURE CAPTIONS</u>

Fig. 1. The corrected γ/π^0 ratio versus P_T, as measured by the Athens-BNL-CERN experiment for four different values of $\sqrt{s}$. The lines are from (12); dashed lines correspond to lowest order QCD calculation, solid lines include the soft gluon correction.

Fig. 2. The rapidity difference between charged tracks and trigger particles a) π^0; b) γ-candidate; c) pure γ. The rapidity difference between π^0 and trigger particles: d) π^0 ; e) γ-candidate; f) pure γ. The rapidity difference between π^0 of $P_T > 0.5$ GeV/c and trigger particle: g) π^0; h) γ-cadidate; i) pure γ. The curves are the expected distributions assuming uncorrelated particles.

Fig. 3. The azimuthal difference between π^0 and single photon candidates with π^0: a) $P_T \leq 0.5$ GeV/c; b) $P_T \gtrsim 0.5$ GeV/c.

Fig. 4. The combined R806 * R807 apparatus.

Fig. 5. Transverse view of: a) π^0 and b) γ event in the R807 drift chamber.

Fig. 6. Definitions of some correlation variables.

Fig. 7. The $\Delta\Phi$ distribution between neutral triggers with $P_T > 4.5$ GeV/c, and charged tracks with $|y| < 0.8$ and $P_T > 1.0$ GeV/c.

Fig. 8. The X_E distribution of charged tracks with $\Delta\Phi > 90^0$, and $|y| < 0.8$ for π^0 and direct photon cadidates with P_T 4.5 GeV/c.

Fig. 9. a) The ratio of the mean number of positive particles to the mean number of negative particles opposite direct photon candidates and π^0's as a function of X_E. Neutral triggers have $P_T > 4.5$ GeV/c and charged tracks have $|y| < 0.8$. b) The same distribution as in fig.9a, but for events satisfying a cut requiring the total momentum of charged tracks in the direction of the neutral trigger to be less than 20% of the trigger momentum.

Fig.10. The charge ratio distribution versus X_E as in Fig.9a, but for particles with $|\Delta\Phi| < 90^\theta$.

Fig.11. The CCOR apparatus.

Fig.12. The ratio of γ/ALL as a function of P_T for two $\sqrt{s}$ values as measured by the CCOR experiment.

Fig.13. The ratio of γ/ALL as a function of $X_T = P_T trig / \sqrt{s}$.

611

Fig.14. The γ/ALL ratio as a function of $P_T$trig for events where
the trigger was: a) accompanied by a particle in the trigger
hemisphere; b) not accompanied by a trigger in the trigger
hemisphere.

Fig.15. The charge ratio R for particles in the hemisphere opposite
to the trigger, as a function of X_E, for those events where
the trigger particle: a) did not convert and was not accompanied by a particle in the trigger hemisphere; b) did convert
and was accompanied by a particle in the trigger hemisphere.

Fig.16. The charge ratio R for $X_E > 0.3$ plotted as a function of
$P_T$trig for direct photons and neutral mesons.

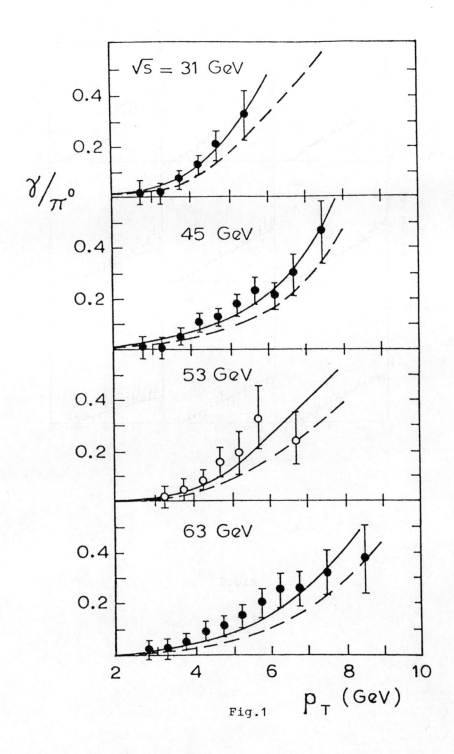

Fig.1

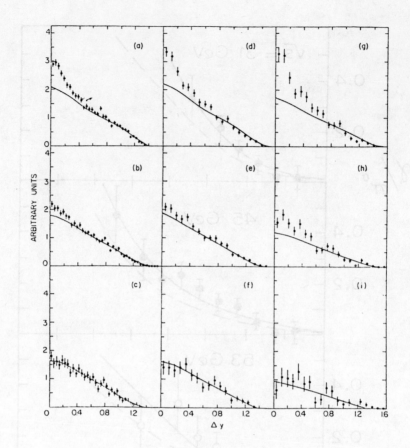

Fig.2

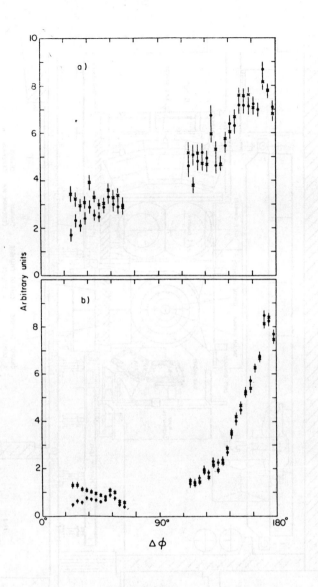

Fig.3

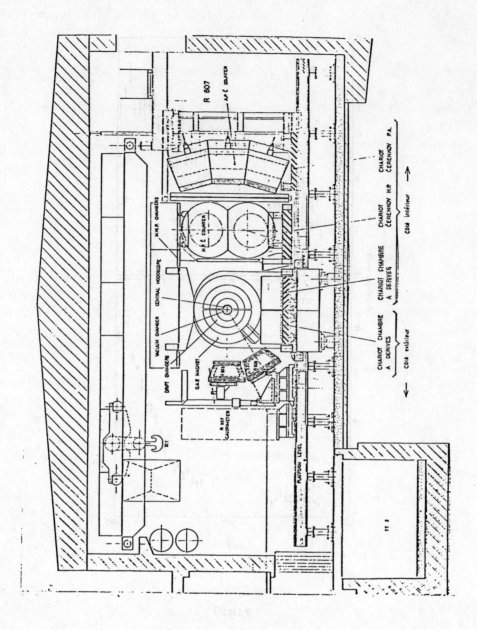

Fig.4

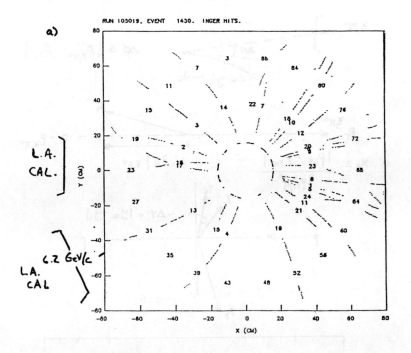

a) RUN 103019. EVENT 1430. INGER HITS.

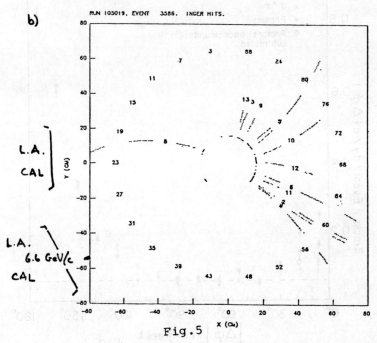

b) RUN 103019. EVENT 3386. INGER HITS.

Fig. 5

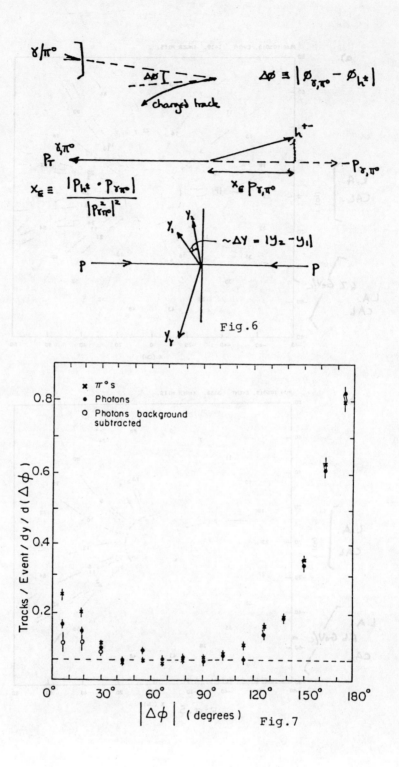

Fig.6

$\Delta\phi \equiv |\phi_{\gamma,\pi^0} - \phi_{h^\pm}|$

$x_E \equiv \dfrac{|P_{h^\pm} \cdot P_{\gamma\pi^0}|}{|P^2_{\gamma\pi^0}|}$

$\sim\Delta Y = |y_2 - y_1|$

Fig.7

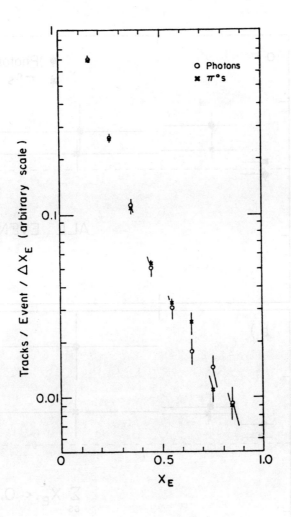

Fig.8

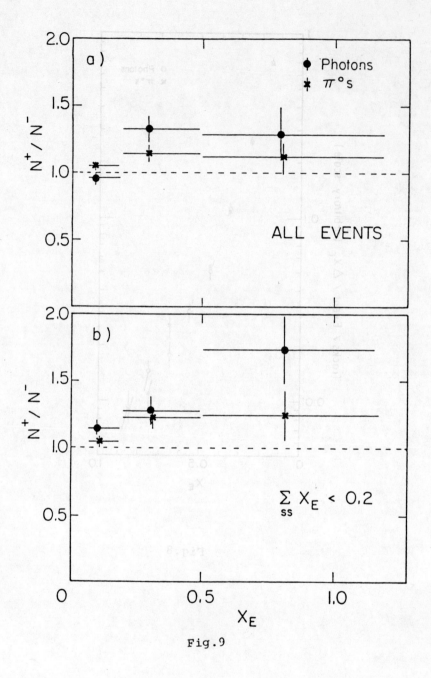

Fig.9

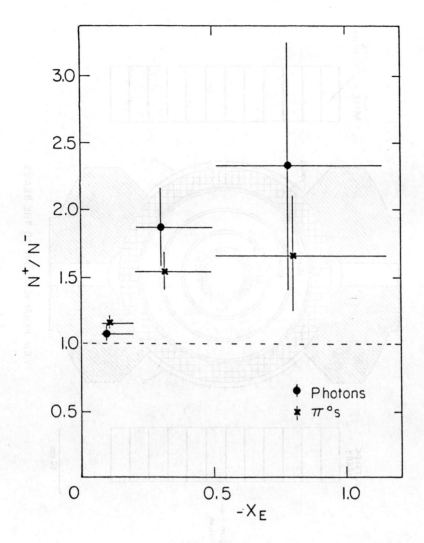

Fig.10

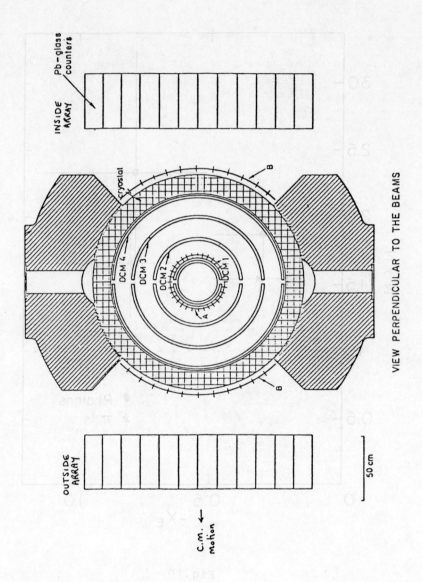

INSIDE ARRAY

Pb-glass counters

cryostat

DCM 4
DCM 3
DCM 2
DCM 1

B

B

A

VIEW PERPENDICULAR TO THE BEAMS

OUTSIDE ARRAY

50 cm

c.m. Motion

Fig.11

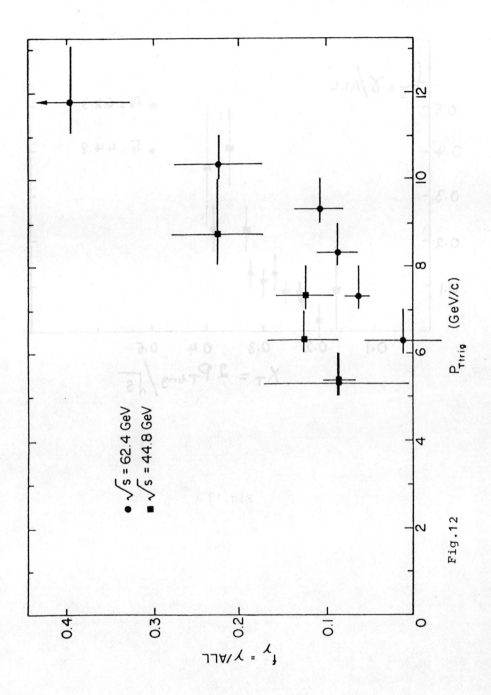

Fig.12

623

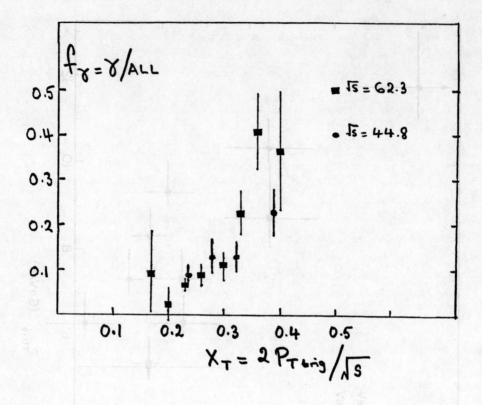

Fig.13

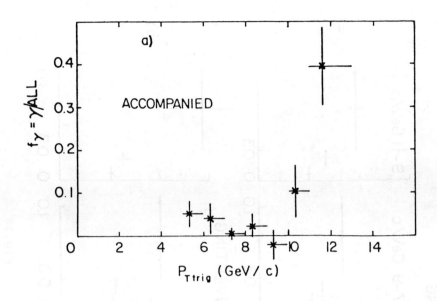

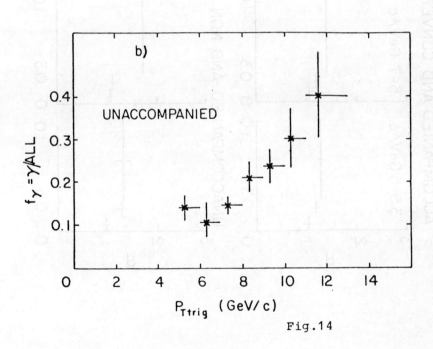

Fig.14

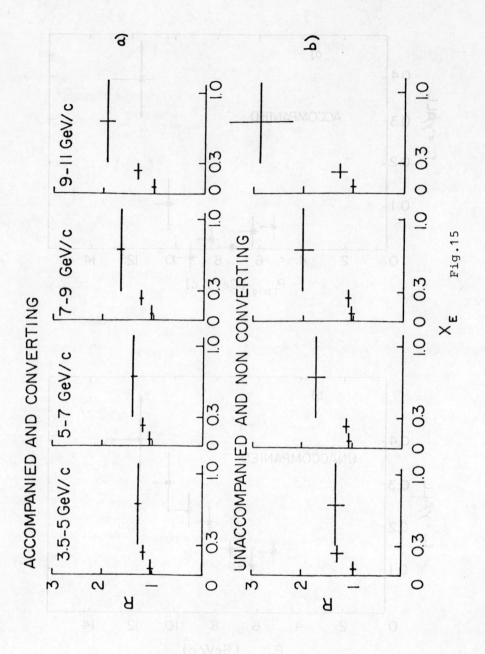

Fig.15

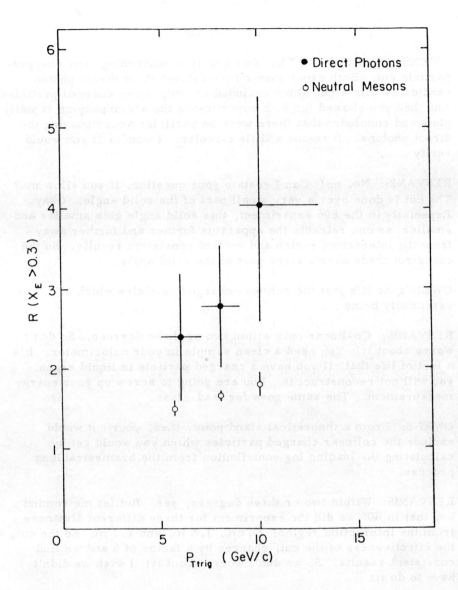

Fig.16

DISCUSSION

OWENS, FLORIDA STATE: One question concerning your charged-particle cut. Both experiments have defined their direct photon candidates with a cut which excluded accompanying charged particles. And then you showed for both experiments the accompanying rapidity plots and concluded that there were no particles accompnaying the direct photons. It seems a little circular. I wonder if you would verify . . . .

RESVANIS: No, no! Can I restate your question, if you allow me? The cut is done over a very small part of the solid angle. Okay. Especially in the 806 experiment, this solid angle gets smaller and smaller as one retracts the apparatus further and further away from the interaction region and we find consistent results. So the cut is not made over a large part of the solid angle.

OWENS: So it's just the colinear charged particles which are essentially being . . . .

RESVANIS: Co-linear only within two or three degrees. So don't worry about it! You need a clean sample in your calorimeter. It's a fact of life that, if you have a charged particle in liquid argon, you will not reconstruct it. You are going to screw up your energy measurement. The same goes for lead-glass.

OWENS: From a theoretical stand point, then, your cut would exclude the colinear charged particles which you would get by calculating the leading log contribution from the bremsstrahlung process.

RESVANIS: Within two or three degrees, yes. But let me remind you that in 806 we did the experiment for three different distances from the interaction region: 60 cm, 1.6 m, and 2.2 m. So the cut, the effectiveness of the cut, changes by a factor of 8 and we find consistent results. So we don't worry about it! I wish we didn't have to do it.

WALKER, DUKE: I just wanted to ask a question about background in the case of the lead-glass experiment. Lead glass is a great detector for the hadrons as well as photons. For example, neutrons or K^o's, several of which could apparently produce a large signal in the lead glass. Is that a possibility in that experiment?

RESVANIS: My personal feeling is to agree with you. That is why we used liquid argon and not lead glass. But it's not my experiment to defend. I think your comment is a good one. It may be there.

GUNION, DAVIS: In the theoretical curve (I think Contogouris will say something about this in his talk) there was substantial disagreement we saw between, say, Jeff Owen's prediction . . . .

RESVANIS: Which theoretical curve? No, its lying on the data. Should I show it to you?

GUNION: Yes, could you put it up again. It looked like they were different.

RESVANIS: (Shows curve and data.) Yes, they are bang on the data.

GUNION: Oh. Okay. The next question is whether or not these include the p_T smearing in the usual on-shell way . . . . The curves look . . . .

RESVANIS: No, no, no! That's two calculations by Contogouris: one with strong glue, the other with soft glue, weak glue. Okay? This is the difference between these.

GUNION: But anyway the claim is that the p_T smearing really doesn't have much effect in this instance.

OWENS: It is much smaller because of the flatter distribution than you have for π^o's.

GUNION: That was my question.

RESVANIS: Remember, Jack, this is a log plot.

Can I add one thing to Walker's question? I believe the CCOR experimenters have made calculations and tests with a model at the PS for the neutron background and find it to be small.

MESTAYER, CHICAGO: Did you use the longitudinal information on the shower development in the liquid argon to define gammas?

RESVANIS: Yes, in fact we used it at two levels. One is at the trigger level. It's a very crude cut but it really buys us a lot of mileage because we reject most hadronic interactions on-line, essentially. And then we do a detailed cut off-line, again using the EGS program. My advice, if you want it, to anybody that works with showers is to use the EGS Monte Carlo program. It's fantastic! [Inaudible question about computer time.] That is a good question. If you have free computer time then it's cheap.

PRELIMINARY RESULTS FOR HIGH X_t π^0 PRODUCTION AND CHARGED PARTICLE CORRELATIONS FOR HADRONIC INTERACTIONS WITH NUCLEAR TARGETS

Bari-Brown-Fermilab-MIT-Warsaw Collaboration
Presented by P. H. Garbincius,
Fermilab, Batavia, Illinois, 60510

INTRODUCTION

At Fermilab energies, the yields of single particles, di-hadrons, or jets produced at high P_t have shown an unexpected and still unexplained dependence on the atomic number of the target nucleus.[1,2,3] These cross sections exhibit a power law dependence on the atomic number, A^α, where this power α increases to values above $\alpha = 1$ at high P_t. This is to be contrasted to the absorption[4] and beam fragmentation[5] cross sections which behave as $A^{3/4}$ and $A^{2/3-1/2}$ respectively. In addition, for total cross section triggers, there is a large increase in the mutliplicity in the target fragmentation region[6] with increasing A.

This experiment was designed to further explore the cause of this anamolous A-dependence by triggering on high P_t π^0's and studying the multiplicities and angular distributions of the correlated charged particles. The hadron jet recoiling against the trigger π^0 also would provide data on the propagation and hadronization of free constituents passing through nuclear matter.

APPARATUS

This experiment was performed in the Fermilab M6 beamline[7] using a 100 GeV π^+, k^+, and p beam. The apparatus is shown in Figure 1. The beam was incident upon a series of carbon, copper, and lead targets. Two thicknesses for each element were used to perform an extrapolation to zero target thickness. The maximum thickness for C, Cu, and Pb was 0.044, 0.045, and 0.019 absorption lengths, respectively. A 20 inch liquid hydrogen target and an empty target holder completed the array.

The π^o's were detected by[8] their two gamma decay mode in a shower detector centered at 77.5^o in the center of mass frame. Lead glass blocks determined the energy, while scintillator hodoscopes at shower maximum determined the position. The trigger consisted of a coincidence between a beam particle and a count in the wedge counter placed at shower maximum which formed an analog P_t signal directly. The major corrections to the raw rate involved the momentum dependent acceptance of the array and the trigger efficiency as a function of P_t.

The associated charged particles were detected in a non-magnetic vertex detector[9] consisting of multi-wire proportional chambers, and scintillator and plastic Cerenkov hodoscopes. The correlation data must be considered very preliminary in that not all hardware and physics corrections have been applied at this time. However, from past experience with the detector, we estimate that the final absolute multiplicities will not change by more than ± 10% from data presented here.

INCLUSIVE π^o DATA

Data extended in P_t out to 3.2 GeV/c which represents $X_t = 0.46$ at 100 GeV. This is well within the kinematic region where $\alpha \geq 1$. Typical cross sections for $h A \to \pi^o X$ are presented in Figure 2. The hydrogen data is well represented by the parametrization of Reference 10 (solid line) or by $\exp(-4P_t)$. The nuclear target data are fit well by $\exp(-5P_t)$. Typical A-dependence is shown in Figure 3 along with fits of the form A^α where the hydrogen point was not included in the fit. The dependence of α as a function of P_t for π^+, k^+, and p beams is presented in Figure 4. The dotted line represents the similar 400 GeV data of Reference 1 for $pA \to \pi^- X$.

CORRELATION DATA

A typical pseudo-rapidity distribution is presented in Figure 5. These show an enhancement (depletion) in the backward (forward) multiplicity with increasing A. In addition, a correlation peak is observed at the trigger π^o pseudo-rapidity.

The total multiplicities for each target are independent of P_t and increase slightly more rapidly

with A or $\bar{\nu}$ than those[5] for total cross section[6] or
fragmentation triggers.[5] Figure 6 shows the backward (n
$\leq$ 1.4) multiplicities. The high P_t triggered
multiplicity increases much more rapidly than that for
the fragmentation or total cross sections showing a
distinct correlation with the α-parameter for these
interactions.

The excess away side correlation in the central
region $1.9 \leq \eta \leq 4.3$, for $|\phi - \phi\pi^0| > 90^\circ$, is presented in
Figure 7 as a function of P_t for different nuclei. The
P_t slopes for each target are consistent with a single
universal curve.

DISCUSSION AND CONCLUSIONS

For high P_t triggers, π^0 production for π^+, k^+, and
p-nucleus interactions behave as A^α with α similar to
that of π^+ or π^-, di-hadron, or jet production[1,2,3]. The
total multiplicities are independent of P_t but show a
strong growth with A in the target fragmentation region
indicating that more than a single nucleon participates
($\alpha > 1$) in the high P_t reaction, either directly through
multiple scattering or indirectly through nuclear
excitation and subsequent break-up. The recoil quark
jet multiplicity increases with P_t but is independent
of A indicating that the characteristic hadronization
length for a free quark is long compared to the
transverse size of the nucleus.

REFERENCES

1. J.W. Cronin, et al., Phys. Rev. $\underline{D11}$, 3105 (1975) and D. Antreasyan, et al., Phys. Rev. $\underline{D19}$, 764 (1979).

2. R.L.McCarthy, et al., Phys. Rev. Lett. $\underline{40}$, 213 (1978).

3. C. Bromberg, et al., Nucl Phys. $\underline{B171}$, 1, (1980) and C. Bromberg, et al., Nucl. Phys. $\underline{B171}$, 38 (1980).

4. A.S. Carroll, et al., Phys. Lett. $\underline{80B}$, 319 (1977).

5. D. Barton, Proc. XI Int. Symp. Multiparticle Dynamics, Bruges, (1980). P.H. Garbincius, A.I.P. Conf. Proc. $\underline{68}$, 77 (1981), (XX ICHEP, Madison, 1980).

6. J.E. Elias, et al., Phys. Rev. $\underline{D22}$, 13 (1980).

7. D.S. Ayres, et al., Phys. Rev. $\underline{D15}$, 3105 (1977).

8. P.H. Garbincius, et al., IEEE $\underline{NS-27}$, 79 (1980).

9. E.F. Anelli, et al., Rev. Sci. Instrum. , $\underline{49}$, 1054 (1978).

10. G. Donaldson, et al., Phys. Rev. Lett. $\underline{36}$, 1110 (1976).

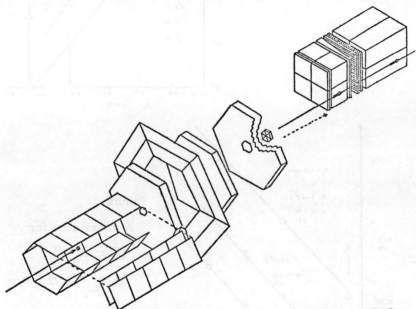

Figure 1. Apparatus consisting of nuclear target, lead glass
scintillator shower detector, and non-magnetic
vertex detector.

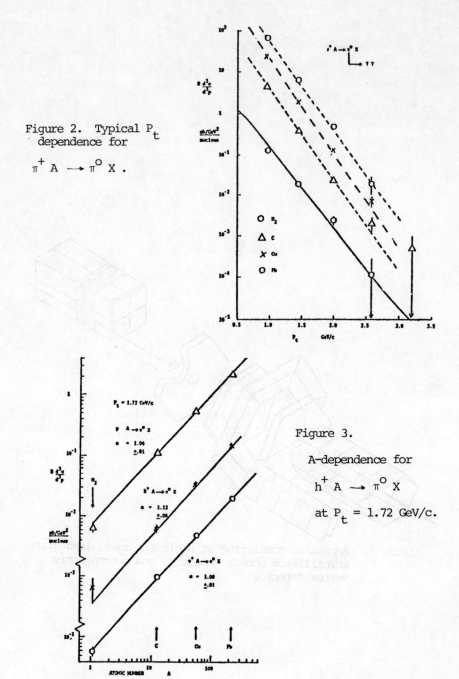

Figure 2. Typical P_t dependence for

$$\pi^+ A \longrightarrow \pi^o X .$$

Figure 3.

A-dependence for

$$h^+ A \longrightarrow \pi^o X$$

at $P_t = 1.72$ GeV/c.

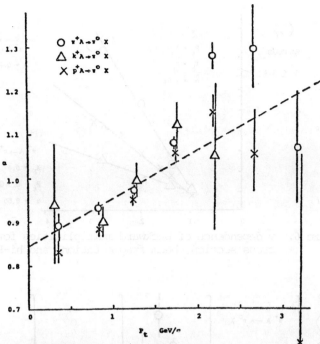

Figure 4. P_t dependence of α parameter.

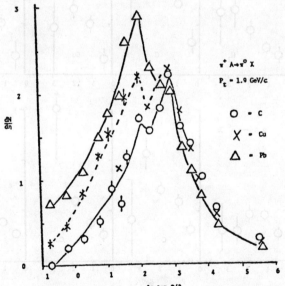

Figure 5. Typical pseudo-rapidity distributions for
$\pi^+ A \longrightarrow \pi^0 X$ at P_t = 1.9 GeV/c.

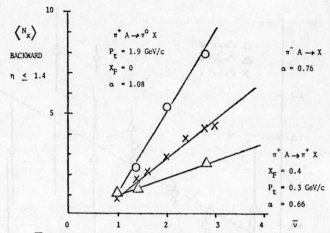

Figure 6. $\bar{\nu}$ dependence of backward multiplicities for total
cross section, beam fragmentation, and hi-P_t triggers.

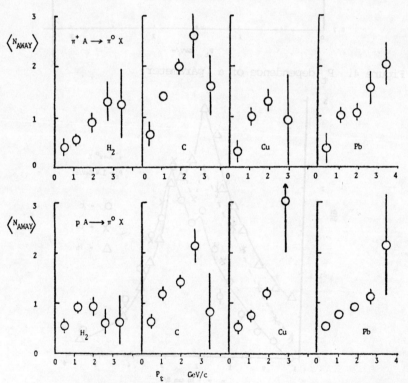

Figure 7. Excess away side multiplicity correlation dependence.

K-FACTORS IN LARGE-p_T DIRECT γ AND $\ell^+\ell^-$ PRODUCTION[(+)]

A.P. Contogouris, S. Papadopoulos and J. Ralston
Department of Physics, McGill University, Montreal, Canada

ABSTRACT

For the QCD subprocess $q + g \rightarrow q + \gamma$ that dominates direct photon and dilepton production at large transverse momenta in p-p collisions, we study perturbative corrections to order α_s^2. The calculation is performed in the soft gluon limit, and results in a large K-factor multiplying the Born term, which significantly improves the agreement with data on $p + p \rightarrow \gamma + X$.

INTRODUCTION

One year ago we discussed the importance of large-p_T direct photon production as a test of QCD[1]. Fig. 1 is a reproduction of the Proceedings[1] and presents data for $p + p \rightarrow \gamma + X$ of the A2BC Collaboration[2] against theoretical predictions. This reaction is known to be dominated by the QCD subprocess

$$q + g \rightarrow q + \gamma \tag{1}$$

When theory uses a weak glue distribution inside the proton the predictions fall utterly below the data (dashed lines). But even when a strong glue, like

$$F_{g/p}(x, Q_o^2) = 0.866 \ (1 + 9x)(1-x)^4 \tag{2}$$

is used, the predictions fall significantly below (solid lines). Several other independent calculations have reached the same conclusions[3][4].

This suggests that a K-factor due to higher order QCD corrections, like that found in the Drell-Yan dilepton cross-section[5], may be in operation here as well. Notice that, apart from the kinematic boundaries of this reaction, most of the K-factor correction arises from effects due to <u>soft</u> gluons (i.e. of 4-momentum $\rightarrow$ 0). As we discuss at the end, the same holds in other reactions, as well.

SOFT GLUON TECHNIQUE AND RESULTS

We have studied $0(\alpha_s^2)$ corrections to (1) due to soft gluon bremsstrahlung and loop contributions[6]. Some of the relevant graphs are shown in Fig. 2 (solid lines: quarks, dashed: hard gluons, dotted: soft gluons, curly: photons). We have used a soft gluon technique, which can be described as follows:

Consider first one of the lowest order terms ($\equiv$Born terms), e.g. corresponding to the graph a of Fig. 2:

$$|a|^2 \sim T_{Born}/[(p_1 + p_2)^2]^2$$

where T_{Born} is a trace. Now consider the contribution to the bremss subprocess $q + g \to q + g + \gamma$ due e.g. to the interference terms of graphs A, B. This has the form

$$Re(AB\star) \sim T(k)/[(p_1 + p_2 - k)^2]^2 (p_1 - k)^2(p_2 - k)^2 \qquad (3)$$

where $T(k)$ is another trace[6]. The contribution of (3) to the cross-section σ ($\equiv q^0 d\sigma/d^3q$) of (1) is obtained by integrating over phase space:

$$\sigma \sim (\frac{\alpha_s}{\pi})^2 \int \frac{d^{n-1}k}{|\vec{k}|} Re(AB\star) \qquad (4)$$

Now, take the soft gluon limit ($k \to 0$): $T(k) \to T(0)$ and, after manipulating Dirac matrices, it turns out that $T(0) \sim s \, T_{Born}$. Moreover:

$$(p_1 + p_2 - k)^2 \to (p_1 + p_2)^2 \equiv s = \text{large} \quad ;$$

the other two propagators in (3) are left unchanged. In this way the analytic structure of the amplitudes at $k \to 0$ (infrared region) remains unaffected. Furthermore, it is clear that for $k \to 0$:

$$Re(A\overset{\star}{B}) \sim s \, |a|^2/(p_1 - k)^2(p_2 - k)^2$$

and the integration in (4) is easy. Finally we obtain a correction

$$\sigma_{brems} \sim \frac{\alpha_s}{2\pi} \sigma_a \, \log^2 s \qquad (5)$$

where σ_a is the contribution of the Born graph a to the cross-section of (1).[a] The factor $\log^2 s$ and the overall structure of (5) are typical of infrared contributions.

The contribution (5) must be added to that of the interference term $Re(A'\overset{\star}{a})$ of the loop graph A' with the graph a (Fig. 2); this is required by the well-known Bloch-Nordsieck procedure of cancellation of infrared divergences. Now

$$Re(A'\overset{\star}{a}) \sim T'(k)/k^2(p_1 - k)^2(p_2 + k)^2[(p_1 + p_2)^2]^2 \qquad (6)$$

and this loop term contributes to the cross-section of (1):

$$\sigma' \sim (\frac{\alpha_s}{\pi})^2 \int d^n k \, Re(A'\overset{*}{a}) \qquad (7)$$

In the soft gluon limit $T'(k) \to T'(0) \sim sT_{Born}$, so that $Re(A'\overset{*}{a})$ becomes also proportional to the Born term $|a|^2$. The remaining integration is easy and finally leads to the correction

$$\sigma_{loop} \sim - \frac{\alpha_s}{2\pi} \sigma_a \, Re \, log^2(-s) \qquad (8)$$

again a factorized form. Now, $Re log^2(-s) = log^2 s - \pi^2$ so that in the sum of (5) and (8) the terms $log^2 s$ cancel but a π^2 remains:

$$\sigma_{brems} + \sigma_{loop} \sim \frac{\alpha_s}{2\pi} \sigma_a \pi^2 \qquad (9)$$

clearly, this gives rise to a large correction[5].

Notice that some of the loop contributions correspond to a box amplitude (Fig. 2). This involves an integral of the typical form

$$\int d^n k \, \frac{N(k)}{k^2(p_1 - k)^2(p_2 + k)^2(p_2 - q + k)^2}$$

In the soft gluon limit, $N(k) \to N(0)$ (= some simple combination of Dirac matrices and external momenta), and $(p_2 - q + k)^2 \to (p_2 - q)^2$; for large transverse momenta q_T, $(p_2 - q)^2$ is large in absolute value. Again, the analytic structure at $k \sim 0$ is preserved and the remaining integration reduces to that of (6).

In the present case there are two sets of graphs that produce π^2 terms: one involving 3-gluon couplings (like A and A'; non-Abelian graphs) and another involving only q - g couplings (like B; QED-like graphs). Taking into account color factors the first set gives a factor $\sim (N_c/2)\pi^2$ and the second $\sim (C_F-N_c/2)\pi^2$, where in color SU(3): $N_c = 3$, $C_F = 4/3$. Adding these contributions together with that of the Born graphs we finally obtain the following total cross-section for (1)[7]:

$$\sigma = (1 + \frac{\alpha_s}{2\pi} (N_c - C_F)\pi^2) \sigma_{Born} \qquad (10)$$

Thus for the transverse momenta of the present data the soft gluon $O(\alpha_s^2)$ effects make a correction of 60 $\sim$ 80%.

In Fig. 1 we present γ/π° calculated on the basis of Eq.(10) (dash-dotted lines). The input glue is of the form (2); for further details, such as partons' intrinsic k_T etc. see Ref. 1. Clearly our correction significantly improves the agreement with the A^2BC data.

Eq. (10) gives also the corresponding K-factor for the

transverse momentum distribution $d\sigma/dq^2 dq_T^2 dy$ for dileptons produced in p-p collisions, where again the QCD subprocess (1) is dominant. Such a K-factor seems also to be required for a better agreement with large-q_T data[8].

CONCLUDING REMARKS

Soft gluon effects are, of course, only part of the $O(\alpha_s^2)$ corrections to (1). However, such effects were often shown to provide the bulk of the correction. Apart from the Drell-Yan cross-section [5], another important case is $e^+e^- \to 3$ jets[9], of which the Born term is $\gamma \to \bar{q} + q + g$; again, in much of the kinematic range, the bulk of the $O(\alpha_s^2)$ correction is a π^2- term, like in Eq. (10)[9][6]. Still another is the unintegrated p_T-distribution for single lepton production in hadron + hadron $\to \ell^\pm + X$[10]. It is of interest to see whether this is the case for the subprocess (1), as well.

On the other hand, these soft gluon corrections have certain important properties. One is that they are expected to occur in any of the usual renormalization schemes[11]. Another is that, as it is argued[12], they can probably be summed to all orders of α_s leading to an exponential factor. If this is true, the soft gluon part of the corrections will be under control.

REFERENCES AND FOOTNOTES

(+) Also supported by the Natural Sciences and Engineering Research Council of Canada and the Quebec Department of Education.
1. Proceedings of XI International Symposium on Multiparticle Dynamics, Bruges (1980), A.P. Contogouris, S. Papadopoulos, C. Papavassiliou p. 497; also Nucl. Phys. B179, 461(1981).
2. M. Diakonou et al (Athens²-Brookhaven-CERN Collaboration), Phys. Lett. B87, 292; B91, 275(1980).
3. M. Kienzle-Focacci, Univ. of Geneva D6-100(1980); O. Benary et al, Tel Aviv TAUP-887-90; J. Owens, these Proceedings (1981).
4. For recent reviews on large-p_T direct photons see L. Resvanis, these Proceedings (1981) and W. Willis CERN-EP.81-45.
5. J. Kubar-André, F. Paige, Phys. Rev. D19, 221(1979); G. Altarelli, R. Ellis, G. Martinelli, Nucl. Phys. B157, 461(1979); B. Humpert, W. Van Neerven, Phys. Lett. B89, 69(1979).
6. A.P. Contogouris, S. Papadopoulos, J. Ralston, Phys. Letters B (in press) and McGill preprint (in preparation). We work in Feynman gauge with massless partons and use dimensional regularization ($n = 4 - 2\epsilon$ dimensions, $\epsilon < 0$).
7. In Eq. (10), in defining the correction we take differences between our subprocesses (Fig. 2) and the corresponding deep inelastic subprocesses related by crossing[6].
8. This follows from calculations like e.g. of Halzen and Scott, Phys. Rev. D18, 3378(1978) if the quark and, in particular, the gluon distribution include scale violations.

9. R. Ellis, D. Ross, A. Terrano, Nucl. Phys. B178, 421(1981).
10. P. Aurenche, J. Lindfors, CERN TH. 2992(1980).
11. W. Celmaster, D. Sivers, Phys. Rev. D23, 227(1981).
12. J. Cornwall, G. Tiktopoulos, Phys. Rev. D13, 3370(1976);
    J. Frenkel, J.C. Taylor, Nucl. Phys. B116, 185(1980); G. Parisi,
    Phys. Lett. B90, 295(1980).

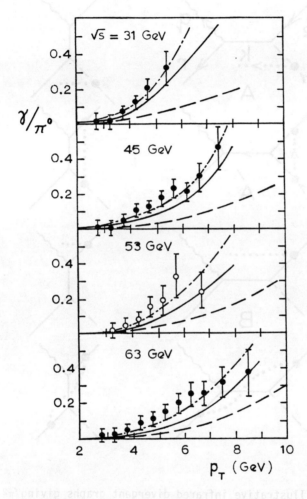

Fig. 1. Predictions for the inclusive γ/π° ratio (θ_{cm} = 90°)
compared with the Athens[2]-Brookhaven-CERN data. Dashed line:
lowest order with weak gluon distribution; Solid line: lowest
order with strong gluon distribution; Dashed-dot: strong glue
including the correction (10). In calculating γ/π° we use π°
data of A[2]BC.

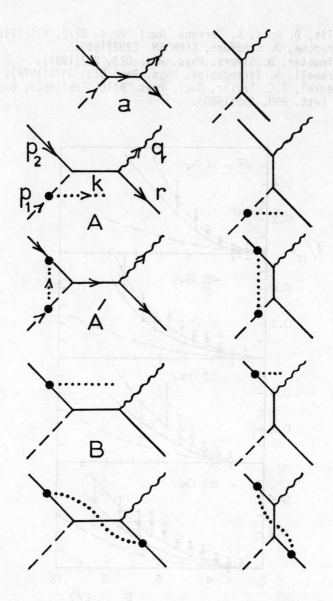

Fig. 2. Illustrative infrared divergent graphs giving π^2 at $O(\alpha_s^2)$ in Feynman gauge. Dashed lines are gluons (dotted in the soft limit).

DISCUSSION

GUNION, DAVIS: I'm still trying to get straight. In the γ/π ratio if you add something to the γ without adding something similar to the pion then of course you increase the ratio. But what about the absolute cross section comparison? Does the k factor make it alright or not?

CONTOGOURIS: We calculate the inclusive production of direct photons either way, soft glue . . . weak glue, soft gluon emission, and so on, and divide by the experimental π^0 result. Now, whether this experimental π^0 comes as a result of soft gluon corrections or other higher order corrections like the ones Owens discussed before, you see these are matters which we leave out completely.

GUNION: So, then, is the statement that the absolute cross section agrees with the absolute direct photon cross section now?

CONTOGOURIS: The inclusive cross section for the process $pp \rightarrow \gamma +$ anything, for example, is going to increase due to soft gluons. This is part of the process that is not going to be cancelled exactly.

GUNION: But the normalization of the cross section, just the cross sections directly, is now comparable to calculation?

CONTOGOURIS: Comparable . . . with the strong gluon.

De GRAND, SANTA BARBARA: You don't have to be a theorist to worry when somebody shows a calculation where the first-order correction is 50% because then you start worrying how big the second-order correction is. What is the status of what people expect will happen when I go to high orders? Does anyone know?

CONTOGOURIS: The reason, in fact, we are very much interested in this particular kind of correction is that there is much evidence that it exponentiates.

De GRAND: How reliable are calculations which claim to exponentiate?

CONTOGOURIS: With respect to Drell-Yan, there is proof including order α^2. There are several proofs. One is by Kripfganz. I have a student at McGill. There are several. Now there are several arguments based essentially on the fact that soft photon effects exponentiate in QED. And there is a paper of Parisi and several other people, you see. I don't really have any complete proof of this part, but it is generally hoped that there is an exponentiation at least of these π^2 terms which I have been discussing and that is why I am interested.

OWENS, FLORIDA STATE: Just a comment to put this talk into appropriate perspective with the results I presented earlier. If in fact the exponentiation of π^2 terms holds, then this part of the correction would appear to be under control, and the comments I made earlier would indicate that the residual higher-order corrections could in fact be made small by appropriate choice of prescription, Q^2 dependence and so on. And in that case, one would have even smaller residual corrections than what I indicated. So, in fact, the two results go hand in hand here.

SESSION F

e^+e^- INTERACTIONS

AND

COMPARISON WITH HADRONIC INTERACTIONS

Thursday, June 25, p. m.

Chairman: S. Ratti
Secretary: J. M. Bishop

RECENT DEVELOPMENTS AT PETRA

by

Sau Lan Wu

Department of Physics, University of Wisconsin, Madison, Wisconsin, USA [+]

and

Deutsches Elektronen Synchrotron DESY, Hamburg, Germany.

Abstract

Recent experimental data are reviewed from the five detectors at PETRA: CELLO, JADE, MARK J, PLUTO, and TASSO. The many interesting new results include, for example, the copious production of baryons, especially Λ and $\bar{\Lambda}$, the study of the properties of the gluon and the observation of high $p_\perp$ jets in photon-photon physics.

+ Supported by the US Department of Energy, Contract EY-76-C-02-0881.

1. INTRODUCTION

At present there are five detectors at PETRA: CELLO, JADE, MARK J, PLUTO and TASSO. All five detectors have been operating well, and four of them - CELLO, JADE, MARK J and TASSO - are taking data at present. The fifth one, PLUTO, will be moved into the beam for the second time in September 1981, equipped with a new forward detector specially built for two-photon physics. There is therefore a great deal of recent data to be reported here, and it is necessary to be brief on each piece of new data. The topics to be covered are:

1. PETRA Status,
2. Baryon Production,
3. Lifetimes of the τ and Bottom Meson,
4. Jet Analysis,
5. Properties of the Gluon,
6. Two-Photon Physics,
7. Electroweak Interaction.

2. PETRA STATUS

PETRA (Positron-Electron Tandem Ring Accelerator) has been operating for nearly three years and performing very well. The most recent improvement was the addition, in March 1981, of focusing quadrupole magnets in so called "mini-beta" sections [1]. This name is derived from the purpose of these quadrupoles to reduce the beta values at the intersection points, because luminosity is inversely proportional to the geometric mean of the horizontal and vertical beta values. Steffen [2] first pointed out how this reduction of β can actually be accomplished at PETRA, and by now the experimental increase in luminosity is in accurate agreement with this theoretical evaluation.

The β values are shown in Table 1, for pre-mini-beta, the present mini-beta scheme, and also a proposed future mini-beta II by Steffen, Voss and Wolf [3]. Theoretical increases in luminosity partially based on the formula luminosity α $(\beta_{vertical}\ \beta_{horizontal})^{-1/2}$ are also shown. The increase in luminosity (nb^{-1} per day) is plotted in Fig. 1 for the period of March and April 1981 when the mini-beta scheme was working for the first time at PETRA. So far the best luminosities are as follows:

luminosity = 1.7×10^{31} cm^{-2} sec^{-1}

integrated luminosity = $\begin{array}{l} 790\ nb^{-1}/day\ (PETRA\ ungated) \\ 600\ nb^{-1}/day\ (TASSO\ gated) \end{array}$

Table 1 - Mini-beta schemes and increases in luminosity

	Pre-mini-beta	Mini-beta	Mini-beta II
$\beta_{vertical}$ (cm)	20	8	1.14
$\beta_{horizontal}$ (cm)	190	120	51
Luminosity ratio	1	3 *	11 *

One of the major problems that had to be solved before the success of the present mini-beta scheme is the reduction of the distance between the interaction point to the nearest quadrupoles. This reduction from 7.5 m to the present 4.45 m is achieved by eliminating the compensating magnets in TASSO and JADE. This is possible because TASSO and JADE can compensate each other.

In mini-beta II [3], the beta values are further reduced to obtain another increase in luminosity of almost a factor of 4. In this

---

* These numbers include the additional improvement due to better control of orbit, vertical dispersion and ring symmetry which gives smaller vertical beam sizes.

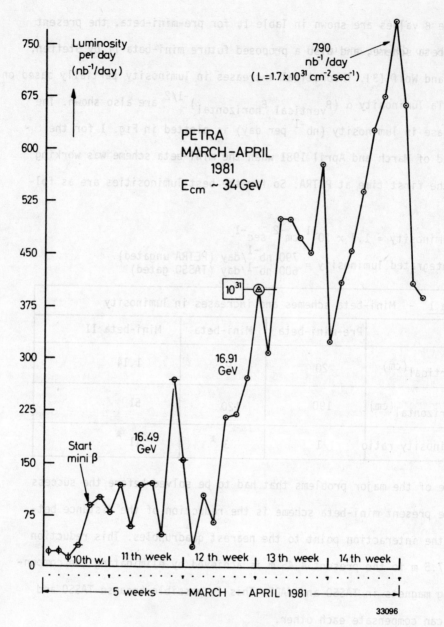

Luminosity per day (nb⁻¹/day)

$\frac{790}{nb^{-1}/day}$ (L = 1.7 × 10³¹ cm⁻² sec⁻¹)

PETRA
MARCH – APRIL
1981
$E_{cm} \sim 34$ GeV

10^{31} ⊕

16.91 GeV

16.49 GeV

Start mini β

10th w. | 11th week | 12th week | 13th week | 14th week

5 weeks —— MARCH – APRIL 1981

33096

Fig. 1 - Luminosity per day at PETRA during the
four weeks period after the start of mini β.

proposed scheme, iron-free superconducting quadrupoles similar to the Fermilab design are used: they are 1.7 m long, 4 cm aperture radius, and 10 cm outer radius, and are placed $\sim$ 1 m away from the interaction point. A major problem for this future design is to allow space for small-angle luminosity monitors and two-photon forward detectors. The scheme is shown [3] in Figure 2, and the target date for its installation is the end of 1982 or early 1983.

The other major future improvement for PETRA is the increase in energy which is underway. It is planned to double the rf power from 4 MW to 8 MW in the spring of 1982 so that the total center-of-mass energy $\sqrt{s}$ is increased from the present 36.7 GeV to about 41 GeV. In 1983, it is further planned to install additional rf cavities in the remaining two straight sections so that $\sqrt{s}$ is increased to 45 GeV. The plan after 1983 is less certain: the general plan is to reach even higher energies by using superconducting rf cavities.

3.   BARYON PRODUCTION - TASSO, JADE

One of the recent excitements at PETRA is the observation of the copious production of baryons. About a year ago, TASSO observed the inclusive production of protons and antiprotons [4]. More recently, JADE [5] confirmed the inclusive antiproton spectrum to about 1 GeV/c and also observed the inclusive anti-Λ spectrum to about 1.4 GeV/c, while TASSO [6] obtained the Λ and $\bar{\Lambda}$ spectrum all the way up 10 GeV/c in momentum.

In JADE, $\bar{p}$ is identified by dE/dx in the drift chamber, while $\bar{\Lambda}$ is identified through the decay mode $\bar{\Lambda} \to \bar{p} \, \pi^{+}$. To discriminate against $K_s \to \pi^{+}\pi^{-}$, the momentum of the π^{+} is required to be less than 40% of the momentum of the $\bar{p}$. In the invariant mass plot, the $\bar{\Lambda}$ peak is seen very clearly. In TASSO, the p and $\bar{p}$ are identified by time-of-flight,

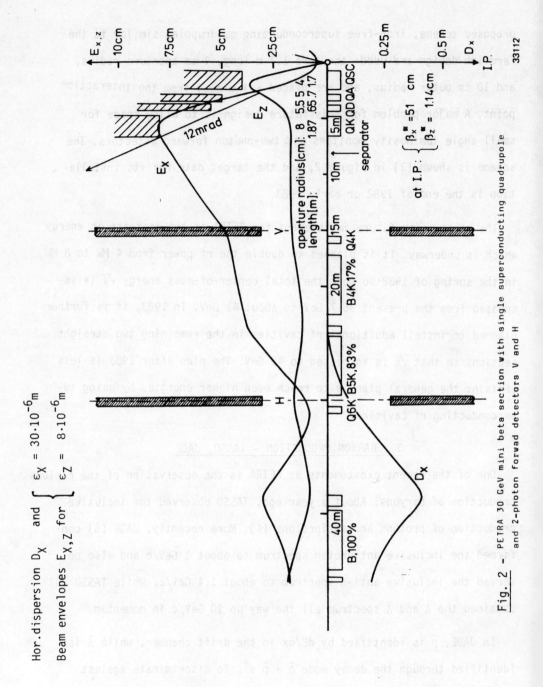

Fig. 2 — PETRA 30 GeV mini beta section with single superconducting quadrupole and 2-photon forwad detectors V and H

while the Λ and Λ̄ are identified by vertex fits to oppositely charged pairs, where the higher-momentum particle is taken to be the p or p̄. Two different methods [6] of analysis are used for Λ and Λ̄, with method 1 emphasizing the lower momentum Λ and Λ̄ and method 2 the higher momenta ones. Invariant mass plots from method 1 are shown in Fig. 3. The experimental results from both TASSO and JADE are summarized in Fig. 4 together with a comparison with the K^0 yield.

In Table 2, the results of JADE and TASSO are compared. While the results on p̄ are barely compatible, those on Λ̄ are in some disagreement. This may be due to the extrapolation procedure used by JADE. For the extrapolation of p̄ and Λ̄ to all momenta the shape predicted by the LUND model [7] has been used by JADE. Note that the observed p and p̄ include those due to the decay of Λ and Λ̄. Another way to express the TASSO results on Λ and Λ̄ is.

$$R_{\Lambda+\bar{\Lambda}} = \frac{\sigma(e^+e^- \to \Lambda x) + \sigma(e^+e^- \to \bar{\Lambda}x)}{\sigma(e^+e^- \to \mu^+\mu^-)} = 1.12 \pm 0.15 \pm 0.17$$

where as usual the first error is statistical while the second one is systematic. Here extrapolation has been carried out to cover also the unobserved range of 0 - 1 GeV/c, which contributes 13% of this value. In this extrapolation, a form of a · exp(-bE) is used in parameterising the invariant cross section $(E/4\pi p^2)$ dσ/dp over the momentum range from 1 to 5 GeV/c.

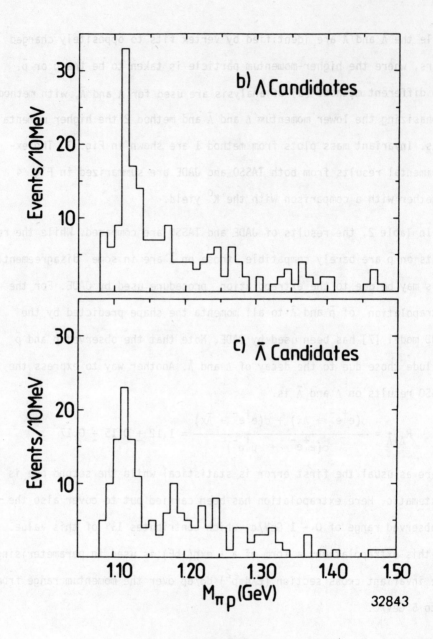

Fig. 3 - a) $M_{p\pi^-}$ spectrum for Λ candidates

b) $M_{\bar{p}\pi^+}$ spectrum for $\bar{\Lambda}$ candidates.

Results from TASSO

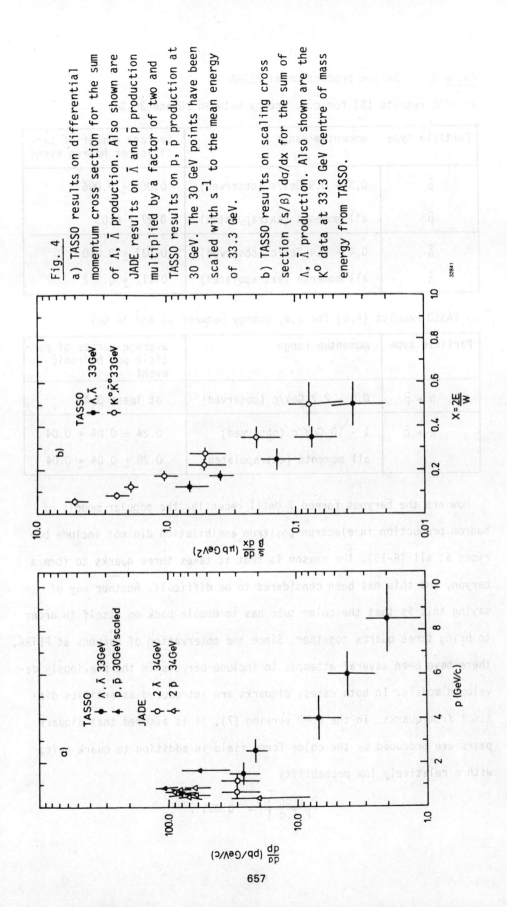

Fig. 4

a) TASSO results on differential momentum cross section for the sum of $\Lambda, \bar{\Lambda}$ production. Also shown are JADE results on $\Lambda, \bar{\Lambda}$ production multiplied by a factor of two and TASSO results on $p, \bar{p}$ production at 30 GeV. The 30 GeV points have been scaled with s^{-1} to the mean energy of 33.3 GeV.

b) TASSO results on scaling cross section $(s/\beta)\, d\sigma/dx$ for the sum of $\Lambda, \bar{\Lambda}$ production. Also shown are the K^0 data at 33.3 GeV centre of mass energy from TASSO.

Table 2 - Baryon production at PETRA

a) JADE results [5] for c.m. energy between 30 and 36 GeV

Particle type	momentum range	average number of particle per hadron event
$\bar{p}$	0.3 - 0.9 GeV/c (observed)	0.062 $\pm$ 0.006
$\bar{p}$	all momenta (extrapolated)	0.27 $\pm$ 0.03
$\bar{\Lambda}$	0.4 - 1.4 GeV/c (observed)	0.037 $\pm$ 0.010
$\bar{\Lambda}$	all momenta (extrapolated)	0.117 $\pm$ 0.032

b) TASSO results [4,6] for c.m. energy between 30 and 36 GeV

Particle type	momentum range	average number of particle per hadronic event
$p + \bar{p}$	0.5 - 2.2 GeV/c (observed)	at least 0.4
$\Lambda + \bar{\Lambda}$	1 - 10 GeV/c (observed)	0.24 $\pm$ 0.04 $\pm$ 0.04
	all momenta (extrapolated)	0.28 $\pm$ 0.04 $\pm$ 0.04

How are the baryons formed ? Until recently, the popular models of hadron production in electron-positron annihilation did not include baryons at all [8-13]. The reason is that it takes three quarks to form a baryon, and this has been considered to be difficult. Another way of saying this is that the color tube has to double back on itself in order to bring three quarks together. Since the observation of baryons at PETRA, there have been several attempts to include baryons in the previously developed models. In both cases, diquarks are introduced as entities distinct from quarks. In the LUND version [7], it is assumed that diquark pairs are produced by the color force field in addition to quark pairs with a relatively low probability

$$\frac{P(qq)}{P(q)} = 0.065 .$$

This ratio was extracted from SPEAR data [14,15] in the 4 GeV center of mass region. In the version of Meyer [16], the Field-Feynman scheme [9] (Fig. 5(a)) is used to include diquark pairs with a similar probability

$$P_{B1} = \frac{P(qq)}{P(q) + P(qq)} = 0.075$$

based on TASSO data [4]. This is shown schematically in Fig. 5(b). The ratios of u, d and s quark pairs are taken as before at 2 : 2 : 1, and the ratio of diquark pairs are taken similarly. The results of these two procedures are shown in Fig. 6. For the solid curve of Meyer in this figure, a second mechanism for diquark pair production has been added as shown in Fig. 5(c). The data are not sensitive enough to support this second mechanism, which in the simplest form tends to increase e^+e^- total hadronic cross section in contradiction with other data [17].

Although the Meyer results agree better with experimental data than the LUND results, a far more interesting and profound point is that these two models produce Λ and $\bar{\Lambda}$ in two qualitatively different ways. In the LUND version, cascades (Ξ^0 and Ξ^-) are hardly produced while the decay $\Sigma^0 \to \Lambda \gamma$ is an important source of Λ. In the Meyer version, the opposite is true: $\Sigma^0 \to \Lambda \gamma$ contributes no more than 20% of the Λ's, while the production of Ξ^0 and Ξ^- is significant. In this way the two models can be distinguished.

We list some of the problems that can perhaps be studied in the near future in connection with baryon production in e^+e^- annihilation.

(i) Short-range baryon-antibaryon correlation. This is universally expected on the basis of most models, and JADE has some preliminary information on this issue [5].

(ii) Long-range baryon-antibaryon correlation.

(iii) Polarization of Λ and $\bar{\Lambda}$. This would give us some information about the source of the Λ's. For example, are Λ's mainly coming from decay of Λ_c's ?

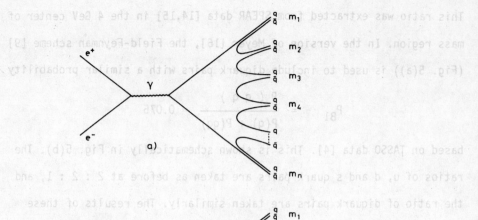

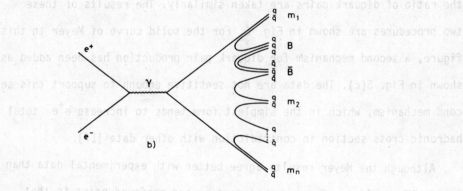

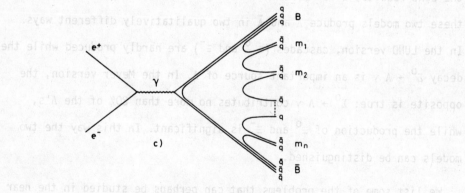

29.7.81   32866

Fig. 5 - a) Diagram for $e^+e^- \to$ mesons
b) Diagram for $e^+e^- \to$ mesons and baryons
c) Diagram for $e^+e^- \to$ mesons and baryons
with leading baryons in each jet.

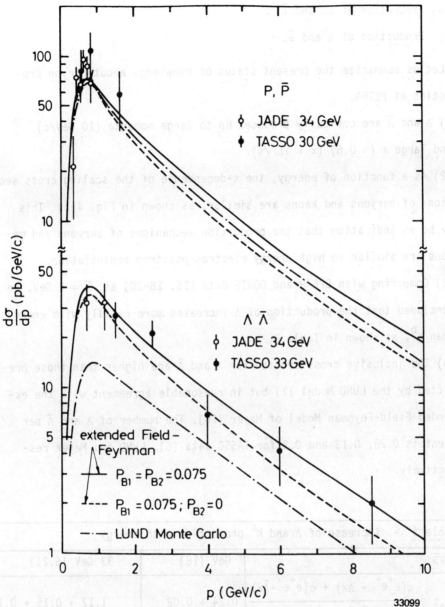

33099

<u>Fig. 6</u> - Inclusive spectra of protons, antiprotons, lambdas, and antilambdas. Data by the TASSO [4, 6] and the JADE[5] Collaborations. (All data points have been scaled with s^{-1} to the energy of 33 GeV).

Solid curves: $P_{B1} = P_{B2} = 0.075$ } Meyer [16]

Dashed curves: $P_{B1} = 0.075$, $P_{B2} = 0.$

Dashed dotted-dashed curve: LUND Model [7] .

661

(iv) Production of Σ^0 and $\bar{\Sigma}^0$.

(v) Production of Ξ and $\bar{\Xi}$.

Let us summarize the present status of knowledge about baryon production at PETRA.

(1) Λ and $\bar{\Lambda}$ are copiously produced up to large momenta (10 GeV/c) and large x (~ 0.6) (x = $2E/\sqrt{s}$).

(2) As a function of energy, the x-dependence of the scaling cross sections of baryons and kaons are similar, as shown in Fig. 4(b). This may be an indication that the production mechanisms of baryons and mesons are similar in high energy electron-positron annihilation.

(3) Comparing with SPEAR and DORIS data [15, 18-20] at $\sqrt{s}$ = 7 GeV, we concluded that the production of Λ increases more rapidly with energy than K^0, as shown in Table 3.

(4) The inclusive cross section for Λ and $\bar{\Lambda}$ are higher than those predicted by the LUND Model [7] but in reasonable agreement with the extended Field-Feynman Model of Meyer [16]. The number of Λ and $\bar{\Lambda}$ per event is 0.28, 0.13 and 0.3 for TASSO data [6], LUND and Meyer respectively.

Table 3 - Increase of Λ and K^0 production with energy		
$\sqrt{s}$	7 GeV [18]	33 GeV [6,21]
$R_{\Lambda+\bar{\Lambda}} = \dfrac{\sigma(e^+e^- \rightarrow \Lambda x) + \sigma(e^+e^- \rightarrow \bar{\Lambda}x)}{\sigma(e^+e^- \rightarrow \mu^+\mu^-)}$	0.24 ± 0.02	$1.12 \pm 0.15 \pm 0.17$
$R_{K^0+\bar{K}^0} = \dfrac{\sigma(e^+e^- \rightarrow K^0x) + \sigma(e^+e^- \rightarrow \bar{K}^0x)}{\sigma(e^+e^- \rightarrow \mu^+\mu^-)}$	2.0 ± 0.2	$5.5 \pm 0.4 \pm 0.8$

4. LIFETIME OF THE τ AND BOTTOM MESON

A. New upper limit for τ lifetime - TASSO

Upper limits on the τ lifetime have previously been given by DELCO [22] and TASSO [23]. They are

$$\text{DELCO} : \tau_\tau < 2.3 \times 10^{-12} \text{ sec (95\% C.L.)}$$
$$\text{TASSO} : \tau_\tau < 1.4 \times 10^{-12} \text{ sec (95\% C.L.)}$$

The latter was based on approximately 3000 nb^{-1} of running at PETRA for $\sqrt{s}$ between 12 and 31.6 GeV.

A new upper limit was recently obtained as [24]

$$\text{TASSO} : \tau_\tau < 5.7 \times 10^{-13} \text{ sec (95\% C.L.)}$$

based on approximately 20,000 nb^{-1} of running for $\sqrt{s}$ between 27 GeV and 37 GeV.

These upper limits may be compared with the theoretical value obtained by scaling the μ lifetime and neglecting the electron mass:

$$\text{Theory} : \quad \tau_\tau = BR(\tau \rightarrow e \; \nu \; \bar{\nu})(m_\mu/m_\tau)^5 \; \tau_\mu$$
$$= (2.7 \pm 0.2) \times 10^{-13} \text{ sec,}$$

where the error is due almost entirely to the uncertainty in the branching ratio $\tau \rightarrow e \; \nu \; \bar{\nu}$.

Recently, MARK II at PEP has obtained a measurement of the τ lifetime being $(4.6 \pm 1.9) \times 10^{-13}$ sec [25].

Briefly, this new upper limit is obtained as follows. Consider events where there are three charged tracks on one side versus only one charged track on the other side such that the total charge is zero. With suitable cuts to reduce background due to radiative Bhabha scattering, there are 150 such events, one of which is shown in Fig. 7. From the side with three charged tracks the position of the τ decay vertex is determined in the

$e^+e^- \to \tau^+\tau^-$

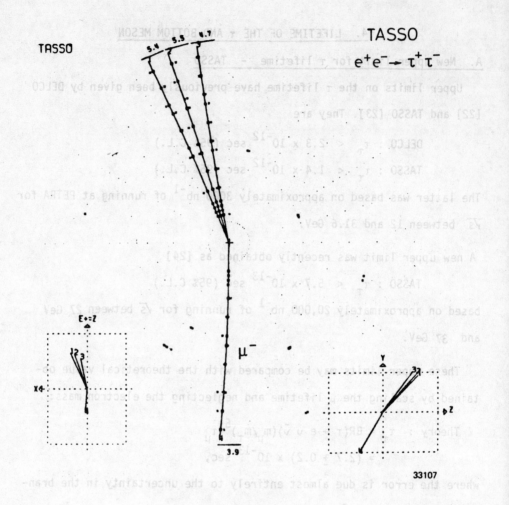

μ^-

33107

TASSO

Fig. 7 - A $e^+e^- \to \tau^+\tau^-$ event from TASSO

transverse plane. The distribution of the resulting transverse distance to the interacting point is shown in Fig. 8. Using the polar angle of the momentum vector of each τ, this transverse distance is converted to a proper time, as shown in Fig. 9. The mean proper time of (-0.25 ± 3.5) x 10^{-13} sec leads to the above mentioned upper limit of 5.7 x 10^{-13} sec at 95% confidence level.

B. New upper limit for the lifetime of botton meson - JADE

JADE [26] has recently obtained an interesting upper limit for the lifetime τ_B of the lightest meson with bottom quantum number. The events used are those hadronic events where there is a single muon with momentum larger than 1.8 GeV/c. There are 73 such events. From each of these events, the closest approach of the μ track to the interaction point is determined in the plane transverse to the e^+, e^- beam directions. Upper limit for this distance d_μ is found to be

$$d_\mu < 0.79 \text{ mm} \qquad (90\% \text{ C.L.})$$

This leads to an upper limit for τ_B

$$\tau_B < 5 \times 10^{-12} \text{ sec } (90\% \text{ C.L.})$$

5. JET ANALYSIS

A. Long-range charge correlation - TASSO, JADE

Since two-jet events in electron-positron annihilation are due to the creation of $q\bar{q}$ pairs, and quarks are charged, it is to be expected that the leading particles in one jet "know" about the charge of the leading particles in the opposite jet. This long-range charge correlation was first observed by TASSO [27,28]. Let y be the rapidity variable defined by

$$y = \frac{1}{2} \ln \frac{E + p_\parallel}{E - p_\parallel}$$

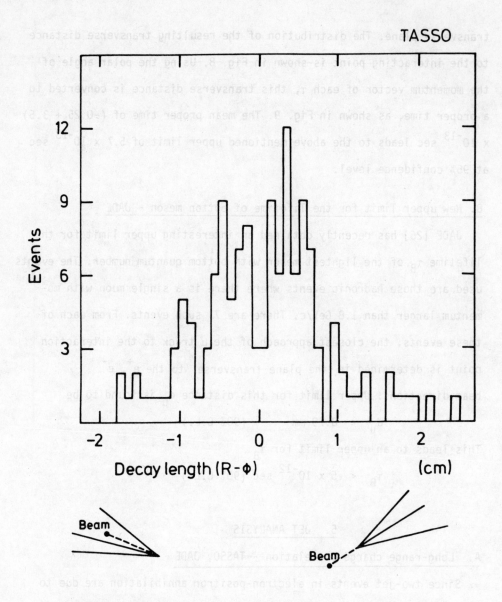

Fig. 8 - Distribution of decay lengths
(distance from the fitted vertex position to the known beam
position on the plane perpendicular to the e^+, e^- beam direction
i.e. R-ø plane) from the tau decay candidates.

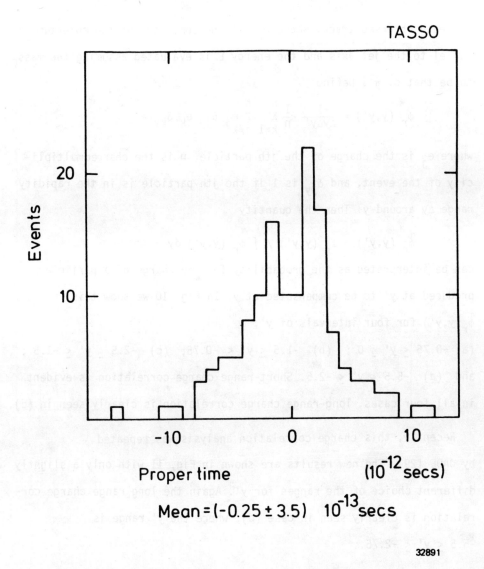

Fig. 9 - τ decay time distribution from TASSO.

for each observed track, where $p_{\parallel}$ is the component of the momentum parallel to the jet axis and the energy E is evaluated assuming the mass to be that of $\pi^{\pm}$. Define

$$\phi_r (y,y') = \frac{-1}{\Delta y \Delta y'} \left\langle \frac{1}{n} \sum_{k=1}^{n} \sum_{j \neq k}^{n} e_j \, \delta_{jy} \, e_k \, \delta_{ky'} \right\rangle$$

where e_j is the charge of the jth particle, n is the chargedmultiplicity of the event, and δ_{jy} is 1 if the jth particle is in the rapidity range Δy around y. Then the quantity

$$\tilde{\phi}_r (y,y') = \phi_r (y,y') \, / \int \phi_r (y,y') \, dy$$

can be interpreted as the probability for the charge of a particle produced at y' to be compensated at y. In Fig. 10 we show this $\tilde{\phi}_r(y,y')$ for four intervals of y':

(a) $-0.75 \leq y' \leq 0$; (b) $-1.5 \leq y' \leq -0.75$; (c) $-2.5 \leq y' \leq -1.5$; and (d) $-5.5 \leq y' \leq -2.5$. Short-range charge correlation is evident in all four cases, long-range charge correlation is clearly seen in (d).

Recently, this charge correlation analysis was repeated by JADE [29]. The new results are shown in Fig. 11 with only a slightly different choice of the ranges for y'. Again the long range charge correlation is clearly seen in case (d), where the y' range is $-5.5 \leq y' \leq -2.75$.

B. Neutral energy - JADE

Both JADE and CELLO can detect photons in 90% of the 4π solid angle, JADE with lead glass counters and CELLO with liquid argon counters. We report here the results from JADE [30].

Two energy fractions are defined as follows: the neutral energy fraction ρ_N is the fraction of energy not carried by charged particles

$$\rho_N = 1 - \sum_j E_j^{ch} \, / \sqrt{s}$$

668

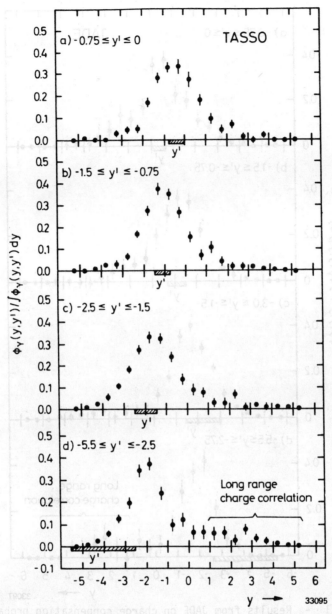

Fig. 10 - Results from TASSO on charge compensation probability $\phi_r(y,y') / \int \phi_r(y,y') \, dy$ as a function of rapidity y for a particle produced at y'.

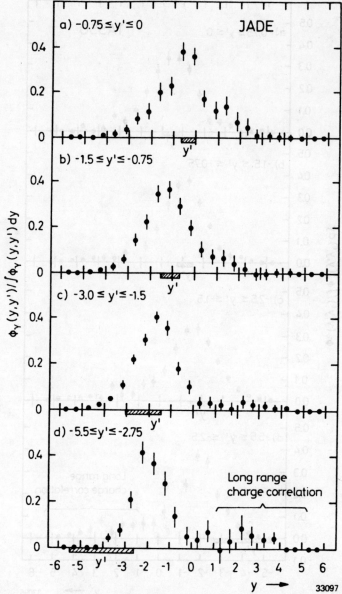

<u>Fig. 11</u> - Results from JADE on charge conpensation probability
$\phi_r(y,y')/\int\phi_r(y,y')\,dy$ as a function of rapidity y for
a particle produced at y'.

and the photon fraction by

$$\rho_\gamma = \sum_j E_j^\gamma / \sqrt{s}.$$

The major contribution to ρ_γ is of course due to π^0. Both quantities have to be carefully corrected. For example, ρ_N includes contributions from K_s and Λ, while for ρ_γ the energy deposited in the lead glass counters by charged particles must be subtracted out.

The difference $\rho_N - \rho_\gamma$ is the fraction of energy carried by long-lived hadrons and neutrinos. The long-lived hadrons include K_L, n, and $\bar{n}$. Using available data at PETRA, these contributions can be estimated and in this way we obtain

ρ_ν = energy fraction carried by neutrinos.

The results are plotted in Fig. 12 and the corrected mean ρ_γ, ρ_N and ρ_ν are listed in Table 4. No difference is seen between two-jet events and planar events.

The most interesting feature of the result is that the fraction of energy carried by neutrinos is quite small. This excludes the Pati-Salam model [31], where multihadron final states arise via quarks of integer charge which decay into hadrons and neutrinos, giving rise to a neutrino energy fraction ρ_ν between 18% and 28%.

Table 4 - The corrected Gamma Ray, Neutral Particle and Neutrino Energy Fractions

$\sqrt{s}$ (GeV)		ρ_γ	ρ_N	Neutrino
12.00	All events	0.25 ± 0.04	0.37 ± 0.04	0.00 ± 0.06
	Planar	0.21 ± 0.05	0.32 ± 0.05	-0.01 ± 0.07
30.30	All events	0.28 ± 0.03	0.36 ± 0.04	-0.03 ± 0.05
	Planar	0.28 ± 0.03	0.36 ± 0.04	-0.03 ± 0.06
34.89	All events	0.30 ± 0.03	0.38 ± 0.04	-0.02 ± 0.05
	Planar	0.32 ± 0.03	0.41 ± 0.04	-0.02 ± 0.06

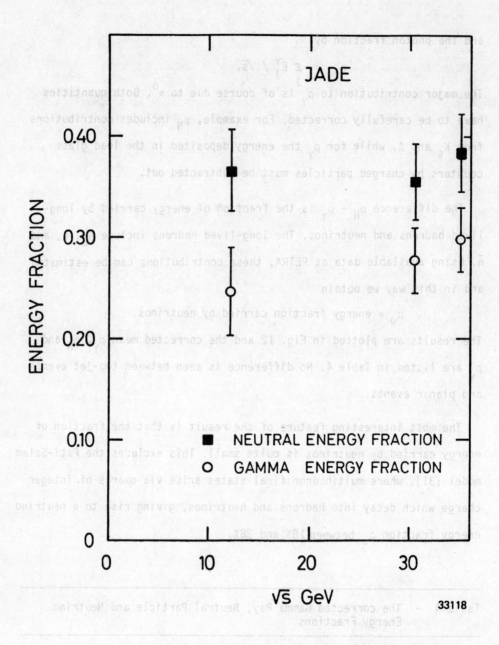

Fig. 12 - Results from JADE on the corrected mean gamma ray
and neutral energy fractions versus center-of-mass
energy.

C. Thrust and energy-flow distributions of photons - CELLO

In an attempt to find out whether neutral and charged particles
are distributed similarly in electron-positron annihilation, CELLO
obtained the thrust using separately the charged particles only and the
photons only, which come mostly from π^0 decay. The result [32] is shown
in Fig. 13, where the histograms represent the experimental data while
the circles with error bars are obtained from a Monte Carlo simulation.
It is seen that the thrust distribution is virtually identical for charged particles and for photons.

Continuing the comparison, CELLO next studied the energy flow distribution [33] which is for many purposesa more sensitive quantity. With
an oblateness [34] cut of ≥ 0.15, the result is shown in Fig. 14. Again
the energy flow pattern for photons alone is the same as that for charged
particles plus photons.

D. Energy flow distribution - MARK J

MARK J has compared their energy-flow pattern with a number of models.
With an oblateness cut of > 0.3, the result [35] is shown in Fig. 15. As
expected, the best agreement is obtained from QCD and the data disagree
with models of $q\bar{q}$, phase space, and a $q\bar{q}$ model with a $\exp(-P_T/650 \text{ MeV})$
fragmentation distribution. This offers one more piece of evidence in
favor of QCD. We shall return to the properties of the gluon in section 6.

E. Jet invariant mass - PLUTO

When jets were first discovered at SPEAR [36], the most useful quantity was the sphericity.At PETRA, a number of quantities have been found
to be useful, including thrust, aplanarity, oblateness, and energy flow.
These quantities all have the character of not being sensitive to missing particles and to the details of fragmentation. As the experiments

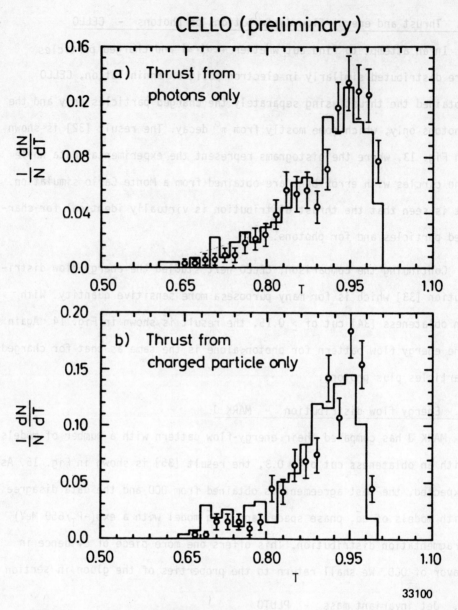

Fig. 13 - CELLO results on

a) thrust distribution reconstructed from photons only;

b) thrust distribution reconstructed from charged particles only.

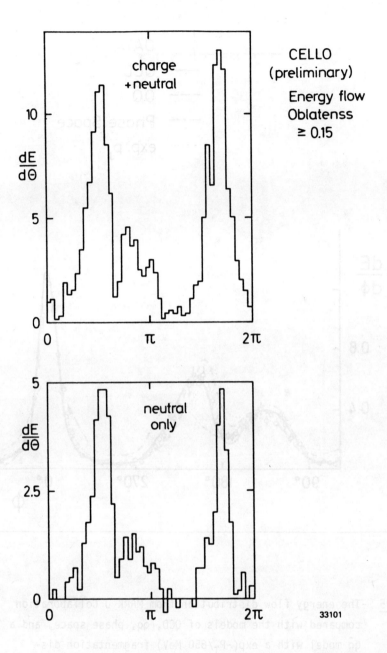

Fig. 14 - CELLO results on energy flow for the events
with oblateness $\geq$ 0.15;
a) for charged and neutral particles
b) for neutral particles only.

675

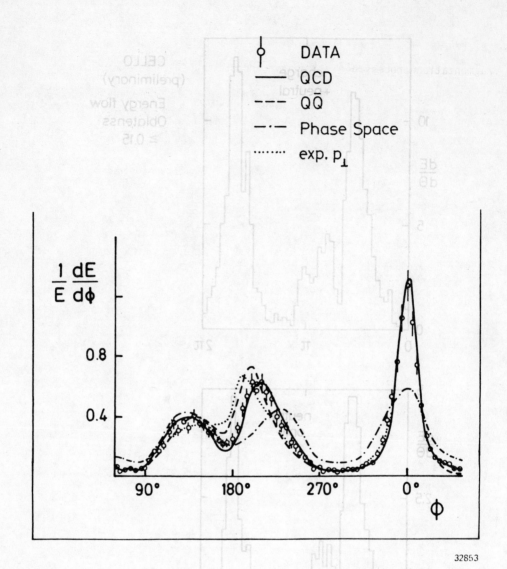

Legend:
- ○ DATA
- —— QCD
- – – – QQ̄
- –·– Phase Space
- ······ exp. $p_\perp$

$\dfrac{1}{E}\dfrac{dE}{d\phi}$

0.8
0.4

90°   180°   270°   0° ϕ

32853

<u>Fig. 15</u> -The energy flow distribution from MARK J Collaboration
        compared with the models of QCD, qq̄, phase space, and a
        qq̄ model with a exp($-P_T$/650 MeV) fragmentation dis-
        tribution.

676

continue to improve, the amount of missing particles decreases and the fragmentation process becomes better understood. Thus this advantage of the presently used quantities becomes less important, and it seems likely that other quantities may be profitably employed to gain a better understanding of the data.

A first step has been taken in this direction recently by PLUTO. They study the jet invariant mass proposed by Clavelli [37]. This is defined as follows for each event. Take all the measured particles, partition them into two sets, and calculate the invariant masses m_1 and m_2 of the two set by minimizing the sum of $(m_1^2 + m_2^2)$ over all partitions. PLUTO has carried out a Monte Carlo study, and it is found that the jet mass distribution for a generated sample of events is considerably degraded after imposing acceptance cuts and detector effects. Accordingly an eleborate correction procedure is necessary. Their result [38] is shown in Fig. 16 together with the QCD results of Hoyer et al. [11] with fragmentation, and leading log approximations [39] without fragmentation. It shows clearly sensitivity to fragmentation. As a first application of the jet invariant mass, it is shown [38] in Fig. 17 the dependence of charged multiplicity on the mass of the heavy jet (M_H) or of the light jet (M_L).

In the near future we expect to see more new and interesting variables used to analyze the experimental data of higher statistics.

6. PROPERTIES OF THE GLUON

A. Spin of the gluon - TASSO, PLUTO

Two years ago gluon bremsstrahlung was first observed at PETRA in the form of three-jet events [40-44]. The determination of the spin of the

677

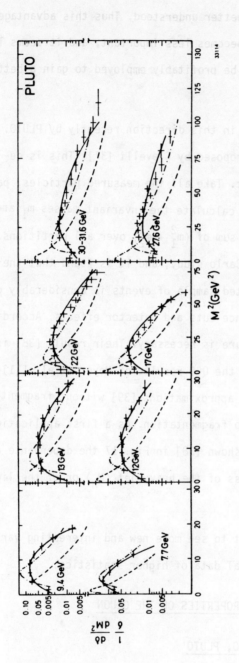

Fig. 16 - Normalized jet invariant mass distributions from 7.7 to 31.6 GeV. The solid curve represents the expectations from Hoyer et al. Monte Carlo. The dotted and dashed dotted curves show the predictions of a LL calculation ref. [39] for two different values of the QCD cut off parameter Λ of 200 MeV and 1.5 GeV respectively.

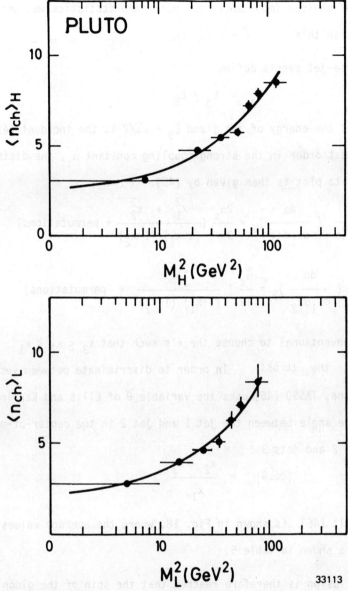

Fig. 17

The mean charged multiplicity of the heavy (M_H) and light (M_L) jets as a function of their masses M for the 27.6 - 31.6 GeV sample. The solid curves show the results of a fit to a QCD inspired form of
$$\langle n_{ch} \rangle = a + b \exp (c \sqrt{\ln M^2/\Lambda^2}) \qquad (M = M_L \text{ or } M_H)$$
with c = 2.4 and Λ = 500 MeV, a = 0.91 $\pm$ 0.19 (0.62 $\pm$ 0.19) and b = 0.019 $\pm$ 0.001 (0.021 $\pm$ 0.001) for the heavy (light) jets.

gluon is of obvious importance, but much more statistics was necessary to accomplish this.

For three-jet events define

$$x_j = E_j / E_B$$

where E_j is the energy of jet j and $E_B = \sqrt{s}/2$ is the incident beam energy. To the lowest order in the strong coupling constant α_s, the distribution in the Dalitz plot is then given by [45]:

$$\text{Vector}: \frac{1}{\sigma_o} \left(\frac{d\sigma}{dx_1 dx_2} \right)_V = \frac{2\alpha_s}{3\pi} \left[\frac{x_1^2 + x_2^2}{(1-x_1)(1-x_2)} + \text{permutations} \right]$$

$$\text{Scalar}: \frac{1}{\sigma_o} \left(\frac{d\sigma}{dx_1 dx_2} \right)_S = \frac{\tilde{\alpha}_s}{3\pi} \left[\frac{x_3^2}{(1-x_1)(1-x_2)} + \text{permutations} \right]$$

It is conventional to choose the x's such that $x_3 \leq x_2 \leq x_1$ and x_1 is the thrust. In order to discriminate between vector and scalar gluons, TASSO [46] uses the variable $\hat{\theta}$ of Ellis and Karliner [47] which is the angle between the jet 1 and jet 2 in the center-of-mass system of jets 2 and jets 3 :

$$|\cos\hat{\theta}| = \frac{x_2 - x_3}{x_1}$$

The result [46] is shown in Fig. 18, where the average values $<|\cos\hat{\theta}|>$ are shown in Table 5.

The conclusion is therefore reached that the spin of the gluon is indeed 1 and the result is insensitive to the details of jet fragmentation as shown in Table 5 and Fig. 18.

More recently PLUTO [48] uses instead the variable $x_\perp$ explained in the upper part of Fig. 19 :

$$x_\perp = \frac{2}{x_1} \left[(1-x_1)(1-x_2)(1-x_3) \right]^{1/2}$$

In terms of this $x_\perp$, the results are shown in Fig. 19. In this figure, PLUTO has used their previous determination of α_s for the case of the vector

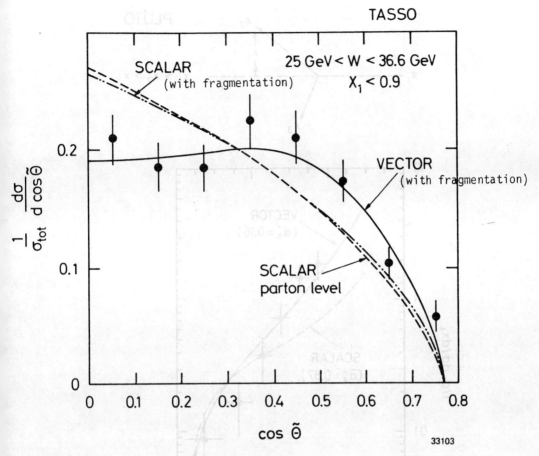

Fig. 18 - Observed distribution of the TASSO data in the region $x_1 < 0.9$ as a function of the cosine of the Ellis-Karliner angle $\tilde{\vartheta}$. The solid line shows the QCD Monte Carlo prediction , dashed line the prediction for the scalar gluons (— ·· — for Monte Carlo scalar model; — — — for scalar model of parton level). All curves are normalized to the number of observed events.

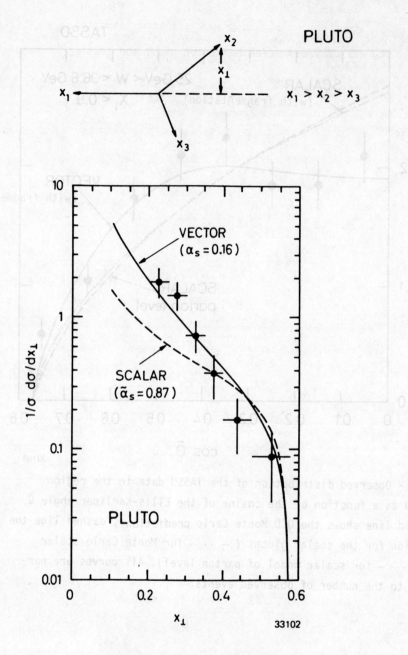

<u>Fig. 19</u> - $x_\perp$ distribution from PLUTO compared with QCD (solid curve) and scalar gluon model (dashed curve).

gluon but the $\tilde{\alpha}_s$ for the scalar gluon is fitted. Thus the χ^2 is 5.7
for six degrees of freedom in the vector case, and 19.7 for five degrees
of freedom in the scalar case. Again the spin of the gluon is determined
to be 1.

Alternative methods of analysis have also been applied to these data
from PLUTO [49] and TASSO [50] leading to the same value of the gluon
spin.

Table 5 - Average value for $|\cos\hat{\theta}|$

	$<|\cos\hat{\theta}|>$	Standard deviation
Data	0.339 ± 0.008	
Vector (Monte Carlo) with acceptance and jet fragmentation	0.341 ± 0.003	$0.2\,\sigma$
Vector (parton level prediction) no acceptance cut no jet fragmentation	0.348	$1.1\,\sigma$
Scalar (Monte Carlo)	0.298 ± 0.003	$4.8\,\sigma$
Scalar (parton level prediction)	0.298	$5.1\,\sigma$

B. Differences between quark and gluon jets - JADE

Ever since the first observation of gluon jets, an intriguing question
is: how do we distinguish a gluon jet from a quark jet ? Although some
differences are expected from QCD, these two kinds of jets turn out to
be remarkably similar. For example, JADE has measured the ratio of
photon energy to charged particle energy separately for the three jets,
and the results are [30] :

683

Jet 1 :   0.47 ± 0.01 ;

Jet 2 :   0.46 ± 0.01 ;

Jet 3 :   0.44 ± 0.02 ;

where the errors are statistical but the systematic errors are expected
to be small.

The first attempt to distinguish these two kinds of jets has very
recently been reported by JADE [51]. At the same jet energy, the average
transverse momentum $<p_\perp>$ of particles with respect to the jet axis is larger
for the gluon jet than for the quark jet. This result is very preliminary, and has not been confirmed by other groups at PETRA. It still
needs a great deal of more study.

JADE starts with 4955 hadronic events in the energy range of $\sqrt{s}$
between 30 to 36 GeV. Planar three-jet events are selected as described
in Reference [52] with the cut $Q_2 - Q_1 \geq 0.07$ and $Q_1 \leq 0.06$ where Q_1,
Q_2 and Q_3 are the normalized eigenvalues of the sphericity tensor with
$0 \leq Q_1 \leq Q_2 \leq Q_3 \leq 1$. To identify 3 jets of particles in a planar event
and to determine the jet direction vectors, the three-jet analysis method
of Reference [53] was used with the modification of maximizing thrust
instead of minimizing sphericity. Events in which one or more of the jets
contain fewer than 4 particles or an energy of less than 2 GeV are rejected. 596 events meet these criteria. The three jets are ordered according to the angles between their direction vectors, projected into the
event plane.

By Monte Carlo calculation using the LUND model it is found, with the
ordering of reconstructed energies: $E_{jet1} > E_{jet2} > E_{jet3}$, that in 51%
of the events jet 3 is the gluon jet; 22% jet 2 is the gluon jet; 12% jet 1
is the gluon jet; and 15% the event is actually a two-jet event without
a gluon jet.

The observed [52] $\langle p_\perp \rangle$ is shown in Fig. 20 for the three jets separately as function of the visible jet energy. It is seen that, for the same visible jet energy, $\langle p_\perp \rangle$ is larger for jet 3 than for jet 1 and 2.

The difference is not present in the Monte Carlo of Hoyer et al.[11], where the fragmentations of gluon and quark jets are identical. It is seen in the LUND Monte Carlo [8,13] where the gluon is associated with a kink in the color force field. Thus the experimental result of JADE agrees with LUND but not with Hoyer et al.

The question has been raised whether the effect is due to the use of the visible jet energy as the variable. Thus a further cut is introduced :

$$|E_{jet} - E_{jet}{}^{visible}| \leq 0.2 \, E_{jet},$$

where E_{jet} is the jet energy calculated from the directions of the three jet axes. The result is shown in Fig. 21.

A great deal more effort is expected to be spent at PETRA on finding differences between gluon and quark jets. It is hoped that this knowledge will deepen our understanding of the physics of electron-positron annihilation.

7. TWO-PHOTON PHYSICS

A. $\gamma\gamma \rightarrow \rho^0\rho^0$ - TASSO, CELLO

A year ago, the TASSO Collaboration observed the $\gamma\gamma \rightarrow \rho^0\rho^0$ process [54] which exhibits a large enhancement near the threshold.

Since each ρ^0 decays into a $\pi^+\pi^-$ pair, experimentally

$$e^+e^- \rightarrow e^+e^- \, \pi^+\pi^+\pi^-\pi^-$$

is observed. It is therefore possible to use the untagged events, and furthermore no extrapolation to $Q^2 = 0$ is needed. The events are identified by two positive and two negative tracks such that

$$|\sum_{j=1}^{4} \vec{p}_{jT}| < 0.15 \text{ GeV/c},$$

685

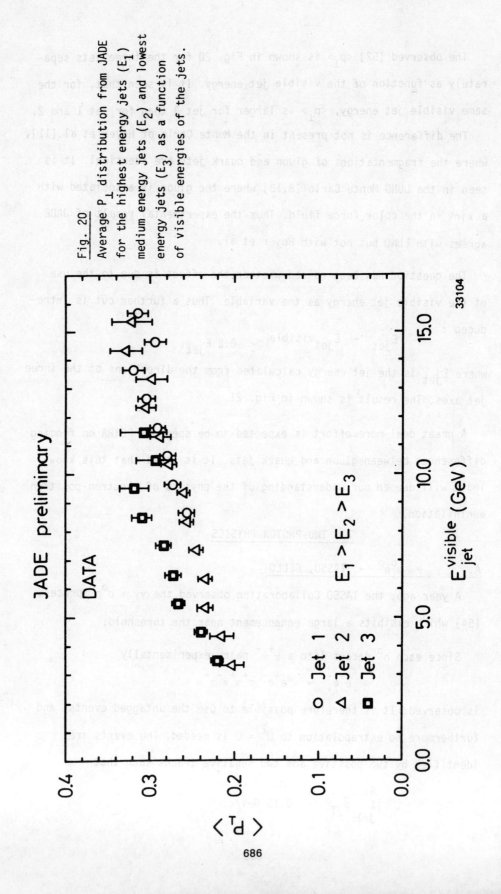

Fig. 20
Average $P_\perp$ distribution from JADE for the highest energy jets (E_1) medium energy jets (E_2) and lowest energy jets (E_3) as a function of visible energies of the jets.

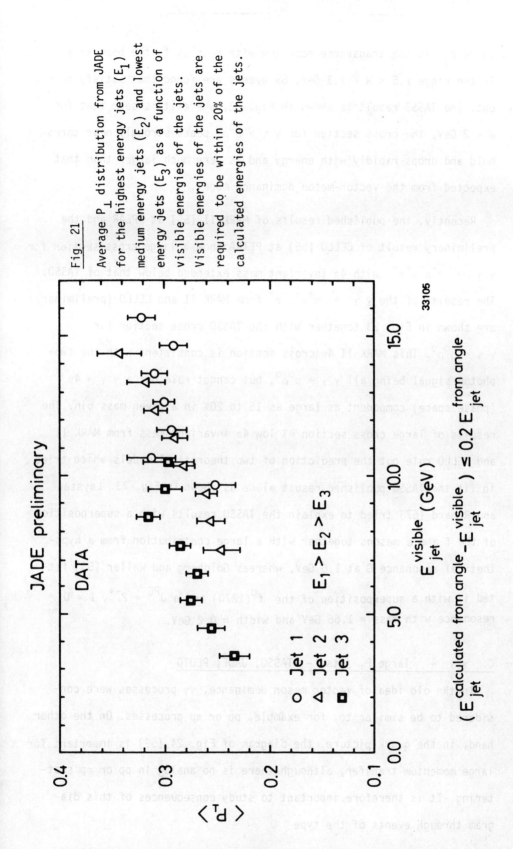

Fig. 21

Average $P_\perp$ distribution from JADE for the highest energy jets (E_1) medium energy jets (E_2) and lowest energy jets (E_3) as a function of visible energies of the jets. Visible energies of the jets are required to be within 20% of the calculated energies of the jets.

JADE preliminary

DATA

○ Jet 1
△ Jet 2 $E_1 > E_2 > E_3$
■ Jet 3

$\langle P_T \rangle$

$E_{jet}^{visible}$ (GeV)

$(E_{jet}^{calculated \; from \; angle} - E_{jet}^{visible}) \leq 0.2 \; E_{jet}^{from \; angle}$

33105

where p_{jT} is the transverse momentum with respect to the beam axis.
In the range $1.5 < W < 2.3$ GeV, 89 events are found that satisfy this
cut. The TASSO result is shown in Fig. 22, where it is seen that for
$W < 2$ GeV, the cross section for $\gamma\gamma \to \rho^0\rho^0$ peaks strongly near threshold and drops rapidly with energy and is very much larger than that
expected from the vector-meson dominance model.

Recently, the published results of MARK II [55] at SPEAR and the
preliminary result of CELLO [56] at PETRA show a large cross section for
$\gamma\gamma \to \pi^+\pi^-\pi^+\pi^-$ with 4π invariant mass extended below that of TASSO.
The results of the $\gamma\gamma \to \pi^+\pi^-\pi^-\pi^-$ from MARK II and CELLO (preliminary)
are shown in Fig. 23 together with the TASSO cross section for
$\gamma\gamma \to \rho^0\rho^0$. This MARK II 4π cross section is consistent with the two-
photon signal being all $\gamma\gamma \to \rho^0\rho^0$, but cannot rule out a $\gamma\gamma \to 4\pi$
(phase space) component as large as 15 to 20% in a given mass bin. The
results of large cross section at low 4π invariant mass from MARK II
and CELLO rule out the prediction of two theoretical models which tried
to fit the TASSO published result alone as shown in Fig. 23. Layssac
and Renard [57] tried to explain the TASSO results with a superposition
of the f and ϵ mesons together with a large contribution from a hypothetical resonance G at 1.6 GeV, whereas Goldberg and Weiler [58] fitted it with a superposition of the $f^0(1270)$ and a $J^{PC} = 2^{-+}$, $I = 0$
resonance with mass $= 1.66$ GeV and width $\simeq 0.2$ GeV.

C. $\gamma\gamma \to$ large P_T jets - TASSO, JADE, PLUTO

In the old idea of vector meson dominance, $\gamma\gamma$ processes were considered to be similar to, for example, pp or πp processes. On the other
hand, in the quark picture, the diagram of Fig. 24 [59] is important for
large momentum transfer, although there is no analog in pp or πp scattering. It is therefore important to study consequences of this diagram through events of the type

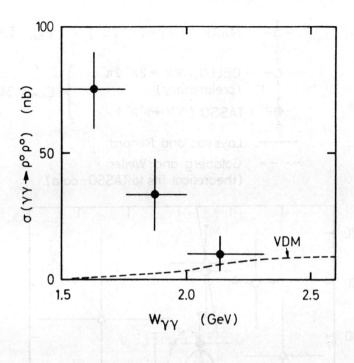

Fig. 22 - Cross section for
$e^+e^- \rightarrow e^+e^- \rho^0\rho^0$ as measured by TASSO.
An asymptotic VMD prediction is shown as
the dotted line.

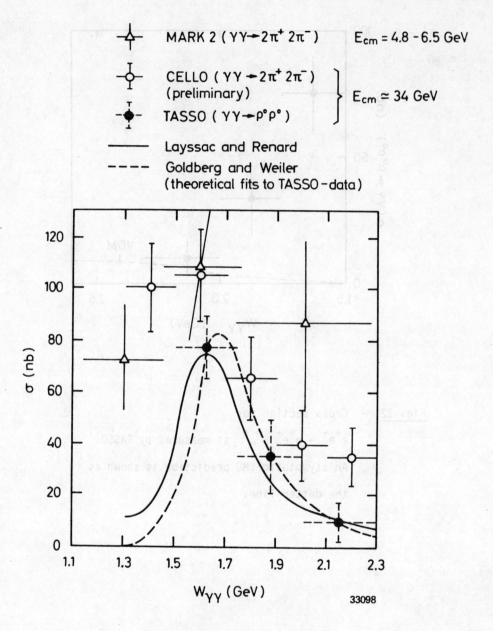

MARK 2 ($\gamma\gamma \to 2\pi^+ 2\pi^-$) $E_{cm} = 4.8 - 6.5$ GeV

CELLO ($\gamma\gamma \to 2\pi^+ 2\pi^-$)
(preliminary)
TASSO ($\gamma\gamma \to \rho^0\rho^0$)   } $E_{cm} \simeq 34$ GeV

——— Layssac and Renard
– – – Goldberg and Weiler
(theoretical fits to TASSO-data)

33098

Fig. 23 – Cross sections of the processes $\gamma\gamma \to \pi^+\pi^-\pi^+\pi^-$
(MARK J and CELLO) and $\gamma\gamma \to \rho^0\rho^0$ (TASSO) in
comparison with two theoretical models.

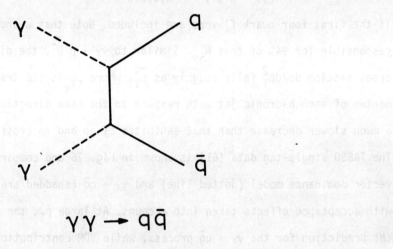

32103

Fig. 24 - Diagram for the creation of a q$\bar{q}$ pair in $\gamma\,\gamma$.

$$e^+e^- \rightarrow e^+e^- + \text{two jets} ,$$

where the jets are not collinear but are coplanar with the beam direction. A single-tagged event of this type from TASSO is shown in Fig. 25.

The main characteristic of the hard scattering diagram of Fig. 24 is the ratio [60]

$$R = \frac{\sigma(\gamma\gamma \rightarrow q\bar{q})}{\sigma(\gamma\gamma \rightarrow \mu^+\mu^-)} = 3\Sigma_{u,d,s,c} Q_i^4 = \frac{34}{27} \quad (Q_i = \text{quark charge})$$

if the first four quark flavors are included. Note that u and c are responsible for 94% of this $R_{\gamma\gamma}$. Similar to $\gamma\gamma \rightarrow \mu^+\mu^-$, the differential cross section $d\sigma/dp_T^2$ falls roughly as p_T^{-4}, where p_T is the transverse momentum of each hadronic jet with respect to the beam direction. This is a much slower decrease than that exhibited by pp and πp cross sections. The TASSO single-tag data [61] is shown in Fig. 26 and compared with the vector dominance model (dotted line) and $\gamma\gamma \rightarrow q\bar{q}$ (shadded area), both with acceptance effects taken into account. At large p_T, the data approach the prediction for the $\gamma\gamma \rightarrow q\bar{q}$ process, while VDM contribution would be much too small to explain the observed differential cross section. Similar results have also been obtained by JADE [62] and PLUTO [63] and those from JADE are shown in Fig. 27. It is interesting to note from both Fig. 26 and Fig. 27 that even at the highest p_T^2 the experimental data tend to be somewhat higher, by a factor of about 1.5 than the expectation evaluation from the diagram of Fig. 24. The reason may be the contributions from QCD corrections, from 3- and 4-jet processes, and from higher twist diagrams [64] which have not yet been incorporated in the calculation.

D. Deep inelastic eγ scattering - PLUTO

Recently, PLUTO [65] obtained interesting first results on the structure function of the photon. This is accomplished by using the large angle tagging. Since the other electron or positron is not tagged, the

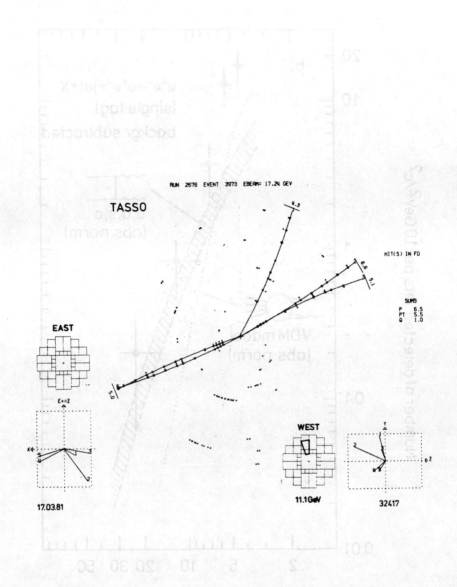

Fig. 25 - $\gamma\gamma \rightarrow 2$ jet event of TASSO at $E_{c.m.} = 34.5$ GeV.
The energy of the single tagged e^- is 11 GeV.

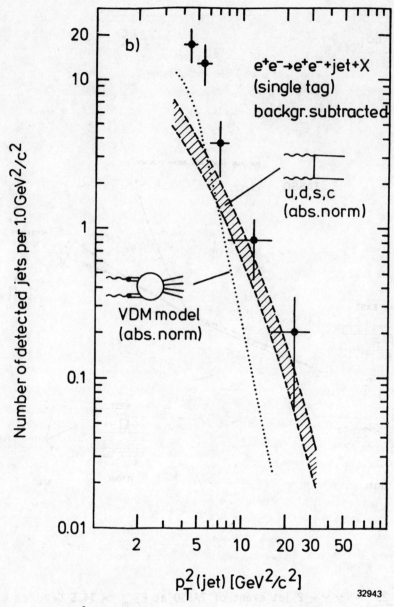

32943

Fig. 26 - The P_T^2(jet) distribution from single tagged data of TASSO for
$e^+e^- \to e^+e^- + jet + X$, compared to a $e^+e^- \to e^+e^- + q\bar{q}$ calculation with four different types of fragmentation model
(shadded band) and to VDM (dotted line).

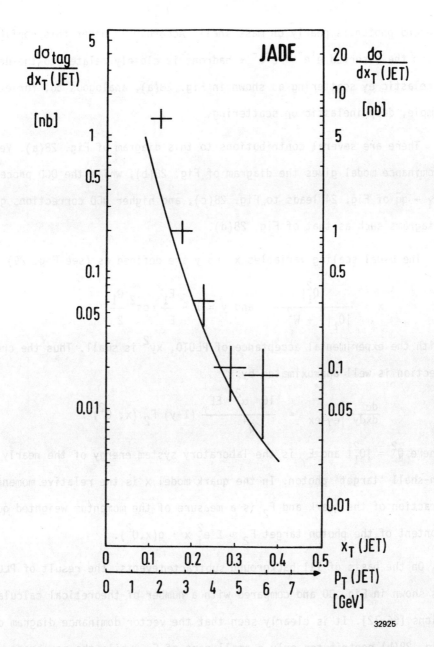

Fig. 27 - Transverse momentum distribution of jets for single tag data of JADE compared with absolute predictions for fractional charged quarks (curve). $x_T(JET) = p_T(JET)/E_{beam}$ is taken with respect to the centre of mass direction of motion. The cross section is given for the JADE tagging condition on the lefthand scale and integrated over all electron angles and energies on the righthand scale.

second photon is nearly on mass shell. Accordingly under this configuration the observed $e^+e^- \rightarrow e^+e^- +$ hadrons is closely related to the deep inelastic $e\gamma$ scattering as shown in Fig. 28(a), analogous to, for example, deep inelastic νp scattering.

There are several contributions to this diagram of Fig. 28(a). Vector dominance model gives the diagram of Fig. 28(b), while the QCD process $\gamma\gamma \rightarrow q\bar{q}$ of Fig. 24 leads to Fig. 28(c), and higher QCD corrections give diagrams such as that of Fig. 28(d).

The usual scaling variables x and y are defined as (see Fig. 29)

$$x = \frac{|Q_1^2|}{|Q_1|^2 + W^2} \quad \text{and} \quad y = 1 - \frac{E_1}{E} \cos^2 \frac{\theta_1}{2}$$

With the experimental acceptance of PLUTO, xy^2 is small. Thus the cross section is well approximated by

$$\frac{d\sigma}{dxdy}\bigg|_{e\gamma \rightarrow ex} = \frac{16\pi \, \alpha^2 \; EE_\gamma}{Q^4} \, (1-y) \, F_2 \, (x, \, Q^2)$$

where $Q^2 = |Q_1^2|$ and E_γ is the laboratory system energy of the nearly on-shell 'target' photon. In the quark model x is the relative momentum fraction of the quark and F_2 is a measure of the momentum weighted quark content of the photon target $F_2 \sim \Sigma \, e_i^2 \, x \cdot q(x,Q^2)$.

On the basis of 111 background subtracted events, the result of PLUTO is shown in Fig. 30 and compared with a number of theoretical calculations [66-72]. It is clearly seen that the vector dominance diagram of Fig. 28(b) contributes only a small part of F_2, while the agreement with QCD is quite good.

QCD also predicts a Q^2 dependence of F_2 at fixed x, equivalent to a strong scale breaking. PLUTO has investigated this effect by plotting in Fig. 31 F_2 versus Q^2, averaged over $0.2 < x_{vis} < 0.8$. This restriction

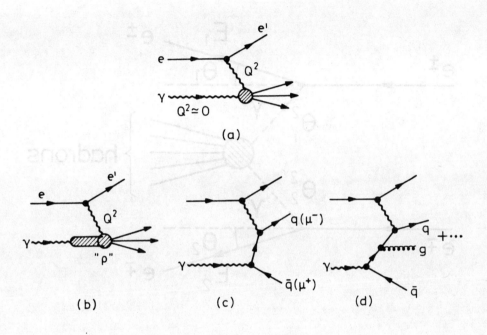

Fig. 28 - Diagrams contributing to e γ → e' hadrons.

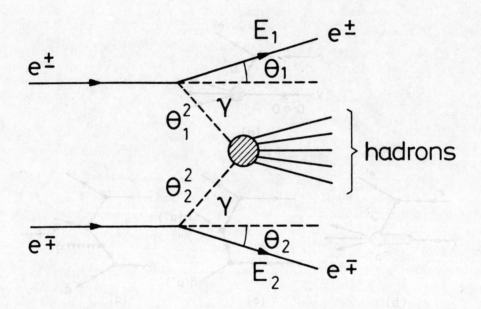

15.12.80   32104

<u>Fig. 29</u> - Diagram for $e^+e^- \rightarrow e^+e^- +$ hadrons.

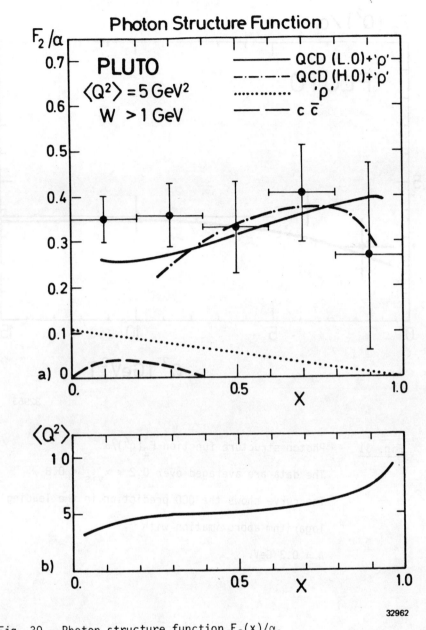

Fig. 30 - Photon structure function $F_2(x)/\alpha$.
The data are averaged over $1 < Q^2 < 15$ GeV^2.

. . . . Hadronic structure function $F_{2,\rho}$ from diagram 28(b)

——— $F_{2,QCD} + F_{2,\rho}$(Λ = 0.3 GeV, u,d,s quarks from diagrams 28(b), (c) and (d) to the leading order)

—·—— $F_{2,QCD} + F_{2,\rho}$ (Λ = 0.3 GeV, u,d,s quards from diagrams 28(b), (c) and (d) including higher order correction)

- - - $F_{2,box}$ for c quarks only (m_q = 1.5 GeV from diagram 28(c)).

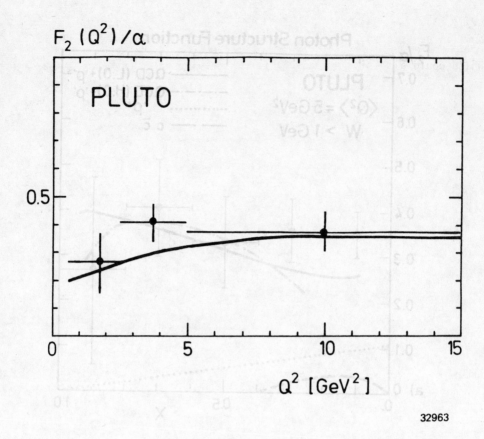

32963

<u>Fig. 31</u> - Photon structure function $F_2(q^2)/\alpha$.
The data are averaged over $0.2 < x_{vis} < 0.8$.
The curve shows the QCD prediction in the leading
logarithm approximation with
$\Lambda = 0.3$ GeV.

on x is necessary in order to avoid Q^2-x correlations (see Fig.30(b)), which could also simulate a scale breaking. The data are clearly consistent with the QCD prediction in the leading order (solid line).

8. ELECTROWEAK INTERACTION

We now come to perhaps the most exciting recent result from PETRA, the observation of weak and electromagnetic interference. Indeed, it was one of the main original purposes of building PETRA to observe such effect.

A. Standard Model

The Standard Model is the elegant model of unified weak and electromagnetic interaction developed by Glashow [73], Weinberg [74], and Salam [75]. In this model, the process

$$e^-e^+ \rightarrow f \bar{f}$$

can proceed not only through one-photon annihilation (Fig. 32a) but also through one-Z^0 annihilation (Fig. 32b), where f can be any fundamental fermion: e^-, μ^-, τ^-, ν_e, ν_μ, ν_τ, u, d, s, c, b or the yet unobserved top quark t. Neglecting radiative corrections, the differential cross section for this process is, in the standard model with the quark mass neglected,

$$
\frac{d\sigma(e^-e^+ \rightarrow f\bar{f})}{d\cos\theta}
$$

$$
= \frac{\pi \alpha^2}{2 s} \left\{ Q_f^2(1+\cos^2\theta) - \frac{2Q_f\, g\, s\, (s/M_Z^2 - 1)[v_e v_f(1+\cos^2\theta) + 2a_e a_f \cos\theta\,]}{(s/M_Z^2 - 1)^2 + \Gamma_Z^2/M_Z^2} \right.
$$

$$
+ \frac{s^2 g^2\, [(v_e^2 + a_e^2)\,(v_f^2 + a_f^2)(1 + \cos^2\theta) + 8v_e a_e\, v_f a_f \cos\theta]}{(s/M_Z^2 - 1)^2 + \Gamma_Z^2/M_Z^2} \tag{8.1}
$$

where M_Z and Γ_Z are the mass and width of Z^0, and

$$
g = \frac{\sqrt{2}\, G_F}{4e^2} = \frac{\sqrt{2}\, G_F}{16\pi\, \alpha} \tag{8.2}
$$

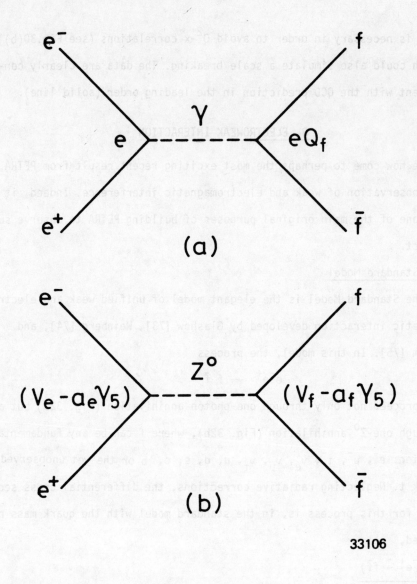

33106

<u>Fig. 32</u> - The process $e^-e^+ \rightarrow f\bar{f}$ in the
Standard Model.

in terms of the Fermi weak coupling constant G_F. The values of v and a in the Standard Model for various fermions are tabulated in Table 6. Integration over θ gives in particular

$$R_f = Q_f^2 - \frac{2Q_f \, g \, s \, (s/M_Z^2 - 1) \, v_e \, v_f - s^2 g^2 \, (v_e^2 + a_e^2)(v_f^2 + a_f^2)}{(s/M_Z^2 - 1)^2 + \Gamma_Z^2/M_Z^2} \qquad (8.3)$$

In both of these formulas, the color factor 3 has not been included when the fermion is a quark.

Table 6 - Values of v_f and a_f in the Standard Model

f	e^-, μ^-, τ^-	d, s, b	u, c, t
Q_f	-1	$-\frac{1}{3}$	$\frac{2}{3}$
v_f	$-1+4\sin^2\theta_W$	$-1+\frac{4}{3}\sin^2\theta_W$	$1-\frac{8}{3}\sin^2\theta_W$
a_f	-1	-1	1

At PETRA energies, Γ_Z is of no consequence. Also, the total cross section is more sensitive to v than to a, while the forward-backward asymmetry is more sensitive to a than to v. This asymmetry has the angular dependence

$$\frac{\cos\theta}{1+\cos^2\theta} \, ,$$

which is largest for $\theta = 0, \pi$. This has the consequence that the experimental observed asymmetry depends significantly on the acceptance of the detector.

B. Determination of $\sin^2\theta_W$ - MARK J, JADE

At the highest PETRA energy of 37 GeV, the dimensionless quantity gs, with g defined by (8.2) is about 0.06. Since this is quite sizable, ac-

703

curate measurements of the various cross sections can be used to determine the Weinberg angle [74] θ_W. This analysis has been carried out by MARK J [76] and JADE [77]. Of course, QCD and QED corrections need to be taken into account. It should be remembered that, in the Standard Model, the mass M_Z also depends on θ_W.

The resulting values of $\sin^2\theta_W$ from JADE and MARK J are

	JADE	MARK J
From hadron events only ($e^-e^+ \rightarrow q\bar{q}$)	0.22 ± 0.10 or 0.56 ± 0.10	$0.27 \begin{array}{c} + 0.34 \\ - 0.08 \end{array}$
From hadron and lepton events ($e^-e^+ \rightarrow q\bar{q}, \mu^-\mu^+, \tau^-\tau^+, e^+e^-$)	0.22 ± 0.08	0.27 ± 0.08

These results provide the first indication that (i) weak neutral current couples to all known quarks, since over 40% of the hadronic events involve a heavy quark, either c or b; (ii) the Standard Model is valid in the time-like region up to 1350 GeV^2.

C. Forward-backward asymmetry

Because of parity non-conservation in weak interactions, a dramatic effect of the interference between the one-photon annihilation of Fig. 32(a) and the one-Z^0 annihilation if Fig. 32(b) is the forward-backward asymmetry in $e^-e^+ \rightarrow f\bar{f}$. This is clearly seen as the $\cos\theta$ terms in eq.(8.1). Very recently, this asymmetry was observed for the first time in the process

$$e^-e^+ \rightarrow \mu^-\mu^+.$$

This gives one of the direct and beautiful experimental verifications of the Standard Model. We give here a detailed description of how this result is obtained.

In Table 7, the observed asymmetries $A_{\mu\mu}$ and $A_{\tau\tau}$ by JADE, MARK J, PLUTO, and TASSO with radiative corrections removed are given for both $e^-e^+ \to \mu^-\mu^+$ and $e^-e^+ \to \tau^-\tau^+$, together with the expected theoretical values from the Standard Model. The differences between the expected theoretical values for the various collaborations are due to the variation of the acceptances of the detectors, as already discussed at the end of section A. Only the value for $A_{\mu\mu}$ from TASSO has been corrected for acceptance. These data were obtained with an average s of

$$<s> \sim 1100 \text{ GeV}^2.$$

Clearly the data on the τ asymmetry is as yet not significant.

Table 7 - Forward-backward asymmetry of $e^-e^+ \to \mu^-\mu^+$ and $e^-e^+ \to \tau^-\tau^+$

	JADE	MARK J	PLUTO	TASSO	combined
$A_{\mu\mu}(\%)$	-11 ± 4	-3 ± 4	7 ± 10	-11.3 ± 5.0	-7.7 ± 2.4
expected	-7.8	-7.1	-5.8	-8.7	-7.8
$A_{\tau\tau}(\%)$		-6 ± 12		0 ± 11	-2 ± 8
expected		-5		-7	-6

Branson [78] has combined these four sets of data on the μ asymmetry to get the value, as given in Table 7, of

$$(- 7.7 \pm 2.4)\%$$

compared with the theoretical expected value of -7.8%. PETRA has therefore observed a forward-backward asymmetry in $e^-e^- \to \mu^-\mu^+$ which is in good agreement with the Standard Model, and is 3.2 standard deviations away from 0. If the preliminary CELLO data of $(-1.3 \, ^{+8}_{-10})\%$ is also included, the result changes by less than a tenth of a standard deviation.

705

In general, combining data from different experiments may be considered problematic. However, in the present case, the situation is favorable for at least two reasons: first all the experiments have been carried out simultaneously on the same accelerator PETRA, and, secondly, there is no known systematic error comparable to the present statistical error in the measurement of the forward-backward asymmetry. Therefore it is perfectly reasonable to think of this asymmetry measurement as one big experiment at PETRA.

There only remains a technical problem of taking into account the different acceptances as reflected in the theoretical expected values. Branson [78] solved this problem by considering only the ratio of the observed and theoretical asymmetries. Let

$$\gamma = \frac{A_{\mu\mu}|_{observed}}{A_{\mu\mu}|_{expected}}$$

then these ratios are

JADE	1.14 ± 0.51
MARK J	0.42 ± 0.56
PLUTO	-1.21 ± 1.72
TASSO	1.30 ± 0.57

These four values are then averaged using as weights the errors to the -2 power:

$$\frac{1.41(0.51)^{-2} + 0.42(0.56)^{-2} - 1.21(1.72)^{-2} + 1.30(0.57)^{-2}}{(0.51)^{-2} + (0.56)^{-2} + (1.72)^{-2} + (0.57)^{-2}} = 0.99.$$

From the point of view of one big experiment, the errors combine in the usual way to give an combined γ of

$$\gamma = 0.99 \pm 0.31$$

It is not crucial what value is used for the combined theoretical expected value. With -7.8%, the combined experimental value of (-7.7 ± 2.4)% is obtained.

Acknowledgements

I wish to thank Professor W.D.Shephard for his hospitality at the 1981 International Symposium on Multiparticle Dynamics. I am grateful to Professors R.Kose, T.Meyer, G.Rudolph, K.Steffen, P.Söding, B.H.Wiik, G.Wolf, and G.Zobernig for helpful and stimulating discussions. Finally I would like to thank the University of Wisconsin and DESY for their support.

707

REFERENCES

1) K.Robinson and G.A.Voss, CEA Report CEA-TM-149 (1965)

2) K.Steffen, Internal Report DESY M-79/07 (1979)

3) K.Steffen, G.A.Voss, and G.Wolf, Internal Report DESY M-81/20 (1981)

4) TASSO Collaboration, R.Brandelik et al., Phys.Lett.94B, 444 (1980)

5) JADE Collaboration, W.Bartel et al., Phys.Lett. 104B, 325 (1981)

6) TASSO Collaboration, R.Brandelik et al., Phys.Lett. 105B, 75 (1981)

7) B.Andersson, G.Gustafson, and T.Sjoestrand, Lund preprint
   LU TP 81-3 (1981)

8) J.Andersson, G.Gustafson, and C.Peterson, Nucl.Phys.B135, 273 (1978)

9) R.D.Field and R.P.Feynman, Nucl.Phys.B136, 1 (1978)

10) T.Meyer of TASSO Collaboration, private communication

11) P.Hoyer, P.Osland, H.G.Sander, and T.F.Walsh,
    Nucl.Phys. B 161, 349 (1979)

12) A.Ali, E.Pietarinen, G.Kramer, and J.Willrodt,
    Phys.Lett. 93 B, 155 (1980)

13) J.B.Andersson, G.Gustafson, and T.Sjoestrand, Phys.Lett. 94B, 211
    (1980)

14) M.Piccolo et al., Phys.Rev.Lett. 39, 1503 (1977)

15) G.S.Abrams et al., Phys.Rev.Lett. 44, 10 (1980)

16) T.Meyer, DESY Report 81/046 (1981)

17) D.Cords, rapporteur talk, Proceedings of the XXth International
    Conference on High Energy Physics, Madison, Wisconsin, USA,
    July 17-23, 1980

18) PLUTO Collaboration, J.Burmeister et al., Phys.Lett. 67B
    367 (1977)

19) SLAC-LBL Collaboration, V.Lüth et al., Phys.Lett.70B, 120 (1977)

20) PLUTO Collaboration, Ch.Berger et al., DESY Report 81/018 (1981)

21) TASSO Collaboration, R.Brandelik et al., Phys.Lett.94B, 91 (1980)

22) DELCO Collaboration, W.Bacino et al., Phys.Rev.Lett.42, 749 (1979)

23) TASSO Collaboration, R.Brandelik et al., Phys.Lett.92B, 199 (1980)

24) TASSO Collaboration (presented by P.L.Woodworth), Proceedings of
European Physical Society International Conference on High Energy
Physics, Lisbon, July 9-15, 1981

25) MARK II Collaboration, to be published.

26) P.Duinker, rapporteur talk, Proceedings of the European Physical
Society International Conference on High Energy Physics,
Lisbon, July 9-15 , 1981

27) TASSO Collaboration (presented by Sau Lan Wu), Proceedings of the
XXth International Conference on High Energy Physics, Madison,
Wisconsin, USA, July 17-23, 1980

28) TASSO Collaboration, R. Brandelik et al., Phys. Lett. 100B, 357 (1981)

29) JADE Collaboration, private communication

30) JADE Collaboration, W.Bartel et al., DESY Report 81/025 (1981)
(to be published)

31) J.C.Pati and Salam, Nucl.Phys. B 144, 445 (1978)

32) CELLO Collaboration, Contribution to the European Physical
Society International Conference on High Energy Physics,
Lisbon, July 9-15, 1981

33) R. Marshall, rapporteur talk, European Physical Society International Conferences on High Energy Physics, Lisbon, July 9-15, 1981

34) For definition of oblateness, see D.P.Barber et al.,
Phys.Rev.Lett. 43, 830 (1979)

35) MARK J Collaboration, D.P.Barber et al.,
MIT-L.N.S. Report 115 (1981)

36) G.Hanson et al., Phys.Rev.Lett. 35, 1609 (1975)

37) L.Clavelli, Phys.Lett. 85B, 111 (1979)
L.Clavelli and D.Wyler, Phys.Lett. 103B, 383 (1981)

38) PLUTO Collaboration, Ch.Berger et al., (to be published)

39) A.V.Smilga, Nucl.Phys. B 161, 449 (1979)

40) B.H.Wiik, Proceedings of the International Neutrino Conference, Bergen, Norway, 18-22 June 1979, p. 113
P.Söding, Proceedings of the European Physical Society International Conference on High Energy Physics, Geneva, Switzerland, 27 June - 4 July 1979, p. 271

41) TASSO Collaboration, R.Brandelik et al., Phys.Lett. 86B, 243 (1979)

42) MARK J Collaboration, D.P.Barber et al., Phys.Rev.Lett. 43, 830 (1979)

43) PLUSO Collaboration, Ch.Berger et al., Phys.Lett.86B, 418 (1979)

44) JADE Collaboration, W.Bartel et al., Phys.Lett. 91B, 142 (1980)

45) J.Ellis, N.K.Gaillard and G.G.Ross,Nucl.Phys. B111, 253 (1976)

46) TASSO Collaboration, R.Brandelik et al., Phys.Lett. 97B, 453 (1980)
and W.Braunschweig, Rapporteur talk, 1981, International Symposium on Lepton and Photon Interactions at High Energies, Bonn, August 24-29, 1981

47) J.Ellis and I.Karliner, Nucl.Phys. B148, 141 (1979)

48) PLUTO Collaboration, private communication

49) PLUTO Collaboration, Ch.Berger et al., Phys.Lett. 97B, 459 (1980)

50) Sau Lan Wu, Proceedings of the Arctic School of Physics, Äkäslompolo, Lapland, Finland (1980)
DESY Report 81/003, January 1981, section 4.3.5

51) JADE Collaboration, to be published

52) JADE Collaboration, Phys.Lett. 101B, 129 (1981)

53) Sau Lan Wu and G.Zobernig, Z.Phys. C2, 107 (1979)

54) TASSO Collaboration, R.Brandelik et al., Phys.Lett. 97B, 448 (1980)

55) D.L.Burke et al., Phys.Lett. 103B, 153 (1981)

56) CELLO Collaboration, contributed paper to the European Physical Society International Conference on High Energy Physics, Lisbon, July 9-15, 1981

57) J.Layssac and F.M.Renard, PM 81/5, Montpellier Report, May 1981

58) H.Goldberg and T.Weiler, Phys.Lett. 102B, 63 (1981)

59) S.M. Berman, J.D. Bjorken and J.B. Kogut, Phys.Rev. D4, 3388 (1971)

60) S.J. Bordsky, T. DeGrand, J. Gunion and J. Weis, Phys. Rev. D19, 1418 (1979)
    and Phys.Rev.Lett. 41, 672 (1978)

61) TASSO Collaboration, R.Brandelik et al., DESY Report 81-053 (1981)

62) JADE Collaboration, W.Bartel et al., DESY Report 81/048 (1981)

63) W.Wagner, Rapporteur talk, Proceedings of the XXth International
    Conference on High Energy Physics, Madison, Wisconsin, USA,
    July 17-23, 1980

64) Stan Brodsky, private communication

65) PLUTO Collaboration, Ch.Berger et al., (to be published)

66) E.Witten, Nucl.Phys. B120, 189 (1977)

67) Ch.Llewellyn Smith, Phys.Lett. 79B, 83 (1978)

68) W.R.Frazer and J.F.Gunion, Phys.Rev. D20, 147 (1979)

69) W.A.Bardeen and A.J.Buras, Phys.Rev. D20, 166 (1979)

70) D.W.Duke and J.F.Owens, Phys.Rev. D22, 2280 (1980)

71) C.Peterson, T.F.Walsh and P.M.Zerwas, Nucl.Phys.B 174, 424 (1980)

72) A.J.Buras and D.W.Duke, unpublished

73) S.L.Glashow, Nucl.Phys. 22, 579 (1961)

74) S.Weinberg, Phys.Rev.Lett. 19, 1264 (1967); Phys. Rev. D5, 1412 (1972)

75) A. Salam, Proceedings of the Eighth Nobel Symposium, May 1968, ed.
    N. Svartholm (Wiley, 1968), p. 367

76) MARK J Collaboration, D.P.Barber et al.,
    Phys.Rev.Lett. 46, 1663 (1981)

77) JADE Collaboration, W.Bartel et al., Phys.Lett. 101B, 361 (1981)

78) J.G.Branson, Rapporteur talk, 1981 International Symposium on
    Lepton and Photon Interactions at High Energies,
    Bonn, August 24-29, 1981

DISCUSSION

DeGRAND, SANTA BARBARA: I have a comment and a question.
The comment is that for the "high low" p_T's which you are seeing
in $\gamma\gamma$ annihilations there are all kinds of "crud" reactions that can
give jets in addition to $\gamma\gamma \to q\bar{q}$. So one should not yet worry too
much that the experiment is above the theory. You worry when it is
below the theory. The question is about the deep inelastic scattering. In your last picture you have a curve of theory versus experiment. Is the theory to which the $F_2^\gamma(x, Q^2)$ data is compared the
QCD calculation, or just the QED (parton model) calculation for the
structure function?

WU: I believe this is just the QED calculation. The gluon is not
included.

DeGRAND: You can't tell us why it dips over at large x then, which
is the way the structure function is supposed to look in the QCD
calculation? Is it some experimental reason?

WU: I think it's some acceptance problem.

GUNION, DAVIS: One thing we have become interested in at this
conference is the source of the increase in the multiplicity (you
know you have rapidly increasing multiplicity in all the PETRA
experiments). . . . where it is coming from in terms of p_T? Is
the rise in multiplicity coming at small p_T or is it related to the
larger-p_T tail that might have something to do with the gluon and
3-jet events?

WU: I skipped that. I thought it was already discussed. Here is
a picture of mean multiplicity as a function of c.m. energy at
PETRA. You can see a very rapid rise at PETRA energy. Now
here is a curve. The lower curve here comes from only the
Field-Feynman $q\bar{q}$ model. The top curve includes the gluon.
Including the gluon does not really change the multiplicity that
much. In fact it only changes about one unit in this region. So the
rapid rise basically comes from the phase space, the volume you
can produce, so therefore it has a lot to do with the low energy.

GUNION: And the low-p_T.

HWA, OREGON: In the proton produced in the jets, the momentum
distribution for different energy showed something peculiar at the
beginning. At higher energy the proton spectrum went down at

712

small x, as the energy went up from 17 to 32 GeV. Is there any change in that with more recent data? Do you have any examples?

WU: For example, this is the antiproton spectrum from JADE and TASSO, but you are talking about the energy variation. I don't have that here.

HWA: Is there any change?

WU: So far there is not much change, but what you have seen is very little statistics. So I think this may change, but we are not yet sure. We actually have very few protons.

HWA: You are saying that there is as much as 10%.

WU: Yes, but still not much.

MESTAYER, CHICAGO: In regard to weak-electromagnetic interference effects, have you measured any forward-backward asymmetry?

WU: Yes but that was . . . . My point here with the weak effect we were talking about was the observation of the Ψ^0 coupled to the quarks, the udscb quarks. For the forward-backward asymmetry, so far we have observed this for μ pair, τ pair, and e pair, so those are for the leptons. However the errors are still very large. We have those events, but it is still not a meaningful result.

MESTAYER: But there are fewer corrections to that kind of analysis aren't there, than our analysis?

WU: Yes, but it is less events also.

ANDERSSON, LUND: Do you have any comment on the angular and multiplicity asymmetry which the JADE group is claiming to see in connection with gluon emission?

WU: No, I have no comment, because we so far haven't been able to reproduce it.

MANN, TUFTS: You showed a plot of $\gamma\gamma$ invariant mass which contained an inclusive π^0 peak. In an untuned Field-Feynman fragmentation model one expects to produce a comparable number of η^0's. Do you observe an inclusive signal for $\eta^0 \to \gamma\gamma$?

WU: No. That is not because we don't agree with the Monte Carlo.

713

It's just that we haven't been able to get a detector to that stage yet. In fact, for the π^0 you see, the efficiency at this moment is not too high. So we are still working on the resolution of the detector. We do expect to see it if it is there.

FRIDMAN, SACLAY-VANDERBILT: You give the value of α_s which is found by the different PETRA groups. The LENA group which studied the Υ direct 3-gluon decay find also an α_s which is exactly the same as yours. But the energy entering in α_s is certainly different for those two cases. Can you comment about this matter?

WU: I do not know in the Υ decay range. I was told that in the Υ decay the higher-order term for α_s can be calculated more exactly, at least at this point. So, as I told you, the α_s we determined is the best number that we can get at this moment. It does not mean that it will be the true value, because we have to wait for these loop diagrams. So it could be smaller.

DeGRAND, SANTA BARBARA: Does any of the groups see a qualitative or quantitative difference between the various jets in a multi-jet event? Do any of them look in any way different from the other ones?

WU: Everybody is trying very hard, but so far there is no very strong statement from any of the groups. There is sort of more indication that they look more the same. However, there may be subtleties that we have not seen.

MORRISON, CERN: You quoted the π:K:p ratio as 55%:35%:10%. Could you or someone in the audience comment on whether these numbers are what one expects?

WU: This is the curve from the Feynman-Field fragmentation for K. So in that sense for kaon and pion there is reasonable agreement. I don't want to comment on proton, because the proton result we have here is with very limited statistics. We have a lot more statistics now and we are working on it.

MORRISON: Really my question is to the theoreticians. Is 10% reasonable? Does it mean you believe in quarks? Does it mean mean diquark-antidiquark production or is this necessary?

WU: Does anyone want to answer that? I also have an answer.

ANDERSSON: Of course it completely depends upon what kind of

714

expectations you have. But if you would blame the so-called SU(3) breaking, that is the fact that you see so many kaons as compared to pions, on an effect similar to this tunneling phenomenon we were talking about before . . . . that is essentially there is a larger mass for the s quark so that they are more difficult to polarize out of a constant force field (like we do in the Lund model and in the color flux tube model) . . . . then you can reasonably try to compute what would be your probability for producing corresponding heavy objects of the type which we have called diquarks. You can just introduce a mass parameter and look at how much there will be. Then you will come up with a number which is somewhere between 5 and 10% in the same scale.

MORRISON: So are you saying that diquarks exist?

ANDERSSON: No, I am not saying that they necessarily exist. I am only saying that it's possible to create masses of that size, evidently, inside a force field of this kind, and that would explain the baryonic production. I am not saying that there is this particular structure of a diquark. I am only saying that it is a part of a baryon which is essential.

WU: Also earlier when I showed the JADE result on $\bar{p}$ distributions, there is a solid curve which comes from the Lund model. The Lund model uses string pictures where you have $q\bar{q}$ and then a string and then a breakup into another $q\bar{q}$ and then they have 2 diquark pairs and from there you can form baryons. So they seem to agree reasonably well with the result. I wanted to go back to answer DeGrand's question on the gluon jet and quark jet. JADE has measured the photon energy versus the charged energy for the event that is a 3-jet candidate. Then you can analyze that into highest-energy jet, meadium-energy jet, and smallest-energy jet. Then the most energetic jets are mainly quark jets, the medium-energy jet is also mainly a quark jet, while the lowest-energy jet is around 60% or so gluon jets, at least from the Monte Carlo. They found that these ratios are not different basically.

GUNION, DAVIS: One more question on the multiplicities. Your Monte Carlos are very successful, including Feynman-Field and the 3-jet events and including the $c\bar{c}$ production and the charm decay and everything. Suppose I just took your Monte Carlos and did it only for the "up" and "down" quarks and then plotted the multiplicity graph you get from your Monte Carlos. How similar would it be to the total multiplicity graph that I see on your curves?

715

WU: I don't have a curve here, but I think it would differ by one or two units.

GUNION: I see, That much. You think that, in other words, the charm decay modes will have a very substantial effect. I think I would be very interested in seeing that graph if you could produce it.

WU: That's my guess.

LOW-p_T PROTON-PROTON COLLISION
AT THE CERN INTERSECTING STORAGE RINGS

M. Basile, G. Cara Romeo, L. Cifarelli, A. Contin, G. D'Ali,
P. Di Cesare, B. Esposito, P. Giusti, T. Massam, R. Nania,
F. Palmonari, V. Rossi, G. Sartorelli, M. Spinetti, G. Susinno,
G. Valenti, L. Votano and A. Zichichi

Presented by G. Valenti

CERN, Geneva, Switzerland
Istituto di Fisica dell'Università di Bologna, Italy
Istituto Nazionale di Fisica Nucleare, Laboratori Nazionali
di Frascati, Italy
Istituto Nazionale di Fisica Nucleare, Sezione di Bologna, Italy
Istituto di Fisica dell'Università di Perugia, Italy
Istituto di Fisica dell'Università di Roma, Italy

ABSTRACT

A study of the properties of multihadron final states in pp collisions at three centre-of-mass energies, $\sqrt{s}$ = 30, 44, and 62 GeV, is presented. Specifically the introduction of the effective hadronic energy E_{had} = $\sqrt{s}/2$ - E_{proton} allows a meaningful comparison with multihadron productions in e^+e^- annihilations.

1. INTRODUCTION

The presence of the leading proton effect[1,2] in proton-proton interactions suggests the use of the effective hadronic energy[3]

$$E_{had} = \sqrt{s}/2 - E_{proton}$$

as a relevant parameter in the description of the multihadron final states resulting from the interaction. In fact the leading proton manifests itself with a nearly flat $x_{Feynman}$ distribution in the interval x_F = 0 to x_F = 1 and therefore takes away a sizeable (on the average, one half) fraction of the incoming proton energy. It is therefore interesting to determine whether the multihadron systems produced in pp interactions, when studied in terms of E_{had}, show properties analogous to the multihadron systems produced in e^+e^- annihilation.

2. DATA TAKING AND ANALYSIS

The experiment was performed at the Split-Field Magnet (SFM) facility[4] of the CERN Intersecting Storage Rings (ISR), and data were collected at three total centre-of-mass (c.m.) energies: $\sqrt{s}$ = 30, 44, and 62 GeV. Two triggers have been used to collect the data:

i) "minimum bias", essentially the requirement of a track anywhere in the apparatus;

ii) "high-x", at least one fast track in a forward cone around the direction of the outgoing beams.

One event is analysed if the following conditions for one hemisphere are both satisfied:

i) at least two tracks fit to a common vertex in the interaction fiducial volume (diamond);

ii) one leading proton is identified, i.e. if the fastest particle in the hemisphere is positively charged and with $0.44 \leq x_F \leq 0.85$ ($x_F = 2p_L/\sqrt{s}$, where p_L is the momentum component along the pp line of flight in the pp c.m. system and $\sqrt{s}$ is the total pp c.m. energy).

Furthermore, only those tracks are retained for which an estimated precision in the momentum determination $\Delta p/p \leq 30\%$ is obtained.

All the data presented have been corrected for acceptance loss determined via Monte Carlo calculation.

3. FRACTIONAL MOMENTUM DISTRIBUTION

The inclusive single-particle fractional momentum distribution is obtained defining, for each accepted particle in the event, the reduced radial momentum fraction in terms of the effective hadronization energy E_{had}:

$$x_R^* = \frac{|\vec{p}|}{E_{had}} .$$

This $(1/N_{ev})(dN_t/dx_R^*)$ is then compared with the equivalent quantity measured in e^+e^-,

$$\frac{1}{2\sigma} \frac{d\sigma}{dx_R} ,$$

where $\sigma \equiv \sigma(e^+e^- \to hadrons) = R\left[\sigma(e^+e^- \to \mu^+\mu^-)\right]$ and $x_R = 2|\vec{p}|/\sqrt{s}_{e^+e^-}$. The comparison is done selecting equal intervals of $\sqrt{s}_{e^+e^-}$ and $2E_{had}$.

Figures 1a,b,c and 2a,b,c show a detailed comparison of $(1/N_{ev})(dN_t/dx_R^*)$ with data obtained in e^+e^- (SLAC, PETRA)[5] in the $2E_{had}$ energy range from 3 to 32 GeV. From this comparison two features emerge:

i) a satisfactory agreement in the inclusive single-particle reduced fractional momentum distribution is obtained at all energies, in particular the strong increase of particle production at small x_R ($x_R \leq 0.1$) is also present in pp interactions for ($x_R^* \leq 0.1$);

ii) data characterized by the same value of $2E_{had}$ obtained at different $\sqrt{s}_{pp}$ are identical within the errors.

718

4. AVERAGE CHARGED MULTIPLICITY

In order to measure the average charged multiplicity $\langle n_{ch} \rangle$ as a function of E_{had} [6], we have dropped the requirement that a track must be determined with a $\Delta p/p \leq 30\%$.

This increases the solid-angle coverage to nearly 4π.

The contamination from $K^0_S \rightarrow \pi^+\pi^-$ has been evaluated using inclusive production data. It turns out to be of the order of 4% and has been subtracted. The effect of misidentification of the leading proton has been studied via Monte Carlo simulation; for our data it is smaller than 2% at all E_{had}.

The average charged multiplicity $\langle n_{ch} \rangle$ [*] versus $2E_{had}$ for data collected at $\sqrt{s_{pp}}$ = 30, 44, and 62 GeV is shown in Fig. 3. The average of these data is compared in Fig. 4 with data from PLUTO [7] at DESY, SPEAR [8] at SLAC, and ADONE [9] at Frascati. Also shown is the inclusive average charge multiplicity measured at the same c.m. energies (standard pp multiplicity).

Notice that the spread of points at all E_{had} is everywhere ≤ 0.4 charged units, i.e. $\leq 0.5\%$. This is comparable to the systematic error.

We have also repeated the analysis for a sample of events where both leading protons are identified. In this case we can define as effective hadronic energy the invariant mass of the system after removal of the two leading protons:

$$M_{12} = \sqrt{(E_{had\,1} + E_{had\,2})^2 - |(\vec{p}_{had\,1} + \vec{p}^{\,*}_{had\,2})|^2} \ .$$

Figure 5 shows the average charge multiplicity as a function of M_{12} for $\sqrt{s_{pp}}$ = 30, 44, and 62 GeV combined data. Again a very satisfactory agreement of the data is found.

5. THE RATIO $\alpha = \langle E_{ch} \rangle / \langle E_{tot} \rangle$

The value of the ratio α in e^+e^- is an interesting characteristic of the multihadron system produced. Its importance is related to the possibility of determining the missing neutral energy; and this, in turn, to the presence of new channels contributing to it. It is therefore of great interest to measure the value of the analogous quantity in pp interactions [10], i.e.

$$\alpha_{pp} = \left. \frac{\langle E_{ch} \rangle}{\langle E_{tot} \rangle} \right|_{pp} \ .$$

Figure 6 shows $\langle \alpha_{pp} \rangle = \langle E_{ch} \rangle / \langle E_{had} \rangle$ versus $2E_{had}$ at $\sqrt{s_{pp}}$ = 30, 44, and 62 GeV; also shown are data points from e^+e^- Mark I (SPEAR) [8] and JADE (PETRA) [11]. For $\langle \alpha_{pp} \rangle$ we obtain the value:

---

*) The experimental multiplicity is multiplied by 2.

719

$$\langle \alpha_{pp} \rangle = 0.57 \pm 0.04 \ .$$

If we consider only events with two identified leading protons, we can study $\langle \bar{\alpha}_{pp} \rangle = \langle E_{ch} \rangle / \langle M_{12} \rangle$ versus the total invariant mass after leading proton subtraction, i.e. M_{12}. Figure 7 shows $\langle \bar{\alpha}_{pp} \rangle$ versus M_{12}. Again we obtain:

$$\langle \bar{\alpha}_{pp} \rangle = 0.57 \pm 0.07 \ .$$

From this value and taking into account the known production of neutral hadrons, we compute the "missing" neutral energy α_{pp}^0 associated with particles other than π^0, K^0, $\bar{K}^0$, n, and $\bar{n}$:

$$\alpha_{pp}^0 = 0.10 \pm 0.10 \ .$$

6. PLANARITY

Details on the structure of the interactions can be inferred by the angular and momentum distributions of the particles inside one event.

In particular the presence of three-jet events at PETRA has been obtained studying planar events. It is therefore interesting to see whether low-p_T pp interactions show a similar jet topology.

The event shape has been studied[12] performing a two-dimensional momentum tensor analysis on the plane orthogonal to the pp line of flight (c.m.s.). For each event one constructs

$$M_{\alpha\beta} = \sum_{j=1}^{N} p_{j\alpha} p_{j\beta} \qquad (\alpha, \ \beta = 1, \ 2) \ ,$$

where N = number of particles in the event, and determines the eigenvectors $\vec{n}_1$, $\vec{n}_2$ and eigenvalues Λ_1, Λ_2 ($\Lambda_1 < \Lambda_2$). With this definition, $\vec{n}_1$ is the direction in which the sum of the square of the momentum components is minimized, and $\vec{n}_2$ together with the pp line of flight defines the event plane (see Fig. 8).

It is then possible to define the average p_T^2 "in" and "out" of the event plane according to

$$\langle p_T^2 \rangle_{out} = \frac{\Lambda_1}{N} = \frac{1}{N} \sum_{j=1}^{N} (\vec{p}_j \cdot \vec{n}_1)$$

$$\langle p_T^2 \rangle_{in} = \frac{\Lambda_2}{N} = \frac{1}{N} \sum_{j=1}^{N} (\vec{p}_j \cdot \vec{n}_2) \ .$$

In Fig. 9, $(1/N_{ev})(dN/d\langle p_T^2 \rangle)$ for the cases "in" and "out" is plotted for $28 \leq 2E_{had} \leq 34$ GeV and is compared with a limited-p_T phase space (LPS) Monte Carlo[13]. While the "out" distribution is well

720

described by the Monte Carlo, the "in" distribution shows a greater number of events with higher $\langle p_T^2 \rangle$, i.e. the multihadron system contains a few percent of "planar events"[*]. Also shown are e^+e^- TASSO data[14] collected at $27.4 \le \sqrt{s}_{e^+e^-} \le 31.6$ GeV.

Figures 10 and 11 show data collected at $\sqrt{s}_{pp}$ = 30, 44, and 62 GeV for two intervals of the effective hadronic energy: $10 \le 2E_{had} \le 16$ GeV and $16 \le 2E_{had} \le 22$ GeV, respectively. Two points should be noted:

i)   the agreement of data taken at different $\sqrt{s}_{pp}$ is very good;

ii) contrary to e^+e^- data, when comparing low E_{had} data with high E_{had} data no substantial increase of the fraction of high $\langle p_T^2 \rangle_{in}$ events is observed.

7. INCLUSIVE p_T^2 DISTRIBUTION

The inclusive p_T^2 distribution has been obtained by determining the p_T of the individual hadrons with respect to the pp beam axis (c.m.s.).

Figures 12a-d show $(1/N_{ev})(dN/dp_T^2)$ for data collected at $\sqrt{s}_{pp}$ = = 30 GeV and for effective hadronic energies between 3 and 13 GeV.

Also shown are fits to e^+e^- data from Mark I at $\sqrt{s}$ = 3, 4.8, and 7.0-7.8 GeV [8] and data points at $\sqrt{s}$ = 12 GeV from TASSO.

Figure 13 shows $(1/N_{ev})(dN/dp_T^2)$ for data collected at $\sqrt{s}_{pp}$ = 62 GeV and for two extreme, effective hadronic energy bands, i.e. $13 \le 2E_{had} \le 17$ GeV and $28 \le 2E_{had} \le 32$ GeV. Also shown are e^+e^- data from TASSO at $\sqrt{s}_{e^+e^-}$ = 12 GeV and 27.4-31.6 GeV [14].

From this comparison one derives that from $2E_{had}$ = 3 to 13 GeV the transverse momentum square distributions obtained in pp interactions agree quite well with e^+e^- data. Above 13 GeV the distribution does not show an increase at high p_T^2 values as marked as that shown by the data of e^+e^-.

8. CONCLUSIONS

The analysis presented covers the main features of the multihadron production in low-p_T pp collisions. The fractional momentum distribution, the average charge multiplicity, and the ratio $\langle E_{ch} \rangle / \langle E_{tot} \rangle$ are very similar to equivalent distributions for hadrons produced in e^+e^- annihilations. Proton-proton data also show a degree of planarity similar to e^+e^- data, however, the $\langle p_T^2 \rangle_{in}$ distribution is nearly constant versus the effective hadronic energy. This feature is confirmed by the inclusive transverse momentum distributions which show no marked energy dependence above $2E_{had}$ = 13 GeV.

The observation of these similarities and differences gives a new perspective to the multihadron production in proton-proton collisions and e^+e^- annihilations, and is certainly bound to be of great importance in the process of understanding the underlying hadronization mechanism at work in these highly inelastic processes.

---

[*] Planar when compared with the LPS Monte Carlo.

REFERENCES

1) M. Basile et al., The "leading" baryon effect in strong and weak interactions, preprint in preparation.
2) M. Basile et al., The "leading" particle effect in hadron physics, preprint in preparation.
3) M. Basile et al., Phys. Lett. 92B, 367 (1980).
4) Details of the apparatus can be found in R. Bouclier et al., Nucl. Instrum. Methods 125, 19 (1975).
   M. Basile et al., Nucl. Instrum. Methods 179, 477 (1981), and 163, 93 (1979).
5) J.L. Siegrist, Thesis SLAC-Report No. 225 (1979).
   TASSO Collaboration (R. Brandelik et al.), Phys. Lett. 89B, 418 (1980).
6) M. Basile et al., Charged particle multiplicities in (pp) interactions and comparison with (e+e-) data, Submitted to the 12th Int. Symp. on Multiparticle Dynamics, Notre Dame, Ind., 1981.
7) PLUTO Collaboration (Ch. Berger et al.), DESY preprint 80/69 (1980).
8) J.L. Siegrist (Ref. 5), and
   V. Lüth et al., Phys. Lett. 70B, 120 (1977).
9) C. Bacci et al., Phys. Lett. 86B, 234 (1979).
10) M. Basile et al., Phys. Lett. 99B, 247 (1981).
11) JADE Collaboration (W. Bartel et al.), DESY Report 80/46 (1980).
12) M. Basile et al., Nuovo Cimento Lett. 29, 491 (1980).
13) D. Drijard, private communication.
14) TASSO Collaboration (R. Brandelik et al.), Phys. Lett. 86B, 243 (1979).

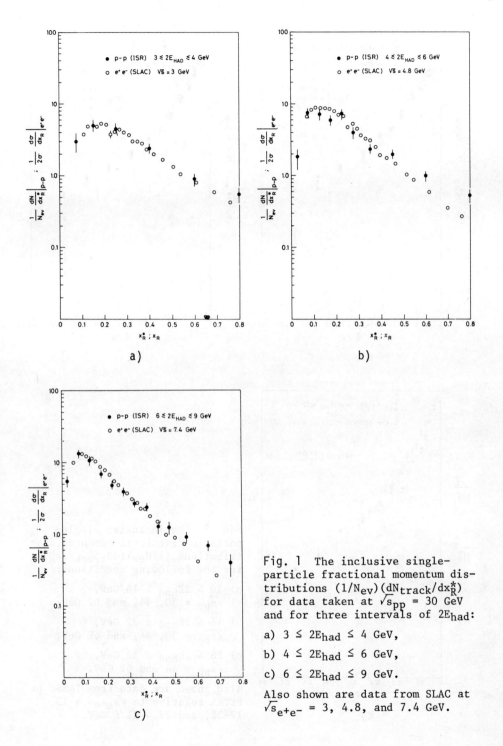

Fig. 1 The inclusive single-particle fractional momentum distributions $(1/N_{ev})(dN_{track}/dx_R^*)$ for data taken at $\sqrt{s}_{pp} = 30$ GeV and for three intervals of $2E_{had}$:

a) $3 \leq 2E_{had} \leq 4$ GeV,

b) $4 \leq 2E_{had} \leq 6$ GeV,

c) $6 \leq 2E_{had} \leq 9$ GeV.

Also shown are data from SLAC at $\sqrt{s}_{e^+e^-} = 3$, 4.8, and 7.4 GeV.

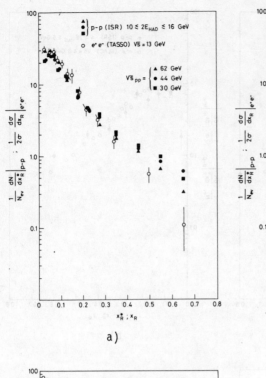

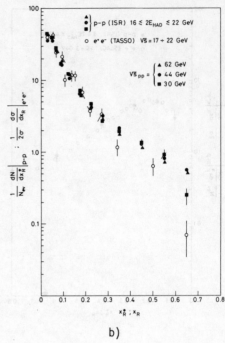

a)   b)

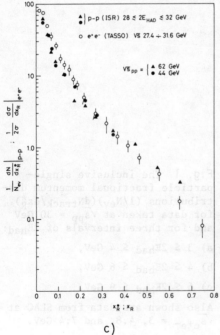

Fig. 2 The inclusive single-particle fractional momentum distributions $(1/N_{ev})(dN_{track}/dx_R^*)$ for the following conditions:

a) $10 \leq 2E_{had} \leq 16$ GeV, $\sqrt{s}_{pp}$ = 30, 44, and 62 GeV;

b) $16 \leq 2E_{had} \leq 22$ GeV, $\sqrt{s}_{pp}$ = 30, 44, and 62 GeV;

c) $28 \leq 2E_{had} \leq 32$ GeV, $\sqrt{s}_{pp}$ = 44 and 62 GeV.

Also shown are data from TASSO at PETRA relative to $\sqrt{s}_{e^+e^-}$ = 13, 17-22, and 27.4-31.6 GeV.

c)

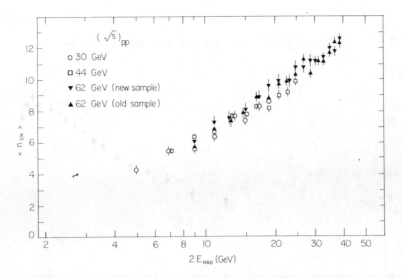

Fig. 3 The average charged multiplicity $\langle n_{ch} \rangle$ plotted versus $2E_{had}$ for pp data collected at $\sqrt{s}_{pp}$ = 30, 44, and 62 GeV

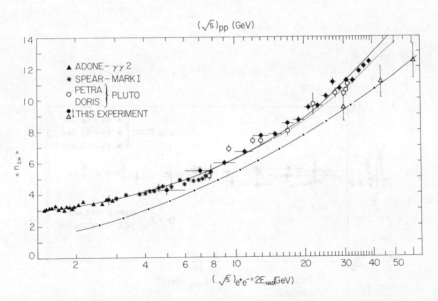

Fig. 4 The combined ($\sqrt{s}_{pp}$ = 30, 44, and 62 GeV) average charged multiplicity for pp data versus $2E_{had}$ compared to e^+e^- data. The dotted-dashed line is a fit to pp standard $\langle n_{ch} \rangle$ measurements.

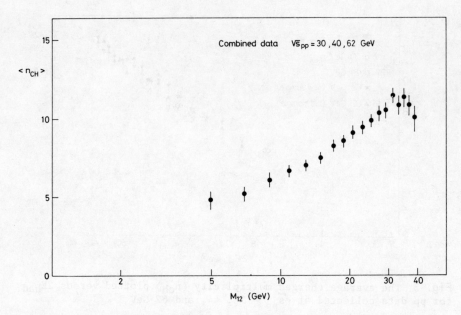

Fig. 5 The combined ($\sqrt{s}_{pp}$ = 30, 44, and 62 GeV) average charged
multiplicity per pp collisions versus M_{12}

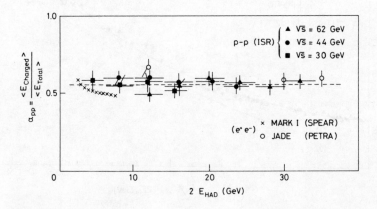

Fig. 6 The charged to total energy ratio obtained in pp collisions,
α_{pp}, plotted versus $2E_{had}$ and compared with e^+e^- data obtained at
SLAC and PETRA

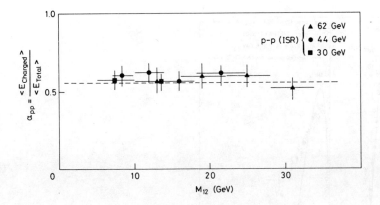

Fig. 7 The charged to total energy ratio $\bar{\alpha}_{pp}$ plotted versus M_{12}

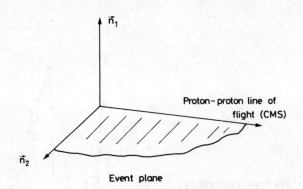

Fig. 8 Event plane definition

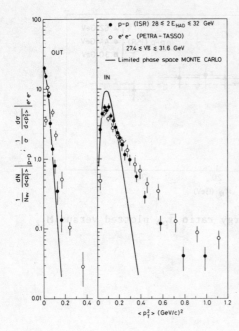

Fig. 9 The average transverse
momentum squared distributions
$(1/N_{ev})(dN_{track}/d\langle p_T^2\rangle)$ for the "in"
and "out" plane for the effective
hadronic energy interval
$28 \leq 2E_{had} \leq 34$ GeV. Also shown
are e^+e^- data obtained at PETRA
(TASSO).

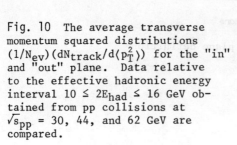

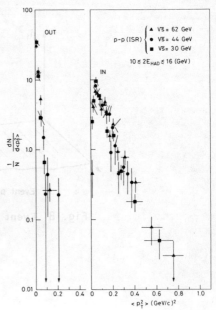

Fig. 10 The average transverse
momentum squared distributions
$(1/N_{ev})(dN_{track}/d\langle p_T^2\rangle)$ for the "in"
and "out" plane. Data relative
to the effective hadronic energy
interval $10 \leq 2E_{had} \leq 16$ GeV obtained from pp collisions at
$\sqrt{s}_{pp} = 30, 44,$ and 62 GeV are
compared.

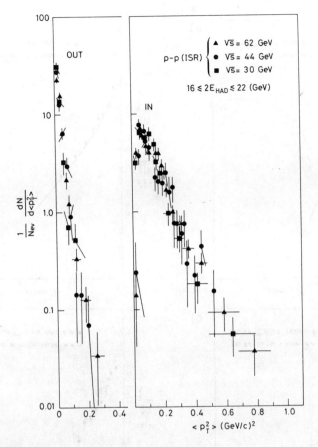

Fig. 11 The average transverse momentum squared distributions
$(1/N_{ev})(dN_{track}/d\langle p_T^2\rangle)$ for the "in" and "out" plane. Data relative
to the effective hadronic energy interval $16 \leq 2E_{had} \leq 22$ GeV obtained from pp collisions at $\sqrt{s}_{pp} = 30$, 44, and 62 GeV are compared.

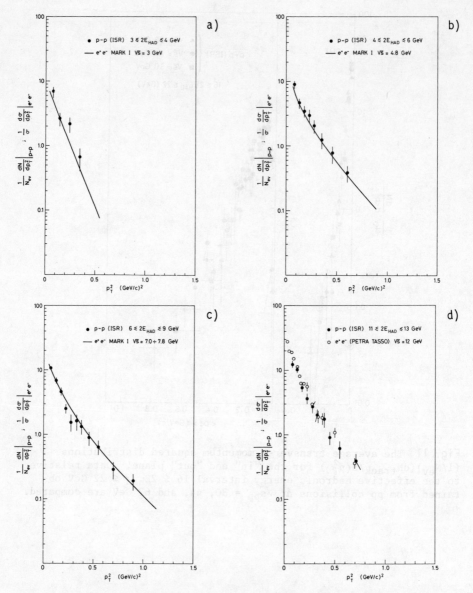

Fig. 12 The inclusive single-particle transverse momentum distributions $(1/N_{ev})(dN_{track}/dp_T^2)$ for data collected at $\sqrt{s}_{pp}$ = 30 GeV, and for the following intervals of the effective hadronic energies:

a) $3 \leq 2E_{had} \leq 4$ GeV,   b) $4 \leq 2E_{had} \leq 6$ GeV,

c) $6 \leq 2E_{had} \leq 9$ GeV,   d) $11 \leq 2E_{had} \leq 13$ GeV.

Also shown are fits to e^+e^- data (a, b, c) and data points obtained at SPEAR (Mark I) and PETRA at $\sqrt{s}$ = 3, 4.8, and 7.0-7.8 GeV and 12 GeV.

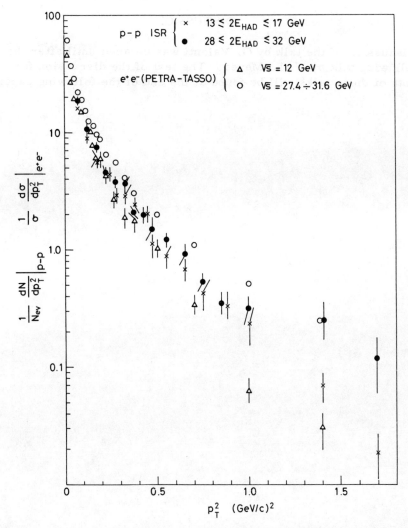

Fig. 13 The inclusive single-particle transverse momentum distributions $(1/N_{ev})(dN_{track}/dp_T^2)$ for data collected at $\sqrt{s}_{pp}$ = 62 GeV and for the two extreme, effective hadronic energy intervals:

i) $13 \leq 2E_{had} \leq 17$ GeV,

ii) $28 \leq 2E_{had} \leq 32$ GeV.

DISCUSSION

Discussion of the talk by G. Valenti was delayed until after the following talk by W. T. Meyer. The text of the discussion for both of these talks will be found at the end of the following paper.

CERN/EP/0481R/TWM/ef
18 August 1981

STUDY OF HADRONIC EVENTS IN pp COLLISIONS AT $\sqrt{s}$ = 62 GeV

AND COMPARISON WITH HADRONIC EVENTS IN e^+e^- COLLISIONS

A. Breakstone, H.B. Crawley, A. Firestone, M. Gorbics,
J.W. Lamsa, W.T. Meyer, and D.L. Parker
Depart. of Phys. and Ames Lab., Iowa State University, Ames, USA

D. Drijard, F. Fabbri[(*)], H.G. Fischer, H. Frehse, P. Hanke,
P.G. Innocenti, and O. Ullaland
CERN, European Organization for Nucl. Res., Geneva, Switzerland

W. Hofmann, M. Panter, K. Rauschnabel, J. Spengler and D. Wegener
Institut für Physik der Universität Dortmund, Dortmund, Germany

W. Geist, M. Heiden, E.E. Kluge, T. Nakada and A. Putzer
Institut für Phys. der Universität Heidelberg, Heidelberg, Germany

K. Doroba, R. Gokieli and R. Sosnowski
Institute for Nuclear Research, Warsaw, Poland

Presented by: W.T. Meyer

ABSTRACT

We present an analysis of minimum bias events from pp
collisions at $\sqrt{s}$ = 62 GeV in the CERN ISR. The effects of both
the leading protons are removed and the B = 0 mesonic residue is
compared to hadronic events of similar energy produced in e^+e^-
collisions. Although the energy dependence of the average sphericity and average charge multiplicity are similar in the two
cases, the transverse momentum distributions with respect to the
sphericity axis show significant differences. These data are
consistent with the predictions of a longitudinal phase-space
model.

Presented at the XII International Symposium
on Multiparticle Dynamics, 21-26 June 1981, Notre Dame, USA

It is of interest to compare the jet structure observed in e^+e^- annihilations into hadrons with possible jet structure in purely hadronic interactions. Such a comparison has been performed using pp collision data taken with a minimum bias trigger in the CERN ISR in which one final state leading proton has been selected [1]. In that work an attempt has been made to account for the observed difference between pp and e^+e^- interactions through the difference in baryon number. Specifically, a leading proton was selected, and the analysis was performed on only those other hadrons in the same centre-of-mass hemisphere as the leading proton. In the present work on pp interactions we insist on a selection of both leading protons, and perform the jet analysis on the entire B = 0 mesonic system remaining after the removal of the two leading protons.

This experiment was performed at the CERN ISR at $\sqrt{s}$ = 62 GeV, using the Split Field Magnet (SFM) detector, a device which allows one to measure the momenta of charged particles in nearly the full solid angle of 4π steradians. Information about the performance of the detector can be found in previous publications [2].

From a sample of 390 000 minimum bias triggers, events with probable protons were selected using the criteria that the particle in each hemisphere with the largest longitudinal momentum with respect to the beam direction should be positively charged, and should have a value of $x = 2p_L^*/\sqrt{s}$ between 0.44 and 0.82 in magnitude. At $x \simeq 0.4$ the π^+ production rate is comparable to that of the protons but as x increases in magnitude, the p/π^+ ratio increases sharply, so the assumption that the leading positive particle is a proton becomes more accurate. The upper limit on x was chosen to remove all diffractive events. In addition, a momentum error cut $\Delta p/p < 0.08$ was required for this leading particle. For this analysis only those events were selected in which there was one leading proton in each of the hemispheres with respect to the beam axis. These form a sample of 3283 events. These data were corrected for geometrical acceptance losses in the detector, losses due to decay and secondary interactions, and inefficiencies of the analysis chain.

In order to compare properly the pp interaction data with data on e^+e^- annihilations, it is necessary not only to remove the two identified protons so that we are comparing B=0 systems, but also to transform all momenta into the rest frame of the mesonic system. This can be done using knowledge of the incident beam momenta and the momenta of the two leading protons without reference to missing neutral particles. In the rest frame of the mesonic system the event appears as sketched in fig. 1. The two beam particles, p_1 and p_2 transfer equal and opposite momenta to the meson system.

The quantity which corres-
ponds to $\sqrt{s}$ in e^+e^- annihilations
is the invariant mass of the
mesonic system, M_0, and this too
can be calculated entirely from
the known beam and proton momenta
without reference to possible
missing neutrals. Due to the
finite range of allowed x over
which protons are selected, there
is a distribution of values for
the invariant mass of the mesonic
system, as shown in fig. 2. Fig. 3

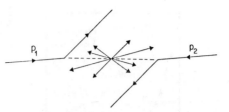

Fig. 1 - Diagram of the
"two- proton" events.

shows the average sphericity as a function of M_0, along with the
e^+e^- annihilation data from the TASSO Collaboration [3]. Although
the two cases are similar, the average sphericity in this experiment is consistently somewhat below that in the e^+e^- case. Also
shown in fig. 3 are the predictions of a longitudinal phase space
model, which agree well with the data of this experiment.

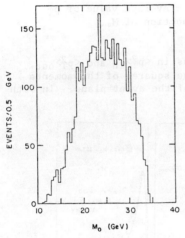

Fig. 2 - Invariant mass of the B=0
mesonic system after the removal
of the two identified protons.

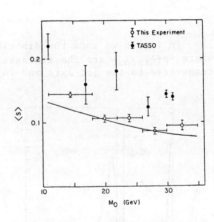

Fig. 3 - Average sphericity
as a function of M_0.

In fig. 4 we show the distributions in p_T^2, where p_T^2 is the
square of the individual hadron's momentum transverse to the jet
axis for events in the lowest and highest regions of M_0. We
observe no significant change in the shape of the p_T^2 distribution
with M_0. Fig. 5 shows the average value of p_T^2 as a function of
M_0. The data of this experiment show no dependence on M_0. In
contrast, the data of the TASSO Collaboration, [3] also shown in
fig. 5 show a pronounced dependence of $\langle p_T^2 \rangle$ on $\sqrt{s}$.

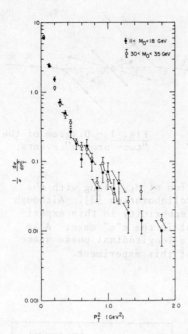

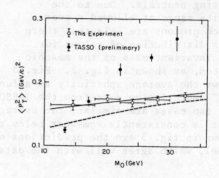

Fig. 4 - Distribution in p_T^2 of
the individual hadrons with respect
to the jet axis for the lowest and
highest regions of M_0.

Fig. 5 - Average value of p_T^2
as a function of M_0.

In fig. 6 we show the distributions in $\langle p_T^2 \rangle_{in}$ and $\langle p_T^2 \rangle_{out}$, where $\langle p_T^2 \rangle_{in/out}$ are the averages of the squares of the momenta transverse to the jet axis and in/out of the event plane. In both

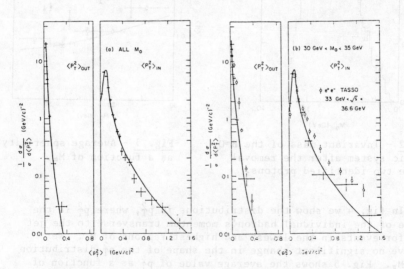

Fig. 6 - Distribution in $\langle p_T^2 \rangle_{in/out}$ with respect to the jet axis in/out of the "event plane": (a) All M_0, and (b) $M_0 > 30$ GeV. In (b) are also shown data from TASSO.

736

cases the data are in agreement with the predictions of the longitudinal phase-space model. This model used as input the observed
charged and neutral multiplicity distribution, and imposed energy,
momentum and charge conservation on the Monte-Carlo events. Individual particles were generated flat in χ and according to
$\exp(-4.5\ M_T)$, where $M_T = \sqrt{p_T^2 + M^2}$
is the transverse mass of the
particle. We observe that there
is no significant broadening of
either the $\langle p_T^2 \rangle_{in}$ or $\langle p_T^2 \rangle_{out}$
distributions with increasing
M_0. This is most clearly
seen in fig. 7, which shows the
distributions in $\langle p_T^2 \rangle_{in}$ for the
lowest and highest regions of M_0.
This result is in disagreement
with that of ref. [5] which claims
such an energy dependence for
$\langle p_T^2 \rangle_{in}$. We also note the generally good agreement with the
predictions of the longitudinal
phase-space model in fig. 6.

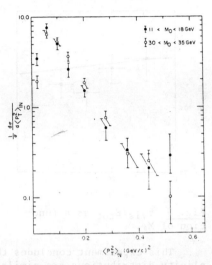

Fig. 7 - Distribution in
$\langle p_T^2 \rangle_{in}$ for the lowest and
highest region of M_0.

In fig. 8 we show the angular distribution of the sphericity
or jet axis with respect to the
direction of the momentum transfers to the B = 0 mesonic system
as shown in fig. 1. The data in
fig. 8 do not show the characteristic $1 + \cos^2\theta$ angular distribution expected from a one-photon
intermediate state and seen in
e^+e^- annihilations, but instead
are sharply peaked in the forward
direction. This is just the behaviour expected if the secondary
particles are produced largely
in the forward-backward direction, as for example in a model
of phase-space with limited
transverse momenta. This distribution does not change significantly with M_0. The dashed
line corresponds to the predictions of the longitudinal phase-
space model.

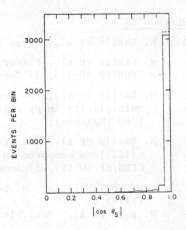

In figs 9 and 10 we show
the quantities E_{vis}/E_{tot} and $\langle n_c \rangle$
as functions of M_0. E_{vis} is the
visible (i.e., charged) energy

Fig. 8 - Angular distrib-
ution of the jet axis.

737

in the overall centre of mass assuming all particles are pions in the $B = 0$ system, while E_{tot} is the total energy in this system obtained from subtracting the energies of the two identified protons from the total c.m. energy.

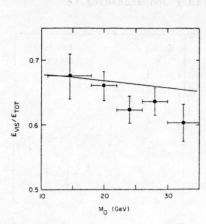

Fig. 9 E_{vis}/E_{tot} as a function of M_0.

Fig. 10 $\langle n_c \rangle$ as a function of M_0.

This experiment concludes that although sphericity and multiplicity distributions are similar for $e^+e^- \rightarrow$ hadrons and purely hadronic interactions, when the effects of the leading proton are removed from the latter, the transverse momentum distribution are significantly different. Furthermore, all the purely hadronic data are consistent with a model of longitudinal phase space.

REFERENCES

[1] M. Basile et al., Phys. Lett. 92B (1980) 367;

     M. Basile et al., Planar jets in pp collisions,
        CERN/EP 80-113, 27 June 1980 (Madison);

     M. Basile et al., The energy dependence of charged particle
        multiplicity in pp collisions, CERN/EP 80-112, 27 June
        1980 (Madison);

     M. Basile et al., The fractional momentum distribution in pp
        collisions compared to e^+e^- annihilation,
        CERN/EP 80-111, 27 June 1980 (Madison).

[2] CCHK Collaboration, M. Della Negra et al., Nucl. Phys. B127
        (1977) 1;
     W. Bell et al., Nucl. Instr. and Meth. 156 (1978) 111.

[3] TASSO Collaboration, R. Brandelik et al., Phys. Lett. 89B
        (1980) 418;
     TASSO Collaboration, R. Brandelik et al., Phys. Lett. 83B
        (1979) 261.

DISCUSSION
(for talks by Valenti and Meyer combined)

VAN HOVE, CERN AND SANTA BARBARA: I have a question which I believe is mainly to the second speaker. It has nothing to do with the comparison with e^+e^- but, using your sample of events in which you know where the protons are and in which you have all the other tracks, have you signs of planarity for these events? I mean the events including the protons.

MEYER: Including the protons? Well, the only planarity plot I can show you looks very similar to the e^+e^-. On a scatter plot like this, you can't get quantitative, but we see most of them clustering down here in the two-jet region and straggling tails out elsewhere.

VAN HOVE: Could you make a similar plot including the proton?

MEYER: Of course we could.

MORRISON, CERN: I make a comment and ask a question. The general comment is that phase space dominates a tremendous number of things, and almost any experiment you can do you can find many effects which are phase space and therefore resemble one another. So by playing little tricks you very often can get agreement between proton, proton-subtract and e^+e^-. That is not a surprise. But it is the duty of every scientist to try to test their data in the most critical manner to look for differences as well as agreements which could come from phase space. And one thing that particularly worries me is that we know a great deal about N^* production. Now maybe I missed it, but I didn't see any graphs of the effects of N^* production and how much Δ^{++} production is there. Can you show me some graph of the $p\pi^+$ mass distribution? What effect does this have?

VALENTI: Okay, as far as I know, in my data we have seen the Δ^{++} signal but I don't have a graph with me. Now on the other hand, when you, make the kind of selection we have done, namely by cutting the x_F between .44 and .85, in a sense you are free from this leading cluster. Instead of the proton coming out of the interaction you have Δ^{++}. Then this will decay into $p\pi^+$. And the x of this particle will be lower. We have simulated the N^* contribution, for example, and we found that it's only a few percent. The events that you will still take in, by mistaking the leading proton for one of the debris of the N^* for example, are very few.

MORRISON: No, that wasn't quite my question. If, instead of simulating things, you take the experimental data and you take your protons as you select by identifying, by finding their x, and you are going to reject them. If you take these protons and you add together the other positive particles do you still observe a peak at the Δ^{++} mass?

VALENTI: We have a peak.

MORRISON: If you have such a peak, will it not inevitably effect your results? This will inevitably influence your results and give you some correlation. It is an effect which does not exist in e^+e^-.

VALENTI: That I agree.

MORRISON: Now I only talk of the Δ^{++} because that is a terribly obvious thing and is quite easy to check. But there are many other N^* isobars.

VALENTI: Okay. That for sure will not change very much the results. It will not change the results as far as multiplicity because of the charge.

MORRISON: But we have known many times that multiplicity is a very poor test. Everything fits multiplicity. That doesn't prove anything. It's phase space that dominates multiplicity.

VALENTI: There are many reactions that have to be taken into account. Unfortunately, the thing becomes more and more complicated because you have all the pionic resonances which you have to look at. You have Σ^+ which will decay. You have the Υ, Υ^*, etc. This we have not done completely, but I think we are confident that the results will not change very much.

MORRISON: So why are you "confident" when you have the data? Shouldn't you test this experimentally before you start publishing?

VALENTI: That's an interesting statement. I don't agree.

MORRISON: This is philosophy of science.

FIAŁKOWSKI, KRAKOW: In line with looking for differences it seems obvious that there is one more difference. Namely, there is a diffractive component in p_T. Did you have any look for the fact that there are some very low multiplicity events left in your sample with leading protons which are certainly not in e^+e^- and which should be counted out?

MEYER: The cut for both experiments on Feynman x of the proton was put at .82 or .84 . . . somewhere in there, specifically to exclude the diffraction peak.

FIAŁKOWSKI: I understand that in your case it excludes because you require protons in both hemispheres. But when you look in only one hemisphere, then for events from the second hemisphere there is a single proton. Then, certainly the proton from diffraction . . .

VALENTI: In the hemisphere where you select the proton. In our case that is the important hemisphere. In our case the analysis of diffraction is only on the hemisphere where the proton has been selected. But in the second hemisphere it is screwed up.

FIAŁKOWSKI: Do you have events in which in the second hemisphere [there] is only a single proton and in the hemisphere where you are investigating is just single diffraction in some cases?

VALENTI: Well, we have events with low multiplicity but we have not investigated . . .

FIAŁKOWSKI: So my guess would be that the dispersion should be very different from e^+e^-. You should have a low multiplicity tail.

RATTI: I might add that, in the International Hybrid Spectrometer Collaboration, we have antiselected the diffractive events, using only non-diffractive events but not subtracting the leadings, and the agreement with the e^+e^- is essentially the same. Namely, subtracting the leading or throwing away the diffractive and taking only the non-diffractive are relatively similar in comparison with the e^+e^-. By chance the sphericity distribution is better for the non-diffractive and the thrust distribution is better for the leading-subtracted events. The overlapping is very small. If you throw away the diffractive you throw away those that contain mostly the leading particle. It is the meaning that might be different. It depends on how you think. Any other question or comment?

PUGH, BERKELEY: If you would really like to test this comparison to the limit you might try it out on the $\alpha\alpha$ interactions from the ISR.

FRIDMAN: I would like to say that I am not particularly surprised by your results, because every time you have hadron hadron which interact you have two jets, more or less. So you have limited p_T, so you have two jets. You see, when you take $p\bar{p}$, which I made at 32 GeV, you see sphericity distribution, thrust distribution. I would like to say that what you are doing is a collision with two particles, so you see also two jets. That is not surprising. You compare it with e^+e^-. So what I think is that in e^+e^- you have two quarks that fragment, so you have two jets. So it should be more or less the same apart from some [differences] with θ distributions. So what I would like only to say is that the important fact is the discovery, I think, that e^+e^- produce hadrons via two quarks and then gives two jets. This is, in my opinion, the important point and not so much that hadron hadron interactions' limited p_T gives two jets.

MEYER: That's essentially the point I was trying to make. That we should not be surprised that we see these two jets.

SESSION G1

PRODUCTION AND DECAY

OF

SYSTEMS INVOLVING HEAVY QUARKS I

Thursday, June 25, p.m.

Chairman: S. Ratti
Secretary: J. M. Bishop

RESULTS FROM C.E.S.R.

M. S. Alam
Vanderbilt University, Nashville, TN 37240

ABSTRACT

On behalf of the CLEO[1] and CUSB[2] collaborations, results on studies of e^+e^- interactions at center-of-mass energies of (9.4-11.8) GeV at CESR, recorded with the CLEO and CUSB detectors between November 1979 and June 1981 are reported. A review of the Υ, Υ', Υ'' and Υ''' resonances and their parameters is presented. Evidence for the production of the B mesons with bare beauty quantum number, their semileptonic decays and strange particle yields is reported. Measurements of inclusive production of pions, kaons and baryons as a function of center-of-mass energy have been made.

INTRODUCTION

The study of e^+e^- interactions has proved to be a powerful tool in particle physics. The coupling of e^+e^- annihilations through the virtual photon to $q\bar{q}$ channels can be used to search for new $q\bar{q}$ thresholds. The discovery[3,4] of charmonium resonances and charmed mesons at SLAC and DORIS has lead to a deeper understanding of quark-antiquark forces from the study of $(c\bar{c})$ charmonium spectroscopy and weak decays of the charmed mesons, besides opening up the study of gluonic annihilations from the measurement of OZI suppressed decay modes of ψ and ψ'. The search for new $q\bar{q}$ thresholds never ends. In this spirit, the Cornell Electron Storage Ring was built to cover the c.m. energy interval (8-15) GeV, not covered by SPEAR-DORIS ($E_{c.m.}$ <8 GeV) or PEP-PETRA ($E_{c.m.}$ >15 GeV) machines.

I now present a list of topics to be covered by me in this report: (i) CESR parameters, (ii) the CLEO and CUSB detectors, (iii) a review of the resonances and their parameters, (iv) hadronic transitions of $\Upsilon(3S)$ and $\Upsilon(2S)$ to $\Upsilon(1S)$, (v) inclusive production of pions, kaons and baryons as a function of center-of-mass energy, (vi) scan above $\Upsilon(4S)$, (vii) evidence for $B\bar{B}$ threshold, (viii) semileptonic and strange particle decays of the B meson and (ix) finally the conclusion.

I. CESR PARAMETERS

The Cornell Electron Storage Ring or CESR[5] is an e^+e^- storage ring with two intersection regions. Between Nov., 1979 and June, 1981,15,000 nb^{-1} of data between c.m. energies of 9.4 to 11.8 GeV have already been recorded with the CLEO detector, operating at the south intersection region. Another detector, built by the CUSB collaboration, has also been operating in the norther intersection region since Nov., 1979. The r.m.s. machine energy resolution is about 4.1 MeV at 10 GeV in the center-of-mass. Since Jan., 1981, CESR has been producing data at around 100 nb^{-1} per day.

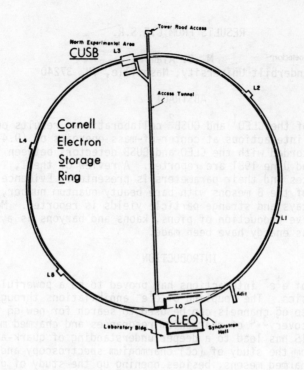

Fig. 1: Layout of the Cornell Electron Storage Ring.

II. THE CLEO DETECTOR

CLEO[6] is a general purpose magnetic detector, consisting of a tracking chamber system inside a 1 meter radius, 3-meter long solenoidal coil, with particle identification and shower detection system outside the coil. The tracking system consists of multi-wire proportional chambers surrounding the beam pipe and a 1 meter radius, 2-meter long, seventeen-layer drift chamber system. The solenoid is conventional and operated at 4.2 kilogauss. The solid angle coverage is $(0.9 \times 4\pi)$ with the spatial resolution measured by a Breit-Wigner function with FWHM of 350 microns. The momentum resolution $\delta p/p = .03p$ is measured with 4.88 GeV/c Bhabha tracks.

Outside the coil, there are eight octants. Radially inmost, there are three layers of outer-z drift chambers. Next follow Cerenkov counters in four octants and dE/dx ionization measurement chambers in the other four octants. The Cerenkov counters, which cover a total of $(.24 \times 4)$ of solid angle and operate at one atmosphere of Freon 12, have a pion threshold of 3 GeV/c. The dE/dx chambers, which cover the same solid angle as the Cerenkov counters, measure the dE/dx along a track

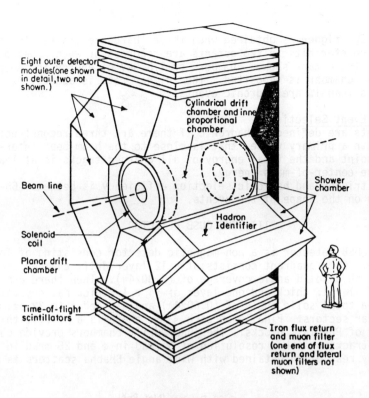

Fig. 2: The CLEO Detector at CESR.

with an r.m.s. resolution of 6% by recording 117 ionization measurements along the track length. The dE/dx chambers operate at 3 atmospheres of (.90/.10) Argonne-methane mixture. A layer of time-of-flight counters measures time of travel from the beam-beam interaction point to the counter with an r.s.m. resolution of 400 ps. The TOF solid angle coverage is about (.5x4π). Pions, kaons and protons below 150, 400, 600 MeV/c respectively range out in the coil and never reach the outer particle identification system. Then there is a 12 r.l. thick MWPC-lead sheet electromagnetic shower detection system in each octant, covering a total solid angle of (0.47x4π), whose energy resolution as measured from Bhabhas and π° is given by σ/E = 0.20/√E. There are shower chambers on both octant ends and magnet pole tips, giving an overall solid angle coverage for shower detection of (0.70x4π). The whole package is inside the muon detection system, which covers a solid angle of (0.85x4π) and consists of (10-14) interaction lengths of steel and a two-layer drift chamber hit-detection system. The muon detection momentum cut-off varies from 1.0 to 1.6 GeV/c depending on the incidence angle.

Trigger:

A CLEO trigger is obtained when three 'hardware tracks' together with TOF counter hits in two octants are recorded or when 2 GeV of energy in the octant shower chambers or 1.8 GeV of energy in the pole-tip shower chamber is deposited. The trigger rate is about 1 Hz of which less than 1% are hadronic events.

Hadronic Event Selection:

Events are defined as hadronic if there are three reconstructed tracks with a primary vertex that is close to the beam-beam inter-section point and the total energy of all charged tracks is at least 15% of the center-of-mass energy.

The trigger and hadronic selection efficiency is between (65-70)% depending on the shape of the events.

III. THE CUSB DETECTOR

The CUSB detector is a non-magnetic detector consisting of four quadrants. Each quadrant consists of a 12 layer drift chamber tracking system, with a solid angle coverage of (0.80x4π). Then there are four layers of ~1 r.l. thick and one layer of ~4 r.l. thick NaI crystals, covering a total solid angle of (0.60x4π) and divided into 32 azimuthal and 2 polar sectors. Finally on each side, there is a ~7 r.l. thick, 8x8 array of Pb-glass blocks. The inner drift chambers provide charged particle tracking with a resolution of 3 mrad in φ and 25 mrad in θ. The energy resolution obtained with wide angle Bhabha scatters is 5.5% FWHM.

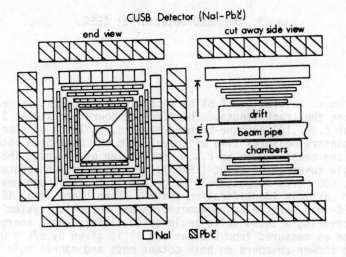

Fig. 3: The CUSB detector at CESR.

<u>Trigger:</u>
      An event is triggered when either 1 GeV or more of energy is
deposited in the outer three NaI layers or at least 700 MeV of energy
is deposited uniformly throughout the detector. The trigger rate is
.1 Hz of which only 5% are hadronic events.

<u>Hadronic Event Selection:</u>
      Events are recognized as hadronic if at least one 'recognized'
minimum ionizing track besides two other charged tracks or neutral
showers are present. The hadronic triggering and selection efficiency
varies from (61-73)%, depending on the shape of the events.

VI. RESONANCES AND THEIR PARAMETERS

      The discovery of two and possibly three long-lived resonances T,
T' and T" at Fermilab[7,8] in 1977 in p + N $\rightarrow$ $\mu^+\mu^-$X and the subsequent
confirmation of the first two in e^+e^- interactions with much better
resolution at DESY[9] in June, 1978 was interpreted as evidence for the
existence of a fifth quark with a new quantum number, named bottom or
beauty.

<u>b$\bar{\text{b}}$ Quarkonium Spectroscopy:</u>
      The success of static potentials with a combination of Coulombic
plus linear terms in correctly predicting the level spacing and
electronic widths in charmonium (c$\bar{\text{c}}$) spectroscopy, leads us to consider the spectroscopy of (b$\bar{\text{b}}$) bottomium states in a similar Coulomb
+ linear static potential. Fig. 4 shows the level scheme for a heavy
Q$\bar{\text{Q}}$ fermion system in any general Coulomb + linear static non-relativistic potential. The spectroscopic notation $2S+1_{n}L_{J}$, J^{PC} is used to
label the levels, where n-1 is the number of radial modes, P=$(-)^{L+1}$
and C=$(-1)^{L+S}$. For each value of L, there are two bands of radial
excitations with opposite charge conjugation depending on where S=0
or 1. The P level will split into one $^1P_1$ level with odd and three
$^3P_{2,1,0}$ levels with even charge conjugation. However, the D level
will split into one $^1D_2$ level with even and three $^3D_{3,2,1}$ levels with
odd charge conjugation. The $^3S_1$ and $^3D_1$ states have the same quantum
numbers as the photon. The wave function of 1^3D_1 state may acquire a
finite value at the origin by mixing with the nearby 2^3D_1 and could
also be produced directly in e^+e^- interactions besides $^3S_1$ states.
      Further, Rosner[10] has shown that for the class of generally
successful potentials with confinement, the number of bound states
below threshold for strong decay increases as the square root of the
quark masses:

$$N_{bound} = 1/4 + \text{const.} * \sqrt{M_Q} \qquad (1)$$

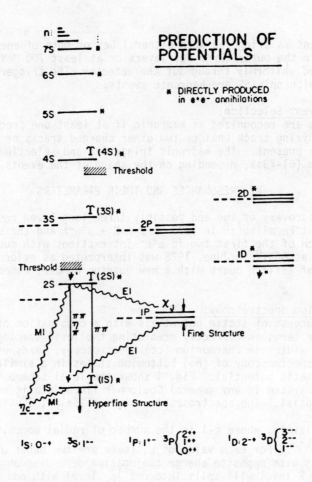

Fig. 4: Spectroscopy of heavy $Q\bar{Q}$ System in a Non-relativistic Coulomb +
Linear Static Potential.

Fixing the value of the constant using the $c\bar{c}$ system, three bound $^3S_1$
states the $B\bar{B}$ threshold are expected for the $(b\bar{b})$ bottomium system.
The mass spectrum of the $(b\bar{b})$ bottomium system will of course
depend on the exact form of the static potential used.

Observation of Resonances:

Both CLEO and CUSB have measured the hadronic yield as a function
of center-of-mass energy. Fig. 5 and Fig. 6 show three distinct
narrow resonances. The first two may be identified as the $\Upsilon(9.4)$ and
$\Upsilon'(10.0)$ resonances, first reported by Herb et al.[7] and later confirmed
at DESY[9]. The third resonance is the first clear confirmation of the
$\Upsilon''(10.4)$ suggested by Ueno et al.[8] from a three gaussian fit to the
background subtracted $\mu^+\mu^-$ mass spectrum. It may be pointed out that
the observed experimental widths of all three resonances are consistent

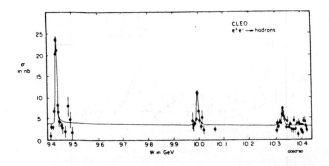

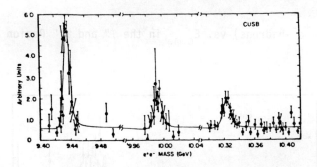

Fig. 5 and 6: The hadronic cross sections at the $\Upsilon(1S)$, $\Upsilon(2S)$, and $\Upsilon(3S)$ for CLEO and CUSB.

with the expected machine resolution. Both CLEO and CUSB have observed a fourth, broader resonance[12,13] at 10.54 GeV. See Figs. 7 and 8. The comparison of the fourth resonance $\Upsilon'''(10.5)$ with the third resonance, $\Upsilon''(10.3)$ show clearly that the observed experimental width of the $\Upsilon'''(10.5)$ is greater than the expected machine resolution.

Resonance Parameters:
    The cross-section for s-channel resonances in e^+e^- interactions is described by a Breit-Wigner function:

$$\sigma_i(W) = \frac{3\pi}{M} \frac{\Gamma_{ee} \cdot \Gamma_i}{(W-M)^2 + \Gamma_t^2/4} \tag{2}$$

where the 'i' refers to the ith-channel $e^+e^- \to \Upsilon \to i$ and Γ_{ee} is partial width into e^+e^-, Γ_i is the partial width to channel i and Γ_t is the total decay width for the s-channel resonance. This will be modified by photon radiation from the inital state and where the natural width of

751

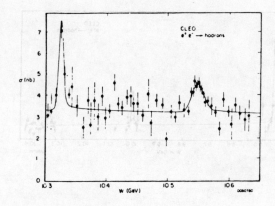

Fig. 7: $\sigma(e^+e^- \to$ hadrons) vs. $E_{c.m.}$ in the Υ'' and Υ''' Region (CLEO).

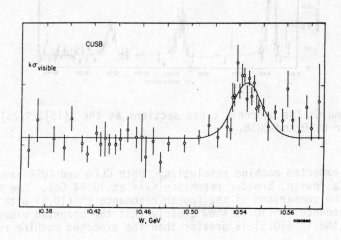

Fig. 8: $\sigma(e^+e^- \to$ hadrons) vs. $E_{c.m.}$ in the Υ''' Region (CUSB).

the resonance is very small compared to machine resolution, the shape of the resonance will be completely dominated by the gaussian machine resolution function.

The masses, experimental widths and the natural area under the resonances are obtained from a global fit to all data with the following representations: (i) the natural width of the narrow resonances are approximated by δ-functions while a gaussian is used for the fourth broad resonance, (ii) a gaussian function with s-dependent width is used for the machine resolution together with initial state radiation corrections and (iii) a 1/s term is used to represent the continuum.

Summing over all final state hadronic channels i and integrating over W, the c.m. energy, we obtain from equation (2):

$$\int \sigma_h(W)\ dW = \frac{6\pi^2}{M^2}\ \frac{\Gamma_{ee}\Gamma_h}{\Gamma_t} \tag{3}$$

where $\sigma_h = \sum_i \sigma_i$ and $\Gamma_h = \sum_i \Gamma_i$. The value of the natural area under the resonance as obtained from the global fit then gives a measure of the reduced electronic width $\Gamma_{ee}' = \Gamma_{ee}\Gamma_h/\Gamma_t$. The true electronic width can be obtained from the reduced electronic width if the branching ratio into dileptons, $BR(\Upsilon \to \ell^+\ell^-)$ is measured independently:

$$\Gamma_{ee} = \Gamma_{ee}'/(1-3B_{\ell^+\ell^-}) \tag{4}$$

where $B_{\ell^+\ell^-} = BR(\Upsilon \to e^+e^-) = BR(\Upsilon \to \mu^+\mu^-) = BR(\Upsilon \to \tau^+\tau^-)$ has been assumed.

Table I shows the results from the global fit to both CLEO and CUSB data. Data from the LENA[14] detector at DORIS are also shown for comparison. There is a 29 MeV difference in the value of the masses from DORIS and CESR and may be attributed to an r.m.s. energy calibration uncertainty of ~30 MeV at CESR and ~10 MeV at DORIS. The mass differences are more meaningful for comparison. Also the ratios $\Gamma'_{ee}(\Upsilon')/\Gamma_{ee}(\Upsilon)$, etc. are more reliable than the widths themselves, since in the experimental measurement the systematic uncertainties and in the theoretical calculation, the higher order corrections tend to cancel out in the numerator and the denominator.

Comparison with Theory:

Theoretical models are based on static, non-relativistic $Q\bar{Q}$ potentials whose (i) short distance behaviour is determined by ideas of asymptotic freedom where single gluon exchange dominates, giving

$$V(r) \underset{r\to 0}{\sim} \frac{1}{r\ln(\Lambda^2/r^2)} \tag{5}$$

and (ii) large distance behaviour comes from the string model and its confirmation in lattice gauge theories, giving $V(r) \underset{r\to\infty}{\sim} kr$ (6)

where Λ is the QCD scale parameter and k is the string tension.

Several approaches to the actual form of the potential exist. Bhanot and Rudaz[15] have suggested a logarithmic interpolaration between the short distance Coulomb behaviour and the long distance linearly confining behaviour. Richardson[16], starting from a single gluon exchange QCD-potential, incorporates the concepts of asymptotic freedom and linear confinement in a unified manner to arrive at a one parameter potential of the form:

TABLE I

EXPERIMENTAL DATA ON UPSILON RESONANCES

	LENA	CLEO	CUSB
M(GeV)			
T	9462 ± .01	9.4330 ± .0003 ± .030	9.4332 ± .0002 ± .03
T'	10.016 ± .02	9.9934 ± .0005 ± .030	9.9938 ± .0002 ± .03
T"		10.3242 ± .0006 ± .030	10.3231 ± .0004 ± .03
T"'		10.5476 ± .0011 ± .032	
M-M$_T$ (MeV)			
T'	553 ± 10	560.4 ± .6 ± 3	560.6 ± .3 ± 3
T"		891.2 ± .7 ± 5	889.9 ± .4
T"'		1140.0 ± 1.1 ± 5.0	1140.0 ± 2.0 ± 5
$\Gamma'_{ee} = \Gamma_{ee}\Gamma_h / \Gamma_{tot}$			
T	1.23 ± 0.10 ± 0.14	0.925 ± 0.06 ± 0.14	
T'	0.55 ± 0.07 ± 0.06	0.468 ± 0.04 ± 0.07	
T"		0.288 ± 0.03 ± 0.05	
T"'		0.221 ± 0.02 ± 0.03	
$\Gamma'_{ee}/\Gamma'_{ee}$ (T)			
T'	0.45 ± 0.06 ± 0.02	0.51 ± 0.05 ± 0.05	0.39 ± 0.06
T"		0.31 ± 0.04 ± 0.03	0.32 ± 0.04
T"'		0.24 ± 0.02 ± 0.03	0.25 ± 0.07
ΔW_{rms} (MeV)			
T	7.3 ± .1	3.14 ± 0.20	
T'		3.53 ± 0.22	4.0 ± .3
T"		3.76 ± 0.24	
T"'		9.00 ± 0.82	8.1 ± 1.7

$$V(q^2) = - \frac{4}{3} \frac{12\pi}{33-2N_f} \cdot \frac{1}{q^2} \frac{1}{\ln(1+q^2/\Lambda^2)} \qquad (7)$$

where q is the three momentum, and Λ may be related to the QCD scale parameter. Fourier transform of eq. (7) gives a short distance behaviour of the form:

$$V(r) \underset{\Lambda r << 1}{\sim} \frac{8\pi}{33-2N_f} \frac{1}{r\ln(1/\Lambda r)} \qquad (8)$$

Buchmüller and Tye[17] have extended the Richardson ansatz, by incorporating one and two loop corrections consistently to arrive at a potential which is completely determined from two parameters only: (i) α', the old Regge slope parameter and (ii) $\Lambda_{\overline{MS}}$, the QCD scale parameter.

Fixing the values of the free parameters from old physics, the potential models are very successful in reproducing the masses and

754

electronic widths of the T' and T" with respect to the T. This leads us to interpret the T, T', T" and T''' as the 1S, 2S, 3S and 4S states respectively of the (bb̄) bottomium system in a static, non-relativistic, flavour-independent potential with a logarithmically modified Coulomb behaviour at short distance and linearly confining behaviour at large distances. However, the success of QCD-like potentials cannot be interpreted as evidence for asymptotic freedom. Martin[18] has shown that a power-law potential of the form $A+Br^{\alpha}$, $\alpha=.14$ can fit the ψ and T spectroscopy just as well.

Charge of the b-Quark:

Quigg, Rosner and Thacker[19] have calculated the electronic widths for the T and T' in a wide variety of potential models. In Fig. 9, they show well defined lower bounds, corresponding to quark charge of 2/3 and 1/3. Both DORIS and CESR data support the value of 1/3 for the b charge.

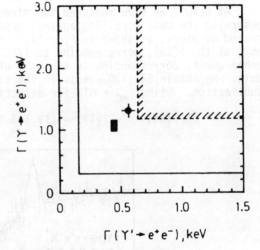

Fig. 9: Predicted lower bound on $\Gamma_{ee}(T)$ and $\Gamma_{ee}(T')$ for q=2/3 (dashed line) and q=1/3 (solid line). Average value of DORIS and CESR data.

V. HADRONIC TRANSITIONS BETWEEN bb̄ STATES

T(3S) and T(2S) are both below threshold for BB̄ decay. However if T(1S), T(2S) and T(3S) are members of the same bb̄ quarkonium system, one might expect hadronic transitions of the T(3S) and T(2S) to the ground -state T(1S). Photon transitions to positive C-parity χ_b and η_b states are also allowed. See Fig. 10. Hadronic transitions of the T(2S) may be expressed as follows:

$$T(2S) \rightarrow \pi\pi T(1S) + \eta°T(1S) \qquad (9)$$

However, the $\eta°T(1S)$ decay mode is expected to negligible due to phase space suppression[20]. The decay T(2S) $\rightarrow$ $\pi\pi T(1S)$ can be searched for in the following ways:

Fig. 10: Decays of Υ(2S).

$$\Upsilon(2S) \to \pi^+\pi^-\Upsilon(1S)$$

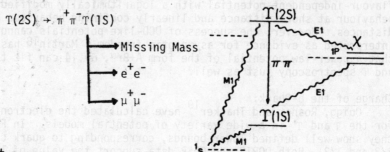

$$\underline{\Upsilon(2S) \to \pi^+\pi^-\Upsilon(1S):}$$

At CLEO[21] we have measured the direct decay Υ(2S) → π⁺π⁻Υ(1S) by observing the mass recoiling against oppositely charged tracks interpreted as pions, as shown in Fig. 11. This data is based on ≈1360 nb⁻¹ at the Υ(2S), corresponding to (7200±300) Υ(2S) decays. A clear enhancement, corresponding to the mass of the Υ(1S) is seen in the data. We obtain 841±130 events in the signal region after background subtraction. Using ε_{MC} = 61% for detection of Υ(2S) → π⁺π⁻X we obtain:

$$BR(\Upsilon(2S) \to \pi^+\pi^-\Upsilon(1)) = (19.1\pm3.1\pm2.9)\% \qquad (10)$$

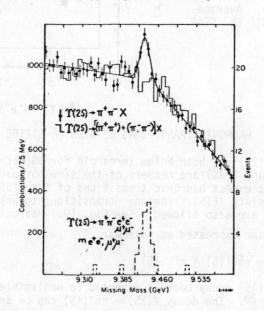

Fig. 11: Data points (histogram) correspond to missing mass spectrum against π⁺π⁻(π±π±) from Υ(2S)→ππX. The curve is a fit to the data. The dashed histogram is mass (ℓ⁺ℓ⁻) for events of the type Υ(2S)→π⁺π⁻ℓ⁺ℓ⁻.

LENA[23] at DORIS has measured this ratio as $(26\pm13)\%$ by comparing track multiplicities in $\Upsilon(2S)$ and $\Upsilon(1S)$ decays. From isospin invariance we expect

$$BR(\Upsilon(2S) \rightarrow \pi^\circ\pi^\circ\Upsilon(1S)) = (1/2)BR(\Upsilon(2S) \rightarrow \pi^+\pi^-\Upsilon(1S))$$

and hence we state:

$$BR(\Upsilon(2S) \rightarrow \pi\pi\Upsilon(1S) = (29\pm5\pm4)\% \tag{11}$$

$\underline{\Upsilon(2S) \rightarrow \pi^+\pi^-e^+e^-}$:

Both CLEO[21] and CUSB[22] have observed the four charged track topology events corresponding to $\Upsilon(2S) \rightarrow \pi^+\pi^-\Upsilon(1S)$ and $\Upsilon(1S)\rightarrow e^+e^-, \mu^+\mu^-$. In Figs. 12(a) and 12(b), we show two characteristic examples of the above decay topology in the CLEO and CUSB detectors respectively.

The electrons are identified by their characteristic shower signatures. The muons are identified by matching hits in the muon chambers behind iron hadron absorbers. In Table II, we give the results of the branching ratio product $BR(\Upsilon(2S)\rightarrow\pi^+\pi^-\Upsilon(1S))*$ $BR(\Upsilon(1S)\rightarrow e^+e^-, \mu^+\mu^-)$ and the corresponding measurement from LENA[23] at DORIS.

TABLE II

$$\Upsilon(2S) \rightarrow \Upsilon(1S) \; \pi^+\pi^-$$
$$\hookrightarrow e^+e^-, \mu^+\mu^-$$

	CLEO	CUSB	LENA
$\Upsilon(2S)$ direct decays	7200	9851	1680
	±300	±335	±110
$\#\Upsilon(2S) \rightarrow \pi^+\pi^-(e^+e^-+\mu^+\mu^-)$	17+8	23	5+3
$BR(\Upsilon(2S) \rightarrow \pi^+\pi^-\Upsilon(1S))$	0.0068	0.0063	0.0061
$* BR(\Upsilon(1S) \rightarrow e^+e^-/\mu^+\mu^-)$	0.0014	±0.0013	±0.0026
$BR(\Upsilon(2S) \rightarrow \pi^+\pi^-\Upsilon(1S))$	$(19.1\pm3.1+2.9)\%$		
$BR(\Upsilon(1S) \rightarrow e^+e^-)$	3.6 ± 0.9	$3.2\pm.8$	$(3.2\pm1.5)\%$

Average $BR(\Upsilon(1S) \rightarrow e^+e^-) = \qquad 3.35\pm.64\%$

DIRECT MEASUREMENT $\Upsilon(1S) \rightarrow \mu^+\mu^-/e^+e^-$

PLUTO	DASP II	LENA
$(3.1\pm1.7)\%$	$(2.9\pm1.3\pm0.5)\%$	$(3.5\pm1.4\pm0.4)\%$

Average $BR(\Upsilon(1S) \rightarrow \ell^+\ell^-) = (3.2\pm.8)\%$ *

New World Average: $\quad BR(\Upsilon(1S) \rightarrow \ell^+\ell^-) = (3.3\pm.5)\%$ *

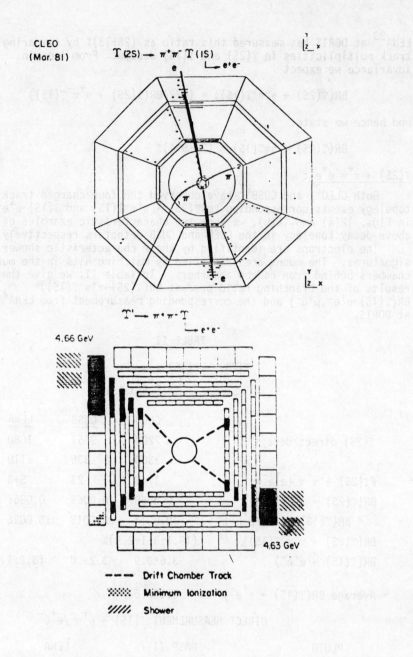

Fig. 12a.(12b): is an example of the decay $T(2S) \rightarrow T(1S)\pi^+\pi^-$ and $T(1S) \rightarrow e^+e^-$ in CLEO (CUSB).

Using the CLEO measurement of $BR(T(2S) \rightarrow \pi^+\pi^- T(1S))=(19.1\pm3.1\pm2.9)\%$ and averaging over the CLEO, CUSB and LENA measurements of the branching ratio product, we obtain $BR(T(1S)\rightarrow e^+e^-) = (3.6\pm.9)\%$, where only the statistical errors are quoted. This is to be compared with the value of $BR(T(1S) \rightarrow \ell^+\ell^-) = (3.2\pm.8)\%$ obtained from the measurements of the direct decay of $T(1S) \rightarrow e^+e^-/\mu^+\mu^-$ as made by PLUTO[24], DASP II[25] and LENA[14] detectors at DORIS[26]. We therefore now quote as the new world average:

$$BR(T(1S) \rightarrow \ell^+\ell^-) = (3.3\pm.5)\% \tag{12}$$

Only statistical errors are quoted. Using this value, we can give a new value for $\Gamma_{ee}(T(1S))$ from CLEO and CUSB measurements of $\bar{\Gamma}_{ee}(T(1S))$.

$$\bar{\Gamma}_{ee} = \Gamma_{ee}(\Gamma_h/\Gamma_{tot}) \rightarrow \Gamma_{ee} = \bar{\Gamma}_{ee}/(1-3B_{\ell\bar{\ell}})$$

We obtain

$$\Gamma_{ee}(T(1S)) = (1.05\pm0.05\pm0.11) \text{ KeV} \tag{13}$$

$T(3S)\rightarrow T(1S)\pi^+\pi^-$ (preliminary):

Both CLEO[27] and CUSB[28] have observed the hadronic transition of the $T(3S)$ to $T(1S)$ in the decay: $T(3S)\rightarrow\pi^+\pi^- T(1S)$ and $T(1S)\rightarrow e^+e^-$, $\mu^+\mu^-$. CLEO and CUSB both see four events, corresponding to the topology $T(3S)\rightarrow\pi^+\pi^- e^+e^-$ and CLEO sees 1 event with $T(3S)\rightarrow\pi^+\pi^-\mu^+\mu^-$. Using the world average of $B_{\mu\mu} = .033\pm.005$, and averaging the CLEO and CUSB measurements, the value for the branching ratio is:

$$BR(T(3S) \rightarrow \pi^+\pi^- T(1S)) = (4.8\pm1.7)\% \tag{14}$$

and from isospin invariance:

$$BR(T(3S) \rightarrow \pi\pi T(1S)) = (7.2\pm2.6)\% \tag{15}$$

$T(3S)\rightarrow T(2S)\pi^+\pi^-$ (preliminary):

No events corresponding to the topology $T(3S)\rightarrow T(2S)\pi^+\pi^-$ and $T(2S)\rightarrow e^+e^-$, $\mu^+\mu^-$ were found. If $BR(T(2S)\rightarrow e^+e^-, \mu^+\mu^-) = 2\%$ is used, a 90% confidence limit of 14% for $BR(T(3S)\rightarrow\pi^+\pi^- T(2S))$ is obtained.

Comparison with Theory:

According to QCD, the hadron transition of the $T(3S)$ and $T(2S)$ to $T(1S)$ proceeds in two steps: (i) The $(b\bar{b})$ in the higher state emits gluons to cascade to the lower state and (ii) the gluons hadronize into $\pi\pi$, η°, etc. Since the gluons emitted are soft, straightforward perturbative QCD cannot

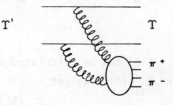

be applied. Gottfried[29] has pointed out that heavy quark systems move slowly and if the wavelength of the radiation emitted is large compared to the dimensions of the source, a multipole expansion of the coupling between the heavy quark system and its color gauge field converges rapidly for sufficiently large heavy quark mass. Huang and Yan[30] have recently calculated the branching ratios for hadronic transitions between the T states and comparison of their predictions with data are shown in Table III[27]:

TABLE III

Model	BR(T(2S)$\to\pi^+\pi^-$T(1S))	BR(T(3S)$\to\pi^+\pi^-$T(1S))
A. Cornell	18.0%	1.3%
B. Bhanot-Rudaz	19.4%	1.9%
C. Buchmüller-Tye	16.7%	1.7%
Experiment:	19.1±3.1	4.8±1.7

VI. CALCULATION OF THE DECAY WIDTHS OF T(1S) and T(2S)

We will abbreviate T(1S) and T(2S) as (1S) and (2S) when they appear as arguments.

T(1S):
The total decay width of the T(1S) may be expressed as:

$$\Gamma_{tot}(1S) = \Gamma_{ggg}(1S) + \Gamma_{\gamma gg}(1S) + (3+R)\Gamma_{ee}(1S)$$

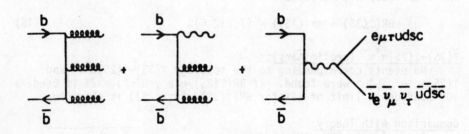

$\Gamma_{tot}(1S)$ can be calculated directly from the measured values of $\Gamma_{ee}(1S)$ and $B_{\mu\mu}(1S)$. We get

$$\Gamma_{tot}(1S) = \frac{\Gamma_{ee}(1S)}{B_{\ell\ell}(1S)} = \frac{1.05\pm0.05}{.033\pm0.005} = (32\pm5) \text{ KeV} \tag{16}$$

Using the value of R = 3.7±.4 as measured at DORIS[31] just below the $\Upsilon(1S)$ and neglecting the $\Gamma_{\gamma gg}(1S)$ term, we get

$$\Gamma_{ggg}(1S) = (25\pm5) \text{ KeV} \tag{17}$$

Only statistical errors are quoted.

$\underline{\Upsilon(2S)}$:
 In this case there is an additional term corresponding to the hadronic transition to the $\Upsilon(1S)$, through emission of a soft gg pair. Thus we express the total decay width as:

$$\Gamma_{tot}(2S) = \Gamma_{ggg}(2S) + (3+R)\Gamma_{ee}(2S) + \Gamma_{\gamma X}(2S) + \Gamma_{\pi\pi\Upsilon}(2S) \tag{18}$$

Here we will make certain approximations,

$$\Gamma_{ee}(2S) = \bar{\Gamma}_{ee}(2S)/(1-3B_{\ell\bar{\ell}}(1S)) = .5\pm.1 \text{ KeV} \tag{19}$$

Further from QCD we expect

$$\frac{\Gamma_{ggg}}{\Gamma_{ee}} \propto \frac{\alpha_s^3 (M_\Upsilon^2)}{\alpha_e^2} \tag{20}$$

If we assume the value of the above constant to be the same at $\Upsilon(1S)$ and $\Upsilon(2S)$, we get

$$\frac{\Gamma_{ggg}(2S)}{\Gamma_{ee}(2S)} = \frac{\Gamma_{ggg}(1S)}{\Gamma_{ee}(1S)} = 23\pm5$$

from which we get

$$\Gamma_{ggg}(2S) = (12\pm3.5) \text{ KeV} \tag{21}$$

From theoretical expectations[32] we expect $\Gamma_{\gamma X}(2S) = (3^{+3}_{-2})$ KeV and expressing $\Gamma_{\pi\pi\Upsilon}(2S)$ as $B_{\pi\pi\Upsilon}(2S)\Gamma_{tot}(2S)$, we get

$$\Gamma_{tot}(2S) = (26^{+7}_{-6}) \text{ KeV} \tag{22}$$

and $B_{\ell\bar{\ell}}(2S) = (1.9\pm.6)\%$ \hfill (23)

and we now obtain the value of $\Gamma_{\pi\pi\Upsilon}(2S)$ as

$$\Upsilon(2S) = B_{\pi\pi\Upsilon}(2S) \ \Gamma_{tot}(2S) = (8\pm2) \text{ KeV} \tag{24}$$

· $\underline{\text{Direct Measurement of } \Upsilon(2S)\rightarrow\mu^+\mu^-}$:
 At CLEO[32a], based on ~780 nb^{-1} of data at the $\Upsilon(2S)$, a direct measurement of $\hat{\Upsilon}(2S)\rightarrow\mu^+\mu^-$, using the muon chambers gives:

$BR(\Upsilon(2S)\to\mu^+\mu^-) = (1.3\pm1.2)\%$ and $BR(\Upsilon(2S)\to\mu^+\mu^-)<2.8\%$ at 90% C.L. The only other published direct measurement for $B_{\mu\mu}$ is by the LENA[32b] collaboration: $BR(\Upsilon(2S)\to\mu^+\mu^-)<3.8\%$ at 90% C.L.

<u>Vector Gluons or Scalar Gluons:</u>

Comparing the decay width of the $\Upsilon(2S)\to\pi\pi\Upsilon(1S)$ to that of $\psi(2S)\to\pi\pi\psi(1S)$, we get

$$\frac{\Gamma_{\pi\pi\psi}}{\Gamma_{\pi\pi\Upsilon}} = \frac{107\pm14}{8\pm2} = 13\pm4 \tag{25}$$

T.M. Yan[33] argues that

$$\frac{\Gamma_{\pi\pi\psi}}{\Gamma_{\pi\pi\Upsilon}} = \frac{<r_\psi>^4}{<r_\Upsilon>^4} \approx \begin{array}{l} 10 \text{ for vector gluons} \\ 1 \text{ for scalar gluons} \end{array} \tag{26}$$

where $<r_\psi>$ and $<r_\Upsilon>$ are the dimensions of the ψ and Υ. Our measurement is clearly consistent with the spin of the gluons being 1.

VII. MEASUREMENT OF THE STRONG INTERACTION COUPLING α_s

In analogy with the coupling of $q\bar{q}$ to the virtual photon characterized by the electromagnetic constant α_{em}, the coupling of $q\bar{q}$ to the virtual gluon is expressed by the strong interaction coupling constant $\alpha_s(\sqrt{s})$ which according to QCD is dependent on s and therefore is a running constant. In lowest order QCD:

$$\alpha_s(\sqrt{s}) = \frac{4\pi}{(11-2/3\ N_f)\ln(s/\Lambda^2)} \tag{27}$$

where N_f = number of flavours accessible at c.m. energy $\sqrt{s}$ and Λ is the unknown QCD scale parameter. For $s\to\infty$, $\alpha_s(\sqrt{s})\to0$ and this is known as asymptotic freedom.

The gluonic and leptonic width of the $\Upsilon(1S)$ provide a sensitive measurement of $\alpha_s(M)$ the running QCD coupling constant at the mass of the $\Upsilon(1S)$. The coupling of the $\Upsilon(1S)$ to e^+e^- proceeds through the virtual photon and is therefore proportional to α_{em}^2, q_b^2 (square of the quark charge) and $|\psi(0)|^2$ (the probability for the $b\bar{b}$ to be at zero separation to annihilate). To the lowest order in QED[34]:

$$\Gamma_{ee} = 16\pi(\alpha^2 q_b^2/M^2)|\psi|0||^2 \tag{28}$$

$$q_b = -1/3$$

Similarly, the annihilation of $b\bar{b}$ in the $\Upsilon(1S)$ to three gluons depends on $\alpha_s^3(M)$ (the running QCD coupling constant) and $|\psi(0)|^2$ (the probability for the $b\bar{b}$ to be at zero separation to annihilate). To the lowest order in QCD[35]:

$$\Gamma_{ggg} = \frac{160}{81} (\pi^2 - 9)(\alpha_s^3/M^2)|\psi|0||^2 \qquad (29)$$

Thus to the lowest order in QCD

$$\frac{\Gamma_{ggg}}{\Gamma_{ee}} = \frac{1-(3+R)+\Gamma_\gamma/\Gamma_{ee})B_{\mu\mu}}{B_{\mu\mu}} = \frac{10(\pi^2-9)}{81\pi q_b^2} \frac{\alpha_s(M)}{\alpha_{em}^2} \qquad (30)$$

where $M = M_{\Upsilon(1S)}$ and $q_b = -1/3$ is the charge of the b-quark. Lepage and Mackenzie[36] have carried out higher order corrections and their result in the $\overline{MS}$ scheme, renormalized at M_Υ is:

$$\frac{\Gamma_{ggg}}{\Gamma_{ee}} = \frac{10(\pi^2-9)}{81\pi q_b^2} \frac{\alpha_s(M_\Upsilon)}{\alpha_{em}^2} [1+(9.1\pm0.5)\frac{\alpha_s(M_\Upsilon)}{\pi}] \qquad (31)$$

Alternately, one may renormalize at $M = .48M_\Upsilon$ where the higher order correction vanishes. Using the world average of $B_{\mu\mu} = .033\pm.005$ and the equation (31), we get[37]:

$$\alpha_s(.48M_\Upsilon)_{\overline{MS}} = 0.158^{+.012}_{-.010} \qquad (32)$$

The higher order parametrization of $\alpha_s(M)$ is expressed as[36]:

$$\alpha_s(M) = \frac{4\pi}{\beta_0 \ln(M^2/\Lambda^2)} - \frac{4\pi\beta_1 \ln[\ln(M^2/\Lambda^2)]}{\beta_0^3 \ln^2(M^2/\Lambda^2)} \qquad (33)$$

where $\beta_0 = 11 - 2/3 N_f$, $\beta_1 = 102 - 38/3 N_f$, $N_f = 4$. Using $\alpha_s(.48M_\Upsilon)_{\overline{MS}} = 0.158^{+.012}_{-.010}$, we get for Λ the QCD scale parameter:

$$\Lambda_{\overline{MS}} = 100^{+34}_{-25} \text{ MeV} \qquad (34)$$

The quoted errors in the $\Lambda_{\overline{MS}}$ are only experimental errors.

The value of α_s obtained from ψ should be higher and has been measured[38] to be:

$$\alpha_s(M_\Upsilon) = 0.19\pm.01$$

and the corresponding value of $\Lambda_{\overline{MS}}$ from the ψ is:

$$\Lambda_{\overline{MS}} = 53^{+16}_{-12} \text{ MeV} \tag{35}$$

VIII. INCLUSIVE PRODUCTION OF HADRONS FROM CLEO

Hadron Identification:

Charged hadrons are identified by either using the time-of-flight counters or dE/dx chambers. See Figs. 13 and 14. A charged track is called a kaon or proton if it is within 2σ of that hypothesis and at least 2.5σ away from the next nearest hypothesis. Because of ranging in the coil at low momenta and confusion with pions at higher momenta, $K^\pm$ and $p(\bar{p})$ are identified only in a limited range:

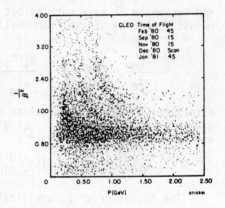

Fig. 13: Plot of $1/\beta^2$ where $\beta = v/c$ measured by TOF is plotted against measured momentum. (CLEO).

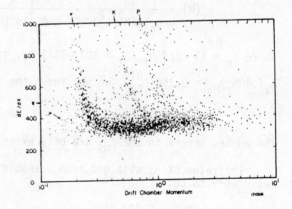

Fig. 14: dE/dx (truncated mean) of charged tracks on $\Upsilon(3S)$ as function of measured momentum (CLEO).

	TOF	dE/dx	
KAONS	.5 - 1.0	.450 - .9	GeV/c
		and >2.0	
PROTONS	.7 - 1.8	.7 - 1.5	GeV/c

Neutral hadrons such as $K^\circ(\bar{K}^\circ)$ and $\Lambda^\circ(\bar{\Lambda}^\circ)$ are detected by reconstructing the secondary decay vertices in the drift chamber. To reduce combinatorial background from tracks originating at the primary vertex, only those secondary vertices which decay at least 1 cm away from the primary vertex in the r-ϕ plane are accepted. $K^\circ(\bar{K}^\circ) \to \pi^+\pi^-$ and $\Lambda^\circ(\bar{\Lambda}^\circ) \to p\pi$ mass spectra are shown in Figs. 15 and 16.

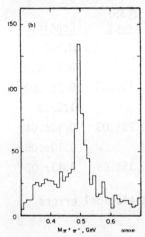

Fig. 15: $\pi^+\pi^-$ Mass Spectrum ($\Upsilon(4S)$).

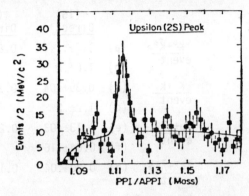

Fig. 16: $(p\pi)^\circ$ Mass Spectrum ($\Upsilon(2S)$).

Direct Resonance Decays of $\Upsilon(1S)$, $\Upsilon(2S)$ and $\Upsilon(3S)$:

The total decay width of the narrow resonances may be expressed as:

$$\Gamma_t = \Gamma_{3g} + \Gamma_{2g} + \Gamma_{\gamma gg} + (3 + R)\Gamma_{ee} \tag{36}$$

Except for the 3 Γ_{ee} component, all other components lead to hadron production in the final states. However, the $R\Gamma_{ee}$ component leads to hadron production by $q\bar{q}$ jets in the same way as hadron production from $q\bar{q}$ jets in the continuum under the resonance peak, as opposed to hadron production from gluon jets which predominate in the $\Gamma_{3g} + \Gamma_{2g} + \Gamma_{\gamma gg}$ component. The hadronic production from gluon jets may be studied from the data taken at the resonances by subtracting the continuum and the $R\Gamma_{ee}$ vacuum polarization contributuion, using data taken off the resonances. The branching ratio for the hadronic contribution from the vacuum polarization is $R\Gamma_{ee}/\Gamma_t = RB_{\mu\mu}$ and turns out to be ~13% at $\Upsilon(1S)$, ~7% at the $\Upsilon(2S)$ and ~6% at the $\Upsilon(3S)$, using R=3.7, $B_{\mu\mu}(1S) = 3.3\%$, $B_{\mu\mu}(1S) = 1.9\%$ and $B_{\mu\mu}(3S)=1.5\%$.

Preliminary data from CLEO[39,40,] [41] on inclusive production of hadrons, $\pi^{\pm}$, $K^{\pm}$, $K^{\circ}(\bar{K}^{\circ})$ and $\Lambda^{\circ}(\bar{\Lambda}^{\circ})$ from the direct decays of the narrow resonances and the continuum region below the $\Upsilon(4S)$ are presented in Table IV. Comparison with data from DASP[42] and PLUTO[43] are also presented. The values are uncorrected for momentum acceptances only.

TABLE IV.

Energy		$\Upsilon(1S)$ Direct	$\Upsilon(2S)$ Direct	$\Upsilon(3S)$ Direct	Contin.
$<K^{\circ}+\bar{K}^{\circ}>$ event	(a)	1.2±.1	0.8±.2		0.8±0.1
	(b)	1.0±.2			0.73±0.16
$<K^{+}+K^{-}>$ event	(c)	0.39±.04	0.39±.04	0.49±.05	0.25±.02
	(d)	1.2±.2			1.3±.3
$<p+\bar{p}>$ event	(e)	0.13±.03	0.22±.02	0.25±.03	0.10±.01
	(f)	0.64±.16			0.10±.06
$<\Lambda^{\circ}+\bar{\Lambda}^{\circ}>$ event	(g)	0.20±.04	0.11±.03	0.15±.05	0.13±.02

Systematic errors are of the order of statistical errors and not included here.

(a) CLEO data 0.3<p<3.0 GeV/c.
(b) PLUTO data 0.3<p<4.5 GeV/c.
(c) CLEO data 0.5<p<0.9 GeV/c.
(d) DASP data 0.3<p<1.5 GeV/c.
(e) CLEO data 0.7<p<1.5 GeV/c.
(f) DASP data 0.3<p<1.5 GeV/c.
(g) CLEO data p>0.4.

Continuum is defined differently for PLUTO, DASP and CLEO:

	Continuum (GeV)
CLEO	10.360 - 10.528
PLUTO	9.30 - 9.44
DASP	around $\Upsilon(1S)$ and $\Upsilon(2S)$

The corrections for momentum acceptance are model dependent; however, results in the same momentum range may be compared. There is weak evidence from CLEO data of enhanced kaon production from the resonances as compared to continuum. DASP shows enhancement of $(p+\bar{p})$/event from the $\Upsilon(1S)$ as compared to that from the continuum, however, the difference is of the order of 3 s.d. CLEO does not see such an enhancement at the $\Upsilon(1S)$; however, there are enhancements at the $\Upsilon(2S)$ and $\Upsilon(3S)$ compared to the continuum. The discrepancy between DASP and CLEO may be statistical or may be attributed to the fact that the data in the two cases correspond to different momentum intervals. CLEO data on $(\Lambda^\circ+\bar{\Lambda}^\circ)$/event shows evidence for an enhancement, but again it is not statistically significant.

Figs. 17-20 show the invariant differential cross-section as a function of particle energies from the $\Upsilon(1S)$, $\Upsilon(2S)$ and $\Upsilon(3S)$ direct decays and the continuum region. The spectra can be approximated by an exponential distribution of the form:

$$\frac{E}{4\pi p^2} \frac{d\sigma}{dp} \propto \exp(-bE) \tag{37}$$

where the slope parameter b is obtained from a least-squares fit to data. The results from the fit to the data are given in Table V.

TABLE V.

$b(GeV^{-1})$	$\Upsilon(1S)$	$\Upsilon(2S)$	Contin.
π	3.8±.4	4.1±.4	3.4±.3
k	3.4±.7	3.3±.7	3.8±.8
p	3.2±.6	2.8±.6	3.6±.7

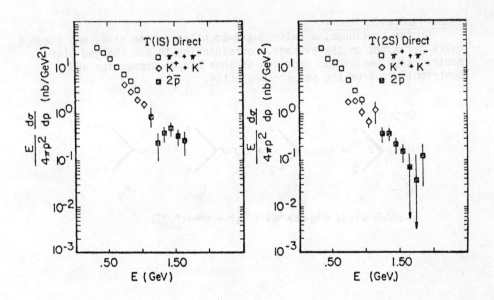

767

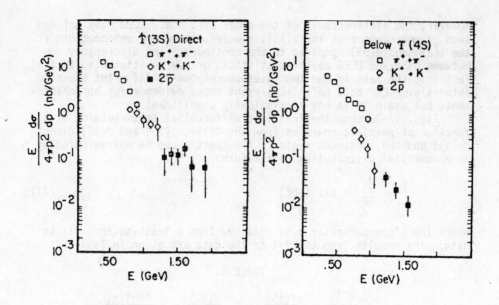

Figs. 17, 18, 19, 20 show the invariant cross-section $E/4\pi p^2$ $d\sigma/dp$ for $\Upsilon(1S)$, $\Upsilon(2S)$, $\Upsilon(3S)$ direct decays and the continuum.

Comparison with Theory:

On the continuum, a naive quark-counting scheme gives one strange quark per event on the average, corresponding to one charged plus neutral kaon per event, provided strange baryon production and contributions from the sea are neglected.

Continuum :

$$e_q^2 \rightarrow \quad 4/9 \qquad 1/9 \qquad 1/9 \qquad 4/9$$

$$(s\text{-quark}/\text{evnt}) = (2 \times 1/9 + 2 \times 4/9)/(10/9) \rightarrow 1 \Rightarrow 1\frac{(K + K^0)}{\text{event}}$$

768

At the $\Upsilon(1S)$, if we assume the gluons couple with equal probability to $u\bar{u}$, $d\bar{d}$ and $s\bar{s}$, then on the average one of the three gluons will couple to $s\bar{s}$, giving two strange quarks per event from $\Upsilon(1S)$ direct decays. This corresponds to two charged plus neutral kaons per event, provided strange baryon production and contributions from the sea are neglected. Hence naively, one would expect enhanced production of strange particles from the direct decays of the resonances as compared to the continuum. The data, however, shows that strange meson production from the direct decays of the resonances is very similar to that from the continuum.

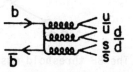

Anderson et al.,[44] have constructed a simple model for baryon production in e^+e^- annihilation events by assuming that diquark-anti-quark pairs can be produced in a color force field in a way similar to quark-antiquark pairs. Suppression of baryon-antibaryon production is achieved on the basis of the larger diquark mass compared to the quark mass. He predicts somewhat enhanced production of baryons from gluon jets as compared to quark jets. Hofman[45] makes a similar pre-diction based on the quark recombination model. He argues that the multiplicity of partons in a parton shower initiated by a primary gluons is 9/4 as large as in quark jets. The enhanced density of partons in phase space makes three parton recombination more likely, yielding more baryons.

IX. THE $B\bar{B}$ THRESHOLD

Total Width of $\Upsilon(4S)$:

According to Table I, the measured experimental r.m.s. width of the $\Upsilon(4S)$ is (9.0 ± 0.8) MeV while the calculated machine resolution at the same center-of-mass energy is $(4.6\pm.3)$ MeV. Subtracting the machine resolution in quadrature from the measured experimental width gives:

$$\Gamma_t(4S) = 19.8\pm5.5\pm5 \text{ MeV} \tag{38}$$

Compared to $\Gamma_t(1S) = (32\pm5)$ KeV and $\Gamma_t(2S) = (26\pm6)$ KeV, we find $\Gamma_t(4S)$ to be three orders of magnitude higher and consistent with the interpretation that the $\Upsilon(4S)$ is above the threshold for a new strong channel. The $\Upsilon(4S)$ may be above the threshold for decay into channels involving bare bottom quantum number, such as:

$$e^+e^- \rightarrow \Upsilon(4S) \rightarrow B\bar{B} + B\bar{B}^* + \bar{B}B^* + B^*\bar{B}^* \tag{39}$$

where $B(b\bar{q})$ and $B^*(b\bar{q})$ are the lowest pseudoscalar and vector bare bottom states

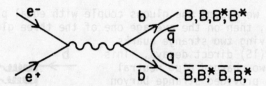

The B$\bar{\text{B}}$ threshold must be between the $\Upsilon(3S)$ and $\Upsilon(4S)$, hence:

$$5.16 < M_B < 5.270 \text{ GeV}$$

The mass of the lowest lying bottom mesons, $B^+(\bar{b}u)$ and $B°(b\bar{d})$ and their charge conjugates may be estimated by scaling from D and D* masses

$$M_B = M_D + m_b - m_c + (3/4)(1-m_c/m_b)(M_{D*}-M_D) = 5.2 \text{ GeV} \quad (40)$$

The mass of the corresponding vector states B^{+*}, $B°^*$ and their charge conjugates may be estimated in a similar manner:

$$M_{B*} - M_B = (m_c/m_b)(M_{D*} - M_D) = 50 \text{ MeV} \quad (41)$$

Martin[47] using very general arguments about flavour independence of $q\bar{q}$ forces and experimental information on strange and charmed meson, arrives on reliable, model-independent bounds on the B, B* masses and their difference:

$$5.208 < M_B < 5.288 \text{ GeV}$$
$$5.285 < M_{B*} < 5.386 \text{ GeV} \quad (42)$$
$$52 < M_{B*}-M_B < 57 \text{ MeV}$$

The width and shape of the $\Upsilon(4S)$ resonance depends on the mass of the $\Upsilon(4S)$, relative to the BB, B$\bar{\text{B}}$* + $\bar{\text{B}}$B* and B*$\bar{\text{B}}$* thresholds. Eichten et al.,[48] using a coupled-channel model which fits the (c$\bar{\text{c}}$) charmonium resonance shapes above D$\bar{\text{D}}$ threshold, predicted the shape of the $\Upsilon(4S)$ resonance for the three cases where $\Delta = M_{4S}-2M_B$ = 50, 70 and 100 MeV. See Fig. 21. A comparison with Figs. 7 and 8 show that the data is compatible with Δ=50 and 100 MeV; the flat-top shape rules out Δ=70 MeV, or intermediate values between 50 and 100 MeV.

If M_{4S} is above B$\bar{\text{B}}$* + $\bar{\text{B}}$B* threshold, then monochromatic photons of ~50 MeV should be observed at the $\Upsilon(4S)$, corresponding to:

$$B* \rightarrow B + \gamma \quad (43)$$

which must account for 100% of B* decays because all hadronic decays are kinematically forbidden.

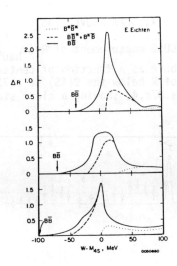

Fig. 21: Shape of the $\Upsilon(4S)$ resonance as function of $W-M_{4S}$ for $\Delta=50$, 70 and 100 MeV.

Photon Spectrum from CUSB:

Fig. 21a shows the inclusive photon spectrum at the $\Upsilon(4S)$ from the CUSB detector. The dashed histogram is the inclusive photon spectrum from $\Upsilon(4S)$ direct decays, obtained by subtracting the continuum contribution. The monochromatic 50 MeV photon line expected from $B^*(\bar{B}^*)\to B(\bar{B})\gamma$ for the case where $\Upsilon(4S)\to B\bar{B}^* + \bar{B}B^*$ occurs 100% of the time, is also shown. Clearly the data is not consistent with this. However, due to uncertainties in the energy resolution and detection efficiencies at very low energies, partial coupling of the $\Upsilon(4S)$ to channels involving B^* cannot be ruled out.

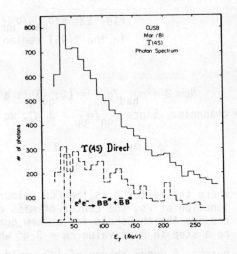

Fig. 21a: Photon Spectrum at $\Upsilon(4S)$ (solid line). Photon spectrum from $\Upsilon(4S)$ direct (dashed line). Photon signal expected for 1 photon per direct $\Upsilon(4S)$ decay. (Dot dashed line.)

X. SCAN ABOVE (4S)

Fig. 22 shows the CUSB measurement of R=$(\sigma_{had}/\sigma_{\mu\mu})$ in the vicinity of Υ(4S) and above as a function of center-of-mass energy.

The average value of R below the Υ(4S) is 3.73±.09±.4 while that above the Υ(4S) is 4.12±.07±.4, showing a clear step of ΔR=.39±.11±.07. See Fig. 22.

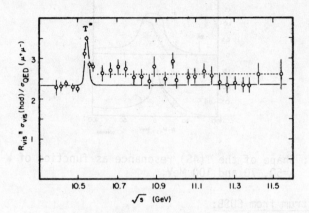

Fig. 22: R=$\sigma_{had}/\sigma_{\mu\mu}$ as a function of $\sqrt{s}$ in the Υ(4S) region and above.

Now R = $\sigma_{had}/\sigma_{\mu\mu}$ = $(\Sigma\sigma_{q\bar{q}})$ is a counter for quark-antiquark channels. Since $\sigma_{qq}/\sigma_{\mu\mu}$ = 3 q_q^2, we have

$$R = \frac{\sigma_h}{\sigma_{\mu\mu}} = 3 \sum_i q_i^2 \qquad (44)$$

where the factor 3 is the QCD colour factor. When R is plotted as function of $\sqrt{s}$, the center-of-mass energy, every time the energy crosses the threshold for a new quark-antiquark channel, there should be a step in R of value ΔR = 3 q^2 where q is the charge of the new quark. A quark charge of 1/3 would give ΔR = 1/3. This is completely consistent with the experimental measurement of ΔR = .4±.1. Hence, this is another indication the charge of the b-quark is 1/3.

It may also be pointed out, there is no evidence for structure above the Υ(4S).

XI. THE STANDARD SIX QUARK MODEL

The Kobayashi-Maskawa[49] model is a natural extension of the Weinberg-Salam-GIM SU(2)xU(1) model to six quarks and six leptons, where quarks and leptons appear in left-handed doublets and right-handed singlets in the weak charged current interactions:

$$\begin{pmatrix} u \\ d' \end{pmatrix}_L \begin{pmatrix} c \\ s' \end{pmatrix}_L \begin{pmatrix} t \\ b' \end{pmatrix}_L \begin{pmatrix} \nu_e \\ e^- \end{pmatrix}_L \begin{pmatrix} \nu_\mu \\ \mu^- \end{pmatrix}_L \begin{pmatrix} \nu_\tau \\ \tau^- \end{pmatrix}_L \qquad \text{Left-Handed Doublets}$$

$$u_R, d_R, c_R, s_R, t_R, b_R, e_R^-, \mu_R^-, \tau_R^- . \qquad \text{Right-Handed Singlets}$$

The d', s' and b' the weak interaction eigenstates are related to d, s and b strong interaction mass eigenstates by the K-M rotation matrix, depending on three generalized Cabibbo angles θ_1, θ_2 and θ_3 and a CP violating phase term δ:

$$\begin{bmatrix} d' \\ s' \\ b' \end{bmatrix} = V \begin{bmatrix} d \\ s \\ b \end{bmatrix} \text{ where } V = \begin{array}{ccc} c_1 & s_1 c_3 & s_1 s_3 \\ -s_1 c_2 & c_1 c_2 c_3 + s_2 s_3 e^{i\delta} & c_1 c_2 s_3 - s_2 c_3 e^{i\delta} \\ -s_1 s_2 & c_1 s_2 c_3 - c_2 s_3 e^{i\delta} & c_1 s_2 s_3 + c_2 c_3 e^{i\delta} \end{array} ,$$

and $c_i = \cos\theta_i$, $s_i = \sin\theta_i$.

The weak charged current interaction is expressed as:

$$J_\mu = (\bar{u} \ \bar{c} \ \bar{t}) \ \gamma_\mu (1-\gamma_5) V \begin{bmatrix} d \\ s \\ b \end{bmatrix} \qquad (45)$$

and the various components in the current may be shown graphically as:

Graphical representation of the weak charged current. Solid lines represent Cabibbo-allowed transitions. Dashed (dotted) lines represent singly (doubly) Cabibbo suppressed transitions.

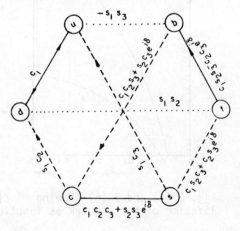

It may be noted that there is natural suppression of flavour changing neutral current and if $m_t > m_b$, the b quark will decay weakly to either a u or c quark. In the graph note that the $b \to cW^-$ transition is proportional to $|c_1 c_2 c_3 + s_2 s_3 e^{i\delta}|^2$ while the $b \to uW^-$ transition is proportional to $|-s_1 s_3|^2$. A measurement of the ratio $\Gamma_{b \to \mu}/\Gamma_{b \to c}$ would be very useful in determining the parameters of the K-M matrix.

Henry Tye[50] has neatly summarized the status of our knowledge of the K-M matrix, based on data from old hadron physics as:

$$s_1^2 = 0.05 \qquad\qquad \text{from } n \to pe^-\bar{\nu}$$

$$0 < s_s < 0.5 \qquad\qquad \text{from } \Lambda \to pe^-\bar{\nu}, \; K \to \pi e\nu$$

$$|c_\delta| \approx 1, c_\delta = \cos\delta = 1 \qquad \text{from CP violation in } K^\circ_L \to \pi^-\pi$$

$$s_2 = 0.2 + 1/2\,(c_\delta + 1/2)\,s_3$$

The only unknowns are s_3 and the sign of c_δ. In Figs. 23(a) and 23(b) from Tye, we show a plot of the ratio $\Gamma_{b \to u}/\Gamma_{b \to c}$ and the lifetime of the b-quark, τ_b as a function of the s_3-parameter for two possible values of $c_\delta = \pm 1$. A value of $\Gamma_{b \to u}/\Gamma_{b \to c}$ greater than .05 would rule out $c_\delta = 1$, as would a value of $\tau_b > 10^{-13}$ sec.

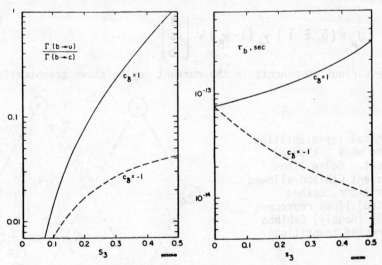

Fig. 23(a) and 23(b) showing $\Gamma(b \to u)$ to $\Gamma(b \to c)$ transition, lifetime of the b-quark as function s_3 for $c_\delta = \pm 1$.

XII. B-LIFETIME

At CLEO a preliminary, indirect estimate of the B-lifetime has been made. If B-mesons have a finite lifetime, they will decay away from the primary vertex and hence if the distance of closest approach to the primary vertex (in the plane transverse to the beam) of all tracks at the $\Upsilon(4S)$ are compared to a similar distribution below the $\Upsilon(4S)$, a Monte Carlo study shows that a mean decay length for B-mesons of 1 mm would appear as a broadening of the distribution at the 2σ-level. In Figs. 24(a) and 24(b), the FWHM of the distance-of-closest approach distributions are measured to be: $\Upsilon(4S)$ = 3.17±.06 mm and Γ(continuum) = 3.07±.09 mm. There is no observed change in the width of the distributions, hence at the 90% c.l., we may state $\lambda_B < 1$ mm. Now

$$\lambda_B \equiv \text{mean decay length} = c\tau_B \sqrt{(\Delta/M_B)} \qquad (46)$$

where $\Delta = M_{\Upsilon(4S)} - 2*M_B$. From the value of $\Gamma_{\Upsilon(4S)}$ = 20 MeV, we infer that $\Delta \cong 40$ MeV. Thus the 90% c.l. upper limit on the B-meson lifetime may be stated as:

$$\tau_B < 3\times10^{-11} \text{ sec} \qquad (47)$$

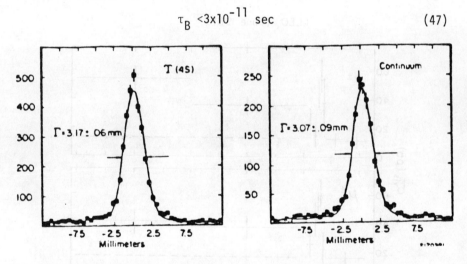

Figs. 24(a) and 24(b): Distance of closest approach to the primary vertex in the plane tranverse to the beam.

Clearly, the measurements have to improve by at least two orders of magnitude before they can be useful in determining the K-M parameters s_3 and c_δ.

XIII. INCLUSIVE PRODUCTION OF LEPTONS NEAR $\Upsilon(4S)$

Both CUSB[51] and CLEO[52] have measured the inclusive electron yield as a function of c.m. energy W. Only electrons above 1 GeV/c are considered, since approximately (75-80)% of the primary electrons from B-decays are expected to be above 1 GeV/c while the contamination of electrons from charm decays above 1 GeV/c is expected to be small. CLEO[53] has also measured the inclusive muon yield as a function of cm-energy W. The cut-off momentum for muons in the 60-cm thick iron hadron absorber is between 1.0 to 1.6 GeV/c depending on the direction of incidence. Preliminary results are presented.

Figs. 25(a) and 25(b) show the inclusive yield of high momentum electrons and muons respectively as function of W as seen by CLEO. Fig. 26 shows the inclusive yield of high momentum electrons as function of W, as seen by CUSB. Both sets of data are uncorrected for acceptances and detection efficiencies. A clear enhancement at the $\Upsilon(4S)$ and a step in going from below the $\Upsilon(4S)$ to above the $\Upsilon(4S)$ in the inclusive lepton yield is observed. This is interpreted as evidence for production of $B\bar{B}$ mesons, which decay weakly with an appreciable semileptonic branching ratio.

CLEO (June 1981)

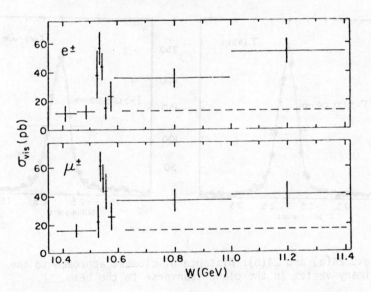

Figs. 25(a) & (b): Inclusive yield of electrons and muons as function of W (CLEO).

776

Momentum Spectrum:

Figs. 27(a) and 27(b) show the continuum subtracted (but uncorrected for efficiencies) electron and muon spectrum respectively from $\Upsilon(4S)$ direct decays in CLEO. Fig. 28 is the corresponding electron spectrum from CUSB and is corrected for all inefficiencies. The solid curve is based on a Monte Carlo model with $B \to D\ell\nu + D^*\ell\nu$, corresponding to 100% $b \to c$. The dashed curve is from a Monte Carlo model with $B \to \pi\ell\nu + \rho\ell\nu$, corresponding to 100% $b \to u$. The data appears to favour $b \to c$, but contributions from $b \to u$ cannot be ruled out. The data is limited in statistics and there are uncertainties in the theoretical models at this point. Data with several times the present statistics will be presented at Bonn[27].

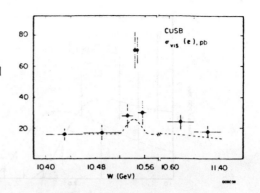

Fig. 26: Inclusive yield of electrons with p>1.0 GeV/c as function of W (CUSB).

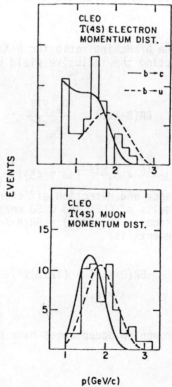

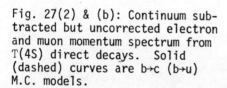

Fig. 27(2) & (b): Continuum subtracted but uncorrected electron and muon momentum spectrum from $\Upsilon(4S)$ direct decays. Solid (dashed) curves are $b \to c$ ($b \to u$) M.C. models.

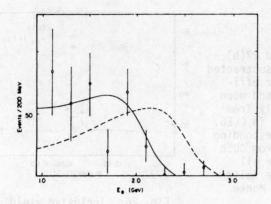

Fig. 28: Momentum spectrum, continuum subtracted and corrected for inefficiency, of electrons from $\Upsilon(4S)$ direct decays (CUSB). Solid (dashed) curve from a $b \to c$ ($b \to u$) M.C. model.

BR($B \to X\ell\nu$):

The branching ratio for $B \to X\ell\nu$ may be measured directly by subtracting the inclusive yield below the $\Upsilon(4S)$ from the on the $\Upsilon(4S)$:

$$ BR(B \to X\ell\nu) = \frac{\Delta\sigma_{e,\mu}^{vis}}{\Delta\sigma_h^{vis}} \cdot \frac{1}{2\varepsilon_{e,\mu}} \cdot \frac{1}{f} \tag{48} $$

where $\Delta\sigma^{vis} = \sigma_i^{vis}$ (on $\Upsilon(4S)$)$-\sigma_i^{vis}$ (below $\Upsilon(4S)$), $\varepsilon_{e,u}$ is the acceptance and detection efficiency and f is the momentum acceptance. The results of CLEO and CUSB measurements are given in Table VI.

Assuming BR($B \to Xe\nu$) = BR($B \to X\mu\nu$), the average of CLEO and CUSB measurements is:

$$ BR(B \to X\ell\nu) = (12\pm2)\% $$
$$ \ell = e,\mu \tag{49} $$

where momentum acceptances have been calculated assuming a $b \to c$ model.

TABLE VI.

$$BR(B{\to}X1\bar{\nu})$$

	Electron		Muon
	CLEO	CUSB	CLEO
Momentum (GeV/c)	1-3	1-3	1-3
ε detection	(24.2±5)%	21%	(26.8±6)%
f(p>1.0 GeV/c)	(75±5)%	(80±5)%	
$\Delta\sigma^{lepton}_{vis}$	26.8±6.6pb	31.5±5	31±8
$\Delta\sigma^{hadron}_{vis}$	520±60	691±45	530±80
BR(B→X $\bar{\nu}$)	(13±3±3)%	(13.6±2.5±3.0)%	(9.4±3.6)%

Average $BR(B{\to}X\ell\bar{\nu}) = (12{\pm}2)\%$
                    $\ell = e,\mu$

$\underline{BR(B{\to}X\ell^{+}\ell^{-}):}$

A search for dilepton events which come from the same B in CLEO data yields no candidates, giving:

$$BR(B{\to}X\ell^{+}\ell^{-}) <1.3\% \text{ (90\% C.L.)} \qquad (50)$$

and

$$\frac{BR(B{\to}X\ell^{+}\ell^{-})}{BR(B-X\ell\bar{\nu})} <0.11 \text{ (80\% C.L.)} \qquad (51)$$

Comparison with Standard Model:

In the spectator model of B-meson decay, the b-quark determines the decay while the light quark acts as a spectator. The b-quark may decay either to a c or u quark while the virtual W^- may couple to $\ell\bar{\nu}$ or $q\bar{q}$ pairs with equal probability except for phase space factors[54]. From naive counting, it is expected that the B will decay semi-leptonically about (15 to 17)% of the time, as shown in the decay diagram above. Also, the semileptonic branching ratio for charged and neutral B's will be the same, since the decay is determined by

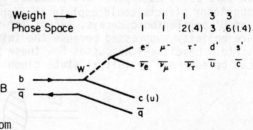

Weight ⟶
Phase Space

| | | | 3 | 3 |
| | | .2(.4) | 3 | .6(1.4) |

e⁻ μ⁻ τ⁻ d' s'
$\bar{\nu}_e$ $\bar{\nu}_\mu$ $\bar{\nu}_\tau$ $\bar{u}$ $\bar{c}$

$b \to c$ $B(B \to X(e/\mu)\nu) = 17\%$
$b \to u$ $= 15\%$

Direct e, μ Momentum Spectrum is hard

the heavier b-quark alone. For charmed mesons, it is now know that
the semileptonic decays of the charged and neutral D mesons are indeed
very different. From DELCO[55]

$$BR(D^+ \to Xe\bar{\nu}) = 22.0^{+4.4}_{-2.2}\% \tag{52}$$

$$BR(D^\circ \to Xe\bar{\nu}) < 4.0\% \text{ (95\% C.L.)}$$

and

$$\frac{\tau_{D^+}}{\tau_{D^\circ}} < 4.3 \text{ (95\% C.L.)} \tag{53}$$

The difference can be explained by introducing non-spectator effects.
The $D^+(c\bar{d})$ can annihilate into a virtual W^+ which can then couple to
$q\bar{q}$ or $\ell\bar{\nu}$ pairs.

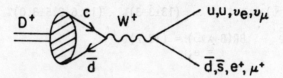

However D° can annihilate only via t-channel exchange:

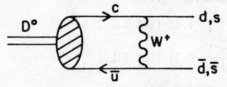

In this case, no leptons are produced in the final state. The non-
spectator effects could explain the apparent differences between the D^+
and D° semileptonic decays, except for the fact that these diagrams
are helicity suppressed because the initial $c\bar{u}$, $c\bar{d}$ are in a spinless
state. Leveille[55] discusses how these helicity-suppressed diagrams
can be enhanced by initial state gluon radiation:

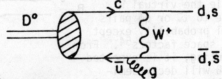

or gluon fluctuations of the energy:

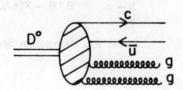

Leveille estimates, when the non-spectator effects are taken into account, the semileptonic branching ratio, $BR(B \to Xl\nu)$ varies between (12.5 to 15)% for $B^\pm$ and (9.5 to 10)% for $B^\circ(\bar{B}^\circ)$ as $\Gamma(b \to u)/\Gamma(b \to c)$ varies from 0 to 1. In the case of B mesons, it appears as though non-spectator effects are less important.

XIV. INCLUSIVE PRODUCTION OF HADRONS NEAR $\Upsilon(4S)$

Kaons:

As discussed in section VIII, charged kaons are identified only between .500 to 1.0 GeV/c while neutral kaons are detected between .300 to 3.0 GeV/c. Preliminary results based on approximately 8000 nb^{-1} of data are presented here. Fig. 29 shows the visible inclusive yield of charged and neutral kaons as a function of W, the center-of-mass energy. Assuming the excess cross-section at $\Upsilon(4S)$ is entirely due to $B\bar{B}$ and subtracting the continuum contribution using the data below the $\Upsilon(4S)$, the inclusive kaon contribution from B decays may be evaluated. The results are given in Table VII. Assuming the charged kaon yield per event is the same as the neutral kaon yield per event, we get (3.6±.4) kaons ($K^+ K^- + K^\circ + \bar{K}^\circ$) per $B\bar{B}$ event and (1.5±.10) kaons ($K^+ + K^- + K^\circ + \bar{K}^\circ$) per event in the continuum below $\Upsilon(4S)$. This is clear evidence of enhanced kaon yield from $\Upsilon(4S)$ direct decays as compared to the continuum below the $\Upsilon(4S)$.

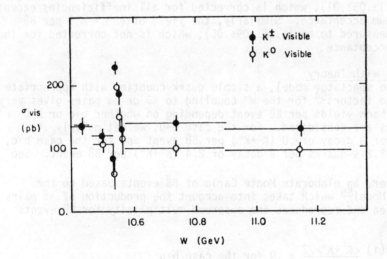

Fig. 29: Inclusive yield of charged and neutral kaons as function of c.m. energy W (CLEO).

781

TABLE VII.

	$\dfrac{\langle K^\circ + \bar{K}^\circ \rangle}{\text{event}}$	$\dfrac{\langle K^+ + K^- \rangle}{\text{event}}$	$\dfrac{(K^\pm + K^\circ)/2}{\text{event}}$
Continuum below Υ(4S)	0.71±0.05	0.90±0.11	0.74±.06
q$\bar{\text{q}}$ jet M.C.			0.95
B$\bar{\text{B}}$ event	1.43±0.25	2.17±0.24	1.80±.17
b→c M.C.			1.6
b→u M.C.			0.9

Only statistical errors are quoted. Systematic errors may be of the order of 15%.

Baryons:

A measurement of baryon production from B decays is of considerable theoretical interest. At CLEO, p($\bar{\text{p}}$)'s are identified only between .700-1.5 GeV/c and Λ°($\bar{\Lambda}^\circ$)'s are identified for momentum greater than .400 GeV/c. The yield of (p+$\bar{\text{p}}$) per B$\bar{\text{B}}$ event is measured to be (0.11±.05±.01), which is corrected for all inefficiencies except the momentum acceptance. Similarly, the yield of (Λ° + $\bar{\Lambda}^\circ$) per B$\bar{\text{B}}$ event is measured to be (-.06±.09±.06), which is not corrected for the momentum acceptance.

Comparison with Theory:

In the spectator model, a simple quark counting with appropriate phase space factors[54] for the W$^-$ coupling to $\ell\bar{\nu}$ or q$\bar{\text{q}}$ pairs gives very different kaon yields per B$\bar{\text{B}}$ event depending on whether b→c or b→u transitions are considered. For the case b→u, we get naively .4 s-quarks per B decay or .8 (K$^\pm$+K$^\circ$) per B$\bar{\text{B}}$ event and for the case b→c, we expect 1.2 s-quarks per B decay or 2.4 (K$^\pm$+K$^\circ$) per B$\bar{\text{B}}$ event. See Fig. 30.

However, an elaborate Monte Carlo of B$\bar{\text{B}}$ events based on the Sjostrand Model[56] which takes into account the production of s$\bar{\text{s}}$ pairs from the sea and reproduces the observed multiplicity for BB events gives:

$$\text{(i)} \quad \frac{\langle K^\pm + K^\circ \rangle / 2}{\text{B}\bar{\text{B}} \text{ event}} = .9 \text{ for the case b→u}$$

$$\text{(ii)} \quad \frac{\langle K^\pm + K^\circ \rangle / 2}{\text{B}\bar{\text{B}}} = 1.6 \text{ for the case b→c} \tag{54}$$

KAON PRODUCTION

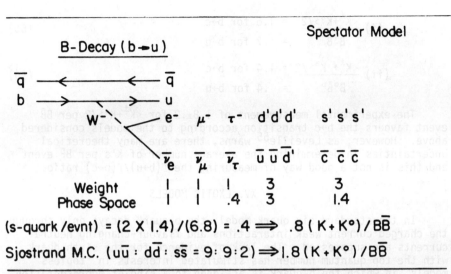

Spectator Model

B-Decay (b→u)

$$\bar{q} \longleftarrow \longleftarrow \bar{q}$$
$$b \longrightarrow \longrightarrow u$$

$$W^- \begin{array}{c} e^- \quad \mu^- \quad \tau^- \quad d'd'd' \quad s's's' \\ \bar{\nu}_e \quad \bar{\nu}_\mu \quad \bar{\nu}_\tau \quad \bar{u}\bar{u}\bar{d}' \quad \bar{c}\bar{c}\bar{c} \end{array}$$

Weight	I	I	I	3	3
Phase Space	I	I	.4	3	1.4

(s-quark/evnt) = (2 X 1.4)/(6.8) = .4 ⟹ .8 (K + K°)/B$\bar{\text{B}}$

Sjostrand M.C. (u$\bar{\text{u}}$: d$\bar{\text{d}}$: s$\bar{\text{s}}$ = 9: 9:2) ⟹ 1.8 (K + K°)/B$\bar{\text{B}}$

B-Decay (b→c)

$$\bar{q} \longleftarrow \longleftarrow \bar{q}$$
$$b \longrightarrow \longrightarrow c$$

$$\begin{array}{c} e^- \quad \mu^- \quad \tau^- \quad d'd'd' \quad s's's' \\ \bar{\nu}_e \quad \bar{\nu}_\mu \quad \bar{\nu}_\tau \quad \bar{u}\bar{u}\bar{u} \quad \bar{c}\bar{c}\bar{c} \end{array}$$

Phase Space	I	I	.2	3	.6

(s-quark/evnt) = (5.2 + 3 X .6) (5.8) = 1.2 ⟶ 2.4 (K + K°)/ B$\bar{\text{B}}$

Sjostrand M.C. (u$\bar{\text{u}}$: d$\bar{\text{d}}$: ss = 9: 9:2) = 3.2 (K + K°)/ B$\bar{\text{B}}$

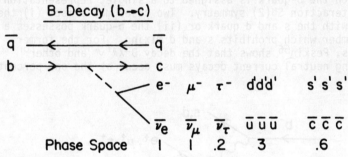

Fig. 30: B-meson decay in the spectator model for the cases (i) b→u and (ii) b→c.

where ($u\bar{u}$; $d\bar{d}$; $s\bar{s}$) = (9:9:2) has been used. However the model is naive and ignores non-spectator and strong interaction effects completely. Leveille[55] has taken these corrections into account and assuming ~20% correction for sea contributions predicts.

(i) $\dfrac{<K^{\pm}+K^{\circ}>/2}{B^+B^-}$ = 1.6 for $b \to c$

                  = .7 for $b \to u$ (55)

(ii) $\dfrac{<K^{\pm}+K^{\circ}>/2}{B^{\circ}\bar{B}^{\circ}}$ = 1.4 for $b \to c$

                  = .4 for $b \to u$ (56)

The experimental measurement of $1.8 \pm .2$ for $<K^{\pm}+K^{\circ}>/2$ per $B\bar{B}$ event favours the $b \to c$ transition according to the models considered above. However, as Leveille[55] warns, there are many theoretical uncertainties in determining the average number of K's per $B\bar{B}$ event and this is not a good way of measuring the $\Gamma(b \to u)/\Gamma(b \to c)$ ratio.

XV. EXOTIC MODELS

In the standard six quark model, the b-quark decays only through the charged current weak interactions and flavour-changing neutral currents are suppressed. The lack of evidence for the sixth quark with the top quantum number has stimulated interests in "topless" models, in which the b-quark is assigned to a singlet representation of the weak interaction SU(2) symmetry. Two cases may arise: (i) the b-quark mixes with the s and d quark or (ii) the b-quark possesses a new quantum number which prohibits s and d mixing. For the former class of models, Peskin[56] shows that the decay $B \to X \ell^+ \nu^-$ and other flavour-changing neutral current decays must occur at the one percent level.

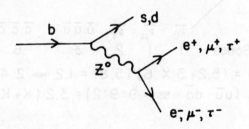

More precisely, it may be shown that:

$$\frac{\Gamma(B \to X \ell^+ \ell^-)}{\Gamma(B \to X \ell \bar{\nu})} \geq 0.125 \qquad (57)$$

and in fact for models with several singlet quarks, the condition is only slightly weaker:

$$\frac{\Gamma(B \to X \ell^+ \ell^-)}{\Gamma(B \to X \ell \bar{\nu})} \geq 0.122 \qquad\qquad (58)$$

The experiment measurement as reported in equation (51) is:

$$\frac{BR(B \to X \ell^+ \ell^-)}{BR(B \to X \ell \bar{\nu})} < 0.11 \text{ at } 80\% \text{ C.L.}$$

and rules out the above class of singlet models at the 80% confidence level limit. The second class of models, in which b mixing with s and d quarks is prohibited by a new quantum number, has been studied by Georgi and Glashow[57]. All b decays have a τ or $\bar{\nu}_\tau$ and the mode $B \to \tau^+ \bar{p}$ is quite large. The experimental measurement on the inclusive baryonic yield from B decays is consistent with zero though the experimental errors are large at this point. Hence, at a preliminary level, the second class of models may also be ruled. The evidence is in favour of the standard model and the top quark will eventually be found.

XVI. B$\bar{B}$ CANDIDATE

The report would be incomplete without showing an example of a reconstructed B°$\bar{B}$° candidate event at $\Upsilon(4S)$. There are two TOF identified kaons in the event which satisfies the energy-momentum constraint besides the following decay hypothesis: (i) $e^+e^- \to B°\bar{B}°$; (ii) $B° \to D°\pi^+\pi^-$, $\bar{B}° \to D^-\pi^+\pi^+\pi^-$; (iii) $D° \to K^-\pi^+\pi^+\pi^-$, $D^- = K^+\pi^-\pi^-$. The masses of the reconstructed B's come out to 5.220±.02 (GeV/c^2).

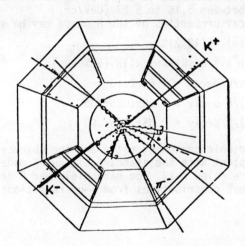

Fig. 31: A B°$\bar{B}$° event candidate (CLEO). Run 8019 rec. 7524.

XVII. SUMMARY AND CONCLUSION

(i) Static non-relativistic potential models with a logarithmically modified Coulomb short distance behaviour and a linear confining large distance behaviour are quite successful in predicting the general features of $c\bar{c}$ and $b\bar{b}$ quarkonium spectroscopy. The T, T', T'', and T''' resonances are identified as the 1^3S_1, 2^3S_1, 3^3S_1, and 4^3S_1, states of the $b\bar{b}$ quarkonium system.

(ii) The hadronic transition rates have been measured:

$$BR(T(2S) \rightarrow T(1S)\pi^+\pi^-) = (19.1 \pm 3.1 \pm 2.9)\%$$

$$BR(T(3S) \rightarrow T(1S)\pi^+\pi^-) = (4.8 \pm 1.7)\%$$

$$BR(T(3S) \rightarrow T(2S)\pi^+\pi^-) < 14\% \ (90\% \ C.L.)$$

(iii) The strong interaction coupling constant α_s has been measured to be:

$$\alpha_s(.48M_T) = 0.158^{+.012}_{-.010}$$

from which the QCD scale parameter $\Lambda_{\overline{MS}}$ is measured to be:

$$\Lambda_{\overline{MS}} = 100^{+34}_{-25} \ \text{MeV}$$

(iv) Inclusive production of hadrons from $T(1S)$, $T(2S)$ and $T(3S)$ direct decays have the same general features as from $q\bar{q}$ jets on the continuum below $T(4S)$.

(v) The $T(4S)$ resonance is above $B\bar{B}$ threshold and the mass of the B meson is between 5.16 to 5.27 (GeV/c².)

(vi) The decay properties of the B meson may be summarized below:

$$BR(X\ell\bar{\nu}) = (12 \pm 2)\%$$

$$BR(X\ell^+\ell^-)/BR(B \rightarrow X\ell\nu) < 11\% \ (90\%C.L.)$$

$$\text{Kaons/B decay} = 1.8 \pm .2 \pm .3$$

$$(p+\bar{p})/B \ \text{decay} = 0.05 \pm .025 \pm .05$$

$$(\Lambda+\bar{\Lambda})/B \ \text{decay} = 0.03 \pm .045 \pm .03$$

The semileptonic branching ratio and the absence of dileptonic decays is consistent with the standard six quark model and most exotic models are ruled out. The kaon yield per B decay favours the $b \rightarrow c$ transition but contributions from $b \rightarrow u$ transitions may also be present.

REFERENCES

1. CLEO Collaboration: D. Andrews, P. Avery, K. Berkelman, R. Cabenda, D. Cassel, J. DeWire, R. Ehrlich, T. Ferguson, M. Gilchriese, B. Gittelman, D. Hartill, D. Herrup, M. Herzlinger, D. Kreinick, N. Mistry, E. Nordberg, R. Perchonok, R. Plunkett, A. Sadoff, K. Shinsky, R. Siemann, P. Stein, S. Stone, R. Talman, G. Thonemann, D. Weber, R. Wilcke (Cornell Univ.); C. Bebek, J. Haggerty, J. Izen, C. Longuemare, W. MacKay, J. Rohlf, W. Tanenbaum, R. Wilson (Harvard Univ.); K. Chadwick, P. Ganci, T. Gentile, H. Kagan, R. Kass, F. Lobkowicz, A. Melissinos, S. Olsen, R. Poling, C. Rosenfeld, G. Rucinski, E. Thorndike (Univ. of Rochester); J. Mueller, D. Potter, F. Sannes, P. Skubic, R. Stone (Rutgers Univ.); D. Bridges, A. Brody, A. Chen, M. Goldberg, N. Horwitz, J. Kandaswamy, H. Kooy, P. Lariccia, G. Moneti (Syracuse Univ.); M. Alam, S. Csorna, R. Panvini, R. Hicks, A. Fridman (Vanderbilt Univ.).

2. CUSB Collaboration: T. Bohringer, F. Constantini, J. Dobbins, P. Franzini, K. Han, S. Herb, L. Lederman, D. Peterson, G. Mageras, E. Rice, J. Yoh (Columbia Univ.); G. Finocchiaro, J. Lee-Franzini, G. Giannini, R. Schamberger, M. Sivertz, L. Spencer, P. Tuts (SUNY at Stony Brook); R. Imlay, G. Levman, W. Metcalf, V. Sreedhar, (Louisiana State Univ.); G. Blanar, H. Dietl, F. Pauss, E. Lorenz, H. Vogel (MPI, Munich).

3. H. Schopper, The Properties of Charmonium and Charm Particles, DESY 77/79, December, 1977.

4. A. Barbaro-Galtieri, Production and Decay of Charmed Particles, LBL-8537, December, 1978.

5. Design Report on CESR, Cornell Univ. report CLNS/360, April, 1977.

6. A. Silverman, The CLEO Detector, Wilson Lab. internal report CBX 81-40, to be published in NIM.

7. S.W. Herb et al., P.R.L., 39, 252 (1977); W.R. Innes et al., P.R.L., 39, 1240 (1977). K. Ueno et al., P.R.L., 42, 486 (1979).

8. Ch. Berger et al., Phys. Lett. 76B, 243 (1978). C.W. Darden et al., Phys. Lett., 76B, 246 (1978).

9. J.L. Rosner, Lectures at Advanced Studies Inst. on Techniques and Concepts of High Energy Physics, St. Croix, Virgin Islands, 1980, p. 2.

10. D. Andrews et al., (CLEO), Phys. Rev. Lett., 44, 1108 (1980).

11. T. Bohringer et al., (CUSB), Phys. Rev. Lett., 44, 1111 (1980).

12. D. Andrews et al., (CLEO), Phys. Rev. Lett., 45, 219 (1980).

13. G. Finocchiaro et al., (CUSB), Phys. Rev. Lett. 45, 222 (1980).

14. B. Niczyporuk et al., (LENA), Phys. Rev. Lett., 46, 92 (1981).

15. G. Bhanot and S. Rudaz, Phys. Lett. 79, 119 (1978).

16. J. Richardson, Phys. Lett., 82B, 272 (1979).

17. W. Buchmuller and H. Tye, Fermilab-Pub-80/94-THY.

18. A. Martin, Phys. Lett. 93B, 338 (1980).

19. J. Rosner, C. Quigg and H. Thacker, Phys. Lett. 74B, 350 (1978).

20. Kurt Gottfried, High Energy e^+e^- Interactions, Vanderbilt, 1980, Edited by S. Csorna and R. Panvini, p. 88.

21. J.J. Mueller et al., (CLEO), Phys. Rev. Lett.,46, 1181 (1981).
22. G. Mageras et al., (CUSB), Phys. Rev. Lett.,46, 1115 (1981).
23. B. Niczyporuk et al., (LENA), Phys. Lett., 100B, 95 (1981).
24. Ch. Berger et al., (PLUTO), Phys. Lett. 93B, 497 (1980).
25. H. Albrecht et al., (DASP2), Phys. Lett. 93B, 500 (1980).
26. H. Schroder, New Results on Υ-System from DORIS, DESY 80/61, June, 1980.
27. M. Alam et al., (CLEO) Observation of $\Upsilon(3S) \to \pi^+\pi^-\Upsilon(1S)$, Submitted to Conf. on Lepton and Photon Interactions at High Energies to be held at Bonn, August 1981.
28. CUSB Internal Report, to be published.
29. K. Gottfried, Phys. Rev. Lett., 40, 598 (1978).
30. Y.P. Kuang and T.M. Yan, Cornell Univ. preprint in preparation.
31. P. Bock et al., Z. Phys. C6, 125 (1980).
32. E. Eichten and F.L. Feinberg, Phys. Rev. Lett., 43, 1205 (1979). T. Sterling, Nuc. Phys., B141, 272 (1978).
32a.Richard Kass, CLEO Internal Memo 7/19/81 to be presented at Bonn Conf., reference 27.
32b.LENA Collab. DESY 80/12, Dec. (1980).
33. T.M. Yan, Phys. Rev., D22, 1652 (1980).
34. R. VanRoyen and V.F. Weisskopf, Nuovo Cimento, 50A, 617 (1967).
35. T. Appelquist and H.D. Politzer, Phys. Rev. D12, 1404 (1978).
36. P. Lepage, P.B. Mackenzie, Cornell Univ. Report, CLNS 80/498 (1981)
37. M. Alam et al., A Determination of α_s from the Leptonic Decay of $\Upsilon(1S)$, Cornell Univ., CBX 81-43, same as ref. 27.
38. A.M. Boyarski et al., Phys. Rev. Lett., 34, 1357 (1975).
39. G. Moneti, H. Kooy, J. Kandaswamy, CLEO Internal Report 7/21/81, to be presented at Bonn, August 1981, see ref. 27.
40. H. Kooy, Syracuse University, Ph.D. thesis, 1981.
41. M. Alam, S. Csorna, R. Panvini, CLEO Internal Report, CBX 81-15, Feb. 20, 1981.
42. H. Albrecht et al., (DASP), DESY 81-001, March (1981).
43. H. Ackermann et al., (PLUTO), DESY 81-018, April (1981).
44. B. Andersson et al., LUND preprint LU 1P 81-3, April (1980).
45. W. Hofman, DESY 81-019, April (1981).
46. E. Eichten et al., Phys. Rev., D17, 3090 (1978); Phys. Rev. D21, 203 (1980).
47. A. Martin, CERN preprint, TH.3068-CERN, April (1981).
48. E. Eichten, Harvard preprint HUTP-80/A027 (1980).
49. M. Kobayashi and T. Maskawa, Prog. Theor. Phys., 49, 652 (1973).
50. H. Tye, CLEO Internal Report, CBX 81-9, Jan. (1981).
51. L.J. Spencer et al., CUSB Internal Report to be published in Phys. Rev. Lett.
52. C. Bebek et al., Phys. Rev. Lett., 46, 84 (1981).
53. K. Chadwick et al., Phys. Rev. Lett. 46, 88 (1981).
54. N. Cabibbo and L. Maiani, Phys. Lett., 87B, 366 (1979).
55. J. Leveille, U. of Michigan (Ann Arbor), UMHE81-18, March (1981).
56. M. Peskin, Cornell Univ., CLEO Report CBX 80-65, October (1980).
57. H. Georgi and S. Glashow, Harvard Univ., Report HUTP-79/A073.

DISCUSSION

ANDERSSON, LUND: Just a small comment in connection with
baryon production on the resonance. I looked up the numbers and
it corresponds in our model for baryon production . . . in which
there is actually more baryon [-antibaryon] production in gluon
jets than in quark jets . . . it turns out that on the Υ resonance we
expect an enhancement of between 8 and 10% as compared to off-
resonance. [The small increase is due to the small amount of
energy available.]

ALAM: We need much higher statistics to confirm things at that
level.

RATTI: Don't you have any example in which the D associated
to the B is not decaying into three particles? You are seeing
beautiful multiparticle decays of this . . . .

ALAM: Actually the answer is no, I don't have.

Results on e^+e^- Interactions in the $\Upsilon(9.46)$ and $\Upsilon'(10.01)$
Energy Region

LENA Collaboration†

A. Fridman*
Vanderbilt University, Nashville, TN 37235

ABSTRACT

Some results on the Υ and Υ' resonance parameters are presented.
Charged hadron production is investigated in the 7-10 GeV c.m. energy
region. The data show KNO scaling and suggest the presence of long
range correlations. The jet structure of the Υ is well described by
the Υ decay into vector gluons and excludes scalar gluons by more
than four standard deviations.

We present results obtained on e^+e^- interactions in a c.m.
energy corresponding to the region of the Υ and Υ' resonances. The data
were obtained by means of the non-magnetic LENA detectors at DORIS.
The detector which is described in detail elsewhere[1] allows us to
measure the energy of the neutral particles and the direction of the
charged tracks produced in the interactions.

The first task of the experiment has consisted in measuring the
resonance parameters of the Υ [1] and Υ' [2]. The integrated hadronic
resonance cross section corrected for radiative effects permits us
to determine $\bar{\Gamma}_{ee} = \Gamma_{ee}\Gamma_h/\Gamma_t$. This quantity is a rather good estimate
of the leptonic width Γ_{ee} as the hadronic width Γ_h is nearly equal to
the total widths Γ_t. A better Γ_{ee} determination can be further
obtained by measuring the branching ratio $B = \Gamma_{ee}/\Gamma_t$ as
$\Gamma_{ee} = \bar{\Gamma}_{ee}/(1-3B)$. The latter formula is derived assuming e,μ,τ
universality. Furthermore the measurement of B allows us to obtain
the total width $\Gamma_t = \Gamma_{ee}/B$. The table presents the resonance
parameters obtained by the LENA collaboration.

Table: Resonance Parameters
(The first error is statistical and the second one
is due to systematics)

	Υ	Υ'
Mass (MeV)	9461.6±0.6±10	10013.6±1.2±10.0
$\bar{\Gamma}_{ee}$ (KeV)	1.10±0.07±0.11	$0.53\pm0.07\,^{+\,0.09}_{-\,0.05}$
B (%)	3.5±1.4±0.4	<3.8
Γ_{ee} (KeV)	1.23±0.10±0.14	—
Γ_t (KeV)	$35\,^{+\,25\,+\,9}_{-\,10\,-\,7}$	$31\,^{+\,16}_{-\,7}$ *

*The $\Gamma_t(\Upsilon')$
has been
estimated by
adding all the
partial width,
the $\Gamma(\Upsilon' \to P\gamma)$
being scaled up
from the
charmonium data
[Phys. Lett.
100B, 45(1981)]

Permanent address: DPHPE, CEN de Saclay, Gif sur Yvette, France

These data can be compared with those obtained at CESR for higher radially excited $\bar{q}q$ states[3,4]. As $B = \Gamma_{ee}/\Gamma_{t}$ was not always measured this comparison can only be made for $\bar{\Gamma}_{ee}$. One notices that $\bar{\Gamma}_{ee}$ decreases with increasing of the principal quantum number k of the $\bar{q}q$ system. This behavior finds a natural explanation in the framework of potential models. In these models the electronic width of a vector meson of Mass M_V can be expressed by the QCD corrected Van Royen-Weisskopf formula[5]

$$\Gamma_{ee} = 16\pi\alpha^2/M_V^2 \ |\psi_k(o)|^2/9[1 - 16\alpha_s/\pi \ ]$$

Here α_s is the QCD strong coupling constant and $|\psi_k(o)|^2$ represents the square of the wave functions at the origin. For monotonic increasing potentials concave downward which are used to describe the Υ and the J/ψ families, it has been argued that[6] $[\psi_k(o)]^2 > [\psi_{k+1}(o)]^2$. This leads through the Van Royen-Weisskopf formula to $\Gamma_{ee}^k > \Gamma_{ee}^{k+1}$ as indeed verified by the data.

Let us now discuss some features concerning the production of charged particles for which one has only directions (no momentum) information. Fig. 1 represents the cosine of the opening angle

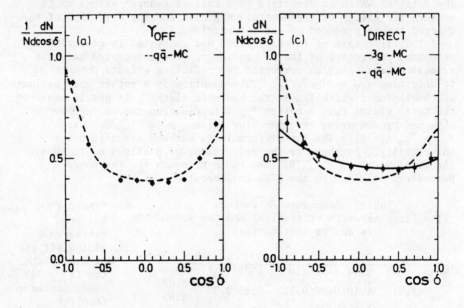

Fig. 1
The distribution of the cosine of the opening angle for the continuum (Υ_{OFF}) and the Υ direct decay data. The experimental points are compared with the two jets $\bar{q}q$ and 3-gluon Monte-Carlo predictions.

distribution for pairs of tracks[7]. These distributions are shown

for the Υ direct decay [8] data and for its nearby continuum. The
continuum is well described by the two jet Monte Carlo a la Field
and Feynman while the direct decay data are at best described by
the 3-gluon predictions. This already shows that direction information of tracks are sensitive to the production mechanism. We
therefore define a thrust like quantity[7] $T' = \max_i \Sigma |\cos \theta_i| /n$

where θ_i are the emission angles of the charged particles (n in total)
defined with respect to an axis which maximalizes $\Sigma |\cos \theta_i|$. The
T' is strongly correlated with the normal thrust variable T_{ch}. This
can be seen from the scatter plot of T' versus T_{ch} (Fig. 2)

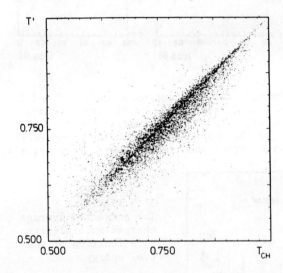

Fig. 2
The scatter plot of T' versus T_{ch} for 3-
gluon Monte Carlo events generated at
9.4 GeV.

obtained from 3-gluon
Monte Carlo events
generated at 9.4 CeV.
Fig. 3 presents the
angular distribution
of the thrust axis
with respect to the
e^+e^- direction,
corrected for our
various efficiencies.
Theoretical calcula-
tions predict that
this angular distri-
bution is of the form
$1 + \bar{\alpha} \cos^2$ where $\bar{\alpha}$
is equal to 0.39 or
-0.995 for vector[9] or
scalar[10] gluons,
respectively. For the
Υ direct decay one
obtains $\bar{\alpha}_\Upsilon = 0.7 \pm 0.3$
excluding thus scalar
gluons by more than
four standard devia-
tions. The distri-
bution for the T'
direct decay[8] is also

shown in Fig. 3 yielding $\bar{\alpha}_{\Upsilon'} = 0.6 \pm 0.4$. For comparison we also
present (Fig. 3a) the thrust angular distribution for the
continuum which for point like partons is $1 + a \cos^2\theta$ with $a = 1$.
The fit of our data gives $a = 0.7 \pm 0.1$, a value somewhat lower
than expected for free quark production.

Using the charged particles produced in the c.m. energy region
of 7.35 to 10.01 CeV we also have investigated the behavior of the
statistical moments deduced from the multiplicity distributions.[7]
To this end we calculated from our data the true number of events
F_n with n charged particles (n = 0,2,...), solving by a least
square method the equations $f_m = \Sigma_n P_{mn} F_n$. Here f_m is the
observed multiplicity (m = 0,1,2,...). The probabilities P_{mn} have
been deduced from a Monte Carlo calculation in which the various
detection efficiencies as well as the photon conversion in the
beam pipe have been incorporated.[11] Fig. 4 presents the average

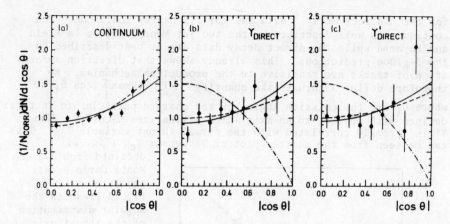

Fig. 3

The thrust angular distributions for the continuum and the Υ and Υ' direct decay data. The full line is obtained by fitting the data with $1 + \alpha\cos^2\theta$. The dashed line in (a) represents the $1 + \cos^2\theta$ distribution. In (b) and (c) the dashed and dotted lines represent the predictions obtained with vector and scalar gluons, respectively.

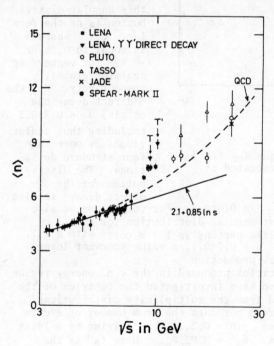

Fig. 4
The corrected average charged multiplicity as a function of the c.m. energy.

charged multiplicity $\langle n \rangle$ as a function of the c.m. energy $\sqrt{s}$. One notices that $\langle n \rangle$ varies faster than $\ln s$ as indeed predicted by QCD models[12]. However for the Υ and Υ' direct decay $\langle n \rangle$ is higher than the values of their nearby continuum.[13] The $\langle n(\Upsilon') \rangle - \langle n(\Upsilon) \rangle = 0.81 \pm 0.28$ can be accounted for by the $\Upsilon \to \Upsilon \pi^+ \pi^-$ decay. In Fig. 5 we present the dispersion $D = \sqrt{\langle n^2 \rangle - \langle n \rangle^2}$ of the charged multiplicity and its second Mueller moment $f_2^- = \langle n_-(n_- -1) \rangle - \langle n_- \rangle^2$ as a function of $\langle n \rangle$. Similarly to hadron interactions D is linearly related to $\langle n \rangle$, namely $D = (0.33 \pm 0.03) \langle n \rangle + (0.23 \pm 0.23)$. The dependance of f_2^- with respect to s (or to $\langle n \rangle$) gives indications about the two particle correlation feature between the produced particles. In particular $f_2^- \propto \ln s$ is a signature of a short range correlation[14]. As the lever arm (in s) is here too small we have plotted the data (Fig. 6) in the KNO form[15]. They manifest the so-called KNO scaling, namely the data points tend to cluster on a curve independent of the c.m. energy. This effect also noted in another experiment[16] means that the reduced moments $c_m = \langle n^m \rangle / \langle n \rangle^m$ are independent of s [14]. This lead s to $f_2 \propto \langle n \rangle^2$. The fact that $\langle n \rangle$ varies faster that $\ln s$ means that f_2^- varies at least as $(\ln s)^2$ This implies a long range correlation between the outgoing particles as advocated by QCD models. [12]

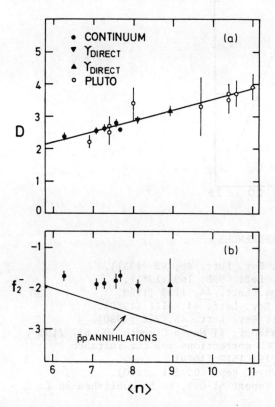

Fig. 5

The dispersion D and the f_2^- moment as a function of $\langle n \rangle$. The full line in (a) is a fit to the data (see text). In (b) the line is obtained by extrapolating low energy $\bar{p}p$ annihilation data (Ref. 7).

†B.Niczyporuk, T. Zeludziewicz, K.W. Chen, R. Hartung, G. Folger, B. Lurz, H. Vogel, U. Volland, H. Wegener, J.K. Bienlein, R. Graumann, J. Kruger, M. Leißner, M. Schmitz, F.H. Heimlich, R. Nernst, A. Schwarz, U. Strohbusch, H.-J. Trost, P. Zschorsch, A. Engler, R.W. Kraemer, F. Messing, C. Rippich, B. Stacey, S. Youssef, A. Fridman, G. Alexander, A. Av-Shalom, G. Bella, Y. Gnat, J. Grunhaus, E. Horber W. Langguth, M. Scheer.

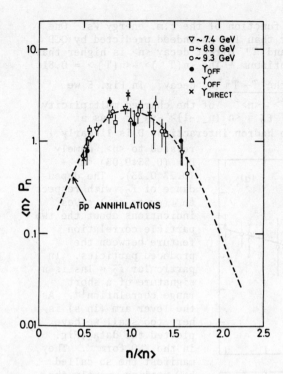

Fig. 6
The multiplicity data
plotted in the KNO form.
Here P_n is the probability
to observe an n prong
event.

REFERENCES

1.   B. Niczyporuk et al, Phys. Rev. Lett. 46, 92 (1981).
2.   B. Niczyporuk et al, Phys. Lett. 99B, 169 (1981).
3.   D. Andrews et al, Phys. Rev. Lett. 44, 1108 (1980.
4.   T. Bohringer et al, Phys. Rev. Lett. 44, 1111 (1980).
     G. Finocchiaro et al, Phys. Rev. Lett. 45, 222 (1980).
5.   R. Van Royen and V.F. Weisskopf, Il Nuovo Cimento 50, 617 (1967)
     and 51, 583 (1967). For QCD corrections see for instance
     W. Celmaster, Phys. Rev. D19, 1517 (1979).
6.   H. Grosse and A. Martin, Phys. Rep. 60, 341 (1980).
7.   B. Niczyporuk et al, DESY Report 81-008, to be published in Z.
     Physik (1981).
8.   The direct decay is obtained by substracting the continuum and
     the vacuum polarization contribution from the resonance data.
     For the Υ one is left with the 3-gluon decay while for the Υ'
     there is in addition other decay channels as Υ' → Υππ, or decays
     via the P state.
9.   K. Koller, H. Krassemann and T.F. Walsh, Z. Physik C1, 71 (1979).
10. K. Koller and H. Krassemann, Phys. Lett. 88B, 119 (1979).
11. Two sets of P_{mn} have been calculated, one for the continuum and
     one for the 3-gluon decay.
12. See for instance K. Konishi, Rutherford Preprint RL-79-035 and
     Proceedings of the XI Internat. Symposium on Multiparticle
     Dynamics, Brügge (1980).

13. The $<n(T)>$ is within the range predicted by QCD calculations, see: K. Koller and T. F. Walsh, Phys. Lett. 72B, 227 (1977) and 73B, 504 (1978); I.I.Y. Bigi and S. Nussinov, Phys. Lett. 82B, 281 (1979).

14. See for instance: A. Fridman, Series of Lecture Notes on High Energy Physics No. 11, published by the Centre de Recherches Nucleaires de Strasbourg and references quoted therein.

15. Z. Koba, H. B. Nielsen and P. Olesen, Nucl. Phys. 1340, 317 (1972).

16. Ch. Berger et al, Phys. Lett. 95B, 313 (1980).

DISCUSSION

DeGRAND, SANTA BARBARA: Do I understand you to be saying that the multiplicity on the Υ is higher than off the Υ? Is this consistent with what earlier experiments have shown?

FRIDMAN: Yes, of course. There is also a theoretical model.

DeGRAND: Never mind theory! . . . for 10 seconds.

FRIDMAN: Of course it is consistent.

DeGRAND: I'm not asking to attack you.

FRIDMAN: Okay, it is consistent but only in this case it is the real multiplicity which is measured, which is really the corrected one, while if you look in the Pluto data there is the uncorrected multiplicity. Anyway, there is a step.

CHARM PRODUCTION AT THE CERN INTERSECTING STORAGE RINGS SPLIT-FIELD MAGNET

M. Basile, G. Cara Romeo, L. Cifarelli, A. Contin, G. D'Ali,
P. Di Cesare, B. Esposito, P. Giusti, T. Massam, F. Palmonari,
G. Sartorelli, G. Valenti and A. Zichichi
CERN, Geneva, Switzerland
Istituto di Fisica dell'Università di Bologna, Italy
Istituto Nazionale di Fisica Nucleare, Bologna, Italy
Istituto Nazionale di Fisica Nucleare, LNF, Frascati, Italy

(Presented by A. Contin)

ABSTRACT

Results on associated charm production in proton-proton interactions at $\sqrt{s} = 62$ GeV are reported. The presence of an electron,
used as trigger for the experiment, was the signature for the semileptonic decay of the anticharmed particles, while the associated
charmed particles, Λ_c^+, D^+ and D^0, were identified via their hadronic
decay modes $\Lambda_c^+ \to pK^-\pi^+$, $D^+ \to K^-\pi^+\pi^+$ and $D^0 \to K^-\pi^+$. Their longitudinal
and transverse momentum distributions have been measured, and esti-
mates have been computed for the associated production cross-sections.

1. INTRODUCTION

The results presented here were obtained by the R415 experiment
using the upgraded Split-Field Magnet (SFM) spectrometer of the CERN
Intersecting Storage Rings (ISR). The main aim of the experiment was
the study of charm production in proton-proton interactions at the
very high c.m. energy of $\sqrt{s} = 62$ GeV.

The charmed particles have a high branching ratio for decaying
semileptonically when compared with the lower-mass hadrons. This
implies that the ratio of charmed/non-charmed events in the final
sample increases when selecting events in which a direct electron is
produced. Moreover, strong interactions conserve charm quantum number, i.e. charmed and anticharmed particles are produced in pairs in
(pp) interactions. The events were thus triggered by an electron
(positron) which possibly comes from the semileptonic decay of an
anticharmed (charmed) particle, and the hadronic decay of the associated charmed (anticharmed) particle was studied by means of the
invariant mass of its decay products.

2. EXPERIMENTAL APPARATUS AND ELECTRON IDENTIFICATION

The main features of the experimental apparatus[1-4], shown in
Fig. 1, were:
i) a powerful electron detector in the 90° region: electromagnetic
   shower detectors (EMSDs), Čerenkov threshold counters, and one
   multiwire proportional chamber (MWPC) with analog read-out
   (dE/dx);

ii) a time-of-flight (TOF) system to perform particle identification up to 1.5 GeV/c (π/K/p separation) or 2.0 GeV/c (πK/p separation);

iii) the SFM MWPC system to measure the momenta of the charged secondaries.

The first analysis of the data consisted in the refinement of the trigger conditions (track reconstruction in the 90° region and Čerenkov pulse-height analysis), in the analysis of the dE/dx signal associated with the trigger track (to reject unresolved and resolved electron pairs from π^0 and η^0 Dalitz decay and from γ conversions), and in the match of the trigger track with an energy cluster in the EMSDs (E > 500 MeV). In the final sample, about 6×10^4 events were retained, corresponding to a total integrated luminosity of about 4.4×10^{36} cm^{-2}, with a charged hadron contamination of the order of 2% and a contamination from neutral hadrons of less than 50%. In order to increase the confidence in the electron charge sign, only electrons with an accuracy in the momentum measurement $\Delta p/p \leq 15\%$ were retained.

3. $\Lambda_c^+ \to pK^-\pi^+$

The associated production of a Λ_c^+ and an anticharmed particle decaying into an e$^-$ was studied. The Λ_c^+ was identified via its decay mode: $\Lambda_c^+ \to pK^-\pi^+$ [5]. The following selection criteria were adopted in order to increase the signal/background ratio in the Λ_c^+ mass region[*]):

  i) p : $|x_L| \geq 0.3$;

 ii) K$^-$: any negative track with a) $|y| \geq 1$ and b) not identified by TOF as π^- or $\bar{p}$;

iii) π^+: any positive track with a) $|y| \geq 1$ and b) not identified by TOF as K$^+$ or p.

All the p, K$^-$, and π^+ tracks were required to have a $\Delta p/p \leq 30\%$ and to fit the event vertex within ±5 cm. Conditions (i), (iia), and (iiia) select forward-produced Λ_c^+'s. The presence of a "leading" system in the rapidity hemisphere opposite to the Λ_c^+ was also requested. Figure 2 shows the final pK$^-\pi^+$ invariant mass spectra for events associated with the e$^-$ (Fig. 2a) and with the e$^+$ (Fig. 2b) triggers. A 4 standard deviations peak is present in the Λ^+ mass region, associated with the e$^-$ trigger. The width of this peak, compatible with the experimental resolution, and the absence of a similar peak in the sample of events triggered by an e$^+$, identify it as a charm signal.

By dividing the invariant mass spectrum of Fig. 2a into a peak region [IN: 2.28 < M(pK$^-\pi^+$) < 2.38 GeV/c^2] and a control region [OUT: 2.18 < M(pK$^-\pi^+$) < 2.28 and 2.38 < M(pK$^-\pi^+$) < 2.48 GeV/c^2], some characteristics of the signal have been studied.

a)   *Transverse momentum distribution*[6]. Figure 3 shows the $(1/p_T)(\Delta N/\Delta p_T)$ experimental distribution of the Λ_c^+ events. The best fit to the data points is:

$$(1/p_T)(\Delta N/\Delta p_T) \propto \exp\left[-(2.5 \pm 0.4)p_T\right].$$

---

[*]) $x_L = 2p_L/\sqrt{s}$, $y = 0.5 \ln\left[(E+p_L)/(E-p_L)\right]$.

b) *Longitudinal momentum distribution*[7]. Figure 4 shows the experimental $\Delta N/\Delta x_L$ distribution of the Λ_c^+ events. The best fit to the data points is

$$(\Delta N/\Delta x_L) \propto (1 - |x_L|)^{0.40 \pm 0.25} .$$

A comparison between the Λ_c^+ and the Λ^0 and $\bar{\Lambda}^0$ $d\sigma/dx_L$ distributions, shown in Fig. 5, indicates that the Λ_c^+ is produced, in the same manner as the Λ^0 and opposite to the $\bar{\Lambda}^0$, in a leading way.

c) *Resonant contributions to the decay channel* $\Lambda_c^+ \to pK^-\pi^+$ [8]. The resonant contributions $\Lambda_c^+ \to \overline{K^{*0}}p$ and $\Lambda_c^+ \to \Delta^{++}K^-$ to the decay channel $\Lambda_c^+ \to pK^-\pi^+$ have been studied. The results are summarized in Fig. 6 and Table I.

$$4. \quad D^+ \to K^-\pi^+\pi^+$$

The reaction studied[9] was:

$$pp \to D^+ + (\text{anticharmed state}) + X ,$$

with the D^+ observed in the decay mode

$$D^+ \to K^-\pi^+\pi^+$$

and the anticharmed state decaying into an e^-.
The events were selected using the following criteria:
i) K^-: identified by TOF (90% confidence level[3]);
ii) π^+: any positive track with $|x_L| < 0.3$ and not identified by TOF as K^+ or p.
The K^- and π^+ tracks were requested to have $\Delta p/p < 30\%$ and to fit the event vertex. Condition (i) limits the K^- momentum to 1.5 GeV/c.
Figure 7 shows the $K^-\pi^+\pi^+$ invariant mass spectrum in events associated with an e^- trigger. A cut in the transverse momentum of the $K^-\pi^+\pi^+$ system, $p_T(K^-\pi^+\pi^+) > 0.7$ GeV/c, was used to increase the signal/background ratio in the D^+ mass region. The result is shown in Fig. 8a, where a D^+ signal of (39 ± 11) combinations is present. The background shape in Figs. 7 and 8a was obtained using the "event mixing" technique. As expected for a real charm signal, no peaks were present when repeating the same analysis using events triggered by an e^+ (Fig. 8b).
The longitudinal and transverse momentum distributions of the D^+ were worked out, using as peak region the mass range $1.8 < M(K^-\pi^+\pi^+) < 2.1$ GeV/c^2, and as a control region the mass ranges $1.5 < M(K^-\pi^+\pi^+) < 1.8$ and $2.1 < M(K^-\pi^+\pi^+) < 2.4$ GeV/c^2. Figure 9 shows the experimental $(1/p_T)(\Delta N/\Delta p_T)$ distribution of the D^+ [10]. The best fit to the data points is

$$(1/p_T)(\Delta N/\Delta p_T) \propto \exp\left[-(2.3 \pm 0.8)p_T\right] .$$

Figure 10 shows the experimental $\Delta N/\Delta x_L$ distribution of the D^+ [11]. The three superimposed curves describe the apparatus acceptance for three different production models of the D^+: central $[E(d\sigma/dx) \propto (1-|x_L|^3)]$; flat in y ($d\sigma/dy = \text{const}$); and flat in x_L ($d\sigma/dx_L = \text{const}$).

Owing to the limited momentum acceptance of the TOF system, nothing can be said about the longitudinal momentum production distribution of the D^+.

5. $D^0 \rightarrow K^- \pi^+$

In order to have a better insight into the production distributions of charmed mesons in (pp) interactions, another reaction has been studied[12]:

$$pp \rightarrow D^0 + \text{(anticharmed state)} + X ,$$

with the D^0 observed in the decay mode,

$$D^0 \rightarrow K^- \pi^+ ,$$

and the associated anticharmed state in the decay mode,

$$\text{(anticharmed state)} \rightarrow e^- + K^+ + \text{anything} .$$

The events were selected as follows:
  i) K^+: identified by TOF (90% confidence level);
 ii) K^-: any negative track not identified by TOF as π^- or $\bar{p}$;
iii) π^+: any positive track with $|x_L| \leq 0.3$ and not identified by
       TOF as K^+ or p.
The K^+, K^-, and π^+ tracks were requested to have $\Delta p/p \leq 30\%$ and to fit the event vertex. As the K^- was not identified by TOF, less limitations on the D^0 acceptance were introduced. The rejection of the non-charmed background events was given by the strong selectivity of the $e^- K^+$ trigger.

Figure 11 shows the $K^- \pi^+$ invariant mass spectrum in events associated with an e^- trigger. As for the D^+ case, a cut on the transverse momentum of the $K^- \pi^+$ system, $p_T(K^- \pi^+) > 0.7$ GeV/c, increases the signal/background ratio in the D^0 mass region. Figure 12 shows the $K^- \pi^+$ invariant mass spectra, with the transverse momentum cut applied, for events triggered by an $e^- K^+$ pair (Fig. 12a), and for events triggered by wrong charge pairs, i.e. $e^- K^-$, $e^+ K^+$, and $e^+ K^-$ (Fig. 12b). The presence of the signal in association only with an $e^- K^+$ trigger identifies the $K^- \pi^+$ peak of Fig. 12a as a charm signal.

Again, a peak region $\left[1.825 < M(K^- \pi^+) < 1.975 \text{ GeV/c}^2 \right]$ and a control region $\left[1.675 < M(K^- \pi^+) < 1.825 \text{ and } 1.975 < M(K^- \pi^+) < 2.125 \text{ GeV/c}^2 \right]$ were defined, to study the transverse and longitudinal momentum distributions of the D^0.

a) *Transverse momentum distribution*[10]. The experimental transverse momentum distribution of the D^0 is shown in Fig. 13. The best fit to the data points is:

$$(1/p_T)(\Delta N/\Delta p_T) \propto \exp\left[-(2.4 \pm 0.5)p_T\right] .$$

b) *Longitudinal momentum distribution*[11]. Figure 14 shows the experimental $\Delta N/\Delta x_L$ distribution of the D^0. Here again, the three superimposed curves represent the apparatus acceptance for three possible production distributions of the D^0. The data indicate that the D^0 is produced in a rather central way $\left[E(d\sigma/dx_L) \propto \propto (1 - |x_L|)^3 \text{ or } d\sigma/dy = \text{const}\right]$, whilst a flat-$x_L$ distribution seems unlikely, despite the low statistics.

6. TOTAL CROSS-SECTIONS

The inclusive cross-section was assumed to be expressed by: $E(d^3\sigma/dp^3) \propto f(y/y_{max}) \exp(-bp_T)$, where the longitudinal and transverse momentum functions will be defined in the following.

In order to estimate the total production cross-sections for the Λ_c^+, D^+, and D^0 signals presented above, the following quantities must be known:

   i) for the associated anticharmed particle decaying semileptonically:
      a) the mass;
      b) the semileptonic branching ratio;
      c) the decay modes;
      d) the decay dynamics;
      e) the production distributions (p_T, x_L);
  ii) for the observed particle:
      a) the decay branching ratio;
      b) the decay dynamics;
      c) the production distributions (p_T, x_L);
iii) the correlation, if any, between the two particles.
Owing to the high number of assumptions, the cross-sections given in the following are only to be taken as indicative.

For all the reactions studied, the associated anticharmed particle was assumed to be a D meson and to have decay branching ratios $(D \rightarrow e^- X)/(D \rightarrow all) = (8.0 \pm 1.0)\%$, $(D \rightarrow Ke^-\nu)/(D \rightarrow e^- X) = 60\%$, and $(D \rightarrow K^*e^-\nu)/(D \rightarrow e^- X) = 40\%$, with a decay dynamics following the $K\ell_3$ matrix. The decay $D \rightarrow e^- K^+ X$, which was observed in association with the D^0 (Section 5), was computed to have a branching ratio of about 4.5%. Three longitudinal momentum production distributions were used to compute the apparatus acceptance: $E(d\sigma/dx_L) \propto (1 - |x_L|)^3$ (model I); $d\sigma/dy = \text{const}$ (model II); and $d\sigma/dx_L = \text{const}$ (model III), with $(d\sigma/dp_T) \propto \exp(-2p_T)$.

The measured branching ratios[13] for the hadronic channels were used: $B(\Lambda_c^+ \rightarrow pK^-\pi^+/\Lambda_c^+ \rightarrow all) = (2.2 \pm 1.0)\%$; $B(D^+ \rightarrow K^-\pi^+\pi^+/D^+ \rightarrow all) = (6.3 \pm 1.5)\%$; $B(D^0 \rightarrow K^-\pi^+/D^0 \rightarrow all) = (3.0 \pm 0.6)\%$; and all decays were assumed to follow the phase space.

The production distributions (I), (II), and (III) defined above were also used for the charmed particles decaying hadronically. Moreover, no correlation was assumed between the two particles.

The results are summarized in Tables II, III, and IV. For the associated $\Lambda_c^+ \bar{D}$ production, the estimates range from 184 ± 75 µb for central $\bar{D}$ and flat-x_L Λ_c^+ productions, up to 1450 ± 580 µb for the extreme model with flat-x_L D and the Λ_c^+ distributed following the observed x_L and p_T behaviours.

For $D\bar{D}$ pair production, the lowest total cross-sections $\left[\sigma(pp \rightarrow D^+\bar{D}X) = 305 \text{ µb and } \sigma(pp \rightarrow D^0\bar{D}X) = 575 \text{ µb with a 50\% error}\right]$ are obtained assuming central production for both D mesons, as indicated by the study of the x_L distribution of the D^0 (Section 5).

803

If flat-y distributions are assumed for both D's, those estimates double.

7. CONCLUSIONS

The charm signals presented here represent the first insight into the mechanisms of charm production in (pp) interactions at $\sqrt{s}$ = 62 GeV. Evidence supports the picture of the charmed baryon Λ_c^+ produced forward with a nearly flat-x_L distribution, and the charmed mesons produced centrally. The total charm production cross-section in (pp) interaction, at this value of $\sqrt{s}$, is of the order of 1 mb, i.e. charm is copiously produced at the ISR.

It is, however, clear that more statistics, with more favourable signal/background ratios, are needed in order to understand charm production in more detail.

Table I. Branching ratios of $p\overline{K^{*0}}$ and $K^-\Delta^{++}$ resonant contributions to the $\Lambda_c^+ \to pK^-\pi^+$ decay channel

		$\overline{K^{*0}}p/pK^-\pi^+$	$\Delta^{++}K^-/pK^-\pi^+$
SPEAR (Mark II)		0.12 ± 0.07	0.17 ± 0.07
ISR (ACCDHW Collab.)		0.41 ± 0.18	0.53 ± 0.22
ISR (this experiment)		0.28 ± 0.16	0.40 ± 0.17
World average	unweighted	0.27 ± 0.08	0.37 ± 0.10
	weighted	0.18 ± 0.06	0.23 ± 0.06

Table II. $\Lambda_c^+\overline{D}$ cross-section estimates obtained with different combinations of production mechanisms

Λ_c^+ model		$\overline{D}$ model		σ_{tot} (μb)
$E(d\sigma/dx_L) \propto (1-x_L)^3$	(I)	$E(d\sigma/dx_L) \propto (1-x_L)^3$	(I)	4200
$d\sigma/dy$ = const	(II)	$d\sigma/dy$ = const	(II)	750
$d\sigma/dx_L$ = const	(III)	$d\sigma/dx_L$ = const	(III)	1125
$d\sigma/dx_L$ = const	(III)	$E(d\sigma/dx_L) \propto (1-x_L)^3$	(I)	184
measured		$E(d\sigma/dx_L) \propto (1-x_L)^3$	(I)	245
$\left\{ \begin{array}{l} d\sigma/dx_L \text{ and} \\ d\sigma/dp_T \end{array} \right\}$		$d\sigma/dy$ = const	(II)	446
		$d\sigma/dx_L$ = const	(III)	1450

804

Table III. $D^+\bar{D}$ cross-section estimates obtained
with different combinations of production mechanisms

D^+ model		$\bar{D}$ model		σ_{tot} (μb)
$E(d\sigma/dx_L) \propto (1-x_L)^3$	(I)	$E(d\sigma/dx_L) \propto (1-x_L)^3$	(I)	305
$d\sigma/dy = \text{const}$	(II)	$d\sigma/dy = \text{const}$	(II)	730
$d\sigma/dx_L = \text{const}$	(III)	$d\sigma/dx_L = \text{const}$	(III)	>5000
$d\sigma/dx_L = \text{const}$	(III)	$E(d\sigma/dx_L) \propto (1-x_L)^3$	(I)	1080

Table IV. $D^0\bar{D}$ cross-section estimates obtained
with the same assumptions as for the $D^+\bar{D}$ production

D^0 model		$\bar{D}$ model		σ_{tot} (μb)
$E(d\sigma/dx_L) \propto (1-x_L)^3$	(I)	$E(d\sigma/dx_L) \propto (1-x_L)^3$	(I)	575
$d\sigma/dy = \text{const}$	(II)	$d\sigma/dy = \text{const}$	(II)	1290
$d\sigma/dx_L = \text{const}$	(III)	$d\sigma/dx_L = \text{const}$	(III)	>5000
$d\sigma/dx_L = \text{const}$	(III)	$E(d\sigma/dx_L) \propto (1-x_L)^3$	(I)	1000

REFERENCES

1. R. Bouclier et al., Nucl. Instrum. Methods 125, 19 (1975).
2. M. Basile et al., Nucl. Instrum. Methods 163, 93 (1979).
3. M. Basile et al., Nucl. Instrum. Methods 179, 477 (1981).
4. H. Frehse et al., Nucl. Instrum. Methods 156, 87 (1978).
   H. Frehse et al., Nucl. Instrum. Methods 156, 97 (1978).
5. M. Basile et al., Nuovo Cimento 63A, 230 (1981).
6. M. Basile et al., Nuovo Cimento Lett. 30, 481 (1981).
7. M. Basile et al., Nuovo Cimento Lett. 30, 487 (1981).
8. M. Basile et al., Nuovo Cimento 62A, 14 (1981).
9. M. Basile et al., Measurement of $(D^+\bar{D})$ charm-meson pair production in (pp) interactions at $\sqrt{s}$ = 62 GeV, to be submitted to Nuovo Cimento (1981).
10. M. Basile et al., The p_T-distribution of charmed mesons produced in (pp) interactions at $\sqrt{s}$ = 62 GeV, to be submitted to Nuovo Cimento (1981).
11. M. Basile et al., The longitudinal momentum distribution of charmed mesons produced in (pp) interactions at $\sqrt{s}$ = 62 GeV, to be submitted to Nuovo Cimento (1981).
12. M. Basile et al., preprint CERN-EP/81-73 (1981).
13. G.A. Trilling, preprint LBL-12283, Feburary 1981, submitted to Physics Reports.

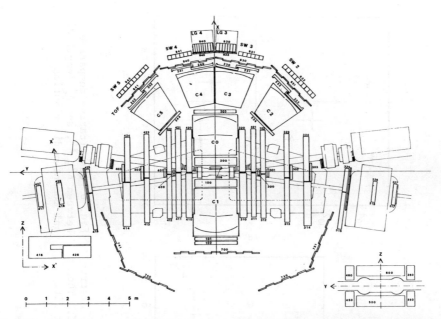

Fig. 1 TOP view of the SFM detector, showing the MWPCs, the time-of-flight system (TOF), the electromagnetic shower detectors (SW, LG), the gas threshold Čerenkov counters (C), and the dE/dx chamber (209).

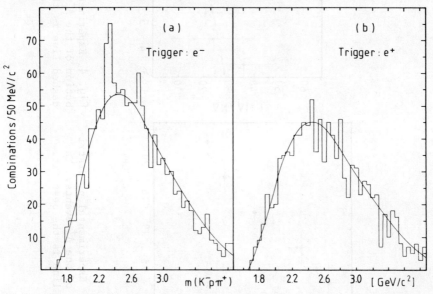

Fig. 2 Mass distribution of pK⁻π⁺ combinations, selected as described in the text, for events triggered by an e⁻ (a) and by an e⁺ (b). The solid line fit represents the background shape as determined by the e⁺ triggered events.

807

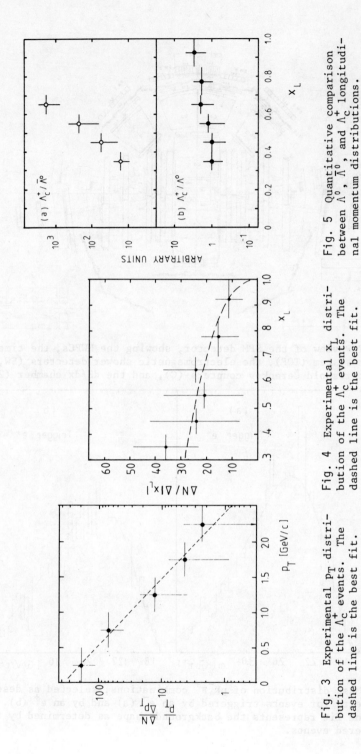

Fig. 3 Experimental p_T distribution of the Λ_c^+ events. The dashed line is the best fit.

Fig. 4 Experimental x_L distribution of the Λ_c^+ events. The dashed line is the best fit.

Fig. 5 Quantitative comparison between Λ^0, $\bar{\Lambda}^0$, and Λ_c^+ longitudinal momentum distributions.

808

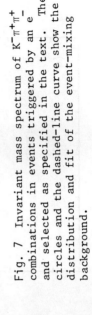

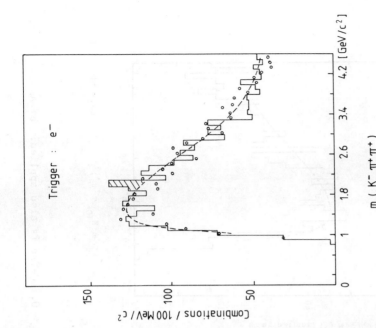

Fig. 7 Invariant mass spectrum of $K^-\pi^+\pi^+$ combinations in events triggered by an e^- and selected as specified in the text. The circles and the dashed-line curve show the distribution and fit of the event-mixing background.

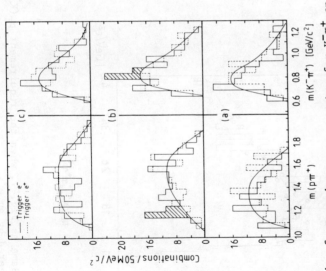

Fig. 6 Invariant mass spectra for $K^-\pi^+$ and $p\pi^+$ combinations as a function of three different cuts in the $pK^-\pi^+$ mass:
a) $2.18 < M(pK^-\pi^+) < 2.28$ GeV/c^2 (below Λ_c^+);
b) $2.28 < M(pK^-\pi^+) < 2.38$ GeV/c^2 (in the Λ_c^+ peak);
c) $2.38 < M(pK^-\pi^+) < 2.48$ GeV/c^2 (above Λ_c^+). The full-line histograms refer to e^- triggers; the dashed-line histograms refer to e^+ triggers. The solid lines are fits to the mass spectra with e^+ triggers.

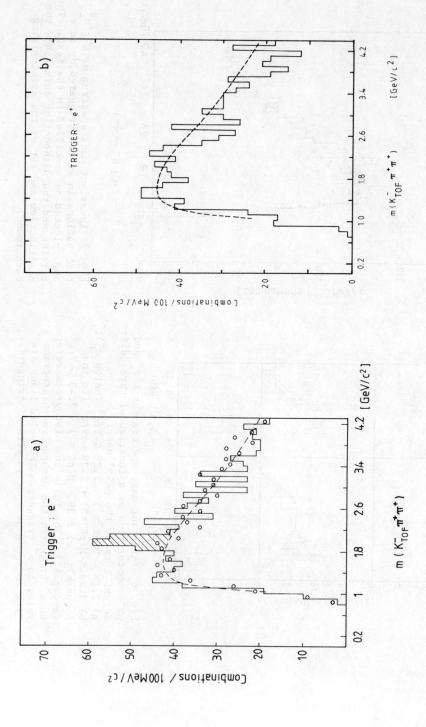

Fig. 8 a) Same as Fig. 7, when the requirement $p_T(K^-\pi^+\pi^+) > 0.7$ GeV/c is also applied. b) As (a) but in events triggered by an e^+.

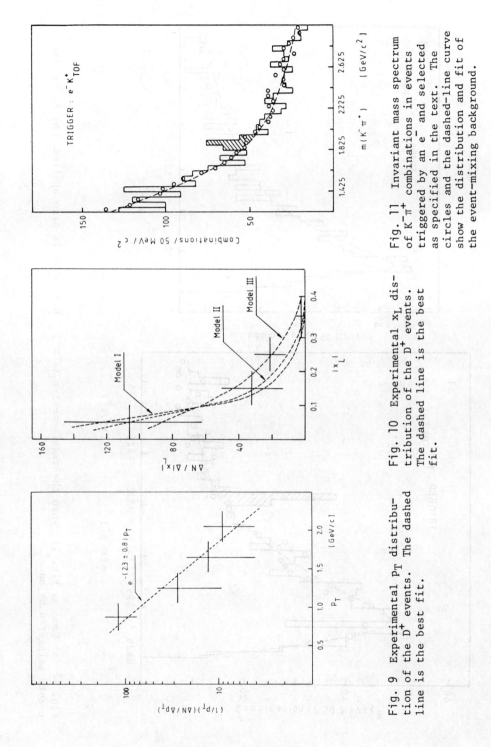

Fig. 9 Experimental p_T distribution of the D^+ events. The dashed line is the best fit.

Fig. 10 Experimental x_L distribution of the D^+ events. The dashed line is the best fit.

Fig. 11 Invariant mass spectrum of $K^-\pi^+$ combinations in events triggered by an e^- and selected as specified in the text. The circles and the dashed-line curve show the distribution and fit of the event-mixing background.

811

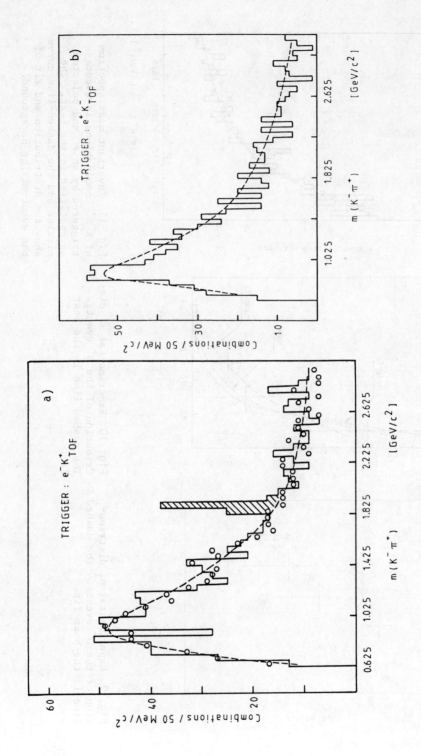

Fig. 12 a) Same as Fig. 11, when the requirement $p_T(K^-\pi^+) > 0.7$ GeV/c is also applied. b) As (a) but in events triggered by an e^+.

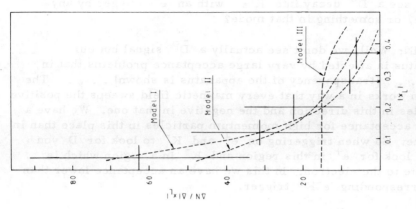

Fig. 14 Experimental x_L distribution of the D^0 events. The dashed line is the best fit.

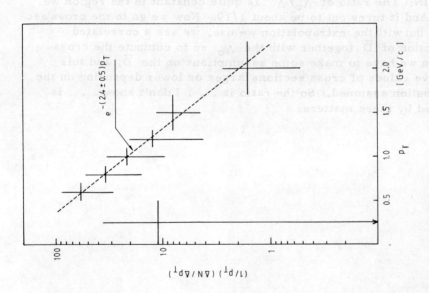

Fig. 13 Experimental p_T distribution of the D^0 events. The dashed line is the best fit.

DISCUSSION

RUCHTI, NOTRE DAME: In your D^o, maybe I missed it, but do you see a $\bar{D}^o$ decay into $K \pi^-$ with an e^+ trigger by any chance, or something in that mode?

CONTIN: Well, we don't see actually a $\bar{D}^o$ signal but our apparatus is affected by very large acceptance problems that in fact . . . . [transparency of the apparatus is shown] . . . . The system works in a way that every magnetic field sweeps the positive particles in this direction and the negative in that one. We have a larger acceptance for high momentum particles in this place than in that one. So when triggering on e^+ and K^- to look for $\bar{D}$ you should look for e^+ in this region and K^- in a region which is opposite to the electron. In this we have an acceptance lower than the corresponding e^-K^+ trigger.

RUCHTI: May I ask one more? What sort of ratio do you have for $(pp \rightarrow \Lambda_c^+ + x)/(pp \rightarrow \Lambda^o + x)$? Is the Λ_c^+ 10% of the Λ^o signal or 1% of a diffractive Λ^o signal? The cross section ratio would be fine. I forget what your $pp \rightarrow \Lambda^o$ is.

CONTIN: The ratio of Λ_c^+/Λ^o is quite constant in the region we see. And it turns out to be about 1/10. Now we go to the cross sections, but with the extrapolation we use, we see a correlated production of $\bar{D}$ together with the Λ_c so to compute the cross section we have to make some assumptions on the $\bar{D}$, and this can give values of cross sections larger or lower depending on the distribution assumed. So the ratio is . . . I don't know . . . is affected by these matters.

BEAUTY PRODUCTION AT THE CERN INTERSECTING STORAGE RINGS
SPLIT-FIELD MAGNET

M. Basile, G. Bonvicini, G. Cara Romeo, L. Cifarelli, A. Contin, G. D'Ali,
P. Di Cesare, B. Esposito, P. Giusti, T. Massam, R. Nania, F. Palmonari,
G. Sartorelli, G. Valenti and A. Zichichi

CERN, Geneva, Switzerland
Istituto di Fisica dell'Università di Bologna, Italy
Istituto Nazionale di Fisica Nucleare, Bologna, Italy
Istituto Nazionale di Fisica Nucleare, LNF, Frascati, Italy

(presented by P. Giusti)

ABSTRACT

The purpose of this talk is to present the evidence for the associated production of states with naked beauty in (pp) collisions at $\sqrt{s}$ = 62 GeV.

The beauty-flavoured state is identified as the first baryon with quark composition (udb), i.e. the Λ_b^0, decaying into a proton, a D^0, and a π^-, while the associated antibeauty-flavoured state is detected via its leptonic decay into a positron plus anything.

The experiment has been performed by the CERN-Bologna-Frascati Collaboration (CBF) at the Split Field Magnet (SFM) of the CERN Intersecting Storage Rings (ISR).

INTRODUCTION

The basic principle of the experiment is shown in Fig. 1. Since the semileptonic decay of heavy quarks is thought to be one of the main sources of the prompt leptons produced in (pp) collisions, the trigger of the experiment was given by the detection of a positron as the signature for the leptonic decay of an antibeauty-flavoured particle. The associated production of a beauty-flavoured particle was then to be investigated with the invariant mass study of its various possible hadronic decays.

The main problem encountered in invariant mass studies is the high combinatorial background due to the following:
  i)   the secondary vertex of the hadronic decay is not known and consequently all possible combinations of all the tracks in the event have to be considered;
 ii)   all possible mass assignments have to be taken into account for the particles which are not specifically identified.
It is therefore necessary to limit the search to particles and decay channels which allow the selection of a phase-space region where the signal-to-background ratio is favourable.

Since our previous studies on the production of the charmed baryon

Λ_c^+ in (pp) collisions[1,2] had shown that this heavy baryon is produced according to a strong leading effect, i.e. with a flat distribution in $x_L = 2p_L/\sqrt{s}$,
and that the identification of a positive particle with large x_L as a proton,
as suggested by the inclusive p/π^+ measurements, is indeed a good criterion
to put in evidence the Λ_c^+ signal, we have applied the same "leading" technique
to the search for naked beauty states produced in (pp) collisions. Accordingly, we have looked for the lowest-mass baryon with a beauty (b) quark, i.e.
Λ_b^0 = (udb) decaying into a fast proton plus other particles, where, since the
b quark is Cabibbo-favoured to decay into a charm (c) quark, one of these
should be a charmed particle, such as, for example, the D^0.

To summarize, the reaction investigated was:

$$pp \longrightarrow \Lambda_b^0 + (\bar{b}\ state) + anything \qquad (1)$$
$$\downarrow \qquad \downarrow e^+ + anything$$
$$\downarrow pD^0\pi^-$$
$$\downarrow K^-\pi^+$$

with "leading" conditions imposed on the baryonic state.

THE EXPERIMENTAL APPARATUS

The experimental apparatus is shown in Fig. 2. The positrons were
detected in the 90° region, by the coincidence ($\check{C}_0\check{C}_3$ or $\check{C}_0\check{C}_4$) of two gas
Čerenkov counters (with a momentum threshold for π's of $\simeq 5.5$ GeV/c) and by
a minimum energy release ($E^e_{min} \geq 500$ MeV) in the electromagnetic shower
detectors (EMSDs) LG3 and LG4 (lead-glass arrays), and SW3 and SW4 (lead/
scintillator hodoscopes)[3]. A multiwire proportional chamber (209) with
analog read-out ("dE/dx" chamber)[4,5] near the interaction region provided
the information needed to reject off-line the unresolved and resolved (e^+e^-)
pairs that are due to external γ conversions and to π^0 and η Dalitz decays.

The multiwire proportional chamber (MWPC) system of the SFM[6] provided
the momentum measurement of the charged secondaries over about 75% of the
total solid angle with a momentum accuracy $\Delta p/p \leq 30\%$.

Finally, a large time-of-flight (TOF) system allowed π, K, p separation
for momenta ≤ 2 GeV/c[7] over about 10% of the total solid angle.

THE POSITRON SELECTION

The total sample of events triggered by a candidate positron ($\simeq 1.5 \times
\times 10^6$), corresponding to an integrated luminosity of 4.39×10^{36} cm^{-2}, was re-
duced to $\simeq 3 \times 10^4$ events by a software filter where a track was searched
for in the solid angle covered by the positron detector. Also
  i) uncorrelated background from γ's and charged hadrons was rejected by
     requiring a reconstructed track through the positron detector ("dE/dx"
     chamber, MWPCs, Čerenkov counter cells, EMSDs) and the matching of the
     impact point of this track with the energy cluster released in the
     EMSDs;
 ii) external γ conversions and π^0 and η Dalitz decays were rejected using
     the "dE/dx" chamber information, either by a pulse-height cut (unresolved

e^+e^- pairs) or by the detection of two close, but resolved, tracks.

After the full reconstruction of the events which had survived this filter, a final check was done to verify that the positron track fitted the interaction vertex.

EVIDENCE FOR $\Lambda_b^0 \rightarrow pD^0\pi^-$

In order to study reaction (1), the following conditions (set α) have been applied to the sample of events selected by the above specified filter:
1) the transverse momentum of the triggering positron, $p_T(e^+)$, was required to be $p_T(e^+) \geq 800$ MeV; the momentum accuracy for the positron track was required to be $\Delta p/p \leq 15\%$;
2) the charged multiplicity of the "anything" in reaction (1), n_{ch}, was required to be $n_{ch} \geq 4$, with at least one particle opposite in direction to the triggering positron;
3) at least one positive particle with $|x_L| = 2|p_L|/\sqrt{s} \geq 0.32$ and momentum accuracy $\Delta p/p \leq 30\%$ was required in the event, and this particle was assumed to be a proton;
4) the rapidity y of the system ($pK^-\pi^+\pi^-$) was required to be $|y| \geq 1.4$.

On the basis of physics intuition, the purpose of conditions (1) and (2) was to select events where it is more likely that the positron is indeed coming from the decay of a heavy mass particle, while conditions (3) and (4) selected events where a leading baryon is also present.

Figure 3 shows the invariant mass spectrum for the system ($K^-\pi^+$) obtained from the events satisfying set α conditions. Only tracks that are fitted to the interaction vertex and have $\Delta p/p \leq 30\%$ are considered, and for those that are not identified by the time of flight, all possible mass assignments have been considered.

In order to select events where a D^0 had been produced, the mass cut

$$1.7 \leq m\ (K^-\pi^+) \leq 2.0\ \text{GeV/c}^2$$

was then applied to the events satisfying the conditions given by set α.

Table 1 shows the number of positron events satisfying the various cuts applied to the data.

Table 1

No conditions	$\simeq 3 \times 10^4$
Condition (1)	14,552
Conditions (1) and (3)	1600
Conditions (1), (3), and (4)	1025
Set α	800
Set α + D^0 cut	208

Figure 4 shows the invariant mass spectrum for the $p(K^-\pi^+)\pi^-$ system, where ($K^-\pi^+$) satisfies the mass cut 1.7-2.0 GeV/c^2, and all tracks fit to the interaction vertex and have $\Delta p/p \leq 30\%$. As previously specified, all mass assignments have been made for all tracks, except for those identified by the time of flight.

A clear enhancement $(29.4 \pm 7.4$ combinations$)^{*)}$ is observed in the mass range $5.35 \leq m[p(K^-\pi^+)\pi^-] \leq 5.5$ GeV/c^2. On the contrary, if the D^0 mass cut condition is released, the invariant mass spectrum of the system $(pK^-\pi^+\pi^-)$ appears as shown in Fig. 5 and no enhancement is observed.

A key point now arises. If the enhancement observed in Fig. 4 is really due to the Λ_b^0 decaying into $pD^0\pi^-$, then a corresponding signal for the D^0 decaying into $K^-\pi^+$ should be observed when a cut in the mass range $5.35 \leq m(pK^-\pi^+\pi^-) \leq 5.5$ GeV/c^2 is applied to the data from which the mass spectrum shown in Fig. 5 is obtained and the $(K^-\pi^+)$ mass spectrum is worked out.

Figure 6a shows the $(K^-\pi^+)$ invariant mass spectrum when the $(pK^-\pi^+\pi^-)$ mass falls inside the above specified mass cut (solid line histogram, IN region) or in two control regions, both 150 MeV/c^2 wide, immediately above and below this mass range (dashed line histogram, OUT regions). Figure 6b shows the bin-to-bin difference of the IN and OUT histograms. An enhancement of (28.5 ± 5.5) combinations is indeed observed for the IN sample at a mass about the D^0 mass, $1.725 \leq m(K^-\pi^+) \leq 1.875$ GeV/c^2.

A second very important cross-check can now be done. The Λ_b^0 signal shown in Fig. 4 is obtained applying a wide cut (1.7-2.0 GeV/c^2) around the nominal D^0 mass, while the D^0 signal observed in Figs. 6a and 6b is narrower: 1.725-1.875 GeV/c^2. If the two observed effects are due to physics (i.e. to the production of Λ_b^0, its decay into $pD^0\pi^-$, with the D^0 decaying into $K^-\pi^+$), then, by applying the narrow, experimentally observed mass cut in the $K^-\pi^+$ spectrum, the signal-to-background ratio for the Λ_b^0 signal should improve without loss of events in the observed enhancement.

The $p(K^-\pi^+)\pi^-$ invariant mass spectrum obtained with the cut 1.725 $\leq m(K^-\pi^+) \leq 1.875$ GeV/c^2 is shown in Fig. 7. Notice that in this case the signal is given by about 30 ± 5 combinations with a 1 to 1 signal-to-background ratio, while for the case of the wide D^0 mass cut this ratio was only 1 to 2.

Λ_b^0 Mass determination

We observe the Λ_b^0 signal in the mass range $5.35 \leq m[p(K^-\pi^+)\pi^-] \leq 5.5$ GeV/c^2 but, at the same time, the $D^0 \rightarrow K^-\pi^+$ enhancement appears at a mass slightly lower than the nominal value: $1.725 \leq m(K^-\pi^+) \leq 1.875$ GeV/c^2. In order to estimate the correct mass range for the Λ_b^0, we have made two hypotheses:

i) the error in the D^0 mass is due to the momentum measurements; by correcting for this effect, the $p(K^-\pi^+)\pi^-$ mass shifts upwards by $\simeq 60$ MeV/c^2;

ii) the error in the D^0 mass is due to the angular measurements; by correcting for this effect the $p(K^-\pi^+)\pi^-$ mass shifts upwards by $\simeq 100$ MeV/c^2.

Accordingly, we conclude that the Λ_b^0 mass is in the range: $5.35 \leq m[p(K^-\pi^+)\pi^-] \leq 5.6$ GeV/c^2, in agreement with theoretical estimates[8,9].

---

*) Note that while 29.4 is the number of combinations above the background, ± 7.4 is the statistical fluctuation of the background level under the observed enhancement. The same applies when discussing the other signals presented in this paper.

Backgrounds

In order to be sure that the observed Λ_b^0 signal is not due to instrumental effects such as, for example, detector acceptance, wire chamber misalignment, etc., we have repeated the same analysis

i) mixing particles from different events;
ii) using as the trigger particle not a positron but a charged hadron;
iii) using as the trigger particle a negative electron.

The results are shown in Fig. 8 for the event-mixing background, in Fig. 9 for the charged hadron background, and in Fig. 10 for the negative electron trigger. No enhancement similar to that obtained with the positron trigger is observed in any of these invariant mass spectra.

Notice that, according to the following process,

$$pp \rightarrow \Lambda_b^0 \quad + \quad (\bar{b}\text{-state}) \quad + \quad \text{anything} \tag{2}$$
$$\quad\; \downarrow pD^0\pi^- \qquad\qquad \downarrow (\bar{c}\text{-state})$$
$$\qquad\; \downarrow K^-\pi^+ \qquad\qquad\quad \downarrow e^- + \text{anything}$$

a Λ_b^0 signal should be observed also with the negative electron trigger. However, Monte Carlo simulation shows that the momentum spectrum of the electrons from the decay of the anticharm particle is such that the acceptance for these electrons is about four times lower than the acceptance for the positrons because of the antibeauty state decay. Consequently, the number of Λ_b^0 signals expected for the e^- trigger is 8 ± 6, which is compatible with what we observed.

Λ_b^0 CROSS-SECTION ESTIMATES AND COMPARISON WITH THE Λ_c^0 CROSS-SECTION

The number of combinations per mass bin in the spectrum shown in Fig. 7 exceeds the number of events by a small amount (25%). In order to work out the cross-section for the observed signal, the number of combinations has, therefore, to be divided by 1.25.

Let us call $\Delta\sigma$ the partial cross-section for the reaction

$$pp \rightarrow \left[p\underbrace{(K^-\pi^+)}\pi^- \right] \quad + \quad e^+ \quad + \quad \text{anything} \tag{3}$$
$$\qquad\qquad D^0$$

with leading baryon conditions — $p_T(e^+) \geq 800$ MeV/c — $n_{ch} \geq 4$

measured under the experimental conditions already specified and synthetically indicated in formula (3). Then

$$\Delta\sigma = N(\Lambda_b^0)/\varepsilon\left[L(pp)\right] ,$$

$N(\Lambda_b^0)$ is the number of observed Λ_b^0 events (25); ε is the total detection efficiency for the final state (3) (11.9 ± 1.2)%; and $L(pp)$ is the total integrated luminosity (4.39×10^{36} cm^{-2}). Therefore

$$\Delta\sigma = (3.8 \pm 1.2) \times 10^{-35} \text{ cm}^2.$$

The value of $\Delta\sigma$ represents the sensitivity of our experimental set-up when illuminated by a final state such as (3).

In order to derive the Λ_b^0 production cross-section it is necessary to know

  i) the nature of the antibeauty state produced in association with the Λ_b^0; this could in fact be either an antibaryon or a meson with an antibeauty quark;

 ii) the longitudinal and transverse momentum production distributions both for the Λ_b^0 and the associated antibeauty state;

iii) the correlation, if any, between these two particles;

 iv) the decay distributions of the Λ_b^0 and the antibeauty state;

  v) the following branching ratios:

$B_1(D^0 \to K^-\pi^+)$ $= (D^0 \to K^-\pi^+)/(D^0 \to \text{all})$

$B_2(\bar{b} \text{ state} \to e^+ + \text{anything}) = (\bar{b} \text{ state} \to e^+ + \text{anything})/(\bar{b} \text{ state} \to \text{all})$

$B_3(\Lambda_b^0 \to pD^0\pi^-)$ $= (\Lambda_b^0 \to pD^0\pi^-)/(\Lambda_b^0 \to \text{all})$.

Since an antibaryon needs three antiquarks while a meson needs only an (antiquark-quark) pair, and the experimental results in (pp) interactions favour the baryon-antimeson associated production with respect to the production of baryon-antibaryon pairs, we have assumed that all the observed Λ_b^0 events originate from the reaction:

$$pp \to \Lambda_b^0 + M_{\bar{b}} + \text{anything}, \tag{4}$$

where $M_{\bar{b}}$ is an antibeauty-flavoured meson.

As suggested by our previous studies of associated charm production[1,10], the transverse momentum production distribution for both the Λ_b^0 and the anti-beauty-flavoured meson was taken to be

$$\frac{d\sigma}{dp_T} \propto p_T \exp(-2.5\ p_T) ,$$

whilst for the longitudinal momentum production distribution we have assumed

$$\left[E\ \frac{d\sigma}{d|x_L|} \right]_{M_{\bar{b}}} \propto (1 - |x_L|)^3$$

for the meson $M_{\bar{b}}$ (i.e. a central production mechanism), and for the baryon Λ_b^0,

$$\frac{d\sigma}{d|x_L|} = \text{const} ,$$

i.e. an x_L production distribution according to the "leading" baryon effect already observed in the Λ_c^+ production studies.

For the decay process of $M_{\bar{b}} \to \bar{D}e^+\nu$, a $K_{\ell 3}$ matrix element was used, whilst for the decay process $\Lambda_b^0 \to pD^0\pi^-$ three models have been considered, using Lorentz-invariant phase-space plus the conditions specified below.

MODEL   I – The "leading" baryon condition. The conditions for Model I
        are: $|x_L|(pD^0\pi^-) \geq 0.32$ and $|y|(pD^0\pi^-) \geq 1.4$.

        Obviously, the result of this model will not be the total
cross-section for reaction (4); it will be the partial cross-
section according to the cuts used for the experimental observation.

MODEL   II – Minimum "leading" baryon condition. The conditions for Model II
        are: $p(\text{proton}) > p(D^0) > p(\pi^-)$.

MODEL III – Isotropic. In Model III the Λ_b^0 decay is purely isotropic with-
        out any leading baryon condition.

Finally we have taken

$$B_1(D^0 \to K^-\pi^+) = (3.0 \pm 0.6)\% \; [\text{Schindler et al.}^{11}]$$

$$B_2(M_b^- \to e^+ + \text{anything}) = (13 \pm 6)\% \; [\text{Bebek et al.}^{12}]$$

whilst of course, $B_3(\Lambda_b^0 \to pD^0\pi^-)$ is unknown.

With the above specified assumptions the results are

$$\Delta\sigma_b(\;\text{I})B_3 = (2.7 \; {}^{+\;1.8}_{-\;1.1}) \times 10^{-30} \; \text{cm}^2 \;,$$

$$\sigma_b(\;\text{II})B_3 = (8.2 \; {}^{+\;6.0}_{-\;2.8}) \times 10^{-30} \; \text{cm}^2 \;,$$

$$\sigma_b(\text{III})B_3 = (27 \; {}^{+\;17}_{-\;11}) \times 10^{-30} \; \text{cm}^2 \;.$$

Note that the large uncertainty of these estimates and their model
dependence does not allow these results to be used as "absolute" measurements. However, they can provide a useful tool for the comparison with
analogous estimates for charm production.

The same models have then been used to compute the Λ_c^+ production cross-
section under conditions identical to those for the Λ_b^0. For the Λ_c^+ the reaction studied was

$$pp \to \Lambda_c^+ \qquad + \qquad \bar{D} \qquad + \text{anything} \qquad\qquad (5)$$
$$\quad\vert\qquad\qquad\qquad\qquad \vert\!\!\to e^- + \text{anything}$$
$$\quad\vert\!\!\to pK^-\pi^+$$

at the same (pp) c.m. energy and using the same apparatus. For the required
branching ratios we have taken

$$B_4(\bar{D} \to e^- + \text{anything}) = (\;8 \pm 1)\% \; [\text{Goldhaber and Weiss}^{13}]$$

$$B_5(\Lambda_c^+ \to pK^-\pi^+) = (2.2 \pm 1)\% \; [\text{Abrams et al.}^{14}] \;.$$

The results are

$$\Delta\sigma_c(\;\text{I}) = (\;77.5 \; {}^{+\;57}_{-\;28}) \times 10^{-30} \; \text{cm}^2 \;,$$

$$\sigma_c(\;\text{II}) = (100 \; {}^{+\;68}_{-\;36}) \times 10^{-30} \; \text{cm}^2 \;,$$

$$\sigma_c(\text{III}) = (190 \; {}^{+\;140}_{-\;68}) \times 10^{-30} \; \text{cm}^2 \;.$$

From the comparison between σ_c and $\sigma_b B_3$ we can now study how the unknown branching ratio $B_3(\Lambda_b^0 \to pD^0\pi^-)$ depends on the ratio of the cross-
sections for charm and beauty. In fact, since theoretical estimates[15,16]

on quasi-diffractive production predict a ratio between heavy-flavour cross-sections of the order of the inverse mass squared of the produced flavours, it is important to check that with this assumption the measured cross-sections produce a reasonable value for $B_3(\Lambda_b^0 \to pD^0\pi^-)$.

The results are shown in Fig. 11 where the correlation between σ_b/σ_c and B_3 is shown for the three models. The 1.5 standard deviation limits are indicated.

Obviously these results are affected by a large uncertainty due to the errors in the various branching ratios and the apparatus acceptances. In addition, also the uncertainties from the various necessary hypotheses concerning the production and decay distributions of the particles involved should be considered. For these reasons the most reliable comparison must be based on Model I, which requires less extreme extrapolations and corresponds to what has been experimentally observed.

The main conclusion, based on Model I, is shown in Fig. 12. Here it is shown that, choosing for σ_b/σ_c the value 1/8 given by the inverse masses squared of the charm and beauty quarks, the value of $B_3(\Lambda_b^0 \to pD^0\pi^-)$ is, within 1.5 standard deviations, compatible with a 2% level.

FURTHER STUDIES ON THE Λ_b^0 SIGNAL

The following studies are important not only because they allow an insight into the production mechanism of beauty-flavoured baryons in (pp) collisions, but also because the observation of any difference between the Λ_b^0 events and events without Λ_b^0 increases the statistical significance and the confidence in the observed signal.

In order to perform these studies, we have compared the results obtained from the IN events, i.e. those events where a $p(K^-\pi^+)\pi^-$ combination falls in the Λ_b^0 mass range $\{5.35 < m[p(K^-\pi^+)\pi^-] < 5.5 \text{ GeV}/c^2\}$, with the results obtained from the OUT events, i.e. those events where the above-mentioned particle combination falls into two control regions, 150 MeV/c^2 below and 150 MeV/c^2 above the Λ_b^0 peak.

The longitudinal momentum distribution of Λ_b^0

Figure 13 shows the x distribution of the Λ_b^0 obtained, as specified above, by making the (IN-OUT) difference. The full line is the best fit to the data,

$$\Delta N / \Delta |x_L| \propto (1 - |x_L|)^\alpha, \text{ with } \alpha = 0.87 \pm 1.26 .$$

Notwithstanding the limited statistics, a flat-x behaviour is clearly shown by the data. The quantities reported in Fig. 14 are the following ratios,

$$\left(\frac{\Delta N}{\Delta|x_L|}\right)_{\Lambda_b^0} \bigg/ \left(\frac{\Delta N}{\Delta|x_L|}\right)_{\Lambda_c^+} = R_{b/c} ,$$

$$\left(\frac{\Delta N}{\Delta|x_L|}\right)_{\Lambda_b^0} \bigg/ \left(\frac{\Delta N}{\Delta|x_L|}\right)_{\Lambda_s^0} = R_{b/s} ,$$

$$\left(\frac{\Delta N}{\Delta|x_L|}\right)_{\Lambda_b^0} \bigg/ \left(\frac{\Delta N}{\Delta|x_L|}\right)_{\overline{\Lambda}_s^0} = R_{b/\bar{s}} ,$$

as a function of x_L. The sharp increase in the ratio $R_{b/\bar{s}}$ together with the flatness of the ratios $R_{b/c}$ and $R_{b/s}$ clearly indicates that despite the large mass difference between the strange (s), the charm (c), and the beauty (b) quarks, the production of these differently flavoured baryonic states shows the same "leading" effect. This can be understood in terms of the fact that all these particles carry two of the original quarks of the incident proton:

$$\left[(ud)u\right] = p \ ,$$
$$\left[(ud)s\right] = \Lambda_s^0 \ ,$$
$$\left[(ud)c\right] = \Lambda_c^\pm \ ,$$
$$\left[(ud)b\right] = \Lambda_b^0 \ .$$

The transverse momentum distribution of the positrons from the antibeauty state decay

Figure 15 shows the transverse momentum spectrum of the positrons associated with the IN Λ_b^0 events with respect to the total positron sample and to the positrons associated with the OUT events. Here again a clear difference is observed, indicating the different origin of the positrons associated with the Λ_b^0 signal.

Possible evidence for $\Sigma_b^\pm \to \Lambda_b^0 \pi^\pm$

We have also investigated the reaction

$$
\begin{aligned}
pp \to \ \Sigma_b^\pm \quad &+ \quad (\bar{b} \text{ state}) \quad + \quad \text{anything} \quad &(6)\\
&\llcorner \Lambda_b^0 \pi^\pm \qquad\qquad \llcorner\!\to e^+ \ + \ \text{anything}\\
&\qquad \llcorner pD^0\pi^-\\
&\qquad\qquad \llcorner K^-\pi^+
\end{aligned}
$$

Figure 16 shows the mass difference $m\left[(pD^0\pi^-)\pi^\pm\right] - m(pD^0\pi^-)$ when the $(pD^0\pi^-)$ mass falls into the Λ_b^0 range (IN region). An enhancement of 14 ± 4.4 events is observed which is not present in Fig. 17, where one sees the same mass difference as that obtained when the $(pD^0\pi^-)$ mass falls into the OUT region.

This result seems to indicate that indeed some of the observed Λ_b^0 come from the strong decay $\Sigma_b^\pm \to \Lambda_b^0 \pi^\pm$.

Study of the decay $\Lambda_b^0 \to \Lambda_c^+ \pi^- \pi^- \pi^+$

We have finally searched for evidence of Λ_b^0 production in (pp) collisions studying the reaction

$$
\begin{aligned}
pp \to \ \Lambda_b^0 \quad &+ \quad (\bar{b} \text{ state}) \quad + \quad \text{anything} \quad &(7)\\
&\llcorner \Lambda_c^+ \pi^- \pi^- \pi^+ \qquad \llcorner\!\to e^+ \ + \ \text{anything}\\
&\qquad \llcorner pK^-\pi^+
\end{aligned}
$$

To the sample of $\simeq 3 \times 10^4$ positron events, we have applied the following cuts:
  i) $p_T(e^+) \geq 800$ MeV/c;
 ii) at least one positive particle with $|x_L| \geq 0.3$, identified as a proton;
iii) $|y|(pK^-\pi^+\pi^-\pi^-\pi^+) \geq 1.6$;
 iv) n_{ch}(anything) ≥ 7 with at least one particle opposite to the positron;
  v) $m(pK^-\pi^+) = m(\Lambda_c^+) \pm 100$ MeV/c^2.

To study the invariant mass spectrum of $pK^-\pi^+\pi^-\pi^-\pi^+$, only tracks with $\Delta p/p \leq 30\%$ and fitted to the interaction vertex have been considered. All mass assignments have been taken into account for those tracks not identified as a given particle by the time of flight.

The $(pK^-\pi^+)\pi^-\pi^-\pi^+$ invariant mass spectrum obtained with the above-mentioned cuts is shown in Fig. 18. An enhancement of 20 ± 5.6 combinations is observed in the mass range $5.35 \leq m[(pK^-\pi^+)\pi^-\pi^-\pi^+] \leq 5.5$ GeV/c^2.

We have then studied the $(pK^-\pi^+)$ mass spectrum as a function of the invariant mass of the system $pK^-\pi^+\pi^-\pi^-\pi^+$ when the condition $m(pK^-\pi^+) = m(\Lambda_c^+) \pm 100$ MeV/c^2 is released. Figure 19 shows the $(pK^-\pi^+)$ mass spectrum when the mass of the system $pK^-\pi^+\pi^-\pi^-\pi^+$ satisfies the condition

$$5.35 \leq m(pK^-\pi^+\pi^-\pi^-\pi^+) \leq 5.5 \text{ GeV/c}^2 \ ,$$

i.e. corresponds to the enhancement observed in Fig. 18 (IN region). An enhancement at a mass corresponding to the Λ_c^+ mass is observed. On the contrary, as shown in Fig. 20, no Λ_c^+ enhancement is observed in the $(pK^-\pi^+)$ mass spectrum when the mass of the $(pK^-\pi^+\pi^-\pi^-\pi^+)$ system corresponds to the two control regions below and above the (5.35-5.5 GeV/c^2) enhancement.

Notwithstanding the limited statistics, it is remarkable that here again, as in the case of $\Lambda_b^0 \to pD^0\pi^-$, a charm signal (the Λ_c^+) is observed in association with a positron, which is forbidden by "charm" physics, only when coupled to an enhancement which can be interpreted as a "beauty" signal.

CONCLUSIONS

The data reported show evidence for an $\simeq 5.5$ standard deviations enhancement in the $p(K^-\pi^+)\pi^-$ invariant mass spectrum in the range $5.35 \leq m[p(K^-\pi^+)\pi^-] \leq 5.5$ GeV/c^2.

The interpretation of the effect is in terms of Λ_b^0 production in (pp) collisions, and is based on two facts:
  i) the association of the signal with a positron, which is the signature for the leptonic decay of an antibeauty state;
 ii) the presence of the D^0 state decaying into $K^-\pi^+$ in association with the mass range where the $pK^-\pi^+\pi^-$ enhancement is observed, whilst, as expected from "charm" physics, no charm signal, associated with a positron, is observed outside this mass range.

Moreover, the reality of the Λ_b^0 signal is strongly supported by
  i) the x_L distribution of the system $pD^0\pi^-$, which is typical of a leading baryon;
 ii) the transverse momentum distribution of the positrons associated with the observed $p(K^-\pi^+)\pi^-$ enhancement, which is different from that obtained when the mass of the above-mentioned system is outside the Λ_b^0 signal;

iii) the indication for the signal $\Sigma_b^\pm \to \Lambda_b^0 \pi^\pm$;

iv) the fact that, even if with smaller statistical significance, an enhancement in the same mass range as for the $pD^0\pi^-$ case is observed in the channel $\Lambda_b^0 \to \Lambda_c^+ \pi^- \pi^- \pi^+$.

From the data obtained with the same apparatus and assuming the same production models, we have estimated the cross-section times the branching ratio $\sigma_b B_3 (\Lambda_b \to pD^0\pi^-)$ for reaction (4) and the cross-section σ_c for reaction (5).

From the comparison of σ_c and $\sigma_b B_3 (\Lambda_b^0 \to pD^0\pi^-)$ as derived from the experimental data with some theoretical predictions for the ratio of σ_b/σ_c, we have determined a value of the branching ratio $B_3 (\Lambda_b \to pD^0\pi^-)$ which is compatible with being as low as a few percent.

*   *   *

REFERENCES

1. M. Basile et al., Nuovo Cimento 63A, 230 (1981).
2. M. Basile et al., Nuovo Cimento Lett. 30, 487 (1981).
3. M. Basile et al., Nucl. Instrum. Methods 163, 93 (1979).
4. H. Frehse et al., Nucl. Instrum. Methods 156, 87 (1978).
5. H. Frehse et al., Nucl. Instrum. Methods 156, 97 (1978).
6. R. Bouclier et al., Nucl. Instrum. Methods 125, 19 (1975).
7. M. Basile et al., Nucl. Instrum. Methods 179, 477 (1981).
8. D. Stanley and R. Robson, Phys. Rev. Lett. B45, 235 (1980).
9. A. Martin, Exact bounds and estimates on the masses of beautiful hadrons, Preprint CERN TH. 3068 (1980).
10. M. Basile et al., Nuovo Cimento Letters 30, 481 (1981).
11. R.H. Schindler et al., SLAC-PUB 2507; LBL 10905, May 1980 (T/E), submitted to Phys. Rev. D.
12. C. Bebek et al., Preprint CLNS-80/4/5 (1980).
13. G. Goldhaber and J.E. Weiss, Preprint LBL 10652, March 1980.
14. G.S. Abrams et al., Phys. Rev. Lett. 44, 10 (1980).
15. G. Gustafson and C. Peterson, Phys. Rev. Lett. B 67, 81 (1977).
16. A. Martin, Preprint CERN TH. 2980 (1980), and private communication.

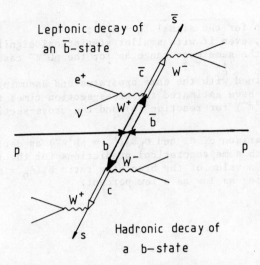

Leptonic decay of
an b̄–state

Hadronic decay of
a b–state

Fig. 1 Schematic diagram showing the principle of the experiment: trigger
on the leptonic decay of an antibeauty state and look for the hadronic
decay of the associated beauty state.

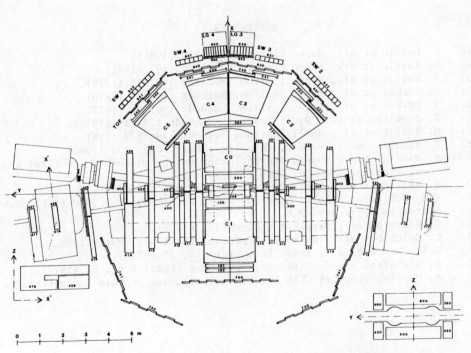

Fig. 2 Top view of the experimental apparatus.

826

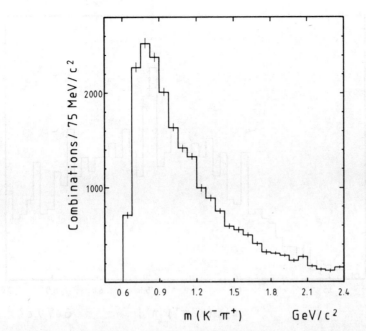

Fig. 3 The (K⁻π⁺) invariant mass spectrum obtained from the events satisfying set α conditions. A mass cut (1.7–2.0 GeV/c²) is used to apply to
these events a software wide D⁰ trigger.

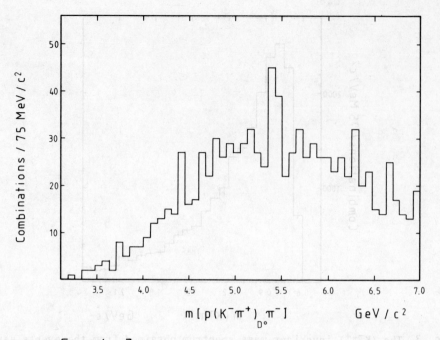

Fig. 4 The $[p(K^-\pi^+)\pi^-]$ invariant mass spectrum obtained from the events satisfying set α conditions and the wide D^0 cut.

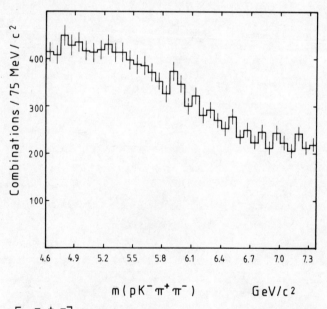

Fig. 5 The $[pK^-\pi^+\pi^-]$ invariant mass spectrum obtained from the events satisfying set α conditions but without the wide D^0 trigger.

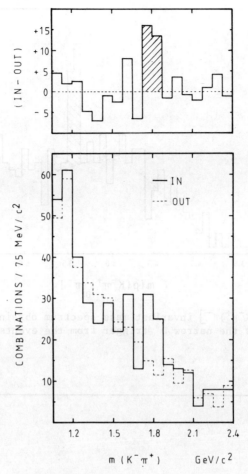

Fig. 6 a) The $(K^-\pi^+)$ invariant mass spectrum when the mass
of the system $\left[p(K^-\pi^+)\pi^-\right]$ falls in the $(5.35\text{-}5.5\ \text{GeV/c}^2)$
mass range (solid line histogram = IN) or in two con-
trol regions above and below this range (dashed line
histogram = OUT).

b) Bin-to-bin difference of the $(K^-\pi^+)$ mass spectra
relative to the IN and OUT regions. A clear D^0 enhancement is observed in the mass range (1.725-1.875
GeV/c^2).

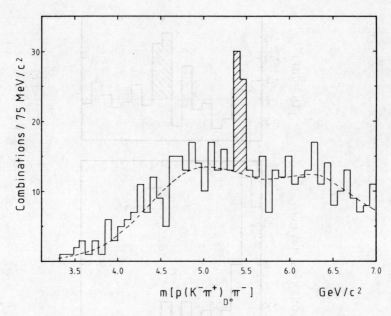

Fig. 7 The $[p(K^-\pi^+)\pi^-]$ invariant mass spectrum obtained when the mass $(K^-\pi^+)$ satisfies the narrow D^0 trigger from the events selected by set α conditions.

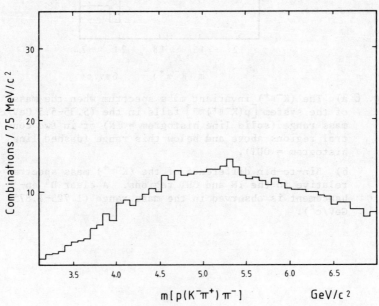

Fig. 8 Background to the $[p(K^-\pi^+)\pi^-]$ mass spectrum from event mixing.

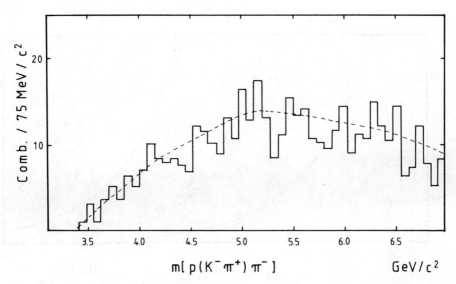

Fig. 9 The $[p(K^-\pi^+)\pi^-]$ invariant mass spectrum obtained when the trigger particle is a charged hadron.

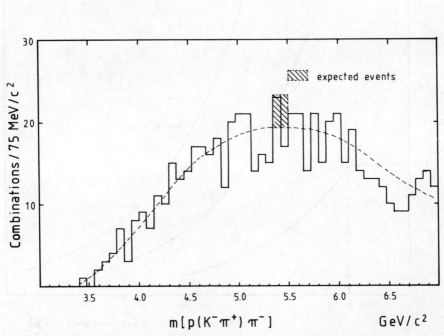

Fig. 10 The $[p(K^-\pi^+)\pi^-]$ invariant mass spectrum obtained when the trigger particle is a negative electron. The expected signal is also shown.

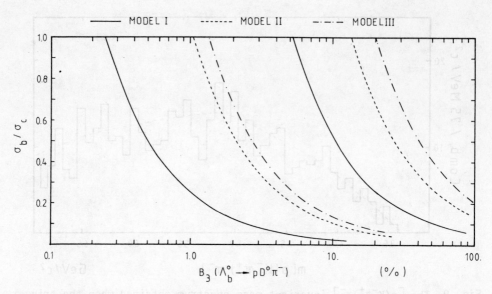

Fig. 11 σ_b/σ_c versus $B_3(\Lambda_b^0 \to pD^0\pi^-)$. The 1.5 standard deviation limits are indicated for the three models described in the text.

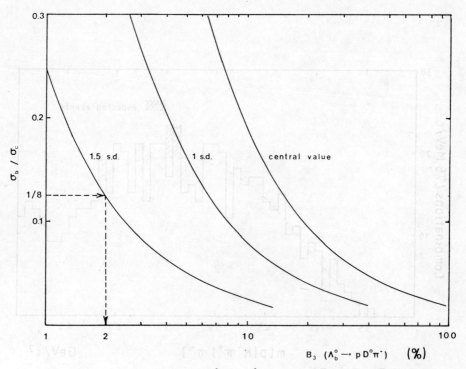

Fig. 12 σ_b/σ_c versus $B_3(\Lambda_b^0 \to pD^0\pi^-)$ relative to Model I.

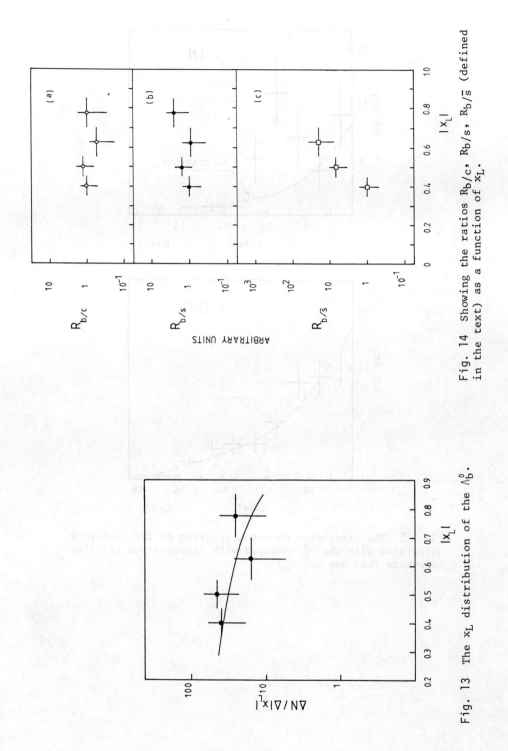

Fig. 14 Showing the ratios $R_{b/c}$, $R_{b/s}$, $R_{b/\bar{s}}$ (defined in the text) as a function of x_L.

Fig. 13 The x_L distribution of the Λ_b^0.

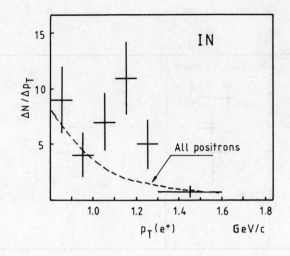

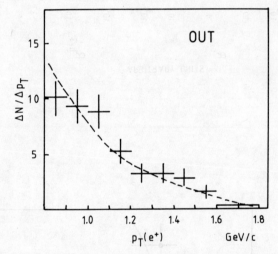

Fig. 15 The transverse momentum spectrum of the positrons associated with the Λ_b^0 compared with the spectrum relative to events that are not Λ_b^0.

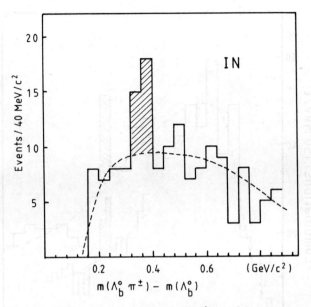

Fig. 16 Possible evidence for the decay $\Sigma_b^\pm \to \Lambda_b^0 \pi^\pm$. The mass difference $\Delta m = m \{[p(K^-\pi^+)\pi^-]\pi^\pm\} - m[p(K^-\pi^+)\pi^-]$ is shown. The histogram corresponds to the case when the $[p(K^-\pi^+)\pi^-]$ mass falls inside the Λ_b^0.

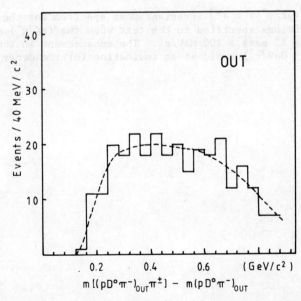

Fig. 17 Showing the same mass difference as in Fig. 16 for the case when the $[p(K^-\pi^+)\pi^-]$ mass falls in two control regions 150 MeV/c^2 below and above the Λ_b^0 peak.

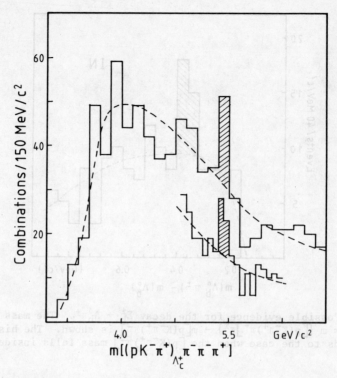

Fig. 18 The $\left[(pK^-\pi^+)\pi^-\pi^-\pi^+\right]$ invariant mass spectrum for the events satisfying the conditions specified in the text when the $(pK^-\pi^+)$ mass corresponds to the Λ_c^+ mass $\pm$ 100 MeV/c^2. The enhancement in the mass range (5.35–5.5 GeV/c^2) could be an indication for the decay $\Lambda_b^0 \rightarrow \Lambda_c^+\pi^-\pi^-\pi^+$.

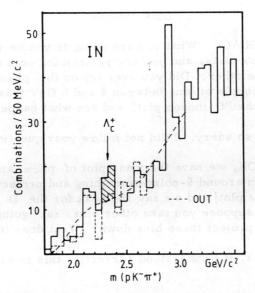

Fig. 19 The (pK⁻π⁺) invariant mass spectrum when the mass of the system
(pK⁻π⁺π⁻π⁻π⁺) corresponds to the possible Λ_b^0 enhancement observed in
Fig. 18.

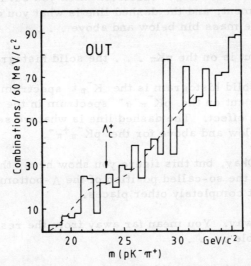

Fig. 20 The (pK⁻π⁺) invariant mass spectrum when the mass of the system
(pK⁻π⁺π⁻π⁻π⁺) corresponds to two control regions 150 MeV/c² above and
below the enhancement shown in Fig. 18.

DISCUSSION

PETERSON, SLAC: What you are doing is you have your two plots, two mass plots, and you are performing sort of a mapping from one to the other.   Did you ever try on the $pK\pi\pi$ mass plot to take say a couple of bins between 4 and 6 GeV mass and project them down to the "D-meson plot" and see what happens?

GIUSTI: I am so sorry.   I did not follow your question.

PETERSON: OK, we have the mass plot of $pK\pi\pi$ and there you show us the bin around 5-point-something and project it down [to a $K\pi$ mass plot] and we saw the peak for the D meson.   My question was, suppose you take other bins, say going from 4.5 to 7 [GeV/c], and project these bins down?   What does it look like?

GIUSTI: I think, if I understood correctly, this is what we have been doing.

RATTI: You mean the projection outside the band of the . . .

PETERSON: No, I mean if you take events within the [$pK\pi\pi$ mass] band and go down and project out the $K\pi$ combinations.

GIUSTI: I don't know if I have understood correctly, but the solid line here is the $K^-\pi^+$ spectrum when you cut in this way, in the enhancement, and the dashed line is what you observe when you cut with the mass bin below and above . . .

RATTI: The cut is on the $pK\pi$ . . . the solid histogram.

GIUSTI: The solid histogram is the $K^-\pi^+$ spectrum that we obtain when we cut on the $pK^-\pi^+\pi^-$ spectrum in the region where we observe the effect.   The dashed line is what we see for $K^-\pi^+$ when we cut below and above for the $pK^-\pi^+\pi^-$.

PETERSON: Okay, but this figure you show here, this is just the cut adjacent to the so-called position of the Λ-bottom, but you haven't tried at completely other places.

RATTI: Far away.   You mean far away from the resonance . . . from the possible . . . .

PETERSON: 6-point-something or whatever.   You never tried that?

838

GIUSTI: No, we have tried but we don't observe the . . . $[K^-\pi^+$ peak].

PETERSON: I see. Oh . . . . . that's so. Okay.

MALHOTRA, TATA INSTITUTE AND FERMILAB: You showed a peak for Σ_b^+ as well as Σ_b^- and then you combined and showed $\Sigma_b^\pm$. How many standard deviations would you say this peak was?

GIUSTI: With the Σ . . . .

MALHOTRA: I guess that . . . .

RATTI: It's 15.4 ± 4.4 if I remember correctly. Is that correct?

GIUSTI: This is, in the case of the negative, you see, 5.5 ± 2.4 . . . .

RATTI: No, the combined. He wants to see the combined distribution.

GIUSTI: The combined. That would be . . . . 14.5 ± 4.4.

ALAM, VANDERBILT UNIVERSITY: Why have you confined yourself only to the $K\pi$ decay mode of the D when there is another whole set of channels which I think you should be easily accessing there?

GIUSTI: Indeed we are still working on this problem. We have actually stopped work on this very nasty project, because there were more serious problems coming up. But one of the reasons we have started with low multiplicity decay channels is that since we don't have the secondary vertex, the decay of the Σ_b, then we must assume all tracks in the event. Then the mass identification [by time of flight] of the particle produced in the event is limited by the acceptance of the time of flight [counters], which is 10% of the total solid angle. So we have lots of combinations. And the more particles we put in the more serious the problems [of combinatorial background] become. That's why we started with as few particles as possible.

ALAM: I appreciate that because we had the same problem at CESR when we looked for B, because the combinations really increase overall geometrically. But still I find from my own experience that you can include actually up to 20% of the branching fraction of the D^o and the D^- and ultimately you really gain if you make tight cuts and things like that. I'm sure you are going to do it.

GIUSTI: We certainly will.

RAJA, FERMILAB: Have you tried looking for the bare B meson by using your D^o's?

GIUSTI: Indeed, we have. I don't have the mass spectrum with me, but we have tried to look for the "beautiful" meson decaying into $D^o \pi^-$ and we observe a signal which is still very preliminary, but it looks nice for the moment.

TAVERNIER, BRUSSELS: Maybe I misunderstood you but were you saying that in what you called Model 1 you assumed that the proton from the decay of your Λ_b was always produced with x-Feynman greater that .3? Is that what you said?

GIUSTI: Yes, that's right.

TAVERNIER: How can that be?

GIUSTI: That implies that the part from Model 1 is not the total cross section. But if you assume the flat production of the Λ_b we lose 35% of this cross section because, by applying this $[x > .3]$ requirement to the proton we obviously imply that Λ_b is also produced at x greater than that [i.e. $x > .3$]. So it is not the total cross section.

TAVERNIER: Okay, then the cross sections which correspond to Model 1 in fact are not correct and they should be bigger than what you quote.

GIUSTI: If you want, maybe they should be bigger by 30%. Yes. But the [reason] we have done things that way is that we wanted to estimate the value for the branching ratio for the process $[\Lambda_b^o \to pD^o \pi^-]$. We felt that that was the more realistic approach and in that case, given the same assumptions for the Λ_b^o and the Λ_c^+, we can do that [estimate the Λ_b^o branching ratio] even if the cross section is not the total one. When we make the ratio the difference disappears.

RUCHTI, NOTRE DAME: My question is in two parts. The first

part is a question of the statistical significance of the signal that you have. When you look at the background drawn underneath the D^o that you see, you quote 29 ± 5.5 or so as the significance. But that's above the background. Am I right or did I misinterpret? In other words, the statistical significance should not only only include the signal channel which you have but also the background underneath the signal which would reduce the significance some. Is that true? Maybe I misunderstood the plot.

GIUSTI: No, in the case of this $[\Lambda_b^o]$ signal when I quoted the σ of 5.5 - 6.2, we have followed this sort of reasoning. We have said: Okay, let us establish the background level in these two mass bins if the effect was not there. And I can do that either by fitting a polynomial curve to the rest of the spectrum or by defining the background level with the two or three bins around the [effect]. So this gives us a background level of 27. Then we have said: Okay, if this is a statistical fluctuation one must ask what is the probability that 27 events goes up to that [peak] level and the error of the 27 is determined to be 5 and something, so this is the statistical significance we have given.

RUCHTI: Okay. My tendency is just to say that you have 30 events in one bin and 26 in the other and I add them together and I get 56. I take the square root of something, which is of order 7-1/2, and then I would have probably drawn the background differently by this viewpoint. The end of my thing is: do you get more running time to pursue these studies further?

GIUSTI: No, sadly not. No.

RATTI: You are asking for more run time probably.

Editor's note: More discussion of this contribution will be found after the following contribution by E. Kluge.

CERN/EP/0463R/EEK/ef
17 July 1981

WAS A BEAUTY BARYON OBSERVED AT THE ISR?

Annecy-CERN-Collège de France-Dortmund-Heidelberg-Warsaw Collaboration

(presented by E.E. Kluge)

D. Drijard, H.G. Fischer, H. Frehse, W. Geist, P. Hanke,
P.G. Innocenti, W.T. Meyer, A. Norton, O. Ullaland and H.D. Wahl[*]
CERN, European Organization for Nucl. Research, Geneva, Switzerland

G. Fontaine, C. Ghesquiere and G. Sajot
Collège de France, Paris, France

W. Hofmann, M. Panter, K. Rauschnabel, J. Spengler and D. Wegener
Institut für Physik der Universität Dortmund, Dortmund, Germany

M. Heiden, E.E. Kluge, T. Nakada and A. Putzer
Institut für Hochenergiephysik der Universität Heidelberg,
Heidelberg, Germany

M. Della Negra and D. Linglin
LAPP, Annecy, France

R. Gokieli and R. Sosnowski
Institute for Nuclear Research, Warsaw, Poland

ABSTRACT

A recent publication [1] claims an evidence for a hadronic
state with beauty flavour, $\Lambda_b \to D^0 p \pi^-$, in p-p collisions at
$\sqrt{s}$ = 63 GeV. We report here a negative result from an experiment which uses the same detector and has a similar sensitivity.
In addition we show that these experiments cannot observe such a
signal with the method proposed lacking two orders of magnitude
in the sample size.

Presented at the XII International Symposium on Multiparticle Dynamics
21-26 June 1981, Notre Dame, Indiana, USA

(*) Now at Institut für Hochenergiephysik, Österr. Akademie der
Wissenschaften, Wien, Austria.

0463R/EEK/ef

1. INTRODUCTION

The observation of a heavy baryonic state was recently report-ed [1] in the channel $p\pi^- K^- \pi^+$ with a mass $\simeq 5.4$ GeV/c^2 produced in p-p collisions at $\sqrt{s} = 63$ GeV using the SFM detector [2] at the ISR. The Λ_b interpretation is supported by a peak in the $K^- \pi^+$ mass distribution about the D^0 mass and an associated e^+ trigger which might originate from a leptonic decay of a B meson produced in a pair with Λ_b. This experiment, later referred to as BCF, eventually estimates the Λ_b signal to be (30 ± 5.0) combinations where $\Lambda_b \rightarrow D^0 p\pi^-$ and $D^0 \rightarrow K^- \pi^+$.

In the course of an experimental program to study the production of heavy flavours [3], we performed a similar experiment as BCF, in the same detector, with a comparable trigger, at the same energy and with an integrated luminosity of the same order of magnitude. We applied to our data selection criteria identical to those of the BCF experiment and found no signal. From the study of sensitivity, we conclude that such a signal should indeed not be observed.

2. EXPERIMENTAL PROCEDURE AND RESULTS

The triggering particle covers a c.m.s. rapidity $|y| < 0.35$ and an azimuthal angle $|\phi| < 9°$ with respect to the beams. A hardware road system guaranties an efficient data taking of events with a track in the trigger region, and a double stage Cerenkov counter selects mostly $e^\pm$ particles. We started with 170 000 e^+ candidates and applied selection criteria to purify the prompt e^+ sample. In particular e^+, e^- pairs of small opening angle were rejected by energy loss measurements in a proportional cham-ber close to the interaction region [2], and most pairs with a large opening angle were rejected using a threshold on effective mass. Similar criteria were used in the BCF experiment leaving still a 50% contamination of not prompt positrons (Dalitz pairs, K_{e3} decays, ...) for triggers with transverse momentum bigger

than 0.8 GeV/c. An electromagnetic shower detector was used in
the BCF experiment, which reduced the hadronic background in the
triggers to a negligible amount. We used these apparatus only as
a test of our efficiency and concluded that 65% of our triggers
with p_T > 0.8 GeV/c are positrons. This conclusion is obtained
from the distribution of the pulse height spectrum of the
electromagnetic shower detector for triggers associated with a
Cerenkov signal (fig. 1(a)) compared to the spectrum for hadron
tracks (fig. 1(b)) for which no Cerenkov signal was required.
The hadron contamination versus trigger transverse momentum is
given in fig. 1(c). Thus 32% of our triggers are prompt e^+,
when the corresponding number for the BCF experiment is 50%.

Fig. 1

Spectrum of the pulse heights
of the electromagnetic shower
counter obtained for:

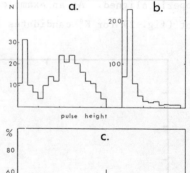

(a)   triggers associated with
      a Cerenkov counter signal,

(b)   particles in the trigger
      region with no Cerenkov
      signal,

(c)   Fraction of hadron conta-
      mination of triggers ver-
      sus trigger transverse
      momentum.

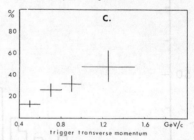

In order to compare our data with those of the BCF
Collaboration, we have used identical selection criteria [1].
They were:

- $(p_T)_{e^+}$ > 0.8 GeV/c and $\Delta p/p$ < 15% gave 18136 events;

- further request of a fast positive particle (assumed to be a
  proton) with x_F > 0.32 and $\Delta p/p$ < 30% gave 4771 events;

- at least 7 additional charged particles are required with
Δp/p < 30% which gave 4149 events.

The detector covers 4π steradians using ∿ 70 000 wires
spread over ∿ 250 planes, giving rise to a global track acceptance ≃ 90%. The BCF Collaboration used our first version of
event reconstruction program, but did not use our later
improvements. These concern mainly a better momentum precision
using vertex information and a better alignment of the planes.
The increased momentum precision is an important matter since the
selection criteria require a good precision, giving in particular
a large effect on the "proton" acceptance. Mass shifts were
reported [1] which we found to be small after chambers are
properly aligned. As an example, we show a $\pi^+-\pi^-$ mass
plot (fig. 2) for K^0 candidates [4].

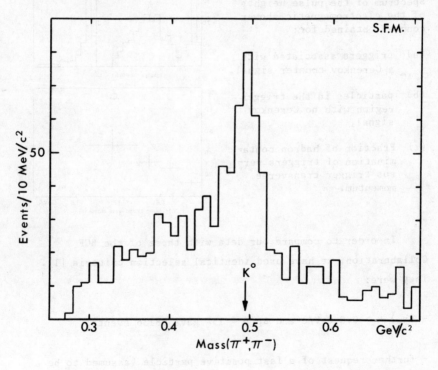

Fig. 2 Mass distribution of π^+, π^- pairs for V^0 candidates.

Our final results are given in fig. 3 where the dotted curve, (144 ± 24) events, represents the signal expectation deduced from the results of the BCF experiment (in particular our hadron contamination in the trigger is accounted for). Our corresponding observation is ≤ (30 ± 30) events. The K^-, π^+ mass is restricted to the range 1.725 to 1.875 GeV/c² as in [1] although no peak is observed. A cut about the nominal D^0 mass 1863 GeV/c² produces the same negative result.

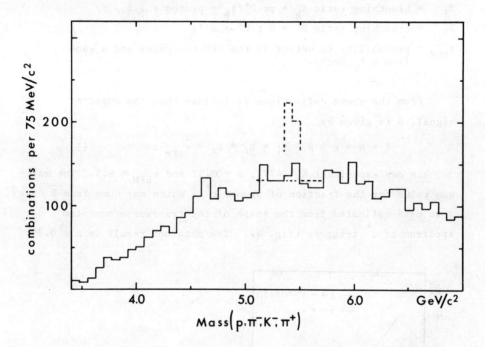

Fig. 3   Mass distribution of $p\pi^-K^-\pi^+$ system when the $K^-\pi^+$
mass is in the region 1.725 - 1.875 GeV/c². The dotted
line indicates the signal expectation deduced from [1].
The smooth background level has been estimated from six
bins adjacent to the 5.35 - 5.50 MeV/c² region.

We conclude that, given conditions mostly identical, the two experiments are not compatible.

3. ESTIMATION OF THE POSSIBLE SIGNAL SIZE

Let us define the following quantities:

N = number of events with an e^+ trigger and with a fast ($x_F > 0.32$) proton,

S = number of events with a $\Lambda_b \to p\pi^- D^0$ associated with a $\bar{B} \to e^+ + \ldots$, $D \to K^-\pi^+$ and a proton with $x_F > 0.32$,

ε = (number of prompt e^+) number of e^+ triggers),

ρ = (number of e^+ from $\bar{B}$)/(number of prompt e^+),

B_1 = (number of Λ_b's)/(number of beauty hadrons produced),

B_2 = branching ratio $\Lambda_b \to p\pi^- D^0/\Lambda_b \to$ proton + $\ldots$,

B_3 = branching ratio $D^0 \to K^- p/D^0 \to$ all,

ε_{SFM} = probability to detect in the SFM two pions and a kaon from a Λ_b decay.

From the above definitions it follows that the expected signal, S is given by

$$S = N * \varepsilon * \rho * B_1 * B_2 * B_3 * \varepsilon_{SFM} . \qquad (1)$$

In our experiment $N = 4771$, $e = 0.32$ and $e_{SFM} = 0.5$. The maximum value for the fraction of prompt e^+'s which may come from $\bar{B}$ decay, ρ we have estimated from the shape of the transverse momentum spectrum of e^+ triggers (fig. 4). The obtained result is $\rho \leqslant 0.25$.

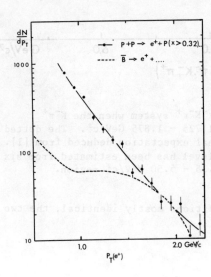

Fig. 4

Transverse momentum spectrum of positron triggers for events with a fast positive particle ($x_F > 0.32$) observed. The dashed line shows the spectrum of positrons from decays of an antibeauty meson.

The branching ratio B_3 has been measured [5] to be
$(3 \pm 0.6)\%$. Let us make an extreme assumption that all beauty
hadrons produced are Λ_b beauty baryons $(B_1 = 1)$ and that Λ_b
decays only via $\Lambda_b \rightarrow p\pi^- D^0$ decay channel $(B_2 = 1)$. We obtain
then an upper bound for the number of Λ_b decays which we could
observe in our sample

$$S \lesssim 5.7 \text{ events.} \tag{2}$$

We conclude, therefore, that a sample of events comparable
in size with that analyzed by us cannot give a statistically
significant signal.

The upper limit for the number of Λ_b decays which could be
seen in the BCF sample is obtained by scaling up our limit (2) by
the factor 5/3 due to the different fraction of prompt positrons in
two triggers. The result has to be renormalized by the ratio of
the number of events used times the probability to observe a
$p\pi^- K^- \pi^+$ system. This factor can be obtained empirically by
comparing the levels of smooth background in the $p\pi^- K^- \pi^+$ mass
plots obtained in two experiments, which directly monitors the
product of the ratio of events and the ratio of the efficiency of
observing the $\pi^- K^- \pi^+$ system. We find this ratio of mass plot
background levels to be $\sim 1:10$, which comes from the ratio of
event samples (1025 for BCF versus 4149 in our sample) and the
ratio of three particle detection efficiency. In the first
approximation this is an efficiency to detect a single particle
raised to the third power. Our recent version of the reconstruction program gives higher precision of the particle momentum and
consequently the requirement that $\Delta p/p < 0.3$ rejects fewer
events. In addition, for our experiment the region opposite to
the trigger was more efficiently filled with proportional chambers, giving in consequence more well-reconstructed tracks.
Accordingly the upper limit for the Λ_b signal reported by the
BCF Collaboration is

$$S \lesssim 5.7 * \frac{5}{3} * \frac{1}{10} \simeq 1.0 \text{ event} \tag{3}$$

The above estimate has been obtained with rather unacceptable extreme assumptions $B_1 = B_2 = 1$. Values such as $B_1 \simeq 0.5$ and $B_2 \lesssim 0.1$ are still high but are more realistic. They lead to upper limits which are lower by a factor of 20 than those given by (2) and (3).

3. CONCLUSIONS

We have shown that in the relatively small sample of events selected, the contribution to $\Lambda_b \rightarrow D^0 p\pi^-$, $D^0 \rightarrow K^-\pi^+$ cannot exceed a few events, which agrees with the negative result of our experiment. This contradicts the interpretation of [1]. In order to raise the upper bound on the signal to a statistically observable level, one would need about hundred times more events. The small branching ratio for $D^0 \rightarrow K^- + \pi^+$ is the main factor which determines the low values of upper limits for a Λ_b in the experiments discussed. It seems, therefore, that future studies of the beauty production require a trigger which more efficiently enriches the event sample with visible decays of charm hadrons.

REFERENCES

[1] M. Basile et al., Evidence for a new particle with a naked
     "beauty" and for its associated production in high-energy
     (pp) interactions, preprint INFN/AE-81/5, and private
     communication.

[2] H. Frehse, F. Lapique, M. Panter and F. Piuz, Nucl. Instr.
     and Meth. 156 (1978) 87;

     H. Frehse, M. Heiden, M. Panter and F. Piuz, Nucl. Instr. and
     Meth. 156 (1978) 97;

     W. Bell et al., Nucl. Instr. and Meth. 156 (1978) 111.

[3] D. Drijard et al., Physics Lett. 81B (1979) 250;

     D. Drijard et al., Physics Lett. 85B (1979) 452.

[4] K. Rauschnabel, V^0 reconstruction in the Split Field
     Magnet spectrometer, CERN, EP Internal Report 81-1, 1981.

[5] A.J. Lankford, Proceedings of the XX Int. Conf. on High
     Energy Phys. in Madison, 1980, American Institute of
     Physics, New-York 1981, p. 373.

851

DISCUSSION

RATTI: Thanks for this Doctoral lecture on statistics. I see already a couple of hands raised. I think that the first one was already up while Kluge was talking so . . . . .

GIUSTI: BOLOGNA: Can I come over there?

RATTI: Do you have a question or a comment?

GIUSTI: A comment. I haven't mentioned but, looking at the [statement that] "one would expect the signal in your case to be 100 events," I don't think this is correct, and in fact I wonder why, the last time I spoke with Sosnowski in Geneva, his statements were quite different from what you are saying, which has changed a bit this week. In fact his conclusions were that his mass spectrum could not exclude the presence of our signal. Another fact which I would like to mention or I would like you to comment on: During this quite frank discussion which we had with Sosnowski, at a certain stage there was a positive fluctuation of 20 events which is what you would expect, not 100, in the exact two mass bins where we observe the signal. And by cutting in that mass range you had a positive fluctuation also in the $K^-\pi^+$ mass spectrum. I would like either you to comment, [or] I give the answer which Sosnowski gave me about this.

KLUGE: Firstly, I tried to be very gentle with you if you have got the message. Secondly, I think that everybody can judge himself what it means, what this spectrum means. I mean there are fluctuations as you can see, but one expects fluctuations of that kind. And whether you can see here a hundred events or not, I leave it to you. I don't believe that. I should tell you something in addition which is the following. If you look carefully, then you see that the number of events [i.e. the] relative number of events, are a factor of three approximately between our experiment and the other one. On the other hand, the number of entries into this $[pK^-\pi^+\pi^-]$ plot is . . . (That is combinations. We are a multi-particle conference.) . . . the number of combinations which you see here is certainly higher than the number of events here. Because you have all sorts of combinations. And in fact the ratio between combinations and events is higher in this [our] case. We think we know why, because this was done in a more advanced program version. So there should be more particles found and therefore the number of combinations is higher. On the other hand, that might cause some more fluctuations than in the other case, which is

852

natural. On the other hand I don't think that we are seeing any
signal just scaled up from the other one. And, I mean, this . . .
if you understood . . . this is not my point. I tried to make clear
that in addition, by looking at reasonable or unreasonable assumptions, one has difficulties understanding that there can be a signal
at all. And this is my main point and you see it even comes also
out from your analysis. I mean, these 30 D^O's which you
observe are exactly the . . . .

RATTI: I think that this point is clear. Your position is clear, as
well as his position.

GIUSTI: There is another point which I have to make about the
statement of the 100 events which one would expect. The
difference . . . our setups were not exactly equal. There is a
small but, we think, a really important difference. Their definition of the lepton is provided by the coincidence of two Cerenkov
counters in series. Our definition of an electron is [the coincidence of] the same two Cerenkov counters and an extra electromagnetic shower detector behind, which, in the range of momenta
which we are interested in, provides a rejection against charged
hadrons of 5%. It reduces the number of charged hadrons to 5%.
You don't have that, so when we start with the same number of
events and we look at the invariant mass spectra, before saying
how many events we expect we should prove that our electron
samples are equal.

RATTI: There is somebody from the floor who would like to ask
a comment.

KLUGE: This is quite a comment and I should certainly answer
this comment first. Right? I don't quite know what you mean to
imply. It is clear that the CERN-Bologna-Frascati collaboration
has just used our setup by adding here some lead glass, and it is
clear that we are measuring momentum in a magnetic field whereas
they have lead-glass information on the other hand. We have been
analyzing the e/π ratio quite carefully and we believe that also,
although we might not have 5% contamination, but 20% contamination . . . .

GIUSTI: No, you have 60%, according to what you say.

KLUGE: It doesn't really matter because . . . .

GIUSTI: No, no . . . .

KLUGE: After all, I mean, I tried to convince you that you can't see anything and therefore, you know, it is hard to . . . .

RATTI: No, I think that now . . . . .

GIUSTI: Let me say that . . . it [the difference in charged hadron contamination] has only the slight effect that one event out of two of our events is a direct-produced electron. In your case that's one out of four. That's all the difference there is.

KLUGE: 60% contamination doesn't mean one out of four. So please . . . . .

GIUSTI: There is one last point that I would like to remark about the internal consistency of the experiments. I would like to ask you if you have evaluated the e/π ratio which you get with the D^+ charm signal. I guess that it comes out to be of order 15×10^{-4} or, if you use the cross section published at the time when the result was found, it is 75×10^{-4}.

KLUGE: This is certainly not true . . . because I don't know where you got these numbers from, and in fact, you see . . . you might be aware or not that in fact your friend [Contin] showed it . . . that our charm cross sections agree quite well. This you said yourself I would guess. We have been triggering on electrons. We see [the same] charm cross section as everybody else sees, and this was the reason why I tried to show you that we were first in the business, that we have seen all sorts of things, and that for this reason we could stand up here and say "Fine, there seem to be funny things going on." And that is what I tried to make clear to you.

RATTI: I hope that we would like to have all of us clear about the point, and I think that everybody has the point very clear that the two results are in contradiction. I think that Randy Ruchti wanted to . . . . .

KLUGE: There is much more to it than this point to be made clear.

RATTI: You will make it clear in the Proceedings, I suppose.

RUCHTI, UNIVERSITY OF NOTRE DAME: You gentlemen have said you have compared what you have done very carefully. I'm curious to know . . . and this is addressed to a slide which you just presented a few moments ago, that beauty will in general

produce leptons which are at fairly large p_T relative to charm
. . . . and I'm curious if you or perhaps your competitors here
have made additional cuts on the transverse momentum of the
electron beyond 800 MeV/c where presumably some substantial
fraction of the B decay should remain but the background should
perhaps go down more rapidly. Has that been done?

KLUGE: No, the only cut was made at 800 MeV/c but there is
one more indication . . . . for the absence of the beauty signal.
We have seen from the CESR results today that the electron
spectrum peaked around 2 GeV. Right? And if you look at the
electron spectrum shown to us by the CERN-Bologna-Frascati
collaboration then the peak was much lower, as I understand. So
clearly this is in accordance with what I tried to tell you: namely,
that there must be quite a few more components in the direct
electrons. And since, in order to explain their signal they need
all electrons to come from beauty, this is a contradiction.

FRIDMAN: Frankly speaking, if I see your mass plot with the
fluctuations, maybe there can be something, maybe there is
nothing. (Also you can say that the way that you count the
statistical deviation, maybe it is a little bit smaller than what you
said.) And I would like only to [remind] you that a lot of bumps
with five and six standard deviations disappear. Remember the
baryonium. And so I think it is really necessary to have at least
two experiments of many standard deviations.

KLUGE: I think you mentioned it. I am going to mention that one
of your peaks which you didn't publish, finally we thought was going
to be the Ψ meson. So on the other hand I agree with our
colleagues from the CERN-Bologna-[Frascati] collaboration that
they are right in publishing [their data]. But on the other hand as
I tried to tell you, if you can close the discussion with that, I
think we should have to be careful! So far it looks as if a few of
these things cannot be understood, and this message I wanted to
make clear!

MORRISON, CERN: Perhaps the main point of Dr. Kluge's talk
is that, when he does the calculation of internal consistency,
instead of having 30 events which is claimed by the BCF you
calculate, rather, 0.4 events. I would like to ask Dr. Giusti . . .
I think we all would like to ask him . . . when he does that
calculation does he also get 0.4? If he does not what number does
he get?

RATTI: He has to give an answer.

GIUSTI: Could you repeat the question?

MORRISON: Dr. Kluge has done a calculation. You say that you observed 30 events. When he makes these same cuts as you, he finds .4 events is the maximum you could possibly get. When you do this calculation do you also get .4 or do you get some other number? If you get some other number what is that number?

GIUSTI: No, we cannot question the exactness of the calculation. And there would be difficulty also to say that it cannot be done, because it can be done and that is the result. That we say. What I intend to say is that the method is incorrect in the sense that anything which we don't know can change that number very easily. In fact if, for an exercise, we repeat the same calculation for the charm signal, we completely saturate the number of K^- and K^+ which we see, which we observe. I have done it.

MORRISON: Sorry, I didn't understand your answer. Do you get 0.4?

GIUSTI: Their argument . . . it looks very strong. Okay, it's very strong. But it is not correct, because . . .

MORRISON: What number do you get when you do a correct calculation?

GIUSTI: It's too late now to show a transparency or anything like that. But we are completely consistent if you allow us to be within 0.5 σ from what we observe, according to our calculation. And when one makes this sort of calculations one should consider that one does not know if the models that he uses are correct. One doesn't know. What else can we say? That is it. Apart from the fact that we like better to make comparisons between beauty and charm cross sections, if you try to extrapolate and, from the obtained cross section [for beauty], we try to calculate the e/π ratio expected from beauty, we get 4.5×10^{-4} which is a couple of times higher than the e/π ratio measured by ISR. Besides, this ratio has not been measured redundantly. There are two measurements, one from Rubbia at low p_T and one from Di Lella at high p_T.

MORRISON: Excuse me. You have talked many subjects. I haven't really understood . . . .

RATTI: Excuse me Douglas, there is also Raja who wants to ask a question or make comments, and, if you don't mind, I would like to give him the floor for a second.

RAJA, FERMILAB: [Question addressed to Kluge.] I don't want to repeat this question, so I will try to speak clearly. We all agree that you don't see the bare beauty signal. Supposing you assume that there is a signal there buried in that $[pK^-\pi^+\pi^-]$ plot and cut there and look at the charm. Do you see the charm?

KLUGE: In fact, I think we didn't even do that. So I couldn't say if it's there. We would have to do it.

RAJA: That is an important check. They have done that. I think you should.

CONTIN, CERN: We have seen the [their] computer output with this check, and we saw that there is a consistent positive fluctuation in the same two bins where we see our D^o. So you have done it, and the result is this.

RATTI: Well, I think that, since the time I was at the Siena conference, and $\triangle S$ was equal to $\triangle Q$ or $\triangle S$ was equal to $-\triangle Q$ and we had beautiful results, after a number of years one of the two speakers was right and the other one was wrong. Obviously, because nature has made its own choice. But I recommend that Bill Shephard be as careful as possible in recording this discussion because it is within the character of this conference to have these kinds of discussions going on. Obviously, the two papers and the two contributions will be included in the Proceedings and I think that we might close the session with this point because we have to be aware of the bus which is going to the banquet and probably Paul Kenney has to give us more announcements.

SESSION G2

PRODUCTION AND DECAY

OF

SYSTEMS INVOLVING HEAVY QUARKS II

Friday, June 26, a.m.

Chairman: V. P. Kenney
Secretary: L. Dauwe

THEORY OF HADRONIC PRODUCTION OF HEAVY QUARKS[*]

C. Peterson[†]
Stanford Linear Accelerator Center
Stanford University, Stanford, California 94305

ABSTRACT

Conventional theoretical predictions for hadronic production of
heavy quarks ($Q\bar{Q}$) are reviewed and confronted with data. Perturbative
hard scattering predictions agree qualitatively well with hidden $Q\bar{Q}$-
production (e.g., ψ,χ,T) whereas for open $Q\bar{Q}$-production (e.g., pp →
Λ_c^+X) additional mechanisms or inputs are needed to explain the forwardly produced Λ_c^+ at ISR. It is suggested that the presence of
$c\bar{c}$-pairs on the 1–2% level in the hadron Fock state decomposition
(intrinsic charm) gives a natural description of the ISR data. The
theoretical foundations of the intrinsic charm hypotheses together
with its consequences for lepton-induced reactions is discussed in
some detail.

1. INTRODUCTION

One of the most important results recently obtained at the ISR
has been the observation of charm with remarkably high cross sections
(0.1–0.5 mb) and with momentum distributions indicating diffraction-
like production mechanisms.[1-8] This experimental fact is in contradiction with what is expected from hard scattering mechanisms which
predict $\sigma_{c\bar{c}} \simeq 50$ μb at ISR energies with soft x_F-spectra for the observed charmed particles.[9] (It should be noted however that at SPS
and FNAL energies these features of the ISR $c\bar{c}$-production has not yet
been settled.) As far as hadronic production of hidden charm is concerned (e.g., pp → ψX) the predictions from hard scattering mechanisms
seem to be confirmed by experiments.[9] In particular the leading subprocess gg → P-state + X has been observed explicitly by the
$\qquad\qquad$ ∟ $\psi+\gamma$
experimental presence of accompanying photons.[10]

Although the nucleon is usually regarded as a three-quark bound
state, its actual Fock state structure in Quantum Chromodynamics is
expected to be more complicated. The proton has a decomposition of
free Hamiltonian eigenstates[11,12]

$$|uud\rangle, \quad |uudg\rangle, \quad |uudq\bar{q}\rangle, \quad \ldots \qquad (1)$$

---

[*]Work supported in part by the Department of Energy under contract
DE-AC03-76SF00515 and by the Swedish National Science Research
Council under contract F-PD8207-101.

[†]On leave of absence from NORDITA, Copenhagen, Denmark.

A nonnegligible $|uudc\bar{c}\rangle$ state in (1) (intrinsic charm) is expected to give fast Λ_c^+'s in hadron-induced reactions since on a long time scale the five constituents should have the same velocity.[13,14] ψ-production from these $|uddc\bar{c}\rangle$ states should however be suppressed due to in particular small wavefunction overlap.[14] A Bag calculation[15] of the $c\bar{c}$-mixture gives $P(|uudc\bar{c}\rangle) \approx 1\%$, which is compatible with the cross section for $pp \rightarrow \Lambda_c^+ X$ observed at ISR. It is of course crucial and of utmost interest to confront the amount of intrinsic charm with data on leptoproduction. Unfortunately measurement on F_2^{charm} from $\mu^\pm N \rightarrow \mu^\pm \mu^\mp X$ has not yet been performed for high enough x_{Bj} to provide a crucial test. Some indirect indications of intrinsic charm exist though: The different values of Λ_{QCD} as measured in μN and νN at high Q^2 can in fact be explained by the onset of charm production which takes place differently in the two reactions.[16] Also for the unexpectedly high rate of observed same sign dimuons in ν-reactions the 1% intrinsic charm contributes substantially in the right direction.[14]

This talk is organized as follows: We first (Sect. 2) review the theoretical expectations for heavy quark production starting with estimates for "soft" production mechanisms and then elaborating more on what is expected from perturbative QCD both with regard to hidden and open heavy flavor production. Comparisons with experimental data on $c\bar{c}$ are found in Sect. 3. In Sect. 4 a general discussion of higher Fock state decomposition of hadronic states is given and in Sect. 5 we argue for the existence of $|uudc\bar{c}\rangle$ or the 1% level and compute the resulting $c(x)$ distributions. Hadronic production of charm is discussed in Sect. 6. In Sect. 7 our model for $c(x)$ is confronted with data from leptoproduction experiments.

2. "CONVENTIONAL" THEORETICAL EXPECTATIONS
FOR HEAVY QUARK PRODUCTION

The production of heavy quarks in hadronic collisions from soft mechanisms is normally expected to be very suppressed. As an example, when considering hadronic productions of particles as a tunneling phenomena one finds the probability to produce a $q\bar{q}$ $(Q\bar{Q})$-pair[17]

$$P(q\bar{q}) \sim \exp(-\frac{\pi}{\kappa} m_\perp^2) \tag{2}$$

where $m_\perp = \sqrt{p_\perp^2 + m_q^2}$ and κ is the string constant ≈ 0.2 GeV^2. Using $m_u = m_d = 0$ MeV, $m_c = 100$ MeV, $m_c = 1500$ MeV and $\langle p_\perp \rangle = 350$ MeV one gets from Eq. (2)

$$u:d:s:c = 1:1:\frac{1}{3}:10^{-10} \tag{3}$$

(The $s\bar{s}$-suppression, 1/3, agrees with data in a cascade picture.)

The reason for the strong suppression of c-quark production is that it is very difficult to localize the energy of a substantial part of a string. Also in other pictures one obtains a strong suppression. For example in the statistical model[18] approach the probability for D-meson production is given by

$$P \sim \exp(-2m_D/160 \text{ MeV}) \tag{4}$$

which gives the ratio $\pi:K:D = 1:0.13:3.10^{-5}$.

However, since large masses are involved one expects that perturbative QCD is applicable. In fact it turns out that the perturbative contribution strongly dominates over the soft one. This is in contrast to, for example, large $p_\perp$-production where perturbative QCD is only responsible for the fall of the spectrum.

We will divide the discussion of the perturbative QCD predictions into two parts--<u>hidden</u> and <u>open</u> heavy quark production:

A. Hidden Heavy Quark Production

Hadronic production of hidden heavy quark pairs, e.g., ψ, can take place through the following hard scattering subprocesses[19] (see Fig. 1)

$$q\bar{q} \to \psi \tag{5a}$$

$$c\bar{c} \to \psi \tag{5b}$$

$$gg \to \text{P-state} \to \chi + \gamma \text{ (or } 3\pi) \tag{5c}$$

$$q\bar{q} \to \text{P-state} \to \psi + \gamma \text{ (or } 3\pi) \tag{5d}$$

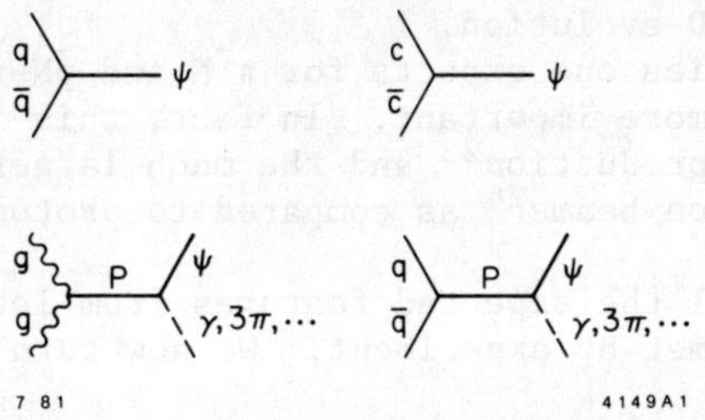

Fig. 1. Lowest order QCD sub-processes for hadron + hadron → ψ + X.

At high energies the gluon-gluon amalgamation process (5c) is the most important due to the abundance of gluons. In the case of $c\bar{c}$-production the P-state in Eq. (5c) is $\chi(3415)$ or $\chi(3555)$. The cross section for, e.g., $pp \to \psi X$ with any of the subprocesses above is given by convolution integrals of the type

$$\sigma(pp \rightarrow \psi X) = F \frac{8\pi^2}{M_\chi^3} \Gamma(\chi \rightarrow gg)$$

$$\iint dx_1 dx_2 \ G_{g/P}(x_1) \ G_{g/P}(x_2) \delta(x_1 x_2 s - M_\chi^2) \qquad (6)$$

where $G_{g/P}(x)$ are the gluon (or quark) distributions respectively and F is an undetermined "fudge"-factor representing uncertainty in color rearrangement, etc.[20]

This hard scattering picture, where the gluon amalgamation process (5c) dominates, is supported by the fact that accompanying photons in connection with ψ-production have been observed (for details see Ref. 21) with the experimental values[10]

$$B(\chi \rightarrow \psi\gamma) \cdot \frac{\sigma_\chi}{\sigma_\psi} = \begin{cases} 0.70 \pm 0.28 \\ 0.48 \pm 0.21 \end{cases} \qquad (7)$$

Furthermore, ψ'-production seems to be suppressed in hadron-hadron collisions[22] which also supports the gg $\rightarrow \chi \rightarrow \psi\gamma$ picture since $m_\chi < m_{\psi'}$. Also $\sigma(s)$ and $d\sigma/dx_F$ from Eq. (6) agrees well with experimental data on pA $\rightarrow \psi X$ when using gluon distribution[9]

$$G_{g/P}(x) \sim \frac{1}{x} (1 - x)^5 \qquad (8)$$

Similarly T production can be fitted by a steeper distribution

$$G_{g/P}(x) \sim \frac{1}{x} (1 - x)^6$$

as expected from QCD-evolution.

At lower energies one expects for $\pi^- N$ and $\bar{p}N$-reactions the $q\bar{q}$-subprocesses to be more important. In fact, this is apparent from the $\bar{p}/p$-ratio in ψ-production[23] and the much larger cross section found for T with pion beams[24] as compared to proton beams at the same energies.

In summary, all the expected features from lowest order graphs Eq. (5) seem to be met by experiment. We now turn to the open charm production.

B.   Open Heavy Quark Production

In the case of open $Q\bar{Q}$ production the following hard scattering processes contribute[25] (see Fig. 2a,b)

$$q\bar{q} \rightarrow Q\bar{Q} \qquad (9a)$$

$$gg \rightarrow Q\bar{Q} \qquad (9b)$$

together with the flavor excitation processes[26] (Fig. 2c)

$$qQ(\bar{Q}) \rightarrow qQ(\bar{Q}) \qquad (10)$$

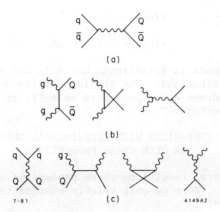

Fig. 2. Lowest order QCD subprocesses for hadron + hadron →
$Q\bar{Q}$ + X.

Predictions from the latter ones are uncertain since the heavy quark
distribution Q(x) is not well known.

Again the gluon amalgamation process (9b) is dominant at high
energies due to the abundance of low-x gluons. The cross section is
given by convolution of distribution functions and the subprocess
cross section ($\hat{\sigma}$)

$$\sigma(h + h \rightarrow Q\bar{Q}X) = \iint_{\substack{x_1 x_2 > x_{min} = \frac{s_{min}}{s}}} dx_1 dx_2 \; G(x_1) \; G(x_2) \hat{\sigma}(x_1, x_2, s) \quad (11)$$

Before comparing resulting cross sections and distributions with data
on charm, we comment on the theoretical uncertainties entering
Eq. (11).

i) The lower limit of Eq. (11). The true kinematical threshold
is $2m_D$ but $2m_c$ is presumably more relevant since the charmed hadrons
are formed in a fragmentation/recombination process, thereby gaining
energy.

ii) The value of m_c. Most authors use m_c = 1.6 GeV. A lower
value like m_c = 1.2 GeV, as obtained from potential calculations,
would increase the cross section by a factor 4.

iii) Higher order graphs are not yet included.

iv) Higher twist contributions. These are unknown and could be
important at such small masses as m_c = 1.6 GeV.

From Eq. (11) the cross section for $c\bar{c}$ and $b\bar{b}$-production in the FNAL
(SPS)-ISR energy range is given by (see Fig. 3)

$$\sigma(c\bar{c}) = 1\text{--}50 \ \mu b$$

$$\sigma(b\bar{b}) = 0.1\text{--}100 \ nb \tag{12}$$

The energy dependence is logarithmic which is due to the $1/x$–behavior of the gluon distributions. The single particle spectrum for the observed charmed hadrons are expected to be soft, reflecting the incoming gluon distributions.

3. COMPARISON WITH EXPERIMENTAL RESULTS ON OPEN CHARM PRODUCTION

The experimental results on charm production are reviewed in detail in Ref. 27. Here we only briefly mention the most important results. They are:

i) At ISR one observes a large cross section (0.1–0.5 mb) for the reaction $pp \rightarrow \Lambda_c^+ X$ (see Fig. 3b).

ii) Moreover, the Λ_c^+ seems to be produced diffractively in the forward region of phase space (see Fig. 4a,b). At least one of the experiments has an explicit diffractive trigger.[2]

iii) At SPS/FNAL experiments the situation is not so clear. One experiment with a diffractive trigger,[28] $\pi^- p \rightarrow D\bar{D}X$, observes a forwardly oriented single particle spectrum. However, indirect information from beam dump experiments are consistent with the soft x_F–spectra typical for hard scattering mechanisms.[27]

iv) Also recently signals from forwardly produced Λ_b at ISR have been reported.[45]

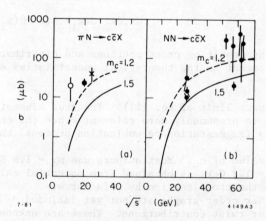

Fig. 3. a) $\sigma(\pi N \rightarrow c\bar{c}X)$ as a function of c.m.s. energy, from Ref. 9.
b) $\sigma(NN \rightarrow c\bar{c}X)$ as a function of c.m.s. energy, from Ref. 9.

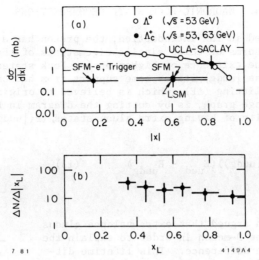

Fig. 4. a) $d\sigma/d|x|$ for Λ_c^+ at 53 and
63 GeV.[1] The smooth curve is a fit
to the Λ° data points. b) Unnormalized
x_L-distribution for Λ_c^+ from ref. 8.

Concerning the <u>large cross sections</u> observed at ISR, one could
imagine to use a very low value for m_c [Eq. (11)] ($m_c = 1.2$ GeV) and
thereby approach the ISR data (see Fig. 3). This is however not an
attractive solution, since the good agreement of the quark-gluon
fusion model with $\gamma p \to c\bar{c}X$ data for $m_c = 1.5$ GeV would then be
destroyed.

The other, and more interesting, discrepancy with the hard
scattering approach is the <u>x_F-spectrum</u> of Λ_c^+ (see Fig. 4): From
general grounds one would expect the Λ_c^+ wave function to favor configurations where the c-quarks have the most momentum (see Fig. 5).

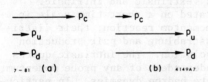

Fig. 5. a) Typical quark momentum configuration in a Λ_c^+.
b) Typical quark momentum configuration after a hard scattering
with a slow c-quark and two fast
valence quarks.

On the other hand, the c-quarks
produced in a hard scattering
process have small x. Hence such
c-quarks would most unlikely end
up in a fast Λ_c^+. It is tempting
to conclude that the only way to
produce fast Λ_c^+ is to have hard
$c(\bar{c})$-quarks initially present in
the proton, i.e., $|uudc\bar{c}\rangle$
states.[13,14] We will discuss this
<u>intrinsic charm</u> hypothesis in some
detail below.

4. HADRONIC FOCK STATE DECOMPOSITIONS

As mentioned in the introduction, the proton has a general decomposition in terms of color singlet eigenstates of the free Hamiltonian. The existence of higher proton Fock states like in Eq. (1) has as far as $|uudg\rangle$ states some support from hadron spectroscopy: The p-Δ mass splitting (ΔE), which is believed to originate from the one gluon exchange graph, is by cutting the diagram in Fig. 6 related to the probability of having extra gluon states, ($P(|uudg\rangle)$), through the relation

$$\Delta E = \sum_{\substack{gluon \\ modes}} P(|uudg\rangle)(E_{uud} - E_{uudg}) \qquad (13)$$

It has also been argued that extra valence gluons are needed in the D-meson in order to explain the D^+ - D° lifetime difference. This lifetime difference can be due to the W-exchange graph possible for D° (but not D^+) provided the helicity suppression is relaxed by the presence of extra valence gluons.[29]

Also in an analysis by Brodsky, Huang and Lepage,[30] it is shown that rigorous constraints from $\pi \to \mu\nu$ and $\pi \to \gamma\gamma$ decays gives a probability <0.25 for having a pion in a pure $q\bar{q}$-state for a large class of wavefunctions.

In the next section we explore the consequences of heavy quark pairs $Q\bar{Q}$ in the Fock state decomposition of the bound state wavefunction of ordinary mesons and baryons. Although proton states such as $|uudc\bar{c}\rangle$ and $|uudb\bar{b}\rangle$ are surely rare, the existence of hidden charm and other heavy quarks within the proton bound state will lead to a number of striking phenomenological consequences.

It is important to distinguish two types of contributions to the hadron quark and gluon distributions: _extrinsic_ and _intrinsic_. Extrinsic quarks and gluons are generated on a short time scale in association with a large transverse momentum reaction; their distributions can be derived from QCD bremsstrahlung and pair production processes and lead to standard QCD evolution. The intrinsic quarks and gluons exist over a time scale independent of any probe momentum, and are associated with the bound state hadron dynamics. In particular, we expect the presence of intrinsic heavy quarks, $c\bar{c}$, $b\bar{b}$, etc., within the proton state by virtue of gluon exchange and vacuum polarization graphs as illustrated in Fig. 7.

The "extrinsic" quarks and gluons correspond to the standard bremsstrahlung and $q\bar{q}$ pair production processes of perturbative QCD. These perturbative contributions yield wavefunctions with minimal power-law fall-off

7-81   4149A5

Fig. 6. One gluon exchange diagram responsible for spin-spin splitting of masses and the existence of higher Fock states containing an extra gluon.

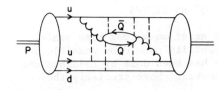

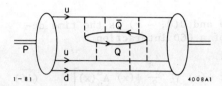

$$\left|\psi(k_{\perp i}, x_i)\right|^2 \sim \frac{1}{k_{\perp i}^2} \qquad (14)$$

Fig. 7. Diagrams which give rise to the intrinsic heavy quarks ($Q\bar{Q}$) within the proton. Curly and dashed lines represent transverse and longitudinal-scalar (instantaneous) gluons, respectively.

and lead to the logarithmic evolution of the structure functions. In contrast, the intrinsic contributions to the quark distribution are associated with the bound state dynamics and necessarily have a faster fall-off in $k_{\perp i}$ ($\psi \sim 1/k_i^2$ or faster[12]). The intrinsic states thus contribute to the initial quark and gluon distributions. A simple illustration of extrinsic and intrinsic $|uudq\bar{q}\rangle$ contributions to the deep inelastic structure functions is shown in Fig. 8a and b. We see that existence of gluon exchange graphs, plus vacuum polarization insertions, automatically yield an intrinsic $|uudq\bar{q}\rangle$ Fock state.

A complete calculation must take into account the binding of the gluon and $q\bar{q}$ constituents inside the hadron (see Fig. 7) so that the analysis is necessarily non-perturbative.

We also note that the normalization of the $|uudq\bar{q}\rangle$ state is not necessarily tied to the normalization of the $|uudg\rangle$ components since the latter only refer to transversely polarized gluons; Fig. 7 shows that $q\bar{q}$-pairs also arise from the longitudinal-scalar (instantaneous) part of the vector potential.

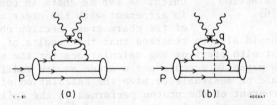

Fig. 8. a) Example with contribution to the deep inelastic structure functions from an _extrinsic_ quark q.
b) Example with contribution to the deep inelastic structure functions from an _intrinsic_ quark q.

5. INTRINSIC HEAVY QUARK STATES

The intrinsic heavy quark states exist on a long time scale. Hence, an estimate of the mixing probability should be possible in the static bag model.[31] Such a study has been done by Donoghue and Golowich[15] in the rest frame of the proton. More precisely they consider states

$$|p\rangle = U(0, -\infty)|p\rangle_0 \qquad (15)$$

where $|p\rangle_0$ is a proton 3-quark state and $U(0, -\infty)$ is the time development operator in the presence of a QCD interaction

$$U(0, -\infty) = T \exp\left[-ig \int_{-\infty}^{0} dt \int d^3x \; \bar{\psi}(x) \; \gamma^\mu \; \frac{\lambda^a}{2} \; \psi(x) \; A_\mu^a(x)\right] \qquad (16)$$

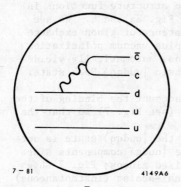

To second order in the coupling constant g, quark pairs are produced according to Eq. (16) via the mechanism in Fig. 9. The probability of a particular $q\bar{q}$-component in the proton is obtained by an overlap of the wavefunction (Eq. (15)) with itself. Using bag wavefunctions[31] as inputs in Eq. (16) and summing over the lowest states the authors of Ref. 15 obtain the result

$$P(|uudu\bar{u}\rangle):P(|uudd\bar{d}\rangle):$$
$$P(|uuds\bar{s}\rangle):P(|uudc\bar{c}\rangle) \qquad (17)$$
$$= 0.20:0.15:0.09:0:0.01$$

7 – 81   4149A6

Fig. 9. A $c\bar{c}$ pair produced by the action of a gluon through the interaction given in Eq. (16).

which, as far as charm is concerned, is in agreement with the order of magnitude of the charm cross section observed at the ISR. It should also be remarked that the results of Eq. (17) are still consistent with previous bag calculations for the static quantities like magnetic moments and average square radii. For our purposes it would be desirable to have the calculation of the intrinsic charm content of the proton performed in the infinite momentum frame. This is presently being investigated.[32]

We now proceed to discuss the c-quark momentum distribution in a $|uudc\bar{c}\rangle$ state. The general form of a Fock state wavefunction is

$$\psi(k_{\perp i}, x_i) = \frac{\Gamma(k_{\perp i}, x_i)}{M^2 - \sum_{i=1}^{n}\left(\frac{m^2 + k_\perp^2}{x}\right)_i} \qquad (18)$$

where Γ is the truncated wavefunction or vertex function. The actual form of Γ must be obtained from the non-perturbative theory, but following Ref. 30 it is reasonable to take Γ as a decreasing function of the off-energy-shell variable

$$\mathscr{E} = M^2 - \sum_{i=1}^{n} \left(\frac{m^2 + k_\perp^2}{x} \right)_i .$$

Independent of the form $\Gamma(\mathscr{E})$, we can read off some general features of the quark distributions:

(1) In the limit of zero binding energy ψ becomes singular and the fractional momentum distributions peak at the values $x_i = m_i/M$. More generally, $\mathscr{E}$ is minimal and the longitudinal momentum distributions are maximal when the constituents with the largest transverse mass $m_\perp = \sqrt{m^2 + k_\perp^2}$ have the largest light-cone fraction x_i. This is equivalent to the statement that constituents in a moving bound state tend to have the same rapidity.

(2) The intrinsic transverse momentum of each quark in a Fock state generally increases with the quark mass. In the case of power law wavefunction $\psi \sim (\mathscr{E})^{-\beta}$ we have $\langle k_\perp^2 \rangle \propto m_Q^2$; for an exponential wavefunction $\psi \sim e^{-\beta \mathscr{E}^{1/2}}$, the dependence is $\langle k_\perp^2 \rangle \propto m_Q$.

In the limit of large $k_\perp$ one can use the operator product expansion near the light cone (or equivalently gluon exchange diagrams) to prove that, modulo logarithms, the Fock state wavefunctions fall off as inverse powers of $k_\perp^2$.[12] For our purpose, which is to illustrate the characteristic shape of the Fock states containing heavy quarks, we will choose a simple power-law form for the Fock state longitudinal momentum distributions

$$P_{(n)}(x_1 \ldots \ldots x_n) = N_{(n)} \frac{\delta\left(1 - \sum_{i=1}^{n} x_i\right)}{\left(M^2 - \sum_{i=1}^{n} \frac{\hat{m}_i^2}{x_i}\right)^2} \tag{19}$$

where the $\hat{m}_i^2$ are identified now as effective transverse masses $\hat{m}_i^2 = m_i^2 + \langle k_\perp^2 \rangle_i$ and the $\langle k_\perp^2 \rangle$ are average transverse momentum. With this choice, single-quark distributions have power-law fall-offs $(1 - x)^2$ and $(1 - x)^3$ for mesons and baryons, respectively.

For example, consider a $|\bar{q}Q\rangle$ state, e.g., a D-meson. Here the momentum distributions of the 2 quarks are according to Eq. (19) given by

$$P(x_1, x_2) = N \frac{\delta(1 - x_1 - x_2)}{\left(m_D^2 - \frac{\hat{m}_c^2}{x_1} - \frac{\hat{m}_u^2}{x_2} \right)^2} \qquad (20)$$

From this expression we obtain the charmed quark distribution

$$P(x_1) = \int_0^{1-x_1} P(x_1, x_2) dx_2 = N' \frac{1}{\left(1 - \frac{1}{x_1} - \frac{\varepsilon}{1-x_1} \right)^2} \qquad (21)$$

where $N' = N/m_D^4$, $\varepsilon = \hat{m}_u^2/m_D^2$ and we take $m_D^2 \approx \hat{m}_c^2$. We see from Fig. 10 that the c-quark tends to carry most of the D-meson momentum ($\langle x_1 \rangle = 0.73$). This leading feature of the c-quark is due to the fact that the quarks should have roughly the same velocity in order for the hadron to "stay together." This can be seen more explicitly by minimizing the off-shellness, i.e., the denominator in Eq. (20)

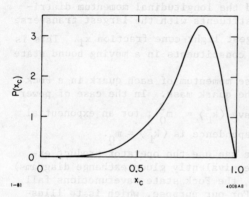

1-81   4008A8

Fig. 10. The x distribution of the charmed quark in a D-meson.

$$\frac{\hat{m}_c^2}{x_1^2} = \frac{\hat{m}_u^2}{x_2^2} \qquad (22)$$

keeping the transverse masses fixed. (A related idea has previously been considered by Bjorken and Suzuki[33] in the context of charm fragmentation into hadrons.)

We now turn to the discussion of $|uudQ\bar{Q}\rangle$ and $|u\bar{d}Q\bar{Q}\rangle$ states. For a $|uudc\bar{c}\rangle$ proton Fock state the momentum distribution is given by

$$P(x_1, \ldots, x_5) = N \frac{\delta\left(1 - \sum_{i=1}^{5} x_i\right)}{\left(m_p^2 - \sum_{i=1}^{5} \frac{\hat{m}_i^2}{x_i} \right)^2} \cdot \qquad (23)$$

In the limit of heavy quarks $\hat{m}_4^2 = \hat{m}_5^2 = \hat{m}_6^2 \gg m_p^2$, $\hat{m}_i^2$ ($i = 1, 2, 3$) we get

872

$$P(x_1, \ldots, x_5) = N_5 \frac{x_4^2 x_5^2}{(x_4 + x_5)^2} \, \delta\!\left(1 - \sum_{i=1}^{5} x_i\right) \tag{24}$$

where $N_5 = 3600\, P_5$ is determined from $\int dx_1 \ldots dx_5 P(x_1, \ldots, x_5) = P_5$, where P_5 is the $|uudc\bar{c}\rangle$ Fock state probability. Integrating over the light quarks (x_1, x_2 and x_3) we get the charmed quark distributions

$$P(x_4, x_5) = \frac{1}{2} N_5 \frac{x_4^2 x_5^2}{(x_4 + x_5)^2} (1 - x_4 - x_5)^2 \quad . \tag{25}$$

By performing one more integration we obtain the charmed quark distribution

$$P(x_5) = \frac{1}{2} N_5 x_5^2 \left[\frac{1}{3}(1 - x_5)(1 + 10 x_5 + x_5^2) - 2 x_5 (1 - x_5) \log \frac{1}{x_5} \right] \tag{26}$$

which has average $\langle x_5 \rangle = 2/7$ and is shown in Fig. 11. This is to be contrasted with the corresponding light quark distribution derived from Eq. (13) and shown in Fig. 12

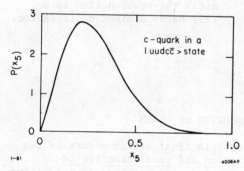

Fig. 11. The x distribution of the charmed quark in a $|uudc\bar{c}\rangle$ state.

$$P(x_1) = 6(1 - x_1)^5 P_5 \quad . \tag{27}$$

The corresponding c- and u-quark distributions in a $|ud c\bar{c}\rangle$ are obtained in the same way.[14] In order to see the contribution of the intrinsic $c\bar{c}$-pairs to the proton structure function we use the value for $P_5 = 0.01$ from the bag model calculations discussed above. The magnitude of the charm cross section at ISR (0.1–0.5 mb)[1] gives for P_5:

$$P_5 = \frac{\sigma_{\Lambda_c}}{2\sigma_{inel}} \approx \frac{250 \text{ μb}}{2.30 \text{ mb}} = 0.004 \tag{28}$$

If the production mechanism is <u>inelastic</u> and

$$P_5 = \frac{\sigma_{\Lambda_c}}{2\sigma_{diff}} \approx \frac{250 \text{ μb}}{2.10 \text{ mb}} = 0.01 \tag{29}$$

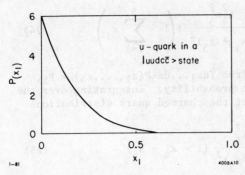

Fig. 12. The x distribution of a light quark in a $|uudc\bar{c}\rangle$ state.

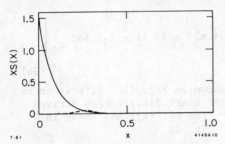

Fig. 13. Comparison of the intrinsic charm sea $xc(x)$ (dashed line) with the total sea at $Q^2 = 5$ GeV^2 as parametrized by Ref. 34.

if it is diffractive. These two possibilities will be discussed in the next section. We conclude that the charm cross section at ISR is compatible with $P_5 = 0.01$.

The charmed quark distribution $c(x) = P(x_5)$ should be measurable in lepto-production for high enough Q^2 and $W^2 > W^2_{th} = 25$ GeV^2. Hence to measure $c(x)$ at, e.g., $x = 0.5$ requires $Q^2 = 25$ GeV^2 ($x = Q^2/(Q^2 + W^2)$). We emphasize that the intrinsic charm sea $c(x)$ is "rare" but not "wee" as is clear from Fig. 13. A discussion on comparing $c(x)$ with lepto-production data is found in Sect. 6. In order to obtain intrinsic u, d and s-distributions ($|uudu\bar{u}\rangle$ states, etc.) the wavefunction in Eq. (19) needs a minor modification.

5. HADRONIC PRODUCTION OF CHARM

Hadronic production of multiparticle final states occurs in two different ways, diffractive dissociation and nondiffractive inelastic production. Although at least one experiment on $\Lambda_c^{\pm}$-production has an explicit diffractive trigger, the situation for charm production is far from settled. We will discuss the two production mechanisms below in the light of intrinsic charm.

A. Diffractive Dissociation

Diffractive production of high M^2-states can be interpreted as a short distance phenomenon due to the large masses involved. Thus perturbative QCD should be applicable to some extent. This idea was first considered in Ref. 35 in the context of charm production. Recently these questions have been studied in more detail for high mass diffraction in general in terms of so-called "transparent states."[36-39] The idea is simple and appealing: when the valence

874

quarks of a hadron are close together the net color extension is almost zero and the hadron does not interact with other hadrons. Hence the absorptive cross section is small and the hadron scatters diffractively off the target which then appears to be transparent. This situation is, as pointed out by Ref. 37, very similar to an analogous process in QED: when e^+e^--pairs are produced in very high energy emulsion experiments, they can only be separated by distances smaller than atomic sizes. The e^+e^- has net charge zero—it is not "seen" by surrounding atoms and hence it does not ionize and give rise to visible tracks. In Ref. 39 the authors explore the knowledge of the pion wavefunction in QCD at short distances in this context and derive interesting results for the jets emerging from the "transparent" target.

As was discussed in connection with Eq. (18) one expects intrinsic heavy quark states to have large $\langle p_\perp \rangle$ and consequently small transverse dimension. It is therefore tempting to assume that the intrinsic heavy quark states scatter diffractively. With that assumption the authors of Ref. 39 obtain in the case of 1% intrinsic charm on a nuclear target

$$\sigma_{charm}^{diff} = 0.01 \cdot \sigma_{el} \approx 0.5 \text{ mb} \cdot A^{2/3} \tag{30}$$

This high value is encouraging as far as production of b- and t-quarks are concerned. A diffractive production mechanism of heavy quarks is also very favorable as far as the combinatorial background is concerned.

For the charm case the $\Lambda_{\bar{c}}$ and D-spectra can be calculated in principle from the strong overlap between the 5-quark and the charmed-hadron state wavefunctions, allowing for decays of excited state, etc. For the purpose of obtaining the x_F-distributions we shall use a simple recombination mechanism for the quarks involved in the states. Neglecting its binding energy, the Λ_c spectrum is given by combining the u, d and c-quark in $|uudc\bar{c}\rangle$ to obtain

$$P(x_{\Lambda_c}) = N_5 \int_0^1 \prod_{i=1}^5 dx_i \, \delta(x_{\Lambda_c} - x_2 - x_3 - x_4) \left(\frac{x_4 x_5}{x_4 + x_5}\right)^2 \delta\left(1 - \sum_{i=1}^5 x_i\right)$$

$$\tag{31a}$$

(see Fig. 14) with $\langle x_{\Lambda_c} \rangle = 1/7 + 1/7 + 2/7 = 4/7$. The ISR data for $d\sigma/dx$ (pp $\to \Lambda_c X$) (see Fig. 4) are consistent with the prediction from Eq. (31) that charmed baryons are produced in the forward fragmentation region, although the existing data are too scarce for a detailed comparison. We expect that the low x region for charm production will be filled in by both perturbative and higher Fock state intrinsic contributions. The corresponding distribution for $D^-(c\bar{d})$ is given by

$$P(x_{D^-}) = N_5 \int_0^1 \prod_{i=1}^5 dx_i \delta(x_{D^-} - x_3 - x_5) \left(\frac{x_4 x_5}{x_4 + x_5}\right)^2 \delta\left(1 - \sum_{i=1}^5 x_i\right)$$

(31b)

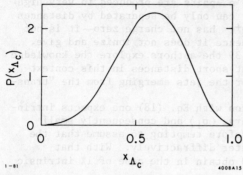

Fig. 14. The x distribution of the Λ_c^+ from the intrinsic charm component of the proton.

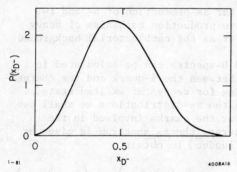

Fig. 15. The x distribution of the D^- from the intrinsic charm component of the proton.

with $\langle x_{D^-}\rangle = 1/7 + 2/7 = 3/7$, and is shown in Fig. 15. The $D^+(c\bar{d})$ distribution would, in principle, be obtained from the $|uudc\bar{c}d\bar{d}\rangle$ Fock state of the proton, where the $d\bar{d}$ could be extrinsic or intrinsic. Assuming that the $\bar{d}$ momentum is small, the D^+ distribution should be close to that of the c-quark shown in Fig. 11. These predictions apply for forward production ($x_F \gtrsim 0.1$), where perturbative contributions and higher Fock state contributions can be neglected. Spectra for pion induced reactions are obtained in the same way.[14]

In addition to charmed mesons and baryons, the J/ψ may also be produced diffractively from the intrinsic charm component of the proton. Compared to the charm production cross section at FNAL energies[27]

$$\sigma(\pi N \to DX) \simeq 20 \ \mu b \ , \qquad (32)$$

J/ψ production data around 200 GeV give[40]

$$\sigma(\pi N \to \psi X) \simeq 100 \ nb \ .$$

Further, the observed x_F-distribution appears to be more strongly peaked near $x \simeq 0$ compared to what would be expected from the intrinsic charm distribution. Evidently most of the ψ production comes from other central production mechanisms such as gluon and $q\bar{q}$ fusion.[19] In order for the intrinsic charm model to be consistent, there must be a large suppression factor for the ψ production from the intrinsic charm compared to the D production

$$\left.\frac{\sigma(\pi N \to \psi X)}{\sigma(\pi N \to DX)}\right|_{\text{intrinsic charm}} \lesssim 5 \times 10^{-5} \ . \qquad (33)$$

In fact, there are a number of factors which act to suppress the production of forward ψ from the intrinsic charm

(1) In the decay of the $|uudc\bar{c}\rangle$ state, the probability that the $\bar{c}$ quark combines to form a $c\bar{c}$ system is about 1/4 (<u>flavor suppression</u>). Similarly, the flavor suppression factor for the $|udc\bar{c}\rangle$ state is about 1/2.

(2) A $c\bar{c}$ system can be formed in either a color octet $c\bar{c}$ or singlet $c\bar{c}$ state. The color octet $c\bar{c}$ state should interact with other colored particles and is most likely to decay into open charm particles such as D's. Therefore, we can take only the color singlet combination of $c\bar{c}$ for ψ production. This occurs only 1/9 of the time (<u>color suppression</u>).

(3) If the color singlet $c\bar{c}$ system has a mass larger than the $D\bar{D}$ threshold, it will decay strongly into charmed particles rather than ψ production. Therefore, we have to require that the invariant mass $M_{c\bar{c}}$ is below the $D\bar{D}$ threshold (<u>mass suppression</u>)

$$2m_c < M_{c\bar{c}} < 2m_D \ . \tag{34}$$

In Ref. 14 the $M_{c\bar{c}}^2$-distribution (see Fig. 16) was calculated from Eq. (25). From that distribution one obtains

$$\int_{4\hat{m}_c^2}^{4m_D^2} dM_{c\bar{c}}^2 \ \frac{dP}{dM_{c\bar{c}}^2} \simeq 10^{-2} \ . \tag{35}$$

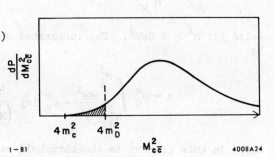

(4) Even if the $c\bar{c}$ system is below $D\bar{D}$ threshold, it may be realized as χ, η_c and ψ' states which do not decay into ψ's. We estimate this suppression factor as 1/3 (<u>channel suppression</u>). If we combine the factors in (1)– (4) we obtain the very rough theoretical estimate

Fig. 16. The $c\bar{c}$ mass spectrum in the intrinsic charm state $|uudc\bar{c}\rangle$. The shaded area corresponds to the χ, η_c, ψ and ψ' production.

$$\left. \frac{\sigma(\pi N \to \psi X)}{\sigma(\pi N \to DX)} \right|_{\text{intrinsic charm}} \simeq 5 \times 10^{-5} \ . \tag{36}$$

Despite these uncertainties, it is clear that although the intrinsic charm model does predict ψ production in the forward fragmentation region, the rate is at a very suppressed level.

B. <u>Inelastic Nondiffractive Production</u>

In Ref. 41 a perturbative analysis was carried out using the graphs of Fig. 2c. For the charm quark distribution $c(x)$ the authors use essentially that from the intrinsic charm.[47] As is seen from Fig. 17, the final charm x-spectra get contributions from both the spectator c and the one participating in the Fig. 2c subprocess.

We end this section by discussing the energy dependence for heavy quark production.

For perturbative heavy quark production mechanisms,[25] the energy dependence of the cross section essentially comes from the lower limit $m_Q/(2\sqrt{s})$ of convolution integrals, and gives rise to a logarithmic energy dependence. To study the energy dependence of the "diffraction" mechanism with "intrinsic" heavy quarks we will use the empirical formula for high mass diffraction[48]

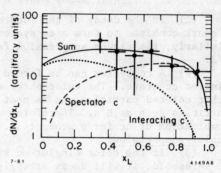

Fig. 17. Longitudinal momentum distributions from Ref. 41 for charm in pp → c$\bar{c}$X based on the diagrams of Fig. 2c. The input charm distribution is essentially that of the intrinsic charm.

$$\frac{d\sigma}{dM^2} = \sigma_0 \frac{1}{M^2} \qquad (37)$$

valid for $M^2 \gtrsim 2 \text{ GeV}^2$. The integrated charm cross section is given by

$$\sigma = \sigma_0^c \int_{M_0^2}^{M_1^2} \frac{dM^2}{M^2} = \sigma_0^c \log \frac{(1 - x_1)s}{\left(M_{\Lambda_c} + M_D\right)^2}, \qquad (38)$$

where in this case M_0^2 is the threshold value for associated production of a pair of hadrons containing charmed quarks. The upper limit M_1^2 is determined from the kinematical relation $M_1^2 = s(1 - x_1)$ where x_1 is the lower fractional momentum cut on the recoiling proton. In the ISR pp → $p_1\Lambda_c$X experiment[2] one triggers on events with $x_1 \geq 0.8$. If we assume that essentially all the charm cross section $\sigma_c \sim 300$ μb is due to diffractive production, then we can determine $\sigma_0 = 77$ μb. From this we predict that at SPS and FNAL energies ($s \cong 400$-600 GeV^2), the total pp → charm cross section should be of the order of 150 μb. Clearly this prediction is larger than present experimental data at SPS/FNAL with both pion and proton beams.[27] The energy dependence thus seems to be stronger than what is implied by Eq. (38).

Concerning production of heavy quarks on nuclear targets one expects an $A^{2/3}$-dependence from the intrinsic charm model. This is in contrast to the perturbative hard scattering cross section, which should be proportional to A.

878

As far as the production of b- and t-quarks are concerned, one can argue on general grounds that the probability of a hadron to contain an intrinsic heavy quark pair should fall as

$$P_{Q\bar{Q}} \propto \frac{\alpha_s^2}{R^2 m_Q^2} \tag{39}$$

where R is a hadron size parameter. Using the same (1-x)-cut as in Eq. (38) and m_t = 20 GeV one obtains the cross sections for b- and t-quark production as shown in Table I.

Table I. Cross section for b- and t-production at ISR and Tevatron energies from Eq. (38) and (39). The numbers in parentheses are the conventional perturbative QCD-predictions.

	ISR ($\sqrt{s}$ = 63 GeV)	Tevatron ($\sqrt{s}$ = 2000 GeV)
b	15 μb (0.5)	70 μb (2)
t (m_t = 20 GeV)	0	3 μb (0.1)

7. THE INTRINSIC CHARM AND LEPTOPRODUCTION EXPERIMENTS

As is clear from Fig. 13, the intrinsic charm sea is very small compared to the total sea. However, it should be visible in experiments explicitly looking for leptoproduction of charm. This is the case in dimuon production (Fig. 18a)

$$\mu^{\pm}N \rightarrow \mu^{\pm}\mu^{\pm}X \tag{40}$$

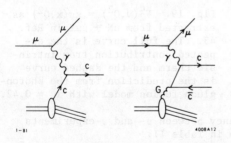

Fig. 18. Lepto-production of charm from the intrinsic charm sea and via the proton-gluon fusion model, respectively.

where one of the final state muons originates from charm decay. However, there exists a competing process with similar experimental signature, photon-gluon fusion[42] (Fig. 18b). The two processes should be additive. Analyses of reaction (40) is somewhat model dependent as far as the charm fragmentation function, $D_c^D(z,k_\perp^2)$ is concerned. The EMC-collaboration has extracted the charm structure function F_2^c from the data[43] and the result is shown in Fig. 19. Unfortunately the data only extends to x_{Bj} = 0.24 so it does not provide a crucial test for intrinsic charm on a 1% level but it is clear from Fig. 19 that the intrinsic charm saturates the highest Q^2 data point. However, at high Q^2 one expects the QCD-evolution to have changed the initial $c(x)$-distribution. In fact it turns out that

$$\langle x(200) \rangle_c = \int_0^1 xc(x, Q^2 = 200 \text{ GeV}^2)dx \approx 0.20 \qquad (41a)$$

as compared to

$$\langle x(Q_0^2) \rangle_c = \int_0^1 xc(x, Q_0^2)dx = \frac{2}{7} \approx 0.29 \qquad (41b)$$

using $Q_0^2 = (1 + 1.5^2)$ GeV with $\Lambda = 0.15$-0.4 GeV.

Although the intrinsic charm contribution to the total structure function F_2 is small globally, it is substantial at large x ($\sim$10% for $x \approx 0.5$). D. P. Roy[16] has used this feature to account for the anomalously small scale breaking ($\Lambda = 0.1$) observed in μN-data without affecting the SLAC-MIT or CDHS results ($\Lambda = 0.3$-0.4 GeV). The idea is the following: The value $\Lambda = 0.4$ GeV corresponds to a 30% decrease of $F_2^{\mu N}$ at large x and $Q^2 = 20$-200 GeV^2. Correspondingly $\Lambda = 0.1$ GeV represents a 20% decrease. Since the charm threshold occurs in μN-reactions for $Q^2 \gtrsim 20$ GeV, the intrinsic charm ($\sim$10% for x = 0.5) increases $F_2^{\mu N}$ in this region thereby lowering the apparent value of Λ. (In the $F_2^{\mu N}$-case this effect is very small since (1) only half of the charm quarks are excited, and (2) the rise occurred already at $Q^2 < 20$ GeV^2). The amount of intrinsic charm required to account for the discrepancy between μ- and ν-experiments for different values of Λ is shown in Table II.

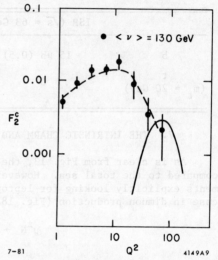

Fig. 19. $F_2^c(\nu, Q^2) = xc(x, Q^2)$ as extracted from $\mu N \rightarrow \mu\mu X$ in Ref. 43. The full curve is the expected contribution from intrinsic charm and the dashed curve is the prediction from the photon-gluon fusion model with $\alpha_s = 0.42$.

A very interesting implication of intrinsic charm for νN and $\bar{\nu}$N charge current reactions is the production of beauty quarks ($\bar{\nu}c \rightarrow \mu^+b$ and $\nu\bar{c} \rightarrow \mu^-\bar{b}$).[14] The subsequent leptonic decay of the b and $\bar{b}$ then leads to same-sign muon pairs (see Fig. 20). The experimentally observed rate of same-sign muon pairs is unexpectedly high, although the different experiments do not all agree.[44] The c $\rightarrow$ b process

880

Table II. QCD induced decrease of $F_2^{\mu N}$ $(x = .5, Q^2)$ over $Q^2 = 20 \to 200$, for various values of Λ. Also shown are the corresponding magnitudes of the intrinsic charm component, required to reproduce a net decrease of 20%, as observed by EMC.

Λ(GeV)	.10	.17	.25	.32	.4
$\dfrac{F_2(20) - F_2(200)}{F_2(20)}$	19.6%	22.4%	25%	27%	29.4%
Required size of the intrinsic charm component	0%	0.5%	1%	1.5%	2%

works in the right direction, but with present limits on the standard left-handed c-b coupling the theoretical prediction from intrinsic charm is below some experimental data. However, in the context of topless models right-handed couplings, $(c,b)_R$, have been suggested which increases the same sign dimuon production from the intrinsic charm.[46]

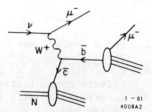

Fig. 20. Same sign dimuon pair production from the intrinsic charm component of nucleons.

8. CONCLUSIONS

We conclude with the following remarks:

- Perturbative QCD with conventional inputs works well for hidden heavy quark production (e.g., ψ, Υ). The dominant subprocess is gluon-gluon fusion.
- For open heavy quark production the predictions from perturbative QCD are in conflict with, in particular, data on $pp \to \Lambda_c^+ X$.
- It is found that higher Fock $c\bar{c}$-states in the proton on the 1% level, as suggested by bag model calculations, gives a natural explanation of the open charm production data.
- These intrinsic charm states, $|uudc\bar{c}\rangle$, are believed to have small transverse extension. Hence they could materialize diffractively in the context of "transparent states."
- Since the intrinsic heavy quark states scale with m_Q^2 one expects nonnegligible cross sections for hadronic production of b- and t-quarks. Furthermore, in diffractive configurations the combinatorial background is less serious than the central collision processes.

Much more theoretical and experimental work is needed, in particular:

- Better understanding of higher Fock states in general; perform refined calculation in the bag model in infinite momentum frame.
- Is charm produced diffractively or not?
- Measurement of $c(x)$ at large x_{Bj} in leptoproduction experiments.
- More experimental study of charm production in the FNAL/SPS energy region.

ACKNOWLEDGEMENT

Part of this work was done in collaboration with S. J. Brodsky, P. Hoyer and N. Sakai. I would like to thank the organizers of this meeting for their kind hospitality. I have also benefitted from conversations with M. Barnhill and C. Heusch.

REFERENCES

1. For a review, see S. Wojcicki, SLAC-PUB-2603; A. Kernan, Proc. Photon-Lepton Conf. at Fermilab, 1979, p. 535.
2. K. L. Giboni et al., Phys. Lett. 85B, 437 (1979) and A. Kernan, private communication.
3. D. Drijard et al., Phys. Lett. 81B, 250 (1979); W. Geist, Proc. of the Topical Workshop on Forward Production of High-Mass Flavors at Collider Energies, College de France, Paris, 1979.
4. W. Lockman et al., Phys. Lett. 85B, 443 (1979).
5. D. Drijard et al., Phys. Lett. 85B, 452 (1979).
6. F. Muller, in High Energy Physics-1980, proceedings of the XXth Int. Conf. on High Energy Physics, Madison, Wisconsin, edited by L. Durand and L. G. Pondrom (AIP, New York), 1981.
7. A. Chilingarov et al., Phys. Lett. 83B, 136 (1979).
8. M. Basile et al., Nuovo Cimento Lett. 30, 481 (1981); ibid 30, 487 (1981).
9. For a review, see R. Phillips, in High Energy Physics-1980, proceedings of the XXth International Conference on High Energy Physics, Madison Wisconsin, edited by L. Durand and L. G. Pondrom (AIP, New York), 1981.
10. T.B.W. Kirk et al., Phys. Rev. Lett. 42, 619 (1979).
11. The states are defined at equal $\tau = t + z$ in the light-cone gauge $A^+ = A^0 + A^3 = 0$.
12. S. J. Brodsky and G. P. Lepage, Phys. Rev. D22, 2157 (1980) and S. J. Brodsky, Y. Frishman, G. P. Lepage, and C. Sachrajda, Phys. Lett. 91B, 239 (1980), and references therein.
13. S. J. Brodsky, P. Hoyer, C. Peterson, and N. Sakai, Phys. Lett. 93B, 451 (1980); P. Hoyer, in High Energy Physics-1980, proceedings of the XXth International Conference, Madison, Wisconsin, edited by L. Durand and L. G. Pondrom (AIP, New York), 1981
14. S. J. Brodsky, C. Peterson and N. Sakai, Phys. Rev. D23, 2745 (1981).

15. J. F. Donoghue and E. Golowich, Phys. Rev. $\underline{D15}$, 3421 (1977).

16. D. P. Roy, Tata Institute Preprint, Phys. Rev.Lett.$\underline{47}$, 213 (1981).

17. For a more detailed discussion, see C. Peterson, proceedings of the Topical Workshop on Forward Production at High-Mass Flavors at Collider Energies, College de France, Paris (1979).

18. See R. Hagedorn, Report No. CERN 71-12.

19. See, e.g., M. B. Einhorn and S. D. Ellis, Phys Rev. $\underline{D12}$, 2007 (1975); C. E. Carlson and R. Suaya, ibid. $\underline{18}$, 760 (1978).

20. See, e.g., Ref. 9.

21. See Raja's talk at this conference.

22. B. C. Brown et al., FERMILAB-77/54-EXP.

23. M. J. Corder et al., Phys. Lett. $\underline{68B}$, 96 (1977).

24. L. Camilleri, proceedings of The 1979 International Symposium on Interactions of Leptons and Photons at High Energies (Fermilab).

25. H. Fritzsch, Phys. Lett. $\underline{67B}$, 217 (1977). F. Halzen, Phys. Lett. $\underline{96B}$, 105 (1977). L. M. Jones and H. W. Wyld, Phys. Rev. $\underline{D17}$, 759, 1782, 2332 (1978). M. L. Gluck and E. Reya, Phys. Lett. $\underline{79B}$, 453 (1978), $\underline{83B}$, 98 (1979). M. Gluck, J. F. Owens, and E. Reya, Phys. Rev. $\underline{D17}$, 2324 (1978). J. Babcock, D. Sivers and S. Wolfram, Phys. Rev. $\underline{D18}$, 162 (1978). C. E. Carlson and R. Suaya, Phys. Rev. $\underline{D18}$, 760 (1978); Phys. Lett. $\underline{81B}$, 329 (1979); H. Georgi et al., Ann. Phys. $\underline{114}$, 273 (1978); K. Hagiwara and T. Yoshino, Phys. Lett. $\underline{80B}$, 282 (1979); J. H. Kuhn, Phys. Lett. $\underline{89B}$, 385 (1980); J. H. Kuhn and R. Ruckl, MPI-PAE/pTH 7/80; V. Barger, W. Y. Keung and R.J.N. Phillips, Phys. Lett. $\underline{91B}$, 253, $\underline{92B}$, 179 (1980); Z. Phys. $\underline{C}$, to be published; Y. Afek, C. Leroy and B. Margolis, Phys. Rev. $\underline{D22}$, 86, 93 (1980); M. Gluck, E. Hoffmann and E. Reya, DO-TH 80/13.

26. B. L. Combridge, Nucl Phys. $\underline{B151}$, 429 (1979).

27. R. Ruchti's talk at this conference.

28. L. J. Koester, in High Energy Physics-1980, proceedings of the XXth International Conference, Madison, Wisconsin, edited by L. Durand and L. G. Pondrom (AIP, New York, 1981); D. E. Bender, Ph.D. thesis, 2980, University of Illinois (unpublished); J. Cooper, proceedings of the XVth Recontre de Moriond, 1981 (unpublished).

29. H. Fritzsch and P. Minkowski, Phys. Lett. $\underline{90B}$, 455 (1980).

30. S. J. Brodsky, T. Huang and G. P. Lepage, SLAC-PUB-2540; T. Huang, in High Energy Physics-1980, proceedings of the XXth International Conference, Madison, Wisconsin, edited by L. Durand and L. G. Pondrom (AIP, New York, 1981).

31. A. Chodos, R. Jaffe, K. Johnson, C. Thorn and V. Weisskopf, Phys. Rev. $\underline{D9}$, 3471 (1974); T. de Grand, R. Jaffe, K. Johnson and J. Kiskis, Phys. Rev. $\underline{D12}$, 2060 (1975).

32. M. Barnhill, private communication.

33. M. Suzuki, Phys. Lett. $\underline{71B}$, 139 (1977); J. D. Bjorken, Phys. Rev. $\underline{D17}$, 171 (1978).

34. A. J. Buras and K.J.F. Gaemers, Nucl. Phys. $\underline{B132}$, 249 (1978).

35. G. Gustafson and C. Peterson, Phys. Lett. $\underline{67B}$, 81 (1977).

36. J. F. Gunion and D. E. Soper, Phys. Rev. $\underline{D15}$, 2617 (1977).

37. J. Pumplin and E. Lehman, Zeitschrift für Physik $\underline{C9}$, 25 (1981).

38. G. Gustafson, LUTP 81-1, talk given at the "IXth International Winter Meeting on Fundamental Physics," Siguenza, Spain, February 1981.

39. G. Bertsch, S. J. Brodsky, A. S. Goldhaber and J. F. Gunion, SLAC-PUB-2748; see also the talk by J. F. Gunion at this conference.

40. J. Badier et al., Proc. Lepton-Photon Conf. at Fermilab, 1979, p. 161; CERN/EP 79-61.

41. V. Barger, F. Halzen and W. Y. Keung, University of Wisconsin preprint, DEO-ER/00881-211.

42. J. P. Leveille and T. Weiler, Nucl. Phys. B147, 147 (1979).

43. C. H. Best, Proc. of XVIth Rencontre de Moriond, 1981.

44. T. Y. Ling, proceedings of ν'81 International Conference on Neutrino Physics and Astro Physics.

45. M. Basile et al., Nuovo Cimento Lett. 31, 97 (1981).

46. V. Barger, W. Y. Keung and R.J.N. Phillips, Phys. Rev. D24, 244 (1981).

47. Note however that in ref. 41 it is hypothesized that the hard $c(x)$-distribution is of pertubative origin and not intrinsic. This is unrealistic since the time-scale involved at $Q^2 \approx m_c^2$ is too short to allow for the produced c-quarks to equalize the velocity with the valence quarks thereby providing a hard $c(x)$-spectrum. (From deep inelastic ep-experiments it is known that long time-scale bound state effects are not important already at $Q^2 \approx 1$ GeV^2).

48. M.G. Albrow et al., Nucl. Phys. B108, 1 (1976).

DISCUSSION

DREMIN, LEBEDEV PHYSICAL INSTITUTE, MOSCOW: Can you predict anything for higher energies? I mean for very high energies . . . 10 TeV or 100 TeV. I mean the total cross sections of charm and bottom production.

PETERSON: As I said, then you need to know what mechanism it is. If it is really a diffractive mechanism and if $1/M^2$ is right (which we observe but our tendency again says $1/M^4$), then I can, of course, use this $1/M^2$ formula to extrapolate both to higher energies and to other flavors. So here you see, for example, what I would expect (and I have a kinematic limit in the upper integral of this integration reflecting a diffractive prejudice) and you see here for bottom production I get 15 microbarns at ISR; at the Tevatron, 70 [microbarns]. For the top production I would get nothing at ISR using this math, and I get 3 microbarns at the Tevatron. And these numbers turn out to be slightly higher than the one from perturbative QCD.

PIETRZYK, CERN: You mentioned that the J/Ψ production is described by QCD. It's a problem of language; it's difficult to say it's described. You have plenty of graphs [describing direct J/Ψ production] and you can play with them and whatever will happen you will not be very much surprised. Raja will tell you in the next talk that nearly 100% of the cross section for J/Ψ production is through decay from other charm particles. I suppose you will be not be surprised and shocked. If it would turn out (which is possible) that J/Ψ is produced in another way, then you would also not be extremely surprised and shocked.

PETERSON: No, but I am encouraged to believe that perturbative QCD, to a large extent, with normal ingredients is responsible for J/Ψ production.

PIETRZYK: But you have large freedom, a factor of five or ten.

PETERSON: Yes, a "fudge factor".

PIETRZYK: This is what I wanted to say.

PETERSON: This is the fact that the widths are not completely known for these χ states.

HWA, OREGON: Could you please explain again why you did not

think that the picture of three clusters, with charm in one of the clusters, would work?

PETERSON: No, I think I said that that is presumably the most appealing picture. The idea of intrinsic charm is no doubt triggered by the observation of Λ_c at large x. I haven't said it's not appealing to start off with 3 quarks and then develop.

HWA: It might work. Do you agree? If you start with the ideas that there could be charm inside a valon and that . . . .

PETERSON: No, then you will never get charm inside the valons.

HWA: Why not? That is what I want to know. Because [each] valon itself has, say, one-third the [proton] momentum on the average. And then the charm [i.e., $c\bar{c}$], being massive, will have most of the momentum of the valon when it is excited. And that [c with roughly half the valon momentum], recombined with the u and d from the other valons will give you very fast [Λ_c].

PETERSON: Okay.

χ meson production in 225 GEV/c π^-p interations[*]

Rajendran Raja
Fermi National Accelerator Laboratory
P.O. Box 500
Batavia, Illinois 60510

ABSTRACT

This paper reports on the preliminary results from E610, which was completed in the Muon Laboratory at Fermilab in August 1980. The experiment used 225 GeV/c negative pions incident on a Beryllium target. By triggering on two muons of opposite charge, we were able to accept a sample of events containing an enhanced number of J/ψ particle decays. Using a lead glass detector it was possible to study the accompanying photons resulting from the decay $\chi \to J/\psi + \gamma$.

Introduction

Due to C-invariance of strong interactions, the J/ψ can couple only to 3 gluons whereas the even C-parity χ states can couple to two gluons. In QCD the hadronic width of the χ is thus calculated to be much larger than the J/ψ. It has therefore been suggested that the χ is produced much more copiously than the J/ψ in hadronic interactions[1] and that a significant fraction of hadronic J/ψ production is due to the electromagnetic decay of $\chi \to J/\psi + \gamma$. Since the branching ratio of the $0^{++} \to J/\psi + \gamma$ is much smaller than the other two,[2], we expect the 1^{++} and the 2^{++} χs to be the dominant source of J/ψ particles.

If the gluon fusion mechanism[1] is dominant then we expect the 2^{++} χ to be produced much more than the $1^{++} \chi$ since the latter does not couple to two zero mass spin 1 gluons due to Yang's theorem. However if quark-antiquark fusion is dominant then we expect[3] the 1^{++} to be more dominant than the $2^{++} \chi$.

Recent results from CERN indicate[4] that both these mechanisms are important at our energies.

---

[*]This is an experiment performed by the Fermilab-Illinois-Pennsylvania-Purdue-Tufts collaboration.

Experimental Set Up

The experiment was performed in the Muon Laboratory in June 1980. Negative pions with momenta 225 GeV/c were incident on a beryllium target placed upstream of the reconstituted Chiago Cyclotron Spectrometer (Figure 1). The momenta of the charged particles were measured by bending them in the large aperture Chicago Cyclotron magnet and tracking them through a set of proportional chambers in the magnet gap and a set of drift chambers downstream of the magnet. Forward - going photons were

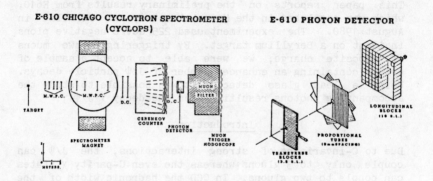

Fig. 1-Experimental set up of the Chicago Cyclotron Magnet Spectrometer.

detected in a lead-glass calorimeter which consisted of three sections: (i) an active converter layer of transverse lead-glass blocks followed by (ii) three stereo layers of proportional tubes to determine the position of the photon, followed by (iii) an array of longitudinal lead-glass blocks to measure the rest of the shower energy. This arrangement allows a position resolution of 0.3 cm and an energy resolution (σ/E) of $6\%/(E)^{1/2}$ for the photons. The experiment trigger was designed to select events in which two oppositely charged muons were produced, by requiring a hit in two diagonally opposite quadrants of the muon hodoscope.

Experimental Results

Dimuon and diphoton mass resolution capabilities of the spectrometer are shown below. The $\mu^+\mu^-$ effective-mass spectrum (Figure 2) shows a clear peak at the J/ψ mass, with FWHM (full width at half maximum) of 150 MeV. The $\gamma\gamma$ invariant mass for those events that have just two photons in the lead glass (Figure 3) similarly shows a clear π^0 peak with FWHM of 30 MeV.

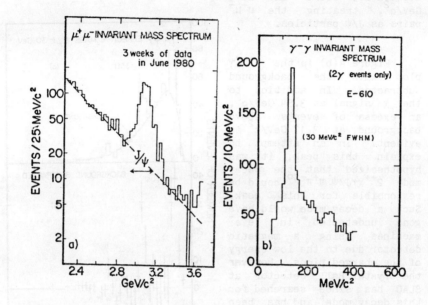

Fig. 2. $\mu^+\mu^-$ effective mass spectrum showing the J/ψ peak.

Fig. 3. $\gamma\gamma$ effective mass spectrum in events having only two γ's showing the π^0 peak.

As previously mentioned, χ states are observed through their decay to J/ψ + γ. Figure 4(a) show the ($\mu^+\mu^-\gamma$) effective mass for γ energies between 1 GeV and 30 GeV, and $\mu^+\mu^-$ pairs with mass between 2.95 and 3.25 GeV/c^2 (the $\mu^+\mu^-$ pairs were constrained to have the J/ψ mass for the plot). To remove photons from π^0 decays, photons from $\gamma\gamma$ pairs with an effective mass between 0.1 GeV/c^2 and 0.2 GeV/c^2 were excluded. A clear peak centered at 3.54 GeV/c^2 is evident.

We interpret this as being due to a mixture of the 1^{++} χ (3510) and the 2^{++} χ (3555) which we are as yet unable to resolve. The 0^{++} χ (3410) is absent from the plot. To estimate the background under the χ, we have used events with $\mu^+\mu^-$ pairs with effective masses between 2.5 GeV/c^2 and 2.75 GeV/c^2, treating the $\mu^+\mu^-$ pairs as J/ψ particles.

Figure 4(b) is the $\mu^+\mu^-\gamma$ plot with the background subtracted. In addition to the χ signal at 3.54 GeV/c^2, an excess of events above background at 3.17 GeV/c^2 is evident. In an attempt to explain this peak, it was hypothesized[5] that the decay mode $2^{++}\chi\rightarrow$J/ψ $\pi^+\pi^-\pi^0$ could be responsible for this peak. Such a decay mode would have gone undetected in e$^+$e$^-$ machines using a magnetic detector due to the low energy of the charged pions. However the Crystal Ball detector at SLAC has since searched for this decay mode[6] and has been able to set an upper limit of 5% for the branching ratio. Also both theoretical and other experimental studies since have been able to demonstrate that the decay $\chi\rightarrow$J/ψ $\pi^+\pi^-\pi^0$ does not occur a significant fraction of the time[7]. We are as yet unable to draw a background that explains the $\psi\gamma$ spectrum in the spectrum in the low mass range.

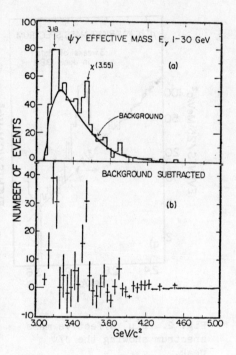

Fig. 4(a) $\psi\gamma$ effective mass spectrum for γ momenta betweem 1 GeV/c and 30 GeV/c. The curve is our best esti- mate of the background. Fig. 4(b) Background sub- tracted.

Conclusion

The 60 events in the $\chi(3.54)$ peak imply that $35\pm10\%$ of the J/ψ's are produced indirectly by the process $\chi\to$J/$\psi+\gamma$. This value agrees quite well with the latest results from the experiment WA11 at CERN.

References

1. Carlson and Suaya, Phys. Rev. D15 (1077) 1416.

2. Particle data group, Rev. Mod. Phys. 52 (1980).

3. J.H. Kuhn, Physics Letters 89B (1979) 385.

4. Y. Lomaigne et al., paper submitted to the XX International Conference, Madison, Wisconsin 1980.

5. R. Raja, Fermilab-Pub-81/32-EXP

6. R. Partridge, Private Communication.

7. C. Flory and I. Hinchliffe, Berkley Report LBL-12703 (May 1981), E.L. Berger and C. Sorensen, ANL-HEP-PR-81-28 (July 1981), R. Barate et al, Paper submitted to the Lisbon Conference (1981).

DISCUSSION

OWENS, FLORIDA STATE: As regards fusion model calculations
for Ψ production there are basically two points of view that one
can take. The first that you have outlined, in a mechanism which
is totally indirect, would involve, order by order in perturbation
theory, involving at the perturbative level the exact constraints
imposed by color. Namely, that you have only color singlets
produced. Another point of view is that you can produce a massive
$c\bar{c}$ pair which can either be a color singlet or a color octet, and
that you calculate that production cross section perturbatively,
that there are then final-state [color] rearrangements which are
inherently soft by assumption, and that out of that $c\bar{c}$ system
come Ψ's. Now this is a very average way of calculating the Ψ's.
In fact you never get the normalization using that technique. But
there is nothing inherently wrong with it. Now at this point you
could just sit back and have a debate, but it turns out that there is
a way that you can actually measure which of these is correct.
This is something that Dennis Duke and I came up with in regards
to photoproduction. If you do Ψ photoproduction with a linearly
polarized photon beam, then you can ask the following question.
If I impose, at the perturbative level, that the $c\bar{c}$ system is
always produced as a color singlet what do I get for the polarization? It turns out that you get zero. The reason is that there are
certain terms which give a dominant polarization effect but which
are ruled out because the $c\bar{c}$ system is produced as an octet.
Then you turn the question around and you ask what happens if I
allow the $c\bar{c}$ system to be produced at the perturbative level in
either a singlet or color octet state? What kind of polarization
will I get? And it turns out to be of the order of 50%. It's an
effect which is large at high energy; the cross sections are large;
it's eminently measurable as soon as they build a linearly
polarized photon beam at Fermilab.

RAJA: Yes, that's going to be done. I agree with you that you
cannot rule that model in or out because its normalization is not
fixed and you can put in any amount of gluon structure function that
you can produce and predict the decays at the right level. However,
there are questions about angular distributions of the χ's which
are produced which go into Ψ + γ. If I can show you the latest
predictions . . . . If you assume that the color singlet model is
correct and there are two ways of producing the χ, either through
$q\bar{q}$ fusion or through gluon-gluon fusion, then you get the following
two angle distributions. [Transparency shown.] The green is the
angle distribution of the photons from χ decay, and the blue is

the lepton pairs from $\chi \rightarrow \Psi + \gamma$ . . . the blue is averaging over the photon direction. So clearly for 2^{++}, for example, if it's quark fusion that is responsible you get a $(1 - (1/3) \cos^2 \theta)$. If gluon fusion is responsible you get $(1 + \cos^2 \theta)$. And if either of these is observed you can have more confidence in the color singlet production level. Whereas, as far as I know, the model where you have color bleaching makes no predictions about angular distributions because you can bleach it with one gluon or you can bleach it with ten gluons. This gives a model which has very little predictive power and therefore you can neither rule it in nor rule it out. It's like Regge theory.

OWENS: But it's got great predictive power for linearly polarized protons. Asymmetries. But there is something I don't understand here. You're demanding that you have the $c\bar{c}$ system be in a color singlet and yet you are talking about quark fusion. That's ruled out at lowest order, because you have a $q\bar{q}$ going into a gluon so it's already a color octet.

RAJA: No, no. $q\bar{q}$ fusing with two gluon exchanges into $c\bar{c}$. That's been calculated by Kuhn. It clearly is not a single gluon exchange.

FRIDMAN, VANDERBILT/SACLAY: I have two small questions which I don't understand. If this gluon-gluon fusion mechanism is the [dominant] one which produces your state you should see the same production if you change the flavor of the incoming beam particle [i.e. the production should be independent of the flavor of the incoming particles].

RAJA: No, it depends on how many gluons there are in the pions as opposed to how many there are in the proton.

FRIDMAN: Okay, but if you take $\pi^+ p$, for instance, you should have the same cross section.

RAJA: Yes, you are right. If you do, pp is different from $p\bar{p}$ at low energies and that means at low energies only gluon-gluon fusion is responsible.

FRIDMAN: So you should see the same thing if you do the experiment with π^+. Now this will be a very good check. So now I have another question which I personally don't understand, which is the following. Why, let's say at the charm level, [does] the production mechanism proceed through gluon-gluon fusion (or you think it is so), [while] in Υ production we know this is not true. For the Υ

production we know that the gluon-gluon fusion is not the dominant mechanism as shown by the CERN . . . .

RAJA: I don't think that has been shown. At threshold you have $q\bar{q}$ production responsible. If you go to high enough energies the Υ production will also be dominated by gg . . . the gluon-gluon fusion. Those are the predictions.

FRIDMAN: You see, at CERN they show very clearly that you have the Υ production cross section increase if you have a valence antiquark in one of the particles.

RAJA: Yes, that is what I'm saying. Let's assume that Υ's also are produced in addition to the χ's . . . the Ψ's by χ production first and then decay. You can get χ production on the Υ also by $q\bar{q}$ fusion and gluon-gluon fusion. Gluon-gluon fusion will dominate as you go to higher and higher energies, for example, ISR. $q\bar{q}$ fusion is definitely important at low energies near threshold. So πp will have a significant fraction of that.

PIETRZYK, CERN: Maybe I will try to answer Fridman's question. You see, when one has an incoming beam of 200 GeV for pions, the center of mass energy is about 19 GeV. So if you produce J/Ψ you need about 3 GeV to produce [one]. If you produce Υ you need 9.6 GeV, and it is very difficult to produce Υ at 200 GeV because you need a lot of energy which, in general, [most] partons don't have. Only valence quarks have [this much energy]. It's a different situation for J/Ψ because gluons at much smaller x can do this, and there are plenty of gluons so you can produce J/Ψ easily through gluon-gluon fusion. But not Υ.

RAJA: Let me answer that a little bit more quantitatively. You have a χ production expression like this (which I didn't go into) and it depends on the τ, which is M_x^2/s, of the producing particle. Now for small τ, which means at high s or small M_x like in the J/Ψ, you will get production by gluons at small x. Where for larger τ, for Υ, valence quarks contribute because they are at larger x. But that contribution falls with [increasing] energy because the τ is going down.

PIETRZYK: So now I have my question. I wonder about the following problem in χ decay (I don't know if its a question to you or to theorists). In χ decay into $\Psi\pi^+\pi^-\pi^0$, phase space is very small. I suppose a clever theorist can tell very fast what this branching ratio between χ decay into $\gamma\Psi$ and χ decaying

$\Psi\pi^+\pi^-\pi^0$ is. Has anybody done it?

RAJA: Do you want me to answer that? Well, I went to the theory group at Fermilab, and I had this idea. I asked them whether they had any estimates for that. And they did not. But that does not mean that in some future day somebody else will not calculate it. It is very difficult to calculate hadron decay fractions in even potential models. But it is true that these decay modes have not been observed, but that may be due to the fact that they are very hard to observe. The η'_c is expected to be at slightly higher mass than the χ and therefore will have a little bit more Q-value to do that. But then it is restricted by an angular momentum barrier. So it all depends on the branching ratios.

RESVANIS, ATHENS: Could you explain to us again what you are doing mixing the dimuons and γ's from 10 events at a time [in estimating your background]? You did say something but you flashed the transparency very fast and I didn't understand it.

RAJA: What we are doing is we're trying to simulate the background. We take all the events [in] that effective mass range and we take sets of ten events at a time and then take combinatorials. We can't take all the events there [at one time], because there would be too many combinatorials. This wastes computer time and space. So we take 10 events at a time and then we take all the combinations, γ from one event and $\mu\mu$ from the other one. I think [it is] similar to what you did.

RESVANIS: Question #2. You said you had a novel photon detector. What is novel about it?

RAJA: As far as I know, maybe you have better ones at CERN. But I gave this talk at CERN and somebody came along to me afterwards and said "I see you guys are quite poor at Fermilab."

RESVANIS: We are working at Fermilab also so we know the story. We agree with you.

RAJA: What is novel about it is that we have these proportional tubes, three layers, which measure the position of the shower as well as its energy. The position accuracy is 3 mm, and that seems to be adequate for the job.

RESVANIS: Last question. If you compare your old result with your new result is [the difference] just statistics or is there a systematic [problem] also?

RAJA: I think that in the old experiment we has about 18 χ events so it is statistics.

MORRISON, CERN: This entire discussion reminds me very much of the problems of production of hidden strangeness or ϕ mesons. You've the same problem. I mean, there is the question of how the Zweig rule operates, how often [there is evidence this occurs frequently] you produce two ϕ's or additional $K\bar{K}$ pairs. The problem of color comes in again, and there are probably some extra gluons emitted. There is much better information, much better statistics on the ϕ. Have you considered the relation between the two?

RAJA: Yes, in fact I proposed that we also look at ϕ production and there is a graduate student who is working on it. We have a Cerenkov counter and we have KK pairs. So, if we have the ϕ's, we would also like to look at ϕ production by indirect means. That is, does it come from $\phi\gamma$ resonances? And the analogy of the χ in $\phi\gamma$ are the A2, and the f and f'. That has to be looked into.

HADRONIC CHARM PRODUCTION

R. C. Ruchti

University of Notre Dame, Notre Dame, Indiana 46556

ABSTRACT

Hadronic interactions are a potentially rich source of heavy particle systems such as charm. However the relatively small charm production cross section and small branching ratios for charm decay have made experimental study difficult. To date neither has the nature of the hadronic production mechanism been firmly established, nor has charm spectroscopy been challenged effectively. In spite of this adversity, a number of recent experiments have been designed to address these difficulties. A survey of some of these experiments is presented.

INTRODUCTION

Fundamental problems in both strong and weak interactions may be addressed through a study of heavy mass particles produced in hadronic reactions. Among the topics of current interest are:

Strong interactions: What is the nature of hadronic pair production - is it central, diffractive, or both? Does the observed spectrum of particles fit into an SU(n) generalization of SU(3)? How do heavy particles hadronize - how do the fundamental constituents, quarks and glue, combine to form heavy systems?

Weak interactions: What are the lifetimes of these particles; are these measurements consistent with decay branching ratios observed in e^+e^- annihilations? What are the Kobayashi-Maskawa mixing angles for heavy quark decay? Is there $D^0\bar{D}^0$ mixing, $B^0\bar{B}^0$ mixing?

Survey: Are there additional leptons or heavy quark systems beyond those currently observed?

This very beautiful experimental program is subject to small production levels:

Reaction	Inelastic cross section 400 GeV/c	Production fraction
$\sigma(H + N \to x)$	~ 30 mb	~ 1
$\sigma(H + N \to C_1 + \bar{C}_2 + x)$	$\gtrsim 20\ \mu b$	$\gtrsim 10^{-3}$
$\sigma(H + N \to B_1 + \bar{B}_2 + x)$	$\lesssim 10$ nb	$\sim 10^{-6}$

(Where H is an incident hadron, and N represents a target nucleon.)

The key to studying heavy flavor production is effective background suppression. A well conceived trigger is essential in this regard, as well as the ability to reconstruct the event topology as completely as possible.

1) Trigger schemes. A number of specialized triggers have been used in experiments to select heavy particle states.

   a) dilepton triggers: $\mu^{\pm}\mu^{\mp}$, $e^{\pm}e^{\mp}$, $\mu^{\pm}e^{\mp}$): Of the three topologies, the μe events are the most restrictive since only weak decays can contribute to a $\mu^{\pm}e^{\mp}$ signal. On the other hand, electromagnetic decays will contribute to the first two topologies. (The decay $\rho \rightarrow \mu^{+}\mu^{-}$ is but one of numerous, annoying examples.) The dilepton trigger schemes are sensitive to the total cross section for charm production since the decay of any charm particle will contribute to the measured rate. The disadvantage is that it is not possible to determine charm particle masses by this technique.

   b) single lepton triggers: ($\mu^{\pm}$ or $e^{\pm}$): If properly executed these triggers allow for a reasonably unbiased study of charm production and decay channels. Since the decay of one state produces a lepton for the trigger, no assumption (at the trigger level) has been made about the decay of the associated state. Of the two lepton triggers, the $\mu^{\pm}$ trigger is the cleanest provided care is taken to minimize the decay length for hadrons. An $e^{\pm}$ trigger is complex because sources of electrons are plentiful (γ conversions in which one of the electrons escapes detection are a prime offender).

   c) topology triggers: These triggers isolate a particular decay mode which has unique kinematical properties: the decay $D^{*} \rightarrow D\pi$ is one of these. The very small Q ($\lesssim 5$ MeV) available constrains the pion to the same direction as the D, and the pion momentum to $p \simeq m_{\pi}p_{D}/m_{D}$. With such trigger schemes, one can study a specific process very accurately.

   d) diffractive triggers: These attempt to isolate larger x production of heavy systems by selecting events which have a slow recoil proton from a target, or by looking for leading particles ($\mu^{\pm}$, $e^{\pm}$, p or $K^{\pm}$) in the final state. In principle these trigger schemes allow access to numerous decay topologies.

2) Observations of Decays-in-Flight. This technique exploits the substantial lifetimes expected for charm states by looking directly for decays-in-flight. The method has been used successfully in the case of neutrino induced reactions in emulsion targets. Use of emulsion targets in hadronic beams is also possible, but requires low beam fluxes. Higher interaction rates can be tolerated if one uses small bubble chambers, streamer chambers, or scintillation crystal targets. The present lifetime range

accessible with these devices is $\lesssim 10^{-13}$ sec (emulsion $\lesssim 10^{-14}$ sec),
with improvements anticipated in the next several years.
3) Combined Trigger and Decays-in-Flight Measurements: This
technique combines the power of both of the above techniques.
Clearly the best way to study the charm system, it may be the
only way to study hadronic beauty production.

There are a number of important experiments recently
completed or currently in progress which employ a number of
the above techniques[1]. Among them are:

I. CERN
    (a) ISR experiments: observation of Λ_c, D^o at large x_F,
                     with large production cross sections.
    (b) NA11 - study of charm production using single electron
            trigger.
    (c) NA16 - study of charm production using LEBC and the
            European Hybrid Spectrometer EHS.

II. Fermilab
    (a) E-515 - study of charm production using a prompt muon
            trigger/passive and active targets.
    (b) E-595 - a beam dump experiment looking for single
            muons and dimuons.
    (c) E-567 - search for the reaction $D^* \rightarrow D\pi$ in the strong
            interactions.
    (d) E-369 - search for charm by triggering on a slow
            recoil proton in conjunction with a forward muon.
    (e) E-490/630 - search for charm decays-in-flight using
                a streamer chamber and muon trigger.

The ISR experiments have received extensive converage in
the literature, and I will not discuss them further except to note
that the quoted cross sections for charm production are very
large: $\sigma \gtrsim 300$ μb/nucleon for $\sqrt{s} \simeq 63$ GeV. Also $d\sigma/dx_F$ is
quite flat, out to large values of x_F ($\gtrsim 0.7$).[1]
    The important experiments NA11, NA16 will not be
releasing results until the Lisbon Conference.
    Hence I will concentrate on the five listed Fermilab experi-
ments:

E-515:   Study of charm production using a prompt muon
              trigger. Northwestern/Carnegie Mellon/Notre Dame/
              Fermilab/Rutgers Collaboration.

This group is studying charm production in 200 GeV/c π^- nucleon interactions via the process:

$$\pi^- + Be \to C_1 + \bar{C}_2 + x$$
$$(200 \text{ GeV/c})$$

$C_1 \to e^-, \mu^-, K^+ \pi^- \ldots$

$\bar{C}_2 \to \mu^+$

The experimental strategy is to trigger on a prompt muon from the semileptonic decay of one of the charm particles (C_1 or $\bar{C}_2$), and to examine the decay of the associated state with open geometry and minimum bias.

The detector (Fig. 1) is a two arm spectrometer. The trigger muon is selected and identified in the upward arm ($\theta_{vert} \geq 40$ mr), consisting of a tungsten and steel absorber and polarized iron (the return yoke of the spectrometer magnet). The arm is instrumented with scintillation hodoscopes and proportional chambers to track the muon and measure its momentum. Particles are detected in this arm if they have $p_T \gtrsim 0.2$ GeV/c and $p \geq 4$ GeV/c (see Fig. 2). By measuring the deflection of the muon in its traversal through the polarized iron, the muon momentum is determined to $\delta p/p \sim 20\%$. Special attention has been paid to minimize the decay length for hadrons (π, K) in this arm. By careful design of the beam and by absorber placement right up to the target, the decay length is $\lesssim 20$ cm (see Fig. 3).

The forward arm of the detector is a conventional magnetic spectrometer consisting of proportional and drift chambers with ancillary detectors for particle identification. Particle momenta are determined in this arm to $\delta p/p \sim 1\%$. Particle identification is provided by a 46 cell, N_2 gas Čerenkov counter for π, K separation in the range $6 \lesssim p \lesssim 22$ GeV/c. Detection of electrons and photons is accomplished using a liquid argon shower detector comprised of 800 separate readout channels. The energy resolution of this device is $\sigma(E)/E \simeq 18\%/\sqrt{E}$. Muon identification is also provided in the forward arm by two hodoscopes placed behind a downstream hadron dump.

Typical acceptances for the apparatus are: in the upward arm, $\sim 5\%$ per muon from charm decay; in the forward arm, $\sim 10\%$ for a $K^- \pi^+$ decay of a D^o meson.

The trigger for the experiment consists of an incident beam particle without halo, plus at least one detected particle which penetrates through the length of the muon arm. The observed trigger rate is $\sim 4 \times 10^{-5}$ per interacting beam pion. By recording events with varying amount of muon-arm absorber removed, it was determined that $\sim 20\%$ of the triggers were due to prompt sources (electromagnetic and weak).

Of the event topologies under study in this experiment, the reaction most sensitive to a small charm production cross section is

$$\pi^- + N \rightarrow C_1 + \bar{C}_2 + x$$
$$\quad \hookrightarrow e^-, \mu^-$$
$$\quad \hookrightarrow \mu^+, e^+$$

in which one looks for $\mu^{\pm} e^{\mp}$ correlations. The muon is detected in the upward arm and the electron is detected in the forward-arm spectrometer and shower detector. The sensitivity to small cross sections is derived from the fact that semileptonic branching ratios for charm states are relatively large and that the experimental acceptance for electrons is good ($\gtrsim 25\%$). Operationally one compares the yield of observed $\mu^{\pm} e^{\mp}$ pairs to the yield of $\mu^{\pm} e^{\pm}$ pairs. (Charm contributes only to the $\mu^{\pm} e^{\mp}$ yield in the absence of $D\bar{D}$ mixing, whereas conventional backgrounds contribute to both topologies.) An excess of opposite-sign pairs over like-sign pairs is then indicative of charm production.

The yield of μe pairs (based on a raw data sample of $\sim 450\,K$ triggers taken during Spring 1980) is presented in Table I. Also shown are the μe yields corrected for charge asymmetry

Table I: μe Yields

Topology	Yield (raw)	Yield (corrected)
$\mu^+ e^-$	1483	1498
$\mu^- e^+$	1303	1511
$\mu^+ e^+$	1251	1251
$\mu^- e^-$	1045	1224

observed in the events due in part to beam/spectrometer misalignment and detection inefficiencies. The excess yield of corrected, opposite-sign pairs is obtained as follows:

$$\text{Excess Yield} = N(\mu^+ e^-) + N(\mu^- e^+) - N(\mu^+ e^+) - N(\mu^- e^-)$$

$$= 534 \pm 74 \text{ events.}$$

This excess is displayed as a function of the momentum of the electron from the μe pair in Fig. 4. The lack of events below 5 GeV/c is due to the acceptance cutoff of the shower detector.

For electron momenta below 10 GeV/c the Čerenkov counter was used as an additional constraint in the electron identification. Since the Čerenkov radiator was N_2 gas at atmospheric pressure, π/e separation could be performed unambiguously below ~ 6 GeV/c and on a statistical basis up to 10 GeV/c. This analysis indicated that 50% of electrons identified in the shower detector were due to hadron "feedthrough". Since the energy resolution of the shower detector improves as the electron energy increases, the 50% "feedthrough" factor should be considered an upperbound over the extended momentum range of 0-40 GeV/c. Thus the yield of opposite sign μe pairs is:

$$267 \pm 52 \text{ events} \leq \text{excess yield} < 534 \pm 74 \text{ events}$$

which translates into a production cross section of

$$16 \, \mu b \leq \sigma_{c\bar{c}} < 32 \, \mu b/\text{nucleon}.$$

Although a detailed model calculation has not yet been performed, the μe data are strongly suggestive of central production for charm. At the present level of analysis, no statistically significant peaks are observed in mass plots relevant to weak hadronic decay modes of charm particles. However the group recently recorded 2×10^6 triggers (a factor of 5 increase in statistics) during Spring 1981. The analysis of this new data is in progress.

Also in Spring 1981 the group tested an active target in conjunction with the spectrometer. The target device (Fig. 5) was a triggerable scintillation camera system developed by D. Potter. Operationally the scintillation light produced in the NaI crystal target is optically focused onto a four-stage imate intensifier whose second stage is quiescently off. When the experimental prompt-muon trigger is satisfied, the second stage is gated on, and the stored image is transferred to film. The dimensions of the NaI target are 1.5 mm thick x 1.8 cm diameter, and the camera system looks into the 1.8 cm face.

Sample events are shown in Fig. 6. The dot size is ~ 30 μm and the dot density is typically 3 - 4 per mm of track length for minimum ionizing particles. The spatial resolution in the plane of best focus is of order 5 μ. The device can operate at substantial interaction rates, $10^4 - 10^5$/sec, and should allow the experimenters to study lifetimes of order $\tau \gtrsim 10^{-13}$ sec. An extensive run of the active target/spectrometer system is planned for early 1982.

E-595: Rochester/Cal Tech/Chicago/Fermilab/Stanford Collaboration

This experiment[2] looks for prompt single muon production in the reaction:

$$p \;+\; Fe \rightarrow charm + x$$
$$(350 \text{ GeV}/c) \qquad\;\; \big\downarrow$$
$$\qquad\qquad\qquad \rightarrow \mu^{\pm}$$

where the interaction occurs in an iron dump. The detector (Fig. 7) consists of a target calorimeter comprised of 49 steel plates (2.4 m of steel) and instrumented with scintillation counters, followed by a forward muon range detector and toroidal magnet spectrometer. The density of the first 1.7 m of the target calorimeter could be varied, allowing data to be taken with three different effective densities (1:2/3:1/2). The density extrapolation technique was then employed to extract the prompt muon signal. Data were taken in two triggering configurations: one required that the produced muon have momentum $p_{min}(\mu) > 8$ GeV/c, the other $p_{min}(\mu) > 20$ GeV/c. The data reported here are from the 20 GeV/c minimum-momentum trigger. Each event had to satisfy the following criteria: a beamline PWC requirement of one and only one incoming hadron within 2% of beam momentum; a hadronic interaction in the upstream 25 cm of the target calorimeter; a requirement that muons originate in the target calorimeter; and that the trigger muon traverse the entire toroid system.

Single muon and dimuon yields are shown in Fig. 8 as a function of inverse density. The data indicate that the prompt $\mu^{\pm}$ rates are equal. After correction for efficiency, the prompt rates for $p_{min}(\mu) > 20$ GeV/c are: $(12.2 \pm 3.8) \times 10^{-6} \mu^{+}$/interacting proton and $(10.1 \pm 2.6) \times 10^{-6} \mu^{-}$/interacting proton.

The prompt single muon distributions as a function of $p(\mu)$ are shown in Fig. 9 and have been compared to two different models of $D\overline{D}$ production, each of which adequately describes the data:

Model A: Assumes D's are produced independently, and are parametrized as:

$$E\frac{d^3\sigma}{dp^3} = (1-x)^\alpha e^{-\beta p_T}$$

with $\alpha = 4.7 \pm 1.0$, $\beta = 2.5$

Model B: Pairs of charm particles with pair mass m are produced as:

$$E\frac{d^3\sigma}{dp^3} = \frac{1}{m^3}(1-x)^\alpha \exp(-\beta p_T - \gamma_m \sqrt{s})$$

with the composite system decaying into $D\bar{D}$.
Here $\alpha = 2 \pm 1.2$, $\beta = 2.5$, $\gamma = 15$.

Shown in Fig. 10 is the rate for producing single muons with momentum $p > p_{min}$, in which the data in Fig. 9 have been corrected for efficiency and the μ^+, μ^- rates have been combined. Models A and B give similar fits (the model A fit is shown). The intercept at $p_{min} = 0$ gives the total prompt single muon rate: $(1.9 \pm 0.4) \times 10^{-4}$ for model A, $(1.8 \pm 0.6) \times 10^{-4}$ for model B. Assuming an 8% average branching ratio and linear A-dependence, these rates correspond to charm production cross sections of 16 ± 4 μb/nucleon (model A) and 15 ± 5 μb/nucleon (model B).

The experimenters conclude that the single muon distribution is adequately described by a central $D\bar{D}$ production model and that the data do not indicate a large component of diffractive charm production of the magnitude reported by ISR experiments.

E567/650: Princeton/Saclay/Torino/BNL Collaboration

This experiment[3] looks specifically for D^* production via the reaction:

$$\pi^- + Be \rightarrow (D^*)^\pm + x$$
$$(200 \text{ GeV/c})$$
$$\downarrow D^0 \pi^\pm$$
$$\downarrow K^+ \pi^\pm$$

The motivation is to exploit both the kinematic constraints which result from the very small Q in $(D^*)^\pm \rightarrow D^0 \pi^\pm$ decay ($\lesssim 5.7$ MeV) and the large branching ratio for this D^* decay mode ($\sim 60\%$). The apparatus (Fig. 11) consists of a V-spectrometer for D^0 detection which includes Čerenkov counters for π, K separation in the range $5 < p < 18$ GeV/c and for p, K separation above 11 GeV/c. Pions of low momentum $1 < p_\pi < 2.5$ GeV/c are detected in two low momentum arms. Momenta were measured to $\delta p/p \sim 1\%$ in all arms.

The trigger requirement consisted of an opposite-sign pair detected in the V-spectrometer with $p > 5$ GeV/c in each arm, accompanied by a pion of momentum $1 < p < 2.5$ GeV/c and angle less than 50 mr. Additionally, only one arm of the V-spectrometer should have a signal in the threshold counter (particle identified as a pion) and the sign of the track in the low momentum arm should match that of the particle identified as a pion in the V-spectrometer. In Fig. 12 is shown the Q-value spectrum for events in the D^0 mass range $1.835 \leq M_{K\pi} \leq 1.875$ GeV/c^2. By fitting a Gaussian with $\sigma = 0.7$ MeV/c^2 above a smooth background (determined from $K\pi$ masses outside the D^0 region) the authors obtained a value of $Q = 5.9 \pm 0.3$ MeV/c^2. The number of D^* events determined by this method, correcting for the number of events outside the $M_{K\pi}$ mass cut, is 71 ± 26 events.

Fig. 13 shows the $K\pi$ invariant mass distribution for events within the Q-value peak shown in Fig. 12. Because the error in Q had large variations depending on the kinematics of particular event, the authors also defined a variable $R = (Q-Q_0)/\delta Q$ where $Q_0 = 5.8$ GeV/c^2 and δQ was calculated for each event. Cuts could therefore be applied on R rather than Q. The events in Fig. 13 have $|R| < 1.4$ or a Q value within $1.4\,\sigma$ of the expected value. The background form was determined by a fit to the $M_{K\pi}$ spectrum with no cut on R. After combining a fit to this background with a Gaussian of $\sigma = 14$ MeV/c^2, and correcting for events outside the cut on R, a signal of 86 ± 27 events was obtained, in agreement with the number obtained from the Q-value peak discussed above. The average of the two analyses, 78 ± 26 events, was then taken as the D^* signal.

Assuming a production cross section of the form $E\,d^3\sigma/dp^3 = A(1 - |x|)^3 \exp(-1.1\,p_T^2)$ and branching ratios for $D^{*+} \rightarrow D^0 \pi^+$ decay and $D^0 \rightarrow K^- \pi^+$ decay of 64% and 2.6% respectively, the total production cross section is calculated to be:

$$\sigma(D^*) = \frac{1}{2}[\sigma(D^{*+}) + \sigma(D^{*-})] = 4.2 \pm 1.4\ \mu b.$$

The corresponding differential cross section is $d\sigma/dy = 1.6 \pm 0.5\,(\pm 0.7)\ \mu b$, which is insensitive to the y-dependence assumed. The first error is statistical. The second is due to normalization uncertainty.

The result of this experiment is in accord with the level of charm production observed in the aforementioned two experiments.

E369: Fermilab/Harvard/Oxford/Illinois/Tufts Collaboration

This experiment looks for diffractive production of charm mesons in the following reaction:

$$\pi^- \; + \; p \to D^+, \, D^o + p$$
$$\text{(217 GeV/c)} \qquad \underset{\longrightarrow \; \mu}{\vphantom{L}}$$

using the Chicago Cyclotron Spectrometer and a liquid hydrogen target. [4] The trigger required that a recoil proton be detected between 60° and 75° relative to the beam and with momentum transfer in the interval $-0.1 \leq t \leq -0.4 \text{ GeV/c}^2$ in coincidence with a muon of $p(\mu) \geq 15 \text{ GeV/c}$ from the semileptonic decay of one of the D mesons. The trigger requirement on the recoil proton insured that the invariant mass of the forward produced system was in the range of twice the D mass. Shown in Fig. 14 is a superposition of all mass spectra $K^{\pm}\pi^+$, $K^{\pm}\pi^{\mp}\pi^{\mp}$. The statistical significance of the peak at the D mass is ~ 8 standard deviations. A peripheral production model yields a diffractive production cross section of $\sigma \sim 12 \; \mu b/\text{nucleon}$. This value is comparable to the central production cross sections observed in this energy range.

E630: Yale/Fermilab Collaboration

This group looks for decays-in-flight of charm particles produced by neutrons in the reaction

$$n + N \to \text{charm} + x$$
$$\text{($\sim$300 GeV/c)} \qquad \underset{\longrightarrow \; \mu, \; \mu\mu}{\vphantom{L}}$$

The apparatus, shown in Fig. 15, consists of a streamer chamber used as a target and vertex detector, followed by a muon spectrometer [5]. The trigger consists of an incident neutral particle, an interaction in the streamer chamber, one or more detected muons in the angular range $30 \text{ mr} < \theta < 120 \text{ mr}$, and no hit in the cone veto. The muon requirement includes muon tracking.

The gas used in the streamer chamber is $Ne/He/CO_2$ in the percentage $.873/.097/.03$ at 600 psi. A high voltage pulse of 150 - 160 kV and 1.3 ns width (FWHM) yields tracks of width 70 - 80 μm. In Fig. 16 sample events are shown. The visible region in the streamer chamber is 4 cm x 3 cm x 0.5 cm.

The group recorded 20 K pictures during Spring 1981 in which they claim 40 detected charm events. This first run represents ~2% of the total data expected when the experiment is completed.

SUMMARY

The charm cross section measured at Fermilab energies $20 \leq \sqrt{s} \leq 28$ GeV is consistent with a level of $\sigma_{c\bar{c}} \simeq 20 \,\mu$b. Diffractive production occurs at the level $\sigma \leq 10 \,\mu$b based on E369 results, and smaller still using the E595 result. These values are at least an order of magnitude smaller than the cross section levels reported from ISR ($\sqrt{s} \simeq 60$ GeV). This abrupt cross section change as a function of energy is different from what one observes for J/Ψ production.

My conclusion is that it is important to measure carefully the energy dependence of $\sigma_{c\bar{c}}$, and to study charm production using both pions and protons. The fixed-target Tevatron program will be important in this study since its coverage falls in the middle of this energy range. Finally, if these heavy mass systems are diffractively produced in a substantive way, it will benefit those looking for decays-in-flight of charm and beauty particles.

The author would like to thank J. Ritchie and R. Coleman (E595), V. Fitch (E567), R. Raja (E369), and J. Slaughter and J. Sandweiss (E630) for helpful discussions and for providing recent results from their experiments.

REFERENCES

1. For a comprehensive summary of charm cross sections observed at Fermilab, SPS, and ISR see S. Wojcicki, "New Flavor Production in γ, μ, ν, and Hadron Beams," High Energy Physics - 1980, XX International Conference, Madison, Wisconsin, L. Durand and L. Pondrom eds., p. 1431. Also see M. Basile et al., CERN-EP/81-22 (23 March 1981) and M. Aguilar-Benitez et al., CERN/EP 81-15 (15 June 1981). The last paper describes the latest LEBC/EHS results.
2. J. L. Ritchie, Prompt Single Muon Production by Protons in Iron, XVI Rencontre de Moriond (1981), and J. L. Ritchie et al., Phys. Rev. Lett. 44, 230 (1980).

3. V. L. Fitch et al., Phys. Rev. Lett. 46, 761 (1981).
4. L. J. Koester, "Diffractive Hadroproduction of Charmed
    D Mesons," High Energy Physics - 1980, XX International
    Conference, Madison, Wisconsin, L. Durand and L.
    Pondrom eds., p. 190, and R. Raja and T. Kirk, private
    communication.
5. This group previously used the streamer chamber and muon
    spectrometer in a beam of protons at 350 GeV/c (E490).
    Results may be found under J. Sandweiss et al., Phys. Rev.
    Lett. 44, 1104 (1980).

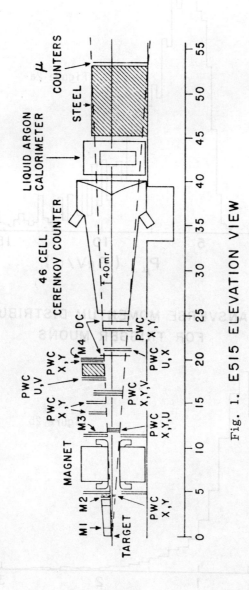

Fig. 1. E515 ELEVATION VIEW

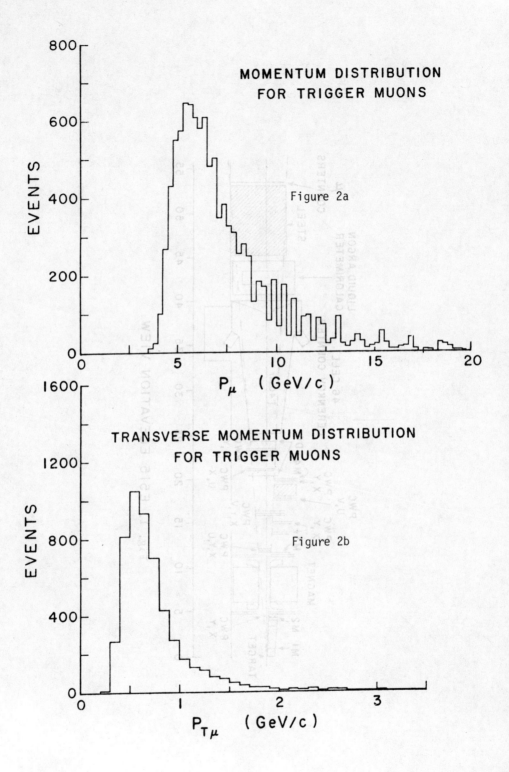

MOMENTUM DISTRIBUTION
FOR TRIGGER MUONS

Figure 2a

P_μ (GeV/c)

TRANSVERSE MOMENTUM DISTRIBUTION
FOR TRIGGER MUONS

Figure 2b

$P_{T\mu}$ (GeV/c)

910

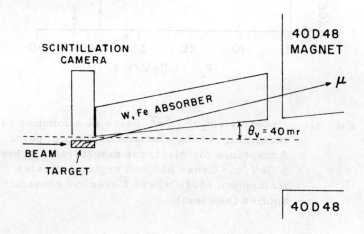

Fig. 3. SCHEMATIC OF THE TARGET REGION
(ELEVATION VIEW)

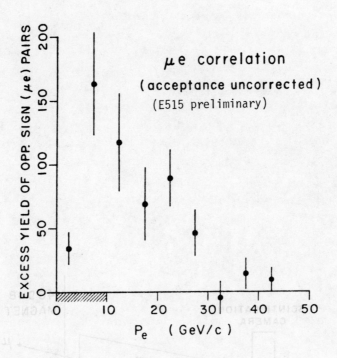

Fig. 4.   Excess yield of $\mu^{\pm}e^{\mp}$ pairs as a function of the momentum of the electron of the μe pair. Acceptance for electrons deteriorates below 5 GeV/c. Cross hatched region indicates momentum range where Cerenkov constraint applies (see text).

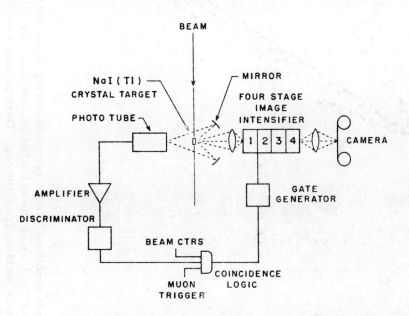

Fig. 5. SCINTILLATION CAMERA SCHEMATIC

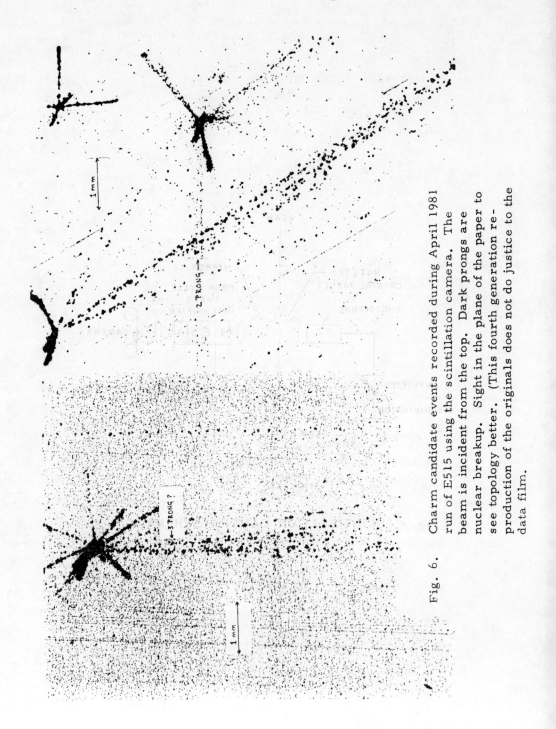

Fig. 6. Charm candidate events recorded during April 1981 run of E515 using the scintillation camera. The beam is incident from the top. Dark prongs are nuclear breakup. Sight in the plane of the paper to see topology better. (This fourth generation reproduction of the originals does not do justice to the data film.

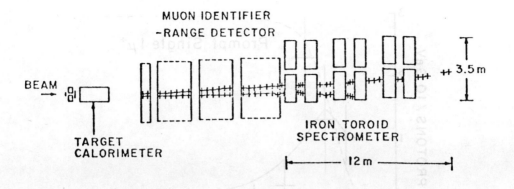

Fig. 7.   Plan view of the E595 apparatus.

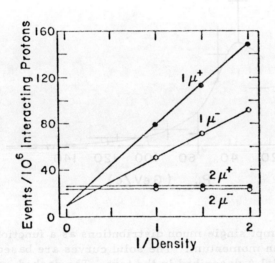

Fig. 8. Event rates versus 1/density.

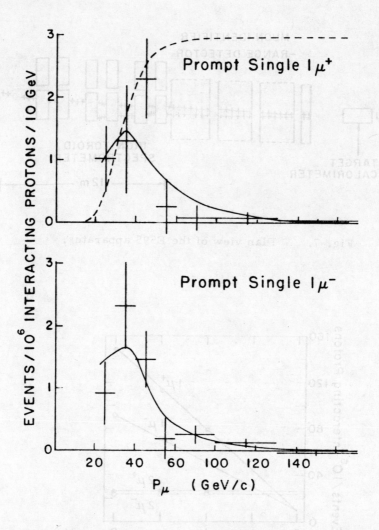

Fig. 9.   Prompt single muon distributions as a function of
          muon momentum. The solid curves are based on
          model A described in the text. The dashed curve
          is the spectrometer acceptance.

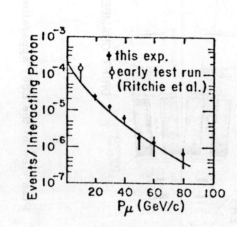

Fig. 10. Total prompt single μ rates (μ^+ and μ^-) with
p(μ) greater than $p_{min}(\mu)$

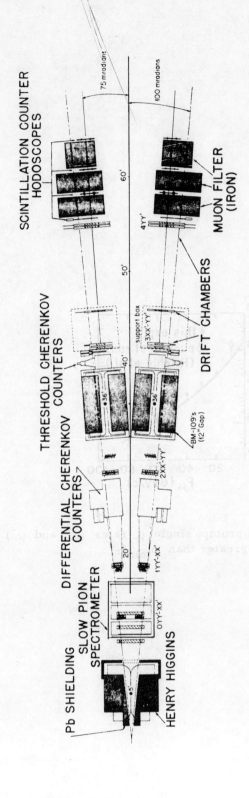

Fig. 11.   Plan view of the E567 apparatus.

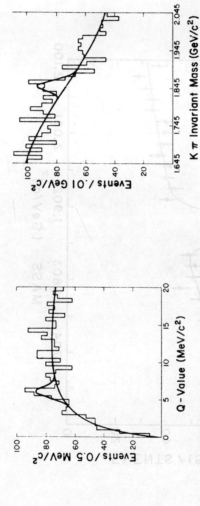

Fig. 12. The Q-value spectrum for events with 1.835 $< M_{K\pi} < 1.875$ GeV/c^2. Both $K^+\pi^-\pi^-$ and $K^-\pi^-\pi^+$ events are included.

Fig. 13. The $K^\pm\pi^\mp$ invariant-mass spectrum for events within the cut on Q value defined in the text.

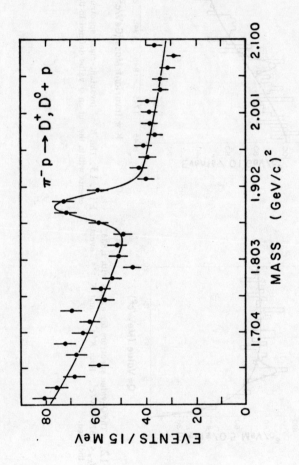

Fig. 14.   Invariant mass distribution combining $K^{\pm}\pi^{\mp}$ and $K^{\pm}\pi^{+}\pi^{+}$ channels from E369.

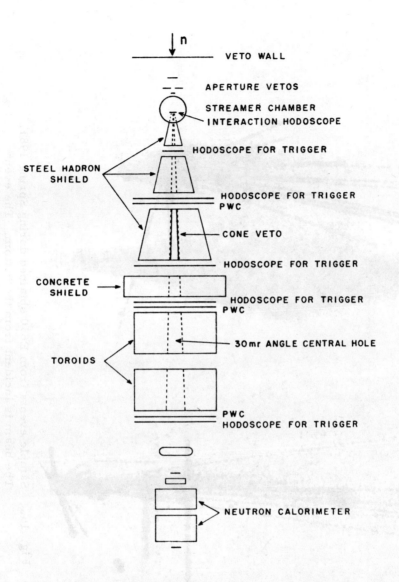

Fig. 15.　Plan view of the E630 apparatus.

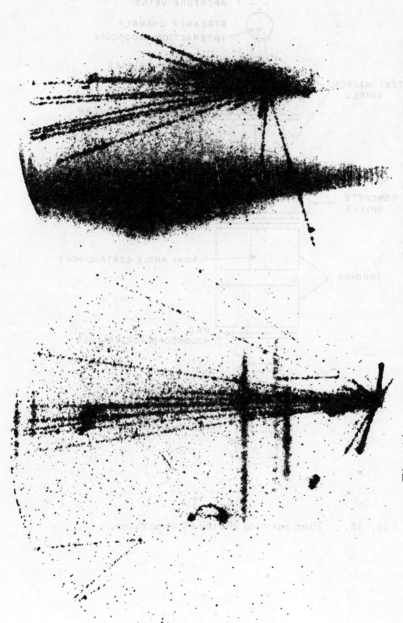

Fig. 16. Sample events from E630 obtained during Spring 1981. The beam is incident from the bottom. The event at the left has a strange particle Vee. The event at right has a kink in a forward prong.

DISCUSSION

TAVERNIER, BRUSSELS: I have one question and one comment. The question is about the experiment which I don't remember the number but I know it by the name of "the Sandweiss Experiment". Can you tell me what is now the size of the streamers and what is the scattering of those streamers around the average track?

RUCHTI: As best as I know it (and I can't find the slide so I can't tell you) the answer is that they appear to be 70 microns. The argument is that bascially they get the illumination from the streamer light. They have ideas of improving that by, in fact, using laser light and then the streamers don't have to form. You don't have to wait as long in order to scatter the light off. So that is one direction they are heading but at the moment, from what I gather from Jean Slaughter, it is 70 to 80 microns.

TAVERNIER: But is that the size of the streamer or the scattering of the streamer about the average value?

[answer from the floor]: That's the scattering.

TAVERNIER: Then I will just say a few words about the NA16 experiment which you haven't talked about. I will just mention that we exist more or less. No more. We have about 20 good candidates. Some of them are absolutely and unambiguously identified, with all decay tracks connected to the observed decay vertex. Some of them are not. We are working very hard but we couldn't get anything ready for this conference. It will be for the Lisbon Conference. So if you want to know more, come to Lisbon.

SESSION H

CLOSING SESSION

Friday, June 26, a.m.

Chairman: V. P. Kenney
Secretary: L. Dauwe

RECENT DEVELOPMENTS IN MULTIPARTICLE DYNAMICS

L. Van Hove
Theoretical Physics Division
CERN
1211 Geneva 23
Switzerland

CONTENTS

1. Introduction.
2. Quark combinatorics for multiplicities.
3. Inclusive distributions and flavour correlations :
   3.1 Fragmentation versus recombination ;
   3.2 Inclusive distribution of heavy flavours.
4. More advanced dynamical considerations.
5. Dense hadronic matter as quark-gluon plasma.
6. Concluding remarks.

1. INTRODUCTION

Rather than summarizing the many topics covered by the Symposium, this talk will concentrate on a selected group of developments and describe some trends which seem to me important for the future. To set the scene, I shall briefly recall the evolution which our topic, the study of multi-hadron processes, went through in the last fifteen years.

The two main changes, obviously related to each other, have been :
 i) the broadening of .the field, which was initially limited to hadron-hadron (h-h) collisions but then extended to cover also lepton-hadron (ℓ-h) collisions and e^+e^- annihilation into hadrons ($e^+e^- \to$ h's),
ii) the change of theoretical basis for the analysis of the phenomena, from a Regge-based description to a parton-based one (the partons being quarks and gluons), which itself is now in the process of becoming a QCD-based description (QCD = quantum chromodynamics).
The present meeting has beautifully illustrated the power and interest of the comparisons between hadron- and lepton-induced processes which are naturally suggested by the parton-based description. It has also illustrated that a comprehensive QCD-based description of multi-hadron phenomena, whether hard or soft, hadron-induced or lepton-induced, cannot escape the necessity to deal with the soft, non-perturbative properties of QCD.

Of course, while the field evolved in this way, certain features of general validity persisted. Diffraction dissociation is one of them, on which interesting new considerations have been presented (see Section 4 below). Another one concerns the space-time development of collisions, in particular impact-parameter and time-dilatation effects, which are very directly relevant in a parton-based description.

It has been known for a long time that many properties of high energy h-h collisions are accounted for by leading particle effects and longitudinal phase space (LPS, i.e., phase space with an exponential cut-off in the transverse momenta p_T of the secondaries). To probe the dynamics one had to go beyond these properties, through the study of non-trivial correlations and flavour dependences. We are confronted today with an extended version of the same situation. Starting with the work of Basile et al. [1], it has become customary to compare various distributions of the hadrons produced in h-h, ℓ-h and $e^+e^- \to$ h's processes. Many striking similarities have been found (the subtraction of leading particles practiced by Basile et al. does not appear to be very important, but diffractive h-h processes must be subtracted, see W. Kittel's report and other contributions to this conference). Is this so-called "jet universality" more significant than the LPS universality of past years ? The answer is probably no, and the most forceful argument, stressed by T. DeGrand, is to note that charm production is minimal in h-h collisions but should make up some 40% of the cross-section in $e^+e^- \to$ h's at centre-of-mass energy $E_{cm} \gtrsim 10$ GeV. In all likelihood, the underlying differences will be more instructive than the superficial similarities, which again are dominated by LPS plus leading particle effects (an interesting finding along the same line is that, except for the leading particle, the Feynman-field parametrization of quark fragmentation is often very similar to mere LPS around the quark axis).

I would like to stress that some differences must exist on very general grounds, and to look for them may offer useful tests of the sensitivity achievable in present experiments. An elementary case is the planarity expected to exist in the final states of h-h and ℓ-h collisions at non-vanishing impact parameter (i.e., in almost all such collisions), the plane being defined by the incident direction and the impact vector. To find visible effects of this planarity, which is a consequence of the extended structure of hadrons, turns out to be very difficult because of the limited multiplicity of secondaries and the non-observation of neutrals. The situation is quite different in $e^+e^- \to$ h's in the one-photon approximation, where planarity, not expected for $e^+e^- \to q\bar{q}$, $e^+e^- \to q\bar{q} + ng$ (n>1, q=quark, g=gluon), occurs for $e^+e^- \to q\bar{q}g$ (three-jet events).

Another important factor in the evolution of multiparticle dynamics has been the discovery of hard processes in h-h collisions, i.e., the high p_T collisions. At the purely hadronic level, the recent progress in the experimental study and theoretical analysis of these processes has been very slow, and the experiment reported by P. Seyboth [2] indicates that much effort is still needed before they will be reliably understood. The main point is the considerable abundance of events which have high transverse energy $E_T = \sum_i (m_i^2 + p_{Ti}^2)^{1/2}$,

i = all secondaries, do not possess a two-jet structure, but may simulate it when the detectors have insufficient coverage in solid angle. As known from the ISR work, very high energy appears to be needed for the emergence of high p_T jets. One may therefore look forward to the results to be obtained at the CERN $p\bar{p}$ collider (E_{cm}=540 GeV), which begins operation this year.

The development has been more satisfactory for the hard h-h processes producing high mass $\mu^+\mu^-$ pairs and large p_T direct photons. Final state leptons and photons, not resulting from the decay of secondary hadrons, have clearly shown to be powerful probes of the structure of incident hadrons. As illustrated in Section 5, they should also provide probes for the dynamical processes occurring during the collision.

We end these general considerations by noting that the field of hadron-nucleus collisions has been consistently recognized as important for the study of aspects of hadron dynamics not accessible in collisions on single hadrons, but progress in obtaining good experimental data has unfortunately remained very slow. On the other hand, as stressed by H. Pugh at this conference, there is a growing realization that relativistic ion collisions may offer a considerable potential of significant research (see also Section 5).

2. QUARK COMBINATORICS FOR MULTIPLICITIES

Our first topic will be the analysis of meson and baryon multiplicities in h-h collisions in terms of combinations of quarks and antiquarks, the valence quarks of the incident hadrons, their sea quarks, and new quarks created by the collision process (the two latter categories to be grouped under the common name of sea quarks). Systematic work on this subject is mainly due to the Leningrad group of Anisovich et al., who pioneered it in 1973 [3] and aptly refer to it as "quark combinatorics" [4]. At high energies, the secondaries produced in the central region of rapidity will mainly result from the combination of sea quarks, whereas the incident valence quarks will mostly emerge (often combined with sea quarks) in secondaries of the two fragmentation regions [*].

Quark combinatorics must take into account that many hadronic resonances occur among the secondaries. The Leningrad group works in the framework of the SU(6) classification. For mesons they analyze the $q\bar{q}$ combinations in the low-lying meson multiplets, $6 \times \bar{6} = 1+35$. The multiplicity ratios of pseudoscalar and vector mesons are well accounted for with dominance of S wave states and a small addition of P wave. In contrast, this method becomes very cumbersome for baryons, as can be understood from the fact that the SU(6) reduction of qqq combinations

$$6 \times 6 \times 6 = 56 + 70 + 70 + 20$$

involves much more than the lowest baryon multiplet 56. A detailed discussion of these problems is given in Ref. 4.

A simpler approach to quark combinatorics for baryons has been developed recently [7,8], with application to the fragmentation of incident protons. The assumption is to use the combinatorial analysis for the lightest baryons of each qqq composition and isospin value, all of which except the Δ are metastable baryons. The method makes use of the fact that a baryon resonance decays into one and only one baryon. It averages over all decay chains leading to the lightest baryons, but is unable to predict baryon resonance multiplicities.

---

[*]   As revealed by the interesting work of the Vienna group [5], however, there are important effects observed in the proton fragmentation region of meson-proton collisions, which depend on the incident meson and can be attributed to annihilation of the valence antiquark of the incident meson with a valence quark of the incident proton. These effects disappear slowly for increasing energy [6].

It has been found [8] to work well for proton fragmentation into protons and hyperons under the simple assumption that the valence quarks of the incident proton emerge in one of the following combinations :
  i) in three mesons (probability a_0),
 ii) in one baryon and two mesons (probability a_1),
iii) in one baryon and one meson (probability a_2),
 iv) in one baryon (probability a_3) :

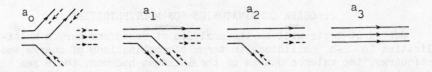

The probabilities a_i verify $\Sigma a_i = 1$, which means that there is a negligible probability for two valence quarks of the incident proton to emerge in two distinct baryons. As shown by Bourquin et al. [7] a second baryon can best be interpreted as due to baryon-antibaryon pair production out of the sea, with opposite flavours for the baryon and the antibaryon in the pair.

As shown in Ref. 8, this scheme puts strong constraints on the proton fragmentation multiplicities into protons and hyperons [defined as $n_f(B) = n(B) - n(\bar{B})$, where the n are the multiplicities of B and $\bar{B}$ in the proton fragmentation centre-of-mass hemisphere]. These constraints follow from the requirement that all $a_i > 0$ and are well satisfied by the data. The analysis of Ref. 8 gives $a_0 \lesssim 0.1$, $a_1 \lesssim 0.3$, $0.2 \lesssim a_3 \lesssim 0.5$, the value of a_2 remaining poorly determined $(0.2 \lesssim a_2 \lesssim 0.8)$. A better determination for a_2 is obtained if one assumes an uncorrelated behaviour of the incident quarks, namely that (except for a small diffractive part δa_3 of a_3) each incident valence quark has the same probability P to emerge in a meson, and $1-P$ in a baryon, without correlations :

$$a_0 = (1 - \delta a_3)P^3, \quad a_1 = 3(1 - \delta a_3)P^2(1 - P),$$
$$a_2 = 3(1 - \delta a_3)P(1 - P)^2, \quad a_3 = (1 - \delta a_3)(1 - P)^3 + \delta a_3$$

The data are compatible with this uncorrelated behaviour but are insufficient to establish it. If it holds, they give $P \sim 0.4$, $\delta a_3 \sim 0.1$, corresponding to

$$a_0 \sim 0.06, \quad a_1 \sim 0.26, \quad a_2 \sim 0.39, \quad a_3 \sim 0.29.$$

In this analysis of proton fragmentation, the flavour of a sea quark or antiquark is s, u, d with probabilities ξ, τ, $\tau = \frac{1}{2}(1-\xi)$, and the baryon multiplicity data give $\xi \sim 0.1$ [8]. The same value was deduced from meson data at large Feynman x (x > 0.5) where resonance effects do not complicate the deduction of ξ from meson multiplicities [*]. From their (anti) hyperon multiplicity data,

----
[*] The determination of ξ from (anti) baryon multiplicity data is only weakly affected by resonance effects because hyperon resonances have rarely K mesons among their decay products.

Bourquin et al. [7] conclude that ξ is probably somewhat larger, $\xi \sim 0.15$, in the central region of rapidity. As mentioned in Section 5, the variation of ξ with rapidity and E_{cm} is of considerable interest.

The above type of combinatorial analysis is able to give, under rather general assumptions, an estimate of what happens to the valence quarks of an incident hadron in a high energy h-h collision. It would be of great interest to have more complete data in order to make it more precise, and to have corresponding data for hadron fragmentation on heavy nuclei in order to determine how the probabilities a_i and ξ change when the fragmentation takes place in nuclear matter. Since the underlying assumptions are likely to hold for more specific dynamical models of hadron fragmentation, the results of quark combinatorics should provide useful guidance for the development of such models.

3. INCLUSIVE DISTRIBUTIONS AND FLAVOUR CORRELATIONS

3.1. Fragmentation versus recombination

Inclusive distributions of hadrons in the Feynman x variable ($x = p_{cm}^{long}/p_{cm}^{inc}$) have been studied in a variety of h-h collisions and have been mainly analyzed in terms of the recombination and the fragmentation models [9]. Despite the fact that they are based on different premises, both models have been remarkably successful in accounting for the data, and this convergence of results has become rather puzzling. Data presented by E. De Wolf at this conference may help to clarify the situation. They concern a comparison between the inclusive distributions of resonant and non-resonant systems of hadrons produced in K^+p interactions at 32 GeV/c [10,11]. The x distributions are found to be different (indicating that different mechanisms are at work), in the sense that the x distribution of a resonant system, e.g., $K^{*+}(892)$ in the K^+ fragmentation region, is found to decrease much less steeply in x than the corresponding non-resonant system, e.g., $K^O\pi^+$ at masses just outside the resonance or in the background below the resonance peak. This is strikingly illustrated by the figures of Ref. 11.

As explained in Ref. 11, this behaviour contradicts the predictions of the recombination model if the valence quarks in a hadron are uncorrelated except for energy-momentum conservation. As we shall now show for the example $K^+ \to K^O\pi^+$ just quoted, the data are compatible with the recombination model if there is a <u>negative correlation between the valence quarks</u> of the K^+, i.e., the type of correlation which is at the basis of the fragmentation model.

We denote by x_u and x_s the x values of the valence quarks u_V and $\bar{s}_V$ of the incident K^+, and by x_O, x_+ those of the outgoing K^O and π^+. In the recombination model, the resonance K^* is mostly composed of the valence $\bar{s}_V$ combined with a u quark from the sea, giving in first approximation $x(K^*) \simeq x_s$ since sea quarks have low x. The K^* distribution therefore probes the distribution of x_s.

For a non-resonant $K^O\pi^+$, the two mesons have in the model the composition $K^O = \bar{s}_V d$, $\pi^+ = u_V \bar{d}$ with d, $\bar{d}$ from the sea, giving $x_O \simeq x_s$, $x_+ \simeq x_u$ and therefore $x(K^O\pi^+) = x_O + x_+ \simeq x_s + x_u$. But the values of x_O, x_+ are constrained by the condition that the mass

931

$m(K^o\pi^+)$ is near the resonant value $m^* = 892$ MeV. At high energies this constraint is expressed by

$$(x_s/x_+)(m_o^T)^2 = \Delta^2 \pm [\Delta^4 - (m_o^T m_+^T)^2]^{\frac{1}{2}}, \quad \Delta^2 = \frac{1}{2}[(m_{K\pi}^T)^2 - (m_o^T)^2 - (m_+^T)^2]$$

where m_o^T, m_+^T and $m_{K\pi}^T$ are the transverse masses of K^o, π^+ and the $K^o\pi^+$ system respectively. Since $x_o \simeq x_s$, $x_+ \simeq x_u$, we see that the region of the (x_s, x_u) plane *) which is probed by the non-resonant $K^o\pi^+$ distribution is given by the previous equations with x_s/x_u replacing x_o/x_+ and the transverse masses ranging over the relevant intervals. Our qualitative interpretation of the steep fall of the $K^o\pi^+$ distribution is therefore that this region is depleted, which means that the two-dimensional probability distribution of u_V and $\bar{s}_V$ in the K^+ is concentrated near the two strips $x_u \ll 1$, $0 < x_v < 1-x_u$ and $x_s \ll 1$, $0 < x_u < 1-x_s$. This is the type of negative correlation between valence quarks which is at the basis of the fragmentation models and can be justified in the dual topological unitarization (DTU) scheme.

The above discussion is admittedly very crude and would have to be redone quantitatively on higher energy data before drawing hard conclusions. Its qualitative result is plausible, however, and may help to unify our picture of hadron fragmentation. The analysis of Ref. 11 has also been carried out for the $\Sigma^*(1385)$ resonance and the conclusion is the same, here in the proton fragmentation region $(x < 0)$. It suggests that in the three-dimensional valence quark distribution of the proton there is a negative correlation between a single quark and a diquark, the single quark having $x \sim 0$. In this case the valence quarks of the diquark can emerge in one baryon or in two separate hadrons, as discussed by U.P. Sukhatme at this conference [12]. Note that they can also emerge together with the third incident valence quark in a single proton (this is forbidden in the DTU scheme but is allowed in gluon-exchange models) ; more generally, all "quark combinatorics" possibilities (Section 2) can occur.

One can ask whether the negative correlation between valence quarks mentioned above pre-exists in the incident hadron. This is not easy to answer, because the collision itself may affect the parton distributions before formation of the outgoing hadrons. The parton sea is undoubtedly excited (see Section 5) and the incident valence quarks are also likely to be affected, as is suggested by the familiar "additive quark model" where one constituent quark of each incident hadron interacts (is "wounded"), while the others are "spectators" [3,4]. It is natural to expect that the valence quark of lowest x is mostly the one belonging to the wounded constituent, the valence quark (or diquark) of higher x belonging to the spectator(s) **). The wounded quark could be pushed to lower x (creating or enhancing the negative correlation), while the spectator quarks can

----

*) We recall that the kinematically allowed area of this plane is the triangle $x_s \geq 0$, $x_u \geq 0$, $x_s + x_u \leq 1$.

**) Also in the DTU scheme the low x valence quark is singled out by the dynamics of the collision.

be expected to have about the same x distribution as in the incident
hadrons. As shown in Ref. 13, the wounding should not affect the
striking similarity (first noted by Ochs and well explained by the recombination model, also in presence of correlations) between the x
distributions of fragmentation mesons at $x \gtrsim 0.4$ and the deep inelastic structure functions of the incident hadron (proton [14] or pion [15]).

   Another striking similarity (first noted by the Lund group,
B. Andersson et al.) exists between these x distributions and parton
fragmentation functions, and it is the aim of the fragmentation models
to explain it. T. Kanki's contribution to the conference [16] shows
that considerable modifications and elaborations are required if the
DTU based model is to reproduce the large increase of the π^+/π^- and
K^+/K^- ratio for $x \to 1$ in proton fragmentation. Nevertheless the
similarity is directly visible, e.g., in the data shown at the conference by A.M. Touchard [17], where the diquark fragmentation functions
deduced from $p \to \pi^{\pm}$ fragmentation in K^-p interactions at 70 GeV/c
agree with those deduced from neutrino and antineutrino–proton interactions. Much could be learned from more accurate data, especially on
the π^+/π^- ratio for $x \gtrsim 0.5$ and on the way the ν and $\bar{\nu}$ data
vary with the x of the quark hit by the weak interaction. Such work
would reveal more about correlations between valence quarks in the
proton, including their possible modification in h–h collisions.

3.2. Inclusive distribution of heavy flavours

   The second topic which we shall mention on inclusive distributions
concerns the production of heavy flavours, i.e., of hadrons containing
the heavy quarks c (for "charm") or b (for "bottom" or "beauty").
Several experiments at the CERN Intersecting Storage Rings have revealed that in pp collisions the production of the charmed baryon
Λ_c^+ (quark composition cdu) is abundant (cross-section of order
0.1-0.5 mb) and is concentrated in the proton fragmentation region,
the x distribution being rather flat as for Λ production [*]. A
recent experiment [18] also reports evidence for production of a beauty
baryon Λ_b (quark composition bdu) at relatively large x, a remarkable result for which confirmation is highly desirable. These
results are very difficult to explain in the conventional models for
particle production which predict smaller cross-sections and more
central production. An attractive explanation has been proposed by
S.J. Brodsky et al. [19]. Since it is extensively covered by the report
of C. Peterson at this conference we shall limit ourselves to a brief
discussion.

   The basic idea is that ordinary hadrons are assumed to contain,
in their wave function in terms of partons, higher Fock states such
as $|uudg\rangle$, $|uudq\bar{q}\rangle$ for the proton (g = gluon, q = quark). In every
Fock state the x distribution of the partons is supposed to be
$\propto \Delta E^{-2}$, with ΔE the energy denominator of old-fashioned perturbation
theory in the infinite momentum frame, varying as $\Delta E \propto m_p^2 - \sum_i (m_{Ti}^2/x_i)$;
the m_{Ti} are the transverse masses of the partons in the state. For
a state $|uudq\bar{q}\rangle$ with q a heavy quark (q = c or b) this implies
$\langle x_q \rangle = \langle x_{\bar{q}} \rangle = 2/7$ for the average x value. Assuming a probability

----

*)
   See various contributions and R.C. Ruchti's review at this conference.

$P_c \sim 1\%$ for the state $|uudc\bar{c}\rangle$ in the proton (a percentage level found by J.F. Donoghue and E. Golowich [20] in the MIT bag model), one obtains qualitative agreement with the Λ_c data. For heavy quarks q, the probability P_q of the state $|uudq\bar{q}\rangle$ is expected to vary as $P_q \propto \left[m_q^{-1}\alpha_s(m_q^2)\right]^2$, with m_q the mass of q and α_s the running coupling strength of QCD. Since $m_b/m_c \sim 3$, the scheme predicts a Λ_b production cross-section around 5-10% of the Λ_c cross-section, with the same shape for the x distribution. The Λ_b evidence presented by M. Basile [18] seems compatible with this prediction.

The higher Fock state proposal of Brodsky et al. [19] has many interesting implications which can be tested experimentally. The most obvious one is that Λ_c would be produced in association with a charmed meson D with an x distribution almost as flat as for Λ_c. The ISR data available so far do not disagree with this but are insufficient to provide a meaningful test. Similar predictions hold for $D\bar{D}$ production by fragmentation of an incident meson. Other tests are provided by charm production in ℓ-h collisions, especially neutrino interactions, where the same higher states of the nucleon must pre-exist. In contrast, the production of hidden charm (the ψ and χ states) out of these states should be weak ; it is suppressed by a colour factor 1/9 and by the low mass of the ψ and χ. The observed production is indeed concentrated in the central region and is reasonably well accounted for by standard QCD processes involving gluons or light quarks (gluon or quark fusion, $gg \to \chi$ or $q\bar{q} \to \psi$ or χ). The same considerations should apply to $b\bar{b}$ mesons.

We end this discussion with two remarks. First, if the proton state contains the state $|uudc\bar{c}\rangle$ at the probability level of 1%, it should contain states such as $|uudg\rangle$; $|uudgg\rangle$ and $|uudq\bar{q}\rangle$ with q = u,d,s at quite a high percentage level. It will therefore be necessary to show how the rules of the model, applied to the structure functions for light quarks and antiquarks and gluons, can be made to reproduce the abundant experimental information now available on these functions. Secondly, even if the model of Brodsky et al. would meet with difficulties on some of the numerous tests it will have to pass, its basic idea can be given different implementations. It states that whenever a heavy quark is present in a hadron, it tends to have a large x value in proportion to its mass. While this was generally accepted to hold when the heavy quark is a valence one, it is now proposed to hold also when it is not. Qualitatively, this is the most natural interpretation of the ISR results on Λ_c production. Good data on D production and associated Λ_c+D production will be needed to decide on the actual shape of the heavy quark components in the hadron wave function. Even more crucial would be the confirmation of Λ_b production and the measurement of its $d\sigma/dx$ in size and shape.

4. MORE ADVANCED DYNAMICAL CONSIDERATIONS

While fragmentation and recombination models have been our most valuable guides in trying to interpret hadron production data, they are strongly phenomenological in character and contain too many arbitrary features (especially arbitrary functions) to qualify as dynamical theories. Much work is going on to give them a more elaborate dynamical basis. Since most of it has been covered in various reports to the conference, our summary can be brief.

With the introduction of the valon as a more precise concept for constituent quarks, R. Hwa and his collaborators have constructed a more elaborate version of the recombination model [21]. Among other things, it links soft processes in a calculable way to the Q^2 dependent hadronic structure functions measured in ℓ-h collisions. This link has been successfully tested on $p \to \pi^{\pm}$ fragmentation in K^+p collisions at 70 GeV/c with the result that the soft processes probe the proton at Q^2 values of 10-20 GeV^2, a surprisingly high range [22].

Fragmentation models have developed in several directions (see A. Capella [9] for a recent review). The Lund group, originator of the model, has worked out in considerable detail a dynamical scheme based on the picture of colour flux tube breaking via $q\bar{q}$ pair creation [23] *) Most other workers on fragmentation start from the DTU scheme, and we refer once more to Ref. 16 for the degree of elaboration needed to fit the data. The DTU based models split again into two distinct versions (Orsay [25] and Austin [26]) for the application to hadron-nucleus collisions.

A very interesting area where phenomenology is forced to become more dynamical has been opened by the discovery at Fermilab of a substantial inclusive polarization in hyperon production at high energy and moderate p_T [27]. This type of polarization effect could be accounted for by means of appropriate dynamical assumptions in the Lund fragmentation model [28] and, more recently and more completely, in the recombination model if one invokes the spin-orbit force resulting from Thomas precession for a scalar potential [29]. The prediction of this approach for Λ polarization in K^+ fragmentation agrees at least in sign with data reported at the conference [30]. This line of work is likely to be a fruitful source of dynamical information, because it can be applied to inclusive production of meson and baryon resonances, for which the density matrices are readily measurable.

The basic aim of all phenomenological studies, of course, should be to bridge the gap between the soft hadronic processes and a fundamental theory of strong interactions, for which QCD is now the best guess. Interesting attempts are being made toward tackling soft hadronic processes directly in terms of QCD. Several were presented at this conference by J.F. Gunion. They analyze soft processes in terms of simple QCD diagrams involving quarks and gluons ; this gives them a perturbative character which is only justified for special final state configurations (e.g., the limit $x \to 1$ in fragmentation). As explained in Gunion's contributions, this approach has been applied to particle production in the central region of h-h collisions (using bremsstrahlung-like diagrams) and to diffraction dissociation. For the latter case two-gluon exchange diagrams are used and the prediction is made that the cross-section $d\sigma/dM$ falls as M^{-4} for large mass M of the system produced by the dissociation, instead of the $d\sigma/dM \propto M^{-2}$ behaviour predicted by Regge-Müller considerations (triple pomeron coupling).

----

*) Tube breaking by $q\bar{q}$ pair creation is a tunnelling process analogous to e^+e^- pair creation in a constant electric field [24].

Another QCD-based approach to diffraction dissociation was developed by J. Pumplin and E. Lehman [31] and by G. Bertsch et al. [32]. The basic idea is that a hadron has coloured constituents with colour charges adding up to zero. When these constituents are very close to each other (e.g., for $q\bar{q}$ pairs with q a heavy quark), they will only interact weakly as they collide with another hadron or a nucleus. This separation-dependent absorption is a source of diffraction dissociation which can be tackled by QCD arguments since small distances are involved. Large cross-sections for diffractive charm production are predicted.

5. DENSE HADRONIC MATTER AS QUARK-GLUON PLASMA

In view of the many successes of the quark-gluon parton model and of QCD as its theoretical basis, it is now very generally believed that dense hadronic matter must take the form of a quark-gluon plasma. If pions could exist as particles in a medium of temperature $T = 1$ GeV (this was the temperature of the Universe at time $t \sim 10^{-6}$ sec of its expansion), there would be of the order of 100 pions overlapping with any single one, so that it would make little sense to imagine them as individual particles, each composed of a quark and an antiquark, and similarly for nucleons if nuclear matter were compressed even at $T = 0$ to a density of 100 times the ordinary nuclear density (~ 0.15 nucleon fm^{-3}). This justifies the great interest which developed among theorists in the last few years for the study of the quark-gluon-plasma as the most probable state for dense hadronic matter, and its transition to an ordinary hadron gas when both the temperature T and the net quark density $\delta n_q = n_q - n_{\bar{q}}$ become small enough (n_q, $n_{\bar{q}}$ are the number densities of quarks and antiquarks respectively). Many authors have discussed the problem of the phase transition (see Ref. 33 for a review and Refs. 34-38 for some recent references). There is considerable agreement concerning the approximate values of the transition temperatures and net densities ; the phase transition probably takes place in the neighbourhood of a line in the T- n_q plane as depicted in the figure, with $T_C \sim 150$-250 MeV and $\delta n_q^C \sim$ ~ 1-2 fm^{-3}. In contrast, it is difficult to assess the nature of the transition with any reliability (crude estimates of the equations of state [*]) in the two phases tend to suggest a first order transition, but this cannot be regarded as significant since better precision is needed if a higher order transition is to be deduced). It has also been argued that percolation aspects could play a rôle in the transition [40].

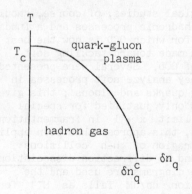

----

[*] It should be noted that the equation of state cannot be derived perturbatively even in the plasma phase, see e.g. Ref. 39.

Is there any hope to observe a quark-gluon plasma (QGP) in high energy collisions ? The answer, although still highly speculative, seems to be yes in two possible ways. Firstly, according to E.V. Shuryak [41], a small QGP could form in the central region of h-h collisions at $E_{cm} \gtrsim 50$ GeV with an initial temperature $T_i \gtrsim 500$ MeV and net quark density $\delta n_q \sim 0$. This QGP would then expand, mostly in longitudinal direction, and every part of it would hadronize (mostly into mesons) as it reaches the transition temperature $T_c \sim 150$-250 MeV, populating the central region of rapidity.

Secondly, according to R. Anishetty et al. [42], a QGP could form in the fragmentation regions of head-on collisions of very heavy nuclei (e.g., uranium) at nucleon-nucleon $E_{cm} \gtrsim 50$ GeV, with an initial $\delta n_q^{\frac{1}{2}} \sim 2$ fm^{-3} (corresponding to a compression factor ~ 4 of nuclear matter) and an initial $T_i \sim 150$ MeV (see also Ref. 43). After expansion and cooling these QGP would hadronize (mostly in nucleons, lighter nuclei and mesons) and populate the fragmentation regions (with in addition a $\delta n_q \simeq 0$ plasma in the central region in between). Since only the latter case was discussed in contributions to the conference, we shall now concentrate on the former one (central region plasma, $\delta n_q \simeq 0$), warning again that the topic is quite speculative.

A crude estimate of the initial temperature T_i of the central region QGP in a h-h collision can be obtained by equating its initial energy density ρ_i to the energy density of a non-interacting gas of massless quarks and gluons at temperature T_i. This gives

$$\rho_i = \frac{\pi^2}{30} g_o T_i^4$$

where g_o is the number of spin states for each boson species ($2 \times 8 = 16$ for the gluons) plus 7/8 times the number of spin states for each fermion species ($2 \times 2 \times 3 \times 7/8 = 10.5$ for each quark flavour, hence 21 for u and d quarks, and 31.5 for u, d, s although the mass of the s quark would make its contribution to ρ_i somewhat smaller). Thus $g_o \simeq 40$ for $T_i \sim T_c$, growing to $g_o \sim 50$ for $T_i \sim 1$ GeV. A rough estimate of ρ_i is obtained as follows (we consider the case of pp collisions). Each proton has in the centre-of-mass frame of the collision a Lorentz contracted volume $v \simeq (4\pi/3) m_\pi^{-3} (2m_p/E_{cm})$. Let $f_o v$ be their overlap in a given collision (f_o depends on the impact parameter and ranges over $0 < f_o < 1$). The energy available for thermalization is at most of order $f_o E_{cm}$, and probably smaller since some of it will go into non-thermal centre-of-mass motion [incident valence quarks flying through with small energy loss and bulk flow of plasma ; note that also $(1-f_o)E_o$ goes into non-thermal motion, mainly incident partons flying through.] Hence $\rho_i \leq f_o E_{cm}/f_o v = E_{cm}/v$ and we shall put $\rho_i = f E_{cm}/v$ with $0 < f < 1$ to take into account this effect, as well as the fact that the volume $f_o v$ may have expanded somewhat as the plasma forms. Putting the numbers together we find

$$T_i \simeq f^{\frac{1}{4}} (E_{cm}/2m_p)^{\frac{1}{2}} \times 100 \text{ MeV}$$

For $E_{cm} = 60$ GeV (CERN Intersecting Storage Rings) and $f = 0.5$ this gives $T_i \simeq 470$ MeV, and for $E_{cm} = 540$ GeV (CERN $p\bar{p}$ collider), $f = 0.25$ one has $T_i \simeq 1.2$ GeV. Even much smaller values of f would still give $T_i > T_c$, especially at the $p\bar{p}$ collider.

One must then consider the expansion of the plasma. It is supposed to be largely longitudinal (non-thermal bulk flow not due to a longitudinal pressure gradient), with a small transverse expansion due to a pressure gradient. As each plasma element expands, it cools down until its local temperature reaches T_c, when it hadronizes. During this expansion phase, the QGP can emit some matter, lepton pairs formed by $q\bar{q}$ annihilation and direct photons (leptons and photons escape easily since they have no strong interaction) and some "early" hadrons. The lepton pairs and direct photons are direct probes of the plasma, the latter being in fact the "excited parton sea" known to exist in h-h collisions [44] and already mentioned earlier.

To describe the space-time expansion of the plasma one can adopt a hydrodynamical scheme similar to Landau's original hydrodynamical model [45]. This has been done by Shuriak [41] who calculated the resulting production of $\mu^+\mu^-$ pairs, direct photons and $c\bar{c}$ pairs during the plasma expansion phase. He finds that the plasma spends most of its time at T near T_c (a feature resulting from general properties of the expansion rather than from specific features of the hydrodynamical equations). For $\mu^+\mu^-$ pair production from the QGP, Shuryak finds that for masses <3 GeV the cross-section is well above the Drell-Yan continuum and is compatible in size and shape with Fermilab data. Turning the argument around, one sees the interest of measuring $\mu^+\mu^-$ pairs in the mass interval 1-3 GeV as a probe for the collision dynamics in the central region (an expanding QGP description could hold even if there is no complete thermalization). Of course one must experimentally correct for the μ's produced after hadronization of the QGP by pion decay or vector meson $\rightarrow \mu^+\mu^-$.

It should be noted that for given E_{cm} the initial temperature of the central QGP will fluctuate from collision to collision, so will its initial volume and total energy (bulk motion + thermal), and so will the energy escaping in the fragmentation regions of the incident hadrons (these quantities are related to the factors f_o, f above). The latter regions are of course mainly populated by hadronization of the incident valence quarks (see Sections 2 and 3) and they receive only a small contribution from the high rapidity tails of the central QGP. These fluctuations may have interesting consequences. In some collisions, for example, most of the energy E_{cm} may go into the plasma (large f_o) and most of the plasma energy may thermalize (large f). Such collisions would undoubtedly have high multiplicity and high total transverse energy $E_T = \sum_j (m_j^2 + p_{Tj}^2)$, with the corresponding depletion of the fragmentation regions. Could they be the class of deep inelastic h-h collisions reported upon by B. Seyboth [2] at this conference and already mentioned in Section 1 ? Is the high E_T entirely due to high central multiplicity at low $<p_T>$, or partly due to larger p_T's (suggesting a stronger transverse expansion in these exceptional collisions) ?

Our final comment concerns an interesting suggestion by Rafelski on the effect of a thermalized QGP on strange particle production [34]. Whereas the u and d quarks have current masses which are $\ll T_c$ and therefore negligible in the QGP, the current mass of the strange quark s is much larger ; values $m_s \sim$ 150-300 MeV are often quoted, and a recent analysis by A. Katz (using QCD and Higgs mass generation at the electroweak level) concludes that $m_S = 497\pm299$ MeV [46]. Treating the plasma as an ideal Fermi gas one easily estimates the ratio $\xi = s/(u+d+s)$ as a function of m_S/T and $\delta n_q/T^3$. At $\delta n_q = 0$ it is 0.25 at $m_S/T = 1.5$ and 0.20 at $m_S/T = 2$. The central QGP plasma hadronizing at T_C will therefore have a higher ξ than the value $\xi = 0.1$ found in proton fragmentation, see Section 2 *). This could be tested experimentally by measuring the production ratios for (anti) baryons of various strangeness in the central region (the K/π ratios are difficult to use for the determination of ξ in the central region, due to problems with meson resonances, esp. the ρ; their variation with E_{cm} would be indicative, however). Rafelski's consideration leads even to more striking effects in a QGP with $\delta n_q > 0$ as could be produced in head-on collisions of heavy nuclei. Due to the chemical potential, the densities of $\bar{u}$ and $\bar{d}$ antiquarks are depressed, but this does not affect the $\bar{s}$ density. Although the transition temperature is now $< T_C$, the value of ξ for antiquarks can therefore be even larger than it is for quarks and antiquarks in the $\delta n_q = 0$ case. This would lead to an abnormally high K^+/π^+ ratio at large x ($x > 0.5$, where resonance effects cause no problem).

6. CONCLUDING REMARKS

Before concluding, I summarize some of the results presented at the Conference.
a) If the findings obtained at 32 GeV/c are confirmed at higher energies, the recombination model analysis of inclusive data implies that there exists a negative correlation between the valence quarks of a hadron, in the sense that at the time of recombination one of the valence quarks has low x value. This is the type of correlation on which the fragmentation model is based. If it holds, the two models become broadly equivalent. In actual detail, however, the recombination and fragmentation models tend to become very elaborate, the latter having different versions (one based on flux tube breaking, the other on dual topological unitarization).
b) In view of the above correlation, nucleon fragmentation in hadron-nucleon collisions is dominantly diquark fragmentation and should resemble diquark fragmentation in deep inelastic lepton-nucleon collisions. Available data exhibit this similarity. Better lepton-nucleon data are needed to test the similarity on such sensitive quantities as the π^+/π^- ratio for $x \gtrsim 0.5$. More generally, there is a vast potential for comparative flavour correlation studies in hadron-hadron, lepton-hadron and e^+e^- collisions. They are likely to be more important for dynamical progress than very global similarities which are often mere consequences of longitudinal phase space corrected for leading particle effects.

----

*) We noted there that Bourquin et al. [7] estimate $\xi \sim 0.15$ in the central region of p nucleon collisions at $E_{cm} = 21$ GeV.

c) <u>Inclusive polarization and alignment effects</u> become a promising field, experimentally (they can be studied on hyperons and on resonances) as well as theoretically (they have already found interesting interpretations).

d) The production of the charmed hyperon Λ_c at large x in proton fragmentation was interpreted as due to the presence of a $uudc\bar{c}$ component in the proton wave function, with probability ~1% and $\langle x_c \rangle \sim \langle x_{\bar{c}} \rangle \sim 2/7$. A similar $|uudb\bar{b}\rangle$ component is predicted with probability ~0.1%, and the first experimental observation of Λ_b was reported, at large x. In addition to giving new insight in hadron structure, this interpretation of <u>heavy flavour production by hadron fragmentation</u> has a number of consequences which can be tested experimentally.

e) On the basis of QCD, new developments were proposed for the theory of <u>diffraction dissociation</u>, with testable predictions on mass spectrum and charm production.

f) The realization that a transition of hadronic matter to a dense <u>quark-gluon plasma</u> could be expected to take place at relatively low temperatures (~200 MeV) has stimulated great interest in the possible production, at attainable energies, of mini-plasmas of this type in hadron-hadron collisions (central rapidity and fragmentation regions, with $\delta n_q \sim 2$ fm^{-3} in the latter). Lepton pairs in the mass range 1-3 GeV, direct photons, and strangeness production rates may offer probes to test these ideas experimentally.

In conclusion, beyond noting the wealth of interesting developments covered by the Conference, I would like to remark on the renewed importance of soft hadronic processes, and of the soft phases involved in all hard hadronic processes, for the study of strong interaction dynamics, i.e., of QCD. Nature does not provide us with any situation controlled by perturbative QCD alone, nor does theory provide us yet with the tools required for non-perturbative QCD. It would be overoptimistic to expect that non-perturbative QCD will be elucidated without the guidance of experiment. With the gradual exhaustion of the possibilities offered by hadron spectroscopy, the resources of hadrodynamics will be increasingly exploited, and this trend has been amply substantiated by the Notre Dame Symposium.

ACKNOWLEDGEMENTS

I wish to express my appreciation to the organizers for their excellent work and their kind hospitality, which made the Symposium such a nice success and such a pleasant experience. I greatly profited from discussions with many participants whom I thank warmly.

REFERENCES

1. M. Basile et al. (Bologna, CERN, Frascati collaboration), Phys.
   Letters 95B (1980) 311 and later publications, also contributions
   to this Conference.
2. C. Favuzzi et al. (Bari-Krakow-Liverpool-MPI Munich-Nijmegen
   collaboration), Deep Inelastic Hadron Scattering with a 2π Calorimeter Trigger, presented at this Conference.
3. V.V. Anisovich and V.M. Shekhter, Nuclear Phys. B55 (1973) 455.
4. V.V. Anisovich, M.N. Kobrinsky and J. Nyiri, Average Multiplicities
   of Secondary Particles in Hadron-Hadron Collisions and Quark Combinatorics, Leningrad Nuclear Physics Institute Preprint 631
   (December 1980).
5. B. Buschbeck, H. Dibon, H.R. Gerhold and W. Kittel, Z.Phys. C3
   (1979) 97.
6. M.M. Schouten et al., Approach to Scaling of π^+/π^- Ratios in
   Target Fragmentation from $\pi^+/K^+/pp$ Interactions at 147 GeV/c,
   Nijmegen Preprint (1981) and paper presented at this Conference.
7. J. Kalinovski, S. Pokorski and L. Van Hove, Z.Phys. C2 (1979) 85 ;
   M. Bourquin et al. (Bristol-Geneva-Heidelberg-Orsay-Rutherford-
   Strasbourg collaboration), Z.Phys. C5 (1980) 275.
8. L. Van Hove, The Rôle of Valence Quarks in Proton Fragmentation,
   CERN Preprint TH. 2997 (November 1980), Z.Phys.C, to appear.
9. For recent work and earlier references, see the contributions of
   R. Hwa (recombination), B. Andersson and T. Kanki (fragmentation)
   and others at this conference. A recent review is found in
   A. Capella, Dynamical Models of Hadron-Hadron and Hadron-Nucleus
   Interactions at Low p_T, Lecture at XVI Rencontre de Moriond,
   Les Arcs (1981), Orsay Preprint LPTHE 81/15 (June 1981).
10. P.V. Chliapnikov et al., Nuclear Phys. B176 (1980) 303.
11. E.A. De Wolf et al. (CERN-Soviet Union collaboration, Brussels-
    Mons-Serpukhov), Production of Resonant and Non-Resonant
    Particle Systems in K^+p Collisions at 32 GeV/c, presented at
    this Conference.
12. U-P. Sukhatme et al., Diquark Fragmentation, Preprint FERMILAB-
    PUB-81/20-THY (January 1981), presented at this Conference.
13. G. Berlad, A. Dar and G. Eilam, Phys.Rev. D22 (1980) 1547.
14. J. Singh et al., Nuclear Phys. B140 (1978) 189.
15. W. Aitkenhead et al. (MIT Cambridge-Bari-Fermilab-Brown Univer-
    sity Providence collaboration), Phys.Rev.Letters 45 (1980) 157.
16. K. Hirose and T. Kanki, Dual Topological Unitarization Approach,
    presented at this Conference.
17. A.M. Touchard et al. (Paris-Rutherford-Saclay collaboration),
    Study of the Longitudinal Distributions in the Fragmentation
    Regions Using the Quark Parton Model in K^-p Interactions at
    70 GeV/c, presented at this Conference.
18. M. Basile et al. (Bologna-CERN-Frascati-Perugia collaboration),
    Lett.Nuovo Cimento 31 (1981) 97, presented at this Conference.
19. S.J. Brodsky et al., Phys.Letters 93B (1980) 451 ; Intrinsic
    Heavy Quark States, Preprint SLAC-PUB-2660 (January 1981), pre-
    sented at this Conference.

20. J.F. Donoghue and E. Golowich, Phys.Rev. D15 (1977) 3421.
21. R.C. Hwa, Hadron Structure and Soft Processes, Lectures at the
    EPS Study Conference on Partons in Soft-Hadronic Processes,
    Erice (8-14 March 1981), Oregon Preprint OITS-165 (May 1981),
    and report to this Conference. See also Xie Qu-bing, Inclusive
    Production of Meson Resonances and the Sea-Quark Distributions
    in the Proton, Oregon Preprint OITS-164 (April 1981), presented
    at this Conference.
22. M. Barth et al. (Brussels-CERN-Genova-Mons-Nijmegen-Serpukhov
    collaboration), Determination of the Internal Structure of
    Valons, presented at this Conference.
23. B. Andersson, G. Gustafson and C. Peterson, Z.Phys. C3 (1980)
    223 and contributions to this Conference.
24. J. Schwinger, Phys.Rev. 82 (1951) 664 ;
    E. Brézin and C. Itzykson, Phys.Rev. D2 (1970) 1191 ;
    A. Casher, H. Neuberger and S. Nussinov, Phys.Rev. D20 (1979) 179.
25. A. Capella and J. Tran Thanh Van, Phys.Letters 93B (1980) 146 and
    Orsay Preprint LPTHE 81/2 ; see also X. Artru, Orsay preprint
    79/8.
26. W.Q. Chao et al., Phys.Rev.Letters 44 (1980) 578 ;
    C.B. Chiu and D.M. Tow, Phys.Letters 97B (1980) 443 ;
    C.B. Chiu, Z. He and D.M. Tow, Austin (Texas) CPT Preprint (1981).
27. G. Bunce et al., Phys.Rev.Letters 36 (1976) 1113.
    Two recent references are K. Rayachauduri et al., Phys.Letters
    90B (1980) 319, and C. Wilkinson et al., Phys.Rev.Letters 46
    (1981) 803.
28. B. Andersson, G. Gustafson and G. Ingelman, Phys.Letters 85B
    (1979) 417.
29. T.A. DeGrand and H.I. Miettinen, Phys.Rev. D23 (1981) 1227 and
    Santa Barbara Preprint UCSB TH-27 (1981).
30. M. Barth et al. (Brussels-CERN-Genova-Mons-Nijmegen-Serpukhov
    collaboration), Inclusive Λ and $\bar{\Lambda}$ Production in K^+p Inter-
    actions at 70 GeV/c, presented at this Conference.
31. J. Pumplin and E. Lehman, A Subtractive Quark Model of the Pomeron,
    Michigan State University Preprint (March 1980), Z.Phys.C., to
    appear.
32. G. Bertsch, S.J. Brodsky, A.S. Goldhaber and J.F. Gunion, Dif-
    fractive Excitation in QCD, Preprint SLAC-PUB 2748 (May 1981).
33. E.V. Shuryak, Physics Reports 61 (1979) 71.
34. J. Rafelski and R. Hagedorn, From Hadron Gas to Quark Matter,
    CERN Preprints TH. 2947 and 2969 (1980), to be published in the
    Proceedings of the International Symposium on Statistical Mechanics
    of Quarks and Hadrons, Bielefeld (August 1980), H. Satz ed.,
    North Holland Publishing Company.
35. L.D. McLerran and B. Svetitsky, Phys.Letters 98B (1981) 195 and
    Santa Barbara Preprint NSF-ITP-81-08 (1981).
36. J. Kuti, J. Polonyi and S. Szlachanyi, Phys.Letters 98B (1981) 199.
37. J. Engels, F. Karsch, I. Montvay and H. Satz, Phys.Letters 101B
    (1981) 89, and 102B (1981) 332.
38. K. Kajantie, C. Montonen and E. Pietarinen, Phase Transition of
    SU(3) Gauge Theory at Finite Temperature, Helsinki Preprint
    HU-TFT-81-8 (1981), Z.Phys.C, to appear.

39. D.J. Gross, R.D. Pisarski and L.G. Yaffe, Revs.Modern Phys. 53 (1981) 43.
40. G. Baym, Physica 96A (1979) 131 ;
    T. Celik, F. Karsch and H. Satz, Phys.Letters 97B (1980) 128.
41. E.V. Shuryak, Phys.Letters 78B (1978) 150, and Yad.Fiz. 28 (1978) 796, transl. Soviet J.Nucl.Phys. 28 (1978) 408.
42. R. Anishetty, P. Koehler and L. McLerran, Phys.Rev. D22 (1980) 2793 and paper presented at this Conference.
43. K. Kajantie and H.I. Miettinen, Temperature Measurement of Quark-Gluon Plasma Formed in High Energy Nucleus-Nucleus Collisions, Helsinki Preprint HU-TFT-81-7 (1981).
44. J.D. Bjorken and H. Weisberg, Phys.Rev. D13 (1976) 1405.
45. L.D. Landau, Izv.Akad.Nauk SSSR, Ser.Fiz. 17 (1953) 51, also Collected Papers of L.D. Landau, ed. D. Ter Haar, Pergamon Press, Oxford (1965) 569.
46. A. Katz, Phys.Rev. D23 (1981) 800.

DISCUSSION

PETERSON, SLAC: One of the very exciting ideas being put forward at this conference is the notion of [color extension and the absorption cross section] having color nonextension and, thereby, no absorption. I would like to call your attention [to] a very nice paper by Pumplin from East Lansing [Michigan State Univ.] in the beginning of 1980 which hasn't gotten the attention that it should have. He really explores this subject.

VAN HOVE: I must say that I knew about that subject from the work of the people on the West coast, so to say, and I was not aware [of the Pumplin paper]. I'm not clear whether this paper has been properly covered in the review talk which was dealing with that. But it is very good that you mentioned it. Thank you very much.

BIAŁAS, KRAKOW: I would like to comment about a "marriage" between Shuryak and McLerran. We [Białas and Czyz, TPJU 12/80] calculated the central region [density] in the collision of two heavy nuclei, and we found that the density in this region, however you try to estimate it, the density of the quarks and gluons and so on can go up to 6 times bigger for uranium than in pp [collisions]. That's probably the maximum you can get. Now compared to, let's say, hadron-proton you get a factor of the order of 2-1/2. So it's not really a very big enhancement, at least in the central region. And since I'm not sure actually that this plasma can really be found in pp (because that requires really that you stop [everything] and that remains to be seen), we can still gain this factor six so then maybe there is still some chance if you get heavy ions.

VAN HOVE: Well, we have a specific way of calculating the central region in nuclei. You draw these color strings, and there is a little bit of an assumption, since all these color strings are on top of each other, on whether you can simply multiply the number of strings in order to find the multiplicity. I believe also that is an open question. And in particular it will be very important to see whether your prediction, which I think is that there is a dip in the middle of the distribution, would be confirmed.

KITTEL, NIJMEGEN: You say that [because of the observed quark correlation] recombination "digs its own grave." But the fact remains that the π^+ distribution in the proton fragmentation [region] looks like a u(x) distribution as measured in deep inelastic [scattering]. Would that then be accidental? [Isn't that too beautiful to be accidental?]

VAN HOVE: No. You see, I made a joke there that shouldn't be taken . . . . One has now to accept that the fragmentation models in the strict austere sense where $x = 0.95$ are not the last word of fragmentation models. One has to do better. I think it was Kanki who tried to develop a quite complex model that was able to reproduce the Ochs analogy. I think all these problems have to be looked at. I would put it this way. If the correlations that are mentioned exist, the two models, so to say, should exist peacefully and then one should be able to have the successes of both coming out of a single scheme. I would rather put it that way.

LIST OF SUBMITTED PAPERS

1. Soft Gluon Corrections to Large-p_T Direct Photon and Dilepton Production
A. P. Contogouris, S. Papadopoulos and J. Ralston
McGill University

2. A Measurement of Direct Electron-Positron Pair Production from Hadronic Bremsstrahlung
A. T. Goshaw et al.
Duke - SLAC

3. Comparison of 147 GeV/c π^-p Low Transverse Momentum Hadron Production with Deep-Inelastic Leptoproduction
V. Kistiakowski et al.
M. I. T. -Brown-Indiana-Oak Ridge-Rutgers-Stevens-Tennessee-Yale Collaboration

4. Diquark Fragmentation
U. P. Sukhatme, K. E. Lassila and R. Orava
Illinois, Chicago Circle - Fermilab

5. Fragmentation of u-Quark, d-Quark and the (ud)-Diquark System - Can It Be Measured in K^-p Reactions by Selection of Fast $\overline{K}^o$'s or Λ's?
B. Buschbeck, H. Dibon
Institut fur Hochenergiephysik, Vienna

6. Study of Hadronic Events in pp Collisions at $\sqrt{s}$ = 62 GeV and Comparison with Hadronic Events in e^+e^- Collisions
W. T. Meyer et al.
Iowa State-CERN-Heidelberg-Dortmund-Warsaw Collaboration

7. Influence of Nuclear Matter on 0^- and 1^+ Resonances of the 3π System
G. Bellini et al.
Dubna-Milano Collaboration

8. On the Recombination Mechanism of Quarks into a Baryon in Proton Fragmentation
C. Iso and S. Iwai
Tokyo Inst. of Technology

947

9. Particle Ratios under Particle Trigger in Baryon
    Fragmentation at Low p_T
    C. Iso
    Tokyo Inst. of Technology

10. Charge Structure of K^-p Interactions
    R. Gottgens et al.
    Aachen-Berlin-CERN-Cracow-London-Vienna-Warsaw
    Collaboration

11. Valence Quark Annihilation and the Total Charge in the
    Forward Hemisphere of Hadron-Hadron Reactions
    M. Szczekowski
    Warsaw

12. Deep Inelastic Hadron Scattering with a 2π Calorimeter
    Trigger
    C. Favuzzi et al.
    Bari-Krakow-Liverpool-MPI Munich-Nijmegen Collabora-
    tion

13. Inclusive and Semi-Inclusive ρ^0 Production in
    $\pi^+/\pi^-/K^+/p$ p Interactions at 147 GeV/c
    M. Schouten et al.
    Nijmegen-Brown-Cambridge-Illinois Inst. of Tech. -
    Indiana-Johns Hopkins-M.I.T. -Mons-Oak Ridge-Pavia-
    Rutgers-Stevens-Technion-Tel-Aviv-Tennessee-
    Weizmann Inst. -Yale Collaboration

14. Dual Topological Unitarization Approach to Single Particle
    Spectra
    K. Hirose and T. Kanki
    Osaka University

15. Properties of Hadrons Associated with Lepton-Pair
    Production
    B. Pietrzyk
    CERN

16. Inclusive Production of Meson Resonances and the Sea-
    Quark Distributions in the Proton
    Xie Qu-Bing
    University of Oregon

17. Inclusive Neutral Kaon Production in 70 GeV/c K^+p Interactions
M. Barth et al.
Brussels-CERN-Genova-Mons-Nijmegen-Serpukhov
Collaboration

18. π^--Neon Collisions at 200 GeV
H.R. Band et al.
Duke University-S.U.N.Y., Albany

19. Observation of ρ^0-Meson Spin Alignment in $\bar{p}p$ Interactions at 22.4 and 5.7 GeV/c
B.V. Batunya et al.
Dubna-Alma Ata-Helsinki-Moscow-Czech. Acad. of Sci.,
Prague-Charles Univ., Prague-Tbilisi Collaboration

20. Diquark Fragmentation in νn and νp Interactions
D. Zieminska et al.
Maryland-Illinois Inst. of Tech.-S.U.N.Y., Stony Brook-
Tohoku-Tufts Collaboration

21. Neutral-Pion Production and Diffraction Dissociation in
High-Energy π^--Nucleon Collisions
H.R. Band et al.
Duke Univ.-S.U.N.Y., Albany

22. Comparison of Strange Antibaryon and Strange Meson
Production in K^+p Interactions at 32 GeV/c
P.V. Chliapnikov et al.
CERN-USSR Collaboration

23. Inclusive $K^{*+}(1430)$, $K^{*0}(1430)$, and $f(1270)$ Production
in K^+p Interactions at 32 GeV/c
P.V. Chliapnikov et al.
CERN-USSR Collaboration

24. Charged Particle and Neutral Strange Particle Production
in 300 GeV/c Proton-Neon Interactions
D. Minette et al.
Univ. of Wisconsin and Uzbec Acad. of Sci.

25. Average Multiplicities of Secondary Particles in Hadron-
Hadron Collisions and Quark Combinatorics
V.V. Anisovich, M.N. Kobrinsky, and J. Nyiri
Acad. Sci. USSR, Leningrad Nucl. Phys. Inst.

26. Intrinsic Heavy Quark States
    S. J. Brodsky, C. Peterson, and N. Sakai
    SLAC-Fermilab

27. Inclusive Λ and $\bar{\Lambda}$ Production in K^+p Interactions at
    70 GeV/c
    M. Barth et al.
    Brussels-CERN-Genova-Mons-Nijmegen-Serpukhov
    Collaboration

28. Jet-Like Properties of Multiparticle Systems Produced in
    K^+p Interactions at 70 GeV/c
    M. Barth et al.
    Brussels-CERN-Genova-Mons-Nijmegen-Serpukhov
    Collaboration

29. Study of the Longitudinal Distributions in the Fragmentation
    Regions Using the Quark Parton Model in K^-p Interactions
    at 70 GeV/c
    Paris-Rutherford-Saclay Collaboration

30. Evidence for a New Particle with Naked "Beauty" and
    For Its Associated Production in High-Energy (pp)
    Interactions
    M. Basile et al.
    CERN-Bologna-Frascati-Perugia Collaboration

31. Cross Section Estimated for Associated "Beauty"
    Production in (pp) Interactions and Comparison with
    "Charm"
    M. Basile et al.
    CERN-Bologna-Frascati-Perugia Collaboration

32. Measurement of Associated Charm Production in pp
    Interactions at $\sqrt{s}$ = 62 GeV
    M. Basile et al.
    Bologna-CERN-Frascati Collaboration

33. A Measurement of Two Resonance Contributions in the
    Λ_c^+ Branching Ratios
    M. Basile et al.
    Bologna-CERN-Frascati Collaboration

34. The Leading Effect in Λ_c^+ Production at $\sqrt{s} = 62$ GeV in Proton-Proton Collisions
M. Basile et al.
CERN-Bologna-Frascati Collaboration

35. Measurement of the Λ_c^+ Transverse Momentum Production Distribution in pp Interactions at $\sqrt{s} = 62$ GeV
M. Basile et al.
CERN-Bologna-Frascati Collaboration

36. The Transverse Momentum Distributions of Particles Produced in pp Reactions and Comparison with e^+e^-
M. Basile et al.
CERN-Bologna-Perugia Collaboration

37. Evidence for Double-Jet Structure in pp Interactions at $\sqrt{s} = 62$ GeV
M. Basile et al.
Bologna-CERN-Frascati-Bari Collaboration

38. Measurements of $< p_T^2 >_{in}$ and $< p_T^2 >_{out}$ Distributions in High-Energy pp Interactions
M. Basile et al.
Bologna-CERN-Frascati-Bari Collaboration

39. The Ratio of Charged-to-Total Energy in High-Energy Proton-Proton Interactions
M. Basile et al.
Bologna-CERN-Frascati-Bari Collaboration

40. Multiparticle Production on Hydrogen, Argon, and Xenon Targets in a Streamer Chamber
C. Favuzzi et al.
Bari-Krakow-Liverpool-MPI Munich-Nijmegen Collaboration

41. Determination of the Internal Structure of Valons
M. Barth et al.
Brussels-CERN-Genova-Mons-Nijmegen-Serpukhov Collaboration

42. Study of Reactions $pp \rightarrow (\omega^0, \rho^0, \eta^0)$ + anything at 300 GeV/c
H. Wald et al.
Tufts-Notre Dame-SUNY (Stony Brook)-Kansas Collaboration

43. Inclusive Meson Resonance Production and Fragmentation
     of u-Quark Jets in High Energy νD Interactions
     C.C. Chang et al.
     Tufts-Illinois Inst. of Tech.-Maryland-SUNY (Stony Brook)-
     Tohoku Collaboration

44. Diquark Fragmentation into Λ Particles in νp and νn
     Interactions
     C.C. Chang et al.
     Tufts-Illinois Inst. of Tech.-Maryland-SUNY (Stony
     Brook-Tohoku Collaboration

45. Production of Vector and Tensor Mesons in K^+p
     Interactions at 70 GeV/c - A Preliminary Analysis
     M. Barth et al.
     Brussels-CERN-Genova-Mons-Nijmegen-Serpukhov
     Collaboration

46. Charged Multiplicities of Hadrons in Hadronic and
     Leptonic Reactions
     M. Bardadin-Otwinowska, M. Szczekowski, and
     A.K. Wroblewski
     Inst. of Exp. Phys., Univ. of Warsaw and Inst. Nucl.
     Research, Warsaw

47. Hadron Multiplicity Distributions in a Dual Model
     K. Fiałkowski and A. Kotanski
     Jagellonian Univ., Krakow

48. A Study of K^{*+} and ρ^0 Production with Meson and
     Baryon Triggers
     I.V. Ajinenko et al.
     Serpukhov-Brussels-Mons Collaboration

49. A Study of Multiparticle Fragmentation in K^+p
     Interactions at 32 GeV/c
     E.A. De Wolf et al.
     CERN-Soviet Union Collaboration

50. QCD and Low-p_T Physics
     J.F. Gunion
     Univ. of California, Davis

51. Higher Twist, QCD, and Inclusive Meson Production at High p_T
J.A. Bagger and J.F. Gunion
Princeton Univ. and Univ. of California, Davis

52. Diffractive Excitation in QCD
G. Bertsch, S.J. Brodsky, A.S. Goldhaber, and J.F. Gunion
Univ. of California, Santa Barbara

53. The Quark-Gluon Plasma and the Little Bang
L. McLerran
Univ. of Washington

54. The Quark-Gluon Plasma and Ultra-Relativistic Nucleus-Nucleus Collisions
L. McLerran
Univ. of Washington

55. Particle Production in High Energy Collisions and the Non-Relativistic Quark Model
V.V. Anisovich and J. Nyiri
Leningrad Nucl. Phys. Inst.

LIST OF PARTICIPANTS

Abramowicz, H.	Warsaw Univ., Poland, & CERN
Alam, M.	Vanderbilt Univ. & Cornell Univ., U.S.A.
Andersson, B.	Univ. of Lund, Sweden
Bardadin-Otwinowska, M.	Univ. of Warsaw, Poland
Beaufays, J.	Universite' de L'Etat a' Mons, Belgium
Białas, A.	Jagellonian Univ., Krakow, Poland
Bishop, J.	Univ. of Notre Dame, U.S.A.
Biswas, N.	Univ. of Notre Dame, U.S.A.
Bruyant, F.	CERN
Buschbeck, B.	Osterr. Akad. d. Wissenschaften, Vienna, Austria
Cason, N.	Univ. of Notre Dame, U.S.A.
Contin, A.	CERN
Contogouris, A.	McGill Univ., Canada
Contri, R.	Ist. di Fisica and Sezione INFN, Genova, Italy
Dauwe, L.	Univ. of Notre Dame, U.S.A.
Davis, T.	Univ. of Notre Dame, U.S.A.
DeGrand, T.	Univ. of California, Santa Barbara, U.S.A.
Devenski, P.	Fermilab, U.S.A.
De Wolf, E.	Universitaire Instelling Antwerpen, Belgium

Dremin, I.	Lebedev Physical Inst., Moscow, U.S.S.R.
Fiałkowski, K.	Jagellonian Univ., Krakow, Poland
Ficenec, J.	Virginia Polytechnic Inst., U.S.A.
Frehse, H.	CERN
Fridman, A.	Vanderbilt Univ./CEN de Saclay, France
Garbincius, P.	Fermilab, U.S.A.
Giusti, P.	INFN Bologna, Italy
Gokieli, R.	CERN
Gunion, J.	Univ. of California, Davis, U.S.A.
Hwa, R.	Univ. of Oregon, U.S.A.
Kalelkar, M.	Rutgers Univ., U.S.A.
Kanki, T.	Osaka Univ., Japan
Kellogg, R.	Univ. of Maryland, U.S.A., & DESY
Kenney, V.P.	Univ. of Notre Dame, U.S.A.
Kistiakowsky, V.	Massachusetts Inst. of Technology, U.S.A.
Kittel, W.	Univ. of Nijmegen, Netherlands
Kluge, E.	Universität Heidelberg, Fed. Rep. of Germany
Kohli, J.	RWTH Aachen & Punjab Univ., India
Lassila, K.	Fermilab, U.S.A.
Malhotra, P.	Tata Inst. of Fundamental Research, India

Mann, W.	Tufts Univ., U.S.A.
McLerran, L.	Univ. of Washington, U.S.A.
Mestayer, M.	Univ. of Chicago, U.S.A.
Meyer, W.T.	Ames Laboratory, Iowa State Univ., U.S.A.
Migneron, R.	SLAC, U.S.A.
Minette, D.	Univ. of Wisconsin, U.S.A.
Morrison, D.	CERN
Orava, R.	Fermilab, U.S.A.
Owens, J.	Florida State Univ., U.S.A.
Palombo, F.	INFN, Univ. di Milano, Italy
Pernegr, J.	INFN, Univ. di Milano, Italy
Peterson, C.	SLAC, U.S.A.
Pietrzyk, B.	CERN
Poirier, J.	Univ. of Notre Dame, U.S.A.
Pugh, H.	Lawrence Berkeley Laboratory, U.S.A.
Raja, R.	Fermilab, U.S.A.
Ratti, S.	IFN Pavia, Italy
Resvanis, L.	Univ. of Athens, Greece
Ross, R.	CERN
Ruchti, R.	Univ. of Notre Dame, U.S.A.
Saarikko, H.	Universität Bonn, Fed. Rep. of Germany

Schmitz, N.	Max-Planck-Institut fur Physik und Astrophysik Munich, Fed. Rep. of Germany
Schubelin, P.	CRN Strasbourg, France
Seyboth, P.	Max-Planck-Instut fur Physik und Astrophysik Munich, Fed. Rep. of Germany
Shephard, W.D.	Univ. of Notre Dame, U.S.A.
Solomon, J.	Univ. of Illinois-Chicago Circle, U.S.A.
Spyropoulou-Stassinaki, M.	CERN
Steiner, R.	Yale Univ., U.S.A.
Suk, M.	Charles Univ., Prague, Czechoslovakia
Sukhatme, U.	Univ. of Illinois-Chicago Circle, U.S.A.
Tavernier, S.	Vrije Universiteit Brussel, Belgium
Touchard, A.	Univ. de Pierre et Marie Curie, Paris, France
Van Hove, L.	CERN & Univ. of California, Santa Barbara
Verbeure, F.	Universitaire Instelling Antwerpen, Belgium
Walker, W.	Duke Univ., U.S.A.
Whitmore, J.	Michigan State Univ., U.S.A.
Wroblewski, A.	Univ. of Warsaw, Poland
Wu. S. L.	DESY & Univ. of Wisconsin, U.S.A.

958

Xie Qu-Bing   Shandong Univ., Peoples Republic
                                of China & Univ. of Oregon

Zieminska, D.   Univ. of Maryland & Univ. of
                                Warsaw, Poland

Zieminski, A.   Univ. of Warsaw, Poland & Univ. of
                                Maryland

960

961

962

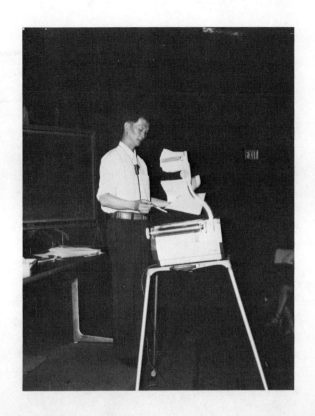